Nuclear Magnetic Resonance

Volume 24

A Specialist Periodical Report

Nuclear Magnetic Resonance

Volume 24

A Review of the Literature Published between June 1993 and May 1994

Senior Reporter

G. A. Webb, *Department of Chemistry, University of Surrey, Guildford*

Reporters

M. W. Anderson, *UMIST, Manchester*
J. R. P. Arnold, *University of Leeds*
J. Barber, *University of Manchester*
A. H. Fawcett, *Queen's University, Belfast*
J. Fisher, *University of Leeds*
H. Fukui, *Kitami Institute of Technology, Japan*
C. J. Jameson, *University of Illinois at Chicago, USA*
K. Kamieńska-Trela, *Polish Academy of Sciences, Warszawa, Poland*
A. Khan, *University of Lund, Sweden*
L. Y. Lian, *University of Leicester*
M. J. W. Prior, *University of Nottingham*
R. R. Sharp, *University of Michigan, USA*
H. Weingärtner, *Universität Karlsruhe, Germany*
S. C. R. Williams, *Institute of Psychiatry, London*
J. Wójcik, *Polish Academy of Sciences, Warszawa, Poland*
B. Wood, *Liverpool John Moores University*
M. Yamaguchi, *Kao Corporation, Japan*

THE ROYAL SOCIETY OF CHEMISTRY

ISBN 0-85404-302-0
ISSN 0305-9804

Published by The Royal Society of Chemistry,
Thomas Graham House, Science Park, Milton Road,
Cambridge CB4 4WF, UK

Printed in Great Britain by Athenaeum Press Ltd, Gateshead, Tyne & Wear

Preface

The format and literature coverage provided by the present volume of Specialist Periodical Reports on Nuclear Magnetic Resonance will be familiar to regular readers of this series. There is the usual combination of annual and biennial reports. Taken together these chapters admirably reflect the quiddity of NMR and its applicability to many areas of science.

It is, perhaps, inevitable that some of the reports tend to be somewhat terse; this merely reflects the nature of the task in hand and is to the reporters credit. I am extremely grateful to all of the contributors for the care taken in manuscript preparation and prompt submission for publication.

Particular thanks go to those who are retiring from the reporting team with this volume, namely Drs J. Barber, B. Wood, and S.C.R. Williams. Finally it is a great pleasure for me to express my thanks to the RSC production team for all of the dedication which they have generously shown to this series for a number of years.

University of Surrey
Guildford
October 1994

G.A. Webb

Contents

Symbols and Abbreviations

These lists contain the symbols and abbreviations most frequently used in this volume, but they are not expected to be exhaustive. Some specialized notation is only defined in the relevant chapter. An attempt has been made to standardize usage throughout the volume as far as is feasible, but it must be borne in mind that the original research literature certainly is not standardized in this way, and some difficulties may arise from this fact. Trivial use of subscripts *etc.* is not always mentioned in the symbols listed below. Some of the other symbols used in the text, *e.g.* for physical constants such as h or π, or for the thermodynamic quantities such as H or S, are not included in the list since they are considered to follow completely accepted usage.

Symbols

a_N	hyperfine (electron–nucleus) interaction constant
A	(i) hyperfine (electron–nucleus) interaction constant (ii) parameter relating to electric field effects on nuclear shielding
B	(i) magnetic induction field (magnetic flux density) (ii) parameter relating to electric field effects on nuclear shielding
B_0	static magnetic field of an n.m.r. or e.s.r. spectrometer
B_1, B_2	r.f. magnetic fields associated with ν_1, ν_2
C_x	spin-rotation coupling constant of nucleus X (used sometimes in tensor form): $C^2 = \frac{1}{3}(C_{\parallel}^2 + 2C_{\perp}^2)$.
$C_{\parallel}$, $C_{\perp}$	components of **C** parallel and perpendicular to a molecular symmetry axis
D	(i) self-diffusion coefficient (ii) zero-field splitting constant
D	rotational diffusion tensor
$D_{\parallel}$, $D_{\perp}$	components of **D** parallel and perpendicular to a molecular symmetry axis
D_{int}	internal diffusion coefficient
D_0	overall isotropic diffusion coefficient
E	electric field
E_n	eigenvalue of $\hat{\mathscr{H}}$ (or a contribution to $\hat{\mathscr{H}}$)
g	nuclear or electronic g-factor
G	magnetic field gradient
H_{ij}	element of matrix representation of $\mathscr{H}$
$\mathscr{H}$	Hamiltonian operator–subscripts indicate its nature
$\boldsymbol{I}_i$	nuclear spin operator for nucleus i
$\boldsymbol{I}_{ix}$, $\boldsymbol{I}_{iy}$, $\boldsymbol{I}_{iz}$	components of $\boldsymbol{I}_i$
I	(i) ionization potential (ii) moment of inertia
^nJ	nuclear spin–spin coupling constant through n bonds (in Hz). Further information may be given by subscripts or in brackets. Brackets are used for indicating the species of nuclei coupled, *e.g.* $J(^{13}C, {}^1H)$ or additionally, the coupling path, *e.g.* J(POCF)
J_r	reduced splitting observed in a double resonance experiment
J	rotational quantum number
^nK	reduced nuclear spin–spin coupling constant (see the notes concerning ^nJ)

m_i	eigenvalue of $\boldsymbol{I}_{iz}$ (magnetic component quantum number)
M_0	equilibrium macroscopic magnetization of a spin system in the presence of B_0
M_x, M_y, M_z	components of macroscopic magnetization
M_n	the number of average mol. wt.
P_A	valence *p* orbital of atom A
P_i	fractional population (or rotamers *etc.*)
P_{uv}	element of bond-order, charge-density matrix
q	electric field gradient
Q	(i) nuclear quadrupole moment (ii) quality factor for an r.f. coil
s_A	valence *s*-orbital of atom A
$S_A^2(0)$	electron density in S_A at nuclear A
S	(i) singlet state (ii) electron (or, occasionally, nuclear spin) *cf. I* (iii) ordering parameter for oriented systems (iv) overlap integral between molecular orbitals
t	elapsed time
T	(i) temperature (ii) triplet state
T_c	coalescence temperature for an n.m.r. spectrum
T_g	the glass transition temperature (of a polymer)
T_1^X	spin–lattice relaxation time of the X nuclei (further subscripts refer to the relaxation mechanism)
T_2^X	spin–spin relaxation time of the X nucleus (further subscripts refer to the relaxation mechanism)
T_2'	inhomogeneity contribution to dephasing time for M_x or M_y
T_2^*	total dephasing time for M_x or M_y; $(T_2^*)^{-1} = T_2^{-1} + (T_2')^{-1}$
T_3	decay time following 90_0–τ–90_{90} pulse sequences
$T_{1\rho}^X, T_{2\rho}^X$	spin–lattice and spin–spin relaxation time of the X nuclei in the frame of reference rotating with B_1
T_{1D}	dipolar spin–lattice relaxation time
X_i	mole fraction of compound *i*
Z_A	atomic number of atom A
α	(i) nuclear spin wavefunction (eigenfunction of $\boldsymbol{I}_z$) for a spin $-\frac{1}{2}$ nucleus (ii) polarizability
β	nuclear spin wavefunction (eigenfunction of $\boldsymbol{I}_z$) for a spin $-\frac{1}{2}$ nucleus
γ_X	magnetogyric ratio of nucleus X
δ_X	chemical shift of a nucleus of element X (positive when the sample resonates to high frequency of the reference). Usually in p.p.m.
δ_{ij}	Kronecker delta (= 1 if $i = j$, and = 0 otherwise)
$\delta(r_{KA})$	Dirac delta operator
Δ	(i) time between field gradient pulses (ii) spectral width
ΔJ	antisotropy in J ($\Delta J = J_\parallel - J_\perp$, for axial symmetry)
Δn	population difference between nuclear states
$\Delta\delta$	change of difference in δ
$\Delta_{\nu\frac{1}{2}}$	full width (in Hz) of a resonance line at half-height
$\Delta\sigma$	(i) anisotropy in $\sigma(\Delta\sigma = \sigma_\parallel - \sigma_\perp$, for axial symmetry) (ii) differences in σ for two different situations
$\Delta\chi$	(i) susceptibility anisotropy ($\Delta_\chi = \chi_\parallel - \chi_\perp$, for axial symmetry (ii) difference in electronegativities
ε_r	relative permittivity
ε_0	permittivity of a vacuum
η	(i) nuclear Overhauser effect (ii) asymmetry factor (*e.g.* in e^2qQ/h) (iii) refractive index (iv) viscosity

μ	magnetic dipole moment
μ_0	permeability of a vacuum
μ_B	Bohr magneton
μ_N	nuclear magneton
ν_i	Larmor precession frequency of nucleus *i* (in Hz)
ν_0	(i) spectrometer operating frequency (ii) Larmor precession frequency (general, or of bare nucleus)
ν_1	frequency of 'observing' r.f. magnetic field
ν_2	frequency of 'irradiating' r.f. magnetic field
σ_i	shielding parameter of nucleus *i* (used sometimes in tensor form). Usually in p.p.m. Subscripts may alternatively indicate contributions to σ.
$\sigma_{\parallel}$, $\sigma_{\perp}$	components of σ parallel and perpendicular to a molecular symmetry axis
σ^d	diamagnetic contribution to σ
σ^p	paramagnetic contribution to σ
τ	(i) pre-exchange lifetime of molecular species (ii) time between r.f. pulses (general symbol)
τ_c	correlation time
τ_{coll}	mean time between molecular collisions in the liquid state
τ_j	angular momentum correlation time
τ_p	pulse duration
τ_t	translational magnetic relaxation correlation time
χ	(i) magnetic susceptibility (ii) electronegativity (iii) nuclear quadrupole coupling constant ($= e^2qQ/h$)
ω	carrier frequency in rad s^{-1}
ω_i, ω_0, ω_1, ω_2	as for ν_i, ν_0, ν_1, ν_2 but in rad s^{-1}
ω_m	modulation angular frequency (in rad s^{-1})
ω_r	sample rotation (rad s^{-1})

Abbreviations

(a) *Physical properties*

a.f.	audiofrequency
a.u.	atomic unit
a.m.	amplitude modulation
b.c.c.	body-centred cubic
c.m.c.	critical micelle concentration
e.d.	electron diffraction
e.f.g.	electric field gradient
e.s.r.	electron spin resonance
f.c.c.	face-centred cubic
f.m.	frequency modulation
h.c.p.	hexagonal close-packed
h.f.	hyperfine
i.d.	inside diameter
i.f.	intermediate frequency
i.r.	infrared
l.c.	liquid crystalline
mol. wt.	molecular weight

n.m.r.	nuclear magnetic resonance
n.q.r.	nuclear quadrupole resonance
o.d.	outside diameter
p.p.m.	parts per million
r.f.	radiofrequency
r.m.s.	root mean square
s.h.f.	super-high frequency
u.h.f.	ultra-high frequency
u.v.	ultraviolet
ADC	analog-to-digital converter
AEE	average excitation energy approximation
AQ	acquire
ARP	adiabatic rapid passage
BIRD	bilinear rotation decoupling
CCPPA	coupled cluster polarization propagator approximation
CH-COSY	carbon-hydrogen correlation spectroscopy
CHESS	chemical shift selection
CHF	coupled Hartree–Fock molecular orbital calculations
CIDEP	chemically induced dynamic electron polarization
CIDNP	chemically induced dynamic nuclear polarization
COSY	correlation spectroscopy
CP	cross polarization
CPMG	Carr–Purcell pulse sequence. Meibom–Gill modification
CSA	chemical shielding anisotropy
CSI	chemical shift imaging
CW	continuous wave
DAC	digital-to-analog converter
DD	dipole-dipole (interaction or relaxation mechanism)
DEPT	distortionless enhancement by polarization transfer
DLB	differential line broadening
DNP	dynamic nuclear polarization
DQ	double quantum
DQF	double quantum filter
ECOSY	exclusive correlation spectroscopy
EHT	extended Hückel molecular orbital theory
ENDOR	electron–nucleus double resonance
EOM	equations of motion
EXSY	exchange spectroscopy
FC	Fermi contact
FID	free induction decay
FLASH	fast low angle shot
FPT	finite perturbation theory
FT	Fourier transform
GIAO	gauge included atomic orbitals
HMQ	heteronuclear multiquantum
HOHAHA	homonuclear Hartman–Hahn
HRPA	higher random phase approximation
IDESS	improved depth selective single surface coil spectroscopy
IGLO	individual gauge for different localized orbitals
INADE-QUATE	incredible natural abundance double quantum transfer experiment
INDO	intermediate neglect of differential overlap
INDO/S	intermediate neglect of differential overlap calculations for spectroscopy
INDOR	internuclear double resonance
INEPT	insensitive nuclei enhanced by polarization transfer
ISIS	image selected *in vivo* spectroscopy
LIS	lanthanide induced shift
LORG	local origin

LSR	lanthanide shift reagent
MASS	magic angle sample spinning
MBPT	many body perturbation theory
MEM	maximum entropy method
MINDO	modified INDO
MQ	multiple quantum
MQC	multiple quantum coherence
MQF	multiple quantum filter
NOE	nuclear Overhauser enhancement
NOESY	nuclear Overhauser enhancement spectroscopy
NQCC	nuclear quadrupole coupling constant
PFG	pulsed field gradient
PRE	proton relaxation enhancement
QF	quadrupole moment/field gradient
QPD	quadrature phase detection
REX	relativistically extended Hückel molecular orbital theory
ROESY	rotating frame Overhauser enhancement spectroscopy
RPA	random phase approximation
SCPT	self consistent perturbation theory
SD	spin dipolar
SECSY	spin echo correlation spectroscopy
SEFT	spin echo Fourier transform
SLITDRESS	slice interleaved depth resolved surface coil spectroscopy
SOPPA	second order polarization propagator approach
SPI	selective population inversion
SPT	selective population transfer
SR	spin rotation (interaction or relaxation mechanism)
TART	tip angle reduced T_1 imaging
TOCSY	total correlation spectroscopy
WAHUHA	Waugh, Huber and Häberlen (cycle of pulses)
ZQ	zero quantum
ZQC	zero quantum coherence

(b) *Chemical species**

acac	acetylacetonato
ACTH	adrenocorticotropic hormone (corticotropin)
ADP	adenosine diphosphate
AMP	adenosine monophosphate
ATP	adenosine triphosphate
BSA	bovine serum albumin
CMP	cytidine monophosphate
cp	cyclopentadienyl
DAP	dodecylammonium propionate
DME	1,2-dimethoxyethane
DMF	dimethylformamide
DML	dimyristoyl-lecithin
DMS	dimethylsiloxane
DMSO	dimethylsulphoxide
DNA	deoxyribonucleic acid
DPG	2,3-diphosphoglycerate
DPI	dipalmitoyl-lecithin
dpm	dipivaloylmethanato
DPPH	diphenylpicrylhydrazyl
DSS	2,2-dimethyl-2-silapentane-5-sulphonate (usually as the sodium salt)
DTBN	di-t-butyl nitroxide

* Lower case initials are used when the species is a ligand.

EBBA	*N*-(*p*-ethoxybenzylidene)-*p*-butylaniline
EDTA	ethylenediaminetetra-acetic acid
EVA	ethylene-vinyl acetate
fod	1,1,1,2,2,3,3-heptafluoro-7,7-dimethyloctane-4,6-dionato
HAB	4,4′-bis(heptyl)azoxybenzene
HMPA	hexamethylphosphoramide
HOAB	*p*-n-heptyloxyazoxybenzene
IHP	inositolhexaphosphate
KDP	potassium dihydrogen phosphate
MBBA	*N*-(*p*-methoxybenzylidene)-*p*-butylaniline
NADH(P)	nicotinamide adenine dinucleotide (phosphate)
NMF	*N*-methylformamide
PAA	*p*-azoxyanisole
PBA	pyrene butyric acid
PBLG	poly(L-benzyl μ-glutamate)
PC	phosphatidyl choline (lecithin)
PCB	polychlorinated biphenyl
PDMS	polydimethylsiloxane
PMA	poly(methacrylic acid)
PMMA	poly(methyl methacrylate)
POM	poly(oxymethylene)
PS	phosphatidylserine
PTFE	polytetrafluoroethylene
PVC	poly(vinyl chloride)
PVF	poly(vinyl fluoride)
PVP	poly(vinyl pyrrolidone)
RNA	ribonucleic acid (tRNA, transfer RNA)
SDS	sodium dodecyl sulphate
TAB	trimethylammonium bromide
TCNQ	tetracyanoquinodimethane
TFA	trifluoroacetic acid
THF	tetrahydrofuran
TMS	tetramethylsilane
UTP	uridine triphosphate

Amino-acid residues

Ala	alanine	Leu	leucine
Arg	arginine	Lys	lysine
Asn	asparagine	Met	methionine
Asp	aspartic acid	Phe	phenylalanine
Cys	cysteine	Pro	proline
Gln	glutamine	Ser	serine
Glu	glutamic acid	Thr	threonin
Gly	glycine	Trp	tryptophan
His	histidine	Tyr	tyrosine
Hyp	hydroxyproline	Val	valine
Ile	isoleucine		

1
NMR Books and Reviews

COMPILED BY G. A. WEBB

This section lists all books and reviews, with NMR as the principal theme, that were known to the compiler, for the period under review. In most cases the source of reference is the 'CA Selects-NMR Chemical Aspects' listings for the period from June 1993 to May 1994. References to technical bulletins, company or institute house journals, special reports, etc., are normally omitted. In general, texts of individual lectures from symposia are not listed nor are NMR sections of general chemistry or spectroscopy textbooks. The compilation is sectionalized for the reader's convenience. The foreign language articles are grouped together according to language and the English title is provided.

(a) **BOOKS**

R1 J.W.Hennel and J. Klinowski,"Fundamentals on NMR",Longman, London, 1993,288pp

R2 C.Jones, B.Mulloy and A.H.Thomas, "Spectroscopic Methods and Analyses",Humana, Totowa, 1993, 395pp

(b) **REGULAR REVIEW SERIES**

Accounts of Chemical Research, 1993, vol.26

R3 H.G. Floss and S.Lee, "Chiral Methyl Groups: Small is Beautiful", p.116

R4 A.Bax and S.Grzesiek, "Methodological Advances in Protein NMR", p.131

R5 D.B. Collum, "Solution Structures of Lithium Dialkylamides and related N-Lithiated Species", p.227

Advances in Chemistry Series, 1993, vol.229

R6 R.A.Wind, G.E.Maciel and R.E.Botto, "Quantitation in ^{13}C NMR Spectroscopy of Carbonaceous Solids", p.3

R7 G.E.Maciel, C.E.Bronnimann and C.F.Ridenour, "^{1}H NMR Spectroscopy: Approaches for Carbonaceous Solids", p.27

R8 R.A.Wind, R.Lewis,H.Lock and G.E.Maciel," Solid Materials research with NMR and Dynamic Nuclear Polarization Spectroscopy", p.45

R9 H.Thomann and M.Bernardo, "Advances in Electron Nuclear Double Resonance Spectroscopy", p.65

R10 M.K.Bowman, "Strategies for Measurement of Electron Spin Relaxation", p.91

R11 R.L.Belford and R.B.Clarkson, "Multifrequency Electron Paramagnetic Resonance Spectroscopy", p.107

R12 Y.Sanada and L.J.Lynch, "High Temperature ESR and NMR Methods Applied to Coal", p.139

R13 W.A.Barton, L.J.Lynch, D.S.Webster and G.Simms, "Diffusion Coupled Relaxation and the Submicroscopic Structure of Bituminous Coals", p.175

R14 A.M.Vassallo, "^{13}C NMR Spectroscopy of Pyridine and Alkylpyridines Sorbed onto Coal", p.201

R15 R.A.Wind, "Distortion Free ^{13}C NMR Spectroscopy in Coal", p.217

R16 R.Sakurovs, L.J.Lynch and W.A.Barton,"^{1}H NMR Thermal Analysis of Argonne Premium Coals", p.229

R17 D.E.Axelson,"Multivariate Statistical Analysis and Pattern Recognition", p.253

R18 Y.Adachi and M.Nakamizo, "Structural Parameters of Argonne Coal Samples: Solid State ^{13}C NMR Spectroscopy", p.269

R19 N.Cyr and N.O.Egiebor," Characterization of Coal Liquefaction Residues by Solid State ^{13}C NMR Spectroscopy", p.281

R20 K.Hayamizu, S.Hayashi, K.Kamiya and M.Kawamura, "^{1}H NMR Spectroscopy and Spin-Lattice Relaxation of Argonne Premium Coals", p.295

R21 J.Hu,L.Li and C.Ye," High Field NMR Studies of Argonne Premium Coals",p.323

R22 J.A.MacPhee, H.Kawashima, Y. Yamashita and Y.Yamada, "Solid State ^{13}C NMR Studies on Coal and Coal Oxidation", p.323

R23 C.Tsiao and R.E.Botto," Measurement of Spin-Lattice Relaxation in Argonne Premium Coal Samples", p.341

R24 L.D.Rosa, M.Pruski and B.Gerstein, "Quantitation of Protons in the Argonne Premium Coals by Solid State ^{1}H NMR Spectroscopy", p.359

R25 J.A.Franz and J.C.Linehan, "Bloch Decay and Cross Polarization-Magic Angle Spinning ^{13}C NMR Study of the Argonne Premium Coals; Effects of High Speed Spinning", p.377

R26 A.Jurkiewicz, C.E.Bronnimann and G.E.Maciel," High Resolution ^{1}H NMR Studies of Argonne Premium Coals", p.401

R27 A.M.Orendt, M.S.Solum, N.K.Sethi, C.D.Hughes, R.J.Pugmire and D.M.Grant, "Measurement of ^{13}C Chemical Shielding Anisotropy in Coal", p.419

R28 B.G.Silbernagel and R.E.Botto," Advanced Magnetic Resonance Techniques Applied to Argonne premium Coals", p.629

Advances in Experimental and Medical Biology, 1992, vol.316

R29 R.Vink,"NMR Spectroscopy and the Study of Tissue Oxygen Metabolism", p.316

Annual Reports on Medicinal Chemistry, 1992, vol.27

R30 J.W.Erickson and S.W.Fesik, "Macromolecular X Ray Crystallography and NMR as Tools for Structure Based Drug Design", p.271

Annual Reports on NMR Spectroscopy, ed. G.A.Webb,1993, vol.27

R31 J.Cavanagh and M.Rance," Sensitivity Enhanced NMR Techniques for the Study of Biomolecules", p.1

R32 Y.Inoue,"NMR Studies of the Structure and Properties of Cyclodextrins and Their Inclusion Complexes", p.59

R33 K.G.Orrell and V.Sik,"Dynamic NMR Spectrosocopy in Inorganic and Organometallic Chemistry", p.103

R34 C.E.Mountford, C.L.Lean, W.B.Mackinnon and P.Russell,"The use of Proton MR in Cancer Pathology", p.173

R35 T.Nose,"Pulsed Field Gradient NMR Studies of the Diffusion of Chain Molecules in Polymer Matrices", p.217

R36 R.H.Contreras and J.C.Facelli, "Advances in Theoretical and Physical Aspects of Spin-Spin Coupling Constants", p.255

1994, vol.28

R37 H.Kawazoe,"Application of NMR Spectroscopy to the Science and Technology of Glasses and Ceramics", p.1

R38 S.Hayashi," High Resolution Solid State NMR Studies on Ceramics", p.29

R39 H.Pfeifer and H.Ernst," NMR Studies of Zeolites", p.91

R40 H.Kurosu, S.Ando, H.Yoshimizu and I.Ando," NMR Studies of Higher Order Structures of Solid Polymers", p.189

R41 F.D.Blum," NMR Studies of Organic Thin Films", p.277

Annual Reports on Progress in Chemistry, Sect.B,1991, vol.87

R42 J.R.Everett and S.A.Redshaw,"NMR Studies on the Structures of Carbohydrates", p.3

1992,vol.88

R43 J.R.Everett and S.A.Redshaw,"NMR Studies on the Structure and Dynamics of Nucleic Acids and Their Complexes with Ions and Other Molecules", p.3

Sect.C,1993, vol.89

R44 G.A.Webb, "NMR Spectroscopy", p.3

Annual Reviews of Biochemistry, 1993, vol.62

R45 J.A.Jaeger, J.Santa Lucia and i.Tinoco," Determination of RNA Structure and Thermodynamics",p.255

Annual Reviews of Biophysics and Biomolecular Structure 1993, vol.22

R46 W,Chiu,"What does Electron Cryomicroscopy Provide that X Ray Crystallography and NMR Spectroscopy Cannot ?", p.233

Biological Magnetic Resonance, ed.L.J.Berliner and J.Reuben, 1992, vol.11
R47 L.Bolinger and R.E.Lenkinski,"Localization in Clinical NMR Spectroscopy", p.1
R48 T.Schleich, G.H.Craines and J.M.Rydzewski, "Off Resonance Rotating Frame Spin-Lattice Relaxation :Theory and *in Vivo* MRS and MRI Applications", p.55
R49 L.H. Chang and T.L.James, "NMR Methods in Studies of Brain Ischemia", P.135
R50 S.K. Miller and G.A. Elgavish, "Shift Reagent Aided ^{23}Na NMR Spectroscopy in Cellular, Tissue and Whole-Organ Systems", p.159
R51 B.S.Selinsky and C.T.Burt, "*In Vivo* ^{19}F NMR", p.241
R52 R.E.London, "*In Vivo* ^{2}H NMR Studies of Cellular Metabolism", p.277
R53 L.J.Berliner and H.Fujii, "Some Applications of ESR to *In Vivo* Animal Studies and EPR Imaging", p.307
1993, vol.12
R54 G.N. La Mar and J.S. de Ropp," NMR Methodology for Paramagnetic Proteins", p.1
R55 L.Banci, "Nuclear Relaxation in Paramagnetic Metalloproteins", p.79
R56 C.C.Lester and R.G.Bryant,"Paramagnetic Relaxation of Water Protons", p.113
R57 F.A.Walker and u.Simonis," Proton NMR Spectroscopy of Model Hemes", p.133
R58 J.D.Satterlee, S.Alam, Q.Yi, J.E.Erman, I.Constantinidis, D.J. Russell and S.J.Moench," Proton NMR Studies of Selected Paramagnetic Heme Proteins", p.275
R59 J.Mispelter, M.Momenteau and J.M. Lhose," Heteronuclear Magnetic Resonance: Applications to Biological and Related Paramagnetic Molecules", p.299
R60 C.Luchinat and S.Ciurli," NMR of Polymetallic Systems in Proteins", p.357
1993, vol.13
R61 B.J.Gaffney and H.J.Silverstone," Simulation of the EMR Spectra of High Spin Iron in Proteins", p.1
R62 P.G.Debrunner," Moessbauer Spectroscopy of Iron Proteins", p.59
R63 R.Basosi, W.E.Antholine and J.S.Hyde," Multifrequency ESR of Copper: Biophysical Applications", p.103
R64 B.M.Hoffman, V.J.DeRose, P.E.Doan, R.J.Gurbiel, A.L.P.Houseman and J.Telser," Metalloenzyme Active Site Structure and Function through Multifrequency CW and Pulsed ENDOR",p.151

R65 J.Huettermann," ENDOR of Randomly Oriented Mononuclear Metalloproteins", p.219

R66 K.Moebius," High Field EPR and ENDOR on Bioorganic Systems", p.253

R67 H.Thomann and M.Bernardo," Pulsed Electron Double and Multiple Resonance Spectroscopy of Metals in Pròteins and Enzymes", p.275

R68 D.C.Thomas, E.M.Ostap, C.L.Berger, S.M.Lewis, P.G.Fajer and J.E.Mahaney," Transient EPR of Spin-Labeled Proteins", p.323

R69 A.Tomasi and A.Iannone,"ESR Spin-Trapping Artifacts in Biological Model Systems", p.353

Bulletin Of Magnetic Resonance, 1993, vol.15

R70 G.C.Leo, R.P.Shank and H.R.Zielke," The Study of Rat Brain Metabolic Pathways by ^{13}C NMR", p.8

R71 A.M.Rossi, D.S.Schramm, O.Chaix and G.Poupeau," Thermal Stability of Trapping Centres in Natural Aragonites", p.14

R72 J.K.Bonesteel, I.Al-Bala'a and R.D.Bates," EPR and NMR Studies of Transient Complex Formation of Chloroform with Nitroxide Free Radicals", p.18

R73 R.K.Harris, D.C.Apperley, N.A.Davies and R.D. Fischer,"High Resolution Solid State NMR of Organometallic Coordination Polymers", p.22

R74 S.G.Kim, M.Salazar and B.R.Ried," Automatic NOE Back Calculation", p.29

R75 T.Takui, Y.Teki, M.Endoh, H.Kitai, Y.Okudo, S.Kita, T.Kinoshita and K.Itoh,"Organic High Spin State ENDOR/TRIPLE Spectroscopy", p.41

R76 A.M.Rossi, N.M.Pinhal and N.V.Vugman,"Evidence of Rotating CO_2^- Radicals in Bioapatites of Calcified Tissue", p.48

R77 D.M.Wang, I.Kovacik, E.J.Reijerse and E.deBoer," Structural Change in $(NH_4)_2[Zn(NH_3)_2(CrO_4)_2]$ and $(NH_4)_2[CdNH_3)_2(CrO_4)_2]$ Doped with Cu(II), Observed with EPR", p.50

R78 R.J.Buist, M.Novak, C.Hudspeth and B.J.Blackburn,"*In Vivo* ^{31}P NMR Spectra of the Liver Region of Mice Infected with *Hymenolepis Microstama* and *Mesocestoides Vogae*", p.55

R79 D.S.Williams J.A.Detre, W.Zhang and A.P.Koretsky," Fast Serial MRI of Perfusion in the Rat Brain Using Spin Inversion of Arterial Water", p.60

R80 A.D.Bain and G.J.Duns,"Equivalence of Pulsed and Continuous Radiation in NMR", p.64

R81 N.J.Clayden,"Morphology in Peek by NMR Relaxation Time Measurements", p.70

R82 C.M.Preston, B.G.Sayer and R.H.Newman,"NMR and Forest Ecosystem Dynamics", p.75

R83 N.Zhou and H.J.Vogel," d($T_8C_4A_8$) Forms a Hairpin with a Bent Stem and an Unstacked Cytosine", p.80

R84 J.M.Aramini and H.J.Vogel," Quadrupolar Metal Ion NMR to Probe Metal Binding to Proteins: ^{27}Al NMR Studies of Ovotransferrin", p.84

R85 L.A.Denys, A.J.Williams and G.Sachs," 2D Hetero and Homonuclear NMR Study of a K^+ Competitive Fluorescent Inhibitor of the H,K-ATPase", p.89

R86 Y.Kuroda, M.Wakita and T.Nakagawa," Interaction Between Dibucaine and Pig Erythrocyte Membranes as Studied by NOESY Experiments in ^{1}H NMR Spectroscopy", p.93

R87 M.Zhang and H.J.Vogel,"Protein Engineering and NMR Studies of Lysine Residues in Mammalian Calmodulin", p.95

R88 H.Fukui, M.Miura and H.Matsuda,"Gauge Invariant Many Body Perturbation Calculations of Nuclear Magnetic Shieldings", p.98

R89 K.Furukawa, T.Matsumura,Y.Teki,T.Kinoshita,T.Takui and K.Itoh," Molecular Design,Synthesis and ESR Detection of a Ground State Tridecet Hydrocarbon as a Model for 2D Organic Ferro- and Superpara-Magnets", p.103

R90 Y.Inagaki, H.Maekawa, T.Yokokawa and S.Shimokawa," High Temperature NMR Studies of Sodium Borates", p.108

R91 A.Diaz, R.Monque and M.Bussolo," Quantitative Determination of Gallium in Zeolites by ^{71}Ga MAS NMR Spectroscopy", p.112

R92 M.R.Morrow, D.Lu, D.Singh and C.W.M.Grant," NMR Studies of Phase Behaviour and Chain Orientational Order in Glycosphingolipid-Phosphatidylcholine Mixtures", p.117

R93 T.Pott and E.J.Dufourc," Cholesterol Inhibits Melittin-Induced Membrane Lysis", p.121

R94 A,Leonard, A.Milon, M.A.Krajewski-Bertrand and E.J.Dufourc," Modulation of Membrane Hydrophobic Thickness by Cholesterol, Cycloartenol and Hopanoid", p.124

R95 R.S.Phadke, S.Arivastava and G.Govil," Antioxidants in Biomembranes: Magnetic Resonance Studies", p.128

R96 V.A.Silaev, K.M.Mysin,O.I.Gnezdilov, E.S. Nefed'ev, L.A. Abrakhmanova and V.G.Khozin," New Results in the Investigations of Polymers by EPR and EPR-Tomography", p.133

R97 L.J.Berliner and J.K.Woodford," Interactions of Hirudin Analogues with Bovine and Human THrombins", p.142

R98 T.Nakai and C.A.McDowell,"The NMR Spectra of Homonuclear Two-Spin Systems Under Magic Angle and Off-Magic Angle Spinning", p.149

R99 M.R.Merrill," Coreflood Experiments on Casper Sandstone", p.154

R100 D.G.Cory and W.E.Maas,"NMR Imaging of Solid Polymers", p.160

R101 R.Ricciolini, A.Miccheli, E.Piccolella, A.Tomassini,C.Manetti, M.E.DiCocco and F.Conti,"Biochemical and Biophysical Effects of Dexamethasone on Human Lymphoblastoid Cell Lines, Studies by ^{31}P and ^{1}H NMR Spectroscopy and Fluorimetry", p.165

R102 T.Aureli, G.Capuani, M.E.Di Cocco, A.Miccheli,N.Corsico, A.Nardone, E.A. Martelli and F.Conti," Peripheral Arteriopathy in the Rat", p.169

R103 P.M.Debenham, P.J.McDonald, A.J.Horsewill and D.Vijayaraghavan," Tunnel Splitting of 2,3 Butanedione with Pressure Studied using Low Field NMR Tunnelling Spectroscopy", p.172

R104 L.C.terBeek and E.E.Burnell,"Spin-Lattice Relaxation for Dideuterium", p.175

R105 C.A.Fyfe, J.Niu, S.Rettig, N.E.Burlinson,CReidsma, D.Wang and M.Poliks," High Resolution ^{13}C and ^{15}N NMR Investigations of the Curing Reaction of Cyanate Resins", p.179

R106 E.Brosio, E.Carnovale,M.Delfini, L.Pizzoferrato and G.Veglia," ^{31}P NMR Study on Phytase Catalysed Hydrolysis of Phytate", p.183

R107 L.J.Mueller and D.P.Weitekamp," Molecular Beam Magnetic Resonance via the Longitudinal Stern Ġerlach Effect", p.187

R108 S.K.Buratto, J.Y.Hwang, N.D.Kurur, D.N.Shykind and D.P.Weitekamp," FT Time Sequenced Optical NMR of Gallium Arsenide", p.190

R109 C.A.Fyfe, P.P.Aroca and Y.Zhang," NMR Characterization of the Homopolymers and Copolymers Formed from Tetraethoxysilane and Methyltriethoxysilane in a Sol-Gel Process", p.195

R110 T.Chandrakumar and E.E.Burnell," Comparison of Short Range Interactions in Zero Electric Field Gradient Mixtures", p.199

R111 S.R.Wassall, S.Ayalasomayajula, G.L.Grffith, M.A.McCabe, W.D.Ehringer and W.Stillwell," ^{2}H NMR Studies of Membrane Stabilization by Vitamin E", p.205

R112 S.J. Hwang and B.C.Gerstein," Spin Counting in Dipolar Coupled Spin 1/2 Systems by Multiple Quantum Coherence", p.211

Chemical Reviews, 1993, vol.93

R113 I.Bertini, P.Turano and A.J.Vila,"NMR of Paramagnetic Metalloproteins", p.2833

Chemical Society Reviews, 1993, vol.22

R114 J.K.M.Sanders,"NMR of Natures Plastics and Spiders Webs: Chemistry, Physics or Biology ?",p.1

R115 D.Reed,"The Role of NMR in Boron Chemistry", p.109

Coordination Chemistry Reviews, 1993,vol.122
R116 E.G.Hope and W.Levason,"Recent Developments in the Coordination Chemistry of Selenoether and Telluroether Ligands", p.109
Magnetic Resonance Reviews, ed.C.P.Poole, 1993.Vol.16
R117 A.V.Ii'Yasov and Y.M.Kargin,"ESR of Electrochemically Generated Radical Ions and Complexes", p.135
R118 S.S.Eaton and G.R.Eaton,"Applications of High Magnetic Fields in EPR Spectroscopy", p.157
R119 M.A.Petrich,"NMR of Amorphous Hydrogenated Silicon", p.183
R120 R.D.Bates,"Dynamic Nuclear Polarization", p.237
NMR Basic Principles and Progress, ed. P.Diehl, S.Fluck , H.Guenther, R.Kosfeld and J.Seelig, 1993, vol.29
R121 D.Freude and J.Haase,"Quadrupole Effects in Solid State NMR", p.1
R122 N.J.Clayden,"Solid State NMR of Synthetic Polymers", p.91
R123 A.Bunn,"Solution NMR of Synthetic Polymers", p.127
1994, vol.30
R124 A.R.Grimmer and B.Bluemich,"Introduction to Solid State NMR", p.1
R125 F.Laupretre,"High Resolution ^{13}C NMR Investigations of Local Dynamics in Bulk Polymers at Temperatures Below and Above the Glass Transition Temperature", p.63
R126 D.Raftery and B.F.Chmelka,"Xenon NMR Spectroscopy", p.111
R127 G.Fleischer and F.Fujara,"NMR as a Generalized Incoherent Scattering Experiment", p.159
R128 P.Bluemler and B.Bluemich,"NMR Imaging of Solids", p.209
Progress in NMR Spectroscopy ed.J.W.Emsley, J.Feeney and L.H.Sutcliffe,1993, vol.25
R129 G.Barbarella,"^{33}S NMR", p.317
R130 P.J.Hore and R.W.Broadhurst,"Photo CIDNP of Biopolymers", p.345
R131 M.S.Searle,"NMR Studies of Dynamics in Nucleic Acids", p.403
R132 A.N.Lane,"NMR Studies of Dynamics in Nucleic Acids", p.481
R133 E.W.Lang and H.D.Luedemann,"Density Dependence of Rotational and Translational Molecular Dynamics in Liquids Studied by High Pressure NMR", p.507
1994, vol.26
R134 J.Jokisaari,"NMR of Noble Gases Dissolved in Isotropic and Anisotropic Liquids", p.1
R135 R.Brueschweiler and D.A.Case,"Characterization of Biomolecular Structure and Dynamics by NMR Cross Relaxation", p.37

R136 A.D.Bain, I.W.Burton and W.F.Reynolds,"Artifacts in 2D NMR", p.59

R137 I.Bertini, C.Luchinat and M.Piccioli,"Copper-Zinc Superoxide Dismutase", p.91

R138 J.Courtieu, J.P.Boyle and B.M.Fung,"Variable Angle Sample Spinning NMR in Liquid Samples", p.141

Progress in Physical Organic Chemistry, 1993, vol.19

R139 U.Edlund and E.Buncel,"Electronic Configuration Structure and Solvation of Phenyl Substituted Group IVA Anions", p.225

Spectroscopic Properties of Inorganic and Organometallic Compounds, ed. G.Davidson, 1993, vol.26

R140 B.E.Mann," NMR Spectroscopy", p.1

(c) **EDITED BOOKS AND SYMPOSIA**

ACS Symposia Series

R141 R.M.Clapperton, B.T.Ingram, R.B.Ottewill and A.R.Rennie,"NMR Spectroscopy and Neutron Scattering Studies on Ammonium Decanoate-Ammonium Perfluorooctanoate Mixtures", 1992,**501**, 268

R142 B.Katz and R.Bersohn,"Hydrogen Isotope Effects in Chemical Reactions and Photodissociations", 1992,**502**, 66

R143 S.B.Liu, J.F.Wu, L.J.Ma, M.W.Lin and T.L.Chen, "Homogeneous Adsorption of Benzene on NaX and NaY Zeolites", 1993,**517**, 272

R144 C.L.Tway, S.J.Bonafede, M.A.Ghazi, C.P.Reed, R.J.De Angelis and T.M.Apple,"Characterization of Ruthenium Trichloride Impregnated NaY Zeolite", 1993,**517**, 372

R145 S.Pang and A.Rudin,"Size Exclusion Chromatographic Assessment of Long Chain Branch Frequency in Polyethylenes", 1993, **521**, 254

R146 O.Meyer, K.Frunzke, J.Tachil and M.Volk," The Bacterial Molybdenum Cofactor", 1993,**535**, 50

R147 R.A.Johnson and S.W.Sholaby," Effect of Structure on Properties of Absorbable Oxalate Polymers", 1994,**540**, 202

R148 H.Weenen, S.B.Tjan, P.J.De Valois, N.Bouter, A.Pos and H.Vonk," Mechanism of PYrazine Formation", 1994,**543**, 142

R149 M.S.Shiao, K.R.Lee, L.J.Lin and C.T.Wang,"Natural Products and Biological Activities of the Chinese Medicinal fungus *Ganoderma Lucidum* ", 1994,**547,** 342

Adenine Nucleotides Cell Energy Transfer Signal Transduction , 1992

R150 H.Munier, E.Krin, A.B.P.Glaser, A.Bouhss, A.Danchin and B.O.Anlaine,"Structural and Functional Organization of the Catalytic Domain of a Bacterial Toxin",p.335

Advances in Lipid Methodology, ed.W.C.Christie,Oily Press, Ayr, 1992

R151 D.Firestone and A.Sheppard,"Determination of *Trans* Fatty Acids", p.273

Amphipathic Helix, ed. R.M.E.Rand, CRC, Boca Raton, 1992

R152 S.J.Opella, J.Gesell and B.Bechinger,"NMR Spectroscopy of Amphipathic Helical Peptides in Membrane Environments", p.87

Buckminsterfullerenes, ed. W.E.Billups and Ciufolini, VCH, New York, 1993

R153 C.Thilgen, F.Diederich and R.L.Whetten,"The Higher Fullerenes", p.59

Cellulose, Sources and Exploitation, 1990

R154 R.D.Short and H.S.Munro,"Some onclusions Drawn from a Study of Cellulose Nitration in Technical Mixed Acids by XPS and ^{13}C NMR", p.263

Chemistry of Acid Derivatives, ed.S.Patai, Wiley, Chichester,1992, vol.2

R155 P.E.Hansen,"NMR of Acids and Acid Derivatives", p.153

Chemistry of Hydroxy, Ether and Peroxide Groups, ed.S.Patai, Wiley, Chichester, 1993

R156 J.Iley and P.G.Taylor,"ESR and NMR Spectroscopy", p.241

Chemistry of Iron, ed.J.Silver, Blackie, Glasgow, 1993

R157 B.W.Fitzsimmons,"Spectroscopic Methods for the Study of Iron Chemistry", p.170

Cholesterol Membrane Models, ed.L.Finegold, CRC, Boca Raton, 1993

R158 J.H.Davis,"Molecular Dynamics, Orientational Order and Thermodynamic Phase Equilibria of Cholesterol/Phosphatidylcholine Mixtures", p.67

Collected Papers of the Second International Symposium on Inovation and Perspectives in Solid State Syntheses, 1992

R159 S.J.Wimalawansa,"Use of Synthetic Peptides in Specific Affinity Chromatography for the Purification of Specific Peptide Receptors", p.111

R160 J.Bahr, H.Mickos, B.Luning and J.Reed,"The 3D Structure of the Arg-Gly-Asp Adhesion Sequence of Fibronectin, Part II", p.337

R161 P.Henklein, M.Beyermann, B.Costisella and R.Sohr,"Activated Intermediates in Solid Phase Synthesis Couplings", p.401

Conference Proceedings of the Italian Physics Society, 1993, vol.43

R162 R.Kimmich,"Self Diffusion and NMR Relaxation in Hydration Shells of Proteins", p.157

R163 I.Muegge and E.W.Knapp,"Protein Water Interactions, Model Studies on BPTI", p.165

Coordination Chemistry of Aluminium, ed. G.H.Robinson, VCH, New York, 1993

R164 J.P.Oliver, R.Kumar and M.Taghiof,"The Chemistry of Alkoxides, Thiolates and theHeavier Group 16 Derivatives of Aluminium and Gallium", p.167

Developments in Carbohydrate Chemistry, ed. R.J.Alexander and H.F.Zobel, American Association of Cereal Chemistry, St.Paul, 1992

R165 M.J.Gidley,"NMR Analysis of Cereal Carbohydraytes", p.163

Experimental Techniques in Glass Science, ed. C.J.Simmons and O.H.El-Bayoumi, American Ceramic Society, Westerville, 1993

R166 J.F.Emerson and P.J.Bray,"NMR Analysis of Glass Structure", p.77

Flavour Science, ed.T.E.Acree and R.Teranishi, ACS, Washington, 1993

R167 C.J.Mussinon,"Instrumental Analysis in the Flavour Industry", p.169

Food Control Research, ed.R.Wittkowski and R.Matissek, Technamic, Lancaster Pa., 1993

R168 G.Takeaka,"Aroma ompounds", p.109

Frontiers in Solid State Science, ed. L.C.Gupta and M.S.Mittani, World Science, Singapore, 1993, vol.1

R169 K.Asayama,"NMR Studies in Highly Correlated Systems", p.585

Glycoconjugates, ed.H.J.Allen and E.S.Kisailus, Dekker, New York, 1992

R170 A.S.Serianni,"NMR Approaches to Oligosaccharide Structure Elucidation", p.71

HPLC Detection, ed.G.Patanay, VCH, New York, 1992

R171 K.Albert and E.Bayer,"HPLC-^{1}H NMR On Line Coupling", p.197

Second International Conference on Electronic, Optical and Acoustic Properties of Polymers, 1990

R172 J.Clements, G.R.Davies and I.M.Ward,"Contribution of Crystallinity Changes to the Piezoelectric Properties of PUDF", p.2

Sixth International Congress on the Technology of Pharmaceuticals, 1992, vol.2

R173 B.Arica, A.Y.Ozer and A.A.Hincol,"Preparation and *In Vitro* Studies on Primaguine Diphosphate Liposomes", p.363

Liquid Crystal Polymers, ed.A.A.Collyer,Elsevier, London, 1992

R174 F.Laupretre,"NMR Studies of Thermotropic Polymers", p.103

Lithium Cells, ed. N.J.Birch, Academic Press, London, 1991

R175 M.S.Hughes,"Intracellular Concentrations of Lithium as Studied by NMR", p.176

Macromolecular Chemistry, Macromolecular Symposia, 1993, vol.67
R176 M.Sawamoto and T.Higashimura,"Principles and Design of Living Cationic Polymerization of Vinyl Monomers", p.299
1993, vol.72
R177 E.G.Brame,"Uncommon Methods for Spectroscopic Analyses of Polymers", p.1
R178 R,Chujo,"NMR Characterization of Biodegradable Polyesters", p.11
Magnetic Resonance, ed.C.L.Khetrapal and G.Govil, Norosa, New Delhi, 1991
R179 N.R.Jagannathan,"^{13}C Shielding Tensors in Organic Molecules", p.280
Magnetic Resonance Microscopy, ed.B.Bluemich and W.Kuhn,VCH, Weinheim,1992
R180 X.Zhou and P.C.Lauterbur,"NMR Microscopy Using Projection Reconstruction", p.3
R181 W.S.Veeman,"High Resolution NMR Imaging of Solids", p.29
R182 D.G.Cory,"Three Dimensional NMR Imaging of Solids", p.49
R183 R.Kimmich,D.E.Demco and S.Hafner,"Slice Selection and Phase Encoding Techniques in Solid State NMR Imaging", p.59
R184 P.Barth, R.Brill, M.staemmler and W.Kuhn,"Relaxation Time Analysis and Parameter Selective Imaging in Materials research", p.85
R185 N.Lugeri, F.De Luca and B.Maraviglia,"Magic Angle Rotating Frame Imaging in Materials", p.103
R186 N.SChuff and P.A.Hornung,"Realization of NMR Microscopy in a Flexible Spectrometer", p.115
R187 M.Ikeya,"Scanning and Computer Tomography ESR Microscopy", p.133
R188 L.J.Berliner,"Applications of EPR Imaging to Materials, Agriculture and Medicine", p.151
R189 B.Bluemich, P.Bluemler, E.Guenther, J.Jansen,G.Schauss and H.W.Spiess,"NMR Imaging of Polymers: Methods and Applications", p.167
R190 J.L.Koenig,"Applications of NMR Imaging to Polymers", p.187
R191 R.A.Komoroski and S.N.Sarkar,"NMR Imaging of Composition and Defects in Industrial Elastomers', P.203
R192 W.Kuhn, E.Koeller and I.Theis,"Investigation of Aging Processes and Cross Linking in Elastomers by Parameter Selective NMR Microscopy", p.217
R193 J.L.Ackerman, L.Garrido, J.R.Moore,B.Pfleiderer and Y.Wu,"Fluid and Solid State MRI of Biological and Nonbiological Ceramics", p.237

R194 B.Tomanek, S.Sagnowski, S.Kwiecinski, A.Jasinski and E.Staszkow,"Magnetic Resonance Imaging of Carbon Implants in the Rabbit Condyle of the Femur", p.261

R195 T.A.Carpenter, L.D.Hall, P.Jezzard, C.J.Wiggins, N.J.Clayden, P.Jackson and N.Walton,"MRI Studies of In-Situ Polymerization Reactions at Elevated Temperatures", p.267

R196 K.Albert,, U.Guenther, M.Ilg, E.Bayer and M.Grossa,"Investigation of Photopolymerization Processes by NMR Imaging", p.277

R197 G.J.Nesbitt, T.W.Fens, J.S. van den Brink and N.Roberts,"Evaluation of Fluid Displacement in Porous Media Using NMR Microscopy", p.287

R198 R.L.Armstrong, A.Tzalmona, M.Menzinger, A.Cross and C.Lemaire,"Imaging the Dynamics of Chemical Waves", p.309

R199 P.T.Callaghan, K.R.Jeffrey and Y.Xia,"Translational Motion Imaging with Pulsed Gradient Spin Echo Methods", p.327

R200 J.Kaerger and H.Pfeifer,"Structure Related Diffusion in Microporous Crystallites", p.349

R201 S.A.Altobelli, A.Caprihan, E.Fukushima and P.D.Majors,"NMR Imaging Studies of Velocity and Concentration Distributions in Flows", p.367

R202 G.C.Chingas, L.Frydman, G.A.Barrall and J.S.Harwood,"Approaches to Flow Statistics, Structure Correlations and 2D Spectroscopy Using Characteristic Functions", p.373

R203 K.P.Datema, J.A. Bolt-Westerhoff, G.J.Nesbitt, P.K.Maarsen, W.Ylstra, P.N.Tutunjian, H.Vinegar and J. Kaerger,"Petrochemical Applications of Pulsed Field Gradient NMR", p.395

R204 M.D.Cockman, F.L.van Nice, C.R.Lindberg and L.W.Jelinski,"Rapid Arterial Flow and Decelaration Immediately Below the Carotid Bifurcation", p.417

R205 R.Bowtell, P.Mansfield, J.C.Sharp, G.D.Brown, M.McJury and P.M.Glover,"NMR Microscopy at 500 MHz", p.427

R206 J.M.Pope,Applications of Chemical Shift Microscopy to Non Invasive Histochemistry of Plant Materials", p.441

R207 V.Sarafis, J.Pope and Y.Sarig,"Some Aspects of NMR Microscopy Applications in Plant Science and Agriculture", p.459

R208 W,E,Hull and K.P.Fichtner,"High Field NMR Imaging of Tumour Models", p.479

R209 G.A.Johnson,H.Benveniste, R.R.Black, G.P.Cofer,S.L.Gewalt, L.W.Hedlund, R.R.Maronpot and S.A.Suddarth,"MR Microscopy of Disease Models", p.501

R210 S.K.Sarkar and R.D.Kapadia,"MRI in Drug Discovery Research", p.513

R211 G.W.Kabalka,G.Q.Cheng and P.Bendel,"^{11}B MRI and MRS", p.533

R212 D.M.Doddrell, I.M.Brereton and S.Crozier,"High Field Localized ^{1}H NMR Spectroscopy", p.547

R213 T.Herrling, N.Groth, K.U.Thiessenhusen, J.Fuchs and U.Ewert,"Spectral Spatial Skin Imaging with Modulated Gradient and Simultaneous Field Scan (MOSS)", p.563

R214 D,Grucker,"Proton Imaging After Dynamic Polarization:, p.573

R215 E.R.Andrew,"Magnetic Resonance Reflections", p.589

Metallothioneins, ed.M.J.Stillman, C.F.Shaw and K.T.Suzuki, VCH, New York,1992

R216 A.H.Robbins and C.D.Stout,"Crystal Structures of Metallothioneins", p.31

Methods in Lignin Chemistry, ed. S.Y.Lin and C.W.Dence,Springer, Berlin, 1992

R217 K.Lundquist,"^{1}H NMR Spectroscopy of Lignin in Solution", p.242

R218 D.Robert,"^{13}C NMR of Lignin in Solution", p.250

Molecular Biology of Atheroscler, ed. M.J.Holpern, Libbey, London, 1992

R219 M.De Buyzene, M.Rosseneu, J.Delanghe, C.Labeur, L.Noens and Y.Benoit,"An Unusual Case of Acquired Bypolipemia and LCAT Deficiency", p.129

Molecular and Cellular Aspects of Basement Membranes, ed.D.H.Rohrbach and R.Timpl,Academic Press, San Diego, 1993

R220 C.L.Haskin, I.L.Cameron and D.H.Rohrbach,"Role of Water of Hydration on the Filtration Function of Proteoglycans of Basement Membranes", p.107

Molecular Conformation and Biological Interactions , ed.P.Balaram and S.Ramaseshan, Indian Institute of Science, Bangalore, 1991

R221 G.C.K.Roberts,"Dynamic Aspects of Protein Specificity", p.401

R222 C.Toniolo and E.Benedetti,"The Fully Extended Polypeptide Conformation", p.511

R223 T.C.Manh,B.Vitoux, R.Demange and T.Marraud,"2D NMR Conformational Analysis of Griselimycin", p.611

National Conference on Advances in Polymer Technology, 1991

R224 T.Narasimhaswamy, N.Thandavarayamurthy, B.S.R.Reddy and S.Rayadurai,"Studies on Beaded Crosslinked Functionalised Phenyl Acrylate Copolymers", p.152

NATO Advanced Study Institute Series, Series A, 1993, vol.246

R225 R.M.A.Knegtel, M.Katahira, J.G.Schilthuis, R.Boelens, D.Eib,P.T. van der Saag and R.Kaptein,"3D Structure of the Human Retinoic Acid Receptor-β DNA Binding Domain", p.153

R226 D.Markman, U.Rao, K.A.Lewis, G.J.Heffron, B.Stec and M.M.Teeter,"Mode of Phospholipid Binding to the Membrane Active Plant Toxin Phoratoxin-A", p.263
Series B, 1993, vol.305

R227 J.Conrad,"Electronic Properties of Alkali Metal Doped C_{60} Fullerene as seen by NMR and ESR and Compared to Graphite Intercalation Compounds", p.243
Series C, 1992, vol.371

R228 P.D.Beer, J.W.Wheeler and C.Moore,"NewApproaches to Anion Coordination Chemistrt, Acyclic Quaternary Polybipyridinium Receptors Designed to Bind Anionic Guest Species", p.105
1992, vol.374

R229 J.J. van der Klink,"Solid Metal Particles Studies by NMR", p.537
1992,vol.379

R230 M.D.Zeidler,"NMR and QNS of Molecular Liquids", p.319

R231 H.D.Burrows,"Structure and Propertiews of Molten Metal Carboxylates", p.333
1992, vol.381

R232 H.Ziffer, Y.Hu and Y.Pu,"Beauveria Sulphuoescens Mediated Oxidation of Dihydroarteniisinin Derivatives", p.361
1993,vol.386

R233 G.A.Webb,"An Overview of Nuclear Shielding Calculations", p.1

R234 T.C.Farrar and J.D.Trudeau,"Experimental and Theoretical Studies of the Chemical Shift Tensors of Pnictides and Fluorophosphates", p.27

R235 H.Eckert, K.Moran, D.Franke and C.Hudalla,"Solid State NMR Chemical Shifts of Chalcogenides and Phosphides", p.49

R236 J.D.Augspurger, C.E.Dykstra, E.Oldfield and J.G.Pearson." Intra- and Inter-molecular Effects on NMR, NQR and IR Spectroscopic Parameters from *Ab Initio* Calculation and Experiment", p.75

R237 C.J.Jameson and A.C. de Dios,"The Nuclear shielding Surface", p.95

R238 A.E.Hansen and T.D.Bouman,"*Ab Initio* Calculation and Analysis of Nuclear Magnetic Shielding Tensors: the LORG and SOLO Approaches", p.117

R239 W.Kutzelnigg, C.v. Wuellen, U.Fleischer, R.Franke and T. v.Mourik," The IGLO Method. Recent Developments", p.141

R240 P.Lazzeretti, M.Malagoli and R.Zanasi,"Electronic Current Density Induced by Magnetic Fields and Magnetic Moments in Molecules", p.163

R241 A.R.Grimmer,"Shielding Tensor Data and Structure: The Bond Related Chemical Shift Concept", p.191

R242 B.L.Phillips, R.J.Kirkpatrick, Y.Xiao and J.G. Thompson,"Investigation of Structural Phase Transitions in Oxide Materials Using High Resolution NMR Spectroscopy", p.203

R243 D.B.Chesnut and C.G.Phung,"A Model Study of Chemical Shielding in a Partially Hydrated Dipeptide", p.221

R244 P.Pulay, J.F.Hinton and K.Wolinski,"Efficient Implementation of the GIAO Method for Magnetic Properties: Theory and Application", p.243

R245 H.Nakatsuji,"Electronic Mechanisms of Metal Chemical Shifts from *Ab Initio* Theory", p.263

R246 J.A.Tossell, " Applications of NMR Shielding Constant Calculations in Minerology and Geochemistry", P.279

R247 R.E.Wasylishen, R.D.Curtis, K.Eichele, M.D.Lumsden, G.H.Penner, W.P.Power and G.Wu,"Applications of Dipolar NMR Spectroscopy in Characterising Nitrogen and Phosphorus Shielding Tensors", p.297

R248 K.W.Zilm and J.C.Duchamp,"Comparisons of Shielding Anisotropies for Different Nuclei and Other Insights into Shielding from an Experimentalists Viewpoint", p.315

R249 J.Geertsen,"An Origin Independent Theory for the Calculation of NMR Shielding Constants", p.335

R250 S.P.A.Sauer and J.Oddershede,"Correlated and Gauge Invariant Calculations of Nuclear Shielding Constants", p.351

R251 `D.M.Grant, J.C.Facelli,D.W.Alderman and M.H.Sherwood,"^{13}C Chemical Shielding Tensors in Sugars", p.367

R252 R.Wolff, C.Vogel and R.Radeglia,"IGLO Calculations of ^{29}Si NMR Chemical Shift Anisotropies in Silicate Models", p.385

R253 W.T.Raynes,"Effects of Isotopic Substitution and Temperature on Nuclear Magnetic Shielding", p.401

R254 R.Dupree, S.C.Kohn, C.M.B.Henderson and A.M.T.Bell,"The Role of NMR Shifts in Structural Studies of Glasses, Ceramics and Minerals", p.421

R255 U.Sternberg,"The Influence of Structure and Geometry on the ^{31}P, ^{29}Si, ^{13}C and ^{1}H Chemical Shifts", p.435

R256 J.Mason,"Nitrogen Shielding Tensors", p.449

R257 M.W.Anderson and J.Klinowski,"^{1}H and ^{13}C Solid State NMR Studies of Catalytic Reactions on Molecular Sieves", p.473

R258 J.H.Iwamiya and G.E.Maciel, "Theoretical Study of Chemical Shielding in Silatrane and its Derivatives", p.495

R259 M.Barfield,"*Ab Initio* IGLO Studies of the Conformational Dependence of the γ Effect in the ^{13}C NMR Spectra of Cyclic Hydrocarbons", p.523

R260 P.D.Ellis, J.D.Odom, A.S.Lipton, Q.Chen and J.M.Gulick, "Gas Phase Measurement and *Ab Initio* Calculations of ^{77}Se and ^{113}Cd Chemical Shifts" p.539

R261 C.J.Jameson,"Overview and Directions for the Future", p.557

1993, vol.392

R262 J.Manna, L.A.Mlinar, R.J.Kulk, R.F.Dallinger,S.J.Geib and M.D.Hopkins,"Methylidyne Complexes", p.75

1993, vol.401

R263 M.Mackowiak and M.Zdanowska-Frzczek,"High Pressure NQR Studies of Hydrogen Bonded Systems", p.299

R264 J.Jonas,"High Pressure NMR Studies of Chemical and Biochemical Systems", p.393

Series H,1992, vol.65

R265 F.J.M.Van de Ven, H,W.van den Hooven, C,W.Hilbers and R.N.H.Konings,"NMR Studies of Lantibiotics: The 3D Structure of Nisin in Aqueous Solution", p.435

1993, vol.76

R266 L.M.G.Heilmeyer, K.Jaquet and K.Karte,"Characterisation of Phosphoproteins by ^{31}P NMR", p.247

Neurotox '91 Papers, ed.I.R.Duce, Elsvier, London,1992

R267 G.C.K.Roberts,"NMR Approaches to Protein Structure and Function", p.223

New Functional Materials, ed.T.Tsuruta, Elsevier, Amsterdam, 1993, vol.B

R268 I.Shimada and Y.Arata,"Stable Isotope Aided NMR Study of Antibody Combining Sites", p.61

Nucleic Acids Symposia Series, 1993,vol.29

R269 S.Yamaguchi, Y.Takai,M.Sawada and K.Tomioka,"Complexation of Nucleotides by Flexible Molecular Tweezers using a π Interaction and an Electrostatic Interaction", p.55

R270 A.Pardi, P.Legault and E.P.Nikonowicz,"Structural Studies of Catalytic RNA's by Heteronuclear Multidimensional NMR", p.117

Nucleic Acid Targeted Drug Design, ed.C.L.Propst and T.J.Perun, Dekker, New York, 1992

R271 A.Wang and H.Robinson,"Structural Studies of Drug-DNA Interactions by X-Ray Crystallography and NMR Spectroscopy", p.17

Papers of the Sixth Annual Symposium of the Protein Society, 1993

R272 S.N.Loh and J.L.Markley,"Measurement of Amide Hydrogen, Deuterium/Hydrogen Fractionation Factors in Proteins by NMR Spectroscopy", p.517

R273 J.Wang, F.O.Sonnichsen, R.Boyko, R.S.Hodges and B.D.Sykes,"Evaluation of NMR Based Structure Determination of Flexible Peptides", p.56
R274 J.Kaufman, L.M.Siegel and L.D.Spicer,"Paramagnetic Proton NMR Methods used in Studying the Hemoprotein Subunit of *E. coli* Sulphite Reductase", p.577
R275 F.Rastinejad, P.Artz and P.Lu," One and Two Dimensional ^{19}F NMR Methods for Proteins and Nucleic Acids", p.585
R276 M.D.Kemple, K.E.Nollet, P.Yuan and F.G.Prendergast,"Active Site Dynamics", p.595
R277 T.S.Harvey and W.F. Van Gunsteren,"The Applications of Chemical Shift Calculations to Protein Structures as Determined by NMR", p.615
R278 M.J.Williams, I.Phan, R.T.Applin,M.Baron and i.D.Campbell,"Towards the Structure of Mosaic Proteins", p.623
Perspectives in Coordination Chemistry, ed.A.F.Williams,C.Floriani and A.E.Merbach, 1992
R279 C.Luchinat, S.Ciurli and F.Copozzi,"Towards an Understanding of the Electronic Structure of Fe_4S_4 High Potential Iron-Sulphur Proteins", p.245
Phospholipids Handbook, ed.G.Cevc, Dekker, New York, 1993
R280 A.Watts,"Magnetic Resonance Studies of Phospholipid-Protein Interactions in Bilayers", p.687
Photosynthetic Reaction Centres, ed.J.Deisenhofer and J.R.Norris, Academic Press, San Diego, 1993, vol.2
R281 A.J.Hoff,"Magnetic Resonance of Bacterial Photosynthetic Reaction Centres", p.331
Physical Chemistry of Food Processes, ed.I.C.Baianu, H.Pessen and T.F.Kumosinski,Van Nostrand, New York, 1993, vol.2
R282 I.C.Baianu and T.F.Kumosinski,"NMR Principles and Applications to the Structure and Hydration of Food Systems with Emphasis on Proteins", p.338
R283 R.L.Dudley,"High Field NMR Instrumentation", p.421
Polyoxides, ed.W.Ando, Wiley, Chichester,1992
R284 B.Plesnicar,"Organic Peroxides", p.479
Proceedings of the Advances in Ion Exchange Meeting '92, 1992
R285 Y.Miyazaki and H.Waki,"^{31}P NMR Spectroscopic Studies on the Complexation Equilibria in the Ion-Exchanger Resin Phase", p.191
Proceedings of the Thirteenth AIRAPT International Conference on High Pressure Science and Technology, 1992
R286 D.Brinkmann,"NMR at Pressures up to 10GPa: Probing Electronic Structures and Diffusion in Alkali Metals", p.857

Proceedings of the Seventh American Physical Society Topics Conference, ed.S.C.Schmidt, North-Holland, Amsterdam, 1992

R287 R.T.Cygan, M.B.Boslaugh and R.J.Kirkpatrick,"NMR Spectroscopy of Experimentally Shocked Silicate Minerals", p.633

Proceedings of the First Biennial International Conference on Frontiers and New Horizons in Amino Acid Research, 1992

R288 K.Wuethrich, T.Szyperski, B.Leiting and G.Otting,"Biosynthetic Pathways of the Common Proteinogenic Amino Acids Investigated by Fractional ^{13}C Labeling and NMR Spectroscopy", p.41

Proceedings of the Seventh International Conference on the Biochemistry and Biophysics of Cytochrome p.450, 1992

R289 V.V.Shumyontseva, M.Y.Pokrovskaya, Y.D.Ivanov, A.A.Izotov, Y.M.Koen and V.Y.Uvorov," Interaction between Organophosphorus Analogues of Biologically Active Compounds and Cytochrome P.450", p.162

Proceedings of the Tenth Coal Testing Conference, 1993

R290 J.T.Riley and S.Burnis,"Rapid Moisture Measurements Using a Magnetic Resonance Method", p.104

Proceedings of the Second International Conferene on Chemical Structure, ed.W.A.Warr, Springer, Berlin,1993

R291 H.Abe, T.Okuyama, K.Yoshida, E.Kouno and S.Sasaki,"A PC Program System for NMR Database Construction", p.15

R292 H.Noeth and E.S.Striedl,"B Base-A Structure Oriented, Numerical, Factual Database for ^{11}B NMR Spectroscopy", p.33

Proceedings of the Fifth International Conference on Advances in Chitin and Chitosan, ed.C.J.Brine, P.A.Sandford and J.P.Zikakis, Elsevier, London,1992

R293 F.Gaili, J.Persson, J.Sugiyama, R.Vuang, S.Tanner and H.Chanzy,"β Chitin From Deep Sea Hydrothermal Vent Worms", p.216

R294 M.Izume, S.Nagae, H.Kawagishi, M.MItsutami and A.Ohtakara," Action Pattern of Bacillus Sp.7.M Chitosanase on Partially N-acetylated Chitosan", p.334

R295 S.Nagae, M.Izume, S.Nishimura, H.Kawagishi and A.Ohtakara,"Preparation of Chitooligosaccharide-Alditols by Hydrogenation with Ruthenium Catalyst", p.378

Proceedings of the Second International Conference and Exhibition of Computer Applications to Materials Molecular Science and Engineering, ed.M.Doyama, North Holland, Amsterdam, 1993

R297 T.Hamada,"Prediction of Macrolactonization Results", p.23

Proceedings of the Eighteenth International Conference on Organic Coating Science and Technology, 1992

R298 D.Stanssens, R.Hermans and H.Wories,"On the Mechanism of the Esterification of a Beta-Hydroxy Alkylamide with a Carboxylic Acid", p.435

Proceedings of an International Conference on Petrol Refining and Petroleum Processes, 1991, vol.3

R299 S.Yin,"Constitutional Analysis and Spectral Identification of the Three Intermediates for the Production of Nylon-12", p.1318

R300 H.Hin, F.Zeng and D.Sun,"Molecular Chain Structure and Crystalline Structure of PB-ET Copolyester", p.1392

Proceedings of the Fourth International Conference on Polyimides, 1993

R301 A.A.Frimer and P.J.Cavano<"Phenylated Polymaleimides as Addition Curing Resins", p.41

R302 T.Takekoshi and J.L.Matson,"Chemistry of PMR-15 Composite Matrix Resin", p.268

R303 G.S.Smith and R.S.Smith,"Methanolysis of 33'44' Benzophenonetetracarboxylic Anhydride", p.281

R304 M.D.Soucek and R.H.Pater,"The Effects of Curing Temperature on Deformation and Fracture Toughness in Thin Films of LaRCRP46", p.388

Proceedings of the Third International Conference on the Role of Formaldehyde in Biological Systems, 1992

R305 G.Blunden, A.Patel, Z.H.Yuan and L.Mathe,"Betaine Distribution in the Labiatae", p.188

Proceedings of the International Conference on Science and Technology of Synthetic Metals, ed.S.Stafstrom, W,R.salaneck, O.Inganas and T.Hjertberg, Elsevier,Amsterdam, 1993

R306 B.Singh, H.K.Sharma,S.P.S.Jadon,M.P.Singh and R.Swarup,"Properties of Si(IV) Conducting Polymeric Crystals", p.618

R307 N.C.Sariciftci,"Overhauser Shift due to Dynamic Nuclear Polarization on the Coduction ESR in Fully Doped Polythiophene at 1.9K", p.624

R308 P.Le Guennec, M.Nechtschein and J.P.Travers,"Thermal Variation of the Spin Susceptibility of Chains of Conducting Polymers", p.630

R309 F.Genoud, M.Nechtschein and C.Santier,"Spin Susceptibility in Polyaniline", p.642

R310 S.M.Long, K.R.Cromack and A.J.Epstein,"High Resolution ESR of Pernigraniline base Solutions", p.648

R311 P.Auban-Senzier, B.Gotschy, A.Farrall, C,Bourbonnais and D.Jerome,"One Dimensional Physics in Organic Conductors $(TMDTDSF)_2X$", p.654

R312 A.Vainrub,I.Heinmaa and E.Lipmaa,"Comparative High Resolution NMR Studies of n- and p-type Doped Polyacetylene Films", p.660

R313 A.M.Kenwright, W.J.Feast, P.Adams, A.J.Milton, A.P.Monkman and B.J.Say,"Solution State NMR Studies of Polyaniline Structure", p.666

R314 P.Le Guennec, J.P.travers and Y.F.Nicolau,"Comparative Study of Spin Dynamics in Polybithiophene and Poly(3-methyl-bithiophene)", p.672

R315 F.L.Pratt, W.Hayes, G.R.Mitchell, B.Rossi, M.S.Kiani, B.D.Malhotra, S.S.Pandey, A.Milton and A.P.Monkman,"Muon Studies of Conducting Polymers", p.677

R316 H.Awano, H.Araki and H.Ohigashi,"ESR Studies of Dynamics of the Phase transition in Wurster's Blue Perchlorate", p.685

R317 P.C.Stein, W.L.Earl and A.Ray,"Two-Dimensional Solid State NMR Studies of Poly(Aniline)", p.702

R318 P.C.Stein and A.Bolobnesi,"NMR Investigation of Doped Poly(3-alkylthiophene)s", p.708

R319 M.Forsyth and M.E.Smith,"Solid State NMR Characterization of polypyrrole", p.714

R320 F.Shimizu, K.Mizoguchi, S.Masubuchi and K.Kume,"Magnetic Resonane Studies of Heavily Doped Polyacetylene", p.720

R321 P.N.Adams,A.P.Monkman and D.C.Apperley,"A Comparison of the molecular Weights of Polyaniline Samples Obtained from Gel Permeation Chromatography and Solid State ^{15}N NMR Spectroscopy", p.725

R322 J.P.Travers,"Polyaniline: Hydration Effects on Spin Dynamics", p.731

Proceedings of the Ninth International Congress on Photosynthesis, ed.N.Murota, Kluwer Dordrecht, 1992, vol.1

R323 J.F.Allen,"Molecular Structural Effects of Protein Phosphorylation in the Regulation of Photosynthesis", p.243

vol.2

R324 Y.Kodera, K.Takura, H.Mino and A.Kawamari,"Pulsed EPR Study of Tyrosine-Z^{+} in Photosystem II", p.57

Vol.3

R325 M.Hebeler, S.Hentrich, A.Mayer, D.Leibfritz and L.H.Grimme,"Phosphate Regulation and Compartmentation in *Chumyclomonas reinhardtii* as Studied by *in vivo* ^{31}P NMR", p.717

Proceedings of the Ninth International Congress on Radiological Research, 1992, vol.2

R326 T.Budinger,"Metabolic Imaging of Tumours using Emission Tomography and NMR", p.841

Proceedings of the Eleventh International Congress on the Theory and Applications of Rheology, 1992, vol.1

R327 G.Chung and J.A.Kornfield,"2D Deuteron Exchange NMR Studies of the Dynamics of Individual Species in Miscible Blends", p.333
Proceedings of the Eighteenth International Symposium on the Controlled Release of Bioactive Materials, 1991

R328 L.J.C.Peachier, J.A.Bouwstra, H.F.Junginger and J.C.Leyte,"Study of Water Mobility in Poly(2-HydroxyethylMethacrylate) Hydrogels by means of the Pulsed Field Gradient NMR Technique", p.136

R329 R.Ek, G.Henriksson, L.Oedberg, H.Nyqvist, S.Bjoerk and G.Dellal,"Self Diffusion Coefficient of Drugs within Cellulose Granules", p.662
Proceedings of the Nineteenth International Symposium on the Controlled Release of Bioactive Materials, 1992

R330 W.Abraham,"Deuterium NMR Study of Structures Formed by Stratum Corneum Lipids *In vitro*", p.135
Proceedings of the Twentieth International Symposium on the Controlled Release of Bioactive Materials, 1993

R331 F.Bezema, E.Maartlin, P.E.H.Roemele, M.A.Salomons,H.J.M. de Groot, J.Brussee, F.Spies and H.E.Bodde,"Evidence for Rapid Isotropic Motion of Ozone in Human Stratum Corneum", p.14

R332 A.R.Rajobi-Siahboomi, R.W.Bowtell, P.Mansfield,M.C.Davies and C.D.Melia,"NMR Microscopy Studies of Swelling and Water Mobility in HPMC Hydrophilic Matrix Systems undergoing Hydration", p.292
Proceedings of the Second International Symposium on Metal Ions in Biology and Medicine, 1992

R333 I.P.Gerothanassis and M Momenteau, "^{17}O NMR Studies of Oxygenated Hemoproteins and Synthetic Model Compounds", p.14

R334 A.Lakavidis, V.Aletras and N.Hadjiliadis, "Ternary Complexes of *cis* and *trans* Platinum II Diammine Dichloride with Amino Acids and Nucleobases", p.59

R335 T.Geyda, B.Henny and J.J.Delpuech,"Complex Formation of Transition Metals with β Alanyl Histamine", p.115
Proceedings of the Fifth International Symposium of the Pakistan-USA Binational Workshop, ed.A.U.Rahman, Harwood, Chur, 1992

R336 B.Zhou,"NMR Assignments and Molecular Modeling", p.413

R337 A.U.Rahman and M.I.Choudhury,"The Use of Modern NMR Techniques in Structure Elucidation of New Natural Products", p.463

Proceedings of the Fourth International Symposium on the Synthesis and Applications of Isotopically Labelled Compounds, ed.E.Buncel and G.W.Kabalka, Elsevier, Amsterdam, 1992

R338 N.E.Mackenzie, P.R.Gooley and D.Zhao,"Structural and Dynamic Differences between Wild Type Rhodobacter Capsulatus Cytochrome C_2 and Point Mutations as Assessed by ^1H-^{15}N HMQC NMR Spectroscopy", p.138

R339 C.O.Meese, M.Eichelbaum, T.Ebner, G.Heinkele, S.Krimmer and P.Fischer,"^{13}C and ^2H NMR as *Ex-vivo* Probe for Monitoring Human Drug Metabolism", p.291

R340 D.W.Boykin,"^{17}O NMR Spectroscopy : Applications in Organic Chemistry", p.343

R341 R.P.Gullapalli, S.Ramaprasad, R.M.Hawk and R.A.Komoroski,"*In vivo* and *In vitro* ^7Li NMR Spectroscopy and Imaging", p.359

R342 G.W.Kabalka, G.Q.Cheng and P.Bendel."*In vivo* ^{11}B MRI in the Rat", p.363

Proceedings of an Intersociety Energy Conversion Conference, 1993

R343 M.Steinbrueck, V.Heinzel, F.Huber, W.Peppler, H.Will and M.Voss,"Investigations of Beta Alumina Solid Electrolytes for Application in AMTEC Cells", p.799

Proceedings of the Second Japanese Symposium on Peptide Chemistry, ed. N.Yanaihara, ESCOM, Leiden, 1993

R344 V.Wray, K.Nokihara and S.Naruse,"Solution Structure of PACAP by ^1H NMR Spectroscopy", p.220

R345 Y.Ohno, M.Segawa, H.Ohishi, M.Doi, M.Inoue and T.Iwashita,"Confformational Characteristics of Dermorphin and Deltorphin II in Aqueous Solution and in Dodecylphosphocholiinemicelles", p.250

R346 N.Kobayashi, S.Endo and E.Munekata,"Conformational Study on the I_gG Binding Domain of Protein G", p.278

R347 S.H.Nakagawa, R.G.Mirmira and H.S.Taper,"Structural Domains and Conformational Adjustments Directing Insulin-Receptor Interactions", p.413

Proceedings of the First Meeting of the Pacific Rim International Conference on Advances in Material Processes, 1993

R348 T.He, J.Zhong, W.Liu, G.Li and M.Ding,"Polyimide Based Composite with better Toughness than PMR-15", p.623

Proceedings of the Thirtysixth OHOLO Conference, 1992

R349 M.Weinstock, M.Razin, I.Ringel, Z.Tashma and M.Charev,"Acetylcholinesterase Inhibition by Novel Carbamates", p.251

Proceedings of the Symposia of the Materials Research Society, 1992, vol.243

R350 R.W.Schwartz, R.A.Assink and T.J.Headley,"Spectroscopic and Microstructural Characterization of Solution Chemistry Effects in PZT Thin Film Processing", p.245

1992, vol.277

R351 G.A.Ozin, C.L.Bowes and M.R.Steele,"Zeolates", p.105

1993, vol.294

R352 B.Little and P.Wagner,"Recent Advances in the Study of Microbiologically Influenced Corosion", p.343

Proceedings of the Seventh Symposium on Salt, 1993, vol.2

R353 G.Siegel, A.Walter, K.Rueckborn, F.Schnalke, E.Buddecke, A.Schmidt and B.Lindman,"Maintenance of the Sodium Ion Distribution in the Arterial Wall", p.395

Proceedings of a Symposium on Solvent Extraction, 1992

R354 T.Sato, S.Ishikawa and K.Sato,"Solvent Extraction of Indium(III) from Hydrochloric Acid Solutions by High Molecular Weight Amines", p.57

Proceedings of the Twentysecond Symposium Superconductivity and Cryoelectronics, 1991

R355 D.Brinkmann,"Static and Dynamic Properties of $YBa_2Cu_4O_8$ as Revealed by Copper NMR and NQR", p.34

Progress in Membrane Biotechnology, ed.J.C.Gomez-Fernandez, D.Chapman and L.Packer, Birkhauser, Basel,1991

R356 D.Chapman and P.I.Haris,"Basic and Technological Studies of biomembrane Components", p.1

Pulsed Magnetic Resonance : NMR, ESR and Optics, ed.D.M.S.Bagguley, Clarendon, Oxford, 1992

R357 D.M.S.Bagguley,"The Echo Phenomenon", p.5

R358 C.P.Slichter, S.E.Barrett, L.Becerra, D.J.Durand, C.A.Klug, J.A.Martindale, C.H.Pennington, J.H.Ross, S.Shore, Z.Wang and D.B.Zax,"Pulsed NMR in Solids- Then and Now", p.54

R359 A.Samoson, B.Q.Sun and A.Pines,"NewAngles in Motional Averaging", p.80

R360 R.R.Ernst,"The Multidimensional Importance of Time Domain Magnetic Resonance", p.95

R361 J.N.S.Evans,"NMR and Enzymes", p.123

R362 J.S.Waugh,"Spin Echoes and Thermodynamics", p.174

R363 M.H.Levitt,"Thermodynamics of Hartmann-Hahn Cross-Polarization", p.184

R364 A.G.Redfield,"On the Possibility of nuclear Self-Spin-Locking in Ferromagnets", p.202

R365 M.Weger,"High-Temperature Superconductivity and an Apparent Breakdown of Causality", p.210

R366 R.Freeman, "Spin Choreography", p.219

R367 R.E.Walstedt,"The Two-Pulse Spin Echo Revisited", p.242

R368 M.S.Conradi, P.A.Fedders and R.E.Norberg,"Deuteron Multiple Echoes", p.260

R369 M.Bloom, C.Morrison, E.Sternin and J.L.Thewalt,"Spin Echoes and the Dynamic Properties of Membranes", p.274

R370 P.Mansfield,"Imaging by NMR", p.317

R371 L.E.Crooks and P.Rothschild,"Clinical Imaging", p.346

R372 H.Thomann and W.B.Mims,"Pulsed Electron-Nuclear Spectroscopy and the Study of Metalloprotein Active Sites", p.362

R373 R.D.Britt and M.P.Klein,"Electron Spin Echo Envelope Modulation Studies of Mixed Valence Manganese Complexes", p.390

R374 A.Henstra and W.T.Weckebach,"Nuclear Orientation via Electron Spin Locking in Si:B", p.411

R375 M.K.Bowman, R.J.Massoth and C.S.Yannoni,"Coherent Raman Beats in Electron Paramagnetic Resonance Spectroscopy", p.423

R376 D.Stehlik and H.M.Vieth,"Time Evolution of Electron-Nuclear Cross-Polarization in r.f. Induced Optical Nuclear Spin Polarization", p.446

R377 M.Mehring, G.Waeckerle and S.Appelt,"Optical Excitation and Detection of Spin Precession", p.478

R378 T.Hashi, Y.Fukuda and M.Tanigawa,"Quantum Beat Echoes", p.492

Recent Advances in Magnetic Properties of Transition Metal Compounds, ed.A.Kotani and N.Suzuki, World Science, Singapore,1993

R379 T.Goto,"Nuclear Magnetic Relaxations in Singlet Ground State and Haldane Gap Systems", p.304

Renal Dispositions in Nephrotoxicity and Xenobiology, ed.M.W.Anders, Academic, San Diego,1993

R380 B.D.Ross and J.C.Bellinger,"NMR: A Tool for the Study of Renal Function and Nephrotoxicity", p.75

Review Program in Quantitative Nondestructive Evaluation, 1992, vol.11A

R381 C.G.Fry, M.J.Lizak and M.S.Conradi,"NMR Imaging of Advanced Composites", p.655

Selected Papers from the Third International Symposium on Guanidino Compounds in Biology and Medicine, 1992

R382 G.Ogimoto, S.Ozawa, T.Maeba, S.Owada and M.Ishida,"^{31}P NMR Study for the Energy Metabolism of Skeletal Muscle in Chronic Hemodialysis Patients", p.281

Twenty Ninth Symposium on Peptide Chemistry, 1992

R383 E.Watanabe, S.Nozaki, M.Akiyama and I.Muramatsu,"CD and ^{1}H NMR Spectra of Steroidogenesis-Activator Polypeptide and its Fragments", p.405

Synthesis, Characterization and Theory of Polymerised Network Gels, ed.S.M.Aharani, Plenum, New York,1992

R384 W,Mormann and B.Brahm,"Novel Polyfunctional Isocyanates for the Synthesis of Model Networks", p.337

Synthesis of Microporous Materials, ed.M.L.Occelli and H.E.Robson, Van Nostrand, New York, 1992

R385 M.L.Occelli, K.Takahama, M.Yokoyama and S.Hirao,"Surface and Cracking Properties of Expanded Clays Dried with a Supercritical Fluid", p.57

Sweet Taste Chemoreceptors, ed.M.Mathlouthi, T.A.Kanters and C.C.Birch, Elsevier, London, 1993

R386 A.Arnoldi, A.Bassoli, L.Martini and E.Ragg,"Progress in Isovanillyl Sweeteners", p.175

Thrombin , ed. L.J.Berliner, Plenum, New York, 1992

R387 F.Ni, K.D.Gibson and H.A.Scheraga,"NMR Studies of Thrombin-Fibrino-Peptide and Thrombin-Hirudin Complexes", p.63

First World Congress on Recent Advances in Cellular and Molecular Biology, 1992, vol.4

R388 C.Ubeaud, B.Valles, A.M.Chauvet-Manges, J.P.Tillement and A.Creval,"Calcium Anti-Ionophoretic Effect of Nicarpidine and Diltiazem", p.183

1992, vol.6

R389 M.Casu, W.A.Gibbons, A.Lai and M.A.Dessi,"^{1}H NMR Study of the Changes in Cellular and Membrane Composition Induced by Carbon Tetrachloride", p.187

(d) REVIEWS IN PERIODICALS

R390 C.Abeygunawardana and A.C.Bush,"Determination of Chemical Structures of Complex Polysaccharides by Heteronuclear NMR Spectroscopy", *Adv. Biophys. Chem.*, 1993,**3**,199

R391 G.E.Adams, J.C.M.Bremner, I.J.Sratford and P.J.Wood,"Can ^{31}P NMR Measurements of Changes in Experimental Tumour Metabolism be Related to Modification of Oxygenation Status?", *BJR Suppl.*, 1992,**24**, 137

R392 S.F.Akber,"Role of Paramagnetic Ions and Water Proton Spin-Lattice Relaxation Times in Biological Systems", *Nuklear Medizin.*,1993,**32**, 52

R393 L.B.Alemanis,"Critical Factors in Obtaining Meaningful Fast MAS NMR Spectra of Non Integral Spin Quadrupolar Nuclei", *Appl. Magn. Reson.*, 1993,**4**, 179

R394 K.Arnold and K.Gawrisch,"Effects of Fusogenic Agents on Membrane Hydration", *Methods Enzymol.*, 1993,**220**, 143

R395 K.Asayama, Y.Kitaoka, H.Nakamura, M.Kyogoku and S.Hiroaka,"NMR Measurements on Cerium and Uranium Compounds", *JJAP Ser.*, 1993,**8**, 205

R396 J.C.Autran,"Protein Analysis of Wheat by Monoclonal Antibodies and NMR", *Mod. Methods Plant Anal.* 1992,**14**, 109

R397 H.Bachelard and R.Badar-Goffer,"NMR Spectroscopy in Neurochemistry",*Neurchem.*, 1993,**61**, 412

R398 L.Banci, L.Bertini, S.Ferretti, C.Luchinat and M.Piccioli," The Structure of Iron-Sulphur Clusters in Proteins as Monitored by NMR, Moessbauer, EPR and Molecular Dynamics", *J. Mol. Struct.*, 1993,**292**, 207

R399 J.Barber, H.F.Cross and J.A.Parkinson, "High Resolution NMR of DNA and Drug-DNA Interactions", *Methods Mol. Biol.*, 1993,**17**, 87

R400 A.R.Bassindale and J.N.Iley,"NMR and ESR of Organosulphur Compounds", *Chem. Sulphur Cont. Funct. Groups*, 1993,245

R401 W.Bauer and P.von R Schleyer,"Recent Results in NMR Spectroscopy of Organolithium Compounds", *Adv. Carbanion. Chem.*, 1992,**1**, 89

R402 J.M.Beale, C.A.Steginsky and B.M.Mamiya,"The Structural Analysis of the Components of Alginic Acid and the Effects of their Metal Ion Complexation as Determined by NMR", *Environ. Sci. Res.*, 1992,**44**, 209

R403 J.D.Bell and P.J.Sadler,"NMR- A Clinical Approach", *Chem. Brit.* 1993,**29**, 597

R404 P.S.Belton,"New Methods for Monitoring Changes in Proteins", *Food Rev. Int.*, 1993,**9**, 551

R405 J.M.Berg,"Zinc Finger Proteins", *Curr. Opin. Struct. Biol.* 1993,**3**, 11

R406 J.C.Bevington,"Structural Irregularities in High Polymers: End Groups Derived from Inhibitors", *Trends Polym. Sci.*, 1993,**1**, 68

R407 S.N.Bhat,"NMR Studies of Local Motions in Fast Protonic Conductors",*Chem. Solid State Mater.*, 1992,**2**, 350

R408 R.Blinc,NMR of Ferroelectric Materials", *J. Mol. Struct.*, 1993,**292**, 269

R409 B.Bluemich and P.Bluemler,"NMR Imaging of Polymer Materials", *Makromol. Chem.* 1993,**194**, 2133

R410 D.Brinkmann,"Probing the Electronic Structure of Yttrium Barium Copper Oxide Superconductors by Cu, O and Ba, NQR and NMR",*Springer Ser. Solid State Sci.,* 1993,**114**, 162

R411 D.Brinkmann and O.Kanert,"NMR Studies of the Quasi One Dimensional Inorganic Complex Salt $K_2Pt(CN)_4Br_{0.3}$ 3.2 H_2O", *Phys. Chem. Mater. Low Dimens. Struct.*,1992,**15**, 229

R412 A.T.Brunger and M.Nilges,"Computational Challenges for Macromolecular Structure Determined by X-Ray Crystallography and Solution NMR Spectroscopy", *Q.Rev. Biophys.*, 1993,**26**, 49

R413 M.Budzynski, J.Sarzynski and R.Wasiewicz,"Application of Hyperfine Interactions to Surface Investigations",*Acta Phys. Pol.,* 1993,**83A**, 643

R414 D.Bullock and K.Moore,"Protein and Fat Determination in Corn",*Mod. Methods Plant Anal.,* 1992,**14**, 181

R415 J.Caro, H.Jobic, M.Bulow, J.Kaerger and B.Zibrowius,"Molecular Mobility Measurements of hydrocarbons by NMR Techniques",*Adv. Catal.*, 1993,**39**, 351

R416 A.F.Casey,"Chiral Discrimination by NMR Spectroscopy", *Trends Anal. Chem.*, 1993,**12**, 185

R417 C.Chachaty,'conformations and Dynamics of Surfactants in Micelles and Liquid Crystals by Nuclear Magnetic Relaxation", *Mol. Eng.*, 1992,**2**, 65

R418 Y.Chen, W.J.Fairbrother and P.E.Wright,"Three Dimensional Structures of the Central Regulatory Proteins of the Bacterial Phosphotransferase System", *J. Cell Biochem.*, 1993,**51**, 75

R419 M.S.Conradi,"Low Temperature NMR Techniques", *Concepts Magn. Reson.,* 1993**5**, 243

R420 I.Dance and G.Lee,"The Chemistry of Cadmium Chalcogenide Clusters", *Spec. Publ, Roy. Soc. Chem.*, 1993,**131**, 87

R421 K.C.Dash,"Mono and Binuclear Dioxomolybdenum (VI) Complexes with Polydentate Schiff Base Ligands", *J. Ind. Chem. Soc.*, 1992,**69**, 501

R422 M.E.Davis and J.B.Higgins,"Zeolites and Molecular Sieves", *Charact. Catal. Mater.*, 1992,129

R423 A.Dell and K.H.Khoo,"Covalent Structure Determination of Glycopolymers", *Curr. Opin. Struct. Biol.,* 1993,**3**, 687

R424 C.E.Dempsey,"Structure and Dynamics of Melittin in Solution and Membranes from Amide Hydrogen Exchange Analysis", *Jerusalem Symp. Quant. Chem. Biochem.*, 1992,**25**, 357

R425 J.W.Dennis and P.Zhuang,"Characterization of O-Glycan Biosynthesis using Aryl-α-D-GalNAC as a Primer *In vivo* ", *Trends Glycosci. Glycotechnol.*, 1993,**5**, 287

R426 C.Dolainsky, T.Koechy, C.Naumann, T.Brumm, S.J.Johnson and T.M.Bayerl,"Structure and Dynamics of Planar and Spherical Supported Phospholipid Bilayers", *Springer Proc. Phys.*, 1992,**66**, 34

R427 M.Duckworth and J.K.Madden,"Determination of Pyruvic Acid in Complex Polysacharides", *Methods Carbohydr. Chem.* , 1993,**9**, 123

R428 G.R.Eaton and S.S.Eaton,"EPR Imaging", *Pract. Spectrosc.*, 1993,**16**, 395

R429 H.Eisenberg,"Combining Experimental Information from Crystal and Solution Studies", *Chemtracts : Biochem. Mol. Biol.*, 1992,**3**, 334

R430 J.Elguero,"The Nature of the Hydrogen Bond and its Influence on the Structure of Inclusion Compounds", *An. Quim.*, 1993,**89**, 43

R431 M.W.Evans,"Review of Optical NMR and ESR", *Speculations Sci. Technol.*, 1993,**16**, 43

R432 A.Favre, P.Laugoa, A.Woisard, P.Clivio and J.L.Fourrey,"Synthesis and photochemistry of Oligonucleotides Containing Thiosubstituted Nucleobases", *Quim. Nova*, 1993,**16**, 304

R433 J.Feeney,"New Dimensions in Biology", *Chem. Br.*, 1993,**29**, 605

R434 S.W.Fesik,"NMR Structure Based Drug Design",*J. Biomol. NMR*, 1993,**3**, 261

R435 S.Forsen, C.Johansson and S.Linse,"Calcium NMR", *Meth. Enzymol.*, 1993,**227**, 107

R436 P.W.Fowler,"Systematics of Fullerenes and Related Clusters", *Philos. Trans. Roy. Soc. London*, 1993,**343A**, 39

R437 R.Freeman,R.K.Haynes, W.A.Loughlin, C.Mitchell and J.V.Stakes,"Synthetic Utilization of Highly Stereoselective Conjugate Addition Reactions of Phosphorus and Sulphur Stabilized Allylic Carbanions", *Pure Appl. Chem.*, 1993, **65**, 647

R438 F.Gaggelli and G.Valensin,"Methods of Single and Double Excitation : Theory and Applications", *Concepts Magn. Reson.*, 1993,**5**, 19

R439 E.A.Gabinski,"Compatible Solutes of Halophilic Eubacteria",*Experientia*, 1993,**49**, 487

R440 Z.Gao and J.C.T.Kwak,"NMR Studies of Interactions between Neutral Polymers and Anionic Surfactants in Aqueous Solution",*Surf. Solution*, 1991,**11**, 261

R441 W.J.Gehring,"Exploring the Homeobox", *Gene*, 1993,**135**, 215

R442 S.J.Glaser,"Developmnet of Made to Measure Multipulse Experiments for NMR of Biological Macromolecules", *GIT Fachz. Lab.*, 1993,**37**, 677

R443 W.Gorecki, P.Donaso and M.Armand,"NMR Spectroscopy in Solid Polymer Electrolytes", *Braz. J. Phys.*, 1992,**22**, 194

R444 J.M.Griffiths and R.G.Griffin,"NMR Methods for Measuring Dipolar Couplings in Rotating Solids", *Anal. Chim. Acta*, 1993,**283**, 1081

R445 A.M.Gronenborn and G.M.Clore,"Protein Structure Determination in Solution using NMR Spectroscopy", *NIDA Res. Monogr.*, 1991,**112**, 78

R446 B.Hall,"NMR Spectroscopy of DNA/RNA Hybrids", *Curr. Opin. Struct. Biol.*, 1993,**3**, 336

R447 B.Halle, P.O.Quist and I.Furo,"Microstructure and Dynamics in Lyotropic Liquid Crystals", *Liq. Cryst.*, 1993,**14**, 227

R448 R.K.Harris,"State of the Art for Solids", *Chem. Br.*, 1993,**29**, 601

R449 K.W.Hart, D.J.Halsall and J.J.Holbrook,"Genetically Inserted Tryptophans in Protein Spectroscopy", *Adv. Spectrosc.*, 1993,**20A**, 195

R450 P.F.Heelis, S.T.Kim, T.Okamura and A.Sancar,"The Photo Repair of Pyrimidine Dimers by DNA Photolyase and Model Systems", *J. Photochem. Photobiol.*, 1993,**17B**, 219

R451 R.W.Hemingway, S.Ohara, J.P.Steynberg, E.V.Brand and D.Ferreira,"C-H HETCOR NMR Studies of Proanthrocyanidins and their Derivatives", *Basic Life Sci.*, 1992,**59**, 321

R452 R.R.Holmes and T.K.Prakasha,"Cyclic Oxyphosphoranes : ^{31}P Chemical Shift Correlations", *Phos. Sulph. Silic. Related Elements,* 1993,**1**, 80

R453 S.W.Homans and T.Rutherford,"Oligosaccharides and Recognition",*Biochem. Soc. Trans.*, 1993,**21**, 449

R454 J.Homer, S.U.Patel and M.J.Howard,"NMR Spectroscopy with Ultrasound", *Spec. Publ. Roy. Soc. Chem.*,1992,**116**, 136

R455 R.F.Howe,"Surface Science in Three Dimensions", *Springer Proc. Phys.*, 1993,**73**, 242

R456 V.J.Hruby, O.Prakash, W.Kazmierski, C.Gehrig and T.O.Matsunaga,"Conformational Analysis of Opioid Receptor-Selective Peptides using NMR and Theoretical Calculations", *NIDA Res. Monogr.*, 1991,**112**,198

R457 J.S.Ingwall and R.G.Weiss,"^{31}P NMR Spectroscopy", *Trends Cardiovasc. Med.*, 1993,**3**, 29

R458 K.Ishihara,"High Pressure Apparatus for the study of Fast Reactions in Solution", *Mem. Sch. Sci. Eng., Waseda Univ.*, 1991,**55**, 307

R459 B.Ittermann, H.Ackermann, E.Diehl, B.Fischer, H.P.Frank and H.J.Soeckmann,"β-Radiation Detected NMR Studies of Point Defects in Semiconductors", *Hyperfine Interact.*, 1993,**79**, 591

R460 J.H.Iwamiya, S.W.Sinton, J.Callahan and G.P.Drobny,"Nondestructive Evaluation of Energetic Materials via NMR Imaging", *Rev. Prog. Quant. Nondestr. Eval.*, 1992,**11A**, 641

R461 H.C.Joao and R.J.P.Williams,"The Anatomy of a Kinase and the Control of Phosphate transfer", *Eur. J. Biochem.*, 1993,**216**, 1

R462 H.Jobic,"Dynamics of Molecules in Zeolites Studied by Neutron Scattering", *Recents Prog. Genie Procedes*, 1991,**5**, 71

R463 J.Jonas,"High Pressure NMR studies of Proteins and Membranes", *Colloq. INSERM*, 1992,**224**, 123

R464 A.A.Jones,"NMR Techniques and Polymer Dynamics", *Polym. Prepr.*, 1992,**33**, 63

R465 C.Jones and B.Mulloy,"Introduction to NMR", *Methods Mol. Biol.*, 1993,**17**, 1

R466 C.Jones and B.Mulloy,"Applications of NMR to Structural Studies on Polysaccharides", *Methods Mol. Biol.*, 1993,**17**, 149

R467 J.Kaerger and H.Pfeifer,"Self Diffusion by NMR Measurements", *Recents Prog. Genie Procedes*, 1991,**5**, 65

R468 T.Kanezaki,"NMR Analysis of Resins in Plastic Products," *Idemitsu Giho*, 1992,**35**, 738

R469 R.Kaptein,"Protein-Nucleic Acid Interactions by NMR", *Curr. Opin. Struct. Biol.*, 1993,**3**, 50

R470 R.A.Kauppinen,^{1}H NMR Spectroscopy Identifies Neural Cell Types : A Promising Step for Neuroimaging?", *Trends Neurosci.*, 1993,**16**, 384

R471 J.Keeler,"Knowing the Unknown", *Chem. Br.*, 1993,**29**, 593

R472 A.P.M.Kentgens, "Off Resonance Nutation NMR Spectroscopy of Half Integer Quadrupolar Nuclei", *J. Magn. Reson.*, 1993,**104A**, 302

R473 J.Klinowski,"Applications of Solid State NMR for the Study of Molecular Sieves", *Anal. Chim. Acta*, 1993,**283**, 929

R474 J.Klinowski,"Solid State NMR Studies of Catalytic reactions on Molecular Sieves",*Top. Inclus. Sci.*, 1992,**4**, 115

R475 Y.A.Knivel, E.V.Vinogradov, A.S.Shashkov, Z.Siodorczyk, A.Rozalski, I.L.Radzejewska and W.Raca," Structural Studies of O-Specific Polysaccharides of Proteins", *J. Carbohyd. Chem.*, 1993,**12**, 379

R476 J.L.Koenig and R.Grinsted,"NMR Imaging of Diffusion in Polymers", *Polym. Prepr.*, 1992,**33**, 763

R477 H.Kolodziez,"^{1}H NMR Studies of Procyanidin Derivatives",*Basic Life Sci.*,1992,**59**, 295

R478 R.A.Komoroski,"Nonmedical Applications of NMR Imaging", *Anal. Chem.*, 1993,**65**, 1068A

R479 J.P.Korb and F.Devraux,"NMR in Gels and Porous Media", *Chem. Solid State Mater.*, 1992,**2**, 432

R480 M.R.laughlin,"^{31}P NMR Spectroscopy in the Study of Mitochondrial Metabolism", *Methods Toxicol.*, 1993,**2**, 146

R481 P.Lauginie, A.Messaoudi and J.Canard,"From Benzene to Fullerenes Through Graphite Intercallation Compounds",*Synth. Meth.* , 1993**56**, 3002

R482 J.T.J.Lecomte and R.C.Matthews,"Unravelling the Mechanism of Protein Folding : New Tricks for an Old Problem", *Protein Eng.*, 1993,**6**,1

R483 W.Levason and D.Pletche,"^{195}Pt NMR Spectroscopy", *Plat. Met. Rev.* , 1993,**37**, 17

R484 L.Y.Lian,"Dynamic and Exchange Processes in Macromolecules studied by NMR", *Methods Mol. Biol.*, 1993,**17**, 149

R485 R.S.H.Liu and L.U.Colmenares,"^{19}F NMR in Studies of Fluorinated Visual Pigment Analogues",*Collloq. INSERM* , 1992,**221**, 251

R486 Z.Liz, H.W.Spiess and J.J.Titman,"Rotor Synchronized MAS 2D Exchange NMR in Solids. Principles and Applications", *Isr. J. Chem.*, 1992,**32**, 145

R487 E.A.C.Lucken,"NQR Studies of Inclusion Compounds",*Topics Inclus. Sci.*, 1992,**4**, 61

R488 G.R.Luckhurst,"Orientational Order in Liquid Crystaks", *Ber. Bunse-Ges. Phys. Chem.*, 1993,**97**, 1169

R489 A.Miijanen, A.Annila, P.Qvintus-Leino and A.Hase,"Use of ^{17}O in the Characterization of Zeolites and Related Structures by NMR", *VTT Tied.*,1993, 1448

R490 A.Makriyannis, D.P.Yang and T.Mavromoustakos,"Combined use of Solid State NMR, Small Angle X-Ray Diffraction and DSC Studies of Cannabinoid -Membrane Interactions", *NIDA Res. Monogr.*, 1991,**112**, 106

R491 D.Malhotra and J.I.Shapiro,"NMR Measurements of Intracellular $_{p}$H", *Concepts Magn. Reson.*, 1993,**5**, 123

R492 X.Marchandise,"NMR and Biomaterials",*Eur. Mater. Res. Soc. Monogr.*, 1992,**3**, 107

R493 L.Margulies, H.Rozen, T.Stern, G.Rytwo, B.Rubin,L.O.Ruzo,S.Nir and L.Cohen,"Photostabilization of pesticides by Clays and Chromophores,"*Arch. Insect Biochem. Physiol.*, 1993,**22**, 467

R494 J.Marko and G.Vermeersch,"Photo CIDNP of Photosensitizing Drugs", *Trends Photochem. Photobiol.*, 1991,**2**,387

R495 G.Martin, G.Remaud and G.J.Martin,"Isotopic Methods for Control of Natural Flavours Authenticity", *Flavour Fragrance J.*, 1993,**8**, 97

R496 G.J.Martin, M.L.Martin and B.L.Zhang,"Site Specific Natural Isotope Fractionation of Hydrogen in Plant Products Studied by NMR",*Plany, Cell Environ.*, 1992,**15**, 1037

R497 R.A.Mashelkar,"Some Excursions in the World of Stimuli-Responsive Polymeric Gels", *J. Indian Inst. Sci.*, 1993,**73**, 193

R498 V.M.Mastikhin,"Characterization of Surface Active Sites of Catalysts with High Resolution Solid State NMR", *Colloids Surf.*, 1993,**78A**, 143

R499 M.A.Mayo, K.M.Brierley and B.A.Goodman,"Developments in the Understanding of the Particle Structure of Tobraviruses", *Biochimie,* 1993,**75**, 639

R500 S.Mazumdar and S.Mitra,"Biomimetic Chemistry of Hemes Inside Aqueous Micelles", *Struct. Bonding,* 1993,**81**, 115

R501 S.F.McCleary,"Pultrusion of PMR 15", *Int. SAMPE Tech. Conf.* , 1991,**23**, 969

R502 M.Mehring,"Introduction to the NMR of High T_C Superconductors", *Springer Ser. Solid State Sci.,* 1993,**114**, 155

R503 D.F.Mierke and H.Kessler,"Improved Molecular Dynamics Simulations for the Determination of Peptide Structures", *Biopolymers*, 1993,**33**, 1003

R504 P.B.Moore,"Recent RNA Structures", *Curr. Opin. Struct. Biol.*, 1993,**3**, 340

R505 D.D.Mota,"Alkali Metal NMR", *Methods Enzymol.*, 1993,**227**, 78

R506 M.A.Navia and D.A.Peattie,"Structure-Based Drug Design", *Trends Pharmacol. Sci.*, 1993,**14**, 189

R507 M.A.Navia and D.A.Peattie,"Structure-Based Drug Design", *Immunol. Today,* 1993,**14**, 296

R508 D.Neuhaus and P.A.Evans,"Structural Studies of Proteins in Solution using ^{1}H NMR",*Methods Mol. Biol.*, 1993,**17**, 15

R509 R.H.Newman and L.J.Porter,"Solid State ^{13}C NMR Studies on Condensed Tannins", *Basic Life Sci.*,1992,**59**, 339

R510 J.L.Nieva and A.Alonso,"On the Mechanism of Phospholipase C-Induced Fusion of Pure Lipid Membranes",*Prog. Membr. Biotechnol.* , 1991, 177

R511 E.Oldfield,"Developments in the NMR Spectroscopy of Lipids and Membranes", *Portland Press Res. Monogr.*, 1992,**3**, 119

R512 G.Otting,"Experimental NMR Techniques for Studies of Protein-Ligand Interactions", *Curr. Opin. Struct. Biol.*, 1993,**3**, 760

R513 G.E.Pake,"NMR in Bulk Matter", *Phys. Today,* 1993,**46**, 46

R514 A.G.Palmer,"Dynamic Properties of Proteins from NMR Spectroscopy", *Curr. Opin. Biotechnol.* 1993,**4**, 385

R515 E.M.Pearce and T.K.Kwei,"Hydrogen Bonding Interactions in Polymer Blends",*Stud. Polym. Sci.* , 1992,**11**, 133

R516 T.Pehk,"The use of 2D NMR Spectroscopy in the Analysis of Isomeric Polyamides", *VTT Symp.*,1992,**133**, 338

R517 J.W.Pritchard,"The Ischemic Penumbra in Stroke : Prospects for Analysis by NMR", *Res. Publ. Assoc. Res. Nerv. Ment. Dis.*, 1993,**71**, 153

R518 R.A.Pike,"Water Based Inorganic Adhesive Primers",*Adhes.'90 Int. Conf.*, 1990, 11

R519 C.M.Preston,"The NMR Users Guide to the Forest", *Canad. J. Appl. Specy.* 1993,**38**, 61

R520 W.S.Price and C.P.Hwang,"Some Recent Developments in NMR Approaches for Studying Liquid Molecular Dynamics and Their Biological Applications", *J. Chin. Chem. Soc.* 1992,**39**, 479

R521 G.K.Radda and D.J.Taylor,"The Study of Bioebergetics *In vivo* Using NMR", *New Comp. Biochem.*, 1992, **23**, 463

R522 S.Ramaprasad, Y.H.Liu, R.K.Pandey, F.Y.Shian and K.M.Smith,"*In vivo* NMR Studies of Deuterium Labeled Photosensitizers in Mice Tumour Models", *Proc. SPIE,* 1993,**1881**, 325

R523 N.F.Ramsey,"Early NMR Experiments", *Phys. Today*, 1993,**46**, 40

R524 J.A.Ripmeester and C.I.Ratcliffe,"Solid State NMR Studies of Host Guest Materials", *Top. Incl. Sci.*, 1992,**4**, 1

R525 K.G.Rice, P.Wu, L.Brand and Y.C.Lee,"Experimental Determination of Oligosaccharide 3D Structure", *Curr. Opin. Struct. Biol.*, 1993,**3**, 669

R526 T.Rizvi,"Pulse Techniques in High Resolution NMR Spectroscopy", *Sci. Int.*, 1992,**4**, 319

R527 J.Ruiz-Cabello and J.S.Cohen."Phospholipid Metabolites as Indicators pf Cancer Cell Function", *NMR Biomed.*, 1992, **5**, 226

R528 J.Ruiz-Cabello and J.S.Cohen," NMR and the Study of the Pathological State in Cells and Tissues", *Int. Rev. Cytol.*, 1993,**145**, 1

R529 U.Schernau, B.Hueser and K.Weber,"Modern Analytical Techniques for Coating and Coating Materials", *ASTM Spec. Tech. Publ.*, 1992,**1119**, 3

R530 F.Scierbaum, S.Radosta, W.Vorweg, V.P.Yuriev, B.B.Braudo and M.L.German,"Characterization of Maltodextrin Gelling by Low Resolution NMR", *Spec. Publ. Roy. Soc. Chem.* 1993,**134**, 278

R531 H.J.Schneider,"NMR Spectroscopy and Molecular Mechanics Calculations in Supramolecular Chemistry", *Rect. Trav. Chim. Pays Bas.,* 1993,**112**, 412

R532 C.A.Scholl,"Nuclear Spin Relaxation and Diffusion", *Diffus. Defect. Data,* 1993,**95A**, 91

R533 A.I.Scott,"Re-routing Nature", *Chem. Brit.,* 1993,**29**, 687

R534 S.A.Sherman and M.E.Johnson,"Derivation of Locally Accurate Spatial Protein Structure from NMR Data", *Prog. Biophys. Mol. Biol.*, 1993,**59**, 285

R535 M.S.Shiao,"Triterpenoid Natural Products in the Fungus *Ganoderma Lucidum* ", *J. Chin. Chem. Soc.* 1992,**39**, 669

R536 R.G.Shulman, A.M.Blamire, D.L.Rothman and G.McCarthy,"NMR Imaging and Spectroscopy of Human Brain Function", *NMR Biomed.*, 1992,**5**, 329

R537 R.L.Silvestri and J.L.Koenig,"Applications of NMR Spectroscopy to Solid Polymers", *Anal. Chim. Acta,* 1993,**283**, 997

R538 K.L.Smith, L.D.Smoot and T.H.Fletcher,"Coal Characteristics, Structure and Reaction Rates", *Coal Sci. Technol.*, 1993,**20**, 131

R539 M.E.Smith,"Applications of ^{27}Al NMR Techniques to Structure Determination in Solids", *Appl. Magn. Reson.*, 1993,**4**, 1

R540 S.O.Smith,"Magic Angle Spinning NMR Methods for Internuclear Distance Measurements", *Curr. Opin. Struct. Biol.*, 1993,**3**, 755

R541 M.L.Snoddy,"The Potential of Process NMR on Flowing Streams", *Spectrosc.(Eugene),* 1993,**8**,41

R542 D.B.Spencer and R.C.Matthews,"Hydrophobic Clustering in Nonnative States of a Protein: Interpretation of Chemical Shifts in the NMR Spectra of Denatured States of Lysozyme", *Biochem. Mol. Biol.*, 1991,**2**, 360

R543 K.Suemer, H.Oja, P.Christjanson and A.Laht," NMR Study of Triazone Compound Formation", *VTT Symp.*, 1992,**133**, 332

R544 J.S.Sun and C.Helene,"Oligonucleoside Directed Triple Helix Formation", *Curr. Opin. Struct. Biol.* , 1993,**3**, 345

R545 K.Suzuki and K.Tohda,"Potentiometric Ion Sensors Based on Natural Carboxylic Polyether Ionophores and their Derivatives", *Trends Anal. Chem.*, 1993,**12**, 287

R546 B.D.Sykes,"Determination of the Conformations of Bound Peptides using NMR Transfered NOE Techniques", *Curr. Opin. Biotechnol.*, 1993,**4**, 392

R547 R.N.E.Thorneley,"An Integrated Interdisciplinary Approach to Understanding the Chemistry of Nitrogenases", *Curr. Plant Sci. Biotech. Agric.* ,1993,**17**, 79

R548 S.R.Thorpe, J.W.Baynes and Z.C.Chroneos,"The Design and Application of Residualizing Labels for Studies of Protein Catabolism", *FASE B J.*, 1993,**7**, 399

R549 E.R.T.Tiekink and G.Winter,"Inorganic Xanthates :A Structural Perspective", *Rev. Inorg. Chem.*, 1992,**12**, 183

R550 A.E.Torda and W.F. van Gunsteren,"Molecular Modeling using NMR Data", *Rev. Comput. Chem.*, 1992,**3**, 143

R551 K.Troev,"Structure of Dialkyl Hydrogen Phosphonates", *Rev. Heteroat. Chem.*, 1993,**8**, 165

R552 B.A.Trofimov and B.A.Shainyan,"Vinyl Sulphides", *Chem. Sulphur Cont. Funct. Groups,* 1993, 659

R553 A.Tsutsumi,"Molecular Dynamics in Biopolymers as Studied by Magnetic Resonance", *Prog. Polym. Sci.*, 1993,**18**, 651

R554 Y.J.Uemura,"MuSR Studies on Superconductivity", *Perspect. Meson Sci.*, 1992, 87

R555 A.S.Ulrich, I.Wallot, M.P.Heyn and A.Watts,"Evidence for a Curved retinal in BR from Solid State ^{2}H NMR", *Colloq. INSERM* , 1992,**221**, 247

R556 W.F. van Gunsteren,"Molecular Dynamics Studies of Proteins",*Curr. Opin. Struct. Biol.*, 1993,**3**, 277

R557 M.van Halbeek," Structural Characterization of Carbohydrate Moieties of Glycoproteins by NMR", *Methods Mol. Biol.*, 1993,**17**, 115

R558 D.Vaux,"The Structure of an Endocytosis Signal", *Trends Cell Biol.*, 1992,**2**, 189

R559 R.Vink,"NMR Characterization of Secondary Mechanisms Following Traumatic Brain Injury", *Mol. Chem. Neuropathol.* ,1993,**18**, 279

R560 G.Wagner,"NMR Relaxation and Protein Mobility", *Curr. Opin. Struct. Biol.*, 1993,**3**, 748

R561 D.A.W. Wendisch,"NMR in Industry", *Appl. Specttrosc. Rev.*, 1993,**28**, 165

R562 M.S.Went and L.W.Jelinski,"Chemical Aspects of NMR Imaging", *Pract. Spectrosc.*, 1993,**16**, 421

R563 W.W. Widmer, P.F.Cancalon and S.Nagy,"Methods for Determining the Adulteration of Citrus Juices", *Trends Food Sci. Technol.*, 1992,**3**, 278

R564 J.R.Williamson,"Guanine Quartets", *Curr. Opin. Struct. Biol.* ,1993,**3**, 357

R565 M.P.Williamson,"Peptide Structure Determination by NMR", *Methods Mol. Biol.*, 1993,**17**, 69

R566 M.P.Williamson,"NMR of Proteins", *Nat. Prod. Rep.*,1993,**10**, 207

R567 C.Wolf, P.J.Quinn and C.Chachaty,"Platelet Activating Factor, a Lipidic Mediator with a Structural Activity in Bilayers", *Prog. Memb. Biotechnol.*, 1991,118

R568 J.R.Wright and L.J.Mathias,"Physical Characterization of Wood and Wood Polymer Composites : An Update", *J. Appl. Polym. Sci.*, 1993,**48**, 2226

R569 W.Xu, H.Wang and S.Yang,"New Developments in the Structural Characterization of Organometallic Complexes by NMR", *Fenzi Cuihua*,1993,**7**, 388

R570 R.Zana,"Methods of Investigation of Mixed Surfactant Systems",*Surfactant Sci. Ser.*, 1992,**46**, 337

(e) **REVIEWS AND BOOKS IN FOREIGN LANGUAGES**

Basque

R571 J.M.Alberdi,"NMR in Organic Solids", *Elhuyar*, 1992,**18**, 94

Chinese

R572 K.Ding, Y.Du and C.Ye,"DANTE Experiments in the High Resolution NMR of Solids", *Bopuxue Zazhi*, 1992,**9**, 313

R573 S,Feng,"Discovery, Preparation, Structure and Properties of Fullerene and its Potential Applications", *Wuli*, 1992,**21**,333

R574 L.A.Gribov,"Expert Systems", *Guangpuxue Yu Guangpu Fenxi*, 1992,**12**, 15

R575 Z.Guo,"Study of the Transformation and Availability of Soil Phosphorus Form with Phosphorus Nuclide", *Turangxue Jinzhan*, 1992,**20**,9

R576 W.Hu, L.Yun and J.Yuan,"Progress in NMR Shielding Effects of Organo-Phosphorus Compounds", *Junshi Yixue Kexueyuan Yuankan*, 1993,**17**, 221

R577 W.Huang, S.Peng and W.Hua,"A New Approach for Drug Design", *Zhangguo Yoake Daxue Xuebao* , 1992,**23**, 317

R578 B.Jiang and J.Yang," Magnetic Properties of C_{60} Compounds", *Huaxue Tongbao*, 1993, 23

R579 H.Jiang and R.Feng,"NMR Investigation of Protein Conformation", *Junshi Yixue Kexueyuan Yuanhan*, 1993,**17**, 152

R580 W.lin and M.Cai,"3D NMR and its Applications",*Huaxue Tongbao*, 1993,57

R581 Y.Qin,"Chemical Shifts in ^{1}H NMR Spectra and the Orientation Effect of Electrophilic Aromatic Substitution",*Huaxue Tongbao*, 1992,22

R582 Z.Shi,"Application of ^{2}H NMR and Isotope Analysis in Food Identification", *Shipin Kexue*, 1993,**157**, 18

R583 P.S.Steyn," Nitrogen Containing Mycotoxins", *Prikl. Biokhim. Mikrobiol.*, 1992,**28**,859

R584 X.Sun, Y.Li and K.Wu,"Synthesis of Large Ring Annulenes and NMR Characterization",*Huaxue Tongbao*, 1993,19

R585 Y.Yang,"^{13}C NMR Study of Structure, Orientation and Molecular Motion of Polymers and their Correlation",*Gongreng Gaofenzi Xuebao*, 1992,**5**, 117

R586 N.Zhang and B.Xie,"Radiation Induced Labeling of Tritiated tracers", *Hejishu*, 1992,**15**, 628

R587 M.Zhou and L.Zhang,"Applications of High resolution ^{13}C NMR in Radiation Chemistry Studies of Polymers", *Fushe Yanjiu Yu Fushe Gangyi Xuebao*, 1992,**10**,193

Dutch

R588 P.Luyten,"NMR Gives a new Look at Disease", *Chem. Mag.*, 1993,171

R589 P.Versloot,"Sulphur Vulcanization of Model Systems", *Kunstst. Rubber*, 1993,21

French

R590 P.Garrigues,"Progress in Molecular Analysis", *Analusis*, 1993,**21**, M24

R591 D.Rouge,"Quality Control of Non Alcoholic Beverages by SNIF NMR", *Bios.*, 1991,**22**,44

German

R592 E.Breitmaier,"From the NMR Spectrum to the Structural Formulae of Organic Compounds", Teubner, Stuttgart, 1992, 261pp

R593 D.Deininger, D.Grzeda and R.Richter,"Applications of Low Resolution ^{1}H NMR in Food Technology and Chemical Analysis",*Wiss. Z. TH Koethen*, 1992,1

R594 Z.Fang,S.Wang,J.Chen,L.Chen and F.Li,"Study of the State of the Methylene Group in Reaction Mixtures for Nitration of Hexamethylenetetramine by NMR", *Int. Ann. Conf. ICT*, 1992,**23**,61

R595 H.Graubaum,"Acylation of Tautomerizable C-Amino Azoles with Amidine or Guanidine Structure", *J. Prakt. Chem.*, 1993,**335**,1

R596 M.Hofman, M.Spraul,R.Streck,I.D.Wilson and A.Rapp,"Liquid Chromatography-NMR Coupling and its Applications", *Labor Praxis*, 1993,**17**,36

R597 J.Kelm,"High Resolution Solid State ^{13}C NMR Spectroscopy of Elastomers, Part 4", *Kautsch Gummi Kunstst.*, 1993,**46**,13

R598 D.Leibfritz, W.Offermann, J.Stelten, F.Koscharek, A.W.H.Jans, W.Kuhn and S.Sohall,"*In vitro* and *In vivo* NMR Spectroscopy", *Nova Acta Leopold*, 1992,**60**, 57

R599 H.Moletta,"NMR and Muon Spin Rotation Methods for Magnetism and Superconductivity Studies", *Twenty Fiurth IFF Ferienkurses*, 1993,71

R600 P.Sohar,"Structural Analysis by NMR", *Nova Acta Leopold*, 1992,**60**, 25

R601 K.Wuethrich,"Structure Determination of Proteins and Nucleic Acids in Solution by means of NMR", *Nova Acta Leopold*, 1992,**60**, 47

Hungarian

R602 P.Hencsei, I.Kovacs and L.Bihatsi,"Silatrane Structure",*Kem. Kozl.*, 1992,**74**, 33

R603 J.Rohonczy,"Total Lineshape Analysis of Dynamic NMR Spectra by an IBM Personal Computer",*Kem. Kozl.*, 1992,**74**, 161

Italian

R604 R.Canese and F.Podo,"*In vivo* NMR Spectroscopy", *Ann. Ist. Super. Sanita*, 1991,**27**, 675

R605 M.J.Glasgow,"Molecular Characterization of Recombinant Proteins", *Biotec (Brescia)*, 1992,**7**, 27

R606 C.Vecchi,"Recent NMR Applications for the Investigation and Characterization of Bitumens", *Riv. Combust.*, 1993,**47**, 117

Japanese

R607 M.Abe,"Development of Inorganic Ion Exchange Materials in View of Advanced Analysis and Assessment", *Zairyo Gijutsu*, 1993,**11**, 84

R608 R.J.Abraham, J.Fisher and P.Loftus,"Introduction to NMR Spectroscopy", Kagaku-Dojin, Kyoto, 1993

R609 T.Akai,"Recent Developments in Solid State NMR and its Applications to Material Science",*Osaka Kagyo Gijutsu Shikensho Kiho*, 1993,**44**, 66

R610 M.Akiba and S.Ito,"Recent Trends in R and D of Thermoplastic Elastomers", *Porima Daijesuto*, 1993,**45**, 17

R611 Y.Arata,"NMR Structural Analyses of Proteins. Scope and Limitations",*Tanpakushitsu Kakusan Koso*, 1993,**38**, 2148

R612 T.Erata,"High Resolution NMR", *Bunseki*, 1993, 264

R613 K.Fukumori,"Deterioration and Properties of Rubber Composites,"*Nippon Gomu Kyokaishi*, 1992,**65**, 445

R614 K.Fukumori,"Principles of Pulsed NMR and its Application to Elastomeric Materials", *Materiaru Raifu*, 1991,**3**, 40

R615 K.Hamada,"Aggregation Behaviour of Azo Dyes Containing Fluorine Atoms in Aqueous Solutions",*Kagaku To Kogyo*, 1993,**67**, 59

R616 H.Hatamizu,"Pharmacokinetic and Metabolic Studies by Using Stable Isotopes", *Yakubutsu Dotai*, 1993,**8**, 379

R617 S,Hayashi and S.Nakata,"Charted Solid NMR of Materials", Kodansha, Tokyo,1993, 190pp
R618 H.Hirota,"Structure Elucidation of Natural Organic Compounds by NMR", *Yuki Gosei Kagaku Kyokaishi*, 1993,**51**, 471
R619 F.Inagaki,"Structural Analysis of Oligosaccharide Chains of Glycolipids by Magnetization Transfer Methods",*Yuki Gosei Kagaku Kyokaishi*, 1993,**51**, 502
R620 F.Inagaki and N.Tate,"Structure Analysis of Sugar Chains by Magnetization Transfer",*Kagaku Zokan (Kyoto)*,1992,**122**, 133
R621 T.Inagaki,"Stable Isotope Aided NMR and Multidimensional NMR Applications to IL-1β", *Saibo*, 1992,**24**, 415
R622 Y.Ishino, H.Maekawa and T.Ohno,"Principles and Applications of Super Conducting FT NMR", *Kagaku To Kogyo*, 1992,**66**, 539
R623 A.Isogai,"General Instrumental Analyses in Pulp and Paper Research. Part II Applications of NMR", *Kami Pa Qikyoshi*, 1992,**46**, 644
R624 Y.Ito and A.Kohno,"Diagnostic Imaging of Adrenal Tumours",*Horumon To Dinsho*, 1993,**41**, 41
R625 T.Kanezaki,"Analysis of the Resin Components of Plastic Products by NMR", *Idemitsu Giho*, 1992,**35**, 738
R626 M.Kanazaki,"Pressure Induced Structural Cahnges in Silicate Glass",*Kobutsugaku Zasshi*, 1992,**21**, 149
R627 K.Kato,"NMR as a Probe for the Study of Diseased State Induced in Mouse Skin", *Fragrence J.*, 1993,**21**, 16
R628 S.Kuroki and I.Ando,"NMR Methods",*Sen'i Gakkaishi*",1993,**49**, 135
R629 H.Kusanagi,"The Variety of Water Structure in Solid Polymers",*Kobunshi*, 1993,**42**, 314
R630 T.Kusumi,"Determination of the Absolute Configuration of Organic Compounds by NMR",*Yuki Gosei Kagaku Kyokaishi*, 1993,**51**, 462
R631 T.Kusumi and I.Ohtani." NMR Chemical Shifts in Structural Formulae", Kodansha, Tokyo, 1993, 188pp
R632 H.Kusuoka, K.Fukuchi and T.Nishimura", *Mol. Med.*, 1993,**30**,1494
R633 H.Maekawa,"Applications of NMR to Silicate Melts",*Nippon Kessho Gakkaishi*, 1992,**34**, 347
R634 Y.Mashima,"Evaluation of Clinical Uses of Human *In vivo* MRS", *Igaku No Ayumi*, 1993,**165**, 889
R635 A.Matsumoto,"High Resolution Solid State NMR", *Kagaku To Kogyo (Osaka)*, 1993,**67**, 507

R636 T.Mori,"Structural Analysis of Isomers of Chemical Products with NMR", *Chemitopiap*, 1993, 12

R637 I.Morishima,"NMR Resolves the Mysteries of Hemoprotein", *Kagaku*, 1993,**48**, 608

R638 E.Nakamura,"Regio and Steriospecific Allylation of Organometallic Reagents",*Yaki Gosei Kagaku Kyokaishi*, 1993,**51**, 985

R639 H.Nakatsuji and M.Sugimoto,"Electronic Mechanisms of Metal NMR Chemical Shifts in Transition Metal Complexes", *Yuki Gosei Kagaku Kenkyusho Koenshu* ,1993,**7**, 109

R640 M.Nomura and S.Murata,"Advances in NMR Spectroscopy and its Applications to Coal Chemistry", *Nippon Enerugi Gakkaishi*, 1992,**71**, 1133

R641 T.Ogata,"ESR Imaging", *Kassei Sanso Furi Rajikaru*, 1992,**3**, 702

R642 Y.Okada,"Study of Cement and Cement Hydrates by ^{29}Si MAS NMR", *Semento Konkurito*, 1992,**548**, 24

R643 S.Okamoto, T.Konishi and T.Nakao,"The Clinical Significance of NMR in HCM", *Igaku No Ayumi*, 1993, 731

R644 T.Okuhara,"Polyoxometalate Chemistry",*Kikan Kagaku Sosetsu*, 1993,**20**, 149

R645 M.Oyabu,"Applications of Field Gradient Pulses in 2D NMR Spectroscopy", *Bunseki*, 1993, 123

R646 H.Saito and A.Naito,"Measuring Distances by High resolution Solid State NMR", *Kagaku*, 1993,**48**, 220

R647 H.Saito, A.Naito and S.Tuzi,"High Resolution Solid State NMR Studies on Proteins", *Seibutsu Butsuri*, 1993,**33**,274

R648 E.Sanuki,MR Spectroscopy", *Nichidai Igaku Zasshi*, 1993,**52**, 587

R649 K.Sato,"MRI", *Kagaku Kogyo*, 1993,**44**,634

R650 Y.Suesishi and N.Nishimura,"High Pressure ESR of Radicals in Solution", *Koatsuryoku No Kagoku To Gijutsu*, 1993,**2**, 139

R651 K.Sugahara,"Investigation of the Sulphated Structure of Heparin", *Yakugaku Kenkyu No Shingo*, 1993,**9**, 83

R652 S,Takeda and T.Atake,"Molecular Motion and Phase Transition in the New Substance C_{60} Fullerene",*Kotai Butsuri*, 1993,**28**, 183

R653 T.Takeuchi,"Method of Determination of Enantiomeric Excess with Chiral Delivering Agents", *Farumashia*, 1993,**29**, 877

R654 H.Tamaoki and Y,Kobayashi,"Structural Analysis of Proteins by NMR",*Saibo Kogaku*, 1993,**12**, 15

R655 T.Tanaka and M.Ikura,"Structural Analysis of Proteins by Multidimensional NMR", *Yuki Gossei Kagaku Kyokaishi*, 1993,**51**, 491

R656 K.Tsukimura,"Importance of Low Temperature Hydrothermal Experiments", *Kobutsugaku Zasshi*, 1992,**21**, 265

R657 T.Watanuki, K.Matsushita and K.Kato,"Response of Antartic Alga to Salt Stress, As Measured by NMR", *Seibutsu Butsuri*, 1993,**33**, 228

R658 A.Yagasaki,"Multinuclear NMR of Polyoxometalates", 1993,**20**, 31

R659 T.Yamada and M.Kikuchi,"Cell Adhesion Mtif Arg-Gly-Asp",*Baisaiensu To Indasutari* , 1993,,**51**, 635

R660 T.Yamamura,"Measuring Methods of Skin Surface Water Content", *Fragrence J.* ,1993,**21**, 35

R661 S.Yamasaki,"Chalcogenide Glass System with Copper", *Kotai Butsuri*, 1992,**27**, 837

R662 T.Yamazaki and K.Nagoyama,"The Progress of Multidimensional NMR", *Bunseki*, 1993, 504

R663 T.Yao,"Computer Software for Protein Engineering", *Kagaku To Sofutowea*, 1993,**15**, 3

R664 I.Yasui," Advances in the Basic Science of Traditional Glasses", *Seramikkusu* , 1993,**28**, 458

R665 K.Yonebayashi,"Modern Instrumental Techniques for Soil Analysis", *Nippon Dojo Hiryogaku* Zasshi, 1993,**64**, 206

R666 T.Yonemoto,"NMR", *Tokyo Joshi Ika Daigaku Zasshi*, 1992,**62**, 1635

Korean

R667 M.W.Chung,"Heavy Hydrogen Isotope Effect on NMR Spectra",*Hwahak Sekye*, 1992,**32**, 1268

R668 J.Liu,"Principles and Applications of NMR Spectroscopy",*Kobunja Kwahak Kwa Kisul*, 1993,**4**, 287

Norwegian

R669 H.Justnes,"NMR in Cement and Concrete Research",*Kjemi*, 1993,**53**, 26

R670 S.Kruse and E.Sletten,"Chemometrics", *Kjemi*, 1993,**53**, 10

Polish

R671 A.Falkowski, P.Latacz and J.Gminski,"Methods Applied in Inositol Phosphates and Measurements",*Ann. Acad. Med. Silensiensis*, 1991,**24**, 41

Romanian

R672 C.Delanu, A.Hirtopeanu, M.T.Caproiu and C.Scotter,"Recent Developments in the EC Research Programs Concerning Applications of NMR Spectroscopy to Food Chemistry",*Rev. Chim.* 1993,**44**, 941

Russian

R673 P.V.Dubovski and L.A.Vasilenko,"Deuterium NMR Studies of Polypeptide Chains and Dynamics in Membranes", *Biol, Membr.*, 1992,**9**, 1189

R674 A.S.Khachaturov and V.P.Ivanova,"Modern Possibilities for NMR Spectroscopy in the Study of Elastomer Structure", *Zh. Vses. Khim. O-Va. im D.I.Mendeleeva*, 1991,**36**, 230

R675 G.A.Kirakosyan,"Multinuclear NMR in the Study of Inorganic Coordination Compounds of Noble Metals", *Koord. Khim.*, 1993,**19**, 507

R676 K.S.Lebedev and I.V.Gritsenko,"Computer Methods for Structure Elucidation by ^{1}H and ^{13}C NMR Spectra Using Data Banks", *Zh. Strukt. Khim.*, 1993,**34**, 51

R677 J.Popelis and E,Lukevics,"NMR Spectra of Furan Derivatives of Group IVA Elements", *Khim. Geterotsikl. Soedin*, 1993, 879

R678 V.D.Shcherbukhin,"Galactomannans of Plants in the Russian Flora",*Prikl. Biokhim. Mikrobiol.*, 1993,**29**, 803

R679 I.Y.Slonim and Y.G.Urman,"Use of NMR in Polymer Chemistry and Technology", *Plast. Massy.*, 1993, 16

Spanish

R680 I.Katime, A.Modoz and J.L.Velada,"Synthesis,Characterization and Polymerization Kinetics of some Itaconic Acid Derivatives",*Third Mem. Simp. Latinoam. Polim.*, 1992, 75

R681 F.Williams,"Unsaturated Polyesters for Powder Paints", *Pint. Acabados. Ind.*, 1993,**35**, 16

2
Theoretical and Physical Aspects of Nuclear Shielding

BY CYNTHIA J. JAMESON

1 Theoretical Aspects of Nuclear Shielding

1.1 General Theory — The nuclear magnetic shielding that we deal with in this chapter, unless otherwise specified, is associated with the linear response of the electrons in a molecule to an external spatially uniform static (i.e., time-independent) magnetic field and the internal perturbation of a nuclear magnetic dipole moment. The non-linear response in the presence of magnetic fields of high intensity (leading to magnetic field dependent chemical shifts) has been reviewed in previous volumes of this series. (Vol. 19) On the other hand, very little is known about the phenomena which could be observed in a non-uniform static magnetic field, which have recently been considered by Lazzeretti and co-workers.[1-3] They point out that the contributions to magnetic properties at a macroscopic scale are very small and hardly measurable, but at the molecular level, it has been shown that there are detectable contributions to the nuclear magnetic shielding: The non-uniformity, over molecular dimensions, of the external magnetic field leads to the well-known "pseudo-contact" shift,[4,5] which can be rationalized via multipole magnetic susceptibilities, according to Buckingham and Stiles.[5] Lazzeretti *et al.* contend that an interpretation of nuclear magnetic shielding in terms of magnetic multipoles of neighboring groups may help in understanding the role played by different domains of the electron distribution in determining NMR chemical shifts. Starting out with the formalism set out by Bloch, they have calculated the nuclear shielding in the presence of a non-uniform magnetic field, in particular, in the presence of a spatially uniform magnetic field gradient, using electronic wavefunctions of increasing accuracy. Test molecules used were H_2O [2] and CH_4 .[3] The magnetic field induced on a nucleus (component along the α direction) by the electrons perturbed by the magnetic field $\mathbf{B}_\beta$ and magnetic field gradient $\mathbf{B}_{\gamma\beta}$ is

$$- \sigma_{\alpha\beta}\, \mathbf{B}_\beta - \sigma_{\alpha,\beta\gamma}\, \mathbf{B}_{\gamma\beta}$$

These authors have calculated the quantities $\sigma_{\alpha,\beta\gamma}$ which they call the magnetic quadrupole contribution to the nuclear magnetic shielding, for the H, O and C nuclei in these molecules.

Recently, the possibility has been raised of observing chemical shifts due to the magnetic field induced at a nucleus by a circularly polarized laser light beam, with tantalizing possible applications to selectively observing nuclei that are spatially close to the optical chromophore in a tumbling chiral molecule.[6] Unfortunately, the possible mechanisms that have been considered for this effect have been shown to lead to chemical shifts that would be too small.[7] A definitive analysis of the induced magnetic field at a nucleus by circularly polarized light and the chemical shifts resulting from this has recently been presented by Buckingham and Parlett.[8] The order of magnitude of this induced field in a right circularly polarized laser beam of 10 W cm^{-2} is estimated to be of the order of 10^{-11} G for protons, giving a shift $\pm 4\times10^{-8}$ Hz. For ^{19}F nuclei the shifts will be of the order of $\pm10^{-5}$ Hz. If the laser is tuned to an optical resonance frequency a resonance enhancement of a factor of 10^3 to 10^6 would be expected. The shifts are reversed by a change in the handedness of the light and depend on the optical frequency, and would be strongly dependent on the local electronic structure.

The shielding tensor components of a nucleus in a molecule immersed in a uniform electric field **F** are given by

$$\sigma_{\alpha\beta} = \sigma_{\alpha\beta}^{(0)} + \sigma_{\alpha\beta\gamma}^{(1)}\mathbf{F}_\gamma + ½\, \sigma_{\alpha\beta\gamma\delta}^{(2)}\mathbf{F}_\gamma\mathbf{F}_\delta + \ldots$$

as given by Buckingham in 1960.[9] In the past few years, there has been a resurgence of interest in the shielding polarizabilities $\sigma_{\alpha\beta\gamma}^{(1)}$ and hyperpolarizabilities $\sigma_{\alpha\beta\gamma\delta}^{(2)}$ which have been reviewed in previous volumes of this series. In the present review period, several important papers dealing with this subject present some very interesting results. The shielding hyperpolarizabilities of the rare gas atoms have been calculated, of which only $\sigma_{xxxx}^{(2)}$ and $\sigma_{yyxx}^{(2)}$ are nonvanishing and unique for an electric field in the x direction, and the mean shielding hyperpolarizability is

$$B_\perp = -(1/6)[\, \sigma_{xxxx}^{(2)} + 2\,\sigma_{yyxx}^{(2)}]$$

The $B_\perp$ values obtained in ppm $/[e/4\pi\varepsilon_0 a_0^2]^2$ were:[10,11]

$B_\perp$	He	Ne	Ar	Kr	Xe
SCF	37.52	312.5	1137	3541	4404
MP2	39.07	402.2	1445	-	-

That is, in every case the rare gas nucleus becomes deshielded when placed in an electric field. There are apparently some contributions from electron correlation, if we use as a guide the difference between the calculations at the SCF level (Coupled Hartree Fock) and the MP2 level. With somewhat smaller basis sets 258.0 and 964.6 ppm au are obtained by Grayson and Raynes for Ne and Ar respectively. In molecules such as N_2, HF, and CO the effect of electron correlation on the $\sigma^{(1)}$ and $\sigma^{(2)}$ are seen to be sufficiently large that higher order than MP2 will be necessary to account for them fully.[12,13] On the other hand H_2 can be well described at the SCF level. The results for shielding polarizabilities and hyperpolarizabilities are even more sensitive to the basis set size than are shieldings in the absence of the electric field. The electric field effects on the shielding of the carbon nucleus in the molecules CH_4, C_2H_6, C_2H_4, C_2H_2, HCN, CH_3F, CH_3CN, and H_2CO have been calculated with large basis set wavefunctions.[14] In choosing basis sets to deal with simultaneous perturbations of a molecule by electric and magnetic fields, it is necessary that the basis be able to produce good values for both the nuclear shielding and the electric polarizabilities. Grayson and Raynes added field derivatives of basis functions to the molecular basis.

Their values for the mean shielding polarizabilities agree with the results obtained by Dykstra *et al.*,[15-18] if the differences in the orientation (of the molecule in the electric field) and in the definition of $B_{\|}$ are taken into account. All the calculations of shielding polarizabilities and hyperpolarizabilities have been carried out with common origin methods. This is true also for the calculations reported in previous years. To the best of our knowledge, a local origin method such as LORG, GIAO, or IGLO has so far never been applied to this problem. The use of local gauge origins may alleviate the problems of basis set dependence.

With the relatively successful application of local origin methods such as IGLO, LORG, and GIAO to the computation of the shielding tensors of nuclei in the first row of the Periodic Table, especially ^{13}C nuclei, in small to medium size molecules, attention has turned to nuclei of the second and third rows and also to the more accurate calculation of shielding in the pathological cases, where it may be assumed that higher level wavefunctions are necessary in order to obtain an adequate description of the shielding tensor. In this review period, different methods have been applied to a common set of molecules CO, N_2, NNO, HCN, F_2 so that the results can be compared with each other. At the correlated level, there are two different strategies for the calculation of σ . One approach , developed by Oddershede and co-workers is based on the polarization propagator ansatz and electron correlation is introduced by means of the second order polarization propagator approximation (SOPPA).[19,20] SOPPA is used with the Hansen and Bouman's LORG[21] method of local gauge origins (second order LORG or SOLO),[22] or with a conventional common gauge origin,[20] or with the truly gauge-origin independent approach of Geertsen[23] which treats the diamagnetic and the paramagnetic part with the same sum-over- states formalism.[24] Another approach is based on analytical derivatives; the shielding is the second derivative of the molecular energy with respect to the simultaneous perturbation of the nuclear magnetic moment and the external magnetic field. Dykstra and Augspurger have used this at the Hartree-Fock level.[16] The analytic derivative approach has been used at the level of second-order many-body perturbation theory otherwise known as second-order Møller-Plesset perturbation theory, MP2. For molecules with only dynamical (weak) electron correlation, MP2 is expected to perform well. Formulations of this in conventional common origin calculations by Bishop and Cybulski,[25] and in the GIAO ansatz by Gauss, so called MP2-GIAO or GIAO-MBPT(2),[26,27] have been used. Another approach is the finite field approach which has been used by Fukui and co-workers at the MP2 level, (so-called FF-MBPT) and also recently at third order Møller-Plesset level (MP3).[28] {A previous version of this method[29,30] has been found inconsistent because σ no longer corresponds to the second derivative of the MBPT(2) energy with respect to the magnetic field and the nuclear moment.}[26] Another approach is to use multiconfigurational SCF wavefunctions (MCSCF) to describe the system in the absence of the external magnetic fields and nuclear magnetic moments, which has been formulated by Helgaker, Jørgensen and co-workers to use gauge-including atomic orbitals (GIAO) which we may call MC-GIAO[31] and formulated by van Wüllen and Kutzelnigg to use individual gauge for localized orbitals (IGLO), which has been called MC-IGLO.[32] The MCSCF response equations are solved for the three components of the magnetic field, using iterative techniques. Finally, sum-over-states perturbation theory combined with Kohn-Sham density functional methodology has been developed with various exchange-correlation potentials for the calculation of shielding tensors.[33]

To compare these various methods of including correlation contributions in the shielding calculations, we consider in Tables 1-4 the molecules to which several of these methods have been applied. For the density functional approach, there is no uncorrelated result corresponding to the CHF value since even the values quoted for the zero order approximation takes into account correlation effects to some extent.[33]

Table 1. *Comparison of shielding values for the CO molecule resulting from calculations at the SCF level (Level 1) and at MP2 or equivalent (Level 2) using the various methods described above.*

Method	Ref.	^{13}C			^{17}O		
		Level 1	Level 2	Diff.	Level 1	Level 2	Diff.
Geertsen GI-RPA/SOPPA	24	-26.14	-21.44	4.7	-88.28	-82.79	5.5
SCF/MP2-GIAO	26	-29.2	11.1	40.3	-95.1	-47.4	47.7
FF-GIAO SCF/MP2	28	-19.8	22.1	41.9	-80.6	-33.5	47.1
CHF/MP2 [a]	25	-25.01	11.55	36.6	-86.77	-44.61	42.2
IGLO/ MC-IGLO	31	-23.4	13.39	36.8	-83.86	-36.66	47.2
SCF-GIAO/ MC-GIAO	31	-23.67	13.18 *8.22* [b]	36.9 *31.9*	-84.25	-36.98 *-38.92* [b]	47.3 *45.3*
DFT	33		9.7			-41.4	
EXPT, with vib. corr.			3.0±0.9 [c]			-40.1 ±17.2 [d]	

[a] These use the conventional common origin at the C or O respectively.
[b] Using what is considered a more balanced CAS rather than full valence CAS.
[c] σ_e from Ref. 34, based on spin-rotation constant for ^{13}C in this molecule. [d] Based on the spin rotation constant for ^{17}O in this molecule[35] and rovibrational corrections.[36]

Third order Møller-Plesset correlation corrections of the shielding have been calculated for N_2, CO, and HCN and for first row hydrides, using modest basis sets in the finite-field many body perturbation theory approach using GIAOs.[28] The results are shown in Table 5. The third order contributions are found to be smaller and opposite in sign to the second order contributions for all nuclei other than hydrogen.

That MP2 calculations generally overestimate the contributions of electron correlation to shielding had been suggested previously.[37] The basis sets used here may not be large enough to say definitely the relative magnitudes of the second and third order corrections for these triply bonded systems, however.

Table 2. *Comparison of shielding values for the N_2 and F_2 molecules resulting from calculations at the SCF level (Level 1) and at MP2 or equivalent (Level 2) using the various methods described above.*

Method	Ref.	N_2			F_2		
		Level 1	Level 2	Diff.	Level 1	Level2	Diff.
Geertsen GI-RPA/SOPPA	24	-109.18	-82.23	27.0			
SCF/MP2-GIAO	26	-128.7	-44.9	83.8			
FF-GIAO SCF/MP2	28	-103.9	-27.2	76.7			
CHF/MP2 [a]	25	-111.54	-39.80	71.7	-170.05	-169.57	0.5
IGLO/ MC-IGLO	31	-109.61	-40.79	68.8	-165.32	-204.32	-39.
SCF-GIAO/ MC-GIAO	31	-110.0	-40.9 *-52.2* [b]	69.1 *57.8*	-167.3	-205.5 *-136.6* [b]	-38.2 *30.7*
DFT	33		-55.5			-202.8	
EXPT w/o vibr. corr.			-61.6 ±0.2 [c]			-232.8 [d]	

[a]These use the conventional common origin at the center of mass. [b] Using an enlarged, more balanced CAS rather than full valence CAS. [c] From gas phase measurements in the zero-pressure limit in the absolute ^{15}N shielding scale based on the NH_3 molecule.[38] [d] From gas phase measurements in the zero-pressure limit converted to absolute shielding based on the HF molecule.[39]

While the isotropic shielding is readily reached even in basis sets of modest size, the experimental anisotropy of the shielding is not so easily reproduced. One such example is ^{31}P in PH_3, the primary reference molecule for the ^{31}P shielding scale. This is one of the simplest molecules of a second row nucleus, so the basis sets that are inadequate for this system are surely inadequate for other molecules containing second row atoms having lone pairs. Earlier calculations of the ^{31}P shielding tensor[40-44] came close to the experimental value for the isotropic shielding but were unable to reproduce the shielding anisotropy obtained from the spin rotation tensor in molecular beam measurements.[47] An extensive basis set and correlation study of ^{31}P in PH_3 has been reported by Pulay and co-workers.[48] The converged values of the isotropic ^{31}P shielding, 582 ppm for SCF and 608 for MP2, were obtained with second

Table 3. *Comparison of shielding values for the HCN molecule resulting from calculations at the SCF level (Level 1) and at MP2 or equivalent (Level 2) using the various methods described above.*

Method	Ref.	^{13}C in HCN			^{15}N in HCN		
		Level 1	Level 2	Diff.	Level 1	Level 2	Diff.
Geertsen GI-RPA/SOPPA	24	69.21	72.13	2.9	-53.23	-35.69	17.5
SCF/MP2-GIAO	26	68.2	87.3	19.1	-56.0	1.0	57.
FF-GIAO SCF/MP2	28	74.3	94.9	20.6	-50.8	3.6	54.4
CHF/MP2 [a]	25	71.07	88.21	17.1	-50.03	1.48	51.5
EXPT w/o vib. corr.			82.1 [b]			-20.4 [c]	

[a]These use the conventional common origin at the C or N respectively. [b]Gas phase measurements in the zero-pressure limit in the ^{13}C absolute shielding scale based on the CO molecule.[34] [c]Gas phase measurements in the ^{15}N absolute shielding scale based on the NH_3 molecule.[38]

Table 4. *Comparison of shielding values for the NNO molecule resulting from calculations at the SCF level (Level 1) and at MP2 or equivalent (Level 2) using the various methods described above.*

Method	Ref.	$^{15}N_{end}$ in NNO			$^{15}N_{cen}$ in NNO			^{17}O		
		Level 1	Level 2	Diff	Level 1	Level 2	Diff	Level 1	Level 2	Diff
LORG/SOLO	45	67	105	38	-21	6	27	169	212	43
SCF/MP2-GIAO	26	51.7	135.1	83.4	-48.2	33.8	82.	149.2	206.2	57.
IGLO/ MC-IGLO	32	64.65	119.43	54.8	-31.97	23.88	55.9	175.73	203.61	27.9
DFT	33		106.9			14.7			186.8	
EXPT w/o vib. corr.			99.5 [a]			11.3 [a]			200.5 ±17.2[b]	

[a] Gas phase measurements in the zero-pressure limit in the ^{15}N absolute shielding scale based on the NH_3 molecule.[38] [b]Gas phase measurements in the absolute shielding scale based on ^{17}O in the CO molecule.[35]

Table 5. *The comparison of second and third order correlation corrections to the shielding with modest basis sets. All values are in ppm.*[28]

	HF	MP2	MP3	Expt.	MP2-HF	MP3-MP2
N_2 [c]	-103.9	-27.2	-57.0	-61.6±0.2 [a]	76.7	-29.8
C in CO [c]	-19.8	22.1	7.4	3.0±0.9 [b]	41.9	-14.7
O in CO [c]	-80.6	-33.5	-54.5	-40.1±17 [b]	47.1	-21.
C in HCN [d]	74.3	94.9	88.0	82.1± 0.9[a]	20.6	-6.9
N in HCN [d]	-50.8	3.6	-21.2	-20.4± 1[a]	54.4	-24.8

[a] Without vibrational corrections. [b] Corrected to the equilibrium distance.
[c] Basis set used was 6-311++G**. [d] Basis set used was 6-311G**.

polarization functions (i.e., *f* orbitals in the basis set for the P atom). Leaving out the *f* orbitals in the SCF and MP2 calculations produces a value of the absolute shielding that is too high by 3 and 5 ppm, and addition of further *d* functions beyond 3 sets contributes 3-4 ppm. At the MP2 level of calculation the correlation contribution to the ^{31}P shielding in PH_3 is 22 - 29 ppm, depending on the basis set used. The anisotropy of the shielding, ($\sigma_{\parallel}$ - $\sigma_{\perp}$), is increased by about 24 ppm with the inclusion of electron correlation. whereas only -4 ppm was found using SOPPA and SOLO. Here again, SOPPA (and SOLO) appear to provide a much smaller correction over the RPA (LORG) level of calculation than is found by other methods such as MP2-GIAO or MC-IGLO. This is consistent with the results shown in Tables 1-4 for other molecules.

Some important conclusions from these papers include the following:

(1) In a direct comparison between GIAO and IGLO at the SCF and MCSCF levels, it has been shown that for smaller basis sets the GIAO method seems superior to IGLO. The GIAO results show better basis set convergence characteristics than the corresponding IGLO results, especially for heavier atoms.[31]

(2) It has been shown that in MCSCF calculations the choice of active space is crucial for an accurate determination of the shieldings. There is a large change in shielding when going from a full valence complete active space (CAS) to an enlarged, more balanced active space. When the active space is enlarged in accordance with the MP2 occupation numbers, large changes are observed in the calculated shieldings.[31]

(3) In the Møller-Plesset expansion, the third order correlation corrections to the shielding are opposite in sign to the second order corrections,[28] and there is some indication that the MP2 may overshoot the true values, in some cases by a large amount.

(4) The basis sets used are of paramount importance to the calculation of accurate shielding values. The basis set dependence has been extensively studied using MCSCF with IGLO and GIAO,[31] using the polarization propagator approach in Geertsen's gauge-invariant method,[24] using MP2-GIAO for the PH_3 molecule, and in the conventional common origin approach.[25] Although not important at the RPA level, second polarization functions (*f* for C, N, O and *d* for H) proved to be important for

correlated calculations.[24] A large set of first and second polarization functions is required to reach the SCF and MP2 basis set convergence limits for the ^{31}P shielding in the PH_3 molecule.[48] This may well be typical for second row atoms with lone pairs. In the gauge-invariant method, the calculation of the diamagnetic part of the shielding by sum-over-states is just as basis-set dependent as the computation of the paramagnetic part. Moreover, some customization of the basis (for example, inclusion of *p* and *d* functions with large exponents) may be necessary to achieve convergence in the diamagnetic part.

(5) SOPPA gives a much smaller correlation contribution (for nuclei other than hydrogen) than the comparable MP2 and multiconfiguration RPA calculations. In comparing with other perturbation theory-based methods, it should be noted that SOPPA includes higher than second order terms.[24]

(6) The small basis sets give a larger correlation correction than one obtains in the large basis sets. Therefore, good agreement with experimental shieldings in small basis sets might be fortuitous, and the correlation correction obtained in a small basis set does not always represent the true correlation correction in the complete basis set limit.[24]

(7) Coupled cluster investigations indicate that orbital relaxation terms are quite important for molecules with triple bonds between atoms with one or more lone pairs. Thus, one would expect difficulties in methods based on a single reference, and the multi-configurational methods could provide better results. The lack of dynamic correlation in the multiconfigurational methods, on the other hand, causes an overestimate of the correlation correction.[24]

It may be too soon to evaluate the methods based on the density functional approach.[33,46] The recent results test different approximations in sum-over-states perturbation theory for the energy difference of the ground and excited states, and use different exchange correlation potentials. Different approximations lead to differences of up to 16 ppm for the shielding of F_2 , and up to 23 ppm in the ^{31}P shielding in the PN molecule. Comparisons of three exchange-correlation potential types using modest basis sets lead to about 11 ppm differences for shielding of ^{19}F in the F_2 molecule. These types of questions are still being investigated. Nevertheless, the touted $O(N^3)$ dependence of the computational time on the number of basis functions, instead of $O(N^5)$ for the MP2 approach, would make for a very efficient procedure to be applied to heavier nuclei, once the methodology is well established. A nice example is shown in Table 6, for ^{77}Se shielding in a few small molecules. In judging the agreement between experiment and calculations in this table, it would be wise to remember the conclusions from basis set dependence studies, that small basis sets can give values that agree fortuitously better with experiment than very large basis sets. Decontraction of basis functions can lead to substantial changes in calculated shieldings.[37] It has been reported that changes of up to 60 ppm (towards more deshielding) were found upon decontraction of the IGLO-II basis set in the case of ^{77}Se shielding in the 3 compounds studied, but not necessarily towards improvement in comparison with experiment.[33]

The ozone molecule was one of the great successes of the MC-IGLO approach, the latter method providing shielding values that are 2000 ppm away from the IGLO results at the SCF level.[51] The density functional method succeeds equally, providing values that are comparable to MC-IGLO, at the same geometry, with the same basis sets, shown in Table 7.[33] The values quoted in the table correspond to the local density approximation used in the other DFT results mentioned in this report, except

Table 6. *^{77}Se shielding values (in ppm) obtained from sum-over-states density functional theory using individual gauges for localized orbitals.*[33]

	Calcd.	Expt.	
		b	*c*
SeH_2	2173.4	2101[a]	2401
$SeHCH_3$	1937.0	1911[d]	2211
$Se(CH_3)_2$	1774.6	1756[d]	2056

[a] $\sigma(^{77}Se)$ in the SeH_2 molecule has been measured in the gas phase in the ^{77}Se absolute shielding scale based on SeF_6(g),[49] (*b*) using non-relativistic diamagnetic shielding for the free atom, or (*c*) using diamagnetic shielding for the free atom with estimated relativistic corrections. [d] The gas phase ^{77}Se chemical shifts between SeH_2 and $SeHCH_3$ and $Se(CH_3)_2$ measured by Ellis *et al.*[50] have been converted to the absolute shielding scale, using the absolute shielding of ^{77}Se in SeH_2.

Table 7. *Results of ^{17}O shielding calculations with electron correlation for the O_3 molecule, in ppm.*

Basis set[e]	Method	central O	terminal O
II	SCF-IGLO[a]	-2928.6	-3054.1
	MC-IGLO[a]	-662.4	-1190.3
	DFT[b]	-742.3	-1229.9
	DFT, zero order[b]	-851.1	-1429.3
III	SCF-IGLO[a]	-2794.3	-2905.8
	MC-IGLO[a]	-663.1	-1176.9
	DFT[b]	-753.0	-1229.9
	DFT, zero order[b]	-856.1	-1420.3
IV	SCF-IGLO[c]	-2730.1	-2816.7
	MC-IGLO[c]	-657.7	-1151.8
	Expt.[d]	-724	-1290

[a] From C. van Wüllen, Ph. D. Thesis, Ruhr-Universität, Bochum, 1992, cited in Ref. 33. [b] From Ref. 33. [c] From Ref. 51. [d] Chemical shifts from Ref. 52 converted to absolute shielding by using $\sigma(^{17}O$, liq. $H_2O) = 307.9$ ppm from Ref. 35. [e] Basis sets are described in Ref. 53.

that we also show the results of the so-called zero-order approximation in this method, which clearly also includes some correlation contributions to the shielding. The very large correlation contributions to the shielding in this molecule make it a good test for a new method of treating correlation in shielding calculations. The experimental value quoted is from liquid phase measurements, and undoubtedly includes intermolecular effects which may be of the order of tens of ppm.

The most commonly used methods to address the gauge problem associated with the calculation of nuclear shielding (and also the magnetizability) GIAO, IGLO, LORG, have been augmented by the new method, individual gauges for atoms in molecules (IGAIM), introduced by Keith and Bader,[54] based on the first order current density distributions, $\boldsymbol{J}^{(1)}(\boldsymbol{r})$. The IGAIM method yields relatively accurate magnetic properties (comparable with the three other methods) because they are determined in terms of the $\boldsymbol{J}^{(1)}(\boldsymbol{r})$ as a sum of separately determined atomic contributions. In this review period, another new method based on $\boldsymbol{J}^{(1)}(\boldsymbol{r})$ has been introduced, which employs a different gauge origin for each point $\boldsymbol{r}$ in real space, that is, performs a separate gauge transformation for each point in space, to calculate the current distribution and the magnetic properties.[55] This new method called CSDGT (continuous set of damped gauge transformations) is based on an empirical expression for the continuous parametric function that chooses the gauge origin for each point in space:

$$\boldsymbol{d}(\boldsymbol{r}) = \boldsymbol{r} - \Sigma_{\mathrm{N}} (\boldsymbol{r} - \boldsymbol{R}_{\mathrm{N}}) \exp[-\alpha_{\mathrm{N}} (\boldsymbol{r} - \boldsymbol{R}_{\mathrm{N}})^4]$$

The use of this expression effectively shifts the gauge origin towards the nucleus N which is nearest the point $\boldsymbol{r}$. The rationale for this choice is as follows: In the region of a given nucleus, the current density is well described with the gauge origin at the nuclear position, which is important for shielding calculations. At a given point in a non-nuclear region of an atom in a molecule, the current density is well described with the gauge origin near the point, as has been demonstrated by these authors in calculations of a property which is insensitive to the current in nuclear regions, the isotropic magnetizability, χ , using $\boldsymbol{d}(\boldsymbol{r}) = \boldsymbol{r}$. The method preserves the advantages of IGAIM in the region of a nucleus, and extends the same quality in the calculated current distribution to all regions of space. The CSDGT method yields first order current density distributions that come closest to obeying the condition of conservation of current over all space which follows from the condition that the divergence vanish everywhere,

$$\nabla \boldsymbol{J}^{(1)}(\boldsymbol{r}) = 0 ,$$

to good numerical accuracy (for example, 10^{-2} to 10^{-3} au or less). Using $\boldsymbol{d}(\boldsymbol{r}) = \boldsymbol{r}$ forces both the exact and the approximate diamagnetic part of $\boldsymbol{J}^{(1)}(\boldsymbol{r})$ to be zero for all $\boldsymbol{r}$, which demands that only the paramagnetic part of the current density must approach the requirement of zero divergence rather than attempting to achieve this result through cancellation of two large terms.[55] The symmetry of the current distribution, preserved in using $\boldsymbol{d}(\boldsymbol{r}) = \boldsymbol{r}$, is guaranteed by assigning equivalent values of the parameter α_{N} for symmetrically equivalent nuclei in the empirical expression above.

The 3-dimensional induced current distributions obtained using CSDGT provide topologically well-behaved displays of the current flow in molecules, which has made possible the topological classification of $\boldsymbol{J}^{(1)}(\boldsymbol{r})$,[56] which is of fundamental interest in itself. An illustrative example is shown in Fig. 1(a) of the current induced in the plane of the nuclei in benzene by a magnetic field applied parallel to the principal symmetry axis, calculated by the CSDGT method described above, with a small basis set. In contrast, when a single gauge origin at the charge centroid is used, the current map produced is shown in Fig. 1(b), which exhibits three-dimensional sources and sinks in the vicinity of each carbon nucleus , thus seriously violating the requirement of zero divergence of $\boldsymbol{J}^{(1)}(\boldsymbol{r})$. In Fig. 1(c), the current is calculated using the IGAIM method, in which one C nucleus serves as the gauge origin. In the region of this carbon nucleus the current density is essentially the same as in Fig. 1(a), both leading to shielding values that are more accurate than that calculated from the Fig. 1(b). On the other hand, the symmetry of the molecule is lost in Fig. 1(c) and this cannot provide accurate values of the magnetizability. Both Fig. 1(b) and 1(c) are physically incorrect descriptions, while Fig. 1(a) possesses the critical points that are predicted for a divergenceless vector field.

From an analysis of the current density plots, such as the one shown here, and also the ones published by Kutzelnigg *et al.*[51] and Lazzeretti *et al.*,[57-59] is it possible to tell whether or not the computed magnetic properties of benzene need to be explained in terms of a ring current? This is subject of a detailed study by Kutzelnigg, Lazzeretti and others.[60] A literal reading of the London ring current model in which the σ electrons do not contribute to the anisotropy of the magnetizability and other such unjustified assumptions, led to the conclusion that the ring current model should not be taken so literally. A detailed comparison of the calculated (IGLO) magnetic properties of benzene and a large number of its isomers led to the following results:[60] The out-of-plane component of the magnetizability is exceptionally high for benzene itself, but for the isomers it is closer to 3 times the value for ethene. The proton shielding tensor has a zz component (z axis is perpendicular to the plane of the molecule) that is about 5 ppm less than the nonconjugated or the conjugated but non-cyclic isomers. The localized MO contributions to σ_{zz} in benzene has a π contribution of -3.17 ppm. This extra deshielding in σ_{zz} for the proton in benzene is taken to be an indication of a ring current effect. That the extra deshielding in σ_{zz} for protons is found also in Hückel annulenes, whereas increased σ_{zz} for protons is found in the anti-Hückel cyclobutadiene, are taken as additional evidence for ring current effects.[60]

The cyclopropene molecule has been of considerable interest as an example of molecular types that may be expected to have substantial antisymmetry in the ^{13}C shielding.[61] The antisymmetric shielding has been measured experimentally for the tetrachlorocyclopropene molecule.[62] Calculations of the shielding in cyclopropene and the tetrachloro derivative by various methods may be compared in Tables 8 and 9.

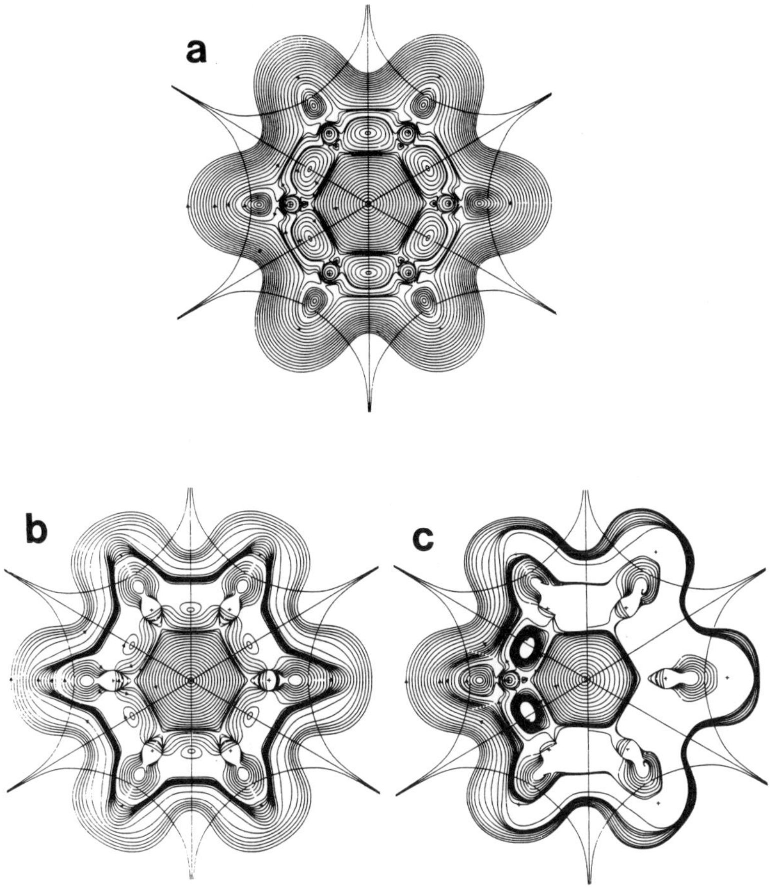

Fig. 1 *(a) Display of the current density induced in the plane of the benzene molecule for a magnetic field applied parallel to the six-fold symmetry axis. The positions of the nuclei are indicated by crosses and the intersections of the interatomic surfaces with the plane are also shown. There is paramagnetic current flow in the ring interior and in the vicinity of each carbon nucleus, on its ring side. This was calculated using the continuous set of damped gauge transformations (CSDGT) method. (b) A similar display calculated using a single gauge origin at the center of symmetry. (c) A similar display calculated with a single gauge origin at the position of a carbon nucleus. This figure is reproduced from Ref. 55 with kind permission of T. A. Keith and R. F. W. Bader and Elsevier Science Publishers.*

Table 8. *The ^{13}C shielding tensor in the olefinic carbon of cyclopropene, C_3H_4, in ppm.*[a]

Method	Basis set	Ref.	σ_{iso} [b]	$\Delta\sigma$ [c]	η [d]	A [e]
IGLO	II[f]	63	73.2	266.0	0.47	104.5
IGLO	III[f]	63	71.4	261.3	0.47	102.6
IGLO	HF limit	63	70.4	262.0	0.46	106.5
MC-IGLO	III[f]	63	96.0	212.0	0.44	84.1
GIAO	97 CGTO	64	73.6	224.4	0.70	-
LORG	6311G**	65	81.4	244.8	0.45	95.3
LORG	[3s3p1d/2s2p]	61	76.7	226.0	0.563	100.4
Expt.	(solid)	66	78.4	196.5	0.565	-

[a] With the convention that $\sigma_{33} \geq \sigma_{22} \geq \sigma_{11}$, the tabulated parameters are defined below:
[b] $\sigma_{iso} = (\sigma_{11} + \sigma_{22} + \sigma_{33})/3$. [c] $\Delta\sigma = (3/2)(\sigma_{iso} - \sigma_{11})$ [d] $\eta = (\sigma_{33} - \sigma_{22})/(\sigma_{iso} - \sigma_{11})$
[e] A or $\xi = (\sigma_{21} - \sigma_{12})/2$ in this case. [f] Basis sets are described in Ref. 53.

The MC-IGLO method has corrected mainly the most deshielding component of the ^{13}C in cyclopropene from -103 to -45 ppm by including pi correlation.[63] The conventional common origin CHF method using basis sets of modest size gave shielding components that are hundreds of ppm away from the HF limit,[65] yet the antisymmetry was relatively stable, within a 25 ppm range throughout all methods and basis set sizes, including the ones shown in Table 8. It has been pointed out by Hansen and Bouman that this striking result is a consequence of the antisymmetry and magnitude of the contributions to the shielding from excitations of A_2 symmetry being truly independent of the gauge choice (being electric dipole forbidden, leads to angular momentum matrix elements that are origin independent).[61] Since the A_2 contribution dominates the xz (in-plane) shielding component and is exactly zero for the zx component, the net antisymmetry, A or $\xi = (\sigma_{xz} - \sigma_{zx})/2$ remains relatively stable throughout gauge origin transformations and to some extent, even major changes in the basis sets used. The same symmetry arguments apply to the tetrachlorocyclopropene derivative. Thus, while each of the principal components of the symmetric part of the shielding tensor in this molecule range over hundreds of ppm depending on method and basis set size,[63,65] the antisymmetry A remains relatively stable within a range of 28 ppm. The sign of the antisymmetry is positive (in Tables 8 and 9) for one of the two olefinic carbons in these molecules, and negative for the other, reflecting the axial or pseudovector nature of the quantity, since the two atoms can be interchanged by a mirror operation in a molecular symmetry plane.

The occasion of these calculations have prompted a clarification in Ref. 63 that the method that has been used in Ref. 65 is not the IGLO method as originally devised by the authors of the method, but is rather a formulation which has been rendered more sensitive to basis set incompleteness by the use of a manipulation that is justified only for a complete basis. It is further noted that to converge to the basis set limit in conventional common origin CHF calculations, steep *d* functions on Cl are necessary whereas these are not necessary for IGLO calculations.

Table 9. *The ^{13}C shielding tensor in the olefinic carbon of tetrachlorocyclopropene, C_3Cl_4, in ppm.*[a]

Method	Basis set	Ref.	σ_{iso}	$\Delta\sigma$	η	*A*
IGLO	II	63	46.0	242.7	0.28	78.8
IGLO	III	63	43.6	236.9	0.27	78.3
IGLO	HF limit	63	42.4	240.0	0.27	79.4
GIAO	6311G** (631G* Cl)	67	64.0	189.0	0.269	71.1
LORG	6311G** (631G* Cl)	65	68.6	184.3	0.26	69.5
Expt.[b]	from T_2/T_1	62	61.8	S = 152±5		51±5

[a] See footnotes to Table 8 and $S^2 = (\Delta\sigma)^2(1+ {}^1/_3\,\eta^2)$ [b] In toluene-d_8 solution.

The use of many basis functions on or in the vicinity of the nucleus whose shielding tensor is desired and fewer functions on centers farther away is a type of mixed basis set approach whereby one emphasizes the spatial region of the "NMR chromophore".[68] Locally dense basis sets have been applied in shielding calculations using the GIAO approach by Chesnut and co-workers.[69] In this review period, detailed comparisons are provided of calculations using locally dense basis sets and balanced basis sets for shielding tensors of hydrogen, carbon, nitrogen, oxygen, fluorine and phosphorus nuclei.[70] It is shown that when chemical functional groups define the appropriate molecular fragment to be taken as locally dense, the shieldings are closer to the results obtained when a balanced set is used for the entire molecule, than if a singly dense basis (only the atom of the observed nucleus has a large number of basis functions) is used for the nucleus whose shielding is being calculated. By selecting the appropriate NMR chromophore, it is possible to investigate systems that would otherwise be inaccessible. Not surprisingly, there is an observed shift of charge toward the NMR chromophore with the higher density of basis functions, and a change in dipole moment may result. Multiple calculations are required if one wants to determine the shielding for many nuclei, that is if more than one NMR chromophore is of interest. If the shieldings of all nuclei in a molecule are desired, then a balanced basis set should be used. The use of locally dense basis functions has been investigated only for the GIAO method where each atomic orbital has its own gauge factor. It has not been tried in the LORG or IGLO approach, where some problems associated with localized molecular orbitals might be anticipated.

1.2 *Ab Initio* Calculations — Various protonated forms usually exhibit different chemical shifts, and the equilibrium concentrations of these various protonated forms determine the pH-dependent chemical shifts. Two examples are ^{31}P shielding in PHO_3^{2-} anion[71] and the ^{13}C and ^{15}N shielding in urea.[72] The *ab initio* calculations on the various protonated forms of urea do not agree very well with the experimentally observed chemical shifts; the ^{15}N shielding calculations were relatively successful with 4 values over a 70 ppm range.

MP2-GIAO calculations have been reported for ^{11}B shielding in boranes and carboranes, where IGLO results at the SCF level could not reproduce experimental chemical shifts. Examples of this are strained systems 1,5-$C_2B_3H_5$ and B_4R_4 (R = CH_3, *t-Bu*, Cl). The large 10-35 ppm discrepancies of the uncorrelated calculations with experiment are resolved when MP2-GIAO is used.[73,74] Remeasurement of the ^{11}B chemical shifts in another molecule, 1,2-$C_2B_3H_7$, led to better agreement with calculations at the uncorrelated level. There is no absolute ^{11}B shielding scale, so all comparisons are for relative shifts only. The *closo*-carborane CB_5H_7 and the isoelectronic $B_6H_7^-$ anion are examples of four-center two-electron bonds. ^{11}B chemical shifts computed at the optimized (MP2/631G*) geometries are found to agree to 1.5 ppm in relative shifts.[75] ^{11}B chemical shifts computed for the Lewis base adducts of BH_3 and small *nido*-boranes B_3H_7 and B_4H_8 with NH_3, CO, PH_3, and PF_3 have been reported for the optimized (MP2/631G*) geometries.[76,77] The solvent effects have of course not been taken into account in the comparisons with solution data. The 10 and 20 ppm difference between the calculated and experimental ^{11}B and ^{15}N shifts in $BH_3{\cdot}NH_3$ are attributed to solvent effects.[77] Two geometric isomers of the iridaborane $(IrB_5H_8)(CO)(PH_3)_2$ have been considered in IGLO calculations of the ^{11}B and ^1H chemical shifts.[78] Qualitative agreement with experimental chemical shifts is reported. It is stated that higher than MP2 level of calculation is required for metallaboranes. *Apropos* what is known in small molecule calculations, these authors should first investigate the basis set problems in such systems. In all of the ^{11}B calculations mentioned here, the experimental chemical shifts are quoted relative to $BF_3{\cdot}OEt_2$, but it is not stated whether the calculated shifts are also relative to the theoretical value for this reference molecule. Similarly, the ^1H shifts are quoted relative to TMS, but it is not stated whether the calculated shieldings are subtracted from theoretical shielding in the TMS molecule.

A large number of ^{13}C shielding calculations have been reported in molecules ranging from onium ions,[79] poly-O-protonated polycations,[80] carbenoid carbons in organo-lithium compounds,[81] phosphanonorbornenes,[82] phenylacetylene,[83] homotropilium cation,[77] cyclobutenyl cation,[84] carbaboranes,[73-75] and a crown ether.[85] These computations used MP2-GIAO,[73,74,79,84] IGLO,[72,75,77,80,81,83] and GIAO[82] methods. What is remarkable about these reports is that the *ab initio* calculations are being carried out on molecules of increasing complexity, invariably comparing with molecules in the solid or in solution, sometimes in very powerful solvents such as superacids, despite the still unanswered questions about shielding in very simple molecules compared to gas phase or molecular beam data, that we have discussed in Section 1.1. The crown ether report[85] is an interesting example of how calculations are helpful even in comparison to solid state data, without correction for intermolecular effects. The ^{13}C NMR spectrum obtained by cross-polarization and magic angle spinning of the C7,10-d_4 -benzo-9-crown-3 ether (which has 10 carbon atoms) shows the expected 10 distinct resonances clearly resolved, with the expected relative intensities based on the X-ray crystallographic structure. In solution there are only five ^{13}C chemical shifts due to conformational averaging. Each of the 5 peaks split up into two, the splittings being 2.3, 8.2, 2.8, 7.9, 2.5 ppm respectively, observed in the CPMAS spectrum. The LORG calculation at the experimental (x-ray diffraction) geometry gives splittings of 2.9, 11, 2.4, 8.2, and 0.1 ppm respectively.[85] The five mean values are 149.9, 123.1, 123.2, 72.1, and 71.6 ppm in the solid, not very different from the conformationally averaged solution peaks, whereas the calculations give 151.45, 135.6, 132.4, 70, and 66 ppm, respectively. The entire spectrum spans 82.8 ppm whereas the predicted spectrum spans 91 ppm. The agreement between calculation and experiment is certainly sufficient to assign every peak. The *ab initio* calculations have been very helpful in this case, and provide a satisfying closure to the experimentalist's observations.

The MC-GIAO method has been applied to the calculation of the ^{15}N shielding in molecules in which the nitrogen is involved in multiple bonds: diazomethane ($H_2C=N=N$), cyanamide ($H_2N-C\equiv N$), carbodiimide (HN=C=NH), and the isomer diazirine (in which C, N, and N form a triangle) and HN=N=N at their experimental geometries.[86] It is expected that there are substantial electron correlation contributions to the ^{15}N shielding in these molecules. Restricted active space SCF wavefunctions were used since a full valence complete active space calculation was considered too large. The results are shown in Table 10. The electron correlation contributions are indeed large for N shielding in these molecules, although they were negligible for the

Table 10. *The nitrogen shielding in small molecules, all in ppm.*[86]

Molecule	Nucleus	SCF-GIAO	MC-GIAO	Expt.	Correl. contrib.
$H-N_a=N_b=N_c$	N_a	196.0	210.3	189.5	14.0
	N_b	-31.3	12.1	-0.9	43.4
	N_c	-29.4	54.4	43.6	83.8
H_2CN_2	N	-275.2^a	-135.2	-87.0	140.0
$H_2C=N_a=N_b$	N_a	-11.6^b	-19.1	-42.0	-7.5
	N_b	-298.8^b	-121.4	-149.0	177.4
$H_2N_a-C\equiv N_b$	N_a	254.3^c	259.1	231.0	4.8
	N_b	28.1^c	71.0	61.0	42.9
HN=C=NH	N	185.5	198.1	142.0	12.6

^{a}The LORG calculation gives -271.45 ppm.[61] ^{b}The IGLO method gives -12.0 and -303 ppm respectively.[87] ^{c}Earlier GIAO calculations give 256.9 and 28.6 ppm, respectively.[88]

^{1}H and ^{13}C shieldings. Some of the discrepancy between experiment and calculation may be due to solvent effects included in the experimental values.

Correlation contributions of the order of 115 ppm were found for the isotropic shielding of ^{15}N in NO_2^- ion in comparing SOLO and LORG calculations using Pople's 6-311G basis augmented with 3 sets of *d* polarization functions.[89] The correlation contributions to the different principal components were not of the same sign; that is, individual tensor components have larger correlation contributions than appears in the isotropic average. Deviations from experiment are rather large however, unless counter ions are included in the computation. The ^{15}N in NO_3^- ion has a smaller correlation contribution, estimated to be about 40 ppm in the isotropic shielding. Surprisingly, in both ions the most shielded component showed large deviations from experiment. The geometry-dependence and the sensitivity of anions to intermolecular effects makes comparison with experiment somewhat difficult.

Since the number of possible isomers increases dramatically with an increase in the number of carbons in higher fullerenes, and since the absence of single crystals does not permit the determination of structure in the usual ways, it would be useful to be able to limit the possible structures on the basis of the solution chemical shifts alone. Semi-direct GIAO calculations of the ^{13}C shielding in the most abundant of the higher fullerenes, C_{84}, is consistent with a 2:1 mixture of two isomers.[90]

Attention has turned to shielding calculations in the second row nuclei ^{27}Al, ^{29}Si, and ^{31}P in recent reports.[42,48,82,91-95] SCF-GIAO and MP2-GIAO methods have been applied to ^{27}Al chemical shifts in fifteen Al(I) compounds spanning about 1000 ppm.[91] There are problems in comparing with experiment since sometimes even the nature of the structures that may be present in solution is not definitely known. It is concluded that both solvation effects and electron correlation have to be included in a study of $AlSiR_3$ compounds.[91] The ^{29}Si and ^{17}O shieldings in model four-ring systems terminated by hydrogens (such as $Si_4O_4H_8$), with counterions ($Si_2Al_2O_4H_8Na_4^{2+}$), and with P (such as $P_2Al_2O_4H_8$) have been calculated as simple models for zeolite and AlPO4-type structures which are of great interest as molecular sieves. These calculations were of the CHF common origin conventional type.[92] ^{29}Si shieldings have been calculated by a more efficient version of Pulay's GIAO code in a variety of compounds for which the experimental values of the absolute shielding have been determined or principal components of the shielding tensor have been reported. Agreement with experiment is very good except in molecules having Si=Si bonds that have biradical character. Higher level calculations will be necessary for these specific systems.[94] ^{31}P shielding in the simplest molecules PH_3, as mentioned in Section 1.1, has been investigated with respect to basis set dependence and electron correlation contributions.[48] In another study the gauge-origin dependence using large polarized basis sets has been carried out for this molecule, and the ions PH_2^- and PH_4^+.[95] Exponents of polarization functions were optimized to maximize the paramagnetic susceptibility. Up to four sets of *d* functions on P and three sets of *p* functions on H were used, and contracted vs. uncontracted sets . Using conventional common-origin CHF calculations at the optimized GAMESS geometry, and at the experimental geometry, these customized basis sets were tested with intrinsic criteria such as sum rules, gauge-invariance conditions and hypervirial theorems. The conclusions are that the degree of origin-independence of the ^{31}P and ^{1}H shielding tensor is fairly good only if very large uncontracted basis sets are used,[95] and even with these very large basis sets, the anisotropy of the P shielding in PH_3 is still too small compared to experiment, when electron correlation is neglected. Phosphorus chemical shifts are very dependent on the structure of the carbon framework in which the P nucleus resides, which aspects have been investigated by GIAO and IGLO calculations in phosphanorbornenes and phosphaalkynes.[42,82] GIAO calculations of all the nuclear shieldings (P, C, and H) in 7-phosphanorbornenes semi-quantitatively reproduce experimental shieldings and such effects as shielding differences of the order of 74 ppm that have been observed between *syn* and *anti* isomers.[82]

The shielding of heavier nuclei received some attention in this review period. For example, ^{73}Ge,[96] ^{77}Se,[97] ^{67}Zn, and ^{113}Cd [98] shieldings were calculated with modest basis set sizes using conventional common-origin methods, and ^{119}Sn shieldings were calculated using DZ or DZ+P basis and the IGLO method.[77] At these levels of calculations, the authors focus on correlations of net atomic charges, ligand substitution, coordination number, and other such quantities with calculated shieldings. For many of the species considered, the actual structures in solution are not well established and there may actually exist several species in fast exchange, resulting in average shifts. In the case of ^{77}Se, the calculated isotropic shieldings in the three molecules shown in Table 6 : 2173.4, 1937.0, 1774.6 ppm [33] for H_2Se, MeHSe, and Me_2Se may be compared with the 2119, 1965, and 1874 ppm respectively calculated by Nakatsuji *et al.*[97]

Proton shieldings have been calculated by IGLO and GIAO methods in model systems for surface hydroxyl groups.[99,100] The largest of the double four-membered silicate ring models studied was $H_8Si_8O_{20}$ in the S_2 and S_6 symmetry with 648 basis functions altogether. Using the semi-direct SCF/GIAO method, the proton shielding obtained was 2.3 ppm referred to TMS by way of a theoretical reference calculation for OH in methanol.[100] This is in good agreement with the 1.8-2.3 ppm shift range observed for terminal silanol groups on silica and zeolites. For C-OH groups, Chauvel and True had found an excellent empirical correlation between the energy of deprotonation as a measure of acidity and the proton chemical shift measured in the dilute gas phase.[101] Calculations of both quantities for 30 molecules containing -XOH groups in which X = H, B, C, N, O, F, Al, Si, P, S, and Cl shows that the correlation is not generally found for X other than C.[99]

1.3 Semiempirical Calculations —With the general availability of efficient software for a*b initio* calculations, calculations that would have been done by semiempirical methods ten years ago are now being done *ab initio*. Only a few papers were found:

FPT-CNDO/INDO has been rediscovered and applied to ^{95}Mo shielding calculations in $[MoO_nS_{4-n}]^{2-}$,[102] the experimental correlation between O-Si-O angle and the anisotropy of the ^{29}Si shielding tensor in silicates is qualitatively reproduced when the CNDO method is used to calculate orbital populations and mean excitation energies,[103] and there is an attempt to relate ^{29}Si chemical shifts in substituted bridgehead polycycloalkanes to Mulliken populations and hybridization of bonds to Si.[104] Shielding of nuclei in large molecular systems are sometimes treated empirically by considering the contributions of atoms not directly bonded to the nucleus in question as if they were contributions from outside the molecule. For example, the Raynes-Buckingham-Bernstein model[105] for intermolecular effects has been applied to the chemical shifts of all C_α protons in 9 proteins such as ubiquitin, BPTI, lysozyme, etc.[106] Ring currents, magnetic anisotropy of the carbonyl group and C'-N bonds, and electric field effects from partial charges assigned to the carbonyl and C'-N are all parametrized (for example each type of ring in the residues is assigned a different intensity parameter) to give a best fit. A somewhat different semiempirical procedure is the SIBFA procedure (sum of interactions between fragments *ab initio* computed) which has been applied to stacked conformers of cyclohexyl-based crotonates and glyoxylates.[107]

2 Physical Aspects of Nuclear Shielding

2.1 Anisotropy of the Shielding Tensor — When comparing two chemical shift tensors (for example those measured in two inequivalent sites, or of the same system in two different laboratories, or for two very similar nuclear sites in related molecules), how does one express how close the two tensors are to each other? Grant and co-workers have examined this problem and have developed a measure of the difference between two tensors by defining the scalar distance between them. Chemical shift tensors are treated as functions whose domain is the surface of the sphere and the mathematical definition of the quadratic distance between two functions is invoked. Expressions for the distance between two chemical shift tensors are developed in the Cartesian representation, in the irreducible spherical representation, and in a new icosahedral representation (wherein the tensor is specified by the shifts when the magnetic field is along 6 directions defined by the vertices of an icosahedron).[108]

Assignments of principal components of the shielding tensor obtained from solid state NMR experiments have generally been assisted by accompanying *ab initio* calculations. It is the full tensor that is much more demanding of accuracy than the isotropic value oberved in solution NMR. Naturally, the intermolecular effects are neglected in making comparisons between (1) a molecule in vacuum for which the calculations are done (usually at some average experimental geometry or an *ab initio* equilibrium geometry) and (2) the actual system surrounded by other molecules and affected by packing effects in the crystal. Correct assignments of tensor components usually do not require that such intermolecular effects be taken into account, nor that the absolute accuracy of the calculated components be better than ±10 or 20 ppm. In a dramatic application of *ab initio* calculations toward the elucidation of the shielding tensor and its dependence on molecular geometry, it has been shown that the departures from D_{2h} symmetry observed in the experimental chemical shift tensor components of naphthalene in the single crystal (leading to the reduced symmetry, C_i) could be accounted for entirely by small geometrical variations that are smaller than the error bars of the x-ray diffraction parameters.[109] This leads to the conclusion that solid state NMR methods can be used for refining structural data, especially in those cases where imperfections such as translational disorder or occlusion of molecular impurities degrade diffraction data while having no effect on chemical shift data. The method of comparison of tensors described above has been used to quantitatively compare model tensors to measured tensors for the 24 different ^{13}C shielding tensors in the two molecules in the unit cell of a single crystal of sucrose.[110] A 3D NMR technique for obtaining the dipolar spectra for individual carbons assisted the assignments.

The shielding tensors associated with the amide fragment in peptides are of great interest because of their potential in the determination of protein secondary structure, conformation, and intramolecular hydrogen bonding. A systematic study of the ^{13}C shielding tensors in 20 amino acids in polycrystalline powder solids has been reported,[111] and also the carboxyl carbons for glycine and oxalate in a single crystal of diglycine oxalate.[112] The ^{13}C static NMR spectra of ^{13}CO-labeled acetanilide exhibit ^{14}N dipolar coupled chemical shift powder line shapes similar to those observed for the amide fragment in peptides.[113] Asymmetric $^{14}N,^{13}C$ dipolar splittings apparent in the δ_{11} region of the powder pattern , symmetrical splittings at δ_{33} and unresolved splittings at δ_{22} in both systems are discussed in detail. Asymmetric dipolar splittings will arise in dipolar-chemical shift powder spectra whenever the dipolar vector is not collinear with one principal component of the shielding tensor of the observed nucleus, provided the dipolar interaction is of comparable magnitude to the chemical shift interaction.

The dipole-CSA cross correlation in spin lattice relaxation provides shielding anisotropies in solution. This method has been used to obtain estimates of anisotropies for ^{13}C shielding in triphenyl methane,[114] and in a pentacyano derivative of pentadienyl anion,[115] as well as for ^{29}Si and ^{119}Sn shielding in Ph_3SiH and Ph_3SnH.[114] The chemical shift anisotropy found for 1,1,3,5,5-penta-1,4-diene-3-ide anion is 370 ppm, much larger than that calculated (IGLO) for the parent $C_5H_7^-$anion. The combination of neutron diffraction and 2D spin diffusion NMR measurements in static powder samples provides an accurate method for positioning shielding tensors in a molecular frame.[116] It is found that the most shielded principal axis of the nearly symmetric ^{13}C tensor in α and β phases of methanol is tilted from the C-O bond in the direction of the hydroxyl group by 7.0°±1.8° and 5.5°± 1.5° respectively. The tilt and the temperature dependence of the structure accounts for apparent differences between NMR and x-ray diffraction results reported earlier. The ^{13}C shielding of carbonyl bonded to metal is not axially symmetric in pentacarbonyl(methyl)rhenium(I) complex[117] and in heme model compounds.[118] This leads to the conclusion that the Fe-C-O unit in heme compounds is bent.

The ^{31}P shielding tensor in dihydrogen phosphate salts are indeed axially symmetric,[119] but strong dipolar couplings can lead to incorrect analysis in MAS NMR spectra of powder samples. Shielding tensors have been determined for ^{31}P nuclei in $Hg(PPh_3)_2(NO_3)_2$,[120] a diphosphirane,[121] phosphoroxidesulfide ($P_4O_6S_n$, n = 0-4),[122] triphenyl phosphite $P(OPh)_3$, and the chromium complex.[123]

Somewhat more accurate shielding tensors for ^{51}V in the metavanadates are reported for (Li, K, NH_4, α- and β-Na)VO_3, and new values for $TlVO_3$ and KV_3O_8.[124] It is interesting to compare these with the density functional theory calculations on $LiVO_3$, KVO_3 and $NaVO_3$ [33] in Table 11. ^{51}V shielding tensors in the divalent metal (Mg, Sr, Ba, Zn, Pb, Ca) vanadates and divanadates have also been reported.[125]

Table 11. *Comparison of ^{51}V chemical shift tensors in vanadates from density functional calculations and experiment, in ppm relative to liquid $VOCl_3$.*

Compound	$\delta_{iso} = (\delta_{11} + \delta_{22} + \delta_{33})/3$		$(3/2)(\delta_{11} - \delta_{iso})$		$\eta = (\delta_{33} - \delta_{22})/(\delta_{iso} - \delta_{11})$	
	Expt.[a]	Calcd.[b]	Expt.[a]	Calcd.[b]	Expt.[a]	Calcd.[b]
$LiVO_3$	-573.4	-615.6	331.5	384.8	0.70	0.694
KVO_3	-552.7		435		0.65	
NH_4VO_3	-569.5		360		0.70	
$TlVO_3$	-529.1		397.5		0.76	
α-$NaVO_3$	-572.7	-623.0	388.5	385.8	0.68	0.666
β-$NaVO_3$	-510.4	-623.0	768	385.8	0.17	0.666
KV_3O_8	-548.1		607.5		0.0	
	-510.0		688.5		0.0	

[a] From Ref. 124, with the convention that $\delta_{33} \le \delta_{22} \le \delta_{11}$, [b] Based on calculations on the $VO_2(OH)^-$ cluster at the V-O geometry of the x-ray structure with the $VOCl_3$ shielding calculated at -1505.4 ppm absolute shielding, Ref. 33.

Shielding tensors for ^{113}Cd in seven Cd complexes were determined for all the sites.[126] ^{133}Cs NMR spectra in a single crystal of Cs_2CrO_4 provide the magnitudes and orientations of the ^{133}Cs shielding and quadrupolar tensors for each of the two crystallographically distinct Cs sites. For each site the principal axis system is oriented along the crystallographic a,b, and c axes.[127] ^{201}Hg and ^{199}Hg NMR spectra in polycrystalline $K_2Hg(CN)_4$ have both been observed. The anisotropy of the Hg shielding in the tetrahedral $Hg(CN)_4^{2+}$ ion is zero. That the electric field gradient at the ^{201}Hg nucleus in this site vanishes allows this first detection of the NMR signal for this nucleus in the solid state.[128] The report of the ^{199}Hg shielding tensor in the Hg_2^{2+} ion in mercurous nitrate dihydrate may be the first instance of a tensor determination of a metallic species in a metal-metal bond. Not surprisingly, the tensor is almost axially symmetric, with a shielding anisotropy of 3203 ppm.[129]

2.2 Rovibrational Averaging and Shielding Surfaces — A correlated calculation of the shielding surfaces for the ^{15}N in ammonia using SOPPA, and rovibrational averaging using nonrigid inverter theory[130] leads to somewhat different thermal average values for the ^{15}N shielding in the NH_3 and ND_3 molecules than was obtained from a LORG surface.[131] The effect of correlation is a small shift and the qualitative behavior of the temperature dependence is the same as that found earlier from a LORG surface, except for a displacement of the whole curve by about 0.91 ppm. However, the rovibrational average calculated in this recent paper gives a much larger contribution to the isotope effect from centrifugal distortion (rotational contributions), leading to a much smaller magnitude of the isotope shift: -0.54 ppm at 293 K. The previous calculation found -2.36 ppm[131] and the experimental value is -1.87 ppm.[132]

The shielding surface of ^{1}H in H_2O has been calculated using the common origin derivative Hartree-Fock method.[133] The results are similar to the earlier calculations on the isolated water molecule.[134] The anisotropy of the ^{1}H shielding shows a significantly greater dependence on the bond length of the other hydrogen than does the isotropic shielding, and the authors find that the deformation of water's structure that occurs in ice is associated with a decrease in anisotropy instead of the experimentally found increase. The increase in anisotropy in going from the isolated water molecule to ice has already been shown to be largely intermolecular in origin, which is well reproduced by a GIAO calculation using a cluster of 17 water molecules.[135]

The ^{19}F shielding surface of FHF^- ion has been calculated using GAMESS and LORG.[136] The change of ^{19}F shielding with bond angle is, of course, symmetrical so that the first derivative is zero at the equilibrium linear configuration. From their figure, it is possible to deduce the second derivative $\partial^2\sigma/\partial\theta^2 = -11\times10^{-3}$ ppm deg^{-2}. The primary shielding change, i.e., the change in the ^{19}F shielding when the bond to it is stretched, is characterized by the derivative $\partial\sigma F_1/\partial r(F_1\text{-}H) = 1450$ ppm/Å, (this is an unusual sign) whereas the secondary shielding change, i.e., the change in the ^{19}F shielding when the other bond is stretched, is characterized by the derivative $\partial\sigma F_1/\partial r(F_2\text{-}H) = -1750$ ppm/Å.

Characterization of the ^{13}C and ^{15}N shielding in the amide environment has been of great interest for some time. In acetanilide and *N*-methylacetanilide, IGLO calculations reproduce the trends in the tensor components but are less successful in determining the absolute magnitudes.[137] Neglect of hydrogen-bonding interactions in the models used in the calculations may be part of the problem. A partial solution is to use point charges or solvent molecules (such as the HF molecule) as part of the model,[139] as in the recent LORG calculations of ^{13}C, ^{15}N, and ^{17}O nuclear shieldings in acetamide and *N*-methylacetamide.[138]

The dependence of the chemical shifts on the dihedral angle has been studied by several workers. The most dramatic results come from the shielding surfaces that are being calculated by de Dios and Oldfield,[139] reported previously in this series, (Vol. 23) but there are other contributions to this subject.[140-144] The ^{13}C chemical shifts of the C_α and C_β of the L-alanine residues in RNaseH in solution and basic pancreatic trypsin inhibitor (BPTI) is simulated by GIAO calculations on the *N*-acetyl-*N'*-methyl-[1-^{13}C]L-alanine model fragment.[140] Unfortunately, the agreement with experiment is rather poor, in part perhaps due to the very small basis set used (4-31G). An IGLO study of the ϕ and ψ dependence of the ^{13}C shieldings in the peptide model *N*-acetyl-*N'*-methylglycinamide[141] using relaxed geometries at each set of angles incorporates, of course, some distance and bond angle dependences into the results as well, so that the resulting surfaces are not the surfaces corresponding to variation in ϕ and ψ alone. In another study, the GIAO method was used at the SCF and the MP2 levels to examine the dependence of the ^{13}C, ^{15}N, and ^{17}O chemical shifts on the dihedral angle in glycine and glycine amide.[145] With the successes of the calculations of Oldfield and co-workers elucidating the ϕ and ψ dependence in proteins, the following questions are posed: Do the SCF-level calculations which had been used constitute a sufficient level of theory, or do correlated calculations at the MP2-GIAO level give different results ? The conclusions are that the absolute shieldings depend strongly on the level of theory at which the geometry and the shieldings are calculated. However, the relative change of the chemical shifts with respect to the dihedral angle ψ is almost the same at all levels of theory.[145] Hence, the authors suggest that in order to obtain a chemical shift surface which has almost MP2 quality, only the shielding of one reference conformation has to be computed at the MP2-GIAO level. The shielding of any other conformation may be obtained by adding the difference between the MP2-GIAO and the SCF-GIAO shielding of that reference to the SCF-GIAO shielding at any conformation.[145] The first detailed study of the effects of varying bond lengths, bond angles, and torsion angles on the ^{13}C shielding of amino acid residues in proteins has been reported. In particular, the ^{13}C shielding of the C^α and C^β sites in model fragments for glycine, alanine, and valine residues in proteins has been calculated using GIAO.[146] The results indicate that ϕ and ψ are the primary determinants of C^α shielding, and that the bond lengths from x-ray structures of proteins are of insufficient accuracy for shielding calculations. The use of shielding surfaces to obtain ϕ and ψ information is proposed.

The ^{13}C shieldings for the unsaturated carbons of *all-trans*-retinal and its analogues have been calculated as a function of the $\phi_{6\text{-}7}$ angle (the $C_5{=}C_6{-}C_7{=}C_8$ dihedral angle)using 4-31G and D95V basis sets and the LORG method.[142] Based on the results, in which the isotropic shielding and two of the shielding components of C_5 and C_8 are expressed in terms of $A + B\cos\phi_{6\text{-}7} + C\cos2\phi_{6\text{-}7}$, the authors assume that the ring-chain conformations of retinals within rhodopsin and bacteriorhodopsin can be estimated from the chemical shift between C_5 and C_8 in these systems. The γ- gauche effect is examined by GIAO calculations in the alkanes from methane to heptane using 4-31G basis sets.[143] The calculated energies were also used to estimate the fractional populations of the conformers and a room temperature average was estimated according to the fractional population weighting. This is, of course, not the proper way to carry out the averaging because the barriers are not so high as to restrict the averaging to the respective bottoms of the conformational wells. A proper torsional average should have been carried out instead. IGLO calculations of ^{13}C chemical shifts at optimized geometries of cyclobutane, substituted cyclobutanes and averaging over the puckering potential lead to semiquantitative agreement with experimental chemical shifts in many cases, but not for the fluoro-substituted ones and the CN carbons.[147] Pucker angle averaging is a relatively small effect, leading to corrections of about 1 ppm at most. Also included were bicyclo[1.1.1]pentanes with the same substituents, and γ- effects for the ring carbons were examined.

The γ- gauche effect in a tetrameric unit of polyisobutylene has been found in the *ab initio* studies using the semi-direct IGLO method in an investigation designed to explain the observed spread of chemical shifts observed in the CH_3 and CH_2 resonances.[144] A six-state rotational isomeric state model was used and for the selected 49 conformations the rest of the molecule were put through geometry optimization using empirical force fields. The shape of the envelope of individual conformer shifts weighted by Monte Carlo-derived conformational statistics resembles the experimental band shapes remarkably well.

2.3 Isotope Shifts —As in previous chapters of this series, the deuterium-induced chemical shifts are denoted by $^{n}\Delta X(^{2/1}H)$ for example, for n bonds separating the substituted atom and the observed nucleus X, that is, consistent with the IUPAC convention for chemical shifts,

$$^{n}\Delta X(^{2/1}H) = \delta^{X}\text{ (deuterio-compound) - }\delta^{X}\text{ (protio-compound)}$$

This is usually a negative chemical shift, that is, the heavy isotopomer is downfrequency (upfield) of the light isotopomer. Where substitution of more than one atom of a equivalent set of atoms is possible, such as in $^{1}\Delta N(^{2/1}H)$ for $^{15}ND_nH_{3-n}$ the isotope shifts are additive or very nearly so. The additivity has been derived theoretically by Jameson and Osten,[148] who also noted that small deviations from additivity can arise due to a contribution proportional to n^2 (i.e., nonlinear in the number of deuteriums) with the proportionality constant involving secondary geometrical isotope effects, leading to a prediction of the ratio of the deviations from additivity 0 : 3 : 4 : 3 : 0 for deuterium substitution at CH_4 molecular types such as $^{119}SnD_nH_{4-n}$, $^{15}ND_nH_{4-n}{}^{+}$ ion, where, indeed these ratios have been observed.[149,150] Extension to NH_3 molecular types leads to ratios of deviations 0 : 1: 1 : 0 ,[132] as are observed in $^{119}SnD_nH_{3-n}{}^{-}$ ion, $^{15}ND_nH_{3-n}$, and $^{31}PD_nH_{3-n}$.[151,132,152] The observed signs of the deviations are *positive* except in $^{119}SnD_nH_{3-n}{}^{-}$ ion and $^{31}PD_nH_{3-n}$. A recent paper suggests that, instead of the secondary geometrical isotope effects (which lead to negative deviations from additivity), the quadratic terms in the dependence of Δr_{CH} and Δr_{CD} on the number of deuteriums in $^{13}CD_nH_{4-n}$ are responsible for (likewise negative) deviations from additivity in $^{13}CD_nH_{4-n}$.[153] The deuterium-induced ^{13}C isotope shifts have been measured in the $^{13}CD_nH_{4-n}$ series by Bennett and Raynes,[154] but unfortunately the $^{13}CD_2H_2$ isotopomer was not included. Nevertheless, the observed deviations from additivity are positive in this series.

The sign of the $^1\Delta C(^{2/1}H)$ isotope shift has been found to be positive for substituted methanes. However, recently two examples of unusual sign have been reported:[155]

$$\delta^C(CDI_3) - \delta^C(CHI_3) = 386.3 \pm 1.0 \text{ ppb}$$

and $$\delta^C(CHDI_2) - \delta^C(CH_2I_2) = 107.3 \pm 0.5 \text{ ppb}$$

whereas $$\delta^C(CD_3I) - \delta^C(CH_3I) = -147.6 \pm 0.1 \text{ ppb} \quad \text{or} \quad -49.3 \text{ ppb/D}.$$

Deuterium isotope effects on ^{13}C shifts in sterically hindered intramolecularly hydrogen-bonded aromatic ketones are found to be unusual: the 2-bond isotope shifts are very large and some show unusual signs.[156] These isotope shifts $^n\Delta C(^{2/1}H)$ in intramolecularly hydrogen-bonded systems, where the OH protons have been exchanged with deuteriums, have been reviewed.[157] The one-bond deuterium isotope effects on nitrogen shielding and (unusual sign) two-bond isotope effects at ^1H shielding have been studied in ammonium ions in complexes with crown ethers and cryptands.[158] The insight gained from LORG calculations on model systems $NH_3D\cdots NH_3$ and $NH_3D\cdots OH_2$ are used in the interpretation of the results. Deuteration at a position not in the N–H···N bond gives rise to a negative $^2\Delta N(^{2/1}H)$, confirming that the (unusual) positive sign is connected to hydrogen-bonding of the deuterium. On the other hand, deuteration at the ammonia molecule has no effect on the ammonium ion. ^{18}O-induced isotope shifts of ^{13}C have been reported for succinic acid and dimethylmaleic anhydride.[159] These two systems illustrate how the ^{18}O-isotope shift may be employed to distinguish between isotopomers that have identical ^{18}O enrichment but different ^{18}O-labeling patterns. In multiply-isotopically labeled mixtures it is found that with spectral editing techniques (using homo- and heteronuclear multiple-quantum filtration to obtain subspectra for the various isotopomers), more precise measurements of the effects of $^{2/1}H$ and $^{18/16}O$ substitution on the nuclear shielding of ^1H, ^{13}C, and ^{31}P can be made.[160]

In the ^{195}Pt spectra of $[Pt^{35}Cl_n{}^{37}Cl_{6-n}]^{2-}$ five distinct signals with n = 2 to 6 exhibiting 0.18 ppm/^{37}Cl isotope shifts of the usual sign were observed, and $\delta^{Pt}\,[PtF_5{}^{37}Cl]^{2-} - \delta^{Pt}\,[PtF_5{}^{35}Cl]^{2-} = -0.32$ ppm.[161] These isotope shifts are rather large for $^{37}Cl/^{35}Cl$ substitution, as might be expected from the very large chemical shift range of the ^{195}Pt nucleus.

2.4 Intermolecular Effects on Nuclear Shielding — In a study of electrostatic effects on ^{19}F shielding, calculations using the conventional common-origin with derivative Hartree-Fock approach of fluorine shielding in CH_3F, CH_3CH_2F, $H_2C{=}CHF$, $HC{\equiv}CF$, $FC{\equiv}CF$, and FC_6H_5 at their experimental geometries were carried out with and without an external ideal dipole.[162] Less contracted valence *p* functions and three sets of polarization functions (except in the case of fluorobenzene) were used in the basis set. The range of shifts due to the external dipole approaching the fluorine center up to 3 Å was found to correlate linearly with the isotropic shielding of the unperturbed molecule. They conclude that the effect of an electrostatic interaction on the shielding is mostly via the paramagnetic contribution.

The ^{15}N shielding in CH_3CN, HCN, and pyridine has an intermolecular effect of unusual sign. While most nuclei suffer deshielding in going from the isolated molecule to the liquid, these nitrogen sites have the opposite intermolecular shifts, they suffer an increased shielding while the ^{13}C nuclei in the same molecules exhibit the usual deshielding.[163] A simple model of association in CH_3CN is the linear head-to-tail dimer and trimer. GIAO calculations at the SCF level have been carried out on these interaction models using rigid monomers at the experimental geometries and optimizing the distances between monomers.[164] The results are in qualitative agreement with the experimental gas-to-liquid shifts. In another example, the ^1H and ^{14}N shielding tensors in solid ammonia are modeled by GIAO calculations at the SCF level.[165] Comparison with ^1H and ^{15}N shielding observed in gaseous NH_3 ,[166] shows that both isotropic shieldings decrease in going from gas to solid, and the ^{14}N shielding anisotropy decreases while the ^1H anisotropy increases.

A model system for the Bronsted acid sites in zeolites with which adsorbed molecules such as NH_3, CH_3OH, and H_2O may interact are small clusters containing the AlO_4 unit. GIAO calculations at the SCF level have been carried out using 3 types of clusters: (1) one AlO_4 unit saturated with hydrogen atoms and the bridging hydroxyl proton, $HO(H)Al(OH)_3$ (2) the 2 oxygen atoms acting as adsorption sites are bound to SiH_3 groups, $H_3SiO(H)Al(OH)_2OSiH_3$ and (3) for interaction with NH_3 molecule, $H_3SiO(H)Al(OSiH_3)_3$. The structure of the complex between the sorbate molecule and the cluster is optimized at the SCF level. Shieldings of the protons of the sorbate and the cluster were calculated.[167] It is concluded that ammonium ions are formed on adsorption of ammonia onto bridging hydroxyl sites in zeolites. This is based on the agreement between the calculated shielding change of 6.8 ppm and the observed proton resonances between 6.5 and 7 ppm in NH_4-Y zeolite. The situation with H_2O and CH_3OH is still not clear. The ^{129}Xe chemical shifts observed when Xe is dissolved in liquid CO_2, NNO, ethane, and propane have been reported.[168] It is found that the gas phase chemical shifts of Xe in CO_2 can be extrapolated to the density of liquid CO_2 and find 63.2 ppm which agrees reasonably well with the measured 58.8 ppm shift. It is concluded that coordination numbers of 7.5-8.5 can be deduced from the ^{129}Xe gas-to- liquid shifts. A review of the effects of the environment on the nuclear shielding treats the so-called van der Waals shifts measured in gases, in gas-to-liquid shifts and in zeolites, primarily of ^{129}Xe.[169]

REFERENCES

1. P. Lazzeretti, *Theor. Chim. Acta*, 1993, **87**, 59.
2. M. C. Caputo, M. B. Ferraro, P. Lazzeretti, M. Malagoli, and R. Zanasi, *J. Mol. Struct. (Theochem)*, 1994, **111**, 89.

3. M. C. Caputo, M. B. Ferraro, P. Lazzeretti, M. Malagoli, and R. Zanasi, *Phys. Rev. A*, 1994, **49**, 3445.
4. H. M. McConnell, *J. Chem. Phys.*, 1957, **27**, 226.
5. A. D. Buckingham and P. J. Stiles, *Mol. Phys.*, 1972, **24**, 99.
6. W. S. Warren, S. Mayr, D. Goswami, and A. P. West, Jr., *Science*, 1992, **255**, 1683.
7. R. A. Harris and I. Tinoco, Jr., *Science*, 1993, **259**, 835.
8. A. D. Buckingham and L. C. Parlett, *Science*, 1994, **264**, 1748.
9. A. D. Buckingham, *Can J. Chem.*, 1960, **38**, 300.
10. D. M. Bishop and S. M. Cybulski, *Chem. Phys. Lett.*, 1993, **211**, 255.
11. D. M. Bishop and S. M. Cybulski, *J. Magn. Reson. Ser. A*, 1994, **107**, 99.
12. D. M. Bishop and S. M. Cybulski, *Mol. Phys.*, 1993, **80**, 199.
13. D. M. Bishop and S. M. Cybulski, *Mol. Phys.*, 1993, **80**, 209.
14. M. Grayson and W. T. Raynes, *Mol. Phys.*, 1994, **81**, 533.
15. J. D. Augspurger, C. E. Dykstra and E. Oldfield, *J. Am. Chem. Soc.*, 1991, **113**, 2447.
16. J. D. Augspurger and C. E. Dykstra, *J. Phys. Chem.*, 1991, **95**, 9230.
17. J. D. Augspurger and C. E. Dykstra, *Mol. Phys.*, 1992, **176**, 229.
18. J. D. Augspurger, J. G. Pearson, E. Oldfield, C. E. Dykstra, K. D. Park, and D. Schwartz, *J. Magn. Reson.*, 1992, **100**, 342.
19. J. Oddershede, P. Jørgensen, and D. L. Yeager, *Comput. Phys. Rep.* , 1984, **2**, 33.
20. J. Oddershede and J. Geertsen, *J. Chem. Phys.*, 1990, **92**, 6036.
21. A. E. Hansen and T. D. Bouman, *J. Chem. Phys.*, 1985, **82**, 5035.
22. T. D. Bouman and A. E. Hansen, *Chem. Phys. Lett.*, 1990, **175**, 292.
23. J. Geertsen, *Chem. Phys. Lett.*, 1991, **179**, 479.
24. S. A. Sauer, I. Paidarova, and J. Oddershede, *Mol. Phys.*, 1994, **81**, 87.
25. S. M. Cybulski and D. M. Bishop, *J. Chem. Phys.*, 1993, **98**, 8057.
26. J. Gauss, *J. Chem. Phys.*, 1993, **99**, 3629.
27. J. Gauss, *Chem. Phys. Lett.*, 1992, **191**, 614.
28. H. Fukui, T. Baba, H. Matsuda, and K. Miura, *J. Chem. Phys.*, 1994,**100**, 6608.
29. H. Fukui, K. Miura, and H. Matsuda, *J. Chem. Phys.*, 1992, **96**, 2039.
30. H. Fukui, K. Miura, and H. Matsuda, *Bull. Magn. Reson.*, 1993, **15**, 98.
31. K. Ruud, T. Helgaker, R. Kobayashi, P. Jørgensen, K. L. Bak, and H. J. A. Jensen, *J. Chem. Phys.*, 1994, **100**, 8178.
32. C. van Wüllen and W. Kutzelnigg, *Chem. Phys. Lett.* 1993, **205**, 563.
33. V. G. Malkin, O. L. Malkina, M. E. Casida, and D. R. Salahub, *J. Am.Chem. Soc.*, 1994, **116**, 5898.
34. A. K. Jameson and C. J. Jameson, *Chem. Phys. Lett.*, 1987, **134**, 461.
35. R. E. Wasylishen, S. Mooibroek, and J. B. Macdonald, *J. Chem. Phys.*, 1984, **81**, 1057.

36. C. J. Jameson, Theoretical and Physical Aspects of Nuclear Shielding, in *Nuclear Magnetic Resonance*, G. A. Webb, ed., Specialist Periodical Reports, Royal Society of Chemistry, 1989, Vol. 18, p. 1.
37. J. A. Tossell, ed., *Nuclear Magnetic Shieldings and Molecular Structure*, NATO ASI Series C Vol. 386, Kluwer, Dordrecht, 1993.
38. C. J. Jameson, A. K. Jameson, D. Oppusunggu, S. Wille, P. M. Burrell,and J. Mason, *J. Chem. Phys.*, 1981, **74**, 81.
39. C. J. Jameson, A. K. Jameson, and P. M. Burell, *J. Chem. Phys.*, 1980, **73**, 6013.
40. C. J. Jameson, A. C. de Dios, and A. K. Jameson, *J. Chem. Phys.*, 1991, **95**, 9042.
41. D. B. Chesnut and C. K. Foley, *J. Chem. Phys.*, 1986, **85**, 2814.
42. U. Fleischer, M. Schindler, and W. Kutzelnigg, *J. Chem. Phys.*, 1987, **86**, 6337.
43. P. Lazzeretti and R. Zanasi, *J. Chem. Phys.*, 1980, **72**, 6768.
44. R. Höller and H. Lischka, *Mol. Phys.*, 1980, **41**, 1041.
45. T. D. Bouman and A. E. Hansen, presented at the June 1990 meeting of the Danish Chemical Society, Odense.
46. V. G. Malkin, O. L. Malkina, and D. R. Salahub, *Chem. Phys. Lett.*, 1993, **204**, 80, 87.
47. P. B. Davies, R. M. Neumann, S. C. Wofsy, and W. Klemperer, *J.Chem. Phys.*, 1971, **55**, 3564.
48. K. Wolinski, C. L. Hsu, J. F. Hinton, and P. Pulay, *J. Chem. Phys.*, 1993, **99**, 7819.
49. C. J. Jameson and A. K. Jameson, *Chem. Phys. Lett..*, 1987, **135**, 254.
50. P. D. Ellis, J. D. Odom, A. S. Lipton, Q. Chen, and J. M. Gulick, in *Nuclear Magnetic Shieldings and Molecular Structure*, ed. J. A. Tossell, NATO ASI Series C Vol. 386, Kluwer, Dordrecht, 1993, p. 539-555.
51. W. Kutzelnigg, C. van Wüllen, U. Fleischer, R. Franke, and T. van Mourik, in *Nuclear Magnetic Shieldings and Molecular Structure*, ed. J. A. Tossell, NATO ASI Series C Vol. 386, Kluwer, Dordrecht, 1993, p. 141-161.
52. I. J. Solomon, J. N. Keith, A. J. Kacmarek, and J. K. Raney, *J. Am. Chem. Soc.*, 1968, **90**, 5408.
53. W. Kutzelnigg, U. Fleischer, and M. Schindler, in *NMR, Basic Principles and Progress*, Vol. 23, Springer-Verlag, Berlin, 1990, p. 165.
54. T. A. Keith and R. F. W. Bader, *Chem. Phys. Lett.*, 1992, **194**, 1.
55. T. A. Keith and R. F. W. Bader, *Chem. Phys. Lett.*, 1993, **210**, 223.
56. T. A. Keith and R. F. W. Bader, *J. Chem. Phys.*, 1993, **99**, 3669.
57. P. Lazzeretti, M. Malagoli, and R. Zanasi, in *Nuclear Magnetic Shieldings and Molecular Structure*, ed. J. A. Tossell, NATO ASI Series C Vol. 386, Kluwer, Dordrecht, 1993, p. 163-190.
58. P. Lazzeretti, E. Rossi, and R. Zanasi, *J. Chem. Phys.* 1982, **77**, 3129.
59. P. Lazzeretti, E. Rossi, and R. Zanasi, *Chem. Phys. Lett.*, 1983, **106**, 67.

60. U. Fleischer, W. Kutzelnigg, P. Lazzeretti, and V. Mühlenkamp, *J. Am. Chem. Soc.*, 1994, **116**, 5298.
61. A. E. Hansen and T. D. Bouman, *J. Chem. Phys.*, 1989, **91**, 3552.
62. F. A. L. Anet, D. J. O'Leary, C. G. Wade, and R. D. Johnson, *Chem. Phys. Lett.*, 1990, **171**, 401.
63. C. van Wüllen, U. Fleischer, and W. Kutzelnigg, *Mol. Phys.*, 1994, **81**, 1373.
64. A. Barszczewicz, M. Jaszunski, K. Kamienska-Trela, T. Helgaker, *Theor. Chim. Acta*, 1993, **87**, 19.
65. C. M. Smith, R. D. Amos, and N. C. Handy, *Mol. Phys.*, 1992, **77**, 381.
66. K. W. Zilm, R. T. Conlin, D. M. Grant, and J. Michl, *J. Am. Chem. Soc.*, 1980, **102**, 6672.
67. M. Jaszunski, K. L. Bak, P. Jørgensen, T. Helgaker, K. Ruud, and H. J. A. Jensen, *Chem. Phys. Lett.*, 1993, **204**, 608.
68. A. E. Hansen and T. D. Bouman, *J. Math. Chem.*, 1992, **10**, 221.
69. D. B. Chesnut and K. D. Moore, *J. Comput. Chem.*, 1989, **10**, 648.
70. D. B. Chesnut, B. E. Rusiloski, K. D. Moore, and D. A. Egolf, *J. Comput. Chem.*, 1993, **14**, 1364.
71. T. C. Farrar, J. L. Schwartz, and S. Rodriguez, *J. Phys. Chem.*, 1993, **97**, 7201.
72. G. Rasul, G. K. S. Prakash, G. A. Olah, *J. Org. Chem.*, 1994, **59**, 2552.
73. P. v. R. Schleyer, J. Gauss, M. Bühl, R. Greatrex, M. A. Fox, *J. Chem. Soc., Chem. Commun.*, **1993**, 1766.
74. M. Bühl, J. Gauss, M. Hofmann, and P. v. R. Schleyer, *J. Am. Chem. Soc.*, 1993, **115**, 12385.
75. M. L. McKee, M. Bühl, O. P. Charkin, and P. v. R. Schleyer, *Inorg. Chem.*, 1993, **32**, 4549.
76. A. M. Mebel, D. G. Musaev, and K. Morokuma, *Chem. Phys. Lett.*, 1993, **214**, 69.
77. D. Cremer, L. Olsson, F. Reichel, and E. Kraka, *Isr. J. Chem.*, 1994, **33**, 369.
78. A. M. Mebel, D. G. Musaev, G. Djamaladdin, N. Koga, and K. Morokuma, *Bull. Chem. Soc. Jpn*, 1993, **66**, 3259.
79. S. Sieber, P. v. R. Schleyer, J. Gauss, *J. Am. Chem. Soc.*, 1993, **115**, 6987.
80. G. A. Olah, J. Bausch, G. Rasul, H. George, and G. K. S. Prakash, *J. Am. Chem. Soc.*, 1993, **115**, 8060.
81. G. Boche, F. Bosold, J. C. W. Lohrenz, A. Opel, P. Zulauf, *Chem. Ber.*, 1993, **126**, 1873.
82. D. B. Chesnut, L. D. Quin, and K. D. Moore, *J. Am. Chem. Soc.*, 1993, **115**, 11984.
83. T. C. Farrar, M. J. Jablonsky, and J. L. Schwartz, *J. Phys. Chem.*, 1994, **98**, 4780.
84. S. Sieber, P. v. R. Schleyer, A. H. Otto, J. Gauss, F. Reichel, and D. Cremer, *J. Phys. Org. Chem.*, 1993, **6**, 445.

85. G. W. Buchanan, A. B. Driega, A. Moghimi, C. Bensimon, R. A. Kirby, and T. D. Bouman, *Can. J. Chem.*, 1993, **71**, 1983.
86. M. Jaszunski, T. Helgaker, K. Ruud, K. L. Bak, and P. Jørgensen, *Chem. Phys. Lett.*, 1994, **220**, 154.
87. M. Schindler, *J. Am. Chem. Soc.*, 1987, **109**, 5950.
88. D. B. Chesnut and D. W. Wright, *J. Comput. Chem.*, 1991, **12**, 546.
89. P. J. Barrie, C. J. Groombridge, J. Mason, and E. A. Moore, *Chem. Phys. Lett.*, 1994, **219**, 491.
90. U. Schneider, S. Richard, M. M. Kappes, and R. Ahlrichs, *Chem. Phys. Lett.*, 1993, **210**, 165.
91. J. Gauss, U. Schneider, R. Ahlrichs, C. Dohmeier, and H. G. Schnöckel, *J. Am. Chem. Soc.*, 1993, **115**, 2402.
92. J. A. Tossell, *Am. Mineral.*, 1993, **78**, 911.
93. L. Olsson and D. Cremer, *Chem. Phys. Lett.*, 1993, **215**, 433.
94. J. F. Hinton, P. L. Guthrie, P. Pulay, and K. Wolinski, *J. Magn. Reson., Ser. A*, 1993, **103**, 188.
95. M. C. Caputo, M. B. Ferraro, R. C. Bochicchio, P. Lazzeretti, M. Malagoli, and R. Zanasi, *J. Mol. Struct. (Theochem)*, 1993, **106**, 77.
96. H. Nakatsuji and T. Nakao, *Int. J. Quantum Chem.*, 1994, **49**, 279.
97. H. Nakatsuji, T. Higashioji, and M. Sugimoto, *Bull. Chem. Soc. Jpn*, 1993, **66**, 3235.
98. J. A. Tossell and D. J. Vaughan, *Geochim. Cosmochim. Acta*, 1993, **57**, 1935.
99. U. Fleischer and W. Kutzelnigg, *Phosphorus, Sulfur Silicon Relat. Elem.*, 1993, **77**, 105.
100. J. Sauer and J. R. Hill, *Chem. Phys. Lett.*, 1994, **218**, 333.
101. J. P. Chauvel, Jr. and N. S. True, *Chem. Phys.* 1985, **95**, 435.
102. X. Xie, Z. Liu, and H. Liu, *J. Magn. Reson., Ser. A*, 1993, **102**, 351.
103. C. Vogel, R. Wolff, and R. Radeglia, *Collect. Czech. Chem. Commun.*, 1993, **58**, 1255.
104. D. Kovacek, Z. B. Maksic, S. Elbel, and J. Kudnig, *J. Mol. Struct. (Theochem)*, 1994, **110**, 247.
105. W. T. Raynes, A. D. Buckingham, and H. J. Bernstein, *J. Chem. Phys.*,1962, **36**, 3481.
106. M. P. Williamson and T. Asakura, *J. Magn. Reson., Ser. B*, 1993, **101**, 63.
107. J. F. Maddaluno, N. Gresh, and C. Giessner-Prettre, *J. Org. Chem.*, 1994, **59**, 793.
108. D. W. Alderman, M. H. Sherwood, and D. M. Grant, *J. Magn. Reson., Ser. A*, 1993, **101**, 188.
109. J. Facelli and D. M. Grant, *Nature*, 1993, **365**, 325.
110. M. H. Sherwood, D. W. Alderman, and D. M. Grant, *J. Magn. Reson., Ser. A*, 1993, **104**, 132.
111. C. Ye, R. Fu, J. Hu, L. Hou, S. Ding, *Magn. Reson. Chem.*, 1993, **31**, 699.
112. R. Q. Fu, J. Z. Hu, C. H. Ye, and M. K. Lu, *Prog. Nat. Sci.*, 1993, **3**, 60.

113. K. Eichele, M. D. Lumsden, and R. E. Wasylishen, *J. Phys. Chem.*, 1993, **97**, 8909.
114. L. Mäler, L. Di Bari, and J. Kowalewski, *J. Phys. Chem.*, 1994, **98**, 6244.
115. J. Bardakji, G. Kontaxis, H. Sterk, and R. Janoschek, *Spectrochim. Acta A*, 1994, **50**, 735.
116. P. Robyr, B. H. Meier, P. Fischer, and R. R. Ernst, *J. Am. Chem. Soc.*, 1994, **116**, 5315.
117. D. Lafleur, Y. Huang, D. F. R. Gilson, and I. S. Butler, *J. Solid State Chem.*, 1994, **108**, 99.
118. I. P. Gerothanassis, M. Momenteau, G. E. Hawkes, and P. J. Barrie, *J. Am. Chem. Soc.*, 1993, **115**, 9796.
119. K. Eichele and R. E. Wasylishen, *J. Phys. Chem.*, 1994, **98**, 3108.
120. G. Wu and R. E. Wasylishen, *J. Chem. Phys.*, 1994, **100**, 4828.
121. R. Challoner, C. A. McDowell, M. Yoshifuji, K. Toyota, and J. A. Tossell, *J. Magn. Reson., Ser. A*, 1993, **104**, 258.
122. M. Jansen, F. Frick, W. Hoffbauer, A. R. Grimmer, W. Kutzelnigg, and U. Fleischer, *Naturwissenchaften*, 1993, **80**, 465.
123. Y. Hua, H. L. Uhm, D. F. R. Gilson, and I. S. Butler, *Inorg. Chem.*, 1994, **33**, 804.
124. J. Skibsted, N. C. Nielsen, H. Bildsoe, and H. J. Jakobsen, *J. Am. Chem. Soc.*, 1993, **115**, 7351.
125. S. Hayakawa, T. Yoko, and S. Sakka, *Bull. Chem. Soc. Jpn.*, 1993, **66**, 3393.
126. J. Sola, P. Gonzalez-Duarte, J. Sanz, I. Casals, T. Alsina, I. Sobrados, A. Alvarez-Larena, J. F. Piniella, and X. Solans, *J. Am. Chem. Soc.*, 1993, **115**, 10018.
127. W. P. Power, S. Mooibroek, R. E. Wasylishen, and T. S. Cameron, *J. Phys. Chem.*, 1994, **98**, 1552.
128. G. Wu and R. E. Wasylishen, *Magn. Reson. Chem.*, 1993, **31**, 537.
129. R. A. Santos and G. S. Harbison, *J. Am. Chem. Soc.*, 1994, **116**, 3075.
130. I. Paidarova, V. Spirko, and J. Oddershede, *J. Mol. Spectrosc.*, 1993, **160**, 311.
131. C. J. Jameson, A. C. de Dios, A. K. Jameson, *J. Chem. Phys.*, 1991, **95**, 1069.
132. R. E. Wasylishen and J. O. Friedrich, *Can. J. Chem.*, 1987, **65**, 2238.
133. J. D. Augspurger and C. E. Dykstra, *Mol. Phys.*, 1993, **80**, 117.
134. P. W. Fowler, G. Riley, and W. T. Raynes, *Mol. Phys.*, 1981, **42**, 1463.
135. J. F. Hinton, P. L. Guthrie, P. Pulay, and K. Wolinski, *J. Am. Chem. Soc.*, 1992, **114**, 1604.
136. E. A. Moore, a. Healy, M. Mortimer, and N.F. Peirson, *J. Mol. Struct. (Theochem)*, 1993, **104**, 229.
137. M. D. Lumsden, R. E. Wasylishen, K. Eichele, M. Schindler, G. H. Penner, W. P. Power, and R. D. Curtis, *J. Am. Chem. Soc.*, 1994, **116**, 1403.
138. P. E. Hansen, J. Abildgaard, Aa. E. Hansen, *Chem. Phys. Lett.*, 1994, **224**, 275.
139. A. C. de Dios, J. G. Pearson, and E. Oldfield, *Science*, 1993, **260**, 1491.

140. N. Asakawa, H. Kurosu, I. Ando, A. Shoji, and T. Ozaki, *J.Mol. Struct.*, 1994, **317**, 119.
141. D. Jiao, M. Barfield, and V. J. Hruby, *J. Am. Chem. Soc.*, 1993, **115**, 10883.
142. M. Wada, M. Sakurai, Y. Inoue, Y. Tamura, and Y. Watanabe, *J. Am. Chem. Soc.*, 1994, **116**, 1537.
143. F. Imashiro, Y. Masuda, M,. Honda, and S. Obara, *J. Chem. Soc., Perkin Trans. 2*, **1993**, 1535.
144. R. Born, H. W. Spiess, W. Kutzelnigg, U. Fleischer, and M. Schindler, *Macromolecules*, 1994, **27**, 1500.
145. H. M. Sulzbach, P. v. R. Schleyer, H. F. Schaefer III, *J. Am. Chem. Soc.*, 1994, **116**, 3967.
146. A. C. de Dios, J. G. Pearson, E.Oldfield, *J. Am. Chem. Soc.*, 1993, **115**, 9768.
147. M. Barfield, *J. Am. Chem. Soc.*, 1993, **115**, 6916.
148. C. J. Jameson and H. J. Osten, *J. Chem. Phys.*, 1984, **81**, 4293
149. K. L. Leighton and R. E. Wasylishen, *Can. J. Chem.*, 1987, **65**, 1469.
150. R. E. Wasylishen and J. O. Friedrich, *J. Chem. Phys.*, 1984, **80**, 585.
151. R. E. Wasylishen and N. Burford, *Can. J. Chem.*, 1987, **65**, 2707.
152. A. K. Jameson and C. J. Jameson, *J. Magn. Reson.* 1978, **32**, 455.
153. N. M. Sergeyev, N. D. Sergeyeva, and W. T. Raynes, *Chem. Phys. Lett.*, 1994, **221**, 385.
154. B. Bennett and W. T. Raynes, *Spectrochim. Acta A*, 1989, **45**, 1267.
155. N. M. Sergeyev, N. D. Sergeyeva, and W. T. Raynes, *J. Chem. Soc. Chem. Commun.*, **1994**, 485.
156. P. E. Hansen, S. N. Ibsen, T. Kristensen, and S. Bolvig, *Magn. Reson. Chem.*, 1994, **32**, 399.
157. P. E. Hansen, *J. Mol. Struct.*, 1994, **321**, 79.
158. P. E. Hansen, Aa. E. Hansen, A. Lycka, and A. Buvari-Barcza, *Acta Chem. Scand.*, 1993, **47**, 777.
159. T. L. Mega and R. L. van Etten, *J. Am. Chem. Soc.*, 1993, **115**, 12056.
160. W. W. Wooten, R. K. Dua, G. D. Dotson, and R. W. Woodard, *J. Magn. Reson., Ser. A*, 1994, **107**, 50.
161. E. Parzich, G. Peters, and W. Preetz, *Z. Naturforsch. B*, 1993, **48**, 1169.
162. J. D. Augspurger and C. E. Dykstra, *J. Am. Chem. Soc.*, 1993, **115**, 12016.
163. C. J. Jameson, A. K. Jameson, D. Oppusunggu, and S. Wille, *J. Chem. Phys.*, 1982, **76**, 152.
164. K. Jackowski, *J. Mol. Struct. (Theochem)*, 1994, **303**, 239.
165. K. Jackowski, A. Barszczewicz, and K. Wozniak, *Solid State Nucl. Magn. Reson.*, 1993, **2**, 265.
166. C. J. Jameson, A. K. Jameson, S. M. Cohen, H. Parker, D. Oppusunggu, P. M. Burrell, and S. Wille, *J. Chem. Phys.*, 1981, **74**, 1608.
167. F. Haase and J. Sauer, *J. Phys. Chem.*, 1994, **98**, 3083.
168. Y. H. Lim, N. E. Nugara, and A. D. King, Jr., *J. Phys. Chem.*, 1993, **97**, 8816.
169. D. M. Bishop, *Int. Rev. Phys. Chem.*, 1994, **13**, 21.

3
Applications of Nuclear Shielding

BY M. YAMAGUCHI

Introduction

The format and coverage of this report remain similar as in the previous years. Various chemical and physical influences to nuclear shieldings are considered in the first section. The shieldings of particular nuclear species are described in the following section according to their position in the periodic table. Since there are huge number of articles on n.m.r. spectroscopy during the period of this review, only widely available and common journals, which are mainly published in English, were examined to give appropriate comments.

1 Various Chemical and Physical Influences to Nuclear Shieldings

1.1 Computer Assisted Structural Assignment

1.1.1 Spectrum simulation and computer assisted assignments

Linear models were developed to predict the ^{13}C n.m.r. chemical shifts of carbons in indole moieties based on a critical review[1]. The mathematical basis and the nature of the recently proposed method for optimised prediction of ^{13}C n.m.r. spectra were discussed[2]. A personal computer based program was developed that permits fast estimation of the ^{13}C chemical shifts of common organic compounds, spectral simulation of these compounds, and shift/structure search in a user-maintained spectral library[3]. Computational neural networks could predict complex mappings between input and output data of ^{13}C n.m.r. shifts[4]. Programs of pattern recognition methods were described on ^{13}C n.m.r. spectroscopy of sesquiterpene skeletal types[5]. Estimation of ^{13}C n.m.r. spectra of organic molecules was discussed with the SpecInfo system which is spectroscopic search system provided with data base of n.m.r., mass, and IR spectra[6]. An automatic system for the interpretation of two-dimensional n.m.r. spectra of proteins, HIPS was presented[7]. Computer-assisted assignment of multidimensional n.m.r. spectra of proteins was reported[8]. A new computational procedure was described for the analysis of heteronuclear multiple-quantum coherence (HMQC)-J n.m.r. spectra[9]. An object-oriented programming approach for computer simulations in magnetic resonance was described[10]. The first successful calculation of ^{19}F n.m.r. spectrum of a protein, the galactose binding protein from Escherichia coli, labelled with [5-^{19}F]tryptophan, was reported[11]. A new method to predict ^{31}P n.m.r. chemical shift based on neural networks was reported[12]. Molecular dynamics simulations were employed to determine the relative ^{1}H n.m.r. chemical shifts in liquid state[13].

The computer-assisted analysis was applied on the structure determination of regular branched polysaccharides containing 2,3-disubstituted rhamnopyranose and mannopyranose residues on the basis of ^{13}C n.m.r[14]. The novel methodology for protein 2D n.m.r. assignment was presented based upon protein spin coupling graph theory analysis, fuzzy graph pattern recognition, and tree searching[15]. 3D TOCSY HECTOR n.m.r. computer aided ^{1}H and ^{13}C resonance assignments of raffinose were reported[16].

1.1.2 Nuclear shielding calculations

Several reviews of recent developments of theoretical calculation methods were published[17,18,19,20,21,22,23,24,25].

Ab initio ^{13}C nuclear shielding calculations of the γ-gauche effect for an *n*-paraffin using the GIAO-CHF procedure were reported[26]. ^{13}C n.m.r. chemical shifts for conformers of alkanes from methane to heptane were calculated by the ab initio GIAO method[27]. The accurate C_{2v} phenonium and benzenonium ion structures were confirmed by correlated GIAO-MP_2 level n.m.r. chemical shift calculations[28]. Computation of the $^{13}C^{\alpha}$ chemical shifts (or shieldings) of glycine, alanine and valine residues in bovine and Drosophila calmodulins and Staphylococcal nuclease, and comparison with experimental values, was reported using a GIAO approach[29]. The ab initio study of ^{13}C n.m.r. chemical shieldings in glycine, alanine, and valine residues was reported to examine the effects of bond lengths, bond angles, and torsion angles on the ^{13}C chemical shielding of C^{α} (and C^{β}) sites in model fragments in protein[30]. Ab initio chemical shielding calculations employing the GIAO method were carried out in a series of phosphorus bicyclic and monocyclic molecules in order to provide an understanding of the unusual chemical shifts seen in 7-phosphanorbornene compounds[31]. The ^{31}P shielding tensor for PH_3 was calculated without electron correlations (SCF) and with electron correlation (2nd-order Mollet-Plesset) using the GIAO method[32]. The GIAO approach was used within the coupled Hartree-Fock (CHF) approximation to compute the ^{1}H and ^{14}N n.m.r. shielding constants in solid ammonia[33]. Correlated ab initio (GIAO-MP2) chemical shift computations gave more reliable ^{11}B and ^{13}C n.m.r. values for the problem carboranes, 1,5-$C_2B_3H_5$ and 1,2-$C_2B_3H_7$, which had given poor IGLO results[34].

The ab initio IGLO method was used to examine the conformational dependencies of the isotropic ^{13}C chemical shifts in the model peptide N-acetyl-N'-methylglycinamide[35]. The structure of the classical 2-norbornyl cation was rigorously defined by ab initio calculation and its best computer (IGLO) ^{13}C chemical shifts were compared with the experimental n.m.r. data[36]. Ab initio MO study of structure, stability and rearrangements of nido-metallaboranes with Ir, Co, and Fe transition metals was performed at HF and correlated MP2 levels, and ^{11}B and ^{1}H chemical shifts calculated by the IGLO method reproduced experimental spectra[37]. IGLO calculations of ^{29}Si n.m.r. chemical shift anisotropies in silicate models were reported[38]. The MC-IGLO method for the calculation of n.m.r. chemical shifts and magnetic susceptibilities with inclusion of electron correlation effects was presented[39].

Ab initio calculations showed that additivity of the intermolecular shielding exists in a model system consisting of fluorobenzene interacting with hydrogen fluoride molecules, C_6H_5F-$(HF)_n$, where n = 1-5[40]. Secondary and tertiary structural effects on protein ^{1}H, ^{13}C, ^{15}N, and ^{19}F n.m.r. chemical shifts were predicted by an ab initio calculations[41]. Ab initio quantum-mechanical chemical shift calculations for the ^{29}Si n.m.r. in a variety of compounds were reported[42]. The geometries, formation energies, and ^{11}B n.m.r. chemical shifts of Lewis base adducts of the B_3H_7

and B_4H_8 nido-baranes, B_3H_7CO, $B_3H_7NH_3$, B_4H_8CO, $B_4H_8PH_3$, and $B_4H_8PF_3$, were investigated by ab initio MO method[43].

The influence of variations in the OSiO bond angle α on the anisotropy of the ^{29}Si shielding tensor of the monosilicate anion SiO_4^{4-} in C_{3v} symmetry was studied by semiempirical CNDO/2 calculations[44]. A simple method was demonstrated for the evaluation of the relative effects of shielding due to electrostatic interactions, such as those which may arise from both inter- and intra-residue interactions in proteins and other biomolecules[45]. The locally dense basis sets were used to calculate n.m.r. shieldings[46]. The ^{19}F n.m.r. chemical shielding tensor in the hydrogen difluoride ion FHF^- and its variation with bond length and angle deformation were calculated using the GAMESS set of SCF programs[47]. The ^{77}Se n.m.r. chemical shifts of nine selenium compounds, SeRR' (R, R' = H, Me, Et, Ph, SiH_3, GeH_3), were theoretically investigated by ab initio Hartree-Fock/finite perturbation method[48]. The influence of structure and geometry on the ^{31}P, ^{29}Si, ^{13}C, and ^{1}H chemical shifts were analysed starting with a description of the molecular wave function using bond orbitals[49]. Theoretical study of the ^{15}N and ^{29}Si chemical shielding tensors of silatrane and a set of derivatives showed sensitivity to substituent-induced structural changes[50]. Quadrupole coupling constants, asymmetry parameters of the electric field gradients, and isotropic chemical shifts of ^{23}Na were determined by computer simulation of the central transition line shapes of the ^{23}Na MAS n.m.r. spectra of crystalline sodium compounds in which sodium has exclusively oxygen atoms as nearest neighbours[51]. Computational studies indicated the origins of fluorine chemical shifts in proteins[52].

1.2 Stereochemical Nuclear Shielding Non-equivalence

The techniques for chiral recognition in n.m.r. were reviewed to applications primarily in the life sciences[53]. A short review of enantiomeric recognition of chiral organic ammonium salts by chiral pyridino-18-crown-6 ligands was reported[54].

1.2.1 Asymmetric determination

The direct determination of both the optical purity and absolute configuration of threo-methylphenidate was accomplished in a simple, specific, and accurate manner by ^{1}H n.m.r[55]. The enantiomeric ratio of chiral NH_4^+ halides or alkali carboxylates was determined by ^{1}H and ^{13}C n.m.r[56]. ^{13}C and 2D n.m.r. were used to determine the absolute configuration of chiral secondary alcohols[57].

Achiral and chiral lanthanide shift reagents were used to study α-ethyl-α-phenylsuccinimide[58]; to study the skeletal muscle relaxant, mephenoxalone[59]; to study 5-methyl-5-phenylhydantoin in polar solvents[60]. The enantiomeric purity of substituted N-phthaloyl-2-cyanoglycine methyl esters obtained by resolution of covalent diastereoisomers was established by measuring the shift separation of signals using $Eu(hfc)_3$[61]. ^{1}H and ^{13}C n.m.r. of racemic drugs were studied with chiral solvating agents and lanthanide shift reagents[62,63,64].

Chiral solvating agents was used for direct determination of enantiomeric excess of the cardiac antiarrhythmic, mexiletine[65]. α-Amino-α-trifluoromethylphenylacetonitrile was used as a potential reagent for ^{19}F n.m.r. determination of enantiomeric purity of acids[66]. Poly-γ-benzyl-L-glutamate (PBLG) and dichloromethane form a lyotropic liquid crystal which can be used as solvent for enantiomeric excess analysis through proton decoupled ^2H n.m.r[67]. Diastereomer discrimination of fluorine-labelled enantiomers in chloroform solutions was studied with and without two chiral solvating agents using ^1H, ^{13}C, and ^{19}F n.m.r and molecular mechanics calculations[68]. Heptakis(2,3,6-tri-O-methyl)-β-cyclodextrin was used as convenient chiral solvating agent to determine the enantiomeric composition of chiral trisubstituted allenes by ^1H n.m.r[69].

Several efficient chiral derivatizing reagents which give very large ^{19}F n.m.r. Δδ values in enantiomeric excess determination were reported[70]. ^{31}P n.m.r. was used to determine the enantiomeric excess of alcohols, amines and amino acid esters by diastereoisomeric ester and amide derivatives of phosphoric acid chloride[71]. (+)- And (-)-α-pinene formed diastereoisomeric complexes with α-,2,3,6-tri-O-methyl-α-,2,3,6-tri-O-methyl-β- and 2,3,6-tri-O-acetyl-β-cyclodextrin which were distinguishable in ^1H and ^{13}C n.m.r[72]. Diastereomeric esters of O-arylated (R)-lactic acids were studied by ^1H n.m.r. with respect to their potential application as reagents for chiral analysis[73]. A facile ^1H n.m.r. spectroscopic method was presented for the determination of the optical purity of α-amino aldehydes, via derivatisation with optically pure semicarbazides (S)-$R^1CHMeNHCONHNH_2$ (R^1 = ph, naphthyl)[74].

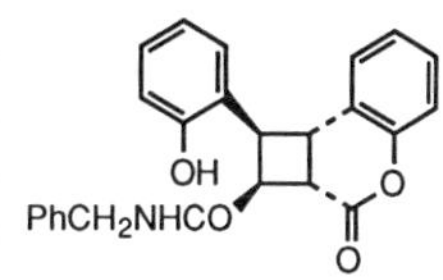

An enatiomerically pure anti head-to-head coumarin dimer derivative was used to determine the enantiomeric excess of chiral amines[75]. 1-Fluoro-2,4-dinitrophenyl-5-(S)-alanine amide (Marfey's reagent) was used as a chiral derivatising agent for benzylic amines in order to determine optical purities[76]. ^{31}P n.m.r. was used to determine the enantiomeric composition of 1-hydroxyalkylphosphonic acids using 1-(1-naphthyl)ethylamine as a chiral agent[77]. The absolute configuration of the stereogenic carbinol centres in nine annonaceous acetogenins was determined by careful ^1H and ^{19}F n.m.r. analysis of (S)- and (R)-Mosher's ester derivatives[78].

The enantioface-differentiating abilities of macrocyclic pseudopeptides were investigated by ^1H n.m.r. in $CDCl_3$ using various (R)- and (S)-ammonium and (R)- and (S)-α-amino acid ester salts[79].

1.2.2 Other stereochemistry determination

A new method and computer program for lanthanide-induced shift (LIS) calculations of highly flexible molecules was developed[80].

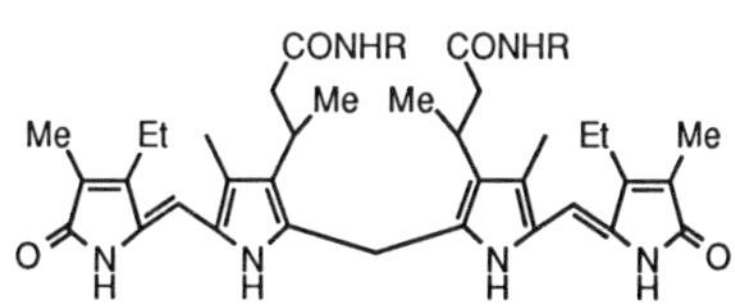

The conformation of the optically active bilirubin diamides (R = H, Me) was stabilised in a ridge-tile shape by intramolecular hydrogen bonding, as detected by ^1H n.m.r. and CD spectroscopy[81]. A simple technique was reported to distinguish between C_s and C_2 diastereomeric diols derived from chiral substrates by the ^{13}C n.m.r. spectra of the easily prepared Mosher's ester of each diol[82]. The lanthanoid-induced shifts (LIS) of ^1H and ^{13}C n.m.r. spectra and difference NOE spectra of benzyl, 1-phenylethyl, and cumyl formates were measured to examine the preferred conformations[83].

1.3 Isotope Effects

One-bond deuterium isotope effects on nitrogen nuclear shielding, $1\Delta N(D)$ and 2-bond isotope effects at ^1H nuclear shielding, $2\Delta H(D)$ were studied in ammonium ions in complexes with crown ethers and cryptands[84] One- and two-bond deuterium isotope effects (1Δ and 2Δ) on ^{13}C n.m.r. chemical shifts for methylpyridines and their N-oxides were investigated[85]. ^2H isotope effects on ^1H and ^{13}C chemical shifts due to deuteration of OH groups in a series of 2,6-dihydroxy-acylaromatic compounds were determined at both ambient and low temperature[86]. ^2H isotope effects on 13C n.m.r. chemical shifts $(^2H/^1H)^n\Delta(^{13}C)$ of 2-methylene-d_2-adamantane, 4-methylene-d_2-2-adamantanone and 6-methylene-d_2-2-adamantanone were reported[87]. One-bond deuterium-induced ^{13}C isotope shifts in several chlorine, bromine and iodine derivatives of methane were measured[88].

^{15}N isotope effects on the acid-base equilibria of amino groups in amino acids were determined by ^{13}C n.m.r[89]. The ^{18}O isotope shifts in ^{13}C n.m.r. were studied for $[^{18}O,1,4\text{-}^{13}C_2]$succinic acid and $[^{18}O]$dimethylmaleic anhydride[90].

The apparent field dependence observed for some isotropic shifts in high resolution solid-state n.m.r. was restudied using the Floquet formalism[91]. N.m.r. isotope shifts and $^3J_{CNCD}$ in ^{13}C n.m.r. spectra of N-neopentyl-4-*tert*-butylpiperidine-1'-d provided evidence for an unusual, eclipsed, syn geometry for the side chain[92].

Intramolecular reversible nondegenerate double proton transfer reaction was evidenced by ^1H and ^{13}C n.m.r. study of kinetic HH/HD/DH/DD isotope effects on the tautomerism of acetylporphyrin[93].

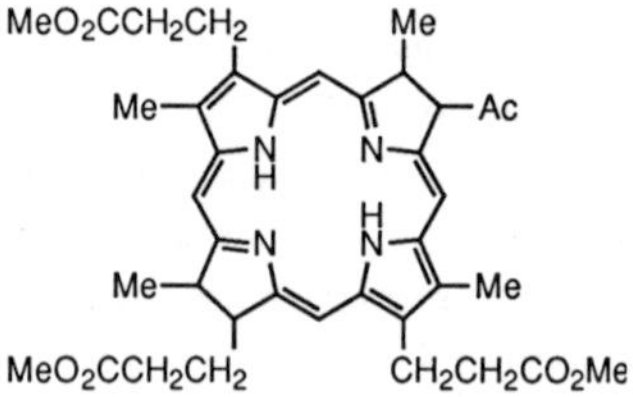

1.4 Substituent Effects

1.4.1 Proton substituent effects

Substituent effects on the ^1H n.m.r. spectra of forty four vicinally substituted 2-, 3-, and 4-aminopyridines[94]; of alkyl-substituted tetramethyl 4H-quinolizine-1,2,3,4-tetracarboxylates[95] were investigated. Substituent effects and chiral discrimination were studied in the complexation of benzoic, 4-methylbenzoic and (RS)-2-phenylpropanoic acids and their conjugate bases by β-cyclodextrin and 6^A-amino-6^A-deoxy-β-cyclodextrin in aqueous solution[96]. The aromatic ^1H n.m.r. shifts of ortho-disubstituted benzenes were used to derive 285 pair-induced chemical shifts (pairwise corrections) for vicinal pairs of the substituent set (F, Cl, Br, I, NH_2, NHCOMe, NO_2, OH, OMe, Me, CHO, COMe, CN, Ph)[97].

The ^1H and ^{13}C n.m.r. spectra of 21 technically important sulphonated 1- and 2-naphthylamines were measured in deuterium oxide, and the substituent chemical shifts of a sulphonic acid group on naphthylamine skeleton were calculated[98]. Substituent-induced chemical shifts along with carbon-carbon double bonds were determined for a series of α-mono- and α,α-disubstituted (1,1,3,3-tetramethyl-2-indanylidene)methanes[99]. A wide variety of push-pull alkenes were studied by means of variable-temperature ^1H and ^{13}C n.m.r. spectroscopy with respect to the configuration/conformation and the barriers to rotation about partial C-C and C-N double bonds[100]. N.m.r. spectroscopy of partially oriented molecules was used to determine the structural substituent effects (changes in bond lengths and angles) in nitrobenzene in liquid crystals[101]. The ^1H and ^{13}C

n.m.r. chemical shift assignments of a series of unsymmetrically N,N-disubstituted (E)- and (Z)-N,N-dialkylacetamides were reported and N-substituent effects were analysed[102].

1.4.2 *Carbon substituent effects*

A novel approach for quantitative studies of the substituent effects on chemical shift values of ^{13}C n.m.r. was reported[103]. Principal component analysis technique was applied to analyse the substituent effect on carbonyl stretching frequency and ^{13}C n.m.r. chemical shifts[104]. Methyl substituent effects in ^{13}C n.m.r. chemical shifts for the indole rings of methyltryptophans were reported[105]. A number of 3- and 4- substituted-2',6'-dimethylazobenzenes and 4-substituted-2,6-dimethylazobenzenes were prepared and their ^{13}C n.m.r. spectra were measured to compare the substituent effect on the ^{13}C n.m.r. chemical shifts for C4' with that for the corresponding azobenzenes[106]. The magnitude of the γ-effect on the ^{13}C chemical shifts of C-4 and C-2 was studied as the function of substitution (Me, Et, benzyl, Ph or p-$NO_2C_6H_4$) at C-2 and C-4, respectively, for several N-unsubstituted and N-methyl-substituted 3,4-dihydro-2H-1,3-benzoxazines and N-unsubstituted 1,2-dihydro-4H-3,1-benzoxazines[107]. ^{13}C n.m.r. spectra of vicinally substituted 2-, 3-, and 4-aminopyridines showed that substituent effects are evidently sensitive to the position of both the substituent and the fixed amino group[108]. Substituent induced chemical shifts (SCS) by the phenyl group in sterically congested styrene derivatives were reported[109].

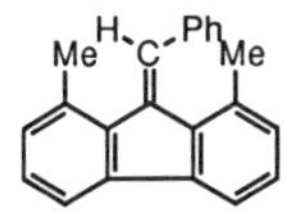

The substituent effects on ^{13}C and ^{17}O n.m.r. chemical shifts of some 5-substituted 2-acetyl thiophenes were analysed by linear free energy relationships[110]. Complete ^1H and ^{13}C n.m.r. assignments and coordination-induced shift [CIS, $\delta_{(complex)} - \delta_{(freeligand)}$] calculations for C σ-bonded ortho-metalated Rh(III) and Ir(III) complexes were described[111]. The ^{13}C n.m.r. chemical shifts for azaferrocenes, phosphaferrocenes and their precursor ligands were used to determine coordination shifts, $\Delta\delta$ (= $\delta_{(complex)} - \delta_{(freeligand)}$) for the complexes[112].

1.4.3 *Heteroatom substituent effects*

The substituent effects on ^{15}N n.m.r. chemical shifts of thirteen m- and p-substituted benzylamines were reported[113]. Correlation of the ^{15}N n.m.r. chemical shifts of the axial nucleotide and corrin ring N22 and N23 was analysed via a multiparameter substituent effect equation[114]. Solvent versus substituent effects on the ^{15}N n.m.r. shielding of the nitro group in substituted benzenes were analysed[115].

Simultaneous determination of primary and secondary thermodynamic isotope effects in tautomeric equilibria was reported[116].

1.5 Intramolecular Hydrogen Bonding Effects and Related Effects

1.5.1 *Proton Shifts*

The temperature and concentration dependence of the ^1H n.m.r. spectra of four anilinomethylenemalonic acid derivatives, 4-$AcC_6H_4NHCH{:}CXY$ (X, Y = CN, CN; CO_2Et, CO_2Et; Ac, CO_2Et; CN, CO_2Et) were determined in $CDCl_3$ and DMSO-d_6 to assign the E-Z isomers and to investigate the formation of intra- and intermolecular hydrogen bonds[117]. Hydrogen bondings in 1-ethyl-2-methyl-4-oxo-1,4-dihydro-pyridin-3-yloxyethanoic acids and 3-(1,2-diethyl-4-oxo-1,4-dihydro-pyridin-3-yloxy)propanoic acids were studied by ^{1}H n.m.r[118]. ^{1}H and ^{13}C n.m.r. were used to study proton transfer in intramolecular hydrogen bonds in substituted 2-diethylaminomethylphenol N-oxides[119].

Intramolecular hydrogen bonding of hydroxy groups to carbanions in indenide- and fluorenide-derivatised alcohols was observed by UV, IR, and ^{1}H n.m.r. in Me_2SO, THF, benzene and toluene[120].

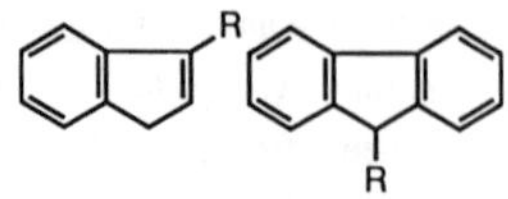

R = $CMe_2CHMeCMe_2OH$

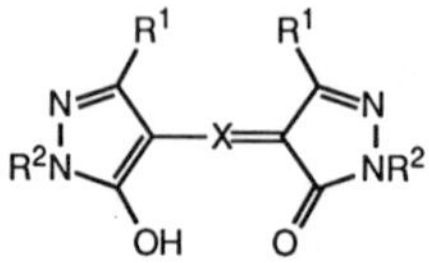

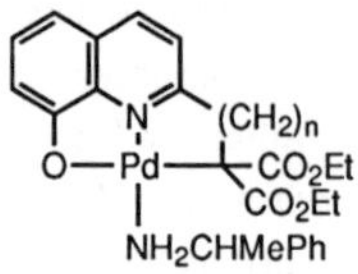

Solution ^{1}H, ^{2}H, ^{13}C, and ^{15}N n.m.r. and solid-state ^{13}C CP/MAS n.m.r. were reported for rubazoic acid derivatives (X = CH, N, R^1 = R^2 = Me; X = CH, R^1 = NH_2, R^2 = Ph; X = CH, R^1 = Me, R^2 = p-$C_6H_4SO_3H$) to evaluate a fast proton transfer equilibrium between two enolic tautomers both in solution and in the solid state[121]. Two novel C-metalated Pd(II) complexes containing diethylmalonate pendants were prepared and their intramolecular hydrogen bondings were analysed by ^{1}H n.m.r. and x-ray crystallography[122].

Free energies of intramolecular hydroxy (donor)-to-ether oxygen (acceptor) hydrogen bonds in hydroxyethers, 10α-hydroxymethyl)-2-oxabicyclo[4.4.0]decane and 4-tert-butyl-2-(hydroxy-methyl)-1-methoxycyclohexane were determined from changes in ^{1}H n.m.r vicinal coupling constants in a broad range of solvents (CCl_4 to D_2O)[123]. The intramolecular hydrogen bond strength between hydroxy and adjacent carbonyl of some sterically crowded substituted phenols was studied by ^{1}H n.m.r[124]. The expression for the limiting chemical shifts for the fully hydrogen bonded state as a function of IR band intensity and chemical shift was derived from the solution studies of N-methylacetamide by n.m.r. and IR spectra[125].

1.5.2 Heteronuclear Shifts

Solid-state ^{13}C and ^{2}H n.m.r. were used to characterise the short hydrogen bonds in salts of dicarboxylic acids[126]. ^{15}N n.m.r. was used to probe hydrogen bonding in ^{15}N-labelled analogues of a self-cDNA fragment, {d[CGTGAATTC(O^{6}Me)GCG]}$_2$ labelled at N1 or N2 position of the O^6-methylguanine residue[127]. ^{17}O n.m.r was used to study intramolecular hydrogen bonding in phenols and salicylaldehydes[128]; in seventeen nitrophenols and four nitrobenzene control compounds[129]; in seven substituted indanones, three fluorenones, and seven 2-methylene-indanones[130].

1.6 Bond Anisotropy, Ring current Effects and Aromaticity

Novel bonding structure of some nonalternant polycyclic systems were reviewed[131]. N.m.r. of π-conjugated polycyclic anions were discussed to show interplay between topology, electronic structure and patterns of charge distribution[132]. The effect of the π-electron ring current on ^{1}H n.m.r. spectra of 1,1-binaphthol derivatives was reported[133]. Syntheses and x-ray crystal structures of the adamantanoparacyclophane and the adamantanopyridinophane, as well as ^{1}H n.m.r. absorption at extremely high field (δ -4.08), unusual conformational behaviour and dienophile addition to the benzene ring of adamantanoparacyclophane were reported[134].

Benzannelated [2.2]paracyclophanes (R = CO_2Me, Ph, etc.) were synthesised and studied by ESR, ENDOR, and n.m.r.[135].

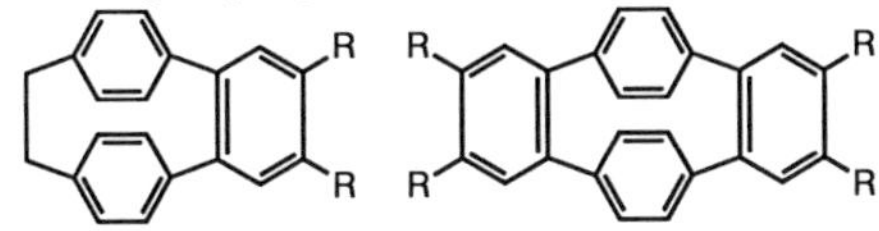

Several methano-bridged annulenes and related compounds were prepared and their π electronic properties were discussed on the basis of ^1H n.m.r. (in $CDCl_3$, CF_3CO_2D, or D_2SO_4), UV (in THF or H_2SO_4) and cyclic voltammetry[136,137,138,139]. Dibenzotrisdehydro[16]annulene-1,4-dione and the related compounds were synthesised and their ^1H n.m.r. spectra were examined to show no ring current effect[140].

21,23-Dithiaporphycene was prepared as the first aromatic sulphur-containing system with porphycene structure and its x-ray crystallographic data, UV/VIS and n.m.r. spectroscopic properties proved the 18π electrons aromaticity of the system, like porphycene[141]. The new 22-π aromatic "pentaplanar" macrocycle, ozaphyrin, was synthesised and characterised by ^1H n.m.r. UV/VIS spectroscopy, cyclic voltammetry, and x-ray crystallography[142].

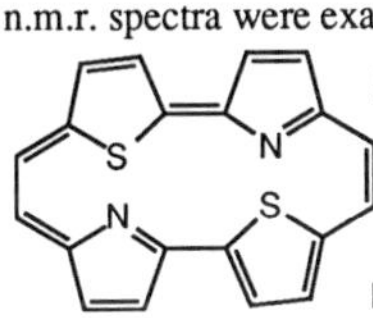

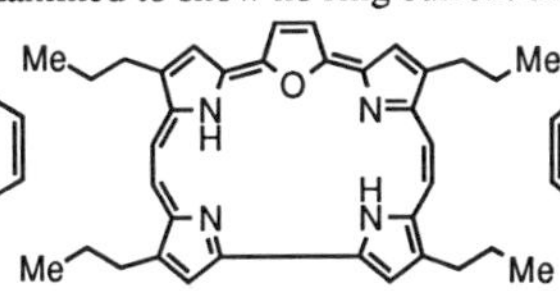

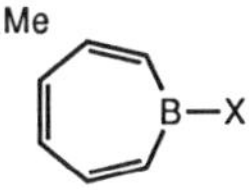

A variety of 1-substituted borepins [X = OH, $O_{1/2}$, OMe, 2,4,6-$Me_3C_6H_2$, $(CH_2)_3NMe_2$, F, NH_2, piperidino, $N(CHMe_2)_2$, H] were prepared and studied by ^1H, ^{11}B, and ^{13}C n.m.r. to evaluate the aromaticity of borepins[143]. The experimental evidence for segregated ring currents in C_{60} was reported[144].

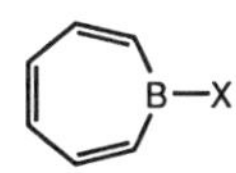

A stable aromatic species containing silicon, 1-tert-butyl-2,3,4,5-tetraphenyl-1-silacyclopentadienide anion, was synthesised and characterised by ^1H, ^{13}C, and ^{29}Si n.m.r. and by the reaction product with Me_2SiHCl[145].

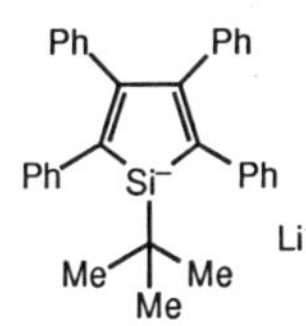

1.7 Inclusion phenomena and Related Effects

Solid-state n.m.r. studies of host-guest materials were reviewed[146]. The role of n.m.r. spectroscopy, with emphasis on shielding variations and nuclear Overhauser effects for structural elucidation of host-guest compounds in solution, was reviewed on the basis of selected cyclophane and cyclodextrin associations[147].

1.7.1 Cyclodextrins

The influence of α-, β- and γ-cyclodextrin (CD) on the ^1H n.m.r. spectrum of the muscarinic antagonist telenzepin dihydrochloride in D_2O was reported[148]. ^{13}C MAS n.m.r. spectra of tropolone included in α- and β-cyclodextrin were observed at various temperatures to study tautomeric proton transfer[149]. Complexation of ephedrine with β-cyclodextrin was studied by ^1H n.m.r. and CD[150]. The interactions between bile acid anions and cyclodextrins in aqueous solution at room temperature were studied in detail by the n.m.r. titration method[151]. ^1H and ^{13}C n.m.r. were used to study complexation of reduced tetracyanoquinodimethane ($TCNQ^-$) with β-cyclodextrin resulted in the formation of a σ-bonded dimer, $(\beta\text{-CD-TCNQ})_2^{2-}$ [152]. Dynamics of ferrocene molecules enclathrated in β-cyclodextrin was studied by variable-temperature one-dimensional SASS (switching-angle sample spinning) n.m.r[153].

α-Cyclodextrin formed inclusion complexes with polyethylene glycol of various molecular weights to give stoichiometric compounds in a crystalline state in high yields and the complexes were characterised by IR, ^{1}H, ^{13}C, ^{13}C CP/MAS n.m.r[154].

1.7.2 Miscellaneous Hosts

The effect of the solvent on the conformations of unsubstituted *p-tert*-butylcalix[4]arene and its methyl ethers were investigated by ^{1}H n.m.r[155]. The cation-π interaction was studied using 9 calix[n]arenes with the different cavity size and the different cavity shape[156]. ^{1}H n.m.r. was used to study the inclusion complexation of ferrocene derivatives by a water-soluble calix[6]arene[157].

High-resolution solid-state ^{13}C n.m.r. and optical microscopy were used to show that *n*-decane molecules migrates from the liquid phase into the tunnel structure of the 1,8-dichlorooctane-urea inclusion compound without loss of single crystal integrity[158]. ^{13}C n.m.r. was used to study poly(L-lactide) in the narrow channels of its inclusion compound with urea[159]. Variable-temperature one-dimensional and two-dimensional ^{13}C n.m.r. were used to investigate conformational properties of monosubstituted cyclohexanes in their thiourea inclusion compounds and in solution[160].

A conformational n.m.r. study of the acid form and Li^{+}, Na^{+}, K^{+}, Rb^{+}, Cs^{+} and NMe_4^{+} neutral 1:1 complexes of carboxylic ionophore cationomycin was carried out in $CDCl_3$ and CD_3OD[161].

A n.m.r. and molecular-interaction energy calculation study of static and dynamic properties of p- and o-xylene guest molecules in Dianin's inclusion compound was reported[162]. The CH-π interaction was proved to play as an important driving force of host-guest complexation in apolar organic media[163]. Functional group interaction of the guest molecules of the formula $X(CH_2)Y$ in channel inclusion compounds of perhydrotriphenylene was studied by ^{13}C CP/MAS n.m.r[164]. A large number of binding studies were reported in aqueous media designed to provide new insights into noncovalent binding interactions, especially the cation-π interactions including 7 different hosts, >70 guests, and >150 new binding constants[165]. The cyclic intermolecular solid state proton transfer involving nitrogen and oxygen atoms in host-guest crystals was reported[166]. A correlation between the mechanism of ^{1}H n.m.r. shift of a chiral guest, e.g., (S)-(-)-$PhCHMeNH_2$, by association with an optically active host compounds, e.g., (R,R)-(-)-bis(hydroxydiphenylmethyl)-2,2-dimethyl-1,3-dioxacyclopentane, in solution and in the solid state was revealed by an x-ray analysis of the host-guest inclusion crystal[167]. ^{13}C Solid-state n.m.r. spectra were measured for γ-valerolactone in the cholic acid host which exhibited efficient optical resolution[168]. N.m.r. measurements of associations between simple organic host and guest molecules bearing alternatively negative charged groups and electroneutral π-moieties showed in aqueous solutions weak but distinct attractions reaching approximately 2kJ mol^{-1} and X^{-}/arene unit[169].

1.8 Intermolecular Hydrogen Bonding Effects and Related Effects

N.m.r. studies of drug-DNA interactions were reviewed[170]. Recent progress in the use of 1D and 2D n.m.r. for the determination of the dynamics and ordering of hydrogen-bonded ferroelectric materials was reviewed[171].

The interaction of different phosphate esters with hydrogen bonding receptor was examined using UV, n.m.r. and x-ray crystallography to show a balance of hydrogen bonding and proton transfer interaction in molecular recognition of phosphate esters[172].

A new type of small molecule-DNA interaction by the monoprotonated form of the water soluble pentapyrrolic sapphyrin was reported and support for the proposed binding mode derived from a variety of spectroscopic and biochemical studies, including visible absorption, solid-state ^{31}P n.m.r. CD, fluorescence and supercoiled DNA unwinding experiments conducted using the calf thymus enzyme, topoisomerase I[173]. Hydrogen bonding of water to chlorophyll a and its derivatives was detected by ^{1}H n.m.r[174].

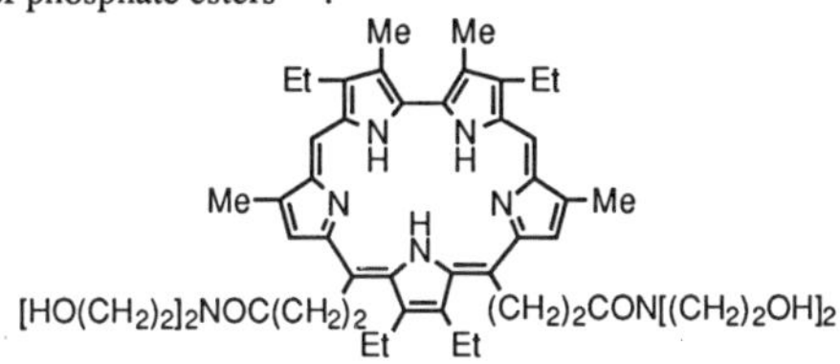

Measurements of ^{1}H n.m.r. chemical shifts and shear viscosities as a function of temperature and concentration for 1,3- and 1,4-dioxane-water systems were carried out to suggest association between water and the dioxane molecules[175]. Solvent effect on the ^{1}H n.m.r. spectra of the α-proton of α,N-diarylnitrones was studied, and comparison studies of chemical shift with corresponding imines were performed[176]. The concentration effects of hydrogen bonding between phenol and various proton acceptors were studied by ^{1}H n.m.r[177].

The structures of supramolecular aggregates based on the cyanuric acid-melamine lattice were characterised using ^{1}H, and ^{13}C n.m.r., UV, gel permeation chromatography, and vapour pressure osmometry to show molecular self-assembly through hydrogen bonding[178]. The aggregation process of peptide GVKGDKGNPGWPGAPY from the triple-helix domain of type IV was investigated by ^{1}H and ^{13}C n.m.r[179]. Self-association of dipyrinones was observed by 2D NOE n.m.r. and dimerization constants were calculated from ^{1}H n.m.r. chemical shifts[180]. The concentration dependent self-association of nicotinamide in solution was studied by ^{1}H and ^{13}C n.m.r[181].

The structure of water in poly(methacrylic acid) gel as a function of the degree of crosslinking of the gel was investigated by experimentally and theoretically calculated ^{1}H n.m.r. shifts[182]. The interaction of water with the silica surface was explored by variable-temperature high-resolution ^{1}H n.m.r[183].

^{15}N n.m.r. shieldings of pyrrole-type nitrogen atoms showed sensitivity to provide an insight into solvent polarity and solute-to-solvent hydrogen bonding effects[184].

1.9 Other Miscellaneous Topics

Proposals for definitions of notation for shifts and shieldings and shift or shielding tensors were presented[185]. The temperature and concentration dependences of the chemical shift of aqueous selenous acid was studied to establish a precise relation between the shift of this reference and that of dimethyl selenide[186]. Systematic explanations of bulbed capillary chemical shifts of benzene, bromoform and chloroform in pure liquids and some binary systems were given using intramolecular and intermolecular effects[187]. The n.m.r chemical shifts of tetramethylsilane (TMS) and sodium 2,2-dimethylsilapentane-5-sulphonate (DSS) were changed considerably as a result of the strong interaction between β- and γ-cyclodextrin in the inclusion complexes[188].

A unified scale for predicting nonspecific solvent polarity employed the physicochemical properties of solutes, such as n.m.r., EPR, electronic transitions, etc., and the equation $\Delta_x = S'P + W$ in order to produce a scale of nonspecific solvating ability[189]. The validity of the bulbed capillary external referencing method with a superconducting magnet n.m.r. was discussed[190].

Solid-state n.m.r. studies of catalytic reactions on molecular sieves were reviewed[191].

2 Shieldings of Particular Nuclear Species

During the period of this review, the n.m.r. spectra of most elements have at least received some chemical investigation. Due to the space limitation, structure determination and related studies will be excluded for the most popular nuclei (^{1}H, $^{10,11}B$, ^{13}C, $^{14,15}N$, ^{19}F, ^{31}P), and each article will be given only brief comments in the following section.

2.1 Group 1 (^{1}H, ^{2}H, ^{3}H, $^{6,7}Li$, ^{23}Na, $^{85,87}Rb$, ^{133}Cs)

2.1.1 Hydrogen (^{1}H)

A huge number of n.m.r. spectra of inorganic and organometallic compounds were collected[192].

Self-organisation of bilayer assemblies in fluorocarbon medium was studied by ^{1}H n.m.r. line broadening with temperature[193]. A n.m.r. method for quantitative determination of the exchange rates of slowly exchanging amide protons in proteins was presented, with the des-[Phe(B25)] mutant of human insulin as an example[194]. The highly selective self-assembly of a chiral bis[2]catenane incorporating two fused cyclobis(paraquat-p-phenylene)rings, each catenated with a bisparaphenylene-34-crown-10 ring, was prepared and ^{1}H and ^{13}C n.m.r. were used to study static and dynamic structure[195]. The ^{1}H and ^{13}C n.m.r. spectra of all $[2_n]$cyclophanes with the same substitution pattern in both benzene rings and of the skew [2.2.2](1,2,4)(1,2,5)cyclophane were assigned by one-dimensional NOE difference and two-dimensional correlation spectroscopy[196]. Chemical shifts of all ^{1}H and ^{13}C atoms were assigned for substituted 7'-aryl-7'-apo-β-carotenes [Ar = C_6H_5, 4-$C_6H_4NO_2$, 4-$C_6H_4OCH_3$, 4-$C_6H_4CO_2CH_3$, C_6F_5 and 2,4,6-$C_6H_2(CH_3)_3$][197,198].

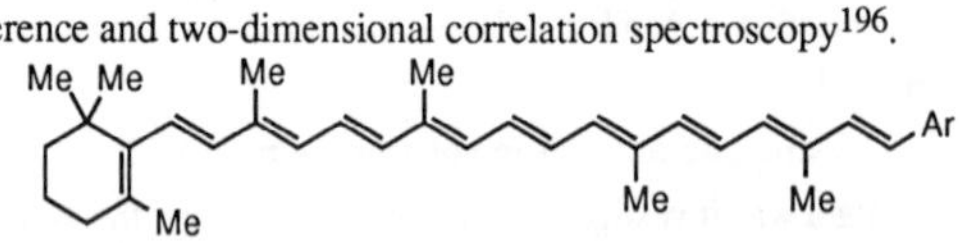

Aluminium electronegativity in molten salts based on 3-butyl-1-methylimidazolium chloride and $AlEtCl_2$, $AlEt_2Cl$, $AlEt_3$, and mixtures of $AlEtCl_2$ and $AlCl_3$ was determined by the modified Daily-Shoolery equation using ^{1}H n.m.r[199]. Mechanism of hydride scrambling in a transition-metal dihydrogen dihydride was studied by solid-state ^{1}H n.m.r[200]. The physicochemical nature of protons generated by the partial reduction of Ag^{+} cations in $Ag_3PW_{12}O_{40}$ with hydrogen was investigated by means of ^{1}H MAS n.m.r[201].

A clear correlation between the ^1H n.m.r. chemical shift and the stretching vibration frequency of hydroxy groups in solids were reported[202]. Hydrogen concentration dependence of the Knight shift of protons in NbH_x ($0.50 \leq x \leq 1.05$) was studied by ^1H MAS n.m.r[203].

High-resolution solid-state ^1H n.m.r spectra were obtained by combining deuterium spin-dilution and high-speed magic-angle spinning[204].

2.1.2 *Deuterium (^2H)*

In vivo ^2H n.m.r. studies of cellular metabolism was reviewed[205].

^2H n.m.r. was used to consider correspondence between existing single-crystal x-ray data for glycosphingolipids and their ceramide backbone conformation in fluid phosphoripid membranes[206]; to study isotope exchange in tritiated alprazolam[207]. The reduction in spectral splitting, or motional narrowing, of the ^2H n.m.r. spectra of D_2O/phospholipid mixtures near the main chain melting phase transition was studied for palmitoyloleoylphosphatidylcholine (POPC), palmioyl-oleoylphosphatidylethanolamine (POPE) and equimolar mixtures of the two at 10% hydration[208]. The effects of 3 terpenoids (cholesterol, cycloartenol, and the hopanoid, bacteriohopaneaminotriol) on dimyristoylphosphatidylcholine membrane structure and dynamics were compared by solid-state ^2H n.m.r[209]. In order to get insights into the effects of cholesterol on protein activity, the lytic power of melittin on 1,2-dipalmitoyl-sn-glycero-3-phosphocholine (DPPC)/cholesterol mixtures was studied by solid-state ^2H and ^{31}P n.m.r[210]. Solid-state ^2H n.m.r. was used to study both equilibrium and dynamic properties of membrane constituents at the molecular level and could yield significant insights regarding the organisation of non-bilayer lipid aggregates[211].

^2H n.m.r. was used to study molecular dynamics and geometry in two classes of onium salts, $(CH_3)_3E^+X^-$ and $C_6H_5M(CH_3)_3{}^+I^-$ [E = O (counter ion BF_4^-), S, Se, Te; M = N, P, As][212]; to demonstrate that the reaction of 1-methylpyrrole with N-chloroimides occurs by addition-elimination[213]. Selective deuterium ^2H,^2H ^{1}D-TOCSY n.m.r. spectroscopy was introduced for signal assignment in perdeuterated compounds[214]. The reaction of cyclopentene and deuterium on silica-supported catalysts were studied with the products of both the exchange and addition reactions being analysed by combined mass spectrometry, gas chromatography, and ^2H n.m.r[215]. The synthesis of the deuterium labelled, endogenously occurring, indolealkylamine hallucinogens N,N-dimethyltryptamine and 5-methoxy-N,N-dimethyltryptamine was described and the compounds were characterised by ^1H, ^2H, and ^{13}C n.m.r[216]. Variable temperature natural abundance ^2H n.m.r. studies of solid-state phase transitions in adamantane, tetrakis(trimethylsilyl)-methane, tetrakis(trimethylsilyl)silane, and 2,2-dimethyl-1,3-propanediol were reported[217]. Solid-state ^2H n.m.r. was used to examine the dynamic behaviour of 18-CD_3 methyl groups in micro-crystalline 6-s-cis-retinoic acid (triclinic) and 6-s-trans-retinoic acid (monoclinic) model compounds, as well as in the membrane protein bacteriorhodopsin regenerated with CD_3-labelled retinal[218]. Hydrogen/deuterium exchange was observed by ^1H and ^2H n.m.r. for the tert-butyl cation ($(CH_3)_2C^+CH_4^-$) in super acid, $DF/SbF_5/SO_2ClF$[219]. ^2H and ^{13}C n.m.r. measurements were reported of the solid and discotic phases of three homologues of the benzenehexa-n-alkanoate series, BHAn, with n (the number of carbons in the alkanoate chains) equal to 6, 7, and 8[220]. ^2H and ^{13}C n.m.r. data for propene adsorbed on high-dispersion, alumina-supported platinum surfaces were obtained after adsorption at various temperatures and thermal treatments[221]. ^{2}H and ^{13}C variable temperature solid-state n.m.r. were used to study the dynamic properties of p-diiodobenzene[222].

^{2}H n.m.r was used to study the effects of the anionic polyelectrolyte poly(sodium styrenesulphonate) (PSSS) on the surface electrostatic charge of polystyrene particles bearing a preformed monolayer of surfactants[223]; to investigate the molecular dynamics of new thermoplastic elastomers based on hydrogen bonding complexes[224]; to study the molecular mobility and dehydration reaction of tert-butyl alcohol, selectively deuterated in the methyl groups (t-BuOH[2-$^{2}H_9$], dTBA), absorbed on H-ZSM-5 zeolite[225].

DSC and ^{2}H n.m.r. were performed to study interaction of phospholipid liposomes with lipid model mixtures for stratum corneum lipids[226].

A novel intermolecular cross-polarisation technique, proton-deuterium CP/MAS n.m.r., was developed for characterisation of polymer blend miscibility[227]. Reproentation of the polystyrene component in a compatible blend of polystyrene and poly(2,6-dimethylphenylene oxide) with 75% polystyrene was observed by two-dimensional solid-state ^{2}H n.m.r. at the glass transition[228].

^{2}H n.m.r. was used to probe the molecular dynamics and structure of the surfactant sodium 4-(1'-heptylnonyl)benzenesulphonate adsorbed from water solution onto alumina[229]; to observe a new sigmatropic liquid crystal, 2-(p-heptylbenzoyloxy)-5-(p-heptylbenzenazo)tropone[230]; to study dodecyltrimethylammonium bromide (DoTAB) and CTAB (D-labelled particles)[231]; to characterise methyl group rotations between their classical and quantum mechanical limits in Cu acetate monohydrate $[Cu(CD_3CO_2)_2]\bullet H_2O$, hexamethylbenzene-$d_{18}$ and acetone-d_6[232]; to study alkyl chain ordering of a series of hexaalkoxytriphenylene solutes dissolved in the calamitic solvent Phase V[233]; to study molecular order and dynamics on pure 8OBCAB-d_2 and on its mixture with compounds of the nTPCHB series[234]; to study dynamics and coexistence of paraelectric and ferroelectric phases in rubidium ammonium-d_4 dideuterium arsenate $(Rb_{0.90}(ND_4)_{0.10}D_2AsO_4)$[235]; to study ND_4ClO_4 single crystal at $\nu_0 = 44$ MHz[236]; to study n-hexane (C_6D_{14}), benzene (C_6D_6), and a mixture of n-hexane and benzene $(C_6D_{14}/C_6H_6$ and $C_6H_{14}/C_6D_6)$ adsorbed on the graphitised carbon black Carbopack B within 300-90 K[237].

^{2}H n.m.r. spectra of solids in natural abundance could often readily be obtained in short accumulation times and with a quality suitable for quantitative analysis if recorded under conditions of MAS[238]. ^{2}H solid echo n.m.r. was used to determine the rate of 180° flips of $^{2}H_2O$ molecules in solid $NdCl_3\bullet 6^{2}H_2O$[239].

The ^{2}H,^{2}H-INADEQUATE experiment was reported[240].

2.1.3 Tritium (^{3}H)

An overview was given of the possibilities of ^{3}H n.m.r. in the characterisation of ^{3}H-labelled compounds[241].

Tributyltin tritide, a versatile tritiation reagent, was synthesised by the reaction of Bu_3SnCl with Li^3H at maximum radioactivity and characterised by ^{1}H, ^{3}H, and ^{119}Sn n.m.r[242].

^3H n.m.r. of specifically tritiated and tritiated/deuterated derivatives of tosylchmotrypsin was examined to show the properties of the tosyl group[243,244].

^3H-Mometasone furoate was synthesised and analysed by ^3H n.m.r to indicate that most of the label (≈ 95%) was located in the 2-position and the remainder in the 1-position[245]. [1',3',4'-$^3H_3$]4-(5',6',7',8'-tetrahydro-5',5',8',8'-tetramethyl-2'-anthracenyl)benzoic acid was synthesised for binding studies of retinoic acid receptors and ^1H and ^3H n.m.r. were used to establish tritium at the 1',3',4'-tetrahydroanthracenyl ring positions[246]. The synthesis of [2-^3H]ATP with specific activity high enough to use for ^3H n.m.r. spectroscopy at micromolar concentrations was accomplished by tritiodehalogenation of 2-bromo-ATP[247].

2.1.4 *Lithium ($^{6,7}Li$)*

Solution structure of lithium dialkylamides and related N-lithiated species was reviewed[248].

^7Li and ^{23}Na n.m.r. were used to study transport and diffusion in liposomes[249].

^6Li and ^{13}C n.m.r. spectra of $Li[C(SiMe_3)_2]_2Li$ revealed a doubly bridged π-structure in aromatic solvents similar to that observed in the crystals[250]. ^{13}C, ^6Li, and ^1H n.m.r., including one- and two-dimensional $^{13}C\{^1H\}\{^6Li\}$ triple resonance experiments, were used to study the solution structures of tert-butyllithium/lithium tert-butoxide mixed aggregates[251].

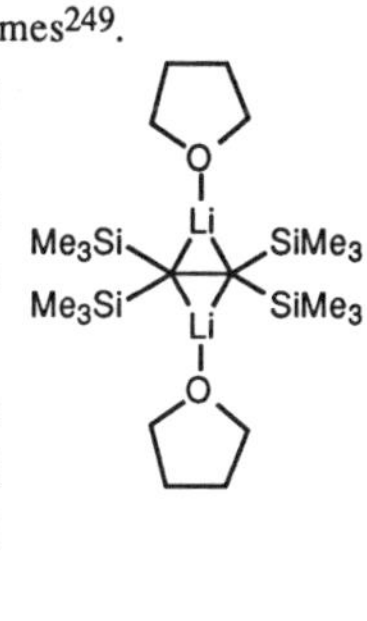

Methyllithium, isolated as the THF solvate, (MeLi•THF), from a cumene/THF solution, was characterised by ^1H, ^7Li, ^{13}C n.m.r. spectra and by x-ray diffraction[252]. ^7Li MAS and static n.m.r. study of mono-, di-, and tetrameric phenyllithium aggregates was reported[253].

Low temperature ^7Li, ^{31}P, ^{13}C, and ^{29}Si n.m.r. were used to study and characterise lithium ion pair structures in THF and THF/HMPA solution and sulphur and silicon substituted carbanions[254,255]. Three examples of simple, nonsolvated, σ-bonded lithium aryls were synthesised and characterised by ^1H, ^{13}C, and ^7Li n.m.r. and x-ray crystallography[256].

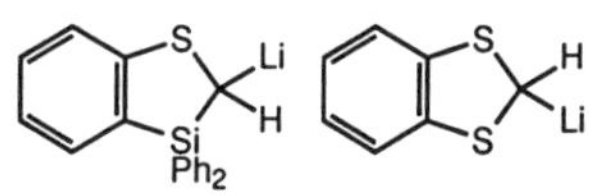

^6Li and ^{15}N n.m.r. spectroscopic studies of ^6Li and ^{15}N isotopically labelled lithium hexamethyldisilazide ([^6Li]LiHMDS and [^6Li, ^{15}N]LiHMDS) confirmed the Kimura and Brown's earlier conclusions[257].

Novel mixed alkali metal organonitrogen ladder, $\{LiNa[N(CH_2Ph)_2]_2{\cdot}OEt_2\}_2$, was synthesised and characterised by ^1H and ^7Li n.m.r. and x-ray crystallography[258]. The trilithium triamide $HC\{(CH_3)_2SiN(Li)\text{-}t\text{-}Bu\}_3$ and its bis-THF solvate $HC\{(CH_3)_2SiN(Li)\text{-}t\text{-}Bu\}_3{\cdot}THF$ were prepared and studied by ^1H and ^7Li n.m.r[259].

^7Li and ^{31}P n.m.r. and x-ray crystallography were used to study dimeric structures of $[Li(Ph_4P_2N_4S_2R){\cdot}THF]_2$ (R = Me, CMe_3, Ph)[260]. ^7Li, ^{13}C and ^{29}Si n.m.r. were used to study the solution structures of tris(trimethylsilyl)methyllithium-tetrahydrofuran(1/2) and -N,N,N',N'-tetramethylethylenediamine(1/1)[261].

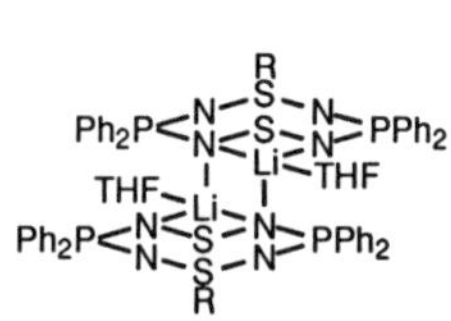

^{13}C n.m.r., ^6Li, ^1H HOESY n.m.r., and IR spectroscopy of α-lithiomethoxyallene and 1-lithio-1-ethoxy-3-tert-butylallene as well as ab initio model calculations on monomeric and dimeric α-lithiohydroxyallene proved the dimeric structure in THF with a nonclassically 1,3-bridged structure[262]. A lithium salt of 2,3-dicarba-nid-o-hexaborate(1-) was synthesised and characterised by ^{1}H, ^{7}Li, ^{11}B and ^{13}C n.m.r. and IR[263]. The ^{13}C and ^{7}Li n.m.r. spectra of the ion-pair structures of pyridine-butyllithium adducts were observed in THF[264].

^{7}Li n.m.r. was used to study the dilithium isoprene polymerisation initiator (1,3-phenylene-bis(3-methyl-1-[methylphenyl]pentylidene)dilithium[265]; to analyse the solutions of methyl α-lithioisobutyrate (a model for living PMMA) in THF with various amounts of LiCl in a wide range of temperatures[266].

^{7}Li and ^{23}Na n.m.r. were used to study the complexation of alkali metal ions by the pendant arm macrocyclic ligand 1,4,7,10-tetrakis(2-hydroxyethyl)-1,4,7,10-tetraazacyclododecane in methanol and DMF[267]. Potentiometric titration and ^{7}Li and ^{23}Na n.m.r. were used to study the complexation of $M^{+} = Li^{+}$, Na^{+}, Ag^{+} and Tl^{+} by the cryptands 4,7,13,16-tetraoxa-1,10-diaza-bicyclo[8.5.5]eicosane (C211) and 4,7,13-tetraoxa-1,10-diazabicyclo[8.5.5]eicosane ($C21C_5$) to form the cryptates $[M.C211]^{+}$ and $[M.C21C_5]^{+}$ in triethyl and tributy phosphate[268]. High-temperature phases of $Li_xV_2O_5$ bronzes ($0.2 \leq x \leq 1$) were studied by ^{6}Li and ^{7}Li n.m.r[269]. ^{7}Li, ^{23}Na, and ^{13}C solid-state n.m.r. spectroscopy under static and magic angle spinning conditions were used to study the salt complexes of lithium chloride, lithium bromide, lithium perchlorate and lithium nitrate with 12-crown-4 and of sodium nitrate with 15-crown-5[270]. ^{7}Li n.m.r. was used to compare the Li^{+} transport in sodium vanadate ($Na_{1+x}V_3O_8$) and lithium vanadate ($Li_{1+x}V_3O_8$)[271]. Variable-temperature ^{7}Li and ambient temperature ^{29}Si n.m.r. were used to study solid-state and solution structures of some lithium salts of tetraphenyldisiloxanediolate(2-) and the lithium-bridged compounds $Li_2[M(OSiPh_2OSiPh_2O)_3{\cdot}3py]$ (py = pyridine, M = Zr, Hf)[272].

2.1.5 Sodium (^{23}Na)

Shift-reagent-aided ^{23}Na n.m.r. in cellular, tissue, and whole-organ systems was reviewed[273].

The use of Tm 1,4,7,10-tetraazacyclododecane-1,4,7,10-tetrakis(methylene phosphonate) as an in vivo ^{23}Na n.m.r. shift reagent for rat liver was evaluated by collecting interleaved ^{23}Na and ^{31}P spectra[274]. Elevation of intracellular sodium ion concentration in human erythrocyte induced by the cardiac glycoside, proscillaridin and its four derivatives was measured by ^{23}Na n.m.r[275]. ^{23}Na n.m.r. was used to determine the intracellular sodium in perfused cancer cells with shift reagent[276]. Double quantum filtered (DQF) ^{23}Na n.m.r. signals were evaluated as a function of $[Na^{+}]$ at constant temperature in two model systems (bovine serum (BSA) and Ficoll 400) and in human red blood cells (RBCs)[277].

The static ^{23}Na n.m.r. spectrum of N,N,N',N'-tetramethylethylenediamine-sodium cyclopenta-dienide revealed a lineshape that could not be unambiguously discussed[278]. ^{23}Na n.m.r. was used to observe and investigate the ionophore-mediated Na^{+} exchange between two populations of liposomes[279]; to study properties of crown ether-alkali metal cation intercalation compounds of transition metal phosphorus sulphides, MPS_3 (M = Mn, Cd, Zn)[280].

^{23}Na n.m.r and DSC were used to characterise a stoichiometric range of sodium salt complexes of amorphous poly[(oxymethylene)oligo(oxyethylene)][281]. The interaction between Na^+ and polymer was studied by ^{23}Na n.m.r. for aqueous 2-hydroxyethylmethacrylate-methacrylic acid copolymer Na salt[282].

The aggregation of $M[Cr_2(L\text{-}tart_2H)(phen)_2]$ (phen = 1,10-phenanthroline, tart = tartrate, M = Na, K) in aqueous solution was studied by measuring ^{23}Na n.m.r. of counterions, ^2H n.m.r. of H_2O molecules, viscosity and electric conductance[283]. Variable-temperature ^{23}Na n.m.r. was used to show complexation of the sodium ion by a pendant-arm tetraaza macrocyclic ligand in aqueous solution[284]. A series of 12 synthetic ionophores containing an amide substituent connected via a methylene bridge to an aza-crown ring evaluated for their ability to transport Na^+ across large unilamellar vesicles using dynamic ^{23}Na n.m.r[285].

Crystalline sodium aluminogermanate hydroxosodalite hydrate $Na_{6+x}[Al_6Ge_6O_{24}](OH)_x{\bullet}H_2O$ ($x \approx 1.6$, $n \approx 3.0$) was synthesised and characterised by simultaneous thermal analysis, DSC, x-ray and neutron diffractions as well as ^1H and ^{23}Na MAS n.m.r. and IR spectroscopy[286]. The water dynamics in $Na_3PMo_{12}O_{40}{\bullet}nH_2O$ and $Na_3PMo_{12}O_{40}{\bullet}nD_2O$ with different n was studied by ^1H, ^2H and, ^{23}Na n.m.r. spectra and relaxation time measurements over a wide temperature range[287]. The adsorption of Na+ on g-alumina at 4 coverages of Na_2CO_3 [5, 10, 15, and 20% (wt./wt.)] was characterised by solid-state ^{23}Na and ^{27}Al n.m.r[288]. The sodium salt of the monoanion $[C_{60}]^{\cdot-}$, $Na^+[C_{60}]^{\cdot-}{\bullet}(THF)_5$ was prepared and the ^{13}C and ^{23}Na solid-state n.m.r. EPR and UV spectra were reported[289].

2.1.6 *Rubidium ($^{85,87}Rb$)*

The use of ^{87}Rb n.m.r. to measure cation fluxes in cells and whole organs was reviewed[290].

Rb_3C_{60} and $Rb_{1.5}K_{1.5}C_{60}$ were studied by ^{87}Rb n.m.r[291]. $^{85,87}Rb$ n.m.r. spectra and relaxation were reported at 4-300 K, for samples of C_{60} intercalated with Rb through a direct-reaction technique at the nominal composition Rb_3C_{60}[292]. The reaction of a stoichiometric amount of Rb with C_{60} in liquid NH_3 provided a rapid and quantitative route to Rb_3C_{60} superconductor and ^{13}C and ^{87}Rb n.m.r. line shapes were reported for Rb_3C_{60}[293]. ^{133}Cs, ^{87}Rb, and ^{13}C n.m.r. measurements on the alkali-metal fullerides CsC_{60} and RbC_{60} were reported[294].

2.1.7 *Caesium (^{133}Cs)*

^{133}Cs n.m.r. chemical shifts and relaxation times were measured for tissue samples in vitro and in vivo from rats which were fed on a high caesium, low potassium diet, which leads to a predominantly intracellular distribution of this ion, similar to that of K^+ [295]. ^{133}Cs n.m.r. was used to study caesium ion complexes of calix[4]arene systems[296].

The high resolution phase diagram of the Cs pentadecafluorooctanoate (CsPFO)/H_2O system was mapped out using mainly ^{133}Cs n.m.r[297]. ^{133}Cs n.m.r. spectra of a single crystal of Cs_2CrO_4 (space group Pnma) were obtained as a function of crystal orientation in an applied magnetic field of 4.70 T and analysed to provide the magnitudes and orientations of the ^{133}Cs chemical shielding and quadrupolar tensors for each of the 2 crystallographically distinct Cs sites[298]. ^{133}Cs n.m.r was performed on $CsCuCl_3$ at 4.2 K under the applied magnetic field parallel to the crystal c-axis[299]. ^{133}Cs n.m.r. was observed in C_{60} superconductors, $Cs_xRb_{3-x}C_{60}$ with x = 1 and 2[300].

2.2 Group 2 (^9Be, ^{25}Mg, ^{43}Ca)

2.2.1 *Beryllium* (^9Be)

Be^{2+} was inserted into the framework of ZSM-5 zeolite by treatment with aqueous NH_4BeF_4 and characterised by ^9Be MAS n.m.r[301]. ^9Be, ^{10}B, and ^{11}B n.m.r. were used to determine the boron substitutional site in boron-doped uranium-beryllium (UBe_{13}-xB_x)[302].

2.2.2 *Magnesium* (^{25}Mg)

Thermodynamic parameters for magnesium binding to a series of DNA molecules of defined sequence were evaluated by ^{25}Mg n.m.r[303]. The essential magnesium-binding site in Escherichia coli ribonuclease H was determined by high field ^1H n.m.r. and the kinetic and thermodynamic parameters were evaluated by ^{25}Mg n.m.r[304]. A novel approach was presented to achieve improved resolution in ^{25}Mg n.m.r. spectrometry[305].

^{25}Mg MAS n.m.r. was used to characterise the reaction sequences in thermal decomposition of magnesite and brucite from 2 geologically different deposits[306]. ^{29}Si, ^{25}Mg, and ^7Li MAS n.m.r were used to study the thermal reactions of synthetic hectorite[307].

2.2.3 *Calcium* (^{43}Ca)

^{43}Ca n.m.r. applications to obtain unique information about biological systems were reviewed[308].

2.3 Group 3 and Lanthanides (^{45}Sc, ^{89}Y, ^{139}La, ^{159}Tb, ^{167}Er, ^{171}Yb)

2.3.1 *Scandium* (^{45}Sc)

The binding of Sc^{3+} to chicken ovotransferrin was investigated by ^{45}Sc and ^{13}C n.m.r[309].

The first order and second order quadrupolar shifts of the ^{45}Sc n.m.r. spectrum were studied at 4-300 K in relation to the structural phase transitions of $Sc(AcO)_3$ and $Sc(CD_3COO)_3$[310].

2.3.2 *Yttrium* (^{89}Y)

^1H, ^{13}C, ^{89}Y n.m.r. were used to study complex formation in solution between the ferrocene-containing cryptand 1,1'(1,4,10-trioxa-7,13-diazacyclopentadecane-7,13-diyldicarbonyl)ferrocene with yttrium perchlorate[311]. Tris(dipivaloylmethanato)yttrium, Y(thd)3 (thdH = dipivaloylmethane), and its monohydrate were synthesised and characterised by elemental analysis, IR, ^1H, ^{13}C, and ^{89}Y n.m.r[312]. The ^{89}Y and the planar ^{17}O n.m.r shifts K and spin-lattice relaxation rates $1/T_1$ in oriented $YBa_2Cu_3O_{7-y}$ powder with $y \approx 0$ and $y = 0.37$ were compared in detail[313].

2.3.3 *Lanthanum* (^{139}La)

Bimetallic lanthanide complexes with lanthanide-transition metal bonds were prepared and characterised by x-ray crystallography and ^{139}La n.m.r[314].

^{139}La n.m.r. was used to study La hydrides in LaH_{2+x} (x = 0.32, 0.38, 0.40, 0.74) at 2.3-303 K[315]; to study metallic $La_2CuO_{4+\delta}$[316]. 2,2'-bipyridine chelates of lanthanum(III) in acetonitrile was characterised by ^{1}H, ^{13}C, ^{17}O and ^{139}La n.m.r[317]. Dy(III)-induced ^{17}O, ligand-induced ^{139}La shift, and Gd(III)-induced ^{13}C relaxation rate enhancement measurements of propionate, acrylate, and propiolate complexes revealed a decreasing extent of inner-sphere complexation across the series[318].

2.3.4 Terbium and Erbium (^{159}Tb and ^{167}Er)

^{159}Tb and ^{167}Er spin echo n.m.r. were measured on $Tb_2Fe_{14}B$, $Er_2Fe_{14}B$ and $(Er_{0.1}Gd_{0.9})_2$-$Fe_{14}B$ compounds[319].

2.3.5 Ytterbium (^{171}Yb)

^{31}P and ^{171}Yb n.m.r. were used to study phosphacyclopentadienyl and arsocyclopentadienyl complexes of ytterbium(II) and samarium(II)[320]. Biphenylene reacted with activated Sm and Yb powders in ethers gave 2,2'-biphenyldiyl-Sm(II) and -Yb(II) complexes, characterised by reactions with electrophiles such as the proton and dimethyldichlorosilane, and by ^{1}H, ^{13}C and ^{171}Yb n.m.r[321].

2.4 Group 4 (^{91}Zr)

2.4.1 Zirconium (^{91}Zr)

^{91}Zr n.m.r. was observed in fcc (∂ phase) and f.c.t. (ε phase) ZrH_x ($1.55 \leq x \leq 2$) at 150-294 K[322].

2.5 Group 5 (^{51}V, ^{93}Nb)

2.5.1 Vanadium (^{51}V)

Vanadium(III) and -(V) alkoxo complexes of the ligands hydridotripyrazolylborate and hydridotris(3,5-dimethylpyrazolyl)borate were prepared and characterised by x-ray diffraction, ^{1}H, ^{13}C, and ^{51}V n.m.r., electrochemistry, optical spectroscopy, and EXAFS[323]. The role of vanadium alkylperoxo complexes in epoxidation of cyclohexene and oxidation of cyclohexane by organic hydroperoxides in the presence of vanadium complexes formed from $VO(acac)_2$ for the catalysts were characterised in situ with ^{51}V and ^{1}H n.m.r. and ESR[324]. ^{51}V n.m.r. was used to establish the equilibrium between mono- and diperoxyvanadates and a number of di- and tripeptides in aqueous solution[325]. The vanadium(III) dinitrogen complex $[(Me_3CCH_2)_3V]_2(\mu\text{-}N_2)$ and its ^{15}N labelled analogue were synthesised and analysed by ^{51}V n.m.r[326]. The new fully oxidised isopolyvanadate, $(Me_4N)_3H_6V_{15}O_{42}{\cdot}2.5H_2O$ was prepared and studied x-ray crystallography and ^{51}V n.m.r[327]. ^{1}H and ^{51}V n.m.r. were used to study the reactions occurring between mono- and diperoxovanadate and a number of amino acids in aqueous solution[328]. By using the ammonium salt of the dimeric coordination compound of 2-ethyl-2-hydroxybutanoic acid and vanadate(V) with known crystal structure as a reference, structures of analogous compounds in aqueous solutions were deduced by ^{1}H, ^{13}C, and ^{51}V n.m.r[329]. Ethanolamine- and glycine-derived vanadium(V) complexes in aqueous solution were examined using a combination of ^{1}H, ^{13}C, ^{51}V, and ^{17}O n.m.r. spectroscopy and IR and UV/VIS spectrophotometries[330]. Three complexes

$(Et_4N)_3[VS_4Cu_4(OC_4H_8dtc)_n(PhS)_{4-n}]$ (n = 0, 1, 2; OC_4H_8dtcH = morpholine-N-dithiocarboxylic acid) were prepared and characterised by ^1H, ^{13}C and ^{51}V n.m.r[331].

Vanadia/silica mixed gel catalysts were characterised by ^{51}V and ^{29}Si CP/MAS solid-state n.m.r[332]. ^{51}V MAS n.m.r. spectra, exhibiting the complete manifold of spinning sidebands from all 7 ^{51}V (I = 7/2) single-quantum transitions, were obtained for metavanadates (MVO_3, M = Li, Na, K, NH_4, Tl) and for KV_3O_8 employing spectral windows up to 2 MHz[333]. ^{51}V static and MAS n.m.r. spectra of crystalline divalent metal vanadates, $M_3(VO_4)_2$ (M = Mg, Sr, Ba, Zn, Pb) and $Mg_2M'(VO_4)_2$ (M' = Sr, Ba, Pb) and divanadates, $M_2V_2O_7$ (M = Mg, Ca, Sr, Ba, Zn, Pb) were measured[334]. The ^{51}V and ^1H solid-state n.m.r. data showed the formation in V_2O_5-TiO_2-SiO_2 catalysts of a number of surface tetrahedral and octahedral V complexes, with their types depending upon the sequence of titanium deposition[335].

2.5.2 *Niobium (^{93}Nb)*

The structure and properties of $LiNbO_2$, including the mobility of Li ions, were determined by Rietveld refinement of the structural model from powder neutron diffraction data and by analyses of ^{93}Nb and ^7Li n.m.r[336]. Solid-state ^1H, ^2H and ^{93}Nb n.m.r. were used to investigate the host-guest bonding and the electric and magnetic environment of these nuclei in layered intercalation compounds of the type $K_x(H_2O)_y[NbS_2]$[337].

2.6 Group 6 (^{95}Mo, ^{183}W)

2.6.1 *Molybdenum (^{95}Mo)*

New possibilities of n.m.r. in mechanistic studies of homogeneous and heterogeneous catalysts were reviewed with several particular examples[338].

^{17}O, ^{95}Mo and ^1H n.m.r. were used to study the composition and reactivity of molybdenum(VI) complexes formed in the catalytic epoxidation of cyclohexene and allyl alcohol with hydrogen peroxide or with ROOH (R = tert-Bu, cumyl) in aprotic solvents[339,340].

^{95}Mo n.m.r. was used to study $(Bu_4N)_3[Mo_6O_{18}(NO)]$, $(Bu_4N)_2[Mo_6O_{17}(OMe)(NO)]$, and $(Bu_4N)_2[Mo_5O_{13}(OMe)_4(NO)\{Na(MeOH)\}\cdot 3MeOH$[341]. ^{17}O and ^{95}Mo n.m.r. were used to study the triangular incomplete cuboidal dimolybdenum-tungsten cluster aqua ion[342]. ^{95}Mo and ^{13}C n.m.r. were used to study some aqueous solutions containing oxomolybdenum(VI) complexes with polyaminocarboxylates at pH values between 2.0 and 9.5[343]. Solid-state ^{95}Mo, ^{31}P, and ^{27}Al n.m.r. were used to investigate surface species on NiO-MoO_3-P-based hydrotreating catalysts supported on γ-Al_2O_3 as a function of calculation temperature and hydration[344].

2.6.2 *Tungsten (^{183}W)*

A Keggin polyoxoanion-supported Ir(1,5-COD)$^+$ complex $(Bu_4N)_4Na_2[(1,5\text{-COD})\text{-}Ir\cdot SiW_9Nb_3O_{40}]$ was synthesised and characterised by elemental analysis, ultracentrifugation solution molecular weight measurements, and ^1H, ^{13}C and ^{183}W n.m.r. and IR spectroscopies in both weakly coordinating and strongly coordinating solvents[345].

^{183}W n.m.r. was used to study nonaqueous associative and dissociative reactions designed to produce $[W_2O_7]^{2-}$ instead revealed the formation of a new polytungstate which contains 5 types of W and reacts with organic substrates to produce functionalised polyoxotungstates[346]; to measure 11 d^3-d^3 face-sharing bioctahedral tungsten-thioether complexes, having the overall formula $[Cl_3W(\mu\text{-}X)_n(\mu\text{-}SR_2)_{3-n}WCl_3]_n^-$ (n = 0, 1) using a horizontal probe to increase sensitivity[347]. $[H_2W_{11}VO_{40}]^{7-}$ and $[H_2W_{11}MoO_{40}]^{6-}$ were identified in aqueous solution by ^{183}W, ^{51}V and ^1H n.m.r[348]. The unambiguous characterisation of A-β-$Si_2W_{18}Ti_6O_{77}{}^{14-}$ was accomplished by elemental analysis, IR, and ^{183}W n.m.r. and a single crystal x-ray structural analysis of the Bu_4N^+/H^+ salt, $(Bu_4N)_7H_7Si_2W_{18}Ti_6O_{77}$[349]. The structures of new heteropolyanions, $Cs_{13}Na_3$-$[(O_3POPO_3)_4W_{12}O_{36}]\cdot 24H_2O$ and $Na_{16}[(O_3PCH_2PO_3)_4W_{12}O_{36}]\cdot 16H_2O$ were shown to have an unusual open structure of approximate D_{2d} symmetry from x-ray crystallography, ^{31}P and ^{183}W n.m.r analysis[350]. The Dawson-type heteropolyoxoanion $P_2W_{15}Nb_3O_{62}{}^{9-}$ was prepared and characterised by a complete elemental analysis, TG/DTA, IR spectroscopy, and ^{31}P and ^{183}W n.m.r[351]. Both the $H_3[PW_{12}O_{40}]\cdot H_2O$ and the H_2WO_4/H_2O_2-H_2O/H_3PO_4 systems showed several ^{31}P and ^{183}W n.m.r. signals which were assigned to new phosphatooxoperoxotungstate species, $[PW_xOy]^{\omega-}$ (x = 1-4)[352]. Three new Dawson-type heteropolyoxotungstate isomers β- and γ-$[As_2W_{18}O_{62}]^{6-}$ and γ-$[P_2W_{18}O_{62}]^{6-}$ were prepared and identified by IR, Raman, UV/VIS, ^{183}W and ^{31}P n.m.r. spectroscopies, and polarographic and kinetic measurements[353].

2.7 Group 7 (^{55}Mn)

2.7.1 Manganese (^{55}Mn)

N.m.r. and n.q.r. signals of ^{55}Mn from site II in β-manganese metal were observed[354]. ^{27}Al, ^{55}Mn and ^{11}B n.m.r. spectra of the quasicrystalline Al-Pd-Mn-B system with the new strong magnetism were reported[355].

2.8 Group 8 (^{57}Fe, ^{99}Ru, ^{187}Os)

2.8.1 Iron (^{57}Fe)

^{57}Fe n.m.r. in $Y_{3-x}Bi_xFe_5O_{12}$ garnet thin films with [111] orientation was measured at liquid helium temperature by the spin echo method[356].

2.8.2 Ruthenium (^{99}Ru)

^1H and ^{99}Ru n.m.r. were used to characterise mer and fac isomers of $[Ru(2,3\text{-dpp})_3][PF_6]_2$ (2,3-dpp = 2,3-bis(2-pyridyl)pyrazine)[357].

2.8.3 Osmium (^{187}Os)

Osmium tetrafluoride dioxide, cis-OsO_2F_4 was prepared and characterised by electron diffraction, vibrational spectroscopy, d. functional theory calculations, ^{19}F and ^{187}Os n.m.r[358].

2.9 Group 9 (^{59}Co, ^{103}Rh)

2.9.1 Cobalt (^{59}Co)

^{59}Co n.m.r. was used to study the correlation between the magnetoresistance ratio and the interface structure, and local strain of cobalt/copper superlattice[359,360]; to study interactions of some cationic cobalt(II) complexes with anions[361]; to study the interactions of cationic or neutral octahedral cobalt(III) complexes such as $[Co(NH_3)_6]^{3+}$, $[Co(en)_3]^{3+}$ (en = ethylenediamine), $[Co(NO_2)(NH_3)_5]^{2+}$, and $[Co(acac)_3]$ (acac = acetylacetone) with surfactants in their nematic liquid crystal solutions[362]. The interface structure of [Cu/Co/Cu/NiFe] superlattices, in which the giant magnetoresistance was induced by a small external field, was studied by ^{59}Co n.m.r. in zero field[363]. The polycrystalline ^{59}Co n.m.r. spectra of the diamagnetic cobalt(III) complexes $Na_3[Co(NO_2)_6]$ and trans-$[Co(en)_2(NO_2)_2]NO_3$ were measured at 7.1 and 9.5 T[364]. The contribution to the ^{59}Co hyperfine fields, $H_{hf}(Co)$, from the Sm atoms in $(Y_{1-x}Sm_x)_2Co_{14}B$ was studied by using ^{59}Co n.m.r[365].

2.9.2 *Rhodium (^{103}Rh)*

^{103}Rh n.m.r. chemical shifts for several enamide complexes of the type [(Z)-ArCH:C(CO_2R')(NHCOR)Rh(L-L)]$^+$, where L-L is a chelating diphosphine, were measured using 2D reverse detection methods via their J-coupled ^{31}P nuclei[366]. A series of indenyl- and $(Cr(CO)_3$indenyl-RhL_2 complexes [L_2 = COD, $(CO)_2$] bearing substituents on both the six- and five-membered ring were synthesised and fully characterised by ^1H, ^{13}C and ^{103}Rh n.m.r[367]. Rhodium(I) and iridium(I) bis(iminophosphoranyl)methanide complexes in which the ligand acts as a σ-N, σ-C chelate were synthesised, characterised by IR and ^{13}C, ^{103}Rh, and solid-state ^{31}P CP/MAS n.m.r., and studied several reactions[368].

2.10 Group 10 (^{195}Pt)

2.10.1 *Platinum (^{195}Pt)*

The homoleptic, cyclometalated complex, cis-bis[2-(2-thienyl)pyridine]platinum(II) underwent thermal or photochemical oxidative addition reactions with a wide range of alkyl halides and the ^{195}Pt n.m.r. was used to characterise the reaction products[369]. Highly reactive cations $[Pt(\eta^4$-$C_4Me_4)(solv)x]^{2+}$ (solv = acetone or nitromethane) could be obtained and the new compounds were characterised by ^1H, ^{13}C and ^{195}Pt n.m.r. and cyclic voltammetry[370]. Ferrocenylamine complexes of platinum(II) were prepared and characterised by ^1H, ^{13}C, and ^{195}Pt n.m.r. and IR spectroscopies[371,372]. Detailed ^{31}P and ^{195}Pt n.m.r. studies on complexes of the type trans-$[MCl(P_3C_2{}^tBu_2)(PR_3)_2]$ (M = Pd, Pt), trans-$[M(P_3C_2{}^tBu_2)_2(PR_3)_2]$, (same M), and cis-$[Pt(P_3C_2{}^tBu_2)_2(dppe)]$ were reported[373].

^{195}Pt n.m.r. was used to study mixed platinum(II)-mercury(II) cytosine nucleobase complexes with metal-metal bonds[374]. Strong metal-metal bonds between trans-bis(amine)platinum(II) and -palladium(II) in heteronuclear complexes of cytosine nucleobases were characterised by ^1H and ^{195}Pt NMR, and x-ray crystallography[375]. ^{195}Pt MAS n.m.r. spectra of the complexes $Pt(en)Cl_x$ (en = 1,2-diaminoethane; x = 2,3 or 4) were reported[376]. ^1H, ^{13}C and ^{195}Pt n.m.r were used to characterise the triangular platinum(II) cluster complexes $[Pt_3(CH_3COO)_4(cdoH)_2(cdoH_2)]$ and $[Pt_3(CH_3COO)_4(dmgH)_2(dmgH_2)]$ ($cdoH_2$ = cyclohexanedionedioxime and $dmgH_2$ = dimethyl-glyoxime)[377]. ^{13}C and ^{195}Pt n.m.r. were used to investigate the hydrogen peroxide oxidation of platinum(II) complexes[378]. Linear triplatinum complexes containing three *cis*-$(Pt(amine)_2)$ units, *cis*-$[\{$*cis*-$PtCl_2(NH_3)(\mu$-$H_2N(CH_2)_4NH_2)\}_2PtCl_2]$ was prepared and characterised by ^1H and ^{195}Pt n.m.r. and fast atom bombardment mass spectroscopy[379]. $[Pt_2(\mu$-L-accys-S$)_2(bpy)_2]$ (L-Haccys =

N-acetyl-L-cysteine, bpy = 2,2'-bipyridine) was isolated and characterised by ^{1}H, ^{13}C, ^{195}Pt n.m.r to study the interaction of platinum with protein sulphur residues[380]. $[PtQCl_3]^-$ (Q = $MeNH_2$, $EtNH_2$, cyclobutylamine, cyclopentylamine, Me_2NH, 1- and 2-adamantanamine, (1-adamantyl-methyl)amine) were synthesised and characterised mainly by ^{195}Pt n.m.r[381]. Deceptively simple Pt complexes of N,N-dialkyl-N'-benzoylthiourea were prepared and their acid-base chemistry in solution was studied by ^{1}H, ^{13}C, ^{195}Pt n.m.r[382]. 1-Methylcytosine and cytidine complexes of platinum(II) stabilised by trimethylphosphine were prepared and studied by ^{1}H, ^{13}C, ^{31}P and ^{195}Pt n.m.r[383]. $[Pt(ER)_2(L\text{-}L)]$ [HER = HSPri, HSPh, HSC_6H_4Me-p, HSePh; L-L = $Ph_2P(CH_2)_n$-PPh_2, n = 1-3)] and $[Pt_2(\mu\text{-}ER)_2(L\text{-}L)][BPh_4]_2$ were prepared and characterised by ^{1}H, ^{31}P, and ^{195}Pt n.m.r[384]. Bis(diphenylphosphino)amine complexes of platinum(II), $[PtCl(PR_3)(dppa)]Cl$ (PR_3 = PMe_3, PMe_2Ph, $PMePh_2$, PBu_3, PPh_3, PEt_3) were prepared and characterised by ^{31}P, ^{195}Pt and ^{1}H n.m.r., microanalysis, and FAB mass spectrometry[385]. ^{195}Pt and ^{15}N n.m.r. were used to study the reaction of potassium trichloroammineplatinate with potassium iodide and amines[386]. ^{31}P and ^{195}Pt n.m.r. were used to study mononuclear, binuclear and trinuclear tris(hydroxymethyl)phosphine-platinum(II) complexes with aquo or hydroxo ligands[387].

The ^{119}Sn and ^{195}Pt Knight shifts of uranium-platinum-tin (UPt_2Sn) were studied[388].

2.11 Group 11 (^{63}Cu, ^{109}Ag)

2.11.1 Copper (^{63}Cu)

^{63}Cu n.m.r. was used to study the ionic diffusion in copper phosphide sulphide bromide (Cu_6PS_5Br)[389]. ^{63}Cu MAS n.m.r. and x-ray diffraction were used to analyse the structure of copper(I) bromide chloride mixed crystal[390]. Solid-state and solution structures of 1,2-bis-(diphenylphosphino)ethane bridged dinuclear copper(I) complexes were investigated by CP/MAS ^{31}P n.m.r., x-ray crystallography, IR, and solution ^{31}P and ^{63}Cu n.m.r[391].

^{63}Cu Knight shifts were measured on $La_{2-x}Sr_xCuO_4$ for x = 0.10, 0.16, and 0.20 in the normal and superconducting states[392]. The characteristic variations of the ^{63}Cu Knight shift in the mechanical alloyed copper-zinc system with Zn content were presented and identified[393]. Knight shift measurements of the ^{63}Cu nuclei in the sandwiching $Cu^{II}O_2$ layers and the middle $Cu^{I}O_2$ layer in the 3-layer bismuth oxide superconductor were performed[394]. ^{63}Cu n.m.r. spectra of [magnetic metal/Cu] superlattices synthesised by the beam sputtering method were measured[395]. ^{63}Cu, ^{207}Pb, and ^{205}Tl n.m.r. line shifts were measured in the cuprates $Tl_{0.5}Pb_{0.5}Sr_2Ca_{1-x}Y_xCu_2O_7$ (x = 0.0-1.0) to study the evolution of the spin susceptibility around the metal-insulator transition, which occurs when the Y content increases up to $x \approx 0.6$[396].

2.11.2 Silver (^{109}Ag)

$[Ag(9\text{-ethylguanine-N7})_2]NO_3 \cdot H_2O$ was prepared and characterised by x-ray crystallography, and ^{109}Ag n.m.r. was applied to study the solution behaviour in DMSO[397].

2.12 Group 12 ($^{111,113}Cd$, $^{199,201}Hg$)

2.12.1 Cadmium ($^{111,113}Cd$)

^{113}Cd n.m.r. applications to metalloproteins were reviewed[398].

^{113}Cd n.m.r. was used to characterise the structure of the divalent cation sites of bacterial phosphotriesterase[399]; to investigate the interaction between a unialgal species (Stichococcus bacillaris) and cadmium in the presence of several different solute[400]. ^{1}H and ^{113}Cd n.m.r. studies indicated that magnesium inhibits formation of 4-calcium-calmodulin-enzyme complex at lower calcium concentration[401].

^{31}P, $^{111,113}Cd$, ^{77}Se n.m.r. data for the systems $Cd(PPh_3)_2(ClO_4)_2$:$Cd(ER)_2$:PPh_3 (ER = $SCHMe_2$, S-c-C_6H_{11}, SPh, SePh, SPr, S-n-C_5H_{11}) provided the evidence for new adamantanoid complexes $[(\mu\text{-ER})_6(CdPPh_3)_n(Cd)_{4-n}]^{2+}$[402]. Mixtures of cadmium salts and benzimidazoles were examined by ^{1}H, ^{13}C, and ^{113}Cd n.m.r. in DMSO and in DMSO-chloroform mixtures[403].

Solid-state ^{113}Cd n.m.r. was used to study cadmium complexes with glycine, L-alanine, and L-cysteine[404]; to study 7 cadmium-sulphur complexes to show shielding tensor-structure correlations[405]. Solid-state ^{113}Cd n.m.r. chemical shift (δ_{iso}) and chemical shift anisotropy (δ_{ii}) data were reported for crystalline samples of 3 different compounds containing the $[S_4Cd_{10}(SPh)_{16}]^{4-}$ cluster[406]. $[Cd(S_2PR_2)_2]$ (R = Et, cyclohexyl) were studied in the solid state by vibrational (IR and Raman) spectroscopy and in CH_2Cl_2 solution by molecular weight measurements and IR and ^{13}C, ^{31}P, ^{113}Cd n.m.r[407].

2.12.2 *Mercury ($^{199,201}Hg$)*

Indirect ^{199}Hg n.m.r. detection was reported for aromatic mercury compounds of biological interest[408]. Organomercury derivatives of 2-phenylpyridine, benzo[h]quinoline, 1-phenylpyrazole and 3,4,5-trimethyl-1-phenylpyrazole were prepared and their ^{199}Hg n.m.r. spectra were reported[409]. Phenyl(perfluoroorgano)mercury derivatives $C_6H_5Hg(Rf)$ [Rf = C_nF_{2n+1} (n = 1-4 or 6) or C_6F_5] were synthesised and characterised by ^{13}C, ^{19}F, and ^{199}Hg n.m.r. and mass spectra[410]. The intermolecular interaction of closo-$B_{10}H_{10}^{2-}$ with [12]-mercuracarborand-4 or its octaetyl derivatives in organic solvents was studied by ^{11}B and ^{199}Hg n.m.r[411].

^{199}Hg n.m.r. chemical shifts of the stable mercury species $HgCl_n^{2-n}$ with n = 0-4 were determined by a least-squares algorithm from ^{199}Hg chemical shifts measured in 50 mM aqueous Hg solution as a function of the Cl concentration at 25°[412]. ^{13}C, ^{19}F, ^{77}Se and ^{199}Hg n.m.r. spectra of mercury(II) trifluoromethane thiolate and selenolate were examined[413]. Alkanethiolato- and arenethiolatomercury(II) polypyrazolylborates were prepared and characterised by IR in the solid-state and ^{1}H, ^{13}C, and ^{199}Hg n.m.r. in solution[414]. The structural nature of optically active ferrocenylphosphine complexes of mercury(II) was studied by ^{199}Hg, ^{31}P one- and two-dimensional solution and solid-state n.m.r[415]. The effects of small deviations from the magic angle on ^{199}Hg solid-state n.m.r. spectra with large chemical shift anisotropy were shown by computation and illustrated with spectra of mercury(II) acetate[416]. Several anisotropic nuclear spin interactions in solid potassium tetracyanomercurate(II), $K_2Hg(CN)_4$, were characterised by ^{13}C, ^{15}N, and ^{199}Hg n.m.r. at 4.70 and 9.40 T[417], and the same authors reported the first observation of ^{201}Hg n.m.r. signals in the solid state for $K_2Hg(CN)_4$[418]. ^{199}Hg n.m.r. measurements of narrow-gap $Hg_{1-x}Cd_xTe$ alloys in the range x = 0.20-0.28 were reported[419]. ^{199}Hg chemical shielding tensor of the mercurous ion in mercurous nitrate dihydrate was reported[420].

2.13 Group 13 (^{11}B, ^{27}Al, $^{69,71}Ga$, ^{115}In, ^{205}Tl)

2.13.1 Boron (^{11}B)

The role of n.m.r. in boron chemistry was briefly reviewed[421].

The ^{11}B n.m.r. chemical shift of certain boron compounds in the presence of cancer cells appears at lower fields in comparison with the chemical shift in the absence of the cells[422]. ^{11}B n.m.r. was used to study peptide boronic acid inhibitor complexes of α-lytic protease[423]. A ^{11}B n.m.r. study of the interaction of representative tertiary amines of variable steric requirements with chlorodicyclohexylborane, Chx_2BCl, suggested that the smaller amines coordinate strongly with Chx_2BCl, while the more bulky amines do not[424].

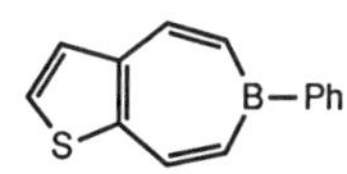

^1H, ^{13}C, and ^{11}B n.m.r. were used to obtain an insight into the electronic state of 1-phenylthieno[2,3-d]borepin[425]. Some $N\text{-}BH_3$, $N\text{-}BF_3$, $N^+\text{-}CH_3$ and $N^+\text{-}H$ imidazole derivatives were prepared and their ^{13}C, ^{11}B, and ^1H n.m.r. spectra were studied[426]. The hydroboration of 1-alkynes was reinvestigated by ^{11}B n.m.r. under optimised conditions (THF, 18h, 0°C) and found to provide trans-vinyl-9-BBN adducts together with variable amounts of 1,1-diborylalkanes depending both upon the excess of alkyne employed and the nature of alkyne substitution[427]. The structure of a new carborane anion [arachno-2,6-$C_2B_6H_{11}$]$^-$, prepared from the reaction of closo-1,6-$C_2B_7H_9$ with tetrabutylammonium fluoride in THF, was determined from a combination of ^{11}B, 2D ^{11}B-^{11}B, and ^{13}C n.m.r. results and the application of ab initio/IGLO/n.m.r. and ab initio/GIAO/n.m.r. methods[428]. The structure of nido-$Se_2B_9H_9$ was determined unambiguously to be the 7,9-isomer by ultrahigh-field (192.48MHz) ^{11}B n.m.r[429].

2.13.2 Aluminium (^{27}Al)

Application of ^{27}Al n.m.r. techniques to structure determination was reviewed[430,431,432,433]. The applications of solid-state n.m.r. in study of structure and catalytic function of zeolitic and aluminophosphate molecular sieves were reviewed[434].

^{27}Al n.m.r. was used to characterise aluminium complexes in tea extract[435]; to probe metal binding to proteins, ovotransferrin was exemplified quadrupolar metal ion n.m.r [436]; to investigate a new room temperature melt containing 1-ethyl-3-methylimidazolium chloride (MEICl) and ethylaluminium dichloride ($EtAlCl_2$) in order to characterise $EtAlCl_2$ dimers and $EtAlCl_3^-$ [437]. ^{27}Al and ^{19}F n.m.r. was applied to the analysis of intact tissues of tea to clarify aluminium kinetics in the tea plant[438]. The reactions of bis(dialkylamino)magnesium [$Mg(NR_2)_2$; R = Et, iso-Pr] with either diethylaluminium chloride or aluminium bromide yielded several products and the products were characterised by elemental analyses, mass spectra, IR, ^1H, ^{13}C, and ^{27}Al n.m.r[439]. [Et_2Al-$P(SiMe_3)_2]_2$, $EtCl_2Al{\cdot}P(SiMe_3)_3$, and iso-$Bu_2ClAl{\cdot}P(SiMe_3)_3$ were prepared and characterised by partial elemental analysis, m.p. data, x-ray crystallography, and ^1H, ^{13}C, ^{31}P and ^{27}Al n.m.r[440]. Field-dependent ^{27}Al n.m.r. studies of the transferrins was reported to show metal ion binding sites in larger proteins[441]. The first five-coordinated simple alkylaluminium chelate complex, $AlMe[O{:}C(OMe)C_6H_4\text{-}o\text{-}O]_2$ was synthesised and characterised by IR, ^1H and ^{27}Al n.m.r., and x-ray crystallography[442].

^{1}H, ^{13}C, and ^{27}Al n.m.r. were used to investigate the isomeric complexes formed between $AlCl_3$ and either Cp_2TiCl_2 or $Cp_2Ti(CH_2SiMe_3)Cl$ in chloroalkane solution[443]. Solutions of aluminium isopropoxide (an initiator to the living polymerisation) in toluene were investigated by ^{13}C and ^{27}Al n.m.r. to study the effect of γ-butyrolactone (nonpolymerisable model, representative of ε-caprolactone)[444].

High-field ^{27}Al MAS n.m.r. was used to study the formation of metakaolinite by flash calcination of kaolinite[445]. ^{27}Al double-rotation n.m.r. spectra of the aluminosilicate polymorph minerals, kyanite, sillimanite and andalusite, showed separately resolved peaks, relating to the crystallographically non-equivalent sites[446]. Mullite ($3Al_2O_3{\cdot}2SiO_2$) was prepared from different precursors by multicomponent metal alkoxide and metal alkoxide-metal nitrate methods and characterised by ^{27}Al and ^{29}Si n.m.r[447]. ^{27}Al and ^{29}Si MAS n.m.r. were used to study reexpansion of collapsed Li-monmorillonites[448].

Synthetic, conformational, and configurational features of group 13 azatranes were analysed by ^{1}H, ^{11}B, ^{13}C, ^{27}Al n.m.r[449]. The system $POCl_3$-$AlCl_3$ was investigated by measuring the Raman and ^{27}Al and ^{31}P n.m.r. spectra as functions of the mole fraction of $AlCl_3$ in $POCl_3$[450]. Alumatrane, $Al(OCH_2CH_2)_3N$, oligomeric behaviour was reinvestigated by ^{1}H, ^{13}C and ^{27}Al n.m.r. and XPS spectroscopy[451]. ^{27}Al CP/MAS n.m.r. spectra of aluminium acetylacetonate showed distinct lineshapes depending on the experimental conditions as the different orientations within the powder cross-polarise with differing efficiencies[452]. Several alumoxanes were characterised by ^{27}Al n.m.r[453]. ^{2}H MAS n.m.r. and echo Fourier ^{27}Al n.m.r. were applied to investigate the Broensted sites in dehydrated hydrogen forms of zeolites[454]. The ^{27}Al n.m.r spectrum of $NaAlO_2$ solutions of pH 9.5-12.0 was consistent with the presence of the $Al(OH)_4^-$ anion[455]. The coordination of Al(III)-aldarate complexes with Ca(II) in aqueous solution was studied using ^{1}H, ^{13}C, ^{17}O and ^{27}Al n.m.r[456].

^{27}Al DOR n.m.r., MAS n.m.r., and x-ray diffraction study of the structure of aluminium molybdate ($Al_2(MoO_4)_3$) were carried out[457]. ^{27}Al DOR n.m.r. was applied to study an ultrastable Y zeolite[458]. The influence of water adsorption in $AlPO_4$-11 was studied by x-ray powder diffraction and ^{27}Al and ^{31}P n.m.r[459]. Highly crystalline samples of the chabazite-related alumino-phosphate $AlPO_4$ 18 (AEI) were prepared and characterised by adsorption measurements and ^{27}Al double rotation n.m.r[460]. Assignment of the ^{27}Al and ^{31}P n.m.r. spectra of the aluminophosphate molecular sieve VPI-5 was reported[461]. ^{23}Na, ^{27}Al and ^{29}Si solid-state n.m.r. were applied to characterise the structure of microporous host-guest systems, especially sodalites with aluminosilicate, aluminate and silicate host-frameworks and a large variety of guest species (inorganic anions and cations, H_2O, OH^-, organic molecules) enclathrated in the sodalite cages[462]. A series of high-area silica gels treated with well-characterised Aluminium solutions, hydrolysed at OH/Al = 0-2, were analysed by ^{27}Al and ^{29}Si solid-state static, MAS, and CP/MAS n.m.r[463]. A series of highly crystalline hydrothermally dealuminated Y zeolites was prepared and characterised by chemical analysis, x-ray diffraction, mid-IR, ^{27}Al MAS n.m.r., analysis of N_2 adsorption isotherms and temperature programmed desorption of NH_3 to determine acid site concentrations[464]. ^{27}Al MAS n.m.r was used to study sodium aluminates and aluminium hydroxides (OH/Al = 5.3-

3.9) formed through varying degrees of acid hydrolysis revealed four- and six-coordinate aluminium species[465]. The ^{27}Al n.m.r. spectra of $AlPO_4$-21, or trialuminium tris(orthophosphate) hydrate with pyrrolidine as a template, was studied[466]. Zeolite β was prepared with different Si/Al ratios by a hydrothermal method and dealuminated by leaching with dilute HCl and characterised by x-ray diffraction and IR and ^{27}Al MAS n.m.r. spectroscopies[467].

2.13.3 Gallium (^{69,71}Ga)

^{71}Ga n.m.r. was used to quantitatively determine the gallium in zeolites[468]; to study mixtures of gallium trichloride and trimethylgallium[469]. High-field ^{71}Ga n.m.r. spectra at 122.03 MHz of aqueous Ga^{3+} solutions containing NCS- ions showed the progressive replacement of the sharp signal of $[Ga(H_2O)_6]^{3+}$ by a broad up-field resonance due a mixture of Ga thiocyanate complexes[470]. ^{71}Ga and ^{31}P n.m.r. were used to study the gallophosphate molecular sieve, cloverite[471]. A linear relation was found to exist between the ^{27}Al and ^{71}Ga n.m.r. chemical shifts of structurally analogous aluminium and gallium compounds having only oxygen in the first metal coordination spheres[472]. Solid-state ^{31}P and ^{69}Ga MAS n.m.r. were used to evaluate gallium phosphide samples quantitatively prepared from rapid solid state metathesis[473].

2.13.4 Indium (^{115}In)

^{115}In n.m.r. studies were reported for the reactions of InX_3 (X = Cl, Br, I) with halides (X = Cl, Br, I) and pseudohalides (NCS, NO_2) in MeCN[474].

2.13.5 Thallium (^{205}Tl)

The binding of Tl^{3+} to human serotransferrin and chicken ovotransferrin in the presence of carbonate and oxalate was examined by ^{205}Tl and ^{13}C n.m.r[475].

^{205}Tl n.m.r. was reported for several zeolite samples (A, X, and Y; hydrated and dehydrated) with Tl(I) exchanged for Na[476]. ^{205}Tl n.m.r. frequency shift and line width due to the magnetic field variation in the mixed state of $Tl_2Ba_2Ca_2Cu_3O_{10}$ were reported[477].

2.14 Group 14 (^{13}C, ^{29}Si, ^{73}Ge, ^{119}Sn, ^{207}Pb)

2.14.1 Carbon (^{13}C)

^{13}C n.m.r. chemical shifts were reported for 39 trimethylsilylated secondary steroid alcohols[478]; for 4-substituted 3-chloro- and 3-bromo-2,6-dimethylpyridine N-oxides in $CDCl_3$[479]; for long chain acids, methyl esters, glycerol esters, wax esters, nitriles, amides, alcohols, and acetates[480]; for a series of 2',3',4'-trimethoxylated flavones were reported[481]; for a variety of phenyl derivatives of boron, silicon, tin, phosphorus, and tellurium fluorides, e.g., $PhBF_2$[482]; for a series of p-phenyl substituted benzoylacetones which undergo a fast, intramolecular proton-transfer reaction between both possible enol tautomers[483]; for 18 abieta-7,13-diene diterpenoids[484]; for a series of acyclic, monocyclic and polycyclic ketones and diketones[485]. The ^{13}C n.m.r. chemical shifts and some ^{1}H-^{13}C coupling constants of pyrazoles and indazoles were reported and discussed[486]. ^{1}H and ^{13}C n.m.r. data were reported for a series of 2-O-, 3-O- and 2,3-di-O-glycosylated methyl α- and β-D-gluco- and β-D-galactopyranosides[487]. ^{1}H and ^{13}C n.m.r chemical shifts of 5 complexes of 1,8-bis(dimethylamino)naphthalene with strong acids were reinvestigated[488].

^{13}C-Labelled celluloses were biosynthesised by Acetobacter xylinum (IFO 13693) from culture media containing D-[1-^{13}C]glucose, D-[6-^{13}C]glucose, or D-[2-^{13}C]glucose as a C source with or without addition of ethanol and their structures were analysed by ^{13}C n.m.r[489]. Preparation and ^{13}C n.m.r. characterisation of hexachlorofullerene ($C_{60}Cl_6$) was reported[490]. A method of generating a thermodynamic acidity function from ^{13}C n.m.r. measurements was presented[491]. ^{13}C n.m.r was used to study ^{13}C-labelled hypervalent (10-C-5) pentacoordinate carbon species[492]. Two-dimensional ^{13}C n.m.r. study of orientational ordering in solid C_{60} was reported between 13.8 and 280 K and the results were compared with various theoretical models[493]. ^{13}C Knight shift of the doped fullerene K_3C_{60} was reported[494].

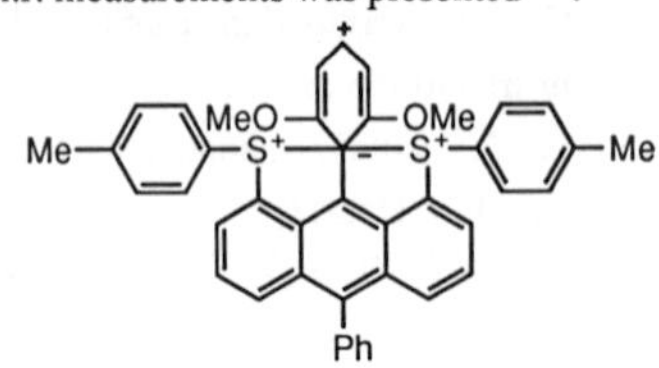

The influence of the antiviral peptide, carbobenzoxy-D-phenylalanyl-L-phenylalanylglycine on the average conformation of phosphatidylcholine in hydrated bilayers was investigated by ^{1}H, ^{2}H, ^{13}C and ^{31}P n.m.r[495]. ^{1}H, ^{13}C, and ^{15}N n.m.r. studies, including the use of the ^{13}C n.m.r. deuterium perturbation method, showed that the interaction between naphthalene or its methylated derivatives and the nitrosonium cation results in the formation of π-complexes[496]. An experimental (^{13}C and ^{15}N n.m.r., x-ray crystallography, and pKa) and theoretical study of N-aminoazoles were carried out in neutral ($CDCl_3$ or DMSO-d_6) and acid (CF_3CO_2H and H_2SO_4) conditions to determine the basicity and protonation site[497].

The ^{13}C chemical shift shielding of carboxylate groups in microcrystalline amino acids was measured and correlated with the protonation state[498]. A new mechanism for metal-catalysed thiophene hydrogenolysis was demonstrated by CP/MAS ^{13}C n.m.r[499]. ^{13}C chemical shift anisotropies of solid amino acids was studied by an approach to spinning sideband separation of protonated and non-protonated carbons in slow spinning conditions via dipolar dephasing[500]. The ^{13}C MAS n.m.r. spectra of several lipid-water systems were reported under a variety of radiofrequency excitation conditions and complex lipid or membrane spectra could be greatly simplified by 'spectral editing' techniques[501]. ^{13}C chemical shift tensors measured in a single crystal of sucrose were assigned and interpreted[502]. The molecular motion of guest molecules in $Cd(mtn)Ni(CN)_4 \cdot 0.5G$ clathrates (mtn = N-methyl-1,3-diaminopropane, G = toluene, nitrobenzene, dichloromethane, and chloroform) were studied by using CP/MAS ^{13}C and ^{2}H n.m.r[503]. ^{13}C and ^{15}N solid-state n.m.r. were employed to characterise the carbonyl carbon and nitrogen chemical shift tensors of the amide fragment of (Z)-acetanilide and (E)-N-methylacetanilide[504]. The reactions of acetone and cyclopentanone on a variety of zeolites with different acidities and pore sizes were investigated in detail by in situ ^{13}C solid-state n.m.r[505].

The reactions of $^{13}CH_3{}^{13}CHO$ on ZSM-5 and the affects of H_2O and or O_2 coadsorption on these reactions was examined by MAS n.m.r[506]. The ^{13}C chemical shift tensor orientations in the

OCH_2CH_2O moieties of poly(ethylene oxide) and poly(ethylene terephthalate) were determined by ^1H-^{13}C separated-local-field n.m.r[507].

The ^{13}C n.m.r. spectra of fullerenes C_{76}, C_{78}, C_{84} and mixtures of C_{86}-C_{102} were reported[508]. ^{13}C n.m.r. was used to study intercalation of oxygen in solid C_{60} fullerene and molecular-rotation hindrance[509]. Solubility was described for 30 aromatic solvents for fullerenes and the efficacy of 1,2-dichlorobenzene and 1-methylnaphthalene as fullerene solvents for ^{13}C n.m.r[510]. ^{13}C n.m.r was used to study C_{60} molecular dynamics in pure C_{60} crystals prepared by sublimation technique and by evaporation of CS_2-solution of C_{60}[511].

Solid-state CP-MAS ^{13}C n.m.r. and ^2H n.m.r. were used to study the extent of charge-transfer and polarisation contributions to the intermolecular binding in the C_6H_6-C_6F_6 complex[512]. Molecular motions and ^{13}C n.m.r. chemical shift anisotropy of acetone adsorbed in H-ZSM-5 zeolite were reported[513]. The paramagnetic rare earth acetates and the analogous deuterated compounds were used as structural probes to measure distance in solid-state ^{13}C n.m.r[514]. Slow spinning ^{13}C CP/MAS n.m.r. was used to study the local geometry and electronic asymmetry of the Fe-C-O unit in heme model compounds[515]. High resolution MAS ^{13}C n.m.r. of potassium fulleride (K_6C_{60}) was reported[516].

2.14.2 Silicon (^{29}Si)

N-silylated diaminosulfanes ($Cl_3SiNR^1SNR^2SiRCl_2$) were prepared and characterised by ^1H, ^{13}C, ^{15}N, ^{29}Si n.m.r. in solution and ^{15}N and ^{29}Si CP/MAS n.m.r. in the solid state and x-ray crystallography[517]. The ^1H, ^{13}C and ^{29}Si n.m.r. spectral data for a series of simple acylsilanes $PhSiCOC_6H_4X$-p and $Me_3SiCoPh$ were given[518]. A New class of supramolecular organometallic compounds, ferrocenyl substituted octakis(dimethylsiloxy)octasilsesquioxanes were synthesised, characterised by ^1H, ^{13}C, and ^{29}Si n.m.r., IR, elemental analysis, and mass spectrometry, and studied electrochemistry[519]. ^{29}Si n.m.r. was used to study the steric effects on ^{29}Si n.m.r. chemical shifts in trimethylsiloxy steroids controlled by the solvent accessibility of the oxygen atom in the molecule[520].

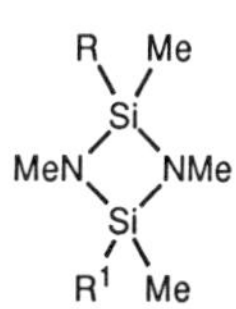

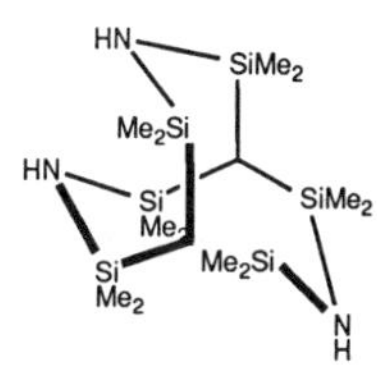

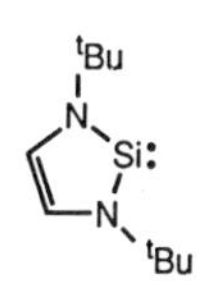

N,N'-Dimethylcyclosilazanes (R = R' = Me, Vi, Et; R = Me, R' = Vi, Et, Ph; Vi = vinyl) were prepared and characterised by ^1H, ^{13}C, ^{29}Si, and ^{15}N n.m.r. and IR spectroscopy and mass spectrometry[521]. The manxane-cage silazane molecule $HC(Me_2SiNHSiMe_2)_3CH$ was synthesised and characterised by ^1H, ^{13}C, and ^{29}Si n.m.r[522]. The first stable two-coordinated silylene was synthesised and its structure was confirmed by gas phase electron diffraction, x-ray crystallography, and ^1H, ^{13}C, ^{29}Si, and ^{15}N n.m.r[523]. Stable silylnitrilium ion, $R_3Si(NC$-$C_3H_7)^+B[3,5$-$(CF_3)_2C_6H_3]_4^-$, was prepared and characterised by ^1H, ^{13}C, ^{11}B, ^{19}F, and ^{29}Si n.m.r[524]. The effect of pentacoordination on ^{29}Si n.m.r. chemical shifts and silicon-hydrogen coupling constants was reported[525]. The x-ray crystallography and ^{29}Si n.m.r. of the tri-iso-propyl silyl species, i-$Pr_3Si(Br_6$-$CB_{11}H_6)$, where the brominated carborane Br_6-$CB_{11}H_6^-$ is perhaps the least nucleophilic anion presently known, revealed the highest degree of silylium cation character

(R_3Si^+)[526]. Ab initio geometry optimisations, ^{29}Si n.m.r. chemical shift calculations, and bond density investigations refuted recent claims about the existence of silylium cations R_3Si^+ (R = Me, Et) in aromatic solvents[527].

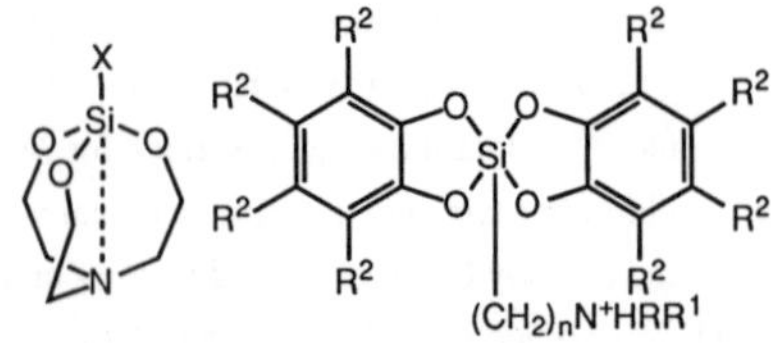

Solid-state ^{13}C, ^{15}N, and ^{29}Si n.m.r. spectra were obtained on silatranes (X = H, Cl, MeO, etc.)[528]. The zwitterionic spirocyclic l5-silicates bis[3,4,5,6-tetrabromo-1,2-benzenediolato(2-)][2-(pyrrolidinio)ethylsilicate [$RR^1 = (CH_2)_4$, R^2 = Br, n = 2, and its monohydrate] and bis[1,2-benzenediolato(2-)][(dimethylammonio)methyl]silicate [R = R^1 = Me, n = 1, R^2 = H) were synthesised and characterised by solid-state n.m.r. spectroscopy (^{29}Si and ^{15}N CP/MAS)[529]. High-resolution solid-state ^{13}C and ^{29}Si n.m.r. were used to investigate the dynamic properties of tetrakis(trimethylsilyl)silane (TTMSS) as a function of temperature[530].

^{29}Si n.m.r. was applied to study the degradation of silicones in vivo[531]. ^{29}Si CP n.m.r. was used to study the interaction between methoxysilanes and $MgCl_2$ when co-milled[532]. ^{29}Si and ^{27}Al MAS n.m.r. were used to study a single crystal of the zeolite chabazite[533].

$NaGeH_n(SiH_3)_{3-n}$ (n = 2,1,0) reacted with $C_4F_9SO_3SiH_3$ to give $H_nGe(SiH_3)_{4-n}$ which were characterised by ^{29}Si n.m.r. and mass spectra[534]. A new monosubstituted, octanuclear silasesqui-oxane with a Si-Co bond, $[Co(CO)_4(H_7Si_8O_{12})]$, was prepared by the reaction of octacarbonyl-dicobalt $[Co_2(CO)_8]$ with octahydrosilasesquioxane $H_8Si_8O_{12}$ in toluene and analysed by ^1H, ^{13}C and ^{29}Si n.m.r. and IR spectroscopies, mass spectrometry and microanalysis[535].

Reversible modification of the silicon environments in the template-free SAPO-34 structure upon hydration-dehydration cycles below ca. 400K was investigated by ^{29}Si MAS n.m.r. and x-ray diffraction[536]. ^{29}Si MAS n.m.r. was used to discriminate phases in EMT/FAU zeolite intergrowths[537]. The bond-polarisation theory was applied to ^{29}Si chemical shifts and its dependence on the molecular geometry of SiO_2 networks[538]. The changes occurring in SAPO-37 upon heating it to 1273 K were studied by ^{29}Si, ^{27}Al, and ^{31}P MAS n.m.r., X-ray diffraction (XRD) and IR spectroscopy[539]. ^{13}C CP/MAS n.m.r., ^{29}Si MAS n.m.r. and XRD were used to study the crystallisation of zeolite ZSM-39 in the tetramethylethylenediamine(TMEDA)-SiO_2-H_2O system[540]. Silica gels were deuterium-exchanged with D2O at 25 and 100° and analysed by ^{29}Si CP/MAS n.m.r[541]. The effect of a paramagnetic Cu^{2+} ion exchange on n.m.r. line broadening was examined by using MAS ^{29}Si n.m.r of CuX zeolite[542]. ^{29}Si n.m.r. was used to quantitatively investigate highly concentrated high-ratio sodium silicate solutions[543]. A beta-zeolite with template was calcined in air at 550° using two different experimental procedures, and IR and ^{29}Si MAS n.m.r. showed that the extent of framework dealumination and the number of Broensted acid sites were greatly affected by the conditions of treatment[544]. Experimental data on chemical shifts in a variety of crystalline and amorphous silicates and ceramics were reviewed and the published correlations between ^{29}Si chemical shifts and structural parameters were critically examined[545]. Five signals from Si(nAl) units, where n = 0-4, were found in ^{29}Si MAS n.m.r. spectra of zeolites X and Y (synthetic faujasites) with different Si/Al ratios[546].

2.14.3 Germanium (^{73}Ge)

^{73}Ge and ^{13}C n.m.r. spectra of several 3-germabicyclo[3.1.0]hexanes (GBCH) were determined[547].

2.14.4 Tin (^{119}Sn)

Organotin(IV) Schiff base complexes of the type (L)SnCl($CH_2CH_2CO_2Me$) (LH_2 = 2-N-salicylideneimino-2-methyl-1-propanol) were prepared and characterised by elemental analyses, IR, ^{1}H, ^{13}C, and ^{119}Sn n.m.r[548]. A convenient method was reported for the synthesis of some novel complexes of composition (R_3SnL_2)ClO_4 (L = pyridine, γ-picoline, 2,2'-bipyridine, 1,10-phenanthroline, R = Bu or Ph), and all of the new ionic complexes were characterised by elemental analysis, molar conductance, IR, ^{1}H, ^{13}C and ^{119}Sn n.m.r., and ^{119}Sn Moessbauer spectroscopies[549].

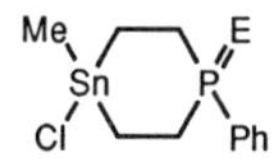

A series of 1-phenyl-1-chalcogeno-4-methyl-4-chloro-1,4-phosphastannanes (E = S, Se, O) was synthesised and studied by ^{119}Sn, ^{31}P and ^{1}H n.m.r. in nonpolar solvents to show a P–E•••Sn coordination which is intramolecular for S and Se and intermolecular for O[550]. Organotin complexes with 1-methyl-2(3H)-imidazolinethione were prepared and characterised by conductance measurements and by IR, Raman, Moessbauer and ^{1}H, ^{13}C and ^{119}Sn n.m.r[551]. The solution structure of trimethyl- and tributylstannyllithium was determined by low temperature ^{7}Li, ^{31}P and ^{119}Sn n.m.r[552]. Allylstannanes of the type $Bu_3Sn(C_3H_3)$ (C_3H_3 = allenyl and propargyl) $Me_3Sn(C_4H_7)$, $Bu_3Sn(C_4H_7)$, $Ph_3Sn(C_4H_7)$, $Me_2Sn(C_4H_7)_2$ and $Bu_2Sn(C_4H_7)_2$ [C_4H_7 = (E)-but-2-enyl, (Z)-but-2-enyl, and but-1-en-3-yl] were prepared via Grignard reactions as isomeric mixtures whose compositions could be directly determined by ^{119}Sn and ^{13}C n.m.r[553]. Seven phenyl-substituted vinylstannanes were prepared and characterised by ^{1}H, ^{13}C, and ^{119}Sn n.m.r[554].

SnR_4 (R = Me, Et, $CHMe_2$, Bu) and Bu_3SnH were reacted with the surface of partially dehydroxylated silica, and interaction between the complexes and the silica surface was followed by IR and ^{13}C and ^{119}Sn MAS n.m.r. spectroscopies and analysis of evolved gases[555]. The ^{119}Sn CP/MAS n.m.r. spectra of the complex $[(Me_2CHSn)_{12}O_{14}(OH_6)]^{2+}$ ion were recorded for two different salts having twelve and four different tin atoms in the asymmetric unit, respectively[556]. Crystalline [K{Sn(CH2Bu-tert)$_3$}(η-$C_6H_5Me)_3$] was prepared and solid-state ^{119}Sn n.m.r. study provided the first observation of direct ^{119}Sn-^{39}K coupling[557]. The reaction of $(n\text{-}C_4H_9)_3SnH$ with the surface of a silica-alumina mixture was investigated by analysis of gases evolved, IR spectroscopy, and by ^{13}C CP/MAS and ^{119}Sn MAS n.m.r[558]. The surface reactions of the well-defined surface organometallic fragment >Si-O-Sn(n-$C_4H_9)_3$ grafted on $silica_{200}$ and on $silica_{500}$ were followed by quantitative measurements of the evolved gases, IR and Moessbauer spectroscopies, ^{13}C CP/MAS and ^{119}Sn n.m.r. spectroscopy, XPS measurements, and electron microscopy (CTEM and STEM EDAX)[559]. Solid-state CP/MAS ^{119}Sn n.m.r. spectra were obtained for six adducts of dimethyl- or diphenyltin dichloride[560].

[NH_4][Sn(cat)$_3$] (H_2cat = catechol or derivative), [SnL_3]X (HL = hydroxyprydinone, X = Cl, Br, I) and [Sn(trop)$_3$]X (Htrop = tropolone) were prepared and characterised by ^{1}H and ^{119}Sn n.m.r., elemental analysis and mass spectroscopy[561]. $Rh(H)\{(\mu\text{-}H)SnR_3\}_2(PPh_3)_2$ (R = Bu, Ph) were prepared and characterised by ^1H, ^{31}P, and ^{119}Sn n.m.r[562]. Several heteroleptic tin(II) isopropoxometallates, $XSn\{Al(OCHMe_2)_4\}$, $XSn\{Zr_2(OCHMe_2)_9\}$, $XSn\{M(OCHMe_2)_6\}$, $\{Al(OCHMe_2)_4Sn\{Zr_2(OCHMe_2)_9\}$, and $\{Nb(OCHMe_2)_6Sn\{Ta(OCHMe_2)_6$ (X = Cl, $OCHMe_2/OCMe_3$; M = Nb/Ta) were synthesised and characterised by elemental analyses, molecular weight measurements, and ^1H ^{13}C, ^{27}Al, ^{119}Sn n.m.r[563]. The reactivity of the azastannatrane $Me_2NSn(NMeCH_2CH_2)_3N$ against protolutic reagents was studied and ^{119}Sn n.m.r. in solution as well as in the solid-state indicated a strong transannular Sn-N bond of the azastannatrane framework[564]. A wide range of novel supramolecular (3-dimensional) heterometallic compounds, synthesised at the University of Hamburg, were studied by ^{13}C, ^{15}N, ^{119}Sn and ^{207}Pb MAS n.m.r[565].

The ^{119}Sn MAS n.m.r. resonances of the paramagnetic stannates $Ln_2Sn_2O_7$ (Ln = Nd, Sm, and $Y_{1.8}Sm_{0.2}$) were found to be extremely sensitive to temperature[566]. The CP/MAS ^{13}C, ^{15}N, and ^{119}Sn n.m.r. spectra of the coordination polymers $[Me_3Sn^{IV})_4M^{II}(CN)_6]_\infty$ [M = Fe, Ru, and Os] were fully determined and optimally assessed[567]. High-resolution solid-state ^{119}Sn n.m.r. spectra of $[H_2B(pz)_2]_2Sn$, $[HB(pz)_3]_2Sn$, $[HB(3,5\text{-}Me_2pz)_3]_2Sn$ and $[H_2B(pz)_2]SnCl$ (pz = pyrazolyl) were measured to correlate the solid and solution structures, and to determine the coordination number of $[HB(pz)_3]_2Sn$[568].

2.14.5 *Lead (^{207}Pb)*

The solid-state CP/MAS ^{207}Pb n.m.r. was used to study molecular routes of lead(II) compounds to lead-based perovskite-phase materials[569].

2.15 Group 15 ($^{14,15}N$, ^{31}P, ^{75}As, $^{121,123}Sb$, ^{209}Bi)

2.15.1 *Nitrogen ($^{14,15}N$)*

Structural studies of peptides and polypeptides in the solid state by ^{15}N n.m.r. were reviewed[570].

The reaction product of isocyanogen (CNCN) with hydrogen chloride was characterised by ^1H, ^{13}C and ^{14}N n.m.r and high resolution mass spectrum[571].

^{15}N n.m.r. was used to determine the structure of active-site histidine-70 adduct formed when β-hydroxydecanoyl thiol ester dehydrase from *E. coli* reacts with the mechanism-based inactivator, S-(3-decynoyl)-N-acetylcysteamine[572]. The binding of netropsin and distamycin to the duplex $[d(CGCGAATTCGCG)]_2$, in which adenine residues A5 or A6 were [3-^{15}N] labelled, was monitored by ^{15}N n.m.r. to probe the sensitivity of these ^{15}N labels to drug binding[573]. The interaction between the oligonucleotide $[d(CGCGAATTCGCG)]_2$ and Hg(II) was studied by ^1H and natural abundance ^{15}N n.m.r[574]. Activation of NAD^+ by uridine nucleotides perturbed the ^{15}N and ^{13}C n.m.r. chemical shifts of selectively enriched NAD^+ bound to the coenzyme site[575]. Solution and solid-state ^{13}C and ^{15}N n.m.r. were used to study 1,8-bis(dimethylamino)naphthalene and its complex with tetrazole[576]. The nitrogen-hydrogen bond lengths of simple peptides in the solid state were determined by natural abundance ^{15}N two-dimensional separated local dipolar field (2D-SLDF) n.m.r[577].

The ^{15}N n.m.r. chemical shifts and $^{1}J(^{15}N, ^{1}H)$ coupling constants were measured for a series of 5-substituted indoles and 4-substituted acetanilides to show the evidence for cross-conjugation effects[578]. ^{15}N, ^{14}N, ^{13}C, and ^{1}H n.m.r. data were used for the structural analysis of seven 1,2,3,4-thiatriazoles[579]. ^{15}N n.m.r. was used to show unusual deshielding of nitrogen in the 7-azabicyclo[2.2.1]heptyl ring system[580]. A marked temperature dependence of the ^{13}C and ^{15}N chemical shifts of 3-hydroxypyridine in D_2O (pD = 7.0) was observed[581]. The individual tautomeric forms of histidine were directly observed in ^{15}N n.m.r. spectra at -55° in ethanol/water solution[582]. ^{15}N, ^{14}N and ^{17}O n.m.r. chemical shifts and ^{15}N, ^{1}H spin coupling constants of N-nitropyrazoles (containing NO_2, CH_3, and CN groups in the pyrazole ring) were measured and a correlation between ^{15}N, ^{14}N and ^{17}O n.m.r. spectral parameters and structure was determined[583]. ^{1}H, ^{13}C, and ^{15}N n.m.r. chemical shifts were measured for N^1-alkyl-N^2-arylthioureas[584]. The ^{15}N n.m.r. spectra of α-, β-, γ-, and δ-carboline and a series of methyl, dihydro and tetrahydro derivatives were reported[585]. ^{13}C and ^{15}N n.m.r. spectra of six nitraminobenzenes and nine nitraminopyridines were measured in DMSO-d_6 and their chemical shifts were assigned in order to clarify the special properties of the aryl-bound nitramino substituent $NRNO_2$ (R = H, CH_3)[586]. The ^{15}N n.m.r. spectra of a series of 1,2,3-thiadiazoles revealed the strong influence of substituents at C-5 on the N-2 resonance[587].

Diamagnetic behaviour in the chemical shift and linewidth of high resolution ^{14}N n.m.r. spectra for paramagnetic chromium(III) diamine complexes was studied in comparison with that for the corresponding diamine dihydrochlorides and the diamagnetic cobalt(III) complexes[588].

The amide ^{15}N and ^{1}H n.m.r. resonances of 9 base-on cobalamines in DMSO-d_6 were observed by inverse-detected ^{1}H,^{15}N HMQC spectroscopy, to evaluate the possibility of the use of ^{15}N n.m.r. as a probe of axial nucleotide ligation in macromolecular complexes of cobalt corrinoids with proteins[589]. A systematic study of sensitivity enhancement in ^{15}N n.m.r. spectra of phosphorus-nitrogen compounds by ^{31}P, ^{15}N polarisation transfer was performed using tris-(dimethylamino)phosphine as a model compound[590]. The Cd(II) and Hg(II) complexes of 2,3-dihydro-2-thioxo-4(1H)-quinazolinone and the free ligand were studied by 1-dimensional and 2-dimensional ^{1}H, ^{13}C, ^{15}N, ^{113}Cd n.m.r. in solution and ^{13}C CP/MAS n.m.r. in the solid state[591].

^{15}N n.m.r. was used to study the nuclear environment of the N as it fills the pores of a microporous zeolite[592]. A novel triple resonance MAS n.m.r. experiment was reported that can be used to probe C-N interatomic distances in polycrystalline and amorphous solids, without the need for ^{15}N isotopic labelling[593]. The 5 solid phases of NH_4NO_3 and transitions between these phases were studied using solid-state ^{15}N MAS and static n.m.r[594].

2.15.2 *Phosphorus (^{31}P)*

An overall survey of ^{31}P n.m.r. chemical shifts versus x-ray data for fifty kinds of cyclo-phosphazenic structures was reviewed to show that n.m.r. in solution is more reliable than crystallographic determinations for predicting how prone cyclophosphazenes are to generate new polymers[595].

^{31}P n.m.r. was used to study the he influence of various parameters on the formation of nonbilayer phases in mixed cardiolipin/phosphatidylcholine liposomes[596]; to observe and elucidate the structures of aged and nonaged phosphorylated cholinesterases[597]; to study phospholipids embedded in lamellar, hexagonal or isotropic phases[598]. The supermolecular organisation of multibilayers formed by aqueous egg phosphatidylcholine-triolein (PC-TO) mixtures was investigated using ^{31}P, ^{13}C, and ^{1}H n.m.r[599]. The simultaneous detection of intracellular $[M_{GT}]/[ATP_T]$ and pH from the three ^{31}P n.m.r. chemical shifts of ATP was demonstrated using two-dimensional calibrations[600].

Protonation constants of four isopropyl esters of clodronic acid were determined by ^{31}P n.m.r. and potentiometric measurements in aqueous solution at 25° and I = 1.0 (Me_4NCl)[601]. ^{31}P n.m.r. appeared to be a very suitable method to follow, directly in the n.m.r. tube, the fate of the different hydrolysis products of phytate[602]. Three phosphorus-nitrogen compounds, tert-butylamino-tert-butylfluorophosphine, tert-butylimino-tert-butylphosphine and of its dimer formed by [1+2] cycloaddition, were studied by ^{1}H, ^{13}C, ^{15}N, ^{19}F, and ^{31}P one- and two-dimensional n.m.r[603]. ^{31}P n.m.r. chemical shifts of phospholipids solubilised in bile salts were studied with respect to variations in the structure of the acyl substituents[604]. The purity of lithium, sodium, and potassium diphenylphosphide, prepared by various methods, was evaluated by ^{31}P n.m.r[605].

Dimethyl methylphosphonate (DMMP) was used as ^{31}P MAS n.m.r. probe of intracellular volume in mammalian cell cultures[606]. ^{31}P CP MAS n.m.r. was used to study correlation of chemical shift tensors and shielding parameters to molecular structure of bis(organothiophosphoryl) disulphides[607]. Solid-state ^{31}P n.m.r., x-ray powder diffraction and ^{1}H, ^{13}C, ^{31}P solution n.m.r. were used to investigate solvation effects of tertiary phosphine derivatives and metal complexes[608].

^{31}P MAS n.m.r. was used to study a series of calcium phosphate glasses[609].

Interactions between phosphines [PPh_nMe_{3-n} ($3 \geq n \geq 1$)] and pentachlorophenol, CF_3COOH and $MeSO_3H$ in EtCl-$CDCl_3$ solutions were examined by ^{1}H and ^{31}P n.m.r[610]. The mechanism of oxidative addition of aryl halides to low-ligated zerovalent palladium species obtained by reduction of their divalent precursor complexes, $Pd^{II}Cl_2(PR_2)$, was investigated on the basis of ^{31}P n.m.r. and electrochemistry[611]. The protonation equilibria of phosphinate, phosphite, and acetate ions in Sephadex gels with different degrees of crosslinking were studied by ^{31}P and ^{13}C n.m.r[612]. Lamellar and hexagonal (H_I) liquid-crystalline mixtures of 1,2-dimyristoyl-sn-glycero-3-phosphocholine ether ($C_{12}E_8$) were studied by solid-state ^{2}H and ^{31}P n.m.r[613]. ^{31}P n.m.r was used to study the comparative hydrolytic breakdown of nickel(II) and cadmium(II) versus zinc(II) bis(O,O-diethyl dithiophosphates) in an aqueous medium[614].

Speciation equilibria and dynamic exchange process occurring in phosphorus-selenium melts above the glass transition temperature were characterised by in-situ static ^{31}P n.m.r. at $25 \leq T \leq 650°$[615]. ^{31}P MAS n.m.r. was used to study the lithium titanium indium phosphate solid solution system ($Li_{1+x}Ti_{2-x}In_xP_3O_{12}$)[616]. A ^{31}P n.m.r. study of powder and single-crystal samples of ammonium dihydrogen phosphate (ADP), $NH_4H_2PO_4$, was undertaken to resolve the apparent discrepancies in the literature concerning the local symmetry at the phosphorus nucleus of the H_2PO_4 moiety[617]. The reaction products of γ-zirconium phosphate with ethylene oxide were analysed by solid-state ^{31}P MAS n.m.r., and ^{31}P and ^{13}C solution n.m.r[618].

2.15.3 Arsenic (^{75}As)

Pressure effects on the ^{31}P and ^{75}As Knight shifts K in CeP and CeAs[619] and temperature dependence of ^{75}As Knight shifts were reported[620]. The microscopic magnetic behaviour of an incipient heavy-electron compound UAs (TN = 126 K) was studied by ^{75}As n.m.r. in the paramagnetic state up to 700 K[621].

2.15.4 Antimony ($^{121,123}Sb$) and Bismuth (^{209}Bi)

The weakly coordinating pnicogen anion series, $M(OTeF_5)_6^-$ (M = As, Sb, Bi), was synthesised as their tetraalkylammonium salts and were structurally characterised by x-ray crystallography, ^{19}F, ^{75}As, $^{121,123}Sb$, ^{125}Te, ^{209}Bi n.m.r. spectroscopy, and Raman spectroscopy[622].

2.16 Group 16 (^{17}O, ^{33}S, ^{77}Se, $^{123,125}Te$)

2.16.1 Oxygen (^{17}O)

^{17}O n.m.r. spectra were measured for some oximes RR^1C:NOH [R = Me, Et, Bu, etc.; R^1 = Me, Me_2CH, (un)substituted Ph, etc.; $RR^1 = (CH_2)_n$ and n = 4, 5, 6, 7] at natural abundance in pyridine and acetonitrile solutions[623]; for cholesterol and 31 other steroid alcohols, esters, ketones, and acids enriched by synthesis with ^{17}O from $H_2{}^{17}O$[624]; for eight β-substituted methylene-indanones obtained at natural abundance in MeCN at 75°[625]; for fifteen 2- and 4-substituted phenols, ten 3- and 5-substituted 2-hydroxybenzaldehydes and eight 3-substituted benzaldehydes, recorded at 75° in acetonitrile[626]; for eleven N-methylbenzamides and eleven N,N-dimethyl-benzamides correlated with Hammett σ^+ constants[627]; for 27 mono-, di-, tri- and tetrahydroxy-bicyclo[2.2.1]heptenes and -heptanes (norbornenes or norbornanes), their methylated derivatives, and some related compounds measured at natural abundance for 0.5M or saturated solutions in MeCN at 65°[628]; for 7-substituted-4-methylcoumarins obtained at natural abundance in acetonitrile at 75°[629]; for disilyl ketone $(Me_3Si)_2CO$ and related compounds to show the strong deshielding by electronic excitation energy[630]. ^{17}O n.m.r. was used to study resonance stabilisation, bond order and excitation energy of carbonyl phosphonium ylides and diazocarbonyl compounds[631]. ^{13}C and ^{17}O n.m.r. of isobenzopyrylium salts were reported[632]. N-Sulphinylanilines were studied by ^{13}C and ^{17}O n.m.r. in chloroform and pyridine[633]. An easy and absolute diagnosis for the coumarin/chromone discrimination by ^{17}O n.m.r. was presented[634].

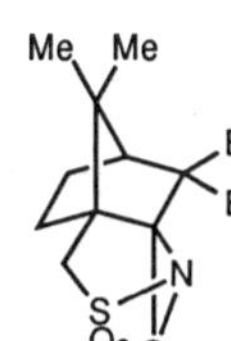

^{13}C, ^{15}N and ^{17}O N.m.r. data for the three ring atoms of various oxaziridines were compared[635]. The mechanism of a novel Baeyer-Villiger rearrangement, characterised by a stereospecific migration of a phosphoryl moiety to a peroxy oxygen, was explored using ^{17}O-enriched hydrogen peroxide (H_2O_2) as the mechanistic probe with ^{17}O n.m.r. as the analysis tool[636]. Seven oxidation products of (+)-3-carene were confirmed by various n.m.r. data including ^{17}O

n.m.r. and other spectra[637]. Three mononuclear 1,3,5-trimethylbenzene (mesitylene) carbonyl transition metal complexes, $(CH_3)_3C_6H_3Cr(CO)_3$, $(CH_3)_3C_6H_3W(CO)_3$, $[(CH_3)_3C_6H_3Mn(CO)_3]BF_4$; and three clusters, $(CH_3)_3C_6H_3Co_4(CO)_9$, $(CH_3)_3C_6H_3Ru_6C(CO)_{14}$ and $Ru_6(CO)_{17}$, were studied by ^1H, ^{13}C, and natural abundance ^{17}O n.m.r[638]. The differentiation of sulphonyl oxygens in cyclic systems was analysed by ^{17}O n.m.r. lanthanide induced shifts[639].

^{17}O MAS n.m.r. of TiO_2-SiO_2 gels formed by hydrolysis of alkoxides with ^{17}O-labelled water provided a novel technique for detecting amorphous or poorly crystalline impurity phases and at scale phase separation[640].

The ^{17}O n.m.r. spectra of sulphonyl- and carbonyl-stabilised diazo compounds ($MeCOCN_2COMe$, $PhCH_2SO_2CHN_2$, etc.) were studied[641]. The tetrameric hydroxo Zr(IV) complex in aqueous solution was studied by ^{17}O and ^1H n.m.r[642]. The structure and dynamics of the lanthanide(III) complexes of the bis(propylamide) of diethylenetriaminepentaacetate in aqueous solution were studied by ^{13}C and ^{17}O n.m.r[643]. $[(LAu)_3(\mu_3\text{-}O)]BF_4$ (L = $PMePh_2$, PMe_2Ph, $PEtPh_2$, $P(CHMe_2)Ph_2$, $P(C_6H_4Cl\text{-}p)_3$, $P(C_6H_4Me\text{-}o)_3$, $P(OEt)Ph_2$, and $P(OMe)_3$) were prepared and ^{17}O n.m.r. data for those gold oxo complexes were reported[644]. The terminal water ligand exchange and substitution by isonicotinamide on the oxo-centred triruthenium(III) complex $[Ru_3(\mu_3\text{-}O)(\mu\text{-}CH_3CO_2)_6(OH_2)_3]^+$ was studied by ^{17}O n.m.r. and UV/VIS spectroscopy in aqueous perchlorate media, I = 1.0M[645]. ^{17}O n.m.r. spectra (in natural abundance) were recorded, at room temperature, for a variety of polynuclear Fe, Ru and Os carbonyl clusters[646]. The rates of the oxygen-exchange reactions of methyl- and butylarsonic acids and dimethylarsinic acid with solvent water were measured in the pH range below 6, and at temperatures between 50 and 90° by using the ^{17}O n.m.r. line-broadening technique[647].

^{17}O n.m.r. was used to study $La_{2-x}Sr_xCuO_4$ in the lightly and heavily doped regions[648]. ^{17}O n.m.r. measurements on triple-layered $Tl_2Ba_2Ca_2Cu_3O_{10-\delta}$ (Tc ≈ 125 K) showed a virtually identical spin susceptibility for both the pyramidal and square Cu-O charge distribution in this system[649]. ^{17}O n.m.r. with high temperature XRD and DTA were used to study the structure of $Ba_2In_2O_5$ and the dynamics of oxygen ion motion between room temperature and 1200° [650].

2.16.2 Sulphur (^{33}S)

^{33}S n.m.r. was reviewed[651].

Tropothione S-oxide was synthesised as the first example of the 6π aromatic stable sulphine and its ^{17}O and ^{33}S n.m.r. data were reported[652].

2.16.3 Selenium (^{77}Se)

The ^{77}Se, ^{13}C and ^1H n.m.r. spectra of 57 mostly new compounds with phenylselenyl groups attached to acyclic alkanes, alkenes and phenylalkanes and of 10 corresponding selenoxides were reported[653]. In order to investigate mechanistic roles of the amino nitrogens located at the active centre of glutathione peroxidase (GPX), a selenium-containing antioxidant enzyme, kinetic analyses and ^{77}Se n.m.r. characterisation of the intermediates of the model reaction (H_2O_2 + 2PhSH → $2H_2O$ + PhSSPh) catalysed by GPX like di-2-(N-cyclohexyl,N-methylamino)methyl)phenyl diselenide were performed[654]. Two dimensional ^{13}C, ^{77}Se correlation spectroscopy was used to assign the selenium and carbon n.m.r. signals in 2,2-bis(phenylseleno)styrene and related compounds[655]. $[M(cod)\{CH_2(PPh_2)(P(Y)R_2)\text{-}P,S\}]^+$ (M = Ir, Rh; Y = S, Se; R = Ph, tert-Bu) were synthesised and characterised by ^{13}C, ^{31}P and ^{77}Se n.m.r[656]. ^1H, ^{13}C, and ^{77}Se chemical shifts of sulphur and selenium analogues of furocumarin (psoralen) were reported[657]. ^{15}N and ^{77}Se n.m.r. were used to study selenium diimides and aminoselenanes[658]. Diarylpolyselanes R_2Se_x (x = 2-7) and diaryldiselenapolysulphanes $(RSe)_2S_y$ (y = 1-15) with R = p-Cl-C_6H_4 were synthesised and characterised by ^{77}Se n.m.r. and UV spectroscopy[659].

A new tetracoordinated cyclic diazaselenurane with amino and ammonio groups as two unsymmetrical apical ligands was prepared and its structure was confirmed by ^{77}Se and ^{15}N n.m.r. using the ^{15}N-enriched aminoammonio-selenurane[660].

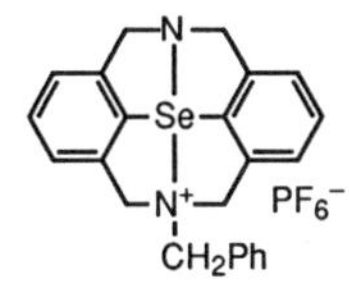

In situ ^{13}C and ^{77}Se magic angle spinning n.m.r. was used to investigate the formation and reactivity of trialkylonium species on zeolite catalysts HZSM-5 and HY[661].

Molten-state equilibria governing the structure of P-Se glasses were studied by high temperature ^{31}P and ^{77}Se n.m.r. over the range 25-550° [662].

Several polyselenides of indium and thallium such as $(Ph_4P)_4[In_2(Se_4)_4(Se_5)]$ were prepared and ^{77}Se n.m.r. spectra of those polyselenides were reported[663].

The reaction of $Ph_2PN_2(SiMe_3)_3$ with organoselenium trichlorides $RSeCl_3$ in acetonitrile at 23° produced heterocycles (R = Me, Et, Ph) , characterised by ^1H, ^{31}P, and ^{77}Se n.m.r., IR, and mass spectroscopy[664,665]. Monomeric selenopalladates and selenoplatinates were prepared and characterised by ^{77}Se and ^{195}Pt n.m.r[666]. The ^{77}Se n.m.r. spectra of $Na_2C_3Se_5$ and $[M(C_3Se_5)_2]^{n-}$ complexes ($C_3Se_5^{2-}$ = 1,3-diselenole-2-selone-4,5-diselenolate; n = 1, M = Au(III); n = 2, M = Zn(II), Cd(II), Ni(II), Pd(II) and Pt(II)) as well as those of $Na_2C_3Se_5$ and $C_3Se_5[C(O)Ph]_2$ were measured[667]. The explosive black solid 3-chloro-1,3,4,2,5-triselena-diazolium chloride, $Se_3N_2Cl_2$, was prepared and the reaction product with SO_2Cl_2 at -78° was characterised by ^{14}N and ^{77}Se n.m.r[668].

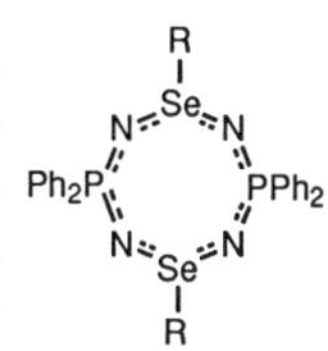

Preparation and ^{31}P and ^{77}Se n.m.r. spectra of platinum and palladium complexes of a $P_2N_4Se_2$ ring were reported. $[NEt_4]_2[\mu\text{-}WSe_4][(SC_4H_3)SeCu]_2$ was synthesised and characterised by ^{77}Se n.m.r[669]. $[ClSeNSeCl]^+[SbCl_6]^-$ was prepared and characterised by elemental analysis and vibrational (IR, Raman), ^{77}Se n.m.r., and mass spectroscopy[670]. Several N-bonded complexes of platinum(II) with phosphorus-nitrogen rings containing three-coordinate chalcogens were prepared and characterised by ^1H, ^{31}P and ^{77}Se n.m.r. spectroscopy and x-ray crystallography[671]. $[Re_2I_2(CO)_6(Se_7)]$ was prepared by the reaction of $[Re_2I_2(CO)_6(THF)_2]$ with Se and characterised by IR, ^{13}C and ^{77}Se n.m.r[672].

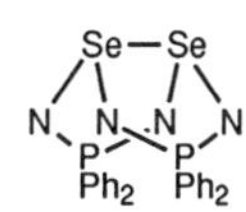

The gas phase ^{77}Se and ^{113}Cd chemical shifts were measured for H_2Se, $HSeCH_3$, $Se(CH_3)_2$, and $Cd(CH_3)_2$ and $Cd(C_2H_5)_2$, respectively and these chemical shifts were calculated by ab initio calculations[673].

2.16.4 Tellurium (^{123,125}Te)

The 2:1 complexes prepared from anilines and tellurium tetrachloride were studied by ^{1}H, ^{13}C, ^{125}Re, and ^{15}N n.m.r. and submitted to chemical degradation[674]. Symmetrical and unsymmetrical diaryl tellurides were synthesised and characterised by ^{125}Te n.m.r[675]. ^{125}Te, ^{1}H, and ^{13}C n.m.r. data were presented for five-membered ring ditellurium derivative[676].

5H, 7H-dibenzo[b,g][1,5]ditellurocin and -telluraselenocin were synthesised and their strong transannular hypervalent interaction between the tellurinyl group and the selenyl group was evidenced by ^{77}Se and ^{125}Te n.m.r[677].

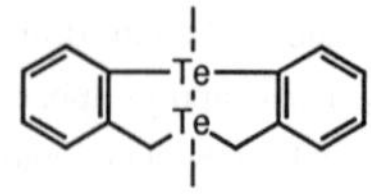

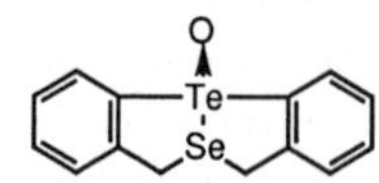

Reaction of 2-(Br_3Te)C_6H_4CH:N-C_6H_4Me-p with BH_4^- resulted in reduction at both the tellurium and double-bond centres to afford intramolecular coordinate bonds characterised by x-ray crystallography, and ^{1}H, ^{13}C, ^{123}Te, ^{125}Te and variable-temperature ^{1}H n.m.r[678]. The first stable telluroketone in solution, 1,1,3,3-tetramethylindantellone, was prepared and its spectroscopic properties of the carbon and tellurium nuclei of the C:Te bond were investigated by ^{13}C and ^{125}Te n.m.r[679].

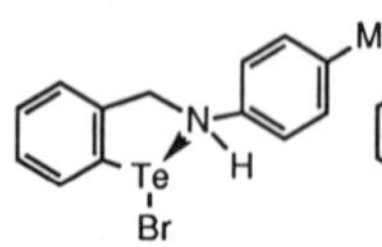

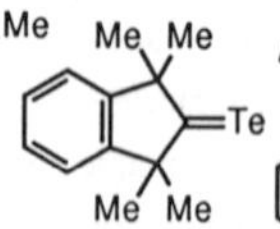

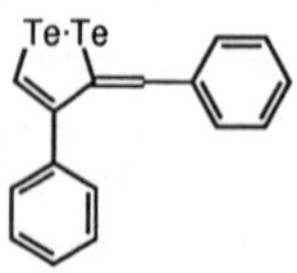

Monomeric terminal chalcogenides, [ETa{$Me_3SiNCH_2CH_2)_3$N}] (E = Se, Te), were prepared via template-induced disilylchalcogenide elimination and characterised by IR, Raman, ^{1}H, ^{13}C, ^{77}Se, and ^{125}Te n.m.r and mass spectra[680]. The TeF_7^- anion was studied by vibrational and ^{19}F and ^{125}Te n.m.r. spectroscopy[681].

X-ray crystallography and solid-state ^{119}Sn and ^{125}Te n.m.r. were used to study 2,2,4,4,6,6-hexakis(trimethylsilylmethyl)cyclotristannatellurane[682]. Bu_4NTeH, Me_4NSeH, and $(Me_4N)_2Te_2$ were synthesised and characterised by ^{123}Te, ^{125}Te, and ^{77}Se solid-state n.m.r. and ^{125}Te Moessbauer spectroscopy[683].

2.17 Group 17 (^{19}F, ^{35}Cl, ^{81}Br)

2.17.1 Fluorine (^{19}F)

The transport of trifluoroacetate (TFA) and difluorophosphate (DFP) into and out of human and sheep erythrocytes was measured by ^{19}F n.m.r[684]. The ^{19}F n.m.r. chemical shifts and coupling constants were given for a series of aryl (Ph, naphthyl and phenanthryl) substituted trifluorovinyls and (E)- and (Z)-perfluoropropenyls[685]. Isomeric ring substituted monofluoro-(E)-3-phenyl-propenals (cinnamaldehydes) were studied to provide an insight into the influence of substituents and the transmission of electronic effects between the aromatic ring and the side chain by ^{1}H, ^{13}C, ^{17}O and ^{19}F n.m.r[686]. The interaction of (R)-1-acetamido-2-(4-fluorophenyl)ethane-1-boronic acid with a-chymotrypsin at pH 4 was studied by a variety of ^{19}F n.m.r. experiments[687]. Conformational differences between complexes of elongation factor Tu were studied by ^{19}F n.m.r[688]. The regioselectivity and rate of the ortho-hydroxylation of 3-fluorophenol by phenol hydroxylase from Trichosporon cutaneum (EC 1.14.13.7) was studied using ^{19}F n.m.r[689]. ^{1}H and ^{19}F n.m.r. chemical shifts, association constants, free association enthalpies, and dissociation barriers were determined for complexes of cryptophan A with CH_4, CHF_2Cl, CF_2Cl_2, CH_2Cl_2, CH_2Br_2, and $CHCl_3$[690].

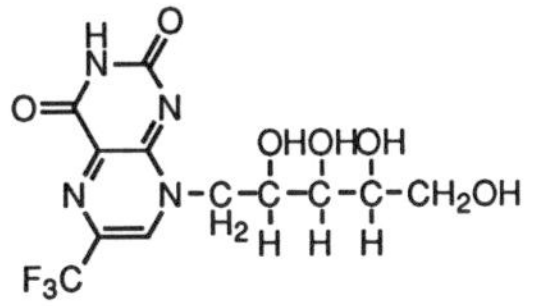

6-(Trifluoromethyl)-8-(D-ribityl)lumazine was synthesised to study its reactivity at C-7 and its binding to riboflavin synthase of Bacillus subtilis by ^{19}F n.m.r[691]. The interaction of 4-fluoro-substituted benzeneboronic acids with subtilisin Carlsberg was studied by ^{19}F n.m.r[692]. New polyfluoroalkyl hypochlorites and fluoroxy compounds containing Cl, H, and Br in the alkyl group were prepared and characterised by ^{19}F and ^{1}H n.m.r. and IR and by their reactions with fluoroalkenes to produce new polyfluoro ethers[693]. The correlation of ^{19}F n.m.r. chemical shielding and chemical non-equivalence was reported[694].

^{19}F n.m.r. was used to study the positional effects of a trifluoromethyl group on the aggregation of azo dyes in aqueous solutions[695]. Distribution and motion of trifluoromethane-sulfonate anions in poly(p-hydroxystyrene) and polystyrene films were quantified by ^{19}F multiple-quantum (MQ) n.m.r[696]. A new technique for probing polymer dynamics on ^{19}F n.m.r. through the refocusing of multiple quantum (MQ) n.m.r. coherences was reported[697].

Solid-state ^{19}F n.m.r. was used to characterise F on diamond powder surfaces after CF_4 and 98% CF_4/2% O_2 radio-frequency plasma treatment[698].

^{1}H, ^{2}H, and ^{19}F n.m.r. analyses of new hydroxide-bridged and water bridged iron and cobalt complexes were reported[699]. ^{19}F n.m.r. was used to study fluorite-type $Sr_{1-x}Bi_xF_{2+x}$ solid solutions ($0 \leq x \leq 0.50$) carried out on samples quenched from 700° to show correlations between short range ordering and ionic conductivity[700]; to observe the molecular motions of a fluorinated ferroelectric liquid crystal, 2,3-difluoro-4-octyloxybiphenyl-4'(-4"-oxycarbonyl-(S)-1-chloro-3-methylbutyl)benzoate[701].

2.17.2 *Chlorine (^{35}Cl)*

Low-temperature ^{7}Li and ^{35}Cl n.m.r. of Gilman cuprate reagents in THF in the presence of Me_3SiCl gave a insight into the potential role of this additive in accelerating cuprate conjugate additions[702].

^{35}Cl n.m.r. spectra and relaxation measurements in the triangular 2-dimensional lattice of VCl_2, with antiferromagnetic interactions V^{2+} $S = 3/2$ ions were presented[703]. The room temperature molten salts formed by combining $AlCl_3$ with 1-ethyl-3-methylimidazolium chloride were investigated by ^{27}Al and ^{35}Cl n.m.r. at 25-100°[704].

2.17.3 *Bromine (^{81}Br)*

The line-broadening of ^{81}Br n.m.r. signals observed in the presence of α-cyclodextrins was applied to evaluate their binding abilities toward inorganic anions in aqueous solutions[705].

2.18 Group 18 (^{3}He, ^{21}Ne, ^{83}Kr, $^{129,131}Xe$)

2.18.1 Helium (^{3}He), Neon (^{21}Ne), Krypton (^{83}Kr), and Xenon (^{131}Xe)

^{3}He n.m.r. chemical shifts in liquid solvents were reported and compared with other noble gases (^{21}Ne, ^{83}Kr, ^{129}Xe)[706]. Precise measurements were presented for the nuclear susceptibility of ^{3}He films adsorbed on exfoliated graphite at mK temperatures as a function of coverage (0.53 - 1.42 layers)[707]. High resolution n.m.r. measurements of the normal ^{3}He Fermi liquid and the bulk superfluid ^{3}He B-phase were reported[708,709]. ^{3}He n.m.r. spectra were presented for the endohedral compounds $^{3}He@C_{60}$ and $^{3}He@C_{70}$ (@ denotes an endohedral compound)[710].

^{21}Ne spin echo n.m.r. was used to determine the self-diffusion coefficient of Ne in water[711].

2.18.2 Xenon (^{129}Xe)

A van der Waals shift of ^{129}Xe n.m.r. resonance shifts for Xe dissolved in polymers was presented which accurately reflects the temperature dependence of ^{129}Xe chemical shifts in polymers above the glass transition temperature[712]. Two-dimensional ^{129}Xe n.m.r. was used to probe microheterogeneity in polymer systems[713].

^{129}Xe n.m.r. was used to study Cu-exchanged Na zeolites having different levels of exchange[714]; to study the distribution of Na metal particles inside the cavity of a NaY zeolite[715]; to study the flexibility of cadmium-rho zeolite[716]; to study the effect of temperature of xenon sorbed in zeolite rho[717]; to study entrapment and controlled release of xenon in cadmium(2+)-exchanged zeolite rho[718]; to study the particle flow between gas and liquid phase[719]; to study faujasite zeolites of varying aluminium content[720]; to monitor the interaction of Xe atoms with monovalent, divalent, and trivalent metal cations and halide anions in aqueous solution[721]; to study the effect of cation substitution on the adsorption of Xe on NaY zeolites[722]. A cell theory of an interacting lattice gas was applied to ^{129}Xe n.m.r. of Xe in zeolites[723]. The chemical shifts of ^{129}Xe dissolved in liquid CO_2, N_2O, C_2H_6, and C_3H_8[724] and in a variety of n-alkanes and their mixtures[725] were measured at 23° using a lightweight sapphire n.m.r. tube-valve assembly designed for this purpose. The ^{129}Xe shift in four very strong acid solvents deviated from its correlation line versus the Rumments refractive index function for forty-two nonacidic solvents[726]. The phenomenon of adsorption and desorption of xenon in an inorganic aluminium/zircon oxide fibre, in silica gel and in an organic polymer blend was compared by ^{129}Xe 2-dimensional-exchange n.m.r[727]. A number of factors in ^{129}Xe n.m.r. were pointed out related to the bulk properties of microporous solids[728]. The porous structure of a number of silica gels was characterised by ^{129}Xe n.m.r. of adsorbed Xe[729].

References

1 M.L. Ranc and P.C. Jurs, *Anal. Chim. Acta*, 1993, **280**, 145
2 L. Chen and W. Robien, *Anal. Chem.*, 1993, **65**, 2282
3 H.N. Cheng and L.J. Kasehagen, *Anal. Chim. Acta*, 1994, **285**, 223
4 J.P. Doucet, A. Panaye, E. Feuilleaubois, and P. Ladd, *J. Chem. Inf. Comput. Sci.*, 1993, **33**, 320
5 V.de Emerenciano, A.C. Bussolini, G.doV. Rodrigues, M. Furlan, and D.L.G. Fromanteau, *Spectroscopy (Amsterdam)*, 1993, **11**, 95
6 D. Redeker, M. Ballabio, D. Borghi, and E. Martinelli, *Chemom. Intell. Lab. Syst.*, 1993, **19**, 233
7 R. Wehrens, C. Lucasius, L. Buydens, and G. Kateman, *Anal. Chim. Acta*, 1993, **277**, 313

8 R. Bernstein, C. Cieslar, A. Ross, H. Oschkinat, J. Freund, and T.A. Holak, *J. Biomol. NMR*, 1993, **3**, 245
9 M.M. Goodgame and S.M. Geer, *J. Magn. Reson., Ser. A*, 1993, **102**, 246
10 S.A. Smith, T.O. Levante, B.H. Meier, and R.R. Ernst, *J. Magn. Reson., Ser. A*, 1994, **106**, 75
11 J.G. Pearson, E. Oldfield, F.S. Lee, and A. Warshel, *J. Am. Chem. Soc.*, 1993, **115**, 6851
12 G.M.J. West, *J. Chem. Inf. Comput. Sci.*, 1993, **33**, 577
13 I.M. Svishchev and P.G. Kusalik, *J. Am. Chem. Soc.*, 1993, **115**, 8270
14 G.M. Lipkind, A.S. Shashkov, N.E. Nifant'ev, and N.K. Kochetkov, *Carbohydr. Res.*, 1992, **237**, 11
15 J. Xu, S.K. Straus, B.C. Sanctuary, and L. Trimble, *J. Chem. Inf. Comput. Sci.*, 1993, **33**, 668
16 G.V.T. Swapna and R. Ramachandran, *J. Magn. Reson., Ser. B*, 1993, **102**, 204
17 H. Eckert, K. Moran, D. Franke, and C. Hudalla, *NATO ASI Ser., Ser. C*, 1993, **386**, 49
18 J.D. Augspurger, C.E. Dykstra, E. Oldfield, and J.G. Pearson, *NATO ASI Ser., Ser. C*, 1993, **386**, 75
19 W. Kutzelnigg, C. van Willen, U. Fleischer, and R. Franke, *NATO ASI Ser., Ser. C*, 1993, **386**, 141
20 P. Pulay, J.F. Hinton, and K. Wolinski, *NATO ASI Ser., Ser. C*, 1993, **386**, 243
21 H. Nakatsuji, *NATO ASI Ser., Ser. C*, 1993, **386**, 263
22 J.A. Tossell, *NATO ASI Ser., Ser. C*, 1993, **386**, 279
23 J. Geertsen, *NATO ASI Ser., Ser. C*, 1993, **386**, 335
24 S.P.A. Sauer and J. Oddershede, *NATO ASI Ser., Ser. C*, 1993, **386**, 351
25 M. Barfield, *NATO ASI Ser., Ser. C*, 1993, **386**, 523
26 H. Kurosu, I. Ando, and G.A. Webb, *Magn. Reson. Chem.*, 1993, **31**, 399
27 F. Imashiro, Y. Masuda, M. Honda, and S. Obara, *J. Chem. Soc., Perkin Trans. 2*, 1993, 1535
28 S. Sieber, P.v.R. Schleyer, and J. Gauss, *J. Am. Chem. Soc.*, 1993, **115**, 6987
29 D.D. Laws, A.C. de Dios, and E. Oldfield, *J. Biomol. NMR*, 1993, **3**, 607
30 A.C. de Dios, J.G. Pearson, and E. Oldfield, *J. Am. Chem. Soc.*, 1993, **115**, 9768
31 D.B. Chesnut, L.D. Quin, and K.D. Moore, *J. Am. Chem. Soc.*, 1993, **115**, 11984
32 K. Wolinski, C.L. Hsu, J.F. Hinton, and P. Pulay, *J. Chem. Phys.*, 1993, **99**, 7819
33` K. Jackowski, A. Barszczewicz, and K. Wozniak, *Solid State Nucl. Magn. Reson.*, 1993, **2**, 265
34 P.v.R. Schleyer, J. Gauss, M. Buehl, R. Greatrex, and M.A. Fox, *J. Chem. Soc., Chem. Commun.*, 1993, 1766
35 D. Jiao, M. Barfield, and V.J. Hruby, *J. Am. Chem. Soc.*, 1993, **115**, 10883
36 P.v.R. Schleyer and S. Sieber, *Angew. Chem. Int. Ed. Engl.*, 1993, **32**, 1606
37 A.M. Mebel, D.G. Musaev, N. Koga, and K. Morokuma, *Bull. Chem. Soc. Jpn.*, 1993, **66**, 3259
38 R. Wolff, C. Vogel, and R. Radeglia, *NATO ASI Ser., Ser. C*, 1993, **386**, 385
39 C. van Wuellen and W. Kutzelnigg, *Chem. Phys. Lett.*, 1993, **205**, 563
40 A.C. de Dios and E. Oldfield, *Chem. Phys. Lett.*, 1993, **205**, 108
41 A.C. de Dios, J.G. Pearson, and E. Oldfield, *Science (Washington, D.C., 1883-)*, 1993, **260**, 1491
42 J.F. Hinton, P.L. Guthrie, and K. Wolinski, *J. Magn. Reson., Ser. A*, 1993, **103**, 188
43 A.M. Mebel, D.G. Musaev, and K. Morokuma, *Chem. Phys. Lett.*, 1993, **214**, 69
44 C. Vogel, R. Wolff, and R. Radeglia, *Collect. Czech. Chem. Commun.*, 1993, **58**, 1255
45 J.D. Augspurger, A.C. deDios, E. Oldfield, and C.E. Dykstra, *Chem. Phys. Lett.*, 1993, **213**, 211
46 D.B. Chesnut, B.E. Rusiloski, K.D. Moore, and D.A. Egolf, *J. Comput. Chem.*, 1993, **14**, 1364
47 E.A. Moore, A. Healy, M. Mortimer, and N.F. Peirson, *THEOCHEM*, 1993, **104**, 229
48 H. Nakatsuji, T. Higashioji, and M. Sugimoto, *Bull. Chem. Soc. Jpn.*, 1993, **66**, 3235
49 U. Sternberg, *NATO ASI Ser., Ser. C*, 1993, **386**, 435
50 J.H. Iwamiya and G.E. Maciel, *NATO ASI Ser., Ser. C*, 1993, **386**, 495
51 H. Koller, G. Engelhardt, A.P.M. Kentgens, and J. Sauer, *J. Phys. Chem.*, 1994, **98**, 1544
52 S.E. Chambers, E.Y. Lau, and J.T. Gerig, *J. Am. Chem. Soc.*, 1994, **116**, 3603

53 A.F. Casy, *Trends Anal. Chem.*, 1993, **12**, 185
54 J.S. Bradshaw, P. Huszthy, C.W. McDaniel, M. Oue, C.Y. Zhu, R.M. Izatt, and S. Lifson, *J. Coord. Chem.*, 1992, **27**, 105
55 G.M. Hanna and C.A. Lau-Cam, *J. Pharm. Biomed. Anal.*, 1993, **11**, 665
56 J. Buddrus, H. Herzog, and K. Risch, *Anal. Chem.*, 1994, **66**, 40
57 T. Pehk, E. Lippmaa, M. Lopp, A. Paju, B.C. Borer, and R.J.K. Taylor, *Tetrahedron: Asymmetry*, 1993, **4**, 1527
58 L. Keller and R. Rothchild, *Spectrosc. Lett.*, 1992, **25**, 1177
59 R. Rothchild, K.S. Venkatasubban, S. Braddock, K. Sanders, and S. Traviglia, *Spectrosc. Lett.*, 1993, **26**, 271
60 R. Rothchild, S. Braddock, S. Traviglia, and K.S. Venkatasubban, *Spectrosc. Lett.*, 1993, **26**, 411
61 P. Hudhomme and G. Duguay, Tetrahedron: Asymmetry, 1993, **4**, 1897
62 K. Roberts, R. Rothchild, and H. Wyss, *Spectrosc. Lett.*, 1993, **26**, 1901
63 R. Rothchild, K. Sanders, and K.S. Venkatasubban, *Spectrosc. Lett.*, 1993, **26**, 1997
64 D. Shields, H. Tran, K.S. Venkatasubban, R. Rothchild, and H. Wyss, *Spectrosc. Lett.*, 1994, **27**, 23
65 H.Y. Aboul-Enein, R. Rothchild, and A. Sinnema, *Spectrosc. Lett.*, 1993, **25**, 1367
66 M. Koos and H.S. Mosher, *Tetrahedron*, 1993, **49**, 1541
67 J.-L. Canet, A. Fadel, J. Salaun, I. Canet-Fresse, and J. Courtieu, *Tetrahedron: Asymmetry*, 1993, **4**, 31
68 B.S. Jursic and Z. Zdravkovski, *J. Org. Chem.*, 1993, **58**, 5245
69 G. Uccello-Barretta, F. Balzano, A.M. Caporusso, and P. Salvadori, *J. Org. Chem.*, 1994, **59**, 836
70 Y. Takeuchi, N. Itoh, T. Satoh, T. Koizumi, and K. Yamaguchi, *J. Org. Chem.*, 1993, **58**, 1812
71 R. Hulst, R.W. Zijlstra, B.L. Feringa, N. Koen de Vries, W. Ten Hoeve and H. Wynberg, *Tetrahedron Lett.*, 1993, **34**, 1339
72 A. Botsi, K. Yannakopoulou, E. Hadjoudis, and B. Perly, *J. Chem. Soc., Chem. Commun.*, 1993, 1085
73 L. Tottie, C. Moberg, and A. Heumann, *Acta Chem. Scand.*, 1993, **47**, 492
74 J. Reiner, R. Dagnino, Jr., E. Goldman, and T.R. Webb, *Tetrahedron, Lett.*, 1993, **34**, 5425
75 Y. Adegawa, T. Kashima, and K. Saigo, *Tetrahedron: Asymmetry*, 1993, **4**, 1421
76 M. Calmes, J. Daunis, A. Hanouneh, and R. Jacqier, *Tetrahedron: Asymmetry*, 1993, **4**, 2437
77 Z. Glowacki, M. Hoffmann, and J. Rachon, *Phosphorus, Sulfur Silicon Relat. Elem.*, 1993, **82**, 39
78 M.J. Rieser, Y.H. Hui, J.K. Rupprecht, J.F. Kozlowski, K.V. Wood, J.L. McLaughlin, P.R. Hanson, Z. Zhuang, and T.R. Hoye, *J. Am. Chem. Soc.*, 1992, **114**, 10203
79 H. Miyake, Y. Kojima, T. Yamashita, and A. Ohsuka, *Makromol. Chem.*, 1993, **194**, 1925
80 J. Lehmann and E. Kleinpeter, *Magn. Reson. Chem.*, 1993, **31**, 367
81 S.E. Boiadjiev, R.V. Person, and D.A. Lightner, *Tetrahedron: Asymmetry*, 1993, **4**, 491
82 M.E. Jung, I.D. Trifunovich, and A.W. Sledeski, *Heterocycles*, 1993, **35**, 273
83 H. Suezawa, A. Mori, M. Sato, R. Ehama, I. Akai, K. Shibahara, M. Hirota, M. Nishio, and Y. Kodama, *J. Phys. Org. Chem.*, 1993, **6**, 399
84 P.E. Hansen, A.E. Hansen, A. Lycka, and A. Buvari-Barcza, *Acta Chem. Scand.*, 1993, **47**, 777
85 Y. Nakashima, M. Fukunaga, K. Suzuki, and K. Takahashi, *Bull. Chem. Soc. Jpn.*, 1993, **66**, 2143
86 P.E. Hansen, M. Christoffersen, and S. Bolvig, *Magn. Reson. Chem.*, 1993, **31**, 893
87 K. Mlinaric-Majerski, V. Vinkovic, L.J. Chyall, and P.G. Gassman, *Magn. Reson. Chem.*, 1993, **31**, 903
88 N.M. Sergeyev, N.D. Sergeyeva, and W.T. Raynes, *J. Chem. Soc., Chem. Commun.*, 1994, 485
89 D.L. Rabenstein and S.V.S. Mariappan, *J. Org. Chem.*, 1993, **58**, 4487
90 T.L. Mega and R.L. Van Etten, *J. Am. Chem. Soc.*, 1993, **115**, 12056
91 M.P. Augustine, K.W. Zilm, and D.B. Zax, *J. Chem. Phys.*, 1993, **98**, 9432

92 D.A. Forsyth and S.M. Johnson, *J. Am. Chem. Soc.*, 1993, **115**, 3364
93 M. Schlabach, H.H. Limbach, E. Bunnenberg, A.Y.L. Shu, B.R. Tolf, and C. Djerassi, *J. Am. Chem. Soc.*, 1993, **115**, 4554
94 D. Rasala, *Spectrosc. Lett.*, 1993, **26**, 227
95 E. Bednarek, K. Kamienska-Trela, and L. Kania, *J. Mol. Struct.*, 1993, **293**, 173
96 S.E. Brown, J.H. Coates, P.A. Duckworth, S.F. Lincoln, C.J. Easton, and B.L. May, *J. Chem. Soc., Faraday Trans.*, 1993, **89**, 1035
97 R.G. Guy, *Magn. Reson. Chem.*, 1993, **31**, 610
98 J. Jirman, *Collect. Czech. Chem. Commun.*, 1993, **58**, 1378
99 R. Knorr, T. von Roman, J. Freudenrich, T.P. Hoang, J. Mehlstaeubl, P. Boehrer, D.S. Stephenson, H. Huber, and B. Schuber, *Magn. Reson. Chem.*, 1993, **31**, 557
100 E. Kleinpeter, S. Thomas, G. Uhlig, and W.D. Rudorf, *Magn. Reson. Chem.*, 1993, **31**, 714
101 P. Streckeisen, O. Muenster, R. Seydoux, and P. Diehl, *J. Mol. Struct.*, 1993, **301**, 137
102 H.G. Bonacorso, M.S.B. Caro, N. Zanatta, and M.A.P. Martins, *Spectrosc. Lett.*, 1993, **26**, 1381
103 L. Chen and W. Robien, *J. Chem. Inf. Comput. Sci.*, 1993, **33**, 441
104 G.F. Fadhil, *Collect. Czech. Chem. Commun.*, 1993, **58**, 385
105 M. Lee and R.S. Phillips, *J. Heterocycl. Chem.*, 1992, **29**, 1181
106 C.J. Byrne and D.A.R. Happer, *Aust. J. Chem.*, 1993, **46**, 887
107 K. Neuvonen and K. Pihlaja, *Acta Chem. Scand.*, 1993, **47**, 695
108 D. Rasala, *J. Phys. Org. Chem.*, 1993, **6**, 307
109 R. Knorr, D.S. Stephenson, P. Boehrer, and T.P. Hoang, *Magn. Reson. Chem.*, 1993, **31**, 388
110 D. Spinelli, L. Lamartina, S. Chimichi, R. Noto, and G. Consiglio, *Acta Chem. Scand.*, 1993, **47**, 160
111 F.O. Garces and R.J. Watts, *Magn. Reson. Chem.*, 1993, **31**, 529
112 A. Houlton, R.M.G. Roberts, J. Silver, and J. Zakrewski, *J. Organomet. Chem.*, 1993, **456**, 107
113 Y. Kusuyama, *Chem. Express*, 1993, **8**, 317
114 K.L. Brown and D.R. Evans, *Inorg. Chem.*, 1994, **33**, 525
115 M. Witanowski, W. Sicinska, Z. Biedrzycka, and G.A. Webb, *Magn. Reson. Chem.*, 1993, **31**, 916
116 B.L. Zhang, F. Mabon, M.L. Martin, *J. Phys. Org. Chem.*, 1993, **6**, 367
117 C. Buenz, D. Heber, U. Holzgrabe, and U. Suerig, *Magn. Reson. Chem.*, 1993, **31**, 371
118 P.S. Dobbin, R.C. Hider, S.K. Rizvi, K.L. Maki, and D. Van der Helm, *J. Chem. Soc., Perkin Trans. 2*, 1993, 451
119 B. Brzezinski, B. Brycki, H. Maciejewska-Urjasz, and G. Zundel, *Magn. Reson. Chem.*, 1993, **31**, 642
120 I. McEwen, M. Roennqvist, *Acta Chem. Scand.*, 1990, **44**, 108
121 A.C. Olivieri, D. Sanz, R.M. Claramunt, and J. Elguero, *J. Chem. Soc., Perkin Trans. 2*, 1993, 1597
122 A. Yoneda, T. Hakushi, G.R. Newkome, Y. Morimoto, and N. Yasuoka, *Chem. Lett.*, 1994, 175
123 C. Beeson, N. Pham, G. Shipps, Jr., and T.A. Dix, *J. Am. Chem. Soc.*, 1993, **115**, 6803
124 F.S. Kamounah and S.R. Salman, *Spectrochim. Acta, Part A*, 1993, **49A**, 601
125 M. Akiyama and T. Ohtani, *Spectrochim. Acta, Part A*, 1994, **50A**, 317
126 N. Kalsbeek, K. Schaumburg, and S. Larsen, *J. Mol. Struct.*, 1993, **299**, 155
127 B. Goswami, B.L. Gaffney, and R.A. Jones, *J. Am. Chem. Soc.*, 1993, **115**, 3832
128 D.W. Boykin, S. Chandrasekaran, and A.L. Baumstark, *Magn. Reson. Chem.*, 1993, **31**, 489
129 D.W. Boykin, *J. Mol. Struct.*, 1993, **295**, 39
130 D.W. Boykin and A. Kumar, *J. Mol. Struct.*, 1993, **298**, 121
131 I. Murata, *Pure Appl. Chem.*, 1993, **65**, 97
132 M. Rabinovitz and A. Ayalon, *Pure Appl. Chem.*, 1993, **65**, 111
133 G. Marin-Puga, V. Horak, and P. Svoronos, *Collect. Czech. Chem. Commun.*, 1993, **58**, 77
134 R. Lemmerz, M. Nieger, and F. Voegtle, *J. Chem. Soc., Chem. Commun.*, 1993, 1168

135 O. Reiser, B. Koenig, K. Meerholz, J. Heinze, T. Wellauer, F. Gerson, R. Frim, M. Rabinovitz, and A. de Meijere, *J. Am. Chem. Soc.*, 1993, **115**, 3511
136 H. Higuchi, K. Asano, K. Nakari, Y. Takai, J. Ojima, and G. Yamamoto, *J. Chem. Soc., Perkin Trans. 1*, 1993, 89
137 H. Higuchi, H. Yamamoto, J. Ojima, G. Yamamoto, *J. Chem. Soc., Perkin Trans. 1*, 1993, 975
138 H. Higuchi, H. Yamamoto, J. Ojima, M. Iyoda, M. Yoshida, G. Yamamoto, *J. Chem. Soc., Perkin Trans. 1*, 1993, 983
139 H. Higuchi, K. Asano, and J. Ojima, *Chem. Lett.*, 1993, 1829
140 H. Higuchi, H. Yamamoto, J. Ojima, G. Yamamoto, *Bull. Chem. Soc. Jpn.*, 1993, **66**, 2323
141 G. De Munno, F. Lucchesini, and R. Neidlein, *Tetrahedron*, 1993, **49**, 6863
142 D.C. Miller, M.R. Johnson, J.J. Becker, and J.A. Ibers, *J. Heterocycl. Chem.*, 1993, **30**, 1485
143 A.J. Ashe, III, W. Klein, and R. Rousseau, *Organometallics*, 1993, **12**, 3225
144 M. Prato, T. Suzuki, F. Wudl, V. Lucchini, and M. Maggini, *J. Am. Chem. Soc.*, 1993, **115**, 7876
145 J.H. Hong, and P. Boudjouk, *J. Am. Chem. Soc.*, 1993, **115**, 5883
146 J.A. Ripmeester and C.I. Ratcliffe, *Top. Inclusion Sci.*, 1992, **4**, 1
147 H.J. Schneider, *Recl. Trav. Chim. Pays-Bas*, 1993, **112**, 412
148 A.F. Casy, *Magn. Reson. Chem.*, 1993, **31**, 416
149 K. Takegoshi and K. Hikichi, *J. Am. Chem. Soc.*, 1993, **115**, 9747
150 T. Ndou, S. Mukundan, Jr., and I.M. Warner, *J. Inclusion Phenom. Mol. Recognit. Chem.*, 1993, **15**, 9
151 Z.J. Tan, X.X. Zhu, and G.R. Brown, *Langmuir*, 1994, **10**, 1034
152 C.J. Hartzell, S.R. Mente, N.L. Easthan, and J.L. Beckett, *J. Phys. Chem.*, 1993, **97**, 4887
153 D. Kawahara, F. Imashiro, and T. Terao, *Chem. Phys. Lett.*, 1993, **204**, 533
154 A. Harada, J. Li, and M. Kamachi, *Macromolecules*, 1993, **26**, 5698
155 L.C. Groenen, E. Steinwender, B.T.G. Lutz, J.H. Van der Maas, and D.N. Reinhoudt, *J. Chem. Soc., Perkin Trans. 2*, 1992, 1893
156 K. Araki, H. Shimizu, and S. Shinkai, *Chem. Lett.*, 1993, 205
157 L. Zhang, A. Macias, T. Lu, J.I. Gordon, G.W. Gokel, and A.E. Kaifer, *J. Chem. Soc., Chem. Commun.*, 1993, 1017
158 A. Mahdyarfar and K.D.M. Harris, *J. Chem. Soc., Chem. Commun.*, 1993, 51
159 C. Howe, S. Sankar, and A.E. Tonelli, *Polymer*, 1993, **34**, 3674
160 A.E. Aliev and K.D.M. Harris, *J. Am. Chem. Soc.*, 1993, **115**, 6369
161 G. Dauphin, G. Jeminet, and F. Vaufrey, *J. Chem. Soc., Perkin Trans. 2*, 1993, 399
162 E. Zaborowski and S. Vega, *Mol. Phys.*, 1993, **78**, 703
163 K. Kobayashi, Y. Asakawa, Y. Kikuchi, H. Toi, and Y. Aoyama, *J. Am. Chem. Soc.*, 1993, **115**, 2648
164 M.D. Hollingsworth and A.R. Palmer, *J. Am. Chem. Soc.*, 1993, **115**, 5881
165 P.C. Kearney, L.S. Mizoue, R.A. Kumpf, J.E. Forman, A. McCurdy, and D.A. Dougherty, *J. Am. Chem. Soc.*, 1993, **115**, 9907
166 F. Toda, K. Tanaka, and C. Foces-Foces, *J. Chem. Soc., Chem. Commun.*, 1993, 1139
167 F. Toda, K. Tanaka, M. Ootani, A. Hayashi, I. Miyahara, and K. Hirotsu, *J. Chem. Soc., Chem. Commun.*, 1993, 1413
168 F. Imashiro, D. Kuwahara, and T. Terao, *J. Chem. Soc., Perkin Trans. 2*, 1993, 1759
169 H.J. Schneider, F. Werner, and T. Blatter, *J. Phys. Org. Chem.*, 1993, **6**, 590
170 M.S. Searle, *Prog. Nucl. Magn. Reson. Spectrosc.*, 1993, **25**, 403
171 R. Blinc, *J. Mol. Struct.*, 1993, **292**, 269
172 S.C. Hirst, P. Tecilla, S.J. Geib, E. Fan, and A.D. Hamilton, *Isr. J. Chem.*, 1992, **32**, 105
173 B.L. Iverson, K. Shreder, V. Kral, and J.L. Sessler, *J. Am. Chem. Soc.*, 1993, **115**, 11022
174 P.H. Hynninen and S. Loetjoenen, *Biochim. Biophys. Acta*, 1993, **1183**, 381
175 D. Lewandowska and T. Podoski, *Spectrosc. Lett.*, 1993, **26**, 693
176 S.Y. Hanna, J.M.A. Al-Rawi, and A.A. Al-Rahman, *Spectrochim. Acta, Part A*, 1993, **49A**, 1022
177 T. Gramstad and L.J. Stangeland, *Acta Chem. Scand.*, 1993, **47**, 605
178 C.T. Seto and G.M. Whitesides, *J. Am. Chem. Soc.*, 1993, **115**, 905

179 V.A. Daragan, E. Ilyina, and K.H. Mayo, *Biopolymers*, 1993, **33**, 521
180 D.F. Nogales, J. Ma, and D.A. Lightner, *Tetrahedron*, 1993, **49**, 2361
181 W.N. Charman, C.S.C. Lai, D.J. Craik, B.C. Finnin, and B.L. Reed, *Aust. J. Chem.*, 1993, **46**, 377
182 H. Yasunaga and I. Ando, *J. Mol. Struct.*, 1993, **301**, 125
183 D.R. Kinney, I.S. Chuang, and G.E. Maciel, *J. Am. Chem. Soc.*, 1993, **115**, 6786
184 M. Witanowski, W. Sicinska, Z. Grabowski, and G.A. Webb, *J. Magn. Reson., Ser. A*, 1993, **104**, 310
185 J. Mason, *Solid State Nucl. Magn. Reson.*, 1993, **2**, 285
186 J. Milne, *Magn. Reson. Chem.*, 1993, **31**, 652
187 K. Momoki and Y. Fukazawa, *Anal. Sci.*, 1993, **9**, 501
188 Z. Li, Q. Guo, T. Ren, X. Zhu, and Y. Liu, *J. Inclusion Phenom. Mol. Recognit. Chem.*, 1993, **15**, 37
189 R.S. Drago, M.S. Hirsch, D.C. Ferris, and C.W. Chronister, *J. Chem. Soc., Perkin Trans. 2*, 1994, 219
190 K. Momoki and Y. Fukazawa, *Anal. Sci.*, 1994, **10**, 53
191 J. Klinowski, *Top. Inclusion Sci.*, 1992, **4**, 115
192 B.E. Mann, *Spectrosc. Prop. Inorg. Organomet. Compd.*, 1992, **25**, 1
193 H. Kuwahara, M. Hamada, Y. Ishikawa, and T. Kunitake, *J. Am. Chem. Soc.*, 1993, **115**, 3002
194 R. Moss, H. Gesmar, and J.J. Led, *J. Am. Chem. Soc.*, 1994, **116**, 747
195 P.R. Ashton, A.S. Reder, N. Spencer, and J.F. Stoddart, *J. Am. Chem. Soc.*, 1993, **115**, 5286
196 L. Ernst, V. Boekelheide, and H. Hopf, *Magn. Reson. Chem.*, 1993, **31**, 669
197 E.S. Hand, K.A. Belmore, and L.D. Kispert, *Magn. Reson. Chem.*, 1993, **31**, 516
198 E.S. Hand, K.A. Belmore, and L.D. Kispert, *Helv. Chim. Acta*, 1993, **76**, 1928
199 Y. Chauvin, F. Di Marco-Van Tiggelen, and H. Olivier, *J. Chem. Soc., Dalton Trans.*, 1993, 1009
200 L.L. Wisniewski, M. Mediati, C.M. Jensen, and K.W. Zilm, *J. Am. Chem. Soc.*, 1993, **115**, 7533
201 T. Baba, M. Nomura, Y. Ono, and Y. Ohno, *J. Phys. Chem.*, 1993, **97**, 12888
202 E. Brunner, H.G. Karge, and H. Pfeifer, *Z. Phys. Chem. (Munich)*, 1992, **176**, 173
203 T. Ueda, S. Hayashi, and K. Hayamizu, *Solid State Commun.*, 1993, **87**, 429
204 L. Zheng, K.W. Fishbein, R.G. Griffin, and J. Herzfeld, *J. Am. Chem. Soc.*, 1993, **115**, 6254
205 R.E. London, *Biol. Magn. Reson.*, 1992, **11**, 277
206 K.S. Hamilton, H.C. Jarrell, K.M. Briere, and C.W.M. Grant, *Biochemistry*, 1993, **32**, 4022
207 W.R. Banks, L.S. Clore, E.J. Dadey, and G.A. Digenis, *Appl. Radiat. Isot.*, 1993, **44**, 1021
208 G. Bryant, J.M. Pope, and J. Wolfe, *Eur. Biophys. J.*, 1992, **21**, 363
209 A. Leonard, A. Milon, M.A. Krajewski-Bertrand, and E.J. Dufourc, *Bull. Magn. Reson.*, 1993, **15**, 124
210 M. Monette, M.-R. Van Calsteren, and M. Lafleur, *Biochim. Biophys. Acta*, 1993, **1149**, 319
211 R.L. Thurmond, G. Lindblom, and M.F. Brown, *Biochemistry*, 1993, **32**, 5394
212 G.H. Penner, J.M. Polson, S.I. Daleman, and K. Reid, *Can. J. Chem.*, 1993, **71**, 417
213 M. De Rosa, R.G. Shadle, and N. Foster, *J. Chem. Soc., Chem. Commun.*, 1993, 1757
214 H.E. Mons, K. Bergander, and H. Guenther, *Magn. Reson. Chem.*, 1993, **31**, 509
215 R. Brown, A.S. Dolan, C. Kemball, and G.S. McDougall, *J. Chem. Soc., Faraday Trans.*, 1993, **89**, 1095
216 P.E. Morris, Jr. and C. Chiao, *J. Labelled Compd. Radiopharm.*, 1993, **33**, 455
217 A.E. Aliev and K.D.M. Harris, *Mendeleev Commun.*, 1993, 153
218 V. Copie, A.E. McDermott, K. Beshah, J.C. Williams, M. Spijker-Assink, R. Gebhard, J. Lugtenburg, J. Herzfeld, and R.G. Griffin, *Biochemistry*, 1994, **33**, 3280
219 G.A. Olah, N. Hartz, G. Rasul, and G.K.S. Prakash, *J. Am. Chem. Soc.*, 1993, **115**, 6985
220 A. Maliniak, S. Greenbaum, R. Poupko, H. Zimmermann, and Z. Luz, *J. Phys. Chem.*, 1993, **97**, 4832

221 J.M. Griffiths, A.T. Bell, and J.A. Reimer, *J. Phys. Chem.*, 1994, **98**, 1918
222 A.E. Aliev, K.D.M. Harris, X. Alcobe, and E. Estop, *J. Chem. Soc., Faraday Trans.*, 1993, **89**, 3797
223 P.M. Macdonald and Y. Yue, *Langmuir*, 1993, **9**, 1206
224 A. Dardin, R. Stadler, C. Boeffel, and H.W. Spiess, *Makromol. Chem.*, 1993, **194**, 3467
225 A.G. Stepanov, A.G. Maryasov, V.N. Romannikov, and K.I. Zamaraev, *Magn. Reson. Chem.*, 1994, **32**, 16
226 A. Blume, M. Jansen, M. Ghyczy, and J. Gareiss, *Int. J. Pharm.*, 1993, **99**, 291
227 N. Zumbulyadis, C.J.T. Landry, and T.E. Long, *Macromolecules*, 1993, **26**, 2647
228 Y.H. Chin, P.T. Inglefield, and A.A. Jones, *Macromolecules*, 1993, **26**, 5372
229 E. Soderlind and F.D. Blum, *J. Colloid Interface Sci.*, 1993, **157**, 172
230 M. Cavazza, C. Forte, G. Galli, M. Geppi, F. Pietra, and C.A. Veracini, *Liq. Cryst.*, 1993, **15**, 275
231 E. Soederlind and P. Stilbs, *Langmuir*, 1993, **9**, 2024
232 U. Werner and W. Mueller-Warmuth, *Z. Phys. B: Condens. Matter*, 1993, **91**, 65
233 D.J. Photinos, Z. Luz, H. Zimmermann, and E. Samulski, *J. Am. Chem. Soc.*, 1993, **115**, 10895
234 D. Imbardelli, B. Wazynska, A. Golemme, G. Chidichimo, and R. Dabrowski, *Mol. Phys.*, 1993, **79**, 1275
235 N.J. Pinto, F.L. Howell, and V.H. Schmidt, *Phys. Rev. B: Condens. Matter*, 1993, **48**, 5983
236 A. Birczynski, Z.T. Lalowicz, L.P. Ingman, M. Punkkinen, and E.E. Ylinen, *Solid State Nucl. Magn. Reson.*, 1993, **1**, 329
237 V. Grundke and B. Boddenberg, *Mol. Phys.*, 1993, **79**, 1215
238 R. Poupko, Z. Olender, D. Reichert, and Z. Luz, *J. Magn. Reson., Ser. A*, 1994, **106**, 113
239 T.-H. Lin, J.A. DiNatale, and R.R. Vold, *J. Am. Chem. Soc.*, 1994, **116**, 2133
240 H.E. Mons, K. Bergander, and H. Guenther, *Helv. Chim. Acta*, 1993, **76**, 1216
241 F.M. Kaspersen, C.W. Funke, J. Vader, and G.N. Wagenaars, *J. Labelled Compd. Radiopharm.*, 1993, **33**, 409
242 D.K. Jaiswal, H. Andres, H. Morimoto, and P.G. Williams, *J. Chem. Soc., Chem. Commun.*, 1993, 907
243 T.M. O'Co el, J.T. Gerig, and P.G. Williams, *J. Am. Chem. Soc.*, 1993, **115**, 3048
244 T.M. O'Connell, P.G. Williams, and J.T. Gerig, *J. Labelled Compd. Radiopharm.*, 1993, **33**, 371
245 D. Hesk, P. Delduca, D. Koharski, P. McNamara, C. Magatti, S. Saluja, and L. Thomas, *J. Labelled Compd. Radiopharm.*, 1993, **33**, 439
246 M.I. Dawson, P.D. Hobbs, S.W. Rhee, H. Morimoto, and P.G. Williams, *J. Labelled Compd. Radiopharm.*, 1993, **33**, 633
247 S. Highsmith, M. Kubinec, D.K. Jaiswal, H. Morimoto, P.G. Williams, and D.E. Wemmer, *J. Biomol. NMR*, 1993, **3**, 325
248 D.B. Collum, *Acc. Chem. Res.*, 1993, **26**, 227
249 A.R. Waldeck, A.J. Lennon, B.E. Chapman, and P.W. Kuchel, *J. Chem. Soc., Faraday Trans.*, 1993, **89**, 2807
250 A. Sekiguchi, M. Ichinohe, T. Nakanishi, and H. Sakurai, *Chem. Lett.*, 1993, 267
251 G.T. DeLong, D.K. Pannell, M.T. Clarke, and R.D. Thomas, *J. Am. Chem. Soc.*, 1993, **115**, 7013
252 C.A. Ogle, B.K. Huckabee, H.C. Johnson, IV, P.F. Sims, S.D. Winslow, and A.A. Pinkerton, *Organometallics*, 1993, **12**, 1960
253 D. Johnels, *J. Organomet. Chem.*, 1993, **445**, 1
254 H.J. Reich, J.P. Borst, R.R. Dykstra, and P.D. Green, *J. Am. Chem. Soc.*, 1993, **115**, 8728
255 H.J. Reich and R.R. Dykstra, *J. Am. Chem. Soc.*, 1993, **115**, 7041
256 K. Ruhlandt-Senge, J.J. Ellison, R.J. Wehmschulte, F. Pauer, and P.P. Power, *J. Am. Chem. Soc.*, 1993, **115**, 11353
257 F.E. Romesberg, M.P. Bernstein, J.H. Gilchrist, A.T. Harrison, D.J. Fuller, and D.B. Collum, *J. Am. Chem. Soc.*, 1993, **115**, 3475
258 D.R. Baker, R.E. Mulvey, W. Clegg, and P.A. O'Neil, *J. Am. Chem. Soc.*, 1993, **115**, 6472

259 L.H. Gade, C. Becker, and J.W. Lauher, *Inorg. Chem.*, 1993, **32**, 2308
260 T. Chivers, M. Edwards, R.W. Hilts, M. Parvez, and R. Vollmerhaus, *J. Chem. Soc., Chem. Commun.*, 1993, 1483
261 A.G. Avent, C. Eaborn, P.B. Hitchcock, G.A. Lawless, P.D. Lickiss, M. Mallien, J.D. Smith, A.D. Webb, and B. Wrackmeyer, *J. Chem. Soc., Dalton Trans.*, 1993, 3259
262 C. Lambert, P.v.R. Schleyer, and E.U. Wuerthwein, *J. Org. Chem.*, 1993, **58**, 6377
263 Y. Wang, H. Zhang, J.A. Maguire, and N.S. Hosmane, *Organometallics*, 1993, **12**, 3781
264 N. Nakamura, A. Yoshino, and K. Takahashi, *Bull. Chem. Soc. Jpn.*, 1994, **67**, 26
265 A.L. Gatzke and D.P. Green, *Macromolecules*, 1994, **27**, 2249
266 J.S. Wang, R. Warin, R. Jerome, and P. Teyssie, *Macromolecules*, 1993, **26**, 6776
267 M.L. Turonek, P. Clarke, G.S. Laurence, S.F. Lincoln, P.A. Pittet, S. Politis, and K.P. Wainwright, *Inorg. Chem.*, 1993, **32**, 2195
268 S.F. Lincoln and T. Rodopoulos, *Inorg. Chim. Acta*, 1993, **205**, 23
269 J. Hirschinger, T. Mongrelet, C. Marichal, P. Granger, J.M. Savariault, E. Deramond, and J. Galy, *J. Phys. Chem.*, 1993, **97**, 10301
270 T. Pietrass and P.K. Burkert, *Magn. Reson. Chem.*, 1993, **31**, 709
271 G. Wang, J. Roos, D. Brinkmann, M. Pasquali, and G. Pistoia, *J. Phys. Chem. Solids*, 1993, **54**, 851
272 M. Motevalli, D. Shah, and A.C. Sullivan, *J. Chem. Soc., Dalton Trans.*, 1993, 2849
273 S.K. Miller and G.A. Elgavish, *Biol. Magn. Reson.*, 1992, **11**, 159
274 N. Bansal, M.J. Germann, V. Seshan, G.T. Shires III, C.R. Malloy, and A.D. Sherry, *Biochemistry*, 1993, **32**, 5638
275 T. Tanase, N. Murakami, A. Nagatsu, and J. Sakakibara, *Biol. Pharm. Bull.*, 1993, **16**, 431
276 L.L. Hansen, J. Rasmussen, E. Friche, and J.W. Jaroszewski, *Anal. Biochem.*, 1993, **214**, 506
277 J.S. Tauskela and E.A. Shoubridge, *Biochim. Biophys. Acta*, 1993, **1158**, 155
278 T. Pietrass and P.K. Burkert, *Inorg. Chim. Acta*, 1993, **207**, 253
279 A.R. Waldeck and P.W. Kuchel, *Biophys. J.*, 1993, **64**, 1445
280 D.S. Gluck, A.R. Brough, P. Mountford, and M.L.H. Malcolm, *Inorg. Chem.*, 1993, **32**, 1893
281 J.P. Lemmon, R.L. Kohnert, and M.M. Lerner, *Macromolecules*, 1993, **26**, 2767
282 J.D. Song and M.S. Jhon, *J. Polym. Sci., Part A: Polym. Chem.*, 1993, **31**, 1687
283 M. Iida, T. Sakai, N. Koine, and S. Kaizaki, *J. Chem. Soc., Faraday Trans.*, 1993, **89**, 1773
284 A.K.W. Stephens and S.F. Lincoln, *J. Chem. Soc., Dalton Trans.*, 1993, 2123
285 Q. Xie, G. Gokel, J. Hernandez, L. Echegoyen, and Y. Li, *J. Am. Chem. Soc.*, 1994, **116**, 690
286 M. Wiebcke, P. Sieger, J. Felsche, G. Engelhardt, P. Behrens, and J. Schefer, *Z. Anorg. Allg. Chem.*, 1993, **619**, 1321
287 S. Ishimaru and N. Nakamura, *Ber. Bunsen-Ges. Phys. Chem.*, 1993, **97**, 777
288 F. Deng, Y. Du. C. Ye, and Y. Kong, *Solid State Nucl. Magn. Reson.*, 1993, **2**, 317
289 R.E. Douthwaite, A.R. Brough, and M.L.H. Green, *J. Chem. Soc., Chem. Commun.*, 1994, 267
290 J.L. Allis, *Annu. Rep. NMR Spectrosc.*, 1993, **26**, 211
291 R.E. Walstedt, D.W. Murphy, and M. Rosseinsky, *Nature (London)*, 1993, **362**, 611
292 M. Corti, *Physica C (Amsterdam)*, 1993, **211**, 299
293 D.R. Buffinger, R.P. Ziebarth, V.A. Stenger, C. Recchia, and C.H. Pennington, *J. Am. Chem. Soc.*, 1993, **115**, 9267
294 R. Tycko, G. Dabbagh, D.W. Murphy, Q. Zhu, and J.E. Fischer, *Phys. Rev. B: Condens. Matter*, 1993, **48**, 9097
295 B.P. Shehan, R.M. Wellard, and W.R. Adam, *Magn. Reson. Med.*, 1993, **30**, 573
296 R. Assmus, V. Boehmer, J.M. Harrowfield, M.I. Ogden, W.R. Richmond, B.W. Skelton, and A.H. White, *J. Chem. Soc., Dalton Trans.*, 1993, 2427
297 N. Boden, K.W. Jolley, and M.H. Smith, *J. Phys. Chem.*, 1993, **979**, 7678
298 W.P. Power, S. Mooibroek, R.E. Wasylishen, and T.S. Cameron, *J. Phys. Chem.*, 1994, **98**, 1552
299 M. Chiba, K. Ohara, Y. Ajiro, and T. Morimoto, *J. Phys. Soc. Jpn.*, 1993, **62**, 4186

300 Y. Maniwa, K. Mizoguchi, K. Kume, K. Kikuchi, I. Ikemoto, S. Suzuki, Y. Achiba, K. Tanigami, T.W. Ebbesen, et al., *Synth. Met.*, 1993, **56**, 3063
301 S. Han, K.D. Schmitt, D.S. Shihabi, and C.D. Chang, *J. Chem. Soc., Chem. Commun.*, 1993, 1287
302 E.T. Ahrens, P.C. Hammel, R.H. Haffner, A.P. Peyes, J.L. Smith, and W.G. Clark, *Phys. Rev. B: Condens. Matter*, 1993, **48**, 6691
303 J.A. Cowan, H.W. Huang, and L.Y. Hsu, *J. Inorg. Biochem.*, 1993, **52**, 121
304 H.-W. Huang and J.A. Cowan, *Eur. J. Biochem.*, 1994, **219**, 253
305 Y.-Y. Lin, N.-H. Ge, and L.-P. Hwang, *J. Magn. Reson., Ser. B*, 1994, **103**, 189
306 K.J.D. Mackenzie and R.H. Meinhold, *J. Mater. Sci. Lett.*, 1993, **12**, 1696
307 K.J.D. MacKenzie and R.H. Meinhold, *Thermochim. Acta*, 1994, **232**, 85
308 S. Forsen, C. Johansson, and S. Linse, *Methods Enzymol.*, 1993, **227**, 107
309 J.M. Aramini and H.J. Vogel, *J. Am. Chem. Soc.*, 1994, **116**, 1988
310 H. Kataoka, S. Takeda, and N. Nakamura, *J. Phys. Soc. Jpn.*, 1993, **62**, 1478
311 C.D. Hall, J.H.R. Tucker, and S.Y.F. Chu, *J. Organomet. Chem.*, 1993, **448**, 175
312 A. Gleizes, S. Sans-Lenain, D. Medus, N. Hovnanian, P. Miele, and J.D. Foulon, *Inorg. Chim. Acta*, 1993, **209**, 47
313 M. Takigawa, W.L. Hults, and J.L. Smith, *Phys. Rev. Lett.*, 1993, **71**, 2650
314 I.P. Beletskaya, A.Z. Voskoboynikov, E.B. Chuklanova, N.I. Kirillova, A.K. Shestakovam I.N. Parshina, A.I. Gusev, and G.K.I. Magomedov, *J. Am. Chem. Soc.*, 1993, **115**, 3156
315 O.J. Zogal, C. Juszczak, A.H. Vuorimaeki, E.E. Ylinen, M. Punkkinen, and H. Drulis, *J. Alloys Compd.*, 1993, **191**, 207
316 P.C. Hammel, A.P. Reyes, S.W. Cheong, Z. Fisk, and J.E. Schirber, *Phys. Rev. Lett.*, 1993, **71**, 440
317 M. Frechette, *Can. J. Chem.*, 1993, **71**, 377
318 J. Huskens, J.A. Peters, and H. van Bekkum, *Radiochim. Acta*, 1993, **61**, 195
319 K. Shimizu, *J. Magn. Magn. Mater.*, 1993, **123**, L9
320 F. Nief, L. Ricard, and F. Mathey, *Polyhedron*, 1993, **12**, 19
321 Y. Chauvin, N. Marchal, H. Olivier, and L. Saussine, *J. Organomet. Chem.*, 1993, **445**, 93
322 K. Niedzwiedz, B. Nowak, and O.J. Zogal, *J. Alloys Compd.*, 1993, **194**, 47
323 C.J. Carrano, M. Mohan, S.M. Holmes, R. de la Rosa, A. Butler, J.M. Charnock, and C.D. Garner, *Inorg. Chem.*, 1994, **33**, 646
324 E.P. Talsi, V.D. Chinakov, V.P. Balbenko, and K.I. Zamaraev, *J. Mol. Catal.*, 1993, **81**, 235
325 J.S. Jaswal and A.S. Tracey, *J. Am. Chem. Soc.*, 1993, **115**, 5600
326 J.K.F. Buijink, A. Meetsma, and J.H. Teuben, *Organometallics*, 1993, **12**, 2004
327 D. Hou, K.S. Hagen, and C.L. Hill, *J. Chem. Soc., Chem. Commun.*, 1993, 426
328 A.S. Tracey and J.S. Jaswal, *Inorg. Chem.*, 1993, **32**, 4235
329 S.E. Harnumg, E. Larsen, and E.J. Pedersen, *Acta, Chem. Scand.*, 1993, **47**, 674
330 D.C. Crans and P.K. Shin, *J. Am. Chem. Soc.*, 1994, **116**, 1305
331 Y. Yang, Q. Liu, L. Huang, D. Wu, B. Kang, and J. Lu, *Inorg. Chem.*, 1993, **32**, 5431
332 K.L. Walther, J. Kuemmerlen, and A. Wokaun, *Ber. Bunsen-Ges. Phys. Chem.*, 1993, **97**, 772
333 J. Skibsted, N.C. Nielsen, H. Bildsoe, and H.J. Jakobsen, *J. Am. Chem. Soc.*, 1993, **115**, 7351
334 S. Hayakawa, T. Yoko, and S. Sakka, *Bull. Chem. Soc. Jpn.*, 1993, **66**, 3393
335 O.B. Lapina, A.V. Nosov, V.M. Mastikhin, K.A. Dubkov, and V.V. Mokrinski, *J. Mol. Catal.*, 1994, **87**, 57
336 M.J. Geselbracht, A.M. Stacy, A.R. Garcia, B.G. Silbernagel, and G.H. Kwei, *J. Phys. Chem.*, 1993, **97**, 7102
337 M. Lobert, W. Mueller-Warmuth, W. Paulus, and R. Schoellhorn, *Colloids Surf., A*, 1993, **72**, 345
338 K. Zamaraev, *J. Mol. Catal.*, 1993, **82**, 275
339 E.P. Talsi, O.V. Klimov, and K.I. Zamaraev, *J. Mol. Catal.*, 1993, **83**, 329
340 E.P. Talsi, K.V. Shalyaev, and K.I. Zamaraev, *J. Mol. Catal.*, 1993, **83**, 347
341 A. Proust, R. Thouvenot, F. Robert, and P. Gouzerh, *Inorg. Chem.*, 1993, **32**, 5299
342 A. Patel, S. Siddiqui, D.T. Richens, M.E. Harman, and M.B. Hursthouse, *J. Chem. Soc., Dalton Trans.*, 1993, 767

343 M.M.C.A. Castro, C.F.G.C. Geraldes, and J.A. Peters, *Inorg. Chim. Acta*, 1993, **208**, 123
344 O.H. Han, C.Y. Lin, N. Sustache, M. McMillan, C.J. Donald, K.W. Zilm, and G.L. Haller, *Appl. Catal., A*, 1993, **98**, 195
345 Y. Lin, K. Nomiya, and R.G. Finke, *Inorg. Chem.*, 1993, **32**, 6040
346 R.J. Errington, M.D. Kerlogue, and D.G. Richards, *J. Chem. Soc., Chem. Commun.*, 1993, 649
347 H.B. Kraatz, J.M. Aramini, X. Gao, P.M. Boorman, and H.J. Vogel, *Inorg. Chem.*, 1993, **32**, 3976
348 J.J. Hastings and O.W. Howarth, *Polyhedron*, 1993, **12**, 847
349 Y. Lin, T.J.R. Weakley, B. Rapko, and R.G. Finke, *Inorg. Chem.*, 1993, **32**, 5095
350 U. Kortz, G.B. Jameson, and M.T. Pope, *J. Am. Chem. Soc.*, 1994, **116**, 2659
351 K. Nomiya, M. Kaneko, N.C. Kasuga, R.G. Finke, and M. Pohl, *Inorg. Chem.*, 1994, **33**, 1469
352 L. Salles, C. Aubry, R. Thouvenot, F. Robert, C. Doremieux-Morin, G. Chottard, H. Ledon, Y. Jeannin, and J.M. Bregeault, *Inorg. Chem.*, 1994, **33**, 871
353 R. Contant and R. Thouvenot, *Inorg. Chim. Acta*, 1993, **212**, 41
354 Y. Kohori, Y. Nobuchi, and T. Kohara, *J. Phys. Soc. Jpn.*, 1993, **62**, 447
355 T. Shinohara, Y. Yokoyama, M. Sato, A. Inoue, and T. Masumoto, *J. Phys.: Condens. Matter*, 1993, **5**, 3673
356 A. Campos, H. Luetgemeier, and P. Novak, *J. Magn. Magn. Mater.*, 1993, **125**, 279
357 G. Predieri, C. Vignali, G. Denti, and S. Serroni, *Inorg. Chim. Acta*, 1993, **205**, 145
358 K.O. Christe, D.A. Dixon, H.G. Mack, H. Oberhammer, A. Pagelot, J.C.P. Sanders, and G.J. Schrobilgen, *J. Am. Chem. Soc.*, 1993, **115**, 11279
359 Y. Saito, K. Inomata, A. Goto, and H. Yasuoka, *J. Phys. Soc. Jpn.*, 1993, **62**, 1450
360 Y. Saito, K. Inomata, A. Goto, and H. Yasuoka, *J. Magn. Magn. Mater.*, 1993, **126**, 466
361 M. Iida, Y. Miyagawa, and S. Kohri, *Bull. Chem. Soc. Jpn.*, 1993, **66**, 2398
362 M. Iida, Y. Miyagawa, S. Kohri, and Y. Ikemoto, *Bull. Chem. Soc. Jpn.*, 1993, **66**, 2840
363 H. Nakamura, T. Suzuki, M. Shiga, H. Yamamoto, and T. Shinjo, *J. Magn. Magn. Mater.*, 1993, **126**, 364
364 S.C. Chung, J.C.C. Chan, S.C.F. Au-Yeung, and X. Xu, *J. Phys. Chem.*, 1993, **97**, 12685
365 F. Maruyama, H. Nagai, and K. Adachi, *J. Phys. Soc. Jpn.*, 1994, **63**, 371
366 B.R. Bender, M. Koller, D. Nanz, and W. von Philipsborn, *J. Am. Chem. Soc.*, 1993, **115**, 5889
367 A. Ceccon, C.J. Elsevier, and J.M. Ernsting, *Inorg. Chim. Acta*, 1993, **204**, 15
368 P. Imhoff, R. van Asselt, J.M. Ernsting, K. Vrieze, C.J. Elsevier, W.J.J. Smeets, A.L. Spek, and A.P.M. Kentgens, *Organometallics*, 1993, **12**, 1523
369 A. von Zelewsky, A.P. Suckling, and H. Stoeckli-Evans, *Inorg. Chem.*, 1993, **32**, 4585
370 G.E. Herberich, U. Englert, and F. Marken, *J. Chem. Soc., Dalton Trans.*, 1993, 1979
371 P.R.R. Ranatunge-Bandarage, B.H. Robinson, and J. Simpson, *Organometallics*, 1994, **13**, 500
372 P.R.R. Ranatunge-Bandarage, J. Simpson, N.W. Duffy, S.M. Johnston, and B.H. Robinson, *Organometallics*, 1994, **13**, 511
373 J.F. Nixon and G.J.D. Sillett, *J. Organomet. Chem.*, 1993, **461**, 237
374 M. Krumm, E. Zangrando, L. Randaccio, S. Menzer, A. Dangrando, D. Holthenrich, and B. Lippert, *Inorg. Chem.*, 1993, **32**, 700
375 M. Krumm, E. Zangrando, L. Randaccio, S. Menzer, A. Danzmann, D. Holthenrich, and B. Lippert, *Inorg. Chem.*, 1993, **32**, 700
376 E.J.W. Austin, P.J. Barrie, and R.J.H. Clark, *J. Chem. Soc., Chem. Commun.*, 1993, 1404
377 T. Yamaguchi, N. Nishimura, and T. Ito, *J. Am. Chem. Soc.*, 1993, **115**, 1612
378 S.O. Dunham, R.D. Larsen, and E.H. Abbott, *Inorg. Chem.*, 1993, **32**, 2049
379 Y. Qu, T.G. Appleton, J.D. Hoeschele, and N. Farrell, *Inorg. Chem.*, 1993, **32**, 2591
380 K.A. Mitchell, K.C. Streveler, and C.M. Jensen, *Inorg. Chem.*, 1993, **32**, 2608
381 F.D. Rochon, M. Doyon, and I.S. Butler, *Inorg. Chem.*, 1993, **32**, 2717
382 A. Irving, K.R. Koch, and M. Matoetoe, *Inorg. Chim. Acta*, 1993, **206**, 193
383 G. Trovo, G. Valle, and B. Longato, *J. Chem. Soc., Dalton Trans.*, 1993, 669

384 V.K. Jain, S. Kannan, R.J. Butcher, and J.P. Jasinski, *J. Chem. Soc., Dalton Trans.*, 1993, 1509
385 P. Bhattacharyya, R.N. Sheppard, A.M.Z. Slawin, D.J. Williams, and J.D. Woollins, *J. Chem. Soc., Dalton Trans.*, 1993, 2393
386 Y. Qu, N. Farrell, M. Valsecchi, L. de Greco, and S. Spinelli, *Magn. Reson. Chem.*, 1993, **31**, 920
387 E. Costa, M. Murray, P.G. Pringle, and M.B. Smith, *Inorg. Chim. Acta*, 1993, **213**, 25
388 K. Hiraoka, Y. Kasamatsu, K. Kojima, and T. Hihara, *Physica B (Amsterdam)*, 1993, **186-188**, 535
389 H. Ohki, K. Harazono, T. Erata, A. Tasaki, and R. Ikeda, *Z. Naturforsch., A: Phys. Sci.*, 1993, **48**, 1005
390 K. Endo, K. Yamamoto, and K. Deguchi, *J. Phys. Chem. Solids*, 1993, **54**, 357
391 N. Vijayashree, A.G. Samuelson, and M. Nethaji, *Curr. Sci.*, 1993, **65**, 57
392 Y.Q. Song, M.A. Kennard, K.R. Poeppelmeier, and W.P. Halperin, *Phys. Rev. Lett.*, 1993, **70**, 3131
393 B. Li and Y. Wang, *Phys. Rev. B: Condens. Matter*, 1993, **47**, 16582
394 B.W. Statt and L.M. Song, *Phys. Rev. B: Condens. Matter*, 1993, **48**, 3536
395 A. Goto, H. Yasuoka, K. Takanashi, K. Saito, and H. Fujimori, *J. Magn. Magn. Mater.*, 1993, **126**, 358
396 A.M. Bogdanovich, Yu.I. Zhadanov, K.N. Mikhalyov, V.V. Lavrentjev, B.A. Aleksashin, S.V. Verkovskij, N. Winzek, P. Gergen, J. Gross, et al., *Physica C (Amsterdam)*, 1993, **215**, 253
397 S. Menzer, E.C. Hillgeris, and B. Lippert, *Inorg. Chim. Acta*, 1993, **210**, 167
398 J.E. Coleman, *Methods Enzymol.*, 1993, **227**, 16
399 G.A. Omburo, L.S. Mullins, and F.M. Raushel, *Biochemistry*, 1993, **32**, 9148
400 W. Zhang and V. Majidi, *Appl. Spectrosc.*, 1993, **47**, 2151
401 S. Ohki, U. Iwamoto, S. Aimoto, M. Yazawa, and K. Hikichi, *J. Biol. Chem.*, 1993, **268**, 12388
402 P.A.W. Dean, N.C. Nicholas, J.J. Vittal, and Y. Yu, *Inorg. Chem.*, 1993, **32**, 4632
403 M.A.R. Dakers, M.N.S. Hill, J.C. Lockhart, and D.J. Rushton, *J. Chem. Soc., Dalton Trans.*, 1994, 209
404 P.J. Barrie, A. Gyani, M. Motevalli, and P. O'Brien, *Inorg. Chem.*, 1993, **32**, 3862
405 J. Sola, P. Gonzalez-Duarte, J. Sanz, I. Casals, T. Alsina, I. Sobrados, A. Alvarez-Larena, J.F. Piniella, and X. Solans, *J. Am. Chem. Soc.*, 1993, **115**, 10018
406 G.S.H. Lee, K.J. Fisher, A.M. Vassallo, J.V. Hanna, and I.G. Dance, *Inorg. Chem.*, 1993, **32**, 66
407 J.S. Casas, M.S. Garcia-Tasende, A. Sanchez, J. Sordo, E.M. Vazquez-Lopez, E.E. Castellano, J. Zukerman-Schpector, *Inorg. Chim. Acta*, 1993, **209**, 137
408 V. Neirnick, D. Robert, and R. Nardin, *Magn. Reson. Chem.*, 1993, **31**, 815
409 D.St.C. Black, G.B. Deacon, G.L. Edwards, and B.M. Gatehouse, *Aust. J. Chem.*, 1993, **46**, 1323
410 H. Layeghi, D. Naumann, and W. Tyrra, *J. Organomet. Chem.*, 1992, **441**, 355
411 X. Yang, C.B. Knobler, and M.F. Hawthorne, *J. Am. Chem. Soc.*, 1993, **115**, 4904
412 G. Klose, F. Volke, G. Peinel, and G. Knobloch, *Magn. Reson. Chem.*, 1993, **31**, 548
413 N.G. Payne, M.E. Peach, and K.J. Schmidt, *J. Fluorine Chem.*, 1993, **64**, 255
414 L.G. Gioia, P. Cecchi, S. Bartolini, C. Pettinari, and A. Cingolani, *Gazz. Chim. Ital.*, 1993, **123**, 641
415 R.E. Blumer, F. Lianza, P.S. Pregosin, H. Ruegger, and A. Togni, *Inorg. Chem.*, 1993, **32**, 2663
416 C.J. Groombridge, *Magn. Reson. Chem.*, 1993, **31**, 380
417 G. Wu and R.E. Wasylishen, *J. Phys. Chem.*, 1993, **97**, 7863
418 G. Wu and R.E. Wasylishen, *Magn. Reson. Chem.*, 1993, **31**, 537
419 J. Shi, M. Wessels, and J.H. Ross, Jr., *Phys. Rev. B: Condens. Mater*, 1993, **48**, 8742
420 R.A. Santos and G.S. Harbison, *J. Am. Chem. Soc.*, 1994, **116**, 3075
421 D. Reed, *Chem. Soc. Rev.*, 1993, **22**, 109
422 Y. Yamamoto, S. Takamatsu, and H. Nakamura, *J. Magn. Reson., Ser. B*, 1993, **101**, 198
423 E. Tsilikounas, C.A. Kettner, and W.W. Bachovchin, *Biochemistry*, 1993, **32**, 12651
424 K. Ganesan and H.C. Brown, *J. Org. Chem.*, 1993, **58**, 7162

425 Y. Sugihara, R. Miyatake, and T. Yagi, *Chem. Lett.*, 1993, 933
426 I.I. Padilla-Martinez, A. Ariza-Castolo, and R. Contreras, *Magn. Reson. Chem.*, 1993, **31**, 189
427 J.C. Colberg, A. Rane, J. Vaquer, and J.A. Soderquist, *J. Am. Chem. Soc.*, 1993, **115**, 6065
428 T. Onak, D. Tan, J. Tseng, M. Diaz, J. Arias, and S. Herrera, *J. Am. Chem. Soc.*, 1993, **115**, 9210
429 O. Ni Dhubhghaill, D. Reed, and T.R. Spalding, *Polyhedron*, 1993, **12**, 1977
430 M.E. Smith, *Appl. Magn. Reson.*, 1993, **4**, 1
431 M.G. Mortuza, R. Dupree, and S.C. Kohn, *Appl. Magn. Reson.*, 1993, **4**, 89
432 L.B. Alemany, *Appl. Magn. Reson.*, 1993, **4**, 179
433 R.J. Kirkpatrick and B.L. Phillips, *Appl. Magn. Reson.*, 1993, **4**, 213
434 J. Klinowski, *Anal. Chim. Acta*, 1993, **283**, 929
435 S.N. Mlhatre, R.K. Iyer, and P.N. Moorthy, *Magn. Reson. Chem.*, 1993, **31**, 169
436 J.M. Aramini and H.J. Vogel, *Bull. Magn. Reson.*, 1993, **15**, 84
437 C.E. Keller, W.R. Carper, and B.J. Piersma, *Inorg. Chim. Acta*, 1993, **209**, 239
438 T. Nagata, M. Hayatsu, and N. Kosuge, *Phytochemistry*, 1993, **32**, 771
439 T.Y. Her, C.C. Chang, J.O. Tsai, Y.Y. Lai, L.K. Liu, H.C. Chang, and J.H. Chen, *Polyhedron*, 1993, **12**, 731
440 R.L. Wells, A.T. McPhail, M.F. Self, and J.A. Laske, *Organometallics*, 1993, **12**, 3333
441 J.M. Aramini, M.W. Germann, and H.J. Vogel, *J. Am. Chem. Soc.*, 1993, **115**, 9750
442 J. Lewinski,J. Zachara, B. Mank, and S. Pasynkiewicz, *J. Organomet. Chem.*, 1993, **454**, 5
443 J.J. Eisch, S.I. Pombrik, and G.X. Zheng, Organometallics, 1993, **12**, 3856
444 N. Ropson, P. Dubois, R. Jerome, and P. Teyssie, *Macromolecules*, 1993, **26**, 6378
445 R.H. Meinhold, R.C.T. Slade, and T.W. Davies, *Appl. Magn. Reson.*, 1993, **4**, 141
446 Z. Xu and B.L. Sherriff, *Appl. Magn. Reson.*, 1993, **4**, 203
447 U. Selvaraj, S. Komarneni, and R. Roy, *J. Solid State Chem.*, 1993, **106**, 73
448 J.M. Trillo, M.D. Alba, R. Alvero, and M.A. Castro, *J. Chem. Soc., Chem. Commun.*, 1993, 1809
449 J. Pinkas, B. Gaul, and J.G. Verkade, *J. Am. Chem. Soc.*, 1993, **115**, 3925
450 F. Birkeneder, B.R. Willestofte, and B.N. Janniksen, *Acta Chem. Scand.*, 1993, **47**, 344
451 J. Pinkas and J.G. Verkade, *Inorg. Chem.*, 1993, **32**, 2711
452 P.J. Barrie, *Chem. Phys. Lett.*, 1993, **208**, 486
453 T. Sugano, K. Matsubara, T. Fujita, and T. Takahashi, *J. Mol. Cat.*, 1993, **82**, 93
454 H. Ernst, D. Freude, and I. Wolf, *Chem. Phys. Lett.*, 1993, **212**, 588
455 L. Mathew, J. Rajaram, V. Ramakrishman, and J.C. Kuriacose, *J. Photochem. Photobiol., A*, 1993, **74**, 109
456 F.R. Venema, J.A. Peters, and H. van Bekkum, *Recl. Trav. Chim. Pays-Bas*, 1993, **112**, 445
457 G.W. Haddix, M. Narayana, S.C. Tang, and Y. Wu, *J. Phys. Chem.*, 1993, **97**, 4624
458 G.W. Haddix, M. Narayana, W.D. Gillespie, M.B. Georgellis, and Y. Wu, *J. Am. Chem. Soc.*, 1994, **116**, 672
459 M.P.J. Peeters, J.W. de Haan, L.J.M. van de Ven, and J.H.C. van Hooff, *J. Phys. Chem.*, 1993, **97**, 5363
460 J. Janchen, M.P.J. Peeters, J.W. de Haan, L.J.M. van de Ven, J.H.C. van Hooff, I. Girnus, and U. Lohse, *J. Phys. Chem.*, 1993, **97**, 12042
461 G. Engehardt and W. Veeman, *J. Chem. Soc., Chem. Commun.*, 1993, 622
462 G. Engelhardt, P. Sieger, and J. Felsche, *Anal. Chim. Acta*, 1993, **283**, 967
463 W.E.E. Stone, G.M.S. El Shafei, J. Sanz, and S.A. Selim, *J. Phys. Chem.*, 1993, **97**, 10127
464 N.P. Rhodes and R. Rudham, *J. Chem. Soc., Faraday Trans.*, 1993, **89**, 2551
465 S.M. Bradley and J.V. Hanna, *J. Chem. Soc., Chem. Commun.*, 1993, 1249
466 A. Samoson, P. Sarv, J.P. van Braam Houckgeest, and B. Kraushaar-Czarnetzki, *Appl. Magn. Reson.*, 1993, **4**, 171
467 I. Kiricsi, C. Flego, G. Pazzuconi, W.O. Parker, Jr., R. Millini, C. Perego, and G. Bellussi, *J. Phys. Chem.*, 1994, **98**, 4627
468 A. Diaz, R. Monque, and M. Bussolo, *Bull. Magn. Reson.*, 1993, **15**, 112

469 Z. Cerny, J. Machacek, J. Fusek, O. Kriz, B. Casensky, and D.G. Tuck, *J. Organomet. Chem.*, 1993, **456**, 25
470 L.J. Baker and M.J. Taylor, *Polyhedron*, 1994, **13**, 251
471 S.M. Bradley, R.F. Howe, and J.V. Hanna, *Solid State Nucl. Magn. Reson.*, 1993, **2**, 37
472 S.M. Bradley, R.F. Howe, and R.A. Kydd, *Magn. Reson. Chem.*, 1993, **31**, 883
473 D. Franke, H. Eckert, R.B. Kaner, and R.E. Treece, *Anal. Chim. Acta*, 1993, **283**, 987
474 M.A. Malyarick and S.P. Petrosyants, *Inorg. Chem.* 1993, **32**, 2265
475 J.M. Aramini, P.H. Krygsman and H.J. Vogel, *Biochemistry*, 1994, **33**, 3304
476 C.J. Groombridge, E.N. Coker, and L.V.C. Rees, *Magn. Reson. Chem.*, 1993, **31**, 1064
477 Y.Q. Song, W.P. Halperin, L. Tonge, T.J. Marks, M. Ledvij, V.G. Kogan, and L.N. Bulaevskii, *Phys. Rev. Lett.*, 1993, **70**, 3127
478 J. Schraml, J. Cermak, and A. Kasal, *Magn. Reson. Chem.*, 1993, **31**, 689
479 H. Ban-Oganowska, *Magn. Reson. Chem.*, 1993, **31**, 1092
480 F.D. Gunstone, *Chem. Phys. Lipids*, 1993, **66**, 189
481 J.C. Wallet and E.M. Gaydou, *Magn. Reson. Chem.*, 1993, **31**, 518
482 C. Wang, Y. Mo, M. Jang, and A.F. Janzen, *Can. J. Chem.*, 1993, **71**, 525
483 R.M. Cravero, M. Gonzalez-Sierra, and A.C. Olivieri, *J. Chem. Soc., Perkin Trans. 2*, 1993, 1067
484 A. San Feliciano, J.M. Miguel del Coral, M. Gordaliza, and M.A. Salinero, *Magn. Reson. Chem.*, 1993, **31**, 841
485 J.E. Gurst, E.M. Schubert, S.E. Boiadjiev, and D.A. Lightner, *Tetrahedron*, 1993, **49**, 9191
486 C. Lopez, R.M. Claramunt, S. Trofimenko, and J. Elgureo, *Can. J. Chem.*, 1993, **71**, 678
487 A.S. Shashkov, N.E. Nifant'ev, E. Nikolay, V.Y. Amochaeva, and N.K. Kochetkov, *Magn. Reson. Chem.*, 1993, **31**, 599
488 Z. Dega-Szafran, B. Nowak-Wydra, and M. Szafran, *Magn. Reson. Chem.*, 1993, **31**, 726
489 T. Arashida, T. Ishino, A. Kai, K. Hatanaka, T. Akaike, K. Matsuzaki, Y. Kaneko, and T. Mimura, *J. Carbohydr. Chem.*, 1993, **12**, 641
490 P.R. Birkett, A.G. Avent, A.D. Darwish, H.W. Kroto, R. Taylor, and D.R.M. Walton, *J. Chem. Soc., Chem. Commun.*, 1993, 1230
491 D. Farcasiu and A. Ghenciu, *J. Am. Chem. Soc.*, 1993, **115**, 10901
492 T.R. Forbus, Jr. and J.C. Martin, *Heteroat. Chem.*, 1993, **4**, 129
493 R. Blinc, J. Seliger, J. Dolinsek, and D. Arcon, *Phys. Rev. B: Condens. Matter*, 1994, **49**, 4993
494 R. Kerkoud, P. Auban-Senzier, D. Jerome, J.M. Lambert, A. Zahab, and P. Bernier, *Europhys. Lett.*, 1994, **25**, 379
495 P.L. Yeagle, A.R. Dentino, F.T. Smith, P. Spooner, and A. Watts, *Biochemistry*, 1993, **32**, 12197
496 G.I. Borodkin, I.R. Elanov, M.M. Shakirov, and V.G. Shubin, *J. Phys. Org. Chem.*, 1993, **6**, 153
497 R.M. Claramunt, D. Sanz, J. Catalan, F. Fabero, N.A. Garcia, C. Foces-Foces, A.L. Llamas-Saiz, and J. Elguero, *J. Chem. Soc., Perkin Trans. 2*, 1993, 1687
498 Z. Gu and A. McDermott, *J. Am. Chem. Soc.*, 1993, **115**, 4282
499 S. Luo, T.B. Rauchfuss, and Z. Gan, *J. Am. Chem. Soc.*, 1993, **115**, 4943
500 C. Ye, R. Fu, J. Hu, L. Hou, and S. Ding, *Magn. Reson. Chem.*, 1993, **31**, 699
501 B. Montez, E. Oldfield, J.A. Urbina, S. Pekerar, C. Husted, and J. Patterson, *Biochim. Biophys. Acta*, 1993, **1152**, 314
502 M.H. Sherwood, D.W. Alderman, and D.M. Grant, *J. Magn. Reson., Ser. A*, 1993, **104**, 132
503 S. Nishikiori, C.L. Ratcliffe, and J.A. Ripmeester, *Can. J. Chem.*, 1993, **71**, 1810
504 M.D. Lumsden, R.E. Wasylishen, K. Eichele, M. Schindler, G.H. Penner, W.P. Power, and R.D. Curtis, *J. Am. Chem. Soc.*, 1994, **116**, 1403
505 T. Xu, E.J. Munson, and J.F. Haw, *J. Am. Chem. Soc.*, 1994, **116**, 1962
506 E.J. Munson and J.F. Haw, *Angew. Chem., Int. Ed. Engl.*, 1993, **32**, 615
507 K. Schmidt-Rohr, M. Wilhelm, A. Johansson, and H.W. Spiess, *Magn. Reson. Chem.*, 1993, **31**, 352
508 R. Taylor, G.J. Langley, A.G. Avent, J.S. Dennis, H.W. Kroto, and D.R.M. Walton, *J. Chem. Soc., Perkin Trans. 2*, 1993, 1029

509 Z. Belahmer, P. Bernier, L. Firlej, J.M. Lambert, and M. Ribet, *Phys. Rev. B: Condens. Matter*, 1993, **47**, 15980
510 W.A. Scrivens and J.M. Tour, *J. Chem. Soc., Chem. Commun.*, 1993, 1207
511 Y. Maniwa, A. Ohi, K. Mizoguchi, and Y. Achiba, *Synth. Met.*, 1993, **56**, 3057
512 M.J. Duer, *J. Chem. Soc., Faraday Trans.*, 1993, **89**, 823
513 A.I. Biaglow, R.J. Gorte, and D. White, *J. Phys. Chem.*, 1993, **97**, 7135
514 A.R. Brough, C.P. Grey, and C.M. Dobson, *J. Am. Chem. Soc.*, 1993, **115**, 7318
515 I.P. Gerothanassis, M. Momenteau, G.E. Hawkes, and P.J. Barrie, *J. Am. Chem. Soc.*, 1993, **115**, 9796
516 F. Rachdi, J. Reichenbach, L. Firlej, P. Bernier, M. Ribet, R. Aznar, G. Zimmer, M. Helmie, and M. Mehring, *Solid State Commun.*, 1993, **87**, 547
517 M. Herberhold, S. Gerstmann, W. Milius, and B. Wrackmeyer, *Z. Naturforsch., B: Chem. Sci.*, 1993, **48**, 249
518 A.G. Brook, R. Kumarathasan, and W. Chatterton, *Organometallics*, 1993, **12**, 4085
519 M. Moran, C.M. Casado, I. Cuadrado, and J. Losada, *Organometallics*, 1993, **12**, 4327
520 J. Schraml, M. Jakoubkova, M. Kvicalova, and A. Kasal, *J. Chem. Soc., Perkin Trans. 2*, 1994, 1
521 E. Duguet, M. Schapacher, and A. Soum, *J. Organomet. Chem.*, 1993, **458**, 9
522 M. Baier, P. Bissinger, and H. Schmidbaur, *Z. Naturforsch., B: Chem. Sci.*, 1993, **48**, 1672
523 M. Denk, R. Lennon, R. Hayashi, R. West, A.V. Belyakov, H.P. Verne, A. Haaland, M. Wagner, and N. Metzler, *J. Am. Chem. Soc.*, 1994, **116**, 2691
524 S.R. Bahr and P. Boudjouk, *J. Am. Chem. Soc.*, 1993, **115**, 4514
525 A.R. Bassindale and J. Jiang, *J. Organomet. Chem.*, 1993, **446**, C3
526 C.A. Reed, Z. Xie, R. Bau, and A. Benesi, *Science (Washington, D.C., 1883-)*, 1993, **262**, 402
527 L. Olsson and D. Cremer, *Chem. Phys. Lett.*, 1993, **215**, 433
528 J.H. Iwamiya and G.E. Maciel, *J. Am. Chem. Soc.*, 1993, **115**, 6835
529 R. Tacke, A. Lopez-Mras, J. Sperlich, C. Strohmann, W.F. Kuhs, G. Mattern, and A. Sebald, *Chem. Ber.*, 1993, **126**, 851
530 A.E. Aliev, K.D.M. Harris, and D.C. Apperley, *J. Chem. Soc., Chem. Commun.*, 1993, 251
531 L. Garrido, B. Pfleiderer, M. Papisov, and J.L. Ackerman, *Magn. Reson. Med.*, 1993, **29**, 839
532 N.J. Clayden and P.J.N. Jones, *J. Polym. Sci., Part A: Polym. Chem.*, 1993, **31**, 2835
533 M. Stocker, *Acta Chem. Scand.*, 1993, **47**, 935
534 T. Lobreyer, H. Oberhammer, and W. Sundermeyer, *Angew. Chem. Int. Ed. Engl.*, 1993, **32**, 586
535 G. Calzaferri, R. Imhof, and K.W. Toernroos, *J. Chem. Soc., Dalton Trans.*, 1993, 3741
536 R. Vomscheid, M. Briend, M.J. Peltre, P. Massiani, P.P. Man, and D. Barthomeuf, *J. Chem. Soc., Chem. Commun.*, 1993, 544
537 J.A. Martens, Y.L. Xiong, E.J.P. Feijen, P.J. Grobet, and P.A. Jacobs, *J. Phys. Chem.*, 1993, **97**, 5132
538 U. Sternberg and W. Priess, *J. Magn. Reson., Ser. A*, 1993, **102**, 160
539 M. Derewinski, M.J. Peltre, M. Briend, D. Barthomeuf, and P.P. Man, *J. Chem. Soc., Faraday Trans.*, 1993, **89**, 1823
540 X. Tang, Y. Sun, T. Wu, L. Wang, L. Fei and Y. Long, *J. Chem. Soc., Faraday Trans.*, 1993, **89**, 1839
541 I.S. Chuang, D.R. Kinney, and G.E. Maciel, *J. Am. Chem. Soc.*, 1993, **115**, 8695
542 J.H. Kwak and R. Ryoo, *J. Phys. Chem.*, 1993, **97**, 11154
543 R.K. Harris, E.F.K. Bahlmann, K. Metcalfe, and E.G. Smith, *Magn. Reson. Chem.*, 1993, **31**, 743
544 C. Jia, P. Massiani, and D. Barthomeuf, *J. Chem. Soc., Faraday Trans.*, 1993, **89**, 3659
545 R. Dupree, S.C. Kohn, C.M.B. Henderson, and A.M.T. Bell, *NATO ASI Ser., Ser. C*, 1993, **386**, 421
546 B. Gil, E. Broclawik, J. Datka, and J. Klinowski, *J. Phys. Chem.*, 1994, **98**, 930
547 Y. Takeuchi, I. Zicmane, G. Manuel, and R. Boukherroub, *Bull. Chem. Soc. Jpn.*, 1993, **66**, 1732

548 S. Gopinathan, M.P. Degaonkar, A.M. Hundekar, and C. Gopinathan, *Appl. Organomet. Chem.*, 1993, **7**, 63
549 T.S. Basu Baul, D. Dey, D.D. Mishra, W.L. Basaiawmoit, and E. Rivarola, *J. Organomet. Chem.*, 1993, **447**, 9
550 H. Weichmann and J. Meunier-Piret, *Organometallics*, 1993, **12**, 4097
551 E.G. Martinez, A.S. Gonzalez, J.S. Casas, J. Sordo, G. Valle, and U. Russo, *J. Organomet. Chem.*, 1993, **453**, 47
552 H.J. Reich, J.P. Borst, and R.R. Dykstra, *Organometallics*, 1994, **13**, 1
553 T. Carofiglio, D. Marton, and F. Lenzmann, *Anal. Chim. Acta*, 1993, **281**, 119
554 J.C. Cochran, H.K. Phillips, S. Tom, A.R. Hurd, and B.S. Bronk, *Organometallics*, 1994, **13**, 947
555 C. Nedez, A. Theolier, F. Lefebvre, A. Choplin, J.M. Basset, and J.F. Joly, *J. Am. Chem. Soc.*, 1993, **115**, 722
556 H. Reuter and A. Sebald, *Z. Naturforsch., B: Chem. Sci.*, 1993, **48**, 195
557 P.B. Hitchcock, M.F. Lappert, G.A. Lawless, and B. Royo, *J. Chem. Soc., Chem. Commun.*, 1993, 554
558 C. Nedez, A. Choplin, F. Lefebvre, and J.-M. Basset, *Inorg. Chem.*, 1994, **33**, 1575
559 C. Nedez, F. Lefebvre, A. Choplin, J.M. Basset, and E. Benazzi, *J. Am. Chem. Soc.*, 1994, **116**, 3039
560 N.A. Davies, K.B. Dillon, R.K. Harris, G.F. Hewitson, and L. Toms, *Polyhedron*, 1994, **13**, 19
561 C.I.F. Denekamp, D.F. Evans, J. Parr, and J.D. Woollins, *J. Chem. Soc., Dalton Trans.*, 1993, 1489
562 L. Carlton and R. Weber, *Inorg. Chem.*, 1993, **32**, 4169
563 S. Mathur, A. Singh, and R.C. Mehrotra, *Polyhedron*, 1993, **12**, 1073
564 W. Plass and J.G. Verkade, *Inorg. Chem.*, 1993, **32**, 5153
565 R.K. Harris, D.C. Apperley, N.A. Davis, and R.D. Fischer, *Bull. Magn. Reson.*, 1993, **15**, 22
566 C.P. Grey, A.K. Cheetham, and C.M. Dobson, *J. Magn. Reson., Ser. A*, 1993, **101**, 299
567 S. Eller, P. Schwarz, A.K. Brimah, R.D. Fischer, and D.C. Apperley, *Organometallics*, 1993, **12**, 3232
568 D.L. Reger, M.F. Huff, S.J. Knox, R.J. Adams, D.C. Apperley, and R.K. Harris, *Inorg. Chem.*, 1993, **32**, 4472
569 A.D. Irwin, C.D. Chandler, R. Assink, and M.J. Hampden-Smith, *Inorg. Chem.*, 1994, **33**, 1005
570 A. Shoji, S. Ando, S. Kuroki, I. Ando, and G.A. Webb, *Annu. Rep. NMR spectrosc.*, 1993, **26**, 55
571 S.J. Goede, B.L.M. van Baar, and F. Bickelhaupt, *J. Org. Chem.*, 1993, **58**, 6930
572 R.R. Annand, J.F. Kozlowski, V.Jo. Davisson, and J.M. Schwab, *J. Am. Chem. Soc.*, 1993, **115**, 1088
573 Y. Rhee, C. Wang, B.L. Gaffney, and R.A. Jones, *J. Am. Chem. Soc.*, 1993, **115**, 8742
574 N.A. Froeystein and E. Sletten, *J. Am. Chem. Soc.*, 1994, **116**, 3240
575 J.R. Burke and P.A. Frey, *Biochemistry*, 1993, **32**, 13220
576 E. Grech, L. Stefaniak, I. Ando, H. Yoshimizu, G.A. Webb, and L. Sobczyk, *Bull. Chem. Soc. Jpn.*, 1990, **63**, 2716
577 A. Naito, S. Tuzi, and H. Saito, *Bull. Chem. Soc. Jpn.*, 1993, **66**, 3309
578 M.S. Morales-Rios, P. Joseph-Nathan, B. Wrackmeyer, and E. Kupce, *Magn. Reson. Chem.*, 1993, **31**, 238
579 J. Jazwinski, and L. Stefaniak, *Magn. Reson. Chem.*, 1993, **31**, 447
580 D. Belkacemi, J.W. Davies, J.R. Malpass, A. Naylar, and C.R. Smith, *Tetrahedron*, 1992, **48**, 10161
581 J. Llor, O. Lopez-Mayorga, and L. Munoz, *Magn. Reson. Chem.*, 1993, **31**, 552
582 S. Farr-Jones, W.Y.L. Wong, W.G. Gutheil, and W.W. Bachovchin, *J. Am. Chem. Soc.*, 1993, **115**, 6813
583 B.L. Ugrak, V.S. Bogdanov, S.A. Shevelev, and I.L. Dalinger, *Mendeleev Commun.*, 1993, 109
584 I. Wawer and V. Koleva, *Magn. Reson. Chem.*, 1993, **31**, 375

585 J.H. Markgraf, D.P. Richardson, and R.C. Livingston, *Magn. Reson. Chem.*, 1993, **31**, 694
586 E. Kolehmainen, K. Laihia, R. Kauppinen, R. Gawinecki, and D. Rasala, *Magn. Reson. Chem.*, 1993, **31**, 659
587 G. L'abbe, P. Delbeke, L. Bastin, W. Dehaen, and S. Toppet, *J. Heterocycl. Chem.*, 1993, **30**, 301
588 T. Fujihara and S. Kaizaki, *J. Chem. Soc., Dalton Trans.*, 1993, 2521
589 K.L. Brown and D.R. Evans, *Inorg. Chem.*, 1993, **32**, 2544
590 D. Gudat, *Magn. Reson. Chem.*, 1993, **31**, 925
591 F.J. Martinez-Martinez, A. Ariza-Castolo, V. Ramos-Nava, N. Barba-Behrens, and R. Contreras, *Magn. Reson. Chem.*, 1993, **31**, 832
592 W.C. Conner, M. Ferrero, J. Bondardet, and J. Fraissard, *J. Chem. Soc., Faraday Trans.*, 1993, **89**, 833
593 C.P. Grey, W.S. Veeman, and A.J. Vega, *J. Chem. Phys.*, 1993, **98**, 7711
594 K.L. Anderson-Altmann and D.M. Grant, *J. Phys. Chem.*, 1993, **97**, 11096
595 M.-C. Labarre and J.-F. Labarre, *J. Mol. Struct.*, 1993, **300**, 593
596 A. Schanck and M. Deleers, *Biochem. Biophys. Res. Commun.*, 1993, **195**, 654
597 Y. Segall, D. Waysbort, D. Barak, N. Ariel, B.P. Doctor, J. Grunwald, and Y. Ashani, *Biochemistry*, 1993, **32**, 13441
598 E.J. Dufourc, J.C. Mailet, T. Pott, and A. Leonard, *Phosphorus, Sulfur Silicon Relat. Elem.*, 1993, **77**, 121
599 K.L. Li, C.A. Tihal, M. Guo, and R.E. Stark, *Biochemistry*, 1993, **32**, 9926
600 G.D. Williams, T.J. Mosher, and M.B. Smith, *Anal. Biochem.*, 1993, **214**, 458
601 H. Ronkkomaki, J. Jokisaari, and L.H.J. Lajunen, *Acta Chem. Scand.*, 1993, **47**, 331
602 E. Brosio, E. Carnovale, M. Delfini, L. Pizzoferrato, and G. Veglia, *Bull. Magn. Reson.*, 1993, **15**, 183
603 B. Wrackmeyer and C. Koehler, *Magn. Reson. Chem.*, 1993, **31**, 573
604 J.M. Pearce and R.A. Komoroski, *Magn. Reson. Med.*, 1993, **29**, 724
605 E.C. Ashby, R. Gurumurthy, and R.W. Ridlehuber, *J. Org. Chem.*, 1993, **58**, 5832
606 J.A. Barry, K.A. McGovern, Y.H.H. Lien, B. Ashmore, and R.J. Gillies, *Biochemistry*, 1993, **32**, 4665
607 M.J. Potrzebowski, *J. Chem. Soc., Perkin Trans. 2*, 1993, 63
608 J.A. Davies and S.G. Dutremez, *Magn. Reson. Chem.*, 1993, **31**, 439
609 J.P. Fletcher, R.J. Kirkpatric, D. Howell, and S.H. Risbud, *J. Chem. Soc., Faraday Trans.*, 1993, **89**, 3297
610 M. Ilczyszyn, *J. Chem. Soc., Faraday Trans.*, 1993, **89**, 2463
611 C. Amatore, A. Jutand, and A. Suarez, *J. Am. Chem. Soc.*, 1993, **115**, 9531
612 Y. Miyazaki and H. Waki, *J. Chem. Soc., Faraday Trans.*, 1993, **89**, 3925
613 R.L. Thurmond, D. Otten, M.F. Brown, and K. Beyer, *J. Phys. Chem.*, 1994, **98**, 972
614 A.J. Burn, S.K. Dewan, I. Gosney, K.G. McKendrick, C.P. Warrens, J.P. Wastle, and C.W. Watson, *J. Chem. Soc., Perkin Trans., 2*, 1994, 373
615 R. Maxwell and H. Eckert, *J. Am. Chem. Soc.*, 1993, **115**, 4747
616 Y. Yue, F. Deng, H. Hu, C. Ye, Z. Lin, and S. Tian, *Chem. Phys. Lett.*, 1993, **208**, 311
617 K. Eichele and R.E. Wasylishen, *J. Phys. Chem.*, 1994, **98**, 3108
618 H. Nakano, T. Ohno, and S. Yamanaka, *Chem. Lett.*, 1994, 9
619 K. Hiraoka, Y. Kasamatsu, K. Kojima, and T. Hihara, *Physica B (Amsterdam)*, 1993, **186-188**, 528
620 K. Hiraoka, Y. Kasamatsu, K. Kojima, and T. Hihara, *Physica B (Amsterdam)*, 1993, **186-188**, 535
621 H. Suzuki, S. Takagi, K. Mattenberger, and O. Vogt, *Physica B (Amsterdam)*, 1993, **186-188**, 755
622 H.P.A. Mercier, J.C.P. Sanders, and G.J. Schrobligen, *J. Am. Chem. Soc.*, 1994, **116**, 2921
623 G. Cerioni and A. Plumitallo, *Magn. Reson. Chem.*, 1993, **31**, 320
624 L.L. Smith, J.E. Herz, and E.L. Ezell, *Steroids*, 1993, **58**, 260
625 A. Kumar and D.W. Boykin, *J. Mol. Struct.*, 1993, **296**, 95
626 D.W. Boykin, S. Chandrasekaran, and A.L. Baumstark, *Magn. Reson. Chem.*, 1993, **31**, 489

627 J. Spychala and D.W. Boykin, *J. Chem. Res., Synop.*, 1993, 426
628 E. Kolehmainen and K. Laihia, *Magn. Reson. Chem.*, 1993, **31**, 763
629 A. Kumar and D.W. Boykin, *Indian J. Heterocycl. Chem.*, 1993, **3**, 1
630 H. Dahn, P. Pechy, and H.J. Bestmann, *J. Chem. Soc., Perkin Trans. 2*, 1993, 1497
631 H. Dahn and P. Pechy, *J. Chem. Soc., Perkin Trans. 2*, 1993, 67
632 A. Saba, F.S. Sib, and J.P. Aycard, *Spectrosc. Lett.*, 1993, **26**, 491
633 G. Cerioni, N. Culeddu, and A. Plumitallo, *Tetrahedron*, 1993, **49**, 2957
634 K.Nagasawa, Y. Higuchi, K. Ito, M. Imanari, and N. Fujii, *Chem. Pharm. Bull.*, 1993, **41**, 211
635 M. Cudic and R. Herrmann, *Magn. Reson. Chem.*, 1993, **31**, 461
636 N.J. Gordon and S.A. Evans, Jr., *J. Org. Chem.*, 1993, **58**, 4516
637 E. Kolehmainen, K. Laihia, M. Heinanen, K. Rissanen, R. Froehlich, J. Korvola, P. Manttari, and R. Kauppinen, *J. Chem. Soc., Perkin Trans. 2*, 1993, 641
638 E. Kolehmainen, K. Laihia, M.I. Rybinskaya, V.S. Kaganovich, and Z.A. Kerzina, *J. Organomet. Chem.*, 1993, **453**, 273
639 G. Barbarella, A. Bongini, V. Cere, C. Paolucci, and S. Pollicino, *J. Org. Chem.*, 1993, **58**, 6499
640 M.E. Smith and H.J. Whitfield, *J. Chem. Soc., Chem. Commun.*, 1994, 723
641 G. Cerioni, N. Culeddu, and A. Saba, *Magn. Reson. Chem.*, 1993, **31**, 829
642 M. Aberg and J. Glaser, *Inorg. Chim. Acta*, 1993, **206**, 53
643 C.F.G.C. Geraldes, A.M. Urbano, M.A. Hoefnagel, and J.A. Peters, *Inorg. Chem.*, 1993, **32**, 2426
644 Y. Yang, V. Ramamoorthy, and P.R. Sharp, *Inorg. Chem.*, 1993, **32**, 1946
645 G. Powell, D.T. Richens, and A.K. Powell, *Inorg. Chim. Acta*, 1993, **213**, 147
646 E. Sappa, D. Cauzzi, G. Predieri, and C. Vignali, *Spectrochim. Acta, Part A*, 1993, **49A**, 1395
647 A. Okumura, M. Fushimi, and M. Yamamoto, *Bull. Chem. Soc. Jpn.*, 1993, **66**, 2814
648 G.-Q. Zheng, T. Kuse, Y. Kitaoka, K. Ishida, S. Ohsugi, K. Asayama, and Y. Yamada, *Physica C (Amsterdam)*, 1992, **208**, 339
649 A.P. Howes, R. Dupree, Z.P. Han, R.S. Liu, and P.P. Edwards, *Phys. Rev. B: Condens. Matter*, 1993, **47**, 11529
650 S.B. Adler, J.A. Reimer, J. Baltisberger, and U. Werner, *J. Am. Chem. Soc.*, 1994, **116**, 675
651 G. Barbarella, *Prog. Nucl. Magn. Reson. Spectrosc.*, 1993, **25**, 317
652 T.Machiguchi, T. Hasegawa, H. Otani, S. Yamabe, and H. Mizuno, *J. Am. Chem. Soc.*, 1994, **116**, 407
653 H. Duddeck, P. Wagner, and B. Rys, *Magn. Reson. Chem.*, 1993, **31**, 736
654 M. Iwaoka and S. Tomoda, *J. Am. Chem. Soc.*, 1994, **116**, 2557
655 T. Faecke, R. Wagner, and S. Berger, *J. Org. Chem.*, 1993, **58**, 5475
656 J. Browning, G.W. Bushnell, K.R. Dixon, and R.W. Hilts, *J. Organomet. Chem.*, 1993, **452**, 205
657 A. Jakobs, G. Llabres, and M. Baiwir, *Magn. Reson. Chem.*, 1993, **31**,786
658 B. Wrackmeyer, B. Distler, S. Gerstmann, and M. Herberhold, *Z. Naturforsch., B: Chem. Sci.*, 1993, **48**, 1307
659 M. Pridoehl, R. Steudel, and F. Baumgart, *Polyhedron*, 1993, **12**, 2577
660 H. Fujihara, H. Mima, T. Erata, and N. Furukawa, *J. Am. Chem. Soc.*, 1993, **115**, 9826
661 E.J. Munson, A.A. Kheir, and J.F. Haw, *J. Phys. Chem.*, 1993, **97**, 7321
662 R. Maxwell and H. Eckert, *J. Am. Chem. Soc.*, 1994, **116**, 682
663 S.S. Dhingra and M.G. Kanatzidis, *Inorg. Chem.*, 1993, **32**, 1350
664 T. Chivers, D.D. Doxsee, and M. Parvez, *Inorg. Chem.*, 1993, **32**, 2238
665 T. Chivers, D.D. Doxsee, J.F. Fait, and M. Parvez, *Inorg. Chem.*, 1993, **32**, 2243
666 J.M. McConnachie, M.A. Ansari, and J.A. Ibers, *Inorg. Chem.*, 1993, **32**, 3250
667 G. Matsubayashi and A. Yokozawa, *Inorg. Chim. Acta*, 1993, **208**, 95
668 J. Siivari, T. Chivers, and R.S. Laitinen, *Inorg. Chem.*, 1993, **32**, 4391
669 C.C. Christuk and J.A. Ibers, *Inorg. Chem.*, 1993, **32**, 5105
670 M. Broschag, T.M. Klapoetke, A. Schulz, and P.S. White, *Inorg. Chem.*, 1993, **32**, 5734
671 T. Chivers, D.D. Doxsee, R.W. Hilts, and M. Parvez, *Can. J. Chem.*, 1993, **71**, 1821

672 A. Bacchi, W. Baratta, F. Calderazzo, F. Marchetti, and G. Pelizzi, *Angew. Chem. Int. Ed. Engl.*, 1994, **33**, 193
673 P.D. Ellis, J.D. Odom, A.S. Lipton, Q. Chen, and J.M. Gulick, *NATO ASI Ser., Ser. C*, 1993, **386**, 539
674 L. Engman and J. Persson, *Organometallics*, 1993, **12**, 1068
675 L. Engman and D. Stern, *Organometallics*, 1993, **12**, 1445
676 L. Stefaniak, B. Kamienski, W. Schilf, S.V. Amosova, and G. Webb, *J. Phys. Org. Chem.*, 1993, **6**, 520
677 H. Fujihara, T. Uehara, T. Erata, and N. Furukawa, *Chem. Lett.*, 1993, 263
678 A.G. Maslakov, W.R. McWhinnie, M.C. Perry, N. Shaikh, S.L.W. McWhinnie, and T.A. Hamor, *J. Chem. Soc., Dalton Trans.*, 1993, 619
679 M. Minoura, T. Kawashima, and R. Okazaki, *J. Am. Chem. Soc.*, 1993, **115**, 7019
680 V. Christou and J. Arnold, *Angew. Chem. Int. Ed. Engl.*, 1993, **32**, 1450
681 K.O. Christe, D.A. Dixon, J.C.P. Sanders, G.J. Schrobilgen, and W.W. wilson, *J. Am. Chem. Soc.*, 1993, **115**, 9461
682 F.W.B. Einstein, I.D.G. Colin, H.W. Jones, A. Riesen, and R.D. Sharma, *Acta Crystallogr., Sect. C: Cryst. Struct. Commun.*, 1993, **C49**, 470
683 R.J. Batchelor, F.W.B. Einstein, I.D. Gay, C.H.W. Jones, and R.D. Sharma, *Inorg. Chem.*, 1993, **32**, 4378
684 A.S.L. Xu and P.W. Kuchel, *Biochim. Biophys. Acta*, 1993, **1150**, 35
685 H. Koroniak, K.W. Palmer, W.R. Dolbier, Jr., and H.Q. Zhang, *Magn. Reson. Chem.*, 1993, **31**, 748
686 K. Laihia, E. Kolehmainen, P. Manttari, and R. Kauppinen, *Magn. Reson. Chem.*, 1993, **31**, 512
687 L.A. Sylvia and J.T. Gerig, *Biochim. Biophys. Acta*, 1993, **1163**, 321
688 J.F. Eccleston, D.P. Molloy, M.G. Hinds, R.W. King, and J. Feeney, *Eur. J. Biochem.*, 1993, **218**, 1041
689 S. Peelen, I.M.C.M. Rietjens, W.J.H. van Berkel, W.A.T. van Workum, and J. Vervoort, *Eur. J. Biochem.*, 1993, **218**, 345
690 L. Garel, J.P. Dutasta, and A. Collet, *Angew. Chem., Int. Ed. Engl.*, 1993, **32**, 1169
691 M. Cushman, H.H. Patel, J. Scheuring, and A. Bacher, *J. Org. Chem.*, 1993, **58**, 4033
692 R.E. London and S.A. Gabel, *J. Am. Chem. Soc.*, 1994, **116**, 2570
693 B.B. Randolph and D.D. DesMarteau, *J. Fluorine Chem.*, 1993, **64**, 129
694 J.D. Augspurger and C.E. Dykstra, *J. Am. Chem. Soc.*, 1993, **115**, 12016
695 K. Hamada, M. Ohira, and K. Miyazaki, *J. Phys. Chem.*, 1993, **97**, 4926
696 S.J. Limb, B.E. Scruggs, and K.K. Gleason, *Macromolecules*, 1993, **26**, 3750
697 D.A. Lathrop and K.K. Gleason, *Macromolecules*, 1993, **26**, 4652
698 B.E. Scruggs and K.K. Gleason, *J. Phys. Chem.*, 1993, **97**, 9187
699 R. Lachicotte, A. Kitaygorodskiy, and K.S. Hagen, *J. Am. Chem. Soc.*, 1993, **115**, 8883
700 S.K. Soo, J. Senegas, J.M. Reau, M. Wahbi, and P. Hagenmuller, *J. Solid State Chem.*, 1993, **104**, 215
701 C. Loubser, P.L. Wessels, J.W. Goodby, and P. Styring, *Liq. Cryst.*, 1993, **15**, 233
702 B.H. Lipshutz, S.H. Dimock, and B. James, *J. Am. Chem. Soc.*, 1993, **115**, 9283
703 F. Tabak, A. Lascialfari, and A. Rigamonti, *J. Phys.: Condens. Matter*, 1993, **5(Suppl. 34B)**, B31
704 S. Takahashi, M.L. Saboungi, R.J. Klingler, M.J. Chen, and J.W. Rathke, *J. Chem. Soc., Faraday Trans.*, 1993, **89**, 3591
705 Y. Yamashoji, M. Fujiwara, and T. Matsushita, *Chem. Lett.*, 1993, 1029
706 R. Seydoux, P. Diehl, R.K. Mazitov, and J. Jokisaari, *J. Magn. Reson., Ser. A*, 1993, **101**, 78
707 R.E. Rapp and H. Godfrin, *Phys. Rev. B: Condens. Matter*, 1993, **47**, 12004
708 T.M. Haard, J.B. Kycia, M.R. Rand, H.H. Hensley, Y. Lee, P.J. Hampt, D.T. Sprague, and W.P. Halperin, *Phys. Rev. Lett.*, 1994, **72**, 860
709 J.B. Kycia, T.M. Haard, M.R. Rand, H.H. Hensley, G.F. Moores, Y. Lee, P.J. Hamot, D.T. Sprague, W.P. Halperin, and E.V. Thuneberg, *Phys. Rev. Lett.*, 1994, **72**, 864
710 M. Saunders, H.A. Jimenez-Vazquez, R.J. Cross, S. Mroczkowski, D.I. Freedberg, and F.A.L. Frank, *Nature (London)*, 1994, **367**, 256

711 M. Hermann, R. Haselmeier, R.K. Mazitov, and H. Weingaertner, *J. Am. Chem. Soc.*, 1994, **116**, 801
712 J.B. Miller, J.H. Walton, and C.M. Roland, *Macromolecules*, 1993, **26**, 5602
713 M. Tomaselli, B.H. Meier, P. Robyr, U.W. Suter, and R.R. Ernst, *Chem. Phys. Lett.*, 1993, **205**, 145
714 A. Gedeon, J.L. Bonardet, and J. Fraissard, *J. Phys. Chem.*, 1993, **97**, 4254
715 E. Trescos, L.C. de Menorval, and F. Rachdi, *J. Phys. Chem.*, 1993, **97**, 6943
716 M.L. Smith, D.R. Corbin, L. Abrams, and C. Dybowski, *J. Phys. Chem.*, 1993, **97**, 7793
717 M.L. Smith, D.R. Corbin, and C. Dybowski, *J. Phys. Chem.*, 1993, **97**, 9045
718 D.R. Corbin, L. Abrams, G.A. Jones, M.L. Smith, C.R. Dybowski, J.A. Hriljac, and J.B. Parise, *J. Chem. Soc., Chem. Commun.*, 1993, 1027
719 M. Tomaselli, B.H. Meier, P. Robyr, U.W. Suter, and R.R. Ernst, *Chem. Phys. Lett.*, 1993, **214**, 1
720 K.J. Chao and D.S. Shy, *J. Chem. Soc., Faraday Trans.*, 1993, **89**, 3841
721 S. McKim and J.F. Hinton, *J. Magn. Reson., Ser. A*, 1993, **104**, 268
722 S.-B. Liu, B.M. Fung, R.-C. Yang, E.-C. Hong, C.-T. Chang, P.-C. Shih, F.-H. Tong, and T.-L. Chen, *J. Phys. Chem.*, 1994, **98**, 4393
723 T.T.P. Cheung, *J. Phys. Chem.*, 1993, **97**, 8993
724 Y.H. Lim, N.E. Nugara, and A.D. King, Jr., *J. Phys. Chem.*, 1993, **97**, 8816
725 Y.H. Lim and A.D. King, Jr., *J. Phys. Chem.*, 1993, **97**, 12173
726 E.M. Arnett and P.C. Wernett, *J. Am. Chem. Soc.*, 1993, **115**, 12187
727 M. Mansfeld and W.S. Veeman, *Chem. Phys. Lett.*, 1993, **213**, 153
728 J.A. Ripmeester and C.I. Ratcliffe, *Anal. Chim. Acta*, 1993, **283**, 1103
729 V.V. Terskikh, I.L. Mudrakovskii, and V.M. Mastikhin, *J. Chem. Soc., Faraday Trans.*, 1993, **89**, 4239

4
Theoretical Aspects of Spin–Spin Couplings

BY H. FUKUI

1 Introduction

Progress in theoretical research for nuclear spin–spin coupling constants is a quite interesting matter for both theoreticians and experimentalists. The aim of this review is to highlight important recent developments in theoretical aspects of spin–spin couplings. The theoretical analysis of spin–spin coupling constants is presently progressing along two different lines, namely accurate calculations where better and more exact techniques are applied, taking advantage of modern computational methods, and simple approximate calculations where the relationships between physical phenomena and experimental data are sought. During the twelve months to the end of May 1994, one theoretical paper,[1] several *ab initio* calculations,[2–8] one semiempirical calculation,[9] and some conformational studies[10–14] have been published in the field of theoretical aspects of spin–spin couplings. These papers will be introduced in the following sections.

2 *Ab initio* Calculations

For last one year we have had the following published papers about nuclear spin–spin coupling calculation, *i.e.*, the relativistic theory for nuclear spin–spin couplings,[1] density functional theory and calculation of coupling constants,[2] SCF calculations of $^3J_{HH}$,[3] theoretical study on $^1J_{CC}$,[4,5] $^1J_{HD}$ in molecular hydrogen complexes,[6] $^1J_{CH}$ and $^2J_{HH}$ surfaces for methane molecule,[7] and *ab initio* Pople–Santry type calculation.[8]

2.1 Relativistic Theory for Spin–Spin Couplings. — Aucar and Oddershede[1] presented a relativistic theory for the nuclear spin–spin coupling tensor within the polarization propagator approach.

2.1.1 *The Definition of Polarization Propagators.* — We start with the definition of time-dependent retarded polarization propagators in the same way as for their nonrelativistic counterparts:[15]

$$\langle\langle P(t); Q(t')\rangle\rangle = -i\theta(t-t')\langle 0|[P(t), Q(t')]|0\rangle, \tag{1}$$

where $\theta(t)$ is the step function, $|0\rangle$ is the reference state, and $P(t)$ and $Q(t')$ are time-dependent property operators in the Heisenberg representation. Namely

$$P(t) = e^{iHt/\hbar} P e^{-iHt/\hbar} \tag{2a}$$

and

$$Q(t') = e^{iHt'/\hbar} Q e^{-iHt'/\hbar}, \tag{2b}$$

in which both P and Q are assumed to be a time–independent Hermitian operator.

We define the propagator in the energy representation by the Fourier transform of $\langle\langle P(t); Q(t')\rangle\rangle$:

$$\begin{aligned}\langle\langle P;Q\rangle\rangle_E &= \hbar^{-1}\int_{-\infty}^{\infty} d(t-t')\langle\langle P(t);Q(t')\rangle\rangle e^{iE(t-t')/\hbar}\\ &= \sum_{m(\neq 0)}\left[\frac{\langle 0|P|m\rangle\langle m|Q|0\rangle}{E-E_m+E_0} - \frac{\langle 0|Q|m\rangle\langle m|P|0\rangle}{E+E_m-E_0}\right].\end{aligned} \tag{3}$$

The energy representation of polarization propagator satisfies the following identity equation,

$$E\langle\langle P;Q\rangle\rangle_E = \langle 0|[P,Q]|0\rangle + \langle\langle P;[H,Q]\rangle\rangle_E. \tag{4}$$

Using Eq. (4) iteratively we can express the energy representation of propagator as

$$\begin{aligned}\langle\langle P;Q\rangle\rangle_E &= E^{-1}\langle 0|[P,Q]|0\rangle + E^{-2}\langle 0|[P,[H,Q]]|0\rangle + E^{-3}\langle 0|[P,[H,[H,Q]]]|0\rangle + \cdots\\ &= \langle 0|[P,(E\hat{1}-\hat{H})^{-1}Q]|0\rangle,\end{aligned} \tag{5}$$

where $\hat{1}$ and $\hat{H}$ are the unit superoperator and the Hamiltonian superoperator, respectively. They are defined for an arbitrary operator Q by

$$\hat{I}Q \equiv Q \tag{6a}$$

and

$$\hat{H}Q \equiv [H,Q]. \tag{6b}$$

In Eq. (5) $(E\hat{1}-\hat{H})^{-1}$ is called the superoperator resolvent. Using a complete set of operators $\{h\}$ we can express $\langle\langle P;Q\rangle\rangle_E$ in a computationally more useful form.[16] We transform Eq. (5) into

$$\langle\langle P;Q\rangle\rangle_E = \langle 0|[P,\tilde{h}]|0\rangle\langle 0|[h^+,(E\hat{1}-\hat{H})\tilde{h}]|0\rangle^{-1}\langle 0|[h^+,Q]|0\rangle. \tag{7}$$

Here, h is a complete set consisting of excitation operators arranged as a column vector and $\tilde{h}$ is its transposed row vector.

For polarization propagators, the operator set is chosen as

$$h = \{h_2, h_4, \cdots\}, \tag{8}$$

where

$$\begin{aligned}h_2 &= \{a_m^+ a_\alpha, a_\alpha^+ a_m\} = \{q^+, q\},\\ h_4 &= \{a_m^+ a_n^+ a_\alpha a_\beta, a_\beta^+ a_\alpha^+ a_n a_m\}\\ &= \{q^+q^+, qq\},\\ &\quad \text{etc.}\end{aligned} \tag{9}$$

In Eq. (9) Roman indices refer to unoccupied orbitals (particle states) and Greek indices to occupied orbitals (hole states). One–electron operators P and Q are written in the second quantized form as

$$P = \sum_{kl} P_{kl} a_k^{+} a_l, \tag{10}$$

in which k and l denote occupied and unoccupied orbitals. If we choose the reference state ($|0\rangle$) as a single configuration Hartree–Fock (HF) wave function ($|\text{HF}\rangle$), only h_2 contributes to $\langle 0|[P,\tilde{h}]|0\rangle$ and $\langle 0|[h^+,Q]|0\rangle$. Then, we have

$$\langle \text{HF}|[P,\tilde{h}_2]|\text{HF}\rangle = (\tilde{P}^*_{m\alpha}, -\tilde{P}_{m\alpha}) \tag{11a}$$

and

$$\langle \text{HF}|[h^+,Q]|\text{HF}\rangle = \begin{pmatrix} Q_{m\alpha} \\ -Q^*_{m\alpha} \end{pmatrix}. \tag{11b}$$

The tilde and non-tilde in Eqs. (11a) and (11b) denote the row vector and column vector, respectively. The second-order correction E^2_{PQ} to the energy, bilinear to the two perturbation energy operators P and Q, can be expressed as

$$E^2_{PQ} = \langle\langle P;Q\rangle\rangle_{E=0}, \tag{12}$$

which is obvious from Eq. (3).

2.1.2 *The Relativistic Hamiltonian.* — To generalize the nonrelativistic expressions for the spin–spin coupling tensor, we must first define the adequate relativistic Hamiltonian. We use the Dirac–Coulomb–Breit (DCB) Hamiltonian[17] as the relativistic Hamiltonian, which is defined by

$$H_{\text{DCB}} = \sum_i h_D(i) + V, \tag{13}$$

$$h_D(i) = c\vec{\alpha}(i)\cdot\vec{p}(i) + (\beta - 1)m_e c^2 + V_{nuc}(\vec{r}_i) + \Lambda_+ U(\vec{r}_i)\Lambda_+, \tag{14}$$

$$V = \frac{1}{2}\sum_{ij} \Lambda_+ g_{ij} \Lambda_+ - \sum_i \Lambda_+ U(\vec{r}_i)\Lambda_+, \tag{15}$$

$$\Lambda_+ = \sum_n |n\rangle\langle n|, \tag{16}$$

where $\vec{\alpha}$ and β are the usual 4×4 Dirac matrices; $V_{nuc}(\vec{r}_i)$, the nuclear potential at position of the i-th electron; $U(\vec{r}_i)$, the Dirac–Hartree–Fock (DHF) average potential; g_{ij}, the two-electron interaction operator including the Breit interaction; and Λ_+, the positive energy projection operator constructed from the whole set of solutions (four-component wave functions) to the Dirac one-electron equation, $h_D|n\rangle = \varepsilon_n|n\rangle$. In the DCB Hamiltonian, the contributions form the negative energy states due to creation of virtual electron–positron pairs are neglected.

In the second-quantization language, the DCB Hamiltonian is expressed in the particle–hole formalism as

$$H_{\text{DCB}} = F + W, \tag{17}$$

$$F = \sum_k \varepsilon_k a_k^{+} a_k, \tag{18}$$

$$W = \frac{1}{2}\sum_{ijkl} (\langle ij|kl\rangle + b_{ijkl})\, a_i^+ a_j^+ a_l a_k - \sum_{kl} u_{kl} a_k^+ a_l, \tag{19}$$

where indices i, j, k, l denote the occupied and unoccupied DHF orbitals. The operators F and W are the second–quantized equivalents of $\sum_i h_D(i)$ and V in Eq. (13), respectively. The matrix elements u_{kl} for the DHF potential are determined as

$$u_{kl} = \sum_{\alpha}^{\mathrm{occ}} \langle k\alpha \| l\alpha\rangle, \tag{20}$$

while b_{ijkl} are the matrix elements of the Breit interaction,

$$b_{ijkl} = -\frac{1}{2}\left(\frac{e^2}{4\pi\varepsilon_0}\right)\langle ij|r_{12}^{-1}\vec{\alpha}(1)\cdot\alpha(\vec{2}) + r_{12}^{-3}\vec{\alpha}(1)\cdot\vec{r}_{12}\vec{\alpha}(2)\cdot\vec{r}_{12}|kl\rangle. \tag{21}$$

Preparatory to later developments, it is convenient to express the field–dependent one–electron relativistic wave equation in two–component form:

$$\begin{bmatrix} V-\varepsilon & c\vec{\sigma}\cdot\vec{\pi} \\ c\vec{\sigma}\cdot\vec{\pi} & V-\varepsilon-2m_e c^2 \end{bmatrix}\begin{bmatrix} \psi^L \\ \psi^S \end{bmatrix} = 0. \tag{22}$$

Here, V denotes all terms in h_D except for the $\vec{\alpha}$ and $(\beta-1)$ dependent terms and $\vec{\sigma}$ is the Pauli spin matrix. We have included the magnetic field $\vec{B} = (\vec{\nabla}\times\vec{A})$ by replacing $\vec{P}$ with $\vec{\pi} = \vec{P} + e\vec{A}$. The symbol ( ) in $(\vec{\nabla}\times\vec{A})$ means $\vec{\nabla}$ acts solely on $\vec{A}$. In Eq. (22) ψ^L and ψ^S are the large and small two–component spinors, respectively. The Pauli spin matrix $\vec{\sigma}$ is defined by

$$\vec{\sigma} = \sigma_x\vec{i} + \sigma_y\vec{j} + \sigma_z\vec{k} \tag{23}$$

and

$$\sigma_x = \begin{bmatrix} 0 & 1 \\ 1 & 0 \end{bmatrix}, \quad \sigma_y = \begin{bmatrix} 0 & -i \\ i & 0 \end{bmatrix}, \quad \sigma_z = \begin{bmatrix} 1 & 0 \\ 0 & -1 \end{bmatrix}. \tag{24}$$

2.1.3 *The Relativistic Tensor* $\hat{J}_{NN'}$. — The second–order correction to the energy due to the indirect, electron–coupled interaction between two nuclear spins, $\vec{I}_N$ and $\vec{I}_{N'}$, can be expressed as

$$E^2_{NN'} = \langle\langle H_N; H_{N'}\rangle\rangle_{E=0}, \tag{25}$$

where H_N is the magnetic–dipole hyperfine interaction Hamiltonian, *i.e.*, the interaction between the magnetic field due to $\vec{I}_N$ and the electrons surrounding the nucleus. This perturbation is

$$H_N = ec\vec{\alpha}\cdot\vec{A}_N, \tag{26}$$

where the vector potential $\vec{A}_N$ represents the dipole field from the pointlike nucleus at the electronic position. Thus, we have

$$\vec{A}_N = g_N\mu_N\left(\frac{\mu_0}{4\pi}\right)\vec{I}_N\times\vec{r}_N/r_N^3, \tag{27}$$

where $\vec{r}_N = \vec{r} - \vec{R}_N$, and $\vec{r}$ and $\vec{R}_N$ are the coordinates of the electron and the nucleus N, respectively. In Eq. (27) μ_N is the nuclear magneton, μ_0 is the magnetic permeability in vacuum, and the other symbols have usual meanings. Substituting Eq. (27) in Eq. (26), we obtain

$$\begin{aligned} H_N &= ecg_N\mu_N\left(\frac{\mu_0}{4\pi}\right)\vec{\alpha}\cdot(\vec{I}_N \times \vec{r}_N/r_N^3) \\ &= ecg_N\mu_N\left(\frac{\mu_0}{4\pi}\right)\sum_{i=x,y,z} I_{Ni}\kappa_{Ni}. \end{aligned} \tag{28}$$

The Cartesian components of the κ operator are defined as

$$\kappa_{Ni} = -\sum_{jk}\varepsilon_{ijk}\alpha_j r_{Nk} r_N^{-3}, \quad i,j,k \in x,y,z, \tag{29}$$

where ε_{ijk} is the Levi–Civita permutation symbol. Conventionally, the interaction energy between nuclear spins $\vec{I}_N$ and $\vec{I}_{N'}$ is expressed as

$$E^2_{NN'} = h\vec{I}_N \cdot \hat{J}_{NN'} \cdot \vec{I}_{N'}. \tag{30}$$

From Eqs. (25), (28), and (30) we see that the components of the $\hat{J}_{NN'}$ tensor are

$$J_{NN',ij} = e^2c^2 g_N g_{N'}\mu_N^2\left(\frac{\mu_0}{4\pi}\right)^2 h^{-1}\langle\langle\kappa_{Ni};\kappa_{N'j}\rangle\rangle_{E=0}. \tag{31}$$

2.1.4 *Relativistic RPA Approximation.* — Aucar and Oddershede gave explicit expressions for the coupling tensor in the random–phase approximation (RPA), neglecting the Breit interaction. At the RPA level

$$|0\rangle = |\text{DHF}\rangle \equiv \text{Dirac–Hartree–Fock ground state}$$

and

$$h = \{h_2\}.$$

In this approximation, all one–electron operators, including κ, may be expanded in the second–quantized form as

$$\kappa = \sum_{m\alpha}(\kappa_{m\alpha}q^+_{m\alpha} + \kappa^*_{m\alpha}q_{m\alpha}), \tag{32}$$

where κ is referred to κ_{Ni} or $\kappa_{N'j}$ and

$$\kappa_{m\alpha} = \langle m|\kappa|\alpha\rangle. \tag{33}$$

We can show that

$$\langle\langle\kappa_{Ni};\kappa_{N'j}\rangle\rangle_E = (\tilde{\kappa}^*_{Ni}, -\tilde{\kappa}_{Ni})\begin{pmatrix} E1 - A & -B^* \\ -B & -E1 - A^* \end{pmatrix}^{-1}\begin{pmatrix} \kappa_{N'j} \\ -\kappa^*_{N'j} \end{pmatrix}, \tag{34}$$

where

$$A_{m\alpha,n\beta} = \delta_{mn}\delta_{\alpha\beta}(\varepsilon_m - \varepsilon_\alpha) + \langle m\beta||\alpha n\rangle \tag{35}$$

and

$$B_{m\alpha,n\beta} = \langle \beta\alpha \| mn \rangle. \tag{36}$$

We use here the conventional notation

$$\langle m\beta \| \alpha n \rangle = \langle m\beta | \alpha n \rangle - \langle m\beta | n\alpha \rangle \tag{37}$$

and

$$\langle ij|kl \rangle = \int \psi_i^*(1)\psi_j^*(2) \frac{e^2}{4\pi\varepsilon_0 r_{12}} \psi_k(1)\psi_l(2)\, d\tau_1 d\tau_2, \tag{38}$$

which means that, in the two–component notation,

$$|\psi_k\rangle = \begin{pmatrix} |\psi_k^L\rangle \\ |\psi_k^S\rangle \end{pmatrix} \tag{39}$$

and

$$\langle ij|kl \rangle = \sum_{T,T'=L,S} \langle \psi_i^T \psi_j^{T'} | \psi_k^T \psi_l^{T'} \rangle. \tag{40}$$

Each $\langle \psi_i^T \psi_j^{T'} | \psi_k^T \psi_l^{T'} \rangle$ integral itself is the sum of four component integrals.

Even though formally Eq. (34) looks very much like the nonrelativistic expression for spin–spin coupling,[18,19] we point out two main differences. In the first place, the A and B matrices are complex and their elements are integrals of DHF orbitals. Each term is composed of up to 16 integrals. Second, we cannot express the operator κ in terms of the usual singlet or triplet tensor operators and, thus, we have only one type of matrix A and B.

The next step is to determine the expressions for the matrix elements of operator κ. The MOs are linear combinations of AOs $\{|r\rangle\}$:

$$|m\rangle = \sum_r C_{rm} |r\rangle, \tag{41}$$

where

$$|r\rangle = \begin{pmatrix} |r^L\rangle \\ |r^S\rangle \end{pmatrix} \tag{42}$$

and $|r^L\rangle$ and $|r^S\rangle$ are two–component spinors. Hence, the matrix elements of κ are

$$\langle m|\kappa|\alpha\rangle = \sum_{rs} C_{rm}^* C_{s\alpha} \langle r|\kappa|s\rangle. \tag{43}$$

From Eq. (29) we see that κ is proportional to α. The effect of the α–matrix on a four–component wave function is to interchange the large and small components:

$$\alpha_j|r\rangle = \begin{pmatrix} 0 & \sigma_j \\ \sigma_j & 0 \end{pmatrix} \begin{pmatrix} |r^L\rangle \\ |r^S\rangle \end{pmatrix} = \sigma_j \begin{pmatrix} |r^S\rangle \\ |r^L\rangle \end{pmatrix}, \quad j \in x, y, z. \tag{44}$$

We can see from Eq. (44) that κ combines only large and small components of the atomic orbital basis set. Thus, we obtain

$$\langle r|\kappa_{Ni}|s\rangle = -\sum_{jk} \varepsilon_{ijk} \left[\langle r^L | \frac{r_{Nk}\sigma_j}{r_N^3} | s^S \rangle + \langle r^S | \frac{r_{Nk}\sigma_j}{r_N^3} | s^L \rangle \right]. \tag{45}$$

As the large and small components of an atomic orbital have opposite parity (symmetry for inversion), one may thus argue that the singlet–triplet splitting is replaced by certain parity constraints in the relativistic case.

The Ramsey's nonrelativistic expressions[20] can be obtained in the limit $c \to \infty$. In this limit we can derive the following relationship between the large and small components of the wave function from Eq. (22):

$$\psi^S = (2m_e c)^{-1}\vec{\sigma} \cdot \vec{\pi}\psi^L. \tag{46}$$

Equation (22) can be written in terms of only the large components as

$$[(2m_e)^{-1}(\vec{\sigma} \cdot \vec{\pi})^2 + V]\psi^L = \varepsilon\psi^L. \tag{47}$$

The large component of the relativistic wave function corresponds to the nonrelativistic wave function. Using the Pauli spin matrices we get

$$(\vec{\sigma} \cdot \vec{\pi})^2 = \vec{\pi}^2 + i\vec{\sigma} \cdot \vec{\pi} \times \vec{\pi}, \tag{48}$$

where

$$\vec{\pi} = -i\hbar\vec{\nabla} + e\vec{A}_N + e\vec{A}_{N'} \tag{49}$$

and

$$\vec{\pi} \times \vec{\pi} = -i\hbar e(\vec{\nabla} \times \vec{A}_N + \vec{\nabla} \times \vec{A}_{N'} + \vec{A}_N \times \vec{\nabla} + \vec{A}_{N'} \times \vec{\nabla}). \tag{50}$$

We have the well-known relationship between the Pauli spin matrix $\vec{\sigma}$ and the electron spin operator $\vec{s}$:

$$\vec{\sigma} = 2\vec{s}. \tag{51}$$

From a second-order perturbation theory, the linear terms in $\vec{A}_N$ or $\vec{A}_{N'}$ produce the FC (Fermi contact), PSO (paramagnetic spin-orbit), and SD (spin-dipolar) terms.[21] The fourth contribution, DSO (diamagnetic spin-orbit) term, arises from the quadratic term in $\vec{A}_N$ and $\vec{A}_{N'}$ using a first-order perturbation theory.

2.2 Density Functional Theory for Spin–Spin Couplings. — Approximate density functional theory (DFT) has over the past decade emerged as a tangible and versatile computational method.[22] Malkin *et al.*[2,23] presented a DFT-based method for spin-spin couplings and obtained a good computational result.

The basic notion in DFT, that the energy of an electronic system can be expressed in terms of its electron density, is almost as old as quantum mechanics and dates back to the early work by Thomas,[24] Fermi,[25] Dirac,[26] and Wigner.[27] In 1951 Slater[28] proposed the X_α method, in which the exchange potential is represented by a function which is proportional to 1/3 power of the electron density. The Thomas–Fermi method and the X_α sheme[28,29] were considered as useful models based on the notion that the energy of an electronic system can be expressed in terms of its electron density. A formal proof of this notion came in 1964 when it was shown by Hohenberg and Kohn[30] that the ground-state energy of an electronic system is uniquely defined by its electron density, although the exact functional dependence of the energy on density remains unknown. Of further importance was the derivation by

Kohn and Sham[31] of a set of one–electron equations from which one in principle could obtain the exact electron density and thus the total energy. The equations by Kohn and Sham have served as a starting point for new approximate DF methods.

2.2.1 *Method.* — By definition the nuclear spin–spin coupling tensor $J_{MN,uv}$ is the second derivative of the total energy of the system with respect to the spins of the nuclei M and N ($M \neq N$):[32]

$$J_{MN,uv} = h^{-1}\left(\frac{\partial^2 E(I_{M,u}, I_{N,v})}{\partial I_{M,u}\partial I_{N,v}}\right)_{I_{M,u}=I_{N,v}=0}, \quad u, v \in x, y, z. \tag{52}$$

The nuclear spin–spin coupling tensor $J_{MN,uv}$ is connected to the reduced coupling tensor $K_{MN,uv}$ by the formula

$$J_{MN,uv} = (\hbar/2\pi)(\mu_0/4\pi)\gamma_M\gamma_N K_{MN,uv}. \tag{53}$$

The reduced coupling tensor $K_{MN,uv}$ is independent of the magnetogyric rations γ_M and γ_N (which could be positive or negative) and depends on the electronic structure of the system only. In many cases the nuclear spin–spin coupling constant J_{MN}, which is the orientationally averaged value of the nuclear spin–spin coupling tensor $J_{MN,uv}$,

$$J_{MN} = \frac{1}{3}(J_{MN,xx} + J_{MN,yy} + J_{MN,zz}) \tag{54}$$

(and an analogous formula for the reduced coupling constant K_{MN}) is the subject of interest.

Malkin *et al.* made a DFT calculation for four contributions to the nuclear spin–spin coupling constants: FC, PSO, SD, and DSO contributions. The FC term is usually the most important of these contributions and the most sentive to geometry changes.[33] They used the finite perturbation theory (FPT) approach to calculate the FC term:

$$K_{MN}^{\rm FC} = \left(\frac{16\pi}{3}\right)\left(\frac{g_e}{2}\right)\mu_B\left(\frac{\partial}{\partial\mu_{M,z}}\left\langle\Psi(\mu_{M,z})\middle|\sum_i\delta(\vec{r}_{iN})s_{iz}\middle|\Psi(\mu_{M,z})\right\rangle\right)_{\mu_{M,z}=0}, \tag{55}$$

where $\vec{r}_{iN} = \vec{r}_i - \vec{R}_N$ is the displacement of electron i relative to nucleus N and g_e is the g–value for the free electron (about 2.0). The matrix elements of the one–electron Hamiltonian for AOs become

$$H_{\nu\sigma}^{\alpha}(\mu_M) = H_{\nu\sigma}(0) + \lambda\langle\chi_\nu|\delta(\vec{r}_M)|\chi_\sigma\rangle \tag{56}$$

for spin α and

$$H_{\nu\sigma}^{\beta}(\mu_M) = H_{\nu\sigma}(0) - \lambda\langle\chi_\nu|\delta(\vec{r}_M)|\chi_\sigma\rangle \tag{57}$$

for spin β, where λ is a perturbation parameter. The unperturbed matrix elements $H_{\nu\sigma}(0)$ are independent of the spin orientation. The final expression for the reduce coupling constant is

$$K_{MN}^{\rm FC} = \left(\frac{8\pi}{3}\right)^2\left(\frac{g_e}{2}\right)^2\mu_B^2\left(\frac{\mu_0}{4\pi}\right)\lambda^{-1}\sum_{\nu\sigma}(P_{\nu\sigma}^{\alpha} - P_{\nu\sigma}^{\beta})\langle\chi_\nu|\delta(\vec{r}_N)|\chi_\sigma\rangle, \tag{58}$$

where $(P^{\alpha}_{\nu\sigma} - P^{\beta}_{\nu\sigma})$ are equal to the spin–density matrix elements. Since Eq. (52) for the nuclear spin–spin coupling tensor is symmetrical with respect to interchange of the indices M and N, the results by FPT method should be invariant, in principle, with respect to such an interchange. However, in reality, the FPT results may be sensitive to the choice of the center of the perturbation.

For calculation of the PSO contribution the Rayleigh–Schrödinger perturbation theory based on density functional theory (RS–DFPT) was used. The RS–DFPT approach for calculation of the PSO contribution leads to the well–know sum–over–states (SOS) equations

$$K^{\rm PSO}_{MN,uv} = -8\mu_B^2\left(\frac{\mu_0}{4\pi}\right)\sum_k^{\rm occ}\sum_a^{\rm unocc}\Delta E^{-1}_{k\to a}\langle\psi_k|L_{M,u}r_M^{-3}|\psi_a\rangle\langle\psi_a|L_{N,v}r_N^{-3}|\psi_k\rangle, \tag{59}$$

where $\vec{L}_M$ is the orbital angular momentum operator at the center M, and ψ_k and ψ_a are occupied and unoccupied Kohn–Sham MOs, respectively. The energy denominator $\Delta E_{k\to a}$ is defined by

$$\Delta E_{k\to a} = \varepsilon_a - \varepsilon_k + \Delta E^{xc}_{k\to a}. \tag{60}$$

Malkin *et al.* used an approximation for $\Delta E^{xc}_{k\to a}$ as

$$\Delta E^{xc}_{k\to a} = \frac{1}{3}C_x\int\left[\frac{1}{2}\rho(\vec{r})\right]^{-2/3}\rho_k(\vec{r})\rho_a(\vec{r})d\tau, \tag{61}$$

$$C_x = \frac{3}{2}\left(\frac{3}{4\pi}\right)^{1/3}, \tag{62}$$

$$\rho_j = \psi_j\psi_j, \quad \rho = 2\sum_{j=1}^{\rm occ}\rho_j. \tag{63}$$

The calculation of the PSO term using the RS–DFPT approach is very quick and yielded good agreement with post–HF results.

The calculation of the DSO contribution is straightforward in the framework of a DFT code. They used a numerical integration procedure for calculation of the following one electron integral ($M \neq N$),

$$K^{\rm DSO}_{MN} = \frac{2}{3}\frac{e^2}{m_e}\left(\frac{\mu_0}{4\pi}\right)\int\rho(\vec{r})[\vec{r}_M\cdot\vec{r}_N/r_M^3r_N^3]d\tau. \tag{64}$$

Since the DSO contribution is relatively small (though it is not negligible[34]), the accuracy of the numerical procedure for calculation of this term is adequate.

The last term is the SD contribution, which usually is also small (at least for long–range coupling). Malkin *et al.* supposed that the importance of the SD contribution is overestimated significantly on the basis of HF calculations. The SD term is usually smaller than the error in the DFT calculations of the FC term. On the other hand, calculation of the SD term would be the most time consuming part of the nuclear spin–spin coupling constant calculations using the DFT approach. Therefore they neglected the SD contribution.

2.2.2 *Results.* — Eriksson *et al.*[35] showed that the DFT calculation of the spin density at nuclei (Fermi contact contribution to the hyperfine constant) other than hydrogen is very sensitive to the type of exchange–correlation functional used (as well as to the quality of basis set and grid for numerical integration). Malkin *et al.*[2] calculated the FC contribution to the

Table 1. All contributions to J_{MN} in comparison with the results of HF and post-HF approaches and experimental data for HF, CO, N_2, and CH_4 molecules. All values are in Hz.

Molecule	Term	RPA[a]	SOPPA[a]	CCSDPPA[a]	MCLR[b]	DFT	Exp.[c]
$^1H^{19}F$	FC	467.3	338.3	329.4		198.1	
	PSO	119.3	195.7	195.7		198.0	
	DSO	−0.1				0.1	
	SD	−12.4	−1.0	−0.6			
	Σ	654.1	532.9	524.4		396.2	529±23
$^{13}C^{17}O$	FC	−8.1	7.3		6.7	13.4	
	PSO	12.2	14.8		13.7	12.4	
	DSO	0.1			0.1	0.1	
	SD	−9.3	−4.0		−4.3		
	Σ	−5.1	18.2		16.1	25.9	16.4±0.1
$^{14}N^{15}N$	FC	−7.7	0.5		−0.2	2.0	
	PSO	0.5	3.3		2.8	2.7	
	DSO				0.0	0.0	
	SD	−8.1	−1.6		−1.9		
	Σ	−15.3	2.2		0.8	4.7	1.8±0.6
$^{13}CH_4$	FC	155.6	126.6	122.1	123.5	122.0	
J_{CH}	PSO	1.4	1.5	1.5	1.5	1.7	
	DSO	0.3			0.3	0.3	
	SD	−0.2	−0.0	−0.0	0.0		
	Σ	157.0	128.3	123.9	125.3	123.9	120.8±0.05
$^{13}CH_4$	FC	−27.6	−16.0	−14.7	−15.7	−11.4	
J_{HH}	PSO	3.6	3.6	3.6	3.6	3.6	
	DSO	−3.5			−3.5	−3.4	
	SD	0.4	0.3	0.3	0.4		
	Σ	−27.1	−15.7	−14.3	−15.3	−11.2	−12.564

[a] Refs. [33,36,37]. [b] Refs. [38,39]. [c] Refs. [33,36,37,38,39].

heteronuclear spin–spin coupling constants for HF, H_2O, NH_3, and CH_4 molecules for different types of exchange–correlation functional, basis sets, grid quality, values of perturbation parameter λ in the FPT approach, and choices of center of the perturbation. They found that the choice of the lighter nucleus (usually hydrogen) as a center of the perturbation in general leads to much better results. The FPT approach was not sensitive to the value of the perturbation parameter λ. A value of this parameter of 0.001 was suitable for all molecules. The quality of the grid was very important for the FC term calculations. However, the extra iteration with an enlarged grid seems to be a reasonable compromise.

The FC contribution strongly depends on the effects of electron correlation and one can expect that this contribution will be sensitive to the proper description of the exchange–correlation effects in DFT as well. Indeed, the dependence of the FC contribution on the type of exchange–correlation functional was dramatic for HF molecule. Fortunately, molecules containing O, N, and especially C and H nuclei were much less sensitive to the type of the exchange–correlation functional employed than the HF molecule. The results were satisfactory, especially for CH_4.

Table 2. J_{MN} for hydrocarbons in comparison with the INDO results and with experimental data. All values are in Hz.

Molecule	J_{MN}	INDO[a]	DFT		Exp.[a]
			F32[b]	F64[b]	
methane	CH	122.9	123.9	121.6	120.78
	HH	−6.1	−11.3	−10.6	−12.564
acetylene	CH	232.7	250.8	249.1	248.7
	C–C–H	2.5	49.0	49.1	49.3
	CC	163.8	185.9	184.3	171.5
	HH	11.0	9.5	9.0	9.5
ethylene	CC	82.1	70.8	61.2	67.6
	CH	156.7	153.9	152.0	156.4
	C–C–H	−11.6	−1.0	−0.7	−2.4
	HH^{gem}	3.2	3.7	4.3	2.5
	HH^{cis}	9.3	10.2	10.1	11.6
	HH^{trans}	25.2	16.9	16.9	19.1
ethane	CH	122.1	123.0	120.9	124.9
	C–C–H	−7.2	−2.8	−2.5	−4.5
	$H\text{–}C\text{–}H^{gem}$	−5.2	−11.5	−10.8	
	$H\text{–}C\text{–}C\text{–}H^{av}$	8.4	7.7	7.7	8.0

[a] The INDO results and experimental data are taken from: refs. [40,41] for methane; ref. [40] for acetylene, ethylene, ethane.

[b] Grid quality.

The results of the DFT calculation for nuclear spin–spin coupling constants are presented in Table 1. Table 1 shows that the results for HF, CO, N_2, and CH_4 molecules are in very good agreement with the results of post–HF approaches (SOPPA, CCSDPPA,[33,36,37] MCLR[38,39]).

A part of their results of J_{CC}, J_{CH}, and J_{HH} calculations for a set of organic molecules are presented in Table 2. The DFT values are compared with the results of INDO calculations because we do not know the results of post–HF calculations with an adequate basis set of the nuclear spin–spin coupling constants of these molecules (except CH_4[37,39]). The DFT approach presented the results which are more or less uniformly in good agreement with experimental data for all the molecules and all the couplings (even though 2, 3, 4, and 5 bonds). In general, the DFT approach led to less accurate results in comparison with post–HF ones, but it has a distinct advantage in the possibility of dealing with large molecules containing more than 40 – 50 atoms and more than 600 – 700 atomic orbitals. The general comparison between J_{CC}, J_{CH}, and J_{HH} constants calculated with the DFT approach for molecules and experimental data are presented visually in Figure 1 for the 64 radial points. Results for the 32 radial points were very slightly different. It is noteworthy that the DFT results seem to be free from the triplet instability which often happens in molecules having multiple bonds (*e.g.*, C_2H_2 and C_2H_4).

2.3 Vicinal Proton–Proton Coupling Constants. — San-Fabian *et al.*[3] carried out an SCF *ab initio* study of the angular dependence and substituent effects upon the vicinal proton–proton coupling constants $^3J_{HH}$ for the molecules CH_3CH_3, CH_2FCH_3, and

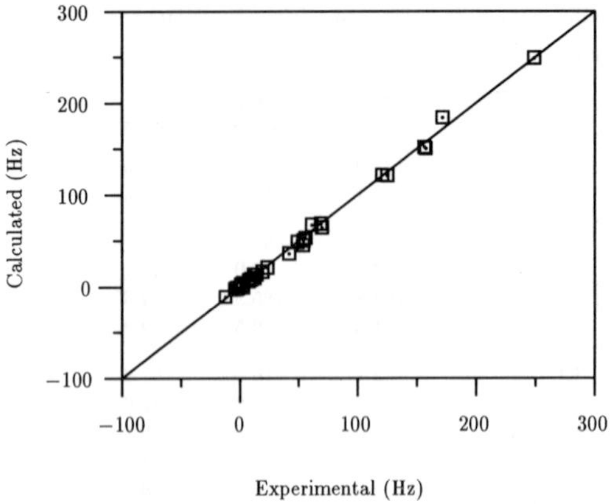

Figure 1. Comparison between J_{MN} calculated with F64 grid and experimental data.

CHF_2CH_3. They computed the four contributions to $^3J_{HH}$, *i.e.*, J^{FC}, J^{SD}, J^{PSO}, and J^{DSO}, using the STO–3G, 6–31G, 6–31G*, and 6–31G** basis sets. The coupling constant $^3J_{HH}$ is the most useful coupling encountered in ^{1}H NMR spectroscopy. The widespread application of this coupling to stereochemistry rests on its strong dependence upon the dihedral angle ϕ between the coupled protons. The second important factor that determines the magnitude of the $^3J_{HH}$ coupling is the effect of substituents attached to the HCCH fragment. An extended Karplus equation including the effect upon $^3J_{HH}$ of both main factors, angle ϕ and substituent electronegativity, has been proposed in their previous paper.[42]

The determination of an accurate equation describing the argular dependence of $^3J_{HH}$ is a cumbersome process that needs to be based on theoretical calculations for ethane derivatives. The semiempirical methods of calculation of J couplings, INDO–FPT[43,44] and EHT–SOS,[45,46] reproduce qualitatively a host of observed trends in the $^3J_{HH}$ coupling due to substituents and stereochemical effects. Both economical methods have been used commonly to calculate the Fermi contribution J^{FC} to $^3J_{HH}$ in ethane derivatives and large molecules. On the contrary, the expensive *ab initio* calculations of $^3J_{HH}$ are scarce and the work reported by San-Fabian *et al.* was the first *ab initio* study of the angular dependence of $^3J_{HH}$ in ethane derivatives.

2.3.1 *Methods and Results.* — The angular dependence of the $^3J_{HH}$ coupling is represented in the substituent effect model as a truncated Fourier series in the torsion angle ϕ of the form

$$^3J_{HH}(\phi) = C_0 + C_1 \cos\phi + C_2 \cos 2\phi + C_3 \cos 3\phi + S_1 \sin\phi + S_2 \sin 2\phi, \qquad (65)$$

which reduces to the Karplus equation

$$^3J_{HH}(\phi) = C_0 + C_1 \cos\phi + C_2 \cos 2\phi, \qquad (66)$$

in the case of the ethane molecule if the term in $\cos 3\phi$ is neglected ($C_3 \cong 0$). The molecular symmetry implies that $S_1 \equiv S_2 \equiv 0$ for ethane.

Table 3. Calculated Fourier coefficients C_n in ethane for the indicated basis sets.

Basis set	J			
	C_0	C_1	C_2	C_3
Exp.[a]	8.02	−1.2(4)	7.0(4)	
STO–3G	6.70	−0.35	5.26	0.07
6–31G	10.42	0.35	8.22	0.01
6–31G*	8.57	−0.10	7.39	−0.05
6–31G**	8.71	−0.05	7.36	−0.06

[a] Ref. [47].

Table 4. Calculated Fourier coefficients ΔK_n in fluoroethane for the indicated basis sets.

Basis set	ΔK_n					
	ΔC_0	ΔC_1	ΔC_2	ΔC_3	ΔS_1	ΔS_2
STO–3G	−0.61	−0.35	−0.57	−0.06	−0.11	0.80
6–31G	−1.57	−0.30	−1.49	−0.13	−0.08	2.46
6–31G*	−1.60	−0.34	−1.50	−0.09	−0.11	2.24
6–31G**	−1.56	−0.30	−1.50	−0.09	−0.08	2.24

The dependence of the $^3J_{HH}$ coupling upon the substituents is embodied into the coefficients K_n (C_n and S_n) in Eq. (65). The individual effect $\Delta K_n^{F_1}$ of a substituent F in position 1 upon the K_n coefficient in Eq. (65) for CH_2FCH_3 may be expressed by the difference

$$\Delta K_n^{F_1} = K_n^{F_1} - K_n^0, \tag{67}$$

where $K_n^{F_1}$ and K_n^0 are the Fourier coefficients for fluoroethane and ethane, respectively. The interaction effect $\delta C_n^{F_1F_2}$ between the two geminal substituents upon the C_n coefficient in Eq. (65) for CHF_2CH_3 is given by the deviation from additivity:

$$\delta C_n^{F_1F_2} = C_n^{F_1F_2} - (C_n^0 + \Delta C_n^{F_1} + \Delta C_n^{F_2}), \tag{68}$$

where $C_n^{F_1F_2}$ is the Fourier coefficient for difluoroethane. In the case of the symmetrical molecule CHF_2CH_3 the S_n coefficients are equal to zero.

San-Fabian *et al.* carried out the *ab initio* calculations for the molecules CH_3CH_3, CH_2FCH_3, and CHF_2CH_3 at the SCF level. A part of the results provided by them are presented in Tables 3, 4, and 5. In their original paper the respective contributions (FC, SD, PSO, and DSO) to K_n, ΔK_n, and δK_n are given. They concluded that the major

Table 5. Calculated Fourier coefficients δK_n in difluoroethane for the indicated basis sets.

Basis set	δC			
	C_0	C_1	C_2	C_3
STO–3G	−0.56	−0.04	−0.17	0.04
6–31G	−1.36	0.10	−0.40	0.15
6–31G*	−1.19	0.09	−0.31	0.13
6–31G**	−1.22	0.07	−0.32	0.12

contributions arise from the FC term and the magnitude of the SD contributions is very small and nearly independent of the size of the basis set. They found that the magnitude of the orbital contributions OR (= PSO + DSO) decreases as the size of the basis set increases.

2.4 One–Bond ^{13}C–^{13}C Couplings. — Since first reported by Karplus[48] in 1959, vicinal spin–spin coupling constants ^3J derived from NMR spectra have been used extensively to elucidate molecular structure due to their dependence on molecular dihedral angles, ϕ. These couplings may be homonuclear (*e.g.*, $^3J_{HH}$ and $^3J_{CC}$) or heteronuclear (*e.g.*, $^3J_{CH}$ and $^3J_{CP}$) and apparently respond to ϕ in a similar fashion (*i.e.*, bell–shaped curves with maxima at $\phi = \sim 0$ and $\sim 180°$ and a minimum at $\phi = \sim 90°$). In addition to ^3J, homonuclear and heteronuclear ^2J values also depend on molecular structure and are applied to problems in organic stereochemistry and conformational analysis.[49]

Despite their application in structure determination in a wide range of molecules, the utility of longer–range spin couplings is sometimes compromised in larger compounds in which NMR line broadening can be substantial, thereby complicating or preventing their measurement. This limitation has led to a renewed interest in the dependence of the considerably larger one–bond spin couplings (*e.g.*, $^1J_{CH}$ and $^1J_{CC}$) on molecular structure and conformation. $^1J_{CH}$ and $^1J_{CC}$ depend on the percent *s*–character of the carbons in CH[50] and CC[51] bonds, respectively. For last one year two *ab initio* calculations of $^1J_{CC}$ have been published.

2.4.1 *Theoretical Study of the $^1J_{CC}$ in Bitetrahedrane.* — In the class of the five convex, tetrahedron, cube, dodecahedron, octahedron, and icosahedron, only the first three members have polycarbocyclic cage analogues of formula C_nH_n ($n = 4, 8, 20$). As yet, cubane (C_8H_8) and dodecahedrane ($C_{20}H_{20}$) have been synthesized and geometrically characterized, whereas only the tetra-*tert*-butyl derivative of tetrahedrane (C_4H_4) has been prepared. These molecules have been the subject of intensive experimental and theoretical investigation.

Recently, the synthesis and X-ray structure of bicubane, $C_{16}H_{14}$ (BC), were reported.[52] An amazing feature of this compound is that the central intercage bond is remarkably short, 1.458 Å. As a matter of fact, it turns out to have the same length as the C–C central bond, made from sp^2 orbitals, in 1,3-butadiene (1.463 Å[53]). Ermer *et al.*[54] suggested bitetrahedrane, C_8H_6 (BT), as a candidate for the molecule incorporating the shortest possible C–C bond in a completely saturated hydrocarbon system. The theoretical study by Xie *et al.*[55–57] confirmed that the intercage distance in BT is very short, namely 1.444 Å(1.434 Å) according to the DZ+P (DZ+P MP2) calculations. Another formidable challenge to organic chemists could be the synthesis of the larger member in the series, bidodecahedrane, $C_{40}H_{38}$ (BD). The structures of BT, BC, and BD bipolyhedra are shown in Scheme 1.

Among the various spectroscopic properties, the NMR coupling constants $^nJ_{CC}$ (*i.e.*, $^nJ(^{13}C,^{13}C)$), in particular $^1J_{CC}$, are faithful mirrors of the electronic distribution in the bonds of alkanes and behave as efficient monitors of alternation in the hybridization of the carbon atoms and, complementarily, in the bond length.

Galasso[4] performed the *ab initio* calculation of the $^nJ_{CC}$ for the theoretical equilibrium geometry[55–57] of BT by the equations–of–motion (EOM) method proposed by Shibuya and McKoy.[58,59] The excitation operator was restricted to all the $1p$–$1h$ pairs. They assert that this method incorporates the main portion of electron correlation. All the four contributions

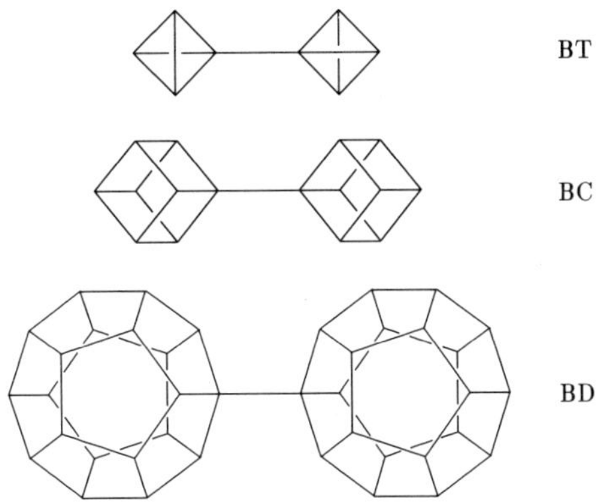

Scheme 1.

to the ${}^{n}J_{CC}$ were considered. The results of the *ab initio* calculation are presented in Table 6. The ultrashort intercage bond of BT was predicted to have an extraordinary large ${}^{1}J_{CC}$, 151 Hz. Thus, it was shown that the theoretical value of BT is as large as that measured for the central bond in 1,3-butadiyne (${}^{1}J_{CC}$ = 154.8 Hz[60]). Furthermore, it lies near the triple–bond region (acetylene, 170.6 Hz[61]). Therfore, the ${}^{1}J_{CC}$ of the bridge bond in BT looks rather similar to that of a bond between *sp*–hybridized carbons. It might be the highest ${}^{1}J_{CC}$ reported for a pure single bond between two saturated, tetravalent carbons.

To get further illustrative insight into the peculiar properties of the central bond in BT, they calculated the central bond ${}^{1}J_{CC}$ for the series of bipolyhedron systems, BT, BC, and BD, by using the self–consistent perturbation theory (SCPT) developed by Blizzard and Santry[62] within the framework of the INDO method.[43] They found the following INDO results for the theoretical ${}^{1}J_{CC}$, the equilibrium central bond distance (r), and the widening of the central bond angle (θ) relative to tetrahedrality ($\Delta\phi = \theta - 109.5°$):

	BT	BC	BD
${}^{1}J_{CC}$ (Hz)	138.0	21.4	26.8
r (Å)	1.444	1.458	1.52
$\Delta\phi$ (°)	35.2	15.8	1.4

The SCPT–INDO semiempirical estimate of the ${}^{1}J_{CC}$ in BT is slightly lower, by 13 Hz, than the EOM *ab initio* value, but still remains unually high for a coupling between two tetravalent carbons. The results indicate that within this series of bipolyhedra no simple correlation can be envisaged between ${}^{1}J_{CC}$ and bond distance or bond angle expressed by $\Delta\phi$. They indicated that application of the empirical relationship between the *s*–character in the carbon hybrid orbitals and ${}^{1}J_{CC}$ values yields unsatisfactory results: 90.6 Hz for BT, 55.2 Hz for BC, and 33.0 Hz for BD. The INDO perturbation theory, instead, gave the reversed order for BC and BD. This striking discrepancy between empirical expectation and theoretical picture unequivocally suggests the extraordinary nature of the intercage bond in the bipolyhedra.

Table 6. *Ab initio* $^nJ_{CC}$ nuclear spin–spin coupling constants of bitetrahedrane (in Hz).

$^1J(C_1-C_{1'})$	FC	151.54	$^1J(C_1-C_2)$	FC	21.74
	DSO	0.38		DSO	0.24
	PSO	−1.39		PSO	−0.29
	SD	0.34		SD	−0.59
	J	150.87		J	21.10
$^1J(C_2-C_3)$	FC	18.93	$^2J(C_1-C_{4'})$	FC	−2.55
	DSO	0.20		DSO	−0.02
	PSO	−0.18		PSO	−0.07
	SD	−0.57		SD	0.10
	J	18.38		J	−2.54
$^3J(C_2-C_{2'})$	FC	2.81	$^3J(C_2-C_{3'})$	FC	−0.02
	DSO	−0.08		DSO	−0.04
	PSO	−0.01		PSO	0.01
	SD	0.13		SD	0.04
	J	2.85		J	−0.01

2.4.2 *The* $^1J_{CC}$ *in Ethylene Glycol.* — Carmichael *et al.*[5] calculated the FC contribution to the $^1J_{CC}$ in conformers of ethylene glycol at the level of sophisticaled QCISD(T) model, recently proposed by Pople *et al.*[63]

One–bond carbon–carbon coupling constants calculated at the SCF level are much greater in magnitude than those observed experimentally, and the second–order Møller–Plesset perturbation theory (MP2) treatments considerably exaggerate the required corrections. A much more thorough treatment of electron correlation is employed in the QCISD(T) method. The effect of amplitudes due to double replacements in the Hartree–Fock self–consistent field wave function is calculated to all orders by the coupled–cluster doubles (CCD) procedure.[64,65] Further, the effect of configurations reached by single replacements in the SCF reference state is also incorpolated in a size–consistent fashion. Lastly, triple excitations are included perturbatively in QCISD(T) method. The magnetic coupling constants are then obtained by finite–field double perturbation theory[32] by performing calculations in the presence of hypothetical nuclear moments (0.002 *au*) placed on the coupled nuclei. The previous study[66] by the QCISD(T) model of electron correlation produced satisfactory results for the indirect nuclear spin–spin couplings in ethane molecule.

Results of the FC contribution to $^1J_{CC}$ in ethanol are presented in Table 7. An analysis of the correlation contributions to the $[5s2p1d|2s]$ calculations for the ethanol conformers is presented in Table 8. The principal importance of the results presented in Table 8 is that correlation components remain roughly constant and independent of conformational changes and that the computed difference at the SCF level is nearly the same as in the more complete correlated QCISD(T) calculations. Table 8 shows that the QCISD(T) results, calculated with the $[5s2p1d|2s]$ basis set, for $^1J_{CC}$ in the two conformers bracket the experimental value (37.6 Hz).

Values of $^1J_{CC}$ in several conformes of ethylene glycol are presented in Table 9. An analysis, similar to that given in Table 8 for ethanol, of the correlation corrections to the SCF value of $^1J_{CC}$ for several conformes of ethylene glycol is presented in Table 10. They

Table 7. Fermi Contact Contribution to the Carbon–Carbon Nuclear Spin–Spin Coupling Constant, $^1J_{CC}$, in Ethanol[a].

	method				
basis	SCF	MP2	CCD	QCISD	QCISD(T)
6–31G	73.1	47.0	48.2	53.6	53.1
6–31G*	61.7	36.2	38.6	43.3	42.3
[4*s*2*p*|2*s*]	79.3	43.2	44.7	52.0	51.5
[4*s*2*p*1*d*|2*s*]	74.6	34.5	38.4	45.9	44.8
[4*s*2*p*1*d*|2*s*1*p*]	73.6	34.1	38.7	46.0	44.7
[5*s*2*p*|2*s*]	71.9	39.0	40.5	46.6	46.1
[5*s*2*p*1*d*|2*s*]	68.8	31.0	34.8	41.8	40.6

[a] Values in hertz from calculations performed at the experimental geometry of the *trans* conformer.[67] The experimental value of $^1J_{CC}$ is 37.6 Hz in aqueous (^2H_2O) solution (present work).

Table 8. Correlation Corrections to the Carbon–Carbon One-Bond Nuclear Spin–Spin Coupling Constant, $^1J_{CC}$, in the *Trans* and *Gauche* Conformers of Ethanol[a].

	conformer	
model (corr)	*trans*	gauche
SCF	68.8	65.3
(doubles)	(−33.9)	(−34.0)
(singles)	(7.0)	(6.6)
(triples)	(−1.1)	(−1.3)
QCISD(T)	40.6	36.6

[a] Contributions (in hertz) from various levels of excitation within the quadratic configuration interaction model. Calculation is performed with the [5*s*2*p*1*d*|2*s*] basis set.

Table 9. Conformational Effects on Calculated Values of $^1J_{CC}$ in Ethylene Glycol[a].

	method	
conformation	SCF	QCISD(T)
tGg′	70.1	40.5
gGg′	67.9	37.7
tGt	75.6	46.2
tGg	71.6	41.5
tTt	77.6	48.7

[a] Fermi contact component calculated with the [5*s*2*p*1*d*|2*s*] basis set for HF/6–31G*-optimized structures. Values reported in hertz.

Table 10. Correlation Corrections to $^1J_{CC}$ in Conformers of Ethylene Glycol[a].

model (corr)	conformer tGg	gGg'	tTt
SCF	71.6	67.9	77.6
(doubles)	(−35.7)	(−35.4)	(−34.8)
(singles)	(6.9)	(6.6)	(7.1)
(triples)	(−1.4)	(−1.5)	(−1.2)
QCISD(T)	41.5	37.6	48.7

[a] Contributions (in hertz) from various levels of excitation within the quadratic configuration interaction model. Calculated for HF/6–31G*-optimized structures with the [5*s*2*p*1*d*|2*s*] basis set.

calculated effects of C–C and C–O torsions on $^1J_{CC}$ in ethylene glycol and found that $^1J_{CC}$ in ethylene glycol fragments depends to an even larger extent on the C–O torsion, reaching a maximum when the hydroxyl proton is anti to a carbon and a minimum in gauche configurations. Thus, in addition to the relative C–C torsion, it is also important to consider the conformational behavior of the C–O bonds. This observation imposes limitations on the use of $^1J_{CC}$ as a structural probe, as in some cases information about C–C and C–O torsions will not be available. In situations where one of these variables is known, $^1J_{CC}$ may be useful to probe the remaining variable.

2.5 The $^1J_{HD}$ in Molecular Hydrogen Complexes. — Since the discovery of the first η^2–H_2 complex, $W(CO)_3(P^iPr_3)_2(\eta^2\text{–}H_2)$ (^{i}Pr = isopropyl), by Kubas *et al.*[68,69] in 1984, the field of molecular hydrogen complexes has expanded rapidly. Neutron and X-ray diffraction studies[68] on this molecule have shown that the hydrogen is bound to the *W* in a sideways. The H_2 bond length, determined from low–temperature neutron diffraction data, is 0.82 Å, *i.e.*, approximately 10% longer than the equilibrium value in "free" H_2 (0.74 Å), indicating a weakening of the H_2 bond.[70,71]

In 1971 Malin and Taube[72] reported the synthesis of the complex $[Os(en)_2H_2]^+$ (en = ethylene diamine) which is now also known to be a molecular hydrogen complex.[73] More recently, Li and Taube[73] have synthesized a series of Os(II) complexes with the general formula $[Os(NH_3)_4L^z(\eta^2\text{–}H_2)]^{(z+2)+}$ where L^z represents a range of ligands. An interesting feature of these complexes is the considerable variation in the hydrogen–deuterium nuclear spin–spin coupling constants $^1J_{HD}$ with the ligand L^z. The observed coupling constant ranges from about 20 Hz when the trans ligand is acetonitrile (CH_3CN) to 4 Hz when it is acetone ($(CH_3)_2CO$). This suggests a possible correlation of $^1J_{HD}$ with the π donor character of the trans ligand (weak and strong respectively in the above example). By comparison, in "free" H_2, J_{HD} = 43 Hz.[74] The observed coupling constants are consistent with the notion that in these osmium complexes the bound H_2 is considerably more stretched than in the complex $W(CO)_3(P^iPr_3)_2(\eta^2\text{–}H_2)$ where $^1J_{HD}$ = 34 Hz,[68] commensurate with a slight increase in the H_2 bond length (from 0.74 to 0.82 Å).

Craw *et al.*[6] calculated the FC contributions to the $^1J_{HD}$ in molecular hydrogen complexes of osmium(II) and compared with the experimental values. They used the FPT method with double perturbations, proposed by Kowalewski *et al.*[32] in which the total en-

Table 11. Calculated and Exprimental Values of the Hydrogen–Deuterium Coupling Constant J_{HD} (Hz)[a] for the $[Os(NH_3)_4L^z(\eta^2\text{-}H_2)]^{(z+2)+}$ Complexes.

L^z	J_{HD}(UHF)	J_{HD}(UMP2)	J_{HD}(expt)[b]
$(CH_3)_2CO$	9.3	−2.0	4.0
H_2O	12.2	−4.6	8.1
CH_3COO^-	1.1	8.6	10.0
Cl^-	3.1	6.4	10.2
H^-	2.7	20.9	
C_5H_5N	18.8	−5.8	19.6
CH_3CN	12.0	−2.9	20.3
H_2^c	54.5	44.6	43.0

[a] Nuclear g values used: 5.5857 (H), 0.8575 (D). [b] Reference [73]. [c] Free H_2 value from ref [74].

ergy is calculated at the unrestricted Hartree–Fock (UHF) level and MP2 level with UHF spin orbitals (UMP2). The results are summarized in Table 11. The underlying assumption in these calculation is that the OR (=PSO+DSO) and SD contributions to $^1J_{HD}$ are likely to be small in comparison with the FC term. Unfortunately this does not seem to be the case. On average, the UHF and UMP2 results are in error by ~6 and ~10 Hz, respectively, *i.e.*, by an order of magnitude. Accordingly, the correlation between theory and experiment is very tenuous. On the other hand, these computations demonstrate that electron correlation is likely to be an important factor in the accurate prediction of the coupling constant in these complexes and that $^1J_{HD}$ is very sensitive to the nature of the complex itself and not likely to depend in a simple way on the H–H distance. This follows from the observation that the Fermi contact contribution to the coupling constant calculated either at the UHF or UMP2 level does not correlate convincingly with the H–H distance, whereas we find that such a treatment appears to be adequate in the case of free H_2 molecule. There is also a strong possibility that in the case of the complexes the level of correlation (UMP2) used is insufficient for the calculation of the coupling constant.

2.6 $^1J_{CH}$ and $^2J_{HH}$ surfaces for CH_4. — Of the two molecular parameters measured in NMR experiments, the chemical shift and the coupling constant, it is the former which is generally considered to be the more important. This is because measurements show that coupling constants are extremely sensitive to molecular structure. Although *ab initio* calculations of coupling constants started to be carried out soon after the first measured values were obtained, the earlier calculations produced results which were less than satisfactory. It is now known that good agreement between calculated nuclear spin–spin coupling constants and their measured values can be achieved only when electronic wave functions incorporating correlation are used, when all four contributions to the coupling — FC, SD, PSO, and DSO — are included, and when the effects of nuclear motion are taken into account.

Nuclear motion must be taken into account because spin–spin coupling constants vary with rovibrational state, and this leads both to the temperature dependence of and the effect of isotropic substitution on the coupling. These are well known experimental phenomena.[75,76] It is usual to adopt the Born–Oppenheimer approximation and to describe the resulting spin–

spin coupling surface (the variation of the coupling with molecular geometry) as a Taylor series expansion in terms of nuclear displacement coordinates. Following this, averages over the nuclear motion are obtained by the use of standard vibration–rotation theory. There have been a considerable number of studies in which the dependence of spin–spin coupling on molecular geometry was calculated. However, to date, calculations of the geometry dependence of the coupling which then proceed to obtain vibrational averages have been confined to the diatomic molecules HD[39,77] and CO and N_2.[38,78] For a polyatomic molecule CH_4, Geertsen *et al.*[7] first presented the calculation of the full coupling surface to be followed by complete rovibrational averaging to any given order.

Geertsen *et al.* used coupled–cluster polarization propagator approximation (CCPPA) to calculate the two surfaces of $^1J_{CH}$ and $^2J_{HH}$ of the methane molecule. To get the coupling surfaces accurate *ab initio* values of the carbon–proton and proton–proton spin–spin coupling constants of the methane molecule were calculated at each of fifty–one distinct geometries. For each geometry the FC, SD, PSO, and DSO contributions were individually calculated. The resulting surfaces were treated as functions of symmetry displacement coordinates. Sixteen independent coefficients are required to describe the carbon–proton coupling surface and twenty–two independent coefficients to describe the proton–proton coupling serface to second order in the symmetry coordinates. Values for these coefficients were reported.[7] The principal features of the results are the dominance of the FC contributions to the two couplings at all geometries, the greater importance of bond stretching for the $^1J_{CH}$, the greater importance of angle–bending for the $^2J_{HH}$, and the almost total cancellation of the PSO and DSO parts of the $^2J_{HH}$ at all geometries. They have reported an unexpected evidence that the FC contribution to the $^2J_{HH}$ becomes more positive as the distance between the coupled protons increases.

The surface data reported are sufficient for calculating the coupling constants J_{CH}, J_{CD}, J_{CT}, J_{HD}, J_{HT}, and J_{DT} in the various isotopomers of methane for any temperature of experimental interest in NMR. It is also possible to determine the individual contributions to each coupling constant of dynamic processes such as bond stretching, angle–bending, and rotational distortion.

2.7 *Ab initio* Calculation by Pople Santry Expression. — Although NMR parameters, chemical shifts (σ_A) and nuclear spin–spin coupling constants (J_{AB}) for magnetic nuclei A and B, potentially provide a wealth of information about molecular structure and dynamics, the theoretical situation with respect to their accurate calculation is still far from satisfactory. Considerable effort has been devoted to calculating coupling constants, but quantitative prediction of J_{AB} remains problematic. *Ab initio* calculation including electron correlation sometimes presents successful results in reproducing experimental couplings in small molecules, but large molecules must still be treated at a semiempirical level, such as the INDO–MO (intermediate neglect of differential overlap molecular orbital) method.[79,80] In short, methods that work well for small molecules must commonly be compromised with semiempirical or INDO–type approximations when applied to larger molecules. Early attempts to calculate J_{AB} with MO–SOS perturbation expression[46] revealed the notorius sensitivity of these expressions to "cancellation of large terms of opposite sign". This sensitivity led to the development of FPT method[44,81] and coupled Hartree–Fock (CHF) method,[82] and to a near–abandonment of SOS techniques. Edison *et al.*[8] examined the theoretical basis of SOS perturbation methods based on the fundamental Ramsey theory[20] and developed a modified SOS approach that incorporates *ab initio* wave functions.

The Ramsey equation for the FC contribution to the J_{AB} between nuclei A and B is given by

$$J_{AB} = -\left(\frac{\hbar}{3\pi}\right)\left(\frac{16\pi}{3}\right)^2\left(\frac{\mu_0}{4\pi}\right)^2 \mu_B^2\gamma_A\gamma_B \times \sum_n \left({}^3E_n - {}^1E_0\right)^{-1}\left\langle 0\left|\sum_j \delta(\vec{r}_{jA})\vec{s}_j\right|n\right\rangle\left\langle n\left|\sum_k \delta(\vec{r}_{kB})\vec{s}_k\right|0\right\rangle. \tag{69}$$

The Ramsey equation, Eq. (69), can be simplified by introducing the Hartree–Fock approximation in two distinct theoretical ways, "uncoupled" or "coupled".[83] In the uncoupled Hartree–Fock (UCHF) treatment of Pople and Santry,[46] the unperturbed Hamiltonian $\mathcal{H}^{(0)}$ is replaced by $\mathcal{H}_{\text{HF}}$, the sum of one–electron Fock operators $f(i)$

$$\mathcal{H}^{(0)} = \mathcal{H}_{\text{HF}} = \sum_i f(i) \tag{70}$$

whose exact eigenfunctions are the complete set of determinantal wave functions D_K constructed from unperturbed Hartree–Fock spin orbitals, with corresponding unperturbed eigenvalues E_K equal to the sum of Hartree–Fock orbital energies ε_r of orbitals occupied in D_K. Namely,

$$\mathcal{H}_{\text{HF}} D_K = E_K E_K, \quad E_K = \sum_r \varepsilon_r. \tag{71}$$

In this case the first–order wave function is a linear combination (formally, infinite in number) of Slater determinants built from unperturbed Hartree–Fock spin orbitals. The exact second–order SOS expression for J_{AB} can then be written in the form first derived by Pople and Santry[46]

$$J_{AB} = -\left(\frac{\hbar}{2\pi}\right)\left(\frac{16\pi}{3}\right)^2\left(\frac{\mu_0}{4\pi}\right)^2 \mu_B^2\gamma_A\gamma_B \times \sum_i^{\text{occ}}\sum_a^{\text{unocc}} (\varepsilon_a - \varepsilon_i)^{-1}\langle\psi_i|\delta(\vec{r}_A)|\psi_a\rangle\langle\psi_a|\delta(\vec{r}_B)|\psi_i\rangle, \tag{72}$$

where ψ_i and ψ_a are occupied and unoccupied molecular orbitals with corresponding orbital energies ε_i and ε_a.

The Pople–Santry equation, Eq. (72), has been employed within the framework of the semiempirical molecular orbital theoretic formalism, but the correlation between calculation and experiment is very tenuous. Edison *et al.* tried to employ the Pople–Santry equation in conjunction with *ab initio* restricted Hartree–Fock (RHF) orbital calculation. In their method the RHF orbitals are evaluated by using standard *ab initio* techniques in a nonempirical manner, and the subsequent evaluation of the SOS summation for J_{AB} is also carried out exactly over the full set of occupied and unoccupied MOs. In this approach, each molecular orbital ψ_i is obtained in standard LCAO (linear combination of atomic orbitals) form as a linear combination of basis functions ϕ_μ, with coefficients $C_{\mu i}$

$$\psi_i = \sum_\mu C_{\mu i}\phi_\mu. \tag{73}$$

The contracted–Gaussian type orbitals (CGTO) are used as AOs, which are, in turn, formed from linear combinations of Gaussian primitive orbitals g_s:

$$\phi_\mu = \sum_s d_{s\mu} g_s. \tag{74}$$

Pople and Santry discarded integrals $\langle\phi_\mu|\delta(\vec{r}_A)|\phi_\nu\rangle$ unless $\phi_\mu = \phi_\nu$ and ϕ_μ is the valence s–type orbital centered at nucleus A. In this reatriction, the non–zero integrals are

$$\langle\phi_\mu|\delta(\vec{r}_A)|\phi_\nu\rangle = |s_A(0)|^2. \tag{75}$$

In contrast, Edison *et al.* included the exact contributions of all basis functions to all integrals by evaluating each $\psi_i(A)$ in the form

$$\psi_i(A) = \sum_\mu C_{\mu i}\phi_\mu(A) \tag{76}$$

and each $\phi_\mu(A)$ exactly in terms of its explicit representation (Eq. (74)) in Gaussian primitives.

We cannot expect quantitative agreement between the theoretical value J_{AB}^{theor}, calculated by Eq. (72), and experimental value J_{AB}^{exp} because the Pople–Santry equation neglects many important effects such as an electron correlation effect. However, we may be able to expect a linear relationship between J_{AB}^{theor} and J_{AB}^{exp}. Edison *et al.* sought empirical proportionality coefficients a and b by which the calculated theoretical values J_{AB}^{theor} can be related to experimental data through an equation of the form

$$J_{AB}^{\text{fit}} = aJ_{AB}^{\text{theor}} + b. \tag{77}$$

Edison *et al.* calculated the FC contribution to ${}^nJ_{AB}$ $(A, B = \mathrm{C}, \mathrm{H})$ in 19 molecules by using the Pople–Santry SOS expression, Eq. (72), in conjunction with modest *ab initio* MO basis sets, 3–21G and 6–31G*. They showed that the common approximation of neglecting basis orbitals not centered on the coupled nuclei is a major source of error in previous calculations by the Pople–Santry equation. They found that calculated values of ${}^3J_{\mathrm{HH}}$, ${}^2J_{\mathrm{HH}}$, and ${}^1J_{\mathrm{CC}}$ show high correlation with experiment ($r^2 \simeq 0.96$–0.99; r^2 is a linear regression coefficient). Calculated ${}^nJ_{\mathrm{CH}}$ values were less highly correlated with experiment, probably resulting from neglect of contributions other than the FC term. It was shown that the increased uncertainty of the convergence in SOS summation with increasing basis set makes results from 6–31G* often less correlated with experiment than those from 3–21G. They concluded that vicinal and geminal proton–proton coupling constants can be estimated with about 1 Hz absolute accuracy by applying their method, based on Eqs. (72) and (77).

3 Semiempirical calculations

The number of papers published in one year, based on the use of semiempirical methods, has been decreasing recently. The reviewer introduces here only one paper written by Domalewski *et al.*[9]

In recent years, the analysis of ${}^1J_{\mathrm{CC}}$ has become a new and important tool in the elucidation of the structure of organic compounds.[49] For the purposes of structural elucidation, long–range substitution effects could be valuable, but they are believed, in general, to be

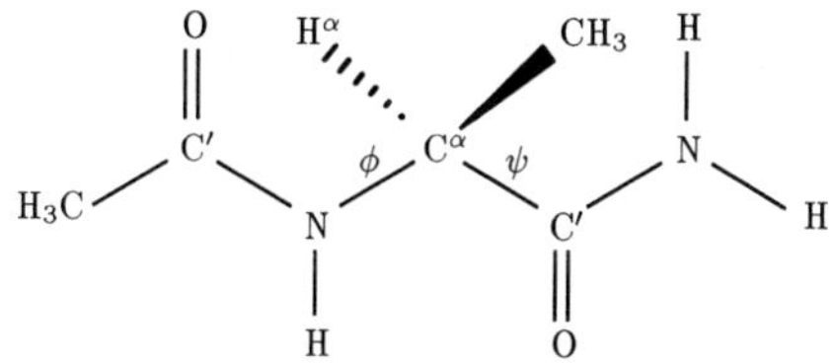

Figure 2. Structure of the model peptide used in the calculations, showing nomenclature used for this study. The angles of rotation are indicated by ϕ and ψ.

negligible when the nearest atom of the coupled carbons and the first atom of the substituent are separated by more than one intervening bond.[84] Domalewski *et al.* studied $^1J_{CC}$ in some substituted nitrobenzenes because of the very strong electron withdrawing character of the nitro group and because the compounds are readily available. The aim of their work was to find systematic changes in the values of $^1J_{CC}$ due to 'close' and 'distant' nitro substituents in the benzene ring with respect to the position of the fixed substituent. They reported the experimentally observed ^{13}C NMR chemical shifts and $^1J_{CC}$ values of 21 substituted nitrobenzenes. They found systematic increases of couplings due to *ortho*, *meta*, and *para* substitution effects. The additivity of these effects was not be confirmed unanimously. They performed calculations of $^1J_{CC}$ by the INDO–SCPT method. The results of calculation were in agreement with the trends observed in the NMR measurements. *Ortho*, *meta*, and *para* substitution effects on $^1J_{CC}$ were significant, but dependent upon the overall structure of the molecule. Practical additivity coefficients obtained are useful for prediction of $^1J_{CC}$ in mono–X-multinitro–substituted benzenes. They reported that *ortho*, *meta*, and *para* NO_2 substitution effects with respect to the position of the substituent X are 10–15 Hz, 1.5–2.5 Hz, and 0.5 Hz, respectively.

4 Conformation Analysis

Edison *et al.*[8] found a good linear relationship between the theoretical coupling constants $J^{\rm theor}$ obtained by using the Pople–Santry SOS expression[46] for the FC contribution, *i.e.*, Eq. (72), within the *ab initio* formalism and the experimentally observed values $J^{\rm exp}$. Subsequently they[10,11] used the same linear scaling approach, Eq. (77), to reproduce the conformational dependence of coupling constants in a model peptide, shown in Figure 2, and to estimate torsion angles in proteins.

Dramatic advances in biomolecular NMR spectroscopy have been brought about by multidimensional NMR methods,[85,86] stable isotropic labeling method,[87] and experimental techniques designed to efficiently correlate atoms along the backbone and side chains of proteins.[88] The ability to measure NMR parameters such as T_1, T_2, and homo– and hetero–nuclear spin–spin coupling constants has been increased. However, although it is now possible to measure many couplings, a satisfactory theoretical basis for their interpretation is lacking.

Edison *et al.*[10] used two separate fitting procedures to find the Fourier series reproducing the conformational dependence of J^{exp}. First, the calculated points (169, at 30° increments) which were obtained by using Eq. (72) at the RHF/3–21G level were fitted to a 2D Fourier series of the two torsion angles ϕ and ψ

$$g(\phi,\psi) = \sum_{i=0}^{M}\sum_{j=0}^{N}\{A_{ij}\cos(i\phi)\cos(j\psi) + B_{ij}\cos(i\phi)\sin(j\psi) + C_{ij}\sin(i\phi)\cos(j\psi) + D_{ij}\sin(i\phi)\sin(j\psi)\} \quad (78)$$

by a least–squares minimization of the differences between the calculated points and $g(\phi,\psi)$. They found that a second–order fit ($M = N = 2$) with 25 coefficients was satisfactory. The absolute magnitudes of the coupling constants obtained from Eq. (72) have little direct resemblance to experimental values. Therefore, they used linear regression analysis to fit J^{calc} ($g(\phi,\psi)$ in its Fourier representation) to J^{exp} (experimental couplings) to give $J^{fit}(\phi,\psi)$, which in turn was used for the construction of the error surfaces.[11]

They presented *ab initio* calculations of the FC term and experimental correlations of six coupling constants, $^3J_{NH,H^\alpha}$, $^1J_{C^\alpha H^\alpha}$, $^2J_{C'H^\alpha}$, $^1J_{C^\alpha N}$, $^2J_{C^\alpha N}$, and $^1J_{C'N}$, in the model peptide as functions of the backbone dihedral angles, ϕ and ψ. Given estimates of experimental uncertainties, they found semiquantitative experimental correlations for $^3J_{NH,H^\alpha}$, $^1J_{C^\alpha N}$, and $^2J_{C^\alpha N}$, qualitative correlations for $^1J_{C^\alpha H^\alpha}$ and $^2J_{C'H^\alpha}$, but no experimental correlations of practical utility for $^1J_{C'N}$, owing to its complex dependence on at least four dihedral angles. Errors in the estimation of experimental dihedral angles from X-ray crystallographic data for proteins, which result from uncertainties in atom–to–atom distances, place substantial limitations on the quantitative reliability of coefficients derived from fitting the calculated coupling constants to such experimental data.

Edison *et al.*[11] used their fitting equations $J^{fit}(\phi,\psi)$, which were determined for the coupling constants ($^3J_{NH,H^\alpha}$, $^1J_{C^\alpha H^\alpha}$, $^2J_{C'H^\alpha}$, $^1J_{C^\alpha N}$, and $^2J_{C^\alpha N}$), to generate error surfaces that can provide estimates of the ϕ and ψ angles in proteins. The error surface of the angles ϕ and ψ is defined by

$$f(\phi,\psi) = \sum_{i=1}^{\text{nexp}} C_i\{J_i^{fit}(\phi,\psi) - J_i^{exp}\}^2, \quad (79)$$

where C_i is a weighting coefficient for coupling type i and nexp is the number of measured couplings for a particular amino acid residue. The most probable set of dihedral angles ϕ and ψ can be found from the minimum of $f(\phi,\psi)$.

They constructed error surfaces for selected residues of the protein staphylococcal nuclease. The residues were chosen to include all those with five experimental couplings, as well as some with four experimental couplings, to demonstrate the relative importance of $^3J_{NH,H^\alpha}$ and $^1J_{C^\alpha H^\alpha}$. For most of the cases, they obtained good agreement between the X-ray structure[89] and the NMR data.

Lit *et al.*[12] used $^3J_{HH}$ values to estimate the changes in conformational equilibria for butanedioic acid as a function of pH, concentration, and nature of the counterion. The $^3J_{HH}$ for butanedioic acid were calculated according to the empirical equation of Haasnoot *et al.*[90] The results supported earlier conclusions that the K_1/K_2 ratio for butanedioic acid is little affected by internal hydrogen bonding in the monoanion.

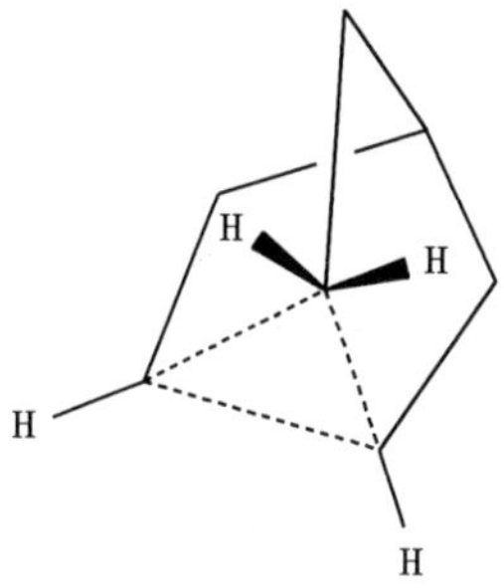

Scheme 2.

Matallana *et al.*[13] studied conformational preferences in hexahydro-2H-azepin-2-ones (caprolactams) by using the calculated and orserved $^3J_{HH}$ values. Calculation of the coupling constants was performed for each of conformations with INDO–FPT method.[91] They found that agreement between calculated and observed coupling constants is acceptable for *tert*-butyl-substituted caprolactams, but not for methyl compound. Some of these compounds showed abnormally small coupling constants for hydrogens near the *tert*-butyl group. Calculations suggested that the effect is related to bond lengthening.

Kelly *et al.*[14] applied the ΔJ equation[92] to polycyclic systems. The ΔJ equation is defined by

$$\Delta J = {}^1J_{CH} \text{ in the alkyl cation} - {}^1J_{CH} \text{ in an appropriate neutral model compound (ketone)},$$

which is a function of the dihedral angle ϕ between the C–H bond and the unoccupied p orbital. This dependence on ϕ is given by

$$\Delta J = A - B\cos^2\phi, \tag{80}$$

where A is the maximum inductive enhancement of J_{CH} (22.5 Hz) and B is the maximum hyperconjugative diminution of J_{CH} (33.1 Hz).[93] They found that for 2-norbornyl cation (scheme 2), exceptionally high values for ΔJ are consistent with a nonclassical σ–bridged structure in superacid.

References

1 G. A. Aucar and J. Oddershede, *Int. J. Quantum Chem.*, 1993, **47**, 425.

2 V. G. Malkin, O. L. Malkina, and D. R. Salahub, *Chem. Phys. Lett.*, 1994, **221**, 91.

3 J. San-Fabian, J. Guilleme, E. Diez, P. Lazzeretti, M. Malagoli, and R. Zanasi, *Chem. Phys. Lett.*, 1993, **206**, 253.

4 V. Galasso, *Chem. Phys.*, 1994, **181**, 363.

5 I. Carmichael, D. M. Chipman, C. A. Podlasek, and A. S. Serianni, *J. Am. Chem. Soc.*, 1993, **115**, 10863.

6 J. S. Craw, G. B. Bacskay, and N. S. Hush, *J. Am. Chem. Soc.*, 1994, **116**, 5937.

7 J. Geertsen, J. Oddershede, W. T. Raynes, and T. L. Marvin, *Mol. Phys.*, 1994, **82**, 29.

8 A. S. Edison, J. L. Markley, and F. Weinhold, *J. Phys. Chem.*, 1993, **97**, 11657.

9 W. Domalewski, L. Stefaniak, and G. A. Webb, *J. Mol. Stract.*, 1993, **295**, 19.

10 A. S. Edison, J. L. Markley, and F. Weinhold, *J. Biomolec. NMR*, 1994, **4**, 519.

11 A. S. Edison, F. Weinhold, W. M. Westler, and J. L. Markley, *J. Biomolec. NMR*, 1994, **4**, 543.

12 E. S. Lit, F. K. Mallon, H. Y. Tsai, and J. D. Roberts, *J. Am. Chem. Soc.*, 1993, **115**, 9563.

13 A. Matallana, A. W. Kruger, and C. A. Kingsbury, *J. Org. Chem.*, 1994, **59**, 3020.

14 D. P. Kelly, K. Aherne, F. Delgado, J. J. Giansiracusa, W. A. Jensen, K. Karavokiros, R. A. Mantello, and M. E. Reum, *J. Am. Chem. Soc.*, 1993, **115**, 12010.

15 J. Oddershede, *Adv. Chem. Phys.*, 1987, **69**, 201.

16 P. Jørgensen and J. Simons, in 'Second Quantization–Based Methods in Quantum Chemistry', Academic Press, New York, 1981, Chapt. 6.

17 J. Sucher, in 'Relativistic, Quantum Electrodynamic and Weak Interaction Effects in Atoms', AIP Conference Proceeding 189, New York, 1989.

18 J. Oddershede and J. R. Sabin, *Int. J. Quantum Chem.*, 1991, **39**, 371.

19 J. Geertsen and J. Oddershede, *Chem. Phys.*, 1984, **90**, 301.

20 N. F. Ramsey, *Phys. Rev.*, 1953, **91**, 303.

21 W. Kutzelnigg, *Theor. Chim. Acta*, 1988, **73**, 173.

22 T. Ziegler, *Chem. Rev.*, 1991, **91**, 651.

23 V. G. Malkin, O. L. Malkina, L. A. Eriksson, and D. R. Salahub, in 'Theoretical and Computational Chemistry, Vol. 1. Density Functional Calculations', *ed. P. Politzer and J. M. Seminario*, Elsevier, Amsterdam, 1994.

24 L. H. Thomas, *Proc. Cambridge Philos. Soc.*, 1927, **23**, 542.

25 E. Fermi, *Z. Phys.*, 1928, **48**, 73.

26 P. A. M. Dirac, *Cambridge Philos. Soc.*, 1930, **26**, 376.

27 E. P. Wigner, *Phys. Rev.*, 1934, **46**, 1002.

28 J. C. Slater, *Phys. Rev.*, 1951, **81**, 385.

29 J. C. Slater, in 'The Self-Consistent Field for Moleculaes and Solids: Quantum Theory of Molecules and Solids', McGraw Hill, New York, 1974, Vol. 4.

30 P. Hohenberg and W. Kohn, *Phys. Rev.*, 1964, **136**, B864.

31 W. Kohn and L. J. Sham, *Phys. Rev.*, 1965, **140**, A1133.

32 J. Kowalewski, A. Laaksonen, B. Roos, and P. Siegbahn, *J. Chem. Phys.*, 1979, **71**, 2896.

33 J. Geertsen, J. Oddershede, and G. E. Scuseria, *J. Chem. Phys.*, 1987, **87**, 2138.

34 G. E. Scuseria, *Chem. Phys.*, 1986, **107**, 417.

35 L. A. Eriksson, V. G. Malkin, O. L. Malkina, and D. R. Salahub, *J. Chem. Phys.*, 1993, **99**, 9756.

36 J. Geertsen, J. Oddershede, and G. E. Scuseria, *Int. J. Quantum Chem. Quantum Chem. Symp.*, 1987, **21**, 475.

37 J. Geertsen, J. Oddershede, W. T. Raynes, and G. E. Scuseria, *J. Magn. Reson.*, 1991, **93**, 458.

38 O. Vahtras, H. Ågren, P. Jørgensen, T. Helgaker, and H. J. Aa. Jensen, *Chem. Phys. Lett.*, 1993, **209**, 201.

39 O. Vahtras, H. Ågren, P. Jørgensen, H. J. Aa. Jensen, S. B. Padkjær, and T. Helgaker, *J. Chem. Phys.*, 1992, **96**, 6120.

40 D. L. Beveridge, in 'Semiempirical Methods of Electronic Structure Calculation', *ed. G. A. Segel*, Plenum Press, New York, 1977, p. 163.

41 Landolt-Börnstein, in 'Numerical Data and Functional Relationships in Science and Technology, Vol. 7. Structure Data of Free Polyatomic Molecules, Group II', *ed. K.-H. Hellwege and A. M. Hellwege*, Springer, Berlin, 1979.

42 E. Diez, J. San-Fabian, J. Guilleme, C. Altona, and L. A. Donders, *Mol. Phys.*, 1989, **68**, 49.

43 J. A. Pople, D. L. Beveridge, and P. A. Dobosh, *J. Chem. Phys.*, 1967, **47**, 2026.

44 J. A. Pople, J. W. McIver, Jr., and N. S. Ostlund, *J. Chem. Phys.*, 1968, **49**, 2960.

45 R. Hoffman, *J. Chem. Phys.*, 1963, **39**, 1397.

46 J. A. Pople and D. P. Santry, *Mol. Phys.*, 1964, **8**, 1.

47 E. Hirota, Y. Endo, S. Saito, and J. L. Dunean, *J. Mol. Spectry.*, 1981, **89**, 285.

48 M. Karplus, *J. Chem. Phys.*, 1959, **30**, 11.

49 J. L. Marshall, in 'Carbon–Carbon and Carbon–Proton NMR Couplings: Applications to Organic Stereochemistry and Conformational Analysis', Verlag Chemie International, Deerfield Beach, FL, 1983.

50 N. Müller and D. E. Pritchard, *J. Chem. Phys.*, 1959, **31**, 768.

51 J. M. Schulman and M. D. Newton, *J. Am. Chem. Soc.*, 1974, **96**, 6295.

52 R. Gilardi, M. Maggini, and P. E. Eaton, *J. Am. Chem. Soc.*, 1988, **110**, 7232.

53 K. Kuchitsu, T. Fukuyama, and T. Morino, *J. Mol. Struct.*, 1968, **1**, 46.

54 O. Ermer, P. Bell, J. Schäfer, and G. Szeimies, *Angew. Chem. Intern. Ed.*, 1989, **28**, 473.

55 Y. Xie and H. F. Schaefer III, *Chem. Phys. Lett.*, 1989, **161**, 516.

56 Y. Xie and H. F. Schaefer III, *Chem. Phys. Lett.*, 1990, **168**, 249.

57 Y. Xie, H. F. Schaefer III, P. Aped, K. Chen, and N. L. Allinger, *Int. J. Quantum Chem.*, 1992, **42**, 953.

58 T. Shibuya and V. McKoy, *Phys. Rev. A*, 1970, **2**, 2208.

59 T. Shibuya, J. Rose, and V. McKoy, *J. Chem. Phys.*, 1973, **58**, 500.

60 K. Kamienska-Trela, *Org. Magn. Reson.*, 1980, **14**, 398.

61 R. M. Lynden-Bell and N. Sheppard, *Proc. Roy. Soc.*, 1962, **A269**, 385.

62 A. C. Blizzard and D. P. Santry, *J. Chem. Phys.*, 1971, **55**, 950.

63 J. A. Pople, M. Head-Gordon, and K. J. Raghavachari, *J. Chem. Phys.*, 1987, **87**, 5968.

64 J. A. Pople, R. Krishnan, H. B. Schlegel, and J. S. Binkley, *Int. J. Quantum Chem.*, 1978, **14**, 545.

65 R. J. Bartlett and G. D. Purvis, *Int. J. Quantum Chem.*, 1978, **14**, 561.

66 I. Carmichael, *J. Phys. Chem.*, 1993, **97**, 1789.

67 M. D. Harmony, V. W. Laurie, R. L. Kuczkowski, R. H. Schwendeman, D. A. Ramsay, F. J. Lovas, W. J. Lafferty, and A. G. Maki, *J. Phys. Chem. Ref. Data*, 1979, **8**, 619.

68 G. J. Kubas, R. R. Ryan, B. I. Swanson, P. J. Vergamini, and H. J. Wasserman, *J. Am. Chem. Soc.*, 1984, **106**, 451.

69 G. J. Kubas, *Acc. Chem. Res.*, 1988, **21**, 120.

70 G. J. Kubas, C. J. Unkefer, B. I. Swanson, and E. Fukushima, *J. Am. Chem. Soc.*, 1986, **108**, 7000.

71 J. Eckert, G. J. Kubas, J. H. Hall, P. J. Hay, and C. M. Boyle, *J. Am. Chem. Soc.*, 1990, **112**, 2324.

72 J. Malin and H. Taube, *Inorg. Chem.*, 1971, **10**, 2403.

73 Z.-W. Li and H. Taube, *J. Am. Chem. Soc.*, 1991, **113**, 8946.

74 T. F. Wimmett, *Phys. Rev.*, 1953, **91**, 476.

75 C. J. Jameson and H.-J. Osten, *J. Am. Chem. Soc.*, 1986, **108**, 2497.

76 B. Bennett, W. T. Raynes, and C. W. Anderson, *Spectrochim. Acta A*, 1989, **45**, 821.

77 J. Oddershede, J. Geertsen, and G. E. Scuseria, *J. Phys. Chem.*, 1988, **92**, 3056.

78 G. E. Scuseria, J. Geertsen, and J. Oddershede, *J. Chem. Phys.*, 1989, **90**, 2338.

79 T. Clark, in 'A Handbook of Computational Chemistry', John Wiley & Sons, New York, 1985.

80 C. N. Cavasotto, C. G. Giribet, M. C. Ruiz de Azua, and R. H. Contreras, *J. Comput. Chem.*, 1991, **12**, 141.

81 J. A. Pople, J. W. McIver, Jr., and N. S. Ostlund, *J. Chem. Phys.*, 1968, **49**, 2965.

82 R. M. Stevens, R. M. Pitzer, and W. N. Lipscomb, *J. Chem. Phys.*, 1963, **38**, 550.

83 P. W. Langhoff, S. T. Epstein, and M. Karplus, *Rev. Mol. Phys.*, 1972, **44**, 602.

84 L. B. Krivdin and G. A. Kalabin, *Prog. Nucl. Magn. Reson. Spectrosc.*, 1989, **21**, 293.

85 R. R. Ernst, G. Bodenhausen, and A. Wokaun, in 'Principles of Nuclear Magnetic Resonance in One and Two Dimensions', Oxford University Press, New York, 1987.

86 C. Griesinger, O. W. Sørensen, and R. R. Ernst, *J. Magn. Reson.*, 1989, **84**, 14.

87 J. L. Markley and M. Kainosho, in 'NMR of Biological Molecules. Practical Approach Series', *ed. G. C. K. Roberts*, Oxford University Press, New York, 1993, p. 101.

88 A. Bax and S. Grzesiek, *Acc. Chem. Res.*, 1993, **26**, 131.

89 P. J. Loll and E. E. Lattman, *Protein Struct. Funct. Genet.*, 1989, **5**, 183.

90 C. A. G. Haasnoot, F. A. A. M. de Leeuw, and C. Altona, *Tetrahedron*, 1980, **36**, 2783.

91 F. A. A. M. de Leeuw, A. A. van Beuzekom, and C. Altona, *J. Comput. Chem.*, 1983, **4**, 438.

92 D. P. Kelly, G. R. Underwood, and P. F. Barron, *J. Am. Chem. Soc.*, 1976, **98**, 3106.

93 D. P. Kelly, G. J. Farquharson, J. J. Giansiracusa, W. A. Jensen, H. M. Hügel, A. P. Porter, I. J. Rainbow, and P. H. Timewell, *J. Am. Chem. Soc.*, 1981, **103**, 3539.

5
Applications of Spin–Spin Coupling Constants

BY KRYSTYNA KAMIEŃSKA-TRELA AND JANEK WÓJCIK

1. Introduction

The number of studies in which spin-spin couplings have been applied is constantly increasing. A particular interest has been observed in the application of spin-spin coupling constants in conformational analysis of biologically important compounds. This concerns not only vicinal proton-proton coupling constants, which have been traditionally used for this purpose, but also couplings across one, two and more bonds between protons and heteronuclei such as ^{13}C, ^{15}N and others. They have become easily available due to the fast development of NMR techniques and computational methods, and the introduction of labelling methods.

A review on the application of coupling constants as restraints in computational structure refinement has been published by Mierke et al.[1] A valuable review on isotope labelling in solution protein assignment and structural analysis has been recently published by LeMaster.[2] The application of multidimensional NMR methods in studies on metal environments in proteins has been reviewed by Canters et al..[3] The NMR results on drug DNA interactions have been collected by Searle,[4] and those on dynamics in nucleic acids by Lane.[5] One chapter in a review by Jaeger et al.[6] has been devoted to NMR application in structural studies of RNA. A review on a conformational study of hydroxymethyl groups in carbohydrates has been published by Bock and Duus.[7]

An extensive review which covers theoretical and physical aspects of spin-spin couplings has been recently published by Contreras and Facelli.[8]

The NMR data including spin-spin coupling constants for azo dyestuffs have been collected by Lycka.[9]

Due to the very rapid relaxation of ^{33}S, coupling constants of this nucleus with other nuclei have been measured only in two cases so far and the corresponding data included in Barbarella's review on ^{33}S NMR.[10] NMR and ESR data of organosulphur compounds including the relevant coupling constants results have been collected by Bassindale and Iley.[11]

It should be noted at this point that the material in the present chapter is arranged as was done previously,[12] i.e. according to (i) increasing number of the nuclei involved, (ii) the number of the bonds separating them. It covers the period from June 1993 to May 1994.

2. Applications of New Methods

A review covering response theory, 1D and 2D FT spectroscopy, extension to higher dimensions, multidimensional FT transformation has been published by Ernst.[13]

In conformational studies on proteins, saccharides and other biologically important compounds the scalar coupling constants play a very important role, and it is especially crucial to know their precise values. However, coupling constants across two-, three- and more bonds are usually small and often obscured by the broad signals in overcrowded spectra, which makes their measurements quite a difficult task. Many 1D, 2D and multidimensional procedures, which allow one to avoid the difficulties encountered, have been already proposed, and still new papers on this topic are appearing.

Several new one-dimensional experiments have been recently proposed by Freeman and his group.[14-18] This includes decoupling experiments, which allow the accurate determination of small coupling constants,[14] measurement of spin couplings by decoupling and reconvolution,[15] and a new type of selective pulse for initiating the measurement and assignment of long-range $J(^{13}C-^{13}C)$ and $J(^{1}H-^{13}C)$ couplings, named after Janus, the Roman god of doors and gates.[16] A new form of correlation spectroscopy based on the application of two simultaneous soft radiofrequency pulses at the chemical-shift frequences of two coupled spins I and S has been designed by this group of authors.[17] The band-selective E-BURP pulses used in this method have proved to be very convenient in practice, and this double-resonance two-spin effect has been called TSETSE. The display is a one-dimensional spectrum containing only responses from the two coupled sites, with the active splitting in an antiphase pattern. Another method designed by Freeman's group[18] employs selective excitation of individual ^{13}C sites under decoupled conditions followed by polarization transfer to protons. The sensitivity of this method was improved by performing eight simultaneous soft-pulse experiments coded according to a Hadamard matrix and later separated by reference to the same matrix. Long-range $^{1}H-^{13}C$ coupling constants in strychnine including those across two bonds have been measured to illustrate its usefullness.Some other one-dimensional experiments have been designed by Adams and Lerner,[19] and Yang and Ye.[20]

A number of new pulse sequences where two- and higher-dimensional techniques have been applied in order to measure accurate spin-spin coupling values have been published under the period reviewed.[21-25]

Soft pulses have been frequently applied in 2D techniques in order to elucidate coupling constants. A novel triple-quantum method that allows one to excite and observe a selected three-spin subsystem coupled to the fourth spin in a molecule has been described by Nicula at al..[26] Simultaneous irradiation at the frequencies of three chemical shifts can be achieved by modulating a radio-frequency carrier by a product of two cosinusoidal audiowaves. The displacements between the two rows in the multiplets in the resulting spectra correspond to the passive coupling constants between one of the active spins and the fourth spin. A detailed discussion of the way of calibration of soft pulses applied to heteronuclei in recording F1-selective 2D correlation spectra has been given by Bernassau.[27] The addition of a bi-selective (BIS) pulse to excite selectively protons coupled to one rare spin-1/2 nucleus prior to the standard HMQC experiment for proton with another rare spin-1/2 nucleus has been proposed by Kupce et al..[28] The improved ^{1}H-detected experiment allows measurement of even very small coupling-constants between rare spin-1/2 nuclei of both types providing that all three spins are coupled. The modifications of the HMQC sequence have been proposed which, together with difference spectroscopy, allow the accurate measurements of long-range $^{1}H-^{31}P$ coupling constants.[29] Bidirectional convolution has appeared to be a very well suited method for improvement of processing the soft-COSY spectra.[30] This new approach allows the stepwise simplification of multiplets.

Wollborn et al.[31] have proposed 2D H,C TOCSY pulse sequence for measuring heteronuclear long-range coupling constants of nonenriched organic molecules. The idea of this method is to use the large chemical-shift dispersion of the C nucleus, which allows to avoid signal superposition.

The INADEQUATE experiment for Li, spin-1, nucleus developed by Guenther and his co-workers has been recently extended to deuteron spin systems of the AX, AA'XX' and AA'MM'X type.[32]

The problem of low sensitivity, which plagued the measurements of the couplings involving ^{13}C and ^{15}N nuclei, has been overcome when, with the introduction of cloning techniques, ^{13}C and ^{15}N enriched proteins have become available. The papers describing new pulse sequences or modifications of the older procedures facilitating measurements of long-range $J(^{13}C-^{13}C)$, $J(^{13}C-^{15}N)$ and $J(^{1}H-^{13}C)$ couplings in enriched samples have been published by Vuister et al.,[33-35] and Madsen et al..[36]

Three- and four-dimensional experiments have already become quite popular in protein studies. Valuable reviews on this topic have been written by Oschkinat et al.,[37] by Bax and Grzesiek,[38] and some new papers published in which this approach has been applied.[39,40] Three-dimensional ^{13}C-^{13}C correlated experiments performed on ^{13}C-enriched oligosaccharides by Chung et al.[41] deserve a special attention.

. The first example of 3D NMR spectroscopy involving ^{1}H,^{13}C and ^{31}P nuclei has been described by Berger and Bast,[42] who adapted the HNCA experiment for a 3D H,C,P correlation. This technique allows one to perform difficult signal assignments in the spectra of organophosphorus compounds, which would otherwise require a performance of several two-dimensional experiments.

A new fast variable-angle spinning (VAS) 2D J-resolved experiment which allows one to measure the indirect spin-spin coupling J between two crystallographically but not magnetically equivalent spins has been described by Wu and Wasylishen.[43,44] A two-bond ^{31}P-^{31}P coupling constant in $Hg(PPh_3)_2(NO_3)_2$, ^{2}J = 250 Hz, has been found by the use of this method. This information is not available from fast MAS NMR spectra or from NMR spectra of isotropic fluids unless a third spin is introduced into the spin system.

A method for determining the relative signs of scalar and dipolar couplings between two spin-1\2 nuclei in a liquid-crystal system has been presented by Tolman and Prestegard.[45] It has been shown by them theoretically that the mixture of dipolar and scalar contributions to the constants which determine the rate of Hartmann-Hahn transfer between two spins in spin-locked experiments and the splitting in directly observed 1D spectra are different. A comparison allows separation and determination of the relative signs of these contribution. A study on the effects of the interaction between spin-3\2 and spin-1\2 nuclei on solid-state MAS NMR spectra of the latter has been performed by Alarcon et al..[46] It concerned such cases where deviations from first order theory were expected. It was shown that, for ^{31}P-$^{63/65}Cu$ and $^{35/37}Cl$-^{119}Sn pairs, the sign of the indirect coupling constants J could be derived.

Finally, it is worthwhile to draw the reader's attention that two-dimensional methods not always lead to the correct results. A comparison has been made by Harris et al.[47] between the classical methods of analysis of 1D spectra and the recently suggested 2D HETCOR method of Beckmann et al.[48] An incorrect absolute $^{2}J(^{13}C-^{31}P)$ value and a wrong sign of this coupling was obtained from 2D analysis performed for *trans*-ethene-1,2-bis(diethyl phosphonate). Some other discrepancies have been also observed between the two set of data obtained by these two methods. It has also been emphasized by Harris et al.[47] that the existence of isotopic chemical shifts cannot be neglected in an analysis of such spectra.

3. One-Bond Couplings to Hydrogen

Unusually large and temperature dependent one-bond proton-proton coupling constants observed in transition metal trihydride complexes L_nMH_3 well deserve a special attention. Recently, *ab initio* calculations have been performed by Jarid et al.[49] in order to explain this unusual phenomenon. In a series of papers Hiller and Harris[50-52] have proposed two theoretical models of the phenomenon which differ in both the predicted sign and the supposed origin of an unusual temperature dependence of these couplings. It must be stressed that the latter are non-magnetic in nature and, as such, they have nothing in common with the ordinary spin-spin couplings except that both lead to identical spectral patterns.

A large $^1J(^1H-^2D)$ coupling constant of 32.2 Hz found in the spectrum of *trans*-[RuH(η2-H_2)(DMPE)$_2$]$^+$ complex, which has an η2-bound H_2 ligand and a σ-bound hydride ligand, has been invoked as evidence for its nonclassical structure.[53]

Similarly to other coupling constants, those across one H-C bond reflect structural changes occurring in an organic molecule, and the number of papers in which they are applied is steadily increasing.

The use of one-bond $^1H-^{13}C$ coupling constants as a probe for protein backbone conformation has been proposed by Vuister et al.[54] The data obtained for free calmodulin and its complex with a 26-amino-acid peptide fragment of myosin light-chain kinase indicate that $^1J(H\alpha-C\alpha)$ couplings depend primarily on ψ and, to a smaller extent, on ϕ, and that they may be useful as a qualitative probe for secondary structure. The $^1J(^1H-^{13}C)$ data have been published for cyclo(Pro-Pro-Gly)$_2$[55] and the $^1J(H_{1'}C_{1'})$ couplings analysed in a study on assignment of anomeric configuration of C-glycopyranosides and C-glycofuranosides.[56]

The one-bond $^1H-^{13}C$ coupling constants have been used to calculate the percent of s-characters of the carbons involved for a series of cycloalkylamine hydrochlorides including 1- and 2-bicyclo[1.1.1]pentyl derivatives. A rough linear relationship was found between these values, and the corresponding pK_a values determined in 50% water-ethanol solution.[57]

The ΔJ equations, which relate the one-bond $^1H-^{13}C$ coupling constant of a group adjacent to a positive charge in a carbocation to the amount of charge and to the dihedral angle between the C-H orbital and the vacant p orbital, have been derived by Kelly and his group and applied recently to some polycyclic systems.[58] An excellent linear correlation between ΔJ and σ^+ was found for 2-arylbicyclo[2.2.1]hept-2-yl cations with no deviation at high electron demand. Kelly's approach has been also applied by Ryu and Shin[59] in their study on the substituted 9-benzonorbornenyl cations.

Recent results obtained by Reich and Dykstra[60] have shown that there is a good correlation between the structure of sulphur and silicon-substituted carbanions and the magnitude of one-bond $^1H-^{13}C$ and $^{13}C-^{29}Si$ couplings. A decrease in $^1J(^1H-^{13}C)$ and $^1J(^{13}C-^{29}Si)$ is observed when the pyramidalized contact ion is formed. A formation of the planar separated ion is accompanied by a dramatic increase in both these couplings. Thus, a small decrease in $^1J(^1H-^{13}C)$ and $^1J(^{13}C-^{29}Si)$ has been observed on going from 2,2-dihydro-3,3-diphenyl-1,3-benzothiasilole, I-H (135 Hz and 51 Hz) to its Li derivative I-Li (114 and 49 Hz), indicating that the contact ions are pyramidal. The solvent-separated ion which is formed upon addition of four equivalents of HMPA to I-Li, I -Li(HMPA)$_4$ evidently has the planar, or nearly planar, structure, since large $^1J(^1H-^{13}C)$ and $^1J(^{13}C-^{29}Si)$ values have been observed for this complex (154 and 95 Hz, respectively).

Extensive studies have been recently performed by Hoffmann et al.[61] on N,N,N′,N′,N‴-

pentamethyldiethylenetriamine-solvated complex of benzylpotassium $C_6H_5CH_2K$. The high $^1J(^1H-^{13}C)$ value of 151.1 Hz found for the CH_2 group in this compound was interpreted in terms of changes occurring in the hybridization of the carbons involved. It has been concluded by the authors that the benzylic carbon atom is approximately $sp^{2.3}$ hybridized, which is in accord with a nearly planar configuration of this fragment observed in the solid-state structures of $PhCH_2M$ (M = K, Rb).

It has been first postulated by Dixon[62] that there should be a correlation between EPR isotropic coupling constants in a radical $R^\cdot$ and NMR isotropic coupling constants in the corresponding molecule RH, since both are dominated by the Fermi contact interaction. Good linear correlations have been already obtained for a series of methanes, silanes, phosphines and aldehydes. Recently, a linear correlation has been found between parallel bridgehead coupling constants observed in the EPR spectra of cubyl and related bridgehead radicals and the one-bond $^1H-^{13}C$ couplings in the NMR spectra of the parent bridgehead alkanes. This relationship:[63]

$$A_{\parallel} = 2.48\ ^1J(^1H-^{13}C) - 147.6; \qquad r = 0.995$$

can be used in the estimation of isotropic coupling constants for the bridgehead radicals, even in those cases where $A_{\perp}$ is unknown.

One-, two- and three-bond proton-carbon coupling constants have been determined for a series of α-mono- and α,α-disubstitued (1,1,3,3-tetramethyl-2-indanylidene)methanes.[64] The $^1J(^1H-^{13}C)$ couplings of the exocyclic C-1′ nucleus with its proton are particularly interesting, since they span the unusually large range 124.3-193.7 Hz. The smallest ^1J values found for trimethylsilyl, dimesitylboryl and trimethylstannyl derivatives (124.3 Hz, 126.3 Hz and 137.6 Hz, respectively) are smaller, by ca. 13 to 20 Hz, than the couplings in analogous simple vinyl compounds. The difference was explained in terms of steric strain which occurs in the compounds studied. The largest ^1J value of 193.7 Hz was found for bromo derivative.

The reactivity order for addition of the phosphinidene complexes $PhPW(CO)_5$ and $MePW(CO)_5$ to exocyclic olefins decreases in the order: methylenecyclopentane > methylenecyclobutane>methylenecyclohexane> methylenecyclopropane. The correlations observed between the relative olefin reactivities and both ^{13}C chemical shifts of the spiro carbon and the $^1J(^1H-^{13}C)$ couplings of the phosphirane CH_2 group of the 1-phosphaspiro[2.n]alkane products have been invoked by the authors as evidence that the observed phosphinidene addition reactivities are product related and therefore solely controlled by thermodynamics.[65]

A set of one-, two-, three- and four-bond $^1H-^{13}C$ coupling constants has been measured for 1-*cis*-methoxy and 1-*trans*-methoxy-1,3-*trans*-butadiene.[66] The results were interpreted in terms of a heavy atom planar conformation in both compounds. They were also used as evidence that the OMe group adopts an *s-syn* conformation in the *trans* compound, and an *s-anti* conformation in the *cis* one. A set of proton-carbon coupling constants including those across one bond has been measured for a series of nitraminobenzenes and -pyridines[67], for some 4-substituted 3-halo-2,6-dimethylpyridine *N*-oxides,[68] for some methyl 3,4-disubstituted benzoates,[69] for a large series of pyrazoles and indazoles,[70,71] for *N*-unsubstituted and *N*-vinyl-4,5,7-trimethyl-4,5,6,7-tetrahydro-pyrrolo[3,2-*c*]pyridines and their nitro derivatives,[72] for a series of 2-thienylethylene derivatives,[73] for 1-substituted 2-thioxo-4(*1H,3H*)-quinazolinones,[74] for some derivatives of phenothiazine,[75] for four fully aromatic isothiazolopyridine isomers,[76] and for a series of pyrazolo[1,5-*a*]pyrimidines.[77] The ^{13}C NMR data obtained in the last paper by Chimichi et al.[77] allowed the authors to assign the correct structure to the final product of the reaction between 3-amino-4-cyanopyrazole and ethyl 3-ethoxymethylene-2,4-dioxopentanoate as being ethyl 3-cyano-7-methylpyrazolo[1,5-

a]pyrimidine-6-oxoacetate and not, as had been formerly claimed, ethyl 3-cyano-7-methyl-6-oxopyrazolo[1,5-*a*][1,3]diazepine-8-carboxylate. Finally, one-bond ^1H-^{13}C coupling constants have been useful in ^{13}C NMR spectral assignment of seven oxidation products of carene.[78]

A review on heterobinuclear and heterotrinuclear transition-metal μ-allenyl complexes which included also $^1J(^1H$-$^{13}C)$ data has been published by Wojcicki.[79] One-bond aromatic ^1H-^{13}C coupling constants have been measured for the tricarbonyl tungsten and chromium complexes of mesitylene, and compared with the values obtained for mesitylene itself. A considerable increase of ca. 10 Hz was interpreted in terms of the increased s-orbital in CH bonds.[80] A one-bond ^1H-^{13}C coupling of 133.6 Hz has been found for the methylene groups of the zirconacyclobutane derivative $(tritox)_2$-$\overline{ZrC(=CH_2)CH_2CH_2}$ (tritox = $(t$-$C_4H_9)_3CO)$, and invoked as evidence of the presence of cyclobutane ring.[81]

^{13}C NMR MAS spectra of urea and thiourea inclusion compounds containing deuteriated guest molecules have been recorded by Aliev et al.[82] at several magnetic fields from 4.7 to 11.7 T under high-power ^1H decoupling condition. Due to the effects of residual dipolar interaction and an indirect spin-spin coupling between ^{13}C and ^2H, the resonances of CD, CD_2 and CD_3 groups appear as triplets, pentets and heptets, respectively. A full analysis of these lineshapes was performed by the authors, which allowed them to determine the scalar spin-spin couplings between carbon and deuterium nuclei for several systems such as $[^2H_{22}]$decane/urea, $[^2H_{11}]$chlorocyclohexane/thiourea and $[^2H_{12}]$cyclohexane/thiourea. The obtained $^1J(^2H$-$^{13}C)$ values of 19 to 23 Hz correspond to the values of $^1J(^1H$-$^{13}C)$ of 124 to 149.5 Hz observed in solutions. These results demonstrate that coupling constants, $^1J(^2H$-$^{13}C)$, can be measured readily from the high-resolution solid-state ^{13}C NMR spectra of the mobile guest molecules in their urea and thiourea inclusion compounds. The data can be subsequently used for calculation of indirect spin-spin couplings providing an indirect method for determining $^1J(^1H$-$^{13}C)$ coupling constants in the solid state.

A multinuclear study has been continued by Jazwinski et al. on mesoionic 1,2,3,4 thiatriazoles[83] and tetrazoles,[84] an interesting group of compounds which can potentially exist in a few mesomeric forms, including chain-ring valence tautomers. The measurements also included one-bond and long-range ^1H-^{15}N couplings.

Basicity and protonation site of a variety of *N*-aminoazoles have been studied by Claramunt et al.[85] The clearest proof that the protonation of *N*-aminoazoles took place on the azole ring came from the observation of $^1J(^1H$-$^{15}N)$ couplings of ca. -104 Hz. These couplings are similar to those observed for protonated pyridines and quinolines, and for protonated 1-methylimidazole, -95 Hz and -101.0 Hz, respectively.

A relationship between $^1J(^1H$-$^{15}N)$ and a second-bond isotope effect at ^1H nuclear shielding has been observed in inclusion complexes with crown ethers and cryptands by Hansen et al.[86]

The effect of pentacoordination on $^1J(^1H$-$^{29}Si)$ coupling constants and ^{29}Si NMR chemical shifts has been studied by Bassindale and Jiang.[87] The coupling values measured for the $R_3SiH{\cdot}Nu$ adducts vary from 210 Hz for Nu = H up to 372 Hz for Nu = 1-methylimidazole. The greatest increase in ^1J was accompanied by the strongest low frequency ^{29}Si NMR chemical shifts. One-bond ^1H-^{29}Si and ^{29}Si-^{29}Si coupling constants have been determined for bromo- and iododisilanes $X_nSi_2H_{6-n}$ (X = Br or I).[88] A relationship between $^1J(^{29}Si$-$^{29}Si)$ and the sum of electronegativities of substituents was observed.

A rather small $^1J(^1H$-$^{29}Si)$ coupling constant of 113.2 Hz has been found in $Cp_2Zr(H)(N$-*t*-

BuSiMe$_2$H) complex. The couplings of similar value from 118.7 to 135.4 Hz have been observed in $Cp_2Zr(X)(N$-*t*-$BuSiMe_2H)$ (X = I, Br, Cl, F). The typical $^1J(^1H-^{29}Si)$ values for silicon hydrides are 160 to 200 Hz. Thus, the observed reduction of the one-bond coupling constant in the zirconium silylamides indicates a significant lengthening and/or rehybridization of the Si-H bond, and has been interpreted by Procopio et al.[89] in terms of the agostic β-Si-H interactions with the metal centre.

The one-bond $^1H-^{125}Te$ coupling constant of 97 to 165 Hz has been measured for TeH^- anion in various solvents. Its value strongly depends on the solvent. The smallest value (97 Hz) has been found in aqueous solution, the largest (160 Hz) in DMF.[90]

4. One-bond Couplings not Involving Hydrogen

One-bond $^6Li-^{13}C$ and $^6Li-^{15}N$ coupling constants have been found to be particularly useful in a study on the aggregation behaviour of some organolithium compounds. Typical multiplet patterns were observed in the ^6Li, ^{13}C and ^{15}N NMR spectra of butyllithium, phenyllithium and lithium diisopropylamide clusters measured in methylal and ethylal. From analysis of these patterns information on the state of aggregation of a given compound could be easily derived.[91]

A splitting of ^{11}B NMR signal observed in the spectrum of $[Ni(C_{12}H_{19})]FB(C_6F_5)_3$, with $^1J(^{11}B-^{19}F)$ = 65 Hz, has provided evidence for the presence of a B-F bond in this compound; $^1J(^{11}B-^{19}F)$ of 57.7 Hz has been found in the $(C_6H_5)_2BF_2^-$ anion.[92]

The largest number of the data presented in this section concerns one-bond $^{13}C-^{13}C$ couplings, which attract an increasing attention of both experimentalists and theoreticians.

The first set of spin-spin coupling constants calculated by the use of the density functional method has been published by Malkin et al.[93] The calculations were performed for acetylene, ethylene, propene, butadiene, benzene and pyridine. The *ab initio* $^1J(^{13}C-^{13}C)$ data have been published for cyclopropene by Barszczewicz et al..[94] The *ab initio* calculations performed by Carmichael et al.[95] at the QCISD(T) level for ethylene glycol have revealed that a $^1J(^{13}C-^{13}C)$ coupling in this compound depends on both the C-C and C-O torsions. The coupling is largest when the hydroxyl substituents are *trans*, and smaller for *gauche* geometries; the coupling reaches a maximum when the hydroxyl proton is *anti* to a carbon and a minimum in *gauche* configurations. Thus, in addition to the relative C-C torsion, one should also take into account the conformational behaviour of the C-O bonds. Studies on one-bond $^1H-^{13}C$ and $^{13}C-^{13}C$ coupling constants have been performed by Zhu and Hu by the use of emiempirical methods.[96, 97]

A new product operator formalism has been developed by Nakai and McDowell[98] which enables one to overcome difficulties encountered in measurements of spin-spin coupling constants in extremely strongly coupled two-spin systems. The outer lines in such AB spectra are very difficult to observe, which often leads to their wrong interpretation (see *infra*). This novel approach allows the time evolution for such systems to be estimated more easily than using the density matrix method, and can be applied to describe, for example, the INADEQUATE experiment. In order to demonstrate its usefulness, the authors used it to determine one-bond and long-range carbon-carbon coupling constants in benzyl bromide, where $\Delta\upsilon/J$ relation was especially unfavourable. The product operator theory has been further extended to ABX systems and applied to describe the INEPT/INADEQUATE and C-relayed H-C INEPT experiments.[99] The same authors have worked up the new refocused INADEQUATE method, which yields in-phase J doublets arising from $^{13}C-^{13}C$ pairs.[100] The designed procedure for obtaining 2D pure absorptive

INADEQUATE spectrum allows one to determine J couplings with the accuracy higher than that in magnitude mode spectra. An advantage of the re-focused method consists also in the fact that it allows one to avoid the accidental cancellation of positive and negative signals in crowded INADEQUATE spectra. Furthermore, Nakai and McDowell have described the spinning spectra of the homonuclear two-spin systems, using the Floquet theory instead of the average Hamiltonian theory.[101] This new approach has allowed them to determine the spin parameters for homonuclear two-spin systems from the magic-angle spinning (MAS), and off-magic-angle spinning (OMAS) spectra. Doubly-$^{13}C_2$ labelled sodium acetate, $^{13}CH_3{}^{13}COONa$, and palmitic acid, $CH_3(CH_2)_{13}{}^{13}CH_2{}^{13}COOH$, were used as model compounds. As a result, $^1J(^{13}C-^{13}C)$ couplings, including their signs (positive ones), were determined for both compounds.

A new DEPT C-C relay sequence for selective detection of CH-C, CH_2-C and CH_3-C connectivities at quaternary carbons with the aid of a spectrum editing technique has been designed by Kawabata and Fukushi.[102] The method allows accurate determination of $^{13}C-^{13}C$ coupling constants and, in order to illustrate its usefulness, it has been applied to measure the couplings in a monoterpene hydrocarbon limonene.

One of the most striking results obtained recently in the field of one-bond $^{13}C-^{13}C$ coupling constants is that obtained by Bocian et al.[103] for a series of mesoionic 3-phenyl-1-thia-2,3,4-triazol-3-ium-5-ylmethanides with various exocyclic groups.

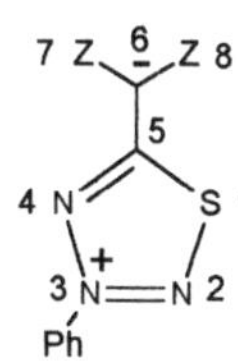

Z= CN, CO_2Et, COMe, COPh

The $^1J(^{13}C-^{13}C)$ coupling values measured across formally single C(5)-C(6), C(6)-C(7) and C(6)-C(8) bonds were unusually large in comparison with $^1J(^{13}C-^{13}C)$ values observed in other compounds. Thus, for example, for Z = CN, $^1J(C_5C_6)$ = 92.7 Hz, $^1J(C_6C_7)$ = 97.2 Hz and $^1J(C_6C_8)$ = 99.1 Hz were found, whereas a typical $^1J(C_{sp3}C{\equiv}N)$ value is of ca. 60 Hz only, and that for C_{sp3}-C_{sp2} bond of 40 Hz. This large increase observed for all the $^{13}C-^{13}C$ couplings in the exocyclic fragment of mesoionic compounds can be interpreted in terms of the increased s character of the central C6 atom, and provides clear evidence for a planar structure around this atom. The latter conclusion is in good agreement with the X-ray results obtained for these compounds.

New examples of a strong $^1J(^{13}C-^{13}C)$ dependence on the electronegativity, E_x, of substituents have been recently published. A coupling of 35.8 Hz has been found for a tetramer of vinyllithium by Dabrowski et al.[104] This result is in agreement with the value published by Bauer et al.[105] (36.3 Hz), who also recorded $^1J(^{13}C-^{13}C)$ of 35 Hz for the dimer.

Substituent effects on $^1J(^{13}C-^{13}C)$ couplings in mono- and disubstituted benzenes have been investigated by Kamieńska-Trela et al.[106] Substituents were varied from Li to F, which resulted in large variations of $^1J(^{13}C-^{13}C)$ values, from 29.5 Hz in phenyl lithium up to 83 Hz in *o*-difluorobenzene. A coupling of 48 Hz only has been reported for $C_6H_5BH_2$ by Hoffmann et al.[61] The $^1J(^{13}C-^{13}C)$ values measured for substituted silyl cyclopropenes also showed a typical ^1J vs. E_x dependence.[107] The coupling across the single bond with a silyl group attached has been found to be significantly smaller (< 3 Hz) than that across the bond bearing an alkyl group (ca. 12 Hz).

A set of 14 $^1J(^{13}C-^{13}C)$ spin-spin coupling constants has been reported by Witanowski and Biedrzycka for 11 five-membered heteroaromatic molecules, including all relevant diazole, triazole, oxazole, oxadiazole, thiazole and thiadiazole systems.[108] The couplings show a clear distinction between carbon-carbon bonds which are formally either double or single. There seems to be no correlation between the couplings and the bond lengths concerned, in spite of some apparent trends.

A strong decrease in $^1J(^{13}C-^{13}C)$ couplings is usually observed upon complexation and, additionally, by analogy to the couplings in free ligands, also the couplings in complexes are governed by the electronegativity of substituent attached to the bond involved. A decrease of a typical value of ca. 23 Hz upon complexation has been observed for $^1J(^{13}C-^{13}C)$ couplings across double bonds in the cuprate complexes of enynes and methyl cinnamate measured by Krause et al.[109]

One-bond $^{13}C-^{13}C$ coupling constants have been measured for (3,4,5,8,9-η^5)tricyclo[5.2.1.0^{2,6}]deca-3,8-dien-5-yl)(η^5-cyclopentadienyl)iron complex.[110] The values obtained for these couplings and for the isotope shifts are comparable to those found for dicyclopentadiene itself.

The ethynyl-bridged complex $W_2(\mu\text{-CCH})(OSiMe_2t\text{-Bu})_5$ reacts in hydrocarbon solvents with xylyl isocyanide, 2,6-$Me_2C_6H_3NC$, to give $W_2(\mu\text{-C(1)C(2)C(3)NHXyl})(OSiMe_2t\text{-Bu})_5(CNXyl)_4$ complex; $^1J(^{13}C-^{13}C)$ of 53 Hz has been found across the C(1)-C(2) bond and that of 36 Hz across the C(2)-C(3) bond.[111]

Removal of two electrons from 1,8-bis[(4-methylphenyl)thio]-9-(2,6-dimethoxyphenyl-1-^{13}C)-10-phenylanthracene has led to a rather unusual (10-C-5) pentacoordinated carbon species. The coupling constants, $^1J(C_{1'}C_{2'})$ of 61.9 Hz and $^1J(C_{1'}C_9)$ of 56.3 Hz, measured for this compound suggest that the carbon C1′ is sp^2 hybridized. The trigonal bipyramidal geometry has been postulated around this carbon centre.[112]

At present, most of the $^{13}C-^{13}C$ coupling studies can be performed using samples with natural abundance of carbon-13 nuclei. These spectra represent AB (or AX) systems and are easy to interpret. However, in many cases the natural ^{13}C level is too low to perform measurements in a reasonable time and ^{13}C enrichment is still required. This concerns in particular many compounds of biological importance.

A sample of totally enriched $^{13}C_{18}$-*meso*-hexestrol, the estrogen receptor ligand (98 at.% of ^{13}C enrichment at every position), has been recently synthesized by Kochanny et al.[113] Its complex ^{13}C NMR spectrum was analyzed by homonuclear decoupling and 1D ^{13}C COSY experiments, and the estimated $^1J(^{13}C-^{13}C)$ values refined via iterative simulation. The authors expect that ^{13}C-enriched samples will be used more frequently in the future in studying receptor-ligand interactions by polarization transfer methods.

The ^{13}C labelling followed by an analysis of the $^{13}C-^{13}C$ coupling patterns has been especially useful in an investigation of the carbon fixation in the phototropic eubacterium *Chloroflexus aurantiacus*.[114] This bacterium does not use any of the known autotrophic CO_2 fixation pathways, and some suggestions have been made that a new cyclic autotrophic pathway in which acetyl-CoA is converted to 3-hydroxypropionate and further to succinate and malate takes place. In order to test this hypothesis the growing cultures of the *C. aurantiacus* have been fed during several generations with 3-hydroxy[1-^{13}C]propionate, [1-^{13}C]acetate, or [2-^{13}C]acetate. Subsequently, a full ^{13}C NMR analysis has been performed for the amino acids and nucleotides isolated from the cells of the bacterium. The data obtained confirm that a novel carbon fixation pathways *via* 3-hydroxypropionate is highly probable.

An extensive use of ^{13}C-^{13}C coupling patterns has been made in a study on the metabolism of [^{13}C]propionate in a wild-type *Escherichia coli* and cells lacking citrate synthase.[115]

A study on 13-enriched saccharides has been continued by Serianni and his group.[116] It includes measurements and analysis of $^1J(^{13}C-^{13}C)$ and $^1J(^1H-^{13}C)$ couplings in ^{13}C-enriched sucrose and some related compounds. The aim of this work was to obtain information on the furanose and glycosidic bond conformations in aqueous solution.

One-bond coupling constants between two carbonyl carbons of 66.9 to 84.0 Hz have been measured for a series of the doubly ^{13}C enriched compounds of a general formula, *p*-$O_2NC_6H_4(CH_2)_2NH$-$^{13}CO^{13}COR$, R = OH, OCH_3, OC_2H_5, $OC_6H_4C_2H_5$ and SC_2H_5.[117] For a compound with R = NHC_3H_7 an evidently incorrect ^1J = 0 Hz has been published. Most probably, the authors overlooked the fact that in this particular compound the two C(O) carbons represented very tightly coupled AB system, and very weak outer lines of this spectrum were difficult to observe. The compounds obtained in this work were further used as the model α-dicarbonyl adducts to study the inactivation of cytochromes P-450 by dichloroacetamide-containing mechanism-based inactivators.

One-bond ^{13}C(O)-^{13}C(O) coupling constants of 73 Hz have been measured for Tl^{3+}/oxalate adducts of the half-molecules of ovotransferrin, found in an avian egg white, and the N-terminal lobe of serotransferrin, a prominent serum protein.[118]

The 1D and 2D INADEQUATE methods have been, as previously, very often applied in order to assign ^{13}C chemical shifts and to elucidate the structure of molecules of medium size. Only a few examples have been included in the present review, with the purpose of showing a variety of structures investigated. Thus, studies have been performed on (*E*)-cyclodec-5-enone,[119] bicyclo[3.1.1]heptane derivatives,[120] a new chiral pentacyclic pinocarvone dimer,[121] the dolabellane, diterpenoid palominol reisolated from the Caribbean gorgonian *Eunicea laciniata*,[122] obscurinervine, the alkaloid isolated from the leaf tissue of *Aspidosperma cruenta*,[123] an unusual cyclization product from the reaction of citreamicin η with sulphene, (1α,3aα,14bβ)-3a,4,14b,15-tetrahydro-14b-hydroxy-11,12-dimethoxy-1,3a-dimethyl-1-[[(methylsulfonyl)oxy]methyl]-18*H*-[1]benzopyrano[3′,2′:5,6]naphth[1,8-*de*][1,2]-(-)-oxathiepino[10,1-*gh*]oxazolo[3,2-*b*]isoquinoline-2(1*H*),8,14,18-tetrone16,16-dioxide,[124] 9,9a-dihydro-2,9a-dimethyl-4-oxo-4H-benz[*f*]indole-3-carboxylic acid Me ester formed from the Friedel-Crafts-based synthesis of benzoylpyrrole calcium channel activators,[125] tricyclohexylmethane,[126] and on poly(vinylchloride).[127]

A review on structure characterization of organometallic complexes by 1D and 2D INADEQUATE techniques has been published by Xu et al.[128]

2-Fluoro- and 4-fluoropyrene and some related compounds dissolved at low temperature in superacids undergo protonation yielding the persistent fluoropyrenium anions. The fluoronium ion character of the resulting delocalized arenium ions has been deduced by Laali and Hansen[129] from the increase in $^1J(^{143}C-^{19}F)$ couplings, deshielding of the fluorine, and reduced deshielding of the fluorine-bearing carbons as compared to position of these carbons in the analogous non-fluorinated pyrenium ions. $^nJ(^{13}C-^{19}F)$ couplings (n = 1-4) have been determined for a series of substituted 2,5-diaryloxazoles and their vinylogues, and for 2-(*ortho*-fluorophenyl)-5-aryloxazoles.[130]

One-bond $^{13}C-^{55}Mn$ coupling constants have been determined by lineshape analysis of the ^{13}CO and $^{13}C(R)$ signals (R = alkyl, acyl or aryl) in σ- and π-organomanganese pentacarbonyl, tetracarbonylphosphine and tricarbonyl complexes. The couplings vary from 29 to 47 Hz for Mn-CO bonds, and from 2 to 28 Hz for Mn-R bonds.[131]

A $^1J(^{13}C-^{59}Co)$ coupling of 117 ± 5 Hz has been determined from the linewidths of the ^{13}C and ^{59}Co resonances of azamacrocyclic compound of cobalt(III) {the tetraphenylborate salt of the six-coordinate [mono(acetone)(5,16-dihydro-16-(*exo*-norborane-2,3-diyl-κC^2)-6,8,15,17-tetramethyldibenzo[*b,i*][1,4,8,11]tetraazadecinato-$\kappa 4N^{5,9,14,18}$)cobalt(III)], and a $^1J(^{13}C-^{103}Rh)$ coupling of 30.4 Hz has been measured for the analogous rhodium complex, hexafluorophosphate salt of the five-coordinate [5,16-dihydro-16-(*exo*-norbornane-2,3-diyl-κC^2)6,8,15,17-tetramethyldibenzo[*b,i*]-[1,4,8,11]tetraazadecinato-$\kappa 4N^{5,9,14,18}$)rhodium(III)].[132]

Two papers by Wrackmeyer et al.[133,134] are devoted to various alkynyltin compounds, which have been characterized by the use of NMR spectroscopy. $^1J(^{13}C-^{119}Sn)$ coupling constants of 1030 to 1175 Hz have been measured for tetra-1-alkynyl compounds, $Sn(C{\equiv}CR)_4$. Significantly larger couplings, up to 1730 Hz, have been found in $Cl_{4-n}Sn(C{\equiv}CR)_n$ compounds reflecting electronegative effect of chlorine. One-bond coupling constants $^1J(^{13}C-^{77}Se)$, $^1J(^{13}C-^{119}Sn)$, $^1J(^{13}C-^{125}Te)$, $^1J(^{77}Se-^{119}Sn)$, and $^1J(^{119}Sn-^{125}Te)$ have been measured for a series of tin (IV) compounds containing ferrocenyl chalcogenate and ferrocenylene dichalcogenate by the same group of authors.[135] The one-bond carbon-tin coupling constants have been measured by Huber et al.[136] for several triorganotin (IV) derivatives of *N*-acetyldipeptides, and by Caruso et al.[137] for the tin complexes of 2,2′,6,6′bipyrimidine [R_2SnCl_2bipym](R = Ph or Bu). The data obtained in these two works were further used in estimation of the C-Sn-C bond angles in the compounds under study.

$^1J(^{13}C-^{199}Hg)_{iso}$ of 1540 Hz ± 2 Hz has been obtained in $K_2Hg(CN)_4$ from ^{13}C and ^{199}Hg MAS NMR spectra of the compound; analysis of the ^{199}Hg NMR spectrum has indicated that this indirect spin-spin interaction is anisotropic.[138] One-bond $^1J(^{13}C-^{205}Tl)$ coupling of 5394 Hz and $^1J(^{13}C-^{203}Tl)$ of 5344 Hz which involved the CN ligand have been observed in the ^{13}C NMR spectrum of *meso*-tetraphenylporhyrinatothallium(III)cyanide at 24°C indicating that the ligand exchange rate in this compound should be small compared with the $^1J(^{13}C-^{203/205}Tl)$ coupling constant.[139]

For the first time, a combination of ^{13}C NMR 2D EXSY and ^{205} Tl 2D EXSY techniques, which rely on very large, i.e. of ca. 8 Kz, $^{13}C-^{205}Tl$ coupling constants, the so-called spy approach, has been applied by Batta et al.[140] to analyze quantitatively the kinetics of cyanide exchange for the thallium (III) cyanide complexes in aqueous solution.

One-bond $^{15}N-^{29}Si$ coupling constants have been measured by Mitzel et al.[141] for benzyl(phenylsilyl)amines, $(PhH_2Si)_x(PhH_2C)_{3-x}N$, and benzyl(p-tolylsilyl)amines, (*para*-$H_3CC_6H_4Si)_x(PhH_2C)_{3-x}N$. A clear relationship was observed between the $^1J(^{15}N-^{29}Si)$ values and the number of silicon atoms attached to nitrogen: ^1J = 7-10 Hz was observed for trisilylamines, ^1J = 13-15 Hz for disilylamines, and ^1J of ca. 21 Hz for monosilylamines. The existence of a linear correlation between $^1J(^{15}N-^{29}Si)$ and the N-Si bond lengths was also suggested for some trisilylamines.

An interesting set of $^1J(^{15}N\text{-}^{29}Si)$ couplings has been published by Herberhold et al.[142] for six-membered cyclic sulfur diimides bearing silyl substituents at the N=S=N fragment, and for their open-chain analogues. The ^{15}N(*amino*)-^{29}Si couplings of 21.1 Hz were found for R(H)NMe$_2$Si(NSN)SiMe$_2$N(H)R compounds with a planar amino nitrogen atom. The $^1J(^{15}N$(*amino*)-$^{29}Si)$ values observed for heterocyclic sulphur diimides, where the amino N atom is a pyramidal one, are significantly smaller, i.e. of 13.1 to 14.6 Hz only. The couplings between the Si nucleus and the nitrogen atom of the N=S=N moiety were not observed. Since rather sensitive techniques were applied, the authors concluded that the couplings in question must be smaller than 1 Hz.

A $^{31}P,^{15}N$ polarization transfer technique has been applied by Gudat[143] in order to measure the one-bond couplings between ^{15}N and ^{31}P across a double bond in substituted *trans*-iminophosphines (R'N=PR); $^1J(^{15}N\text{-}^{31}P)$ of 77.2 to 108 Hz were found depending on the nature of substituents attached. The 2D $^{31}P/^1H_{Me}$ HEED-HETCOR experiment has been performed for dimethyl-(*N*-pyrrolyl)phosphane providing the first experimental proof of $^1J(^{15}N\text{-}^{31}P) > 0$ for a pyrrole-type ^{15}N nucleus.[144] The HEED experiment has been also successfully applied by Wrackmeyer's group[145] in order to measure the triplet pattern in the ^{31}P NMR spectrum arising from the $^{15}N\text{-}^{31}P\text{-}^{15}N$ isotopomer. This was a rather challenging task, since the combined intensity of the triplet amounts only to 1.4 x 10^{-5} compared with that of the parent signal (= 1). The compound chosen was 2-*tert*-butyl-1,3-dimethyl-1,3,2-diazaphospholidine, for which positive $^1J(^{15}N\text{-}^{31}P)$ = 47.8 Hz was found. The same coupling was measured for some other similar compounds, such as 2-chloro-1,3-dimethyl-1,3,2-diazaphospholidine, bis(diethylamino)chlorophosphane and a bis(diethylamino)phosphenium anion (62.3 Hz, 61.7 Hz and 92.2 Hz, respectively).

The largest $^1J(^{15}N\text{-}^{77}Se)$ values of ca. 160 Hz reported so far have been observed in the selenium diimides, RN=Se=NR. The corresponding couplings in aminoselenanes are considerably smaller, i.e. of ca.100 Hz only.[146]

The presence of one-bond $^{15}N\text{-}^{89}Y$ coupling of ca. 14 Hz observed in the ^{15}N NMR spectrum of yttrium (II) isothiocyanate complexes has been invoked as evidence for the bond formation at the anion nitrogen atom.[147]

One-bond coupling constants between ^{15}N and ^{183}W nuclei have been determined for the tungsten complexes of the type W=NPh. The coupling of 102 Hz has been found for W(NPh)Cl(CH$_2$SiMe$_3$)$_3$, and that of 109 Hz for W(NPh)Cl$_4$(OEt$_2$).[148] One-bond $^1J(^{15}N\text{-}^{195}Pt)$ coupling constants of 254 to 329 Hz have been collected for the series $[PtX_n(NH_3)_{4-n}]^{(2-n)+}$ (with n and X(Cl,I) systematically varied) and correlated with $\delta(^{195}Pt)$ and $\delta(^{15}N)$ chemical shifts.[149]

A one-bond $^{19}F\text{-}^{129}Xe$ coupling constant of 5644.2 Hz has been found in xenon difluoride. The ^{129}Xe chemical shift in this compound was very sensitive towards temperature, whereas the coupling was temperature insensitive over the range from 243 to 303 K.[150] The $[PtF_nCl_{n-6}](PPN)_2$ complexes have been characterised by one-bond $^1J(^{19}F\text{-}^{195}Pt)$ coupling constants.[151] The averaged couplings of ca. 1890 Hz have been found for symmetric F-Pt-F axes; those for asymmetric axes F-Pt-Cl are considerably smaller, i.e. of 1296 Hz only. These data reflect the *trans* influence of the ligands.

The ^{29}Si-INEPT-INADEQUATE technique has been applied by Jenkner al.[152] in order to confirm the structure of octadecamethylbicyclo[4.4.0]decasilane - the only bicyclic permethyloligosilane known at present - and its possible structural isomer bi(undecamethylcyclopentasilanyl). It was proved that only the *trans* isomer was formed in the case of decasilane. By use of one- and two-bond couplings the exact assignments of single Si atoms were possible for both compounds. One-bond ^{29}Si-^{29}Si coupling constants observed in solution (ca. 57 Hz) have facilitated the assignment of Si resonances in the two products of the reaction of $(MeSi_3)_3SiLi$ with acetone, $H(Me_3Si)_2Si$-$(Me_3Si)_2Si$-CMe_2SiMe_3 and $(Me_3Si)_3Si$-$(Me_3Si)_2Si$-CMe_2SiMe_3.[153]

One-bond ^{29}Si-^{29}Si coupling constants have been measured by Schenzel and Hassler[154] for a series of variously substituted hexamethyltrisilanes and pentamethyltrisilanes, and compared with SiSi force constants calculated by the use of normal coordinate analysis. The couplings measured varied from 70.5 Hz to 96.8 Hz.

One-bond ^{29}Si-^{183}W coupling constants of ca. 100 Hz have been invoked as evidence for partial silylene character in the chlorosilyl derivatives of mono- and bis(silyl)complexes of tungsten.[155]

The experiments performed by Szalontai et al.[156] have confirmed the presence of two different P atoms (AB system) in the CP MAS spectrum of the strained bis(diphenylphosphino)methane. In the spectra of the more flexible compounds with a linear alkyl chain only one P resonance was observed.

An analysis of the second-order ^{31}P and ^{195}Pt NMR spectra of fluxional η^1- ligated 1,2,4-triphosphacyclopentadienyl platinum complexes of the type $[PtX(R_2C_2P_3)(PEt_3)_2]$, has given one-bond ^{31}P-^{31}P coupling constants of ca. 500 Hz and one-bond ^{31}P-^{195}Pt couplings, which varied from 1860 Hz to almost 3000 Hz. Also two- and three-bond couplings obtained from this analysis provided information on the behaviour of these compounds.[157]

The indirect $^1J(^{31}P$-$^{69/71}Ga)$ coupling constant has been determined for the GaP-included sodalite. Its presence was attributed to $[GaZn_3P]^{6+}$ cage, in which gallium is tetrahedrally coordinated to one phosphorus atom and three oxygen atoms.[158]

The first reported values of indirect one-bond spin-spin coupling constants between ^{31}P and $^{99/101}Ru$ nuclei have been measured by Corrigan et al.[159] for the cluster $(\mu$-$H)Ru_4(CO)_{10}(\mu$-$PPh_2)[\mu_4$-$\eta^1(P),\eta^1(P)\eta^1(P),\eta^1,\eta^2$-$\{C_6H_4\}PPh]$, which contains a five coordinate phosphido-bridge and an orthometallated η^2 arene ring. The obtained $^1J(^{31}P$-$^{99/101}Ru)$ values of 129-174 Hz from the ^{31}P CP MAS NMR spectrum of the solid complex[159,160] are comparable with the range of ^{31}P-^{57}Fe and ^{31}P-^{187}Os couplings of 26-147 Hz[161] and 71-317 Hz, respectively.[162]

$^1J(^{31}P$-$^{103}Rh)$ couplings found for organorhodoxime $[Rh(dimethylglyoxime)_2(PPh)_3(L)R$ varied from 54.9 Hz for R = cyclohexyl up to 80.0 Hz for R = C≡CPh, and have been used as a measure of the *trans* influence of the organo group in these complexes.[163]

Well resolved splittings due to coupling with $^{107/109}Ag$ have been observed in ^{31}P CP MAS NMR spectra of the mononuclear $[AgBr(PPh_3)_2]$ and binuclear $[Ag_2X_2(PPh_3)_4]$ complexes; ($^1J(^{31}P$-$^{107/109}Ag)$ vary from 360 to 423 Hz).[164] In dimers, two P atoms are non-equivalent, which has allowed to measure $^2J(^{31}P$-$^{31}P)$ coupling constants between them (^2J vary from 102 to 131 Hz).

Measurements of indirect spin-spin coupling constants in solution involving quadrupolar nuclei with spin larger than 1/2 usually encounter great difficulties. On the one hand, efficient quadrupolar relaxation causes a broadening of the corresponding peaks. On the other, the NMR signals belonging to neighbouring spin 1/2 nuclei very often do not exhibit resolvable splittings -

caused by spin-spin coupling with quadrupolar nuclei - because of the short relaxation times of the latter. These problems can be overcome by carrying out NMR studies in a solid state. The first observation of splittings in ^{31}P NMR spectrum which arise from coupling between ^{31}P (spin 1/2) and ^{115}In (spin 9/2) has been recently reported by Wasylishen et al.,[165] who recorded the ^{31}P CP MAS NMR spectrum of solid tribromo(tris(4-methoxyphenyl)phosphineindium(III), $Br_3In{\cdot}P(4\text{-}(CH_3O)C_6H_4)_3$; $^1J(^{31}P\text{-}^{115}In)$ of 1109 ± 9 Hz has been found for this compound. The presence of the large anisotropic $J(^{31}P\text{-}^{115}In)$ tensor clearly indicates that the mechanisms other than the Fermi-contact mechanism play an important role in the coupling between ^{31}P and ^{115}In.

One-bond $^{31}P\text{-}^{195}Pt$ coupling constants from 1100 to 2500 Hz have been determined for a large series of recently synthesized luminescent diphosphine dithiolate complexes of platinum (II),[166] and the couplings from ca. 2800 to 3600 Hz have been found for the amphiphilic bis(thiolato)platinum (II) copmlexes with diphosphine or phosphinoarsine ligands.[167]

The first observation of one-bond $^{39}K\text{-}^{119}Sn$ coupling constant has been reported by Hitchcock et al.,[168] ^1J of 289 Hz was measured in the solid-state CP MAS NMR spectrum of $[K\{Sn(CH_2\text{-}t\text{-}Bu)_3\}(\eta^6\text{-}C_6H_5Me)_3]$. The coupling was not observed in solution, probably due to the low site symmetry at the quadrupolar ^{39}K nucleus.

One-bond $^{77}Se\text{-}^{199}Hg$ coupling constant of 24 Hz has been observed in mercury (II) trifluoromethane selenolate, $Hg(SeCF_3)_2$.[169]

The magnitude of the $^1J(^{195}Pt\text{-}^{199}Hg)$ coupling constants found in *trans*-$[(CH_3NH_2)_2Pt(L)_2Hg(NO_3)_2{\cdot}0.5H_2O$, 2783 Hz for L = 1-methyl and 2850 Hz for L = 1,5-dimethylcysteine, provides a clear indication that a direct Pt-Hg bond is formed in these two compounds, despite the somewhat long distances between the two metals.[170]

5. Two-Bond Couplings Involving Hydrogen

A careful analysis of the second order ^1H NMR spectra has been performed for several multibridged $[2_n]$cyclophanes by Ernst et al..[171] It gave geminal and vicinal coupling constants between the bridge protons of this interesting group of compounds. The results obtained were interpreted with regard to molecular geometries around the bridge.

Two- and three-bond coupling constants have been determined for the ketone and epoxide synthetic precursors of β-chamigrene, a component of the essential oil from the leaves of *Chamaecyparis taiwanesis* Masam, and used in a conformational analysis thereof.[172]

The pH dependence of $\delta(^1H)$ chemical shift of the CH_2 group and $^2J(^1H\text{-}^1H)$ coupling constant has been studied for citrate by Moore et al.[173] in order to get information if these parameters change significantly over the physiological pH range, and if they could potentially be used as a noninvasive pH monitor for *in vivo* studies. It was shown earlier that the concentration of citrate is reduced markedly in prostate cancer tissue in comparison to that in the normal prostate. The observed $^2J(^1H\text{-}^1H)$ changes were from 15.9 Hz at pH 1 to 15.2 Hz at pH 6-12.

Proton-proton coupling constants across two and more bonds have been measured for feruloylated arabinoxylan tetrasaccharide containing the β-D-Xyl*p*(1→2)-α-L-Ara*f*,[174] for proliferine, a new bicyclic sesterterpene isolated from *Fusarium proliferatum*,[175] for the four reduced α(2→8)-linked sialic acid compounds (monomer, dimer, trimer and polymer (n>25),[176] Geminal and vicinal proton-proton coupling constants have been measured for novel phenolic organogels obtained from a variety of phenols and the twin-tailed, anionic surfactant sodium bis(2-

ethylhexyl)sulphosuccinate in various solvents.[177] It has followed from an analysis of these data that gel formation does not significantly affect the surfactant equilibrium rotamer distribution, but decreases the energy barrier for interconversion.

A ring opening of the antitumor prodrug temozolomide in deuteriated phosphate buffer has been studied by Wheelhouse and Stevens.[178] An observation of two-bond ^1H-^2H and three-bond ^1H-^{31}P coupling constants in the spectra of the products allowed the authors to follow the deuterium incorporation into the methyl group transferred from the reactive species 5-(3-methyltriazen-1-yl)imidazole-44-carboxamide.

(*Z*)-2-Lithio-1-(*o*-lithiophenyl)ethene in ether solution forms a static dimer stable even at room temperature. The ^{6}Li NMR spectrum of this compound has provided the first example where the assignment of chemically non-equivalent Li atoms in an organolithium cluster could be based solely on ^1H-^6Li coupling constants; $^2J(^1H_2{}^6Li_2)$ = 0.75 Hz, $^2J(^1H_2{}^6Li_{2'})$ = 0.69 Hz and $^3J(^1H_1{}^6Li_2)$ = 0.80 Hz have been measured for this compound.[179]

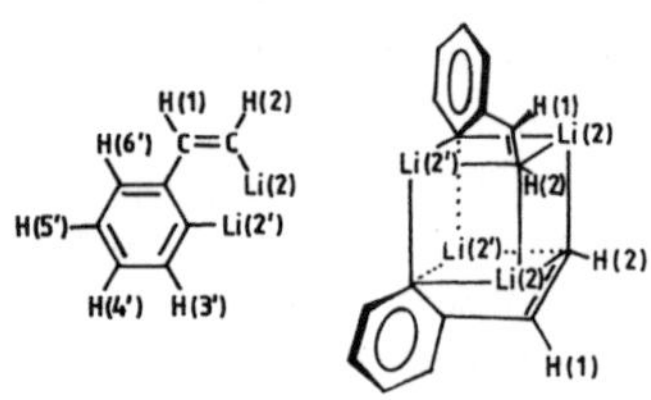

So far, the application of geminal ^1H-^{13}C scalar coupling constants in conformational analysis has been less popular than that of vicinal ones. Without the knowledge of $^2J(^1H$-$^{13}C)$ signs (they can be either negative or positive) an interpretation of their small values could often lead to misleading results. At present, the modern 3D and 4D techniques make possible not only measurements of even very small J values but also facilitate their sign determination. Recently, Hines et al.[180] have shown that the sign of $^2J(^1H$-$^{13}C)$ is exceptionally useful for RNA structure determination. They have reported a novel, independent method for the stereospecific assignment of the H5′(*pro-S*) and H5″(*pro-R*) protons which is based only on the sign of the proton-carbon two-bond scalar couplings.

A careful analysis of two-bond ^1H-^{13}C coupling constants measured for three ^{13}C enriched deoxyribonucleosides, 2′-deoxyadenosine, 2′-deoxycytidine and thymidine, has been performed by Serianni and his group.[181] The results obtained suggest that the difference $|^2J(C_{1'}$-$H_{2'S})$ - $^2J(C_{1'}$-$H_{2'R})|$ may be a useful probe of N/S equilibria in 2′-deoxyribonucleosides in ^2H_2O solution.

A new ^1H-detected 3D NMR experiment allowing to measure two- and three-bond ^1H-^{13}C coupling constants in proteins with selectively ^{13}C-enriched methyl sites has been described by Vuister et al.[34] The method was demonstrated for staphylococcal nuclease selectively [5,5 ^{13}C]-labelled in all 11 leucine positions and ligated with thymidine 3,5-biphosphate and Ca^{2+}. The $^2J(^1H$-$^{13}C)$ and $^3J(^1H$-$^{13}C)$ values combined with $^3J(^{13}C$-$^{13}C)$ coupling constants allowed the elucidation of the χ_2-angles and the stereospecific assignments of the Cδ-Me group with respect to the prochiral β-protons. The sign of $^2J(^1H-^{13}C)$ can also be used to determine the torsion angle γ and the sugar conformation.

A set of two- and one-bond $^1H-^{13}C$ coupling constants has been measured and analysed by Osborne for some quinolone derivatives,[182] for 2,4-dihalogenoquinolines,[183] for 2-alkoxy-4-halogenoquinolines,[184] and by Kolehmainen for all eleven chlorinated derivatives of methyl propanoate.[185]

Geminal and vicinal $^1H-^{15}N$ couplings have been measured for a large series of substituted *N*-methylpyrrazoles by Ugrak et al.[186] The $^2J(H_3N_2)$ couplings were considerably larger than the $^2J(H_1N_5)$ ones, i.e. 11.8-12.7 Hz and 3.0-4.5 Hz, respectively. The difference was explained in terms of the lone pair influence.

Geminal and vicinal coupling constants, $J(^1H\text{-}X)$ ($X = {}^{13}C, {}^{15}N, {}^{31}P$), have been measured for *t*-butylamino-*t*-butylfluorophosphane and for *t*-butylimino-*t*-butylphosphane and its dimer; some erroneous literature data have been corrected.[187]

A full NMR characteristic which included also two-, three- and four-bond $^1H-^{29}Si$ coupling constants has been obtained for β-silylated silyl enol ethers.[188] The couplings were measured by the use of modified selective polarization transfer experiments, and their values were found to be related to the geometry of the coupling path involved.

The first systematic study on two-bond proton-phosphorus coupling constants across the central metal atom, $^2J(^1H\text{-}M\text{-}^{31}P)$, has been carried out by Bampos et al.[189] The couplings including their signs were measured for a series of octahedral hydridometal phosphine complexes [$MRH(PP_3)$; $PP_3 = P(CH_2CH_2CH_2P(CH_3)_2)_3$, M= Ru, Fe; R = H, Cl, 1-pentenyl, cyclopentenyl, 2-methyl-1-butenyl, 3,3-dimethyl-1-butenyl] using a selective 2D NMR experiment. Their magnitude varied from 12.1 to 95.6 Hz depending on the relative disposition of the coupled nuclei, as well on the nature of the metal.

A two-bond $^1H-^{77}Se$ coupling constant combined with $^{13}C-^{77}Se$ couplings have allowed the absolute assignment of the regioisomers in the ^1H and ^{13}C spectra of phenylselenenyl azide and chloride addition products of methylenecyclohexane and cyclohexene.[190]

Very small $^1H-^{103}Rh$ coupling constants of ca. 1 to 3 Hz or smaller have been measured for a series of [$\eta 5\text{-}C_5H_4Cl)Rh(diolefin)$] complexes and the results interpreted in terms of the variation in charge density within the substituted cyclopentadienyl ring.[191]

A two-bond coupling constant, $^2J(^1H-^{119}Sn)$, observed in $[Cl(CH_3)_2Sn]_3N$ (^2J = 70.2 Hz) has been found to be considerably larger than that observed in $[(CH_3)_3Sn]_3N$ (^2J = 56.3 Hz).[192] The difference observed can be easily explained in terms of the *-I* effect of the chlorine substituent in the former compound. It may be also partly due to the changed geometry at the tin atoms. The authors offered a similar explanation for the changes observed between the remaining couplings measured in these two compounds. Geminal and vicinal proton-tin couplings have been measured for the parent tetravinyl tin and some related chlorovinyltin compounds; *geminal* and *cis* couplings are of the similar magnitude through the series studied, whereas the *trans* couplings are ca. two times larger than the two remaining ones.[193]

Two-bond $^1H-^{125}Te$ coupling constants of 20 to 28 Hz have been measured for allyl-, cyanomethyl- and some other organotelluronium salts.[194]

Two-bond $^1H-^{195}Pt$ coupling constants have been reported for *trans* (53 Hz) and *cis* $[PtCl_2(^{15}NH_3)_2]$ (63 Hz) complexes and for *trans* $[PtCl_2(H_2O)_2(^{15}NH_3)_2]^{2+}$ complex (49 Hz).[195]

6. Two-Bond Couplings not Involving Hydrogen

The progressive coordination of four HMPA molecules to the lithium cation has been followed by Reich et al.[196] in the ^{7}Li and ^{31}P NMR spectra through characteristic chemical shifts and $^2J(^7Li-^{31}P)$ coupling constants. The two-bond lithium-phosphorus coupling constants measured for the contact and separated lithium ions fall in a narrow range of 7.4-11.5 Hz, with the smallest values observed for separated $Li(HMPA)_4^+$, and the largest for separated $Li(HMPA)_1^+$. The compounds studied were FlLi (Fl = fluorenyl), Ph_3CLi, Ph_3HgLi, LiCl, LiBr, LiI, MeSeLi, MeLi, and MeSLi.

For the first time HEED pulse sequences have been applied by Wrackmeyer and Kehr[197] to measure two-bond coupling constants $^2J(^{13}C_R{}^{15}N)$ in the 2-substituted pyridines 2-R-C_6H_4N (R = CH_3, CH=CH_2, C≡CH, C(O)H, C(O)CH_3). The couplings were negative and their absolute values varied very little, i.e. from 8.6 to 14.1 Hz, being independent of the hybridization of the carbon atom. This indicates that the presence and the orientation of the lone pair of electrons at the nitrogen atom dominates the two-bond ^{13}C-^{15}N coupling, the nature of the carbon atom and its substituents being of minor importance.

The ^{13}C-^{19}F coupling constants across two, three and more bonds are a useful guide in the assignment of the ^{13}C NMR chemical shifts, since they usually correlate clearly with the distance between the coupled nuclei. Recently, they have been applied by Laihia et al.[198] in order to assign the ^{13}C NMR spectra of several isomeric ring substituted monofluoro-(*E*)-3-phenylpropenals, where a typical trend $^1J > ^2J > ^3J$ was observed for the ring carbons. Only the side-chain carbons in the *ortho*-substituted compounds showed an exception to this rule.

The first example of two-bond ^{13}C-^{89}Y coupling constant has been published by Hall et al.[199] The $^2J(^{13}C-^{89}Y)$ coupling of 3.0 Hz was observed between the ^{89}Y nucleus and the carbons of the amide carbonyl groups in the spectra of the complexes formed between the ferrocene-containing cryptand 1,1'(1,4,10-trioxa-7,13-diazacyclopentadecane-7,13-diyldicarbonyl)ferrocene and yttrium perchlorate.

Two-bond coupling constants, $^2J(^{15}N-M)$ for M = ^{29}Si, ^{119}Sn, ^{209}Pb, including their signs, have been measured for organometallic-2-substituted pyridines by Hahn-Echo Extended (HEED) pulse sequences.[200] An application of these pulse sequences has shown that the couplings can be measured much faster from the M NMR spectra than from the ^{15}N NMR spectra. The fairly large absolute magnitude of $^2J(^{15}N-M)$ couplings has been ascribed to the influence of the lone pair of electrons at the nitrogen atom.

The first report of $^2J(^{14/15}N-^{199}Hg)$ in the solid state obtained from an analysis of ^{199}Hg MAS NMR spectra has been published by Wu and Wasylishen:[138] $^2J(^{14}N-^{199}Hg)_{iso}$ = 20.6 ± 2.0 Hz and $^2J(^{15}N-^{199}Hg)$ = 29.0 ± 2.0 Hz were measured in $K_2Hg(CN)_4$.

A correlation between geminal ^{19}F-^{19}F coupling constants and Hammett σ constants in β,β-difluoro-α-(trifluoromethyl)styrenes has been reported by Morken and Burton:[201]

$$\sigma = -0.13\ ^2J(^{19}F-^{19}F) + 1.6; \qquad r = 0.94,\ n = 8$$

Geminal ^{29}Si-^{119}Sn coupling constants have been measured by Wrackmeyer et al.[202] for a series of the alkenyl(alkynyl)dimethylsilanes and for two *gem*-silyl,stannyl substituted alkenes bearing additionally boryl substituent; the $^2J(^{29}Si-^{119}Sn)$ values varied from 88.5 Hz up to 120.6 Hz.

An analysis of ^{31}P NMR spectra recorded for some bis(phosphinoamines) at ambient and low

(-90°C) temperatures has yielded $^2J(^{31}P\text{-}N\text{-}^{31}P)$ couplings of 12.2-13.4 Hz and 12.2-20.8 Hz, respectively.[203] These rather small $^2J(^{31}P\text{-}N\text{-}^{31}P)$ values clearly indicate that the compounds under study exist mostly in the *trans* C_s conformation at both temperatures. Large, i.e. of ca. 150-660 Hz, $^2J(^{31}P\text{-}N\text{-}^{31}P)$ values would have been expected if the *cis* C_{2V} forms had been present. Two-bond $^{31}P\text{-}^{31}P$ coupling constants of 3.4-10.6 Hz have been measured for some new *trans* and *cis* dithiadiphosphetanes (dimeric perthiophosphoric acid anhydrides).[204] Two-bond $^{31}P\text{-}^{31}P$ coupling constants have been measured for a variety of complexes in order to get insight into their structure. This included diastereomeric olefin complexes of Rh, [(enamide)Rh(L-L)]ClO_4,[205] dirhodium and platinum-rhodium complexes of bis(diphenylphosphino)methane,[206] transition metal complexes of (diphenylphosphino)(diphenylphosphinoselenoyl)phenylamine, and (diphenylphosphino)(diphenylphosphinothioyl)phenylamine, with ^2J of 53 to 112 Hz.[207] The $^2J(^{31}P\text{-}^{31}P)$ couplings of 300 to 450 Hz have been observed in the ^{31}P CP MAS spectra of some linear mono-, bi-, and tri-metallic bis(tri-*n*-butylphosphine) nickel and palladium complexes having 1,4-tetrafluorophenylene bridges; these couplings were of magnitude typical of P-metal-P pairs in a *trans* configuration.[208] A hindered rotation caused by the presence of the bulky xylyl substituent has been observed around the C-NiBr$(PMe_3)_2$ bond in the cyclic complex *trans*-$(Me_3P)_2$BrNi(C(NXy)CH_2-*o*-C_6H_4)NiBr(PMe_3). This, in turn, caused inequivalency of the *trans* phosphine ligands and allowed to measure two-bond coupling between them, $^2J(^{31}P\text{-}^{31}P)$ = 267 Hz.[209]

The ^{119}Sn NMR spectra of tributyl and triphenylstannyl perchlorates recorded in hexamethylphosphoroamide-d_{18} solution have revealed the presence of triplets due to the coupling between ^{119}Sn and ^{31}P nuclei, $^2J(^{31}P\text{-}^{119}Sn)$ of 146 Hz and 163 Hz, respectively. This provided direct spectroscopic evidence for solvated trioorganostannyl cations of bipyramidal form in which the solvent molecules (HMPA) occupy the two axial positions of the five-coordinate trigonal bipyramid.[210]

A two-bond coupling constant between $^{77}Se(sp^2)$ and $^{77}Se(sp^3)$ atoms of 100 Hz has been observed in $C_3Se_5[C(O)Ph]_2$ compound, and two complexes, $[PPh_4]_2[Zn(C_3Se_5)_2]_2$ and $[NBu^n{}_4]_2[Pt(C_3Se_5)_2]$ ($C_3Se_5{}^{2-}$ = 1,3-diselenole-2-selone-4,5-diselenolate).[211] The large splittings of ca. 160 Hz observed in the ^{77}Se solid-state NMR spectra of $[Me_4N]_2Zn(Se_4)_2]$ complex have been attributed to two-bond $^{77}Se\text{-}^{77}Se$ couplings under assumption that such coupling values for selenium rings and chains are greater than those across one bond.[212]

Unusually large coupling constants, $^3J(^{119}Sn\text{-}^{117}Sn)$ = 1514 Hz and $^2J(^{119}Sn\text{-}^{207}Pb)$ = 1145 Hz have been observed in the spectra of the chalcogenide Sn (II) anions, $Sn_2Se_3{}^{2-}$ and $SnPbSe_3{}^{2-}$, respectively.[213] These values are consistent with predominantly p-bonded cages if relativistic effects, which are significant in the heavy-element spin-spin couplings, are taken into account.

Two-bond $^{183}W\text{-}O\text{-}^{183}W$ couplings have been observed in the spectra of $[GeW_9V_3O_{40}]^{7-}$ anion confirming its A-type structure.[214]

7. Three-Bond Couplings Between Hydrogens

There is no doubt that, because of their angle dependence, vicinal proton-proton coupling constants belong to the most widely applied NMR parameters in conformational analysis of peptides, proteins, saccharides, DNA, RNA and other biologically important molecules.

These couplings have been measured for literally dozens of peptides and proteins thus being a source of information that is complementary to the information obtained from NOE experiments.

The knowledge of $^3J(H_NH\alpha)$ and $^3J(H\alpha H\beta)$ coupling constant values allows one, *via* the Karplus dependence, to determine the local conformation of the main chain of protein and that of the side chain, respectively. The values of $^3J(H\alpha H\beta)$ coupling constants also provide the stereospecific assignment of protons.

The 3D NMR techniques applied to the measurements of $^3J(H_NH\alpha)$ and $^3J(H\alpha H\beta)$ coupling constants have been shortly summarized by Oschkinat et al.[37] A new E.COSY-type 2D method, the selective version of XLOC, allowing measurements of $^3J(^1H-^1H)$ for peptides and proteins has been introduced by Soerensen et al.[215] It uses a large heteronuclear J coupling for displacement in the $\omega 1$ dimension. For small peptides, the natural abundance heteroatom level is sufficient to obtain $^3J(^1H-^1H)$ value with the standard deviation of 0.3 Hz. The improved pulse sequence for measuring $^3J(H_NH\alpha)$ coupling constants in $^{13}C/^{15}N$ labelled proteins has been reported by Madsen et al.[36] The technique is a simple modification of E.COSY, in which advantage of $^{13}C\alpha$ is taken with intraresidue $\{^{13}C\alpha,^1H_N\}$ peak in the $\omega 2$-$\omega 3$ plane. A separation of the peak components by the large $^1J(C\alpha H\alpha)$ coupling in $\omega 2$ dimension makes it possible to determine accurately small $^3J(H_NH\alpha)$ coupling in $\omega 3$. The methods based on a transfer of in-phase magnetization during a doubly selective irradiation period which allow one to obtain selected high-resolution multiplets where all signals have positive pure absorption peak shapes, have been reported by Vincent et al.[216] The HNHA method, which relies on quantitative analysis of the diagonal-peak to cross-peak intensity ratio in a ^{15}N-separated HN-Hα homonuclear J correlation experiment has been proposed by Vuister and Bax.[217] The method provides the values of $^3J(H_NH\alpha)$ and intraresidue correlation information in ^{15}N-enriched proteins. Applied to staphylococcal nuclease it yielded 0.73 Hz root mean-square deviation from the Karplus curve based on 86 backbone angles known with good precision from the crystal structure. A method for estimating $^3J(H\alpha H\beta)$ from the shape and fine structure of fingerprint-region crosspeaks of COSY spectra has been devised and applied to the analysis of spectra of hen lysozyme by Bartik and Redfield.[218]

The wealth of NMR data available, new designed algorithms, and the access to powerful computers facilitate now the study of the dynamic properties of quite large peptides. Several new approaches in this respect have been proposed and cyclic peptides such as antamanide have been used as model compounds. The CHARMM 22.0.β atomic force field parameter set has been used for Langevin dynamics yielding a nearly quantitative agreement between calculated $^3J(^1H-^1H)$ coupling constants and those experimentally measured for antamanide.[219] A novel multiconformational search algorithm which uses homonuclear $^3J(^1H-^1H)$ coupling constants as constraints in order to characterize dynamic properties of that peptide has been applied by Blackledge et al.[220] A novel approach which combines NMR parameters with conformational energy calculations including Monte Carlo sampling and statistical evaluation of the results has been proposed by Nikiforovich et al.[221] The statistical weight estimations for low energy backbone conformations, which are consistent with all of the NMR data, were obtained by this method for dermenkephalin (DRE, H-Tyr-D-Met-Phe-His-Leu-Met-Asp-NH_2), a natural peptide found in frog skin.

In several papers published recently all possible vicinal spin-spin coupling constants across the common bond have been used directly as the restraints to remove ambiguity from the MD calculations. It has been done by Mierke and Kessler for several cyclic peptides,[222] by Blackledge et al. for antamanide,[220] by Mierke et al. for cyclo[-D-Pro-Ala_4-],[223] and by Mierke et al. for the side chain rotamers of phenylalanine in a short peptide, Ala-Phe-Ala.[224]

By analogy to the distance restraints, also spin-spin coupling constants can be treated as the time-averaged ones in molecular dynamics simulations. This approach has been recently applied to antamanide by Torda at al.[225]

Analysis of the vicinal coupling constants in the ^1H NMR spectra of the Ac-D-Phe-Pro-boroArg-OH, a potent, competitive inhibitor of thrombin shows that Phe χ_1 is not fixed and the phenyl side chain possesses various degrees of internal motion.[226] Well-defined conformers have been determined for two cyclic analogues of angiotensin (Asp-Arg-Val-Tyr-Val/Ile-His-Pro-Phe, AT) cyclo[Sar^1,Cys^3,Mpt^5]-AT and cyclo[Sar^1,HCys^3,Mpt^5]-AT.[227] These conformers were also present among low energy ones obtained from energy calculations.

Homonuclear coupling constants have been used to elucidate the structure of three tripeptides containing a phosphodiester linkage mimicking a disulphide bridge,[228] of L-pGlu-L-His-Gly, for which two stable conformations have been identified,[229] of cyclic hexapeptides, for which a new structural motif - the pseudo-βI-turn - has been identified,[230] of 14-membered peptide rings and side chains of DPDPE, a conformationally restricted synthetic pentapeptide with the sequence H-Tyr^1-D-Pen^2-Gly^3-Phe^4-D-Pen^5-OH.[231] Homonuclear coupling constants have been also used to check the influence of a saccharide residue in position *i* of a standard β-turn on the formation of reverse turns in S-glycosylated cyclic hexapeptide cyclo(-D-Pro^1-Phe^2-Cys^3(tetra-O-acetyl-β-D-galactopyranosyl)-Trp^4-Lys(Z)^5-Phe^6-).[232]

Analysis of vicinal coupling constants measured in a function of pH for three and tetrapeptides of thymopoietin-type has provided evidence for the presence of non-statistical orientations of the functional groups.[233]

The conformations of four active antagonists of bradykinin (9-residue peptide hormone),[234] the solution structure of the cyclic Ac-Pen-Arg-Gly-Asp-Cys-OH with 17-membered ring, a potent fibrinogen receptor antagonist,[235] and the conformations of BQ123 cyclo(Leu-D-Trp-D-Asp-Pro-D-Val), Ac-D-Trp-Leu-Asp-Ile-Ile-Trp and BQ610 hexahydroazepinocarbonyl-Leu-D-Trp(CHO)-D-Trp[236] have been established with the help of vicinal $^3J(^1H-^1H)$ coupling constants. The latter three peptides are endothelin antagonists.

Homonuclear vicinal coupling constants have been also utilized in the secondary and tertiary structure determination of longer polypeptide chains, and of cyclic peptides. Some indication of nascent helix between residues 5 and 12 has been observed in water for a peptide corresponding to the amino acid region 71-93 of the plasma protein transthyretin, for which averaged coupling constants have been reported.[237] An antibiotic peptide, trichosporin B-V, has been found to be predominantly helical in solution on the basis of small values of $^3J(H_NH\alpha)$ coupling constants.[238] With the use of vicinal homonuclear coupling constants the differences between the solution structures of protonated and unprotonated forms of oxytocin have been found to be due to a change in the conformation of the disulphide bridge.[239] The structure of RP 71955, a new 21 amino acid tricyclic peptide active against HIV-1 virus,[240] and the structure of a monocyclic analogue of endothelin [1,15 Aba]-ET-1 in aqueous acetonitrile solution[241] have been determined with the help of $^3J(H_NH\alpha)$ coupling constants. A set of 10 vicinal coupling constants has been also used in the determination of the solution structure of the Ga(III) complex of pyoverdin GM-II.[242]

The vicinal homonuclear coupling constants of the polypeptide backbone have been used to determine elements of the secondary structure in a number of proteins. The $^3J(H_NH\alpha)$ coupling constants have been measured for N-terminal lipoyl domain (79 residues) of the dihydrolipoyl transacetylase component of the pyruvate dehydrogenase complex from *Azotobacter vinelandii*, and used for identification of two antiparallel β-sheets.[243] The same coupling constants have been used in identifying the secondary structure elements of 99 residue long polypeptide containing the rat LFB1/HNF1 transcription factor in positions 10-90,[244] of uniformly ^{15}N-enriched oxidized *Desulfovibrio vulgaris* flavodoxin (148 residues),[245] of barstar, 89-residue polypeptide inhibitor of barnase,[246] of the glucose transporter of *Escherichia coli,*[247] of capsicein, 98 residues long, 10.2 kDa α-elicitin,[248] of uniformly ^{15}N/^{13}C-labelled 62-residue N-terminal DNA-binding domain of GAL4, a transcriptional activator protein from *Saccharomyces cerevisiae,*[249] of modules corresponding to the fibronectin type 1 and epidermal growth factor-like (residues 1-91),[250] and of the uniformly ^{15}N-labelled reduced Cu(I) form of the blue copper protein azurin (129 residues, 14 kDa) from *Alcaligenes denitrificans.*[251] The Greek key arrangement has been found in calcium loaded *Myxococcus xanthus* protein S (173 residues) with the help of $^3J(H_NH\alpha)$ coupling constants.[252]

The homonuclear proton-proton coupling constants have been frequently used as supporting data in the three-dimensional structure solving of larger peptides and proteins. They have been classified as small, medium and large for that purpose. The backbone vicinal homonuclear proton-proton coupling constants have been helpful in the structure solving of the 27 aminoacid residue polypeptide ω-conotoxin GVIA from the venom of the cone shell *Conus geographus,*[253] of a 43-residue active, synthetic peptide encompassing the peripheral subunit-binding domain of dihydrolipoamide acetyltransferase from the pyruvate dehydrogenase multienzyme complex of *Bacillus stearothermophilus,*[254] of ^{15}N-enriched human type-α transforming growth factor (hTGFα) containing 50 residues and 3 disulfide bonds,[255] of a soluble, glycosylated form of the human complement regulatory protein CD59 of 77 residues,[256] of uniformly ^{15}N-labelled N-terminal 10 kDa fragment of the *Escherichia coli* Ada protein,[257] and of N-terminal DNA-binding domain of the salmonella phage *P22 c2* repressor (residues 1-76).[258]

Further examples of 3D structure determination include the kringle domain from urokinase-type plasminogen activator corresponding to the Ser^{47}-Lys^{135} segment of the u-PA,[259] the fourth and fifth type 1 module pair from the N-terminus of human fibronectin ($^4F_1$.$^5F_1$),[260] the variant-1 (CSE-V1) neurotoxin from the venom of the New World scorpion *Centruroides sculpturatus* Ewing,[261] the GDP form of a truncated version of the rasp21 protein (residues 1-166),[40] a hybrid consisting of chymotrypsin inhibitor 2 (CI2) where part of the active loop was replaced by the nonapeptide and where the 19 structureless residues of the N-terminus were removed,[262] cardiotoxin CTX IIb from *Naja mossambica mossambica,*[263] the glutaredoxin (C14S) from *Eschericha coli* forming mixed disulphide with glutathione,[264] the recombinant mutant α-amylase inhibitor [R19L]Tendamistat,[265] three mutant proteins of calbindin D_{9k},[266] and uniformly ^{15}N/^{13}C-labelled periplasmic cyclophilin type *cis-trans* peptidyl-prolyl isomerase from *Escherichia coli* (167 residues, 18.2 kDa).[267]

In several cases the improved experimental techniques have allowed measurement of larger numbers of coupling constants, which facilitated high resolution structure calculation or refinement. The solution structure of the glucocorticoid receptor DNA-binding domain[268] has been refined. New experimental input has provided about 30% more of dihedral angle constraints in a refinement of the previously determined structure of acyl-coenzyme A binding protein from bovine liver.[269] A set of 29 structures has given an atomic r.m.s. deviation of 0.44Å for the N and Cα atoms in the four α-helices. 123 dihedral angle constraints have been identified in NMR spectra of a 57 residue long protein, dendrotoxin K (Toxin K) in solution.[270] With the other NMR data they lead to the solution structure of the protein with the average r.m.s. deviation of 0.31Å for all backbone atoms. 42 dihedral angle constraints have been used in determination of high-resolution 3D structures of bovine pancreatic ribonuclease A in aqueous solution,[271] and 174 in the case of uniformly ^{15}N-enriched calcium-loaded calbindin-D_{9k}.[272]

In some cases a set of calculated NMR structures has been compared with the X-ray structure of the protein. This involved the NMR structures of cardiotoxin III, a membrane toxin from the venom of Taiwan cobra *Naja naja atra* (60 residues). They were compared with the X-ray crystal structure of cardiotoxin V_4^{II} from *Naja mosambica mosambica.*[273] For the 129-residue protein, turkey egg-white lysozyme, 110 $^3J(H_NH\alpha)$ coupling constants have been measured and compared with the experimental values of $^3J(H_NH\alpha)$ obtained earlier for hen egg-white lysozyme.[274] The values have been also compared with three sets of $^3J(H_NH\alpha)$ computed from known X-ray structures of turkey protein of different resolutions. R.m.s. differences between calculated and experimental data were 1.60 Hz for 0.25 nm resolution, 1.69 Hz for 0.22 nm resolution and 0.80 Hz for 0.15 nm resolution.

Sometimes vicinal coupling constants have been converted to dihedral angle constraints, and subsequently to distance restraints for use in DG calculations, as, for example, in the structure calculation of ω-conotoxin GVIA (27 residues), a peptide antagonist of the N-type neuronal voltage-sensitive calcium channel.[275]

An increasing attention has been paid to elucidation of the conformation of the side chains. The values of homonuclear proton-proton coupling constants have been used in a conformational analysis of the valine side chains of ribonuclease T_1,[276] in solution structure determination of the 129-residue protein hen lysozyme,[277] and an analysis of a side chain mobility has been performed for bovine pancreatic ribonuclease A.[271]

$^3J(^1H-^1H)$ coupling constants have been back-calculated for 1296 different conformations (in 10° steps) of the heme propionates of soybean carbonmonoxy leghemoglobin and compared with experimental data in order to identify the best-fitting propionates conformation.[278]

The structures of proteins which form associates have been also studied. Vicinal coupling constants have been measured and used to define constraints for the dihedral angles in the structure elucidation of a dimeric conformer of gramicidin A, a 15 residue polypeptide,[279] of monomers of the symmetrical dimer of 19.4 kDa of the M13 gene V protein Y41H mutant,[280] and of the isolated V_L domain of the anti-digoxin antibody 26-10, which was proved by the analytical centrifugation to be weakly associating (K_d = 2.5 mM).[281] With the use of $^3J(H_NH\alpha)$ coupling constants the secondary structure of FK506 binding protein when bound to ascomycin has been determined.[282] The 3D structure has been solved for the DNA-bound *Antennapedia* homeodomain in 18 kDa 1:1 protein-DNA complex,[283] and for 86 residues long acyl-coenzyme A binding protein in *holo* form and in complex with palmitoyl-coenzyme A.[284]

The structure of a novel glycopeptidolipid (GPL X-IIb composed of a lipotetrapeptide C-12-Ser-OMe-Ser-Phe-aThr-OMe, of a sugar 3OMe-α-L-6dTal*p*, and of a disaccharide α-L-Rha*p*- (1→3)-2-O-Lau-α-L-Rha*p*), isolated from *Mycobacterium xenopi* has been elucidated with the help of $^3J(^1H-^1H)$ coupling constants.[285]

In the case of nucleosides and nucleotides, $^3J(^1H-^1H)$ coupling constants provide crucial information allowing to distinguish between N and S conformations of furanose nonplanar five-member rings, and to determine populations of the *g+*, *gt* and *g-* staggered rotamers about the C(4′)-C(5′) bond. In many papers a two-state dynamic model with a rapid interconversion between S- and N-type sugar pucker has been considered in order to rationalize the averaged $^3J(^1H-^1H)$ couplings observed.

An effort to determine three-bond coupling constants quantitatively has been undertaken by Emsley et al. in their study of DNA samples.[286] The selective two-dimensional NMR correlation experiments have been demonstrated to be particularly well suited for obtaining coupling constants between sugar protons, which have been then used to determine structural parameters of DNA. The applicability of experimental $^3J(^1H-^1H)$ coupling constants to the determination of sugar conformation of B-DNA form has been discussed by Chuprina et al.[287] The experimental values have been compared with those calculated on the basis of highly resolved B-DNA crystal structures. The simulated vicinal proton coupling constants for the structures generated with a new method, a restrained Monte Carlo calculation, have been compared with the experimental deoxyribose ring coupling constants of the octamer duplex d(GTA-TAATG)·d(CATTATAC).[288] Molecular dynamics calculations have been performed with the program AMBER4 with weighted time-averaged restraints for the DNA octamer d(GTATAATG)·d(CATATTAC).[289] The significant improvement for simulated structures was achieved relative to rMD-derived structures with J-r.m.s. of 0.65 Hz for $^3J(H_1H_2)$ and $^3J(H_2H_3)$ coupling constants.

The North conformer has been found to be predominant in two *C1'*-branched β-D-nucleosides on the basis of NMR data.[290] An analysis of $^3J(^1H-^1H)$ coupling constants has shown that *g+* and *gt* staggered rotamers are preferred in these compounds. A conformational analysis of the series of the 2',3'-*cis*-fused pyrrolidino-β-D-nucleosides has yielded North or South-type conformation and the *g+* conformation preference.[291] A vicinal coupling constant has been critical in the assignment of absolute configuration to the N^6-deoxyadenosine adducts with phenanthrene 9,10-oxide.[292] A conformational analysis of the hitherto unknown product of the photolysis of thymidylyl-(3'→5')-thymidine, *trans-syn* II diasteromer in solution has been performed with the use of three-bond $^3J(^1H-^1H)$.[293] Deoxyribose ring conformations in two *trans-syn* cyclobutane photodimers of thymidylyl (3'-5')deoxyuridine have been calculated from proton-proton coupling constants by pseudorotational analysis.[294] The results of J-coupling analysis have been used as constraints to refine the structure of defluorinated 5-5 photoadduct of the dinucleoside monophosphate, deoxy-m^{5}HO^6-uridylyl(5-5)(3'-5')deoxyuridine, d-T5p5U.[295] The solution conformation of the potent HIV inhibitor 4'-azidothymidine has been determined from NMR coupling constants in D_2O.[296] The configurations of the glucuronide linkage of a series of glucuronides of 5-fluorocytosine and 5-fluorouracil have been assigned on the basis of $^3J(H_{1'}H_{2'})$ of the sugar.[297]

A detailed analysis of the coupling constants of the novel bicyclic and tricyclic thymidine derivatives has shown that their pentose sugars exist in 2'-*endo*-3'-*exo* conformation.[298] The conformation of fused six membered ring has been also established. Vicinal coupling constants have been of use in tracing the significant changes in the percentage of S conformer in 2'-deoxyadenylyl-(3'→5')-2'-deoxyadenosine with pH.[299]

The vicinal $^3J(^1H-^1H)$ coupling constants have been used in conformational studies of 3'-azido and 3'-amino-terminated thymidine dimers containing sulphonate and sulphonamide linkages,[300] of the lariat RNA tetramer and pentamer,[301] of *cis*- and *trans*-DDP adducts with the d(GpTpG) trinucleotide,[302] and of a series of 3'-fluorinated A(2'-5')A(2'-5')A fragments containing 9-(3'-fluoro-3'-deoxy-β-D-xylofuranosyl)-adenine or 3'-fluoro-3'-deoxyadenosine residues at different positions.[303]

A novel conformation having characteristics of both A- and B-form DNA has been found for the activating transcription factor region binding site within the E2A promoter region of adenovirus.[304] Coupling constants have been used to determine sugar puckers. The $^3J(^1H-^1H)$ couplings have been also used in determination of sugar pucker of d[C(TTCG)G] DNA hairpin and its RNA analogue,[305] of the thrombin-binding DNA aptomer d(GGTTGGTGTGGTTGG),[306] of a tetradecamer d(GCTGTTGACAATTA)·d(TAA- TTGTCAACAGC),[307] of a 14-base model of the native TAR hairpin 5′-GAGC[CUGGGA]-GCUC3′,[308] of E73, a 29-nucleotide RNA that has the sarcin/ricin loop sequence from 28S ribosomal RNA,[309] of the DNA oligonucleotide d($G_4T_4G_4$)(Oxy-1.5), which contains the *Oxytricha* telomere repeat T_4G_4,[310] and of a 19-nucleotide RNA containing the CUGGGA loop sequence corresponding to nucleotides 30-35 of the HIV-1 trans-activation response element (TAR).[311]

The J-coupling constants have been used in the assessment of the solution structure of several 5′-phenazinum-tethered DNA duplexes,[312] of the 2′,3′-dideoxy-β-D-gluco-pyranosyl rings of the 'homo-DNA' oligonucleotide ddGlc(A_5-T_5),[313] of sugars in DNA·RNA hybrid (GTCACATG).(caugugac) duplex,[314] of sugar rings of a self-complementary 14-mer DNA duplex d(GGATTGGCCAATCC) containing the GCAAT recognition motif of several transcription factors,[315] of the sugars in the DNA·RNA hybrid d(GTGAACTT)·r(AAGUUCAC),[316] of the duplex $(rGGCGAGCC)_2$, which contains GA mismatches,[317] and of a trisdecamer duplex, d(GGCAGAG-GTGAAA)·d(TTTCACCTCTGCC) with the sequence comprising the 11-base-pair direct repeat sequence found in the hepetitis B viral genome.[318]

Previously estimated torsion angles for sugar rings of a trisdecamer DNA duplex sequence d(AGCTTGCCTTGAG)·d(CTCAAGGCAAGCT) from a conserved region of HIV-1 genome's long terminal repeat have been employed for structure refinement.[319] The structure is basically in the B-DNA family.

A small shift in the conformational equilibrium has been observed for the ribose ring of 3′-GMP in its complex with barnase versus its free form.[320] The solution conformations of a free DNA oligonucleotide d(TCGCGTTTTCGCGA) and its complexes with the anticancer drug actinomycin D have been calculated with the help of 70 torsion angles found from coupling constants estimated only approximately due to the spectral overlap.[321] It was found that the drug intercalated exclusively at the single d(GC) step.

A number of studies have been devoted to mono- and polysaccharides, and some related compounds.

An extensive use of $^3J(^1H-^1H)$ coupling constants has been made by Perrin and Armstrong[322] in order to elucidate the proportions of axial anomers of glucopyranosylamine and some of its *N*-alkyl, tetra-*O*-acetyl, and 4,6-benzylidene derivatives. The data obtained indicated that there is no need to invoke the reverse anomeric effect in order to explain the small shift in the position of equilibrium which occurs upon N-protonation of these compounds, since all changes observed can be easily interpreted in terms of steric effects and a small normal anomeric effect. Therefore, the authors have come to the conclusion that the reverse anomeric effect with any cationic nitrogen substituent does not exist, and that the conclusions drawn in the earlier studies where the presence of an inverse anomeric effect was suggested require re-consideration.

Vicinal coupling constants between the hydroxyl protons and the vicinal ring protons have been measured for a series of carbohydrates by Adams and Lerner.[323] These couplings are easily observed provided that the exchange is slow enough and does not cause broadening of the OH signals. Their values varied from 4 to 8 Hz. $^3J(H_1H_2)$ and $^3J(H_4H_5)$ coupling constants have been examined as a convenient tool for the assignment of anomeric configuration of C-glycopyranosides and C-glycofuranosides.[56] A maximum-entropy method, which makes use of vicinal $^3J(H_5H_6)$ coupling constants, has been employed for calculation of rotameric distribution about the C(5)-C(6) bond in β-gentiobiose.[324]

Vicinal coupling constants have been measured and served for elucidation of ring conformation in a series of di-*O*-isopropylidene-*myo*-inositol derivatives.[325]

A set of $^3J(^1H-^1H)$ coupling constants has been measured for a series of 4-thiomaltose and its nitrogen analog,[326] for some 2-acetamido-2-deoxy monosaccharides and their acetylated derivatives,[327] for the synthetic branched trisaccharide β-D-ManNAc-(1→4)-[α-D-Glc-(1→3)]-L-Rha, which represents the repeating trisaccharide unit of the O-antigenic polysaccharide from *Aeromonas salmonicida*,[328] and for a pentasaccharide glycoside isolated from *Agave Americana*.[329] $^3J(^1H-^1H)$ coupling constants have proved to be useful in a study on the conformation of α-D-Fuc-(1→4)-β-D-GlcNAc-OMe synthetic disaccharide,[330] and of methyl 2-acetamido-2-deoxy- -3-O-(-β-D-glucopyranosyluronic acid)-β-D-glucopyranoside, which stands for a repeating unit of polysaccharide hyaluronan.[331] Preferred solution conformations have been determined for 1,1′-linked C-disaccharides: α,α-, α,β- and β,β-C-trehaloses, on the basis of their vicinal coupling constants.[332] The dependence of the L-arabinofuranosyl ring $^3J(^1H-^1H)$ coupling constants on substitution and linkage position has been characterized for α-L-arabinofuranosyl-(1→2)- and (1→3)-β-D-xylopyranosides.[333] The conformations of the pyranoid and furanoid rings of sucralose, a trichlorilated derivative of sucrose, have been calculated with the help of homonuclear vicinal proton-proton coupling constants.[334] The conformational analysis of the carbohydrate portion of T and TN haptenes in different solvents has been performed with the help of homonuclear proton-proton vicinal coupling constants.[335]

Accurate $^3J(H_3H_4)$ and $^3J(H_4H_5)$ coupling constants have been obtained by simulation of the 600 MHz ^1D NMR spectra of inulin related oligosaccharide, β-D-fructofuranosyl-(2→1)-β-D-fructofuranosyl-(2→1)-β-D-fructofuranosyl-(2→1)-β-D-fructofuranosyl-(2→1)-α-D-glucopyranoside, which allowed the assignment of $^4T_3$ conformation to the fructofuranosyl rings.[336] Molecular mechanics calculations combined with an analysis of the experimental data have been applied to elucidate the structure of the cyclic (1→2)-β-D-glucan secreted by *Rhizobium tropici* CIAT 899.[337] Vicinal coupling constants have been measured for a series of synthesized phytoalexin-elicitor-active oligoglicosides,[338] and for the O-specific polysaccharide of *Salmonella arizonae* O45.[339]

The vicinal proton-proton coupling-constants, $^3J(^1H-^1H)$, also remain the most useful parameter in the study of various organic molecules of different sizes. A new method for quantitative analysis of the conformation of six-membered ring compounds in solution which uses the endocyclic vicinal NMR proton-proton coupling constants has been presented by Haasnoot.[340] The method involves a combination of the generalized Karplus equation and the recently developed Truncated Fourier formalism;[341] it allows one to derive the relevant conformational parameters.

Vicinal proton-proton coupling constants have been applied in a conformational analysis of cyclic compounds such tricyclo[4.3.1.1^{2,5}]undec-3-en-10-one,[342] phenyl-substituted bicyclo[3.3.1]nonenes,[343] and substituted dibenzo-14-crown-4 ethers.[344] The couplings have been used to elucidate the structure of isomers of 3-phenyl-2-selenobicyclo[4.4.0]decane-5-one,[345] the conformation of A and B rings in 2-, 4-, and 6-bromosubstituted steroidal 4-en-3-ones,[346] and the structure of the first steroidal macrocycle, "tetra-acetoxycholaphane", has been calculated with their help.[347]

The proton spectrum of α-pinene has been re-measured and vicinal proton-proton coupling constants re-determined for it by Smith[348] in order to explain the discrepancies existing in the literature. A new set of the ^1H NMR data has been obtained for this compound.

The vicinal proton-proton coupling constants have been also applied to elucidate the correct relative and absolute stereochemistry of natural products and drugs. These included the sesquiterpene ether (-)kessane isolated from the essential oil of *Valeriana officinalis*,[349] nervosin, one of many natural products isolated from the leaves of ***Rabdosia nervosa***, which has antiphlogistic and antibacterial effects,[350] and antipain [(1-carboxy-2-phenylethyl)carbamoyl-L-arginyl-L-valyl-DL-arginal], a naturally occurring peptidyl aldehyde isolated from the culture filtrates of various species of *Actinomycetes* bacteria.[351] A study has been performed on the conformations of the 14-membered lactone ring of major metabolites of macrolide antibiotics roxithromycin and erythromycin A in $CDCl_3$ solution.[352] A combined use has been made of $^3J(^1H-^1H)$ couplings and *ab initio* HF MO STO-3G calculations in an analysis of conformational distribution in several 3-(substituted furan-2-yl) and 3-substituted thien-2-yl) derivatives of γ -aminobutyric acid, an inhibitory neurotransmitter responsible for the regulation of many physiological mechanisms.[353]

Based on the ^{13}C chemical shifts and $^3J(H_CH_N)$ coupling constants, the ratio of the keto-amine and phenol-imine tautomers in six Schiff bases formed from 2-hydroxy-1-naphthaldehyde and substituted anilines have been derived by Salman et al.[354] Vicinal proton-proton coupling constants have been used to estimate the changes in conformational equilibrium for butanedioic acid as a function of pH, concentration, and nature of the counterion.[355]

The vicinal proton-proton coupling constants provide an unequivocal test for the determination of the stereochemistry of the double bond. They have been used to assign the structure of all-*trans*-(9′Z,11′Z)-pyrrhoxanthin, an acetylenic C_{37} - skeletal nor-carotenoid butenolide, isolated from a natural bloom of dinoflagellates (mainly *Ceratium* spp.).[356] However, in the case of chemically equivalent protons this parameter is not easily accessible. Recently, Radeglia and his co-workers have developed the method which allows one to determine the vicinal proton-proton couplings across the symmetrically substituted $-CH_2CH=CHCH_2-$ fragments from the higher order ^{13}C NMR multiplets observed in ^1H continuous-wave off-resonance ^{13}C NMR spectra. The couplings measured by the use of this method have been applied in order to characterize the geometry of a series of *E* and *Z* cycloalkenes (C_5-C_{12})[357] and the four *Z-E* isomers of cyclodeca-1,5,9-triene.[358] The theoretical background of "anomalous" multiplets in off-resonance decoupled ^{13}C NMR spectra of vinylic moieties has been presented by the same group of authors. The method has been further applied to determine the vicinal proton-proton and geminal proton-carbon coupling constants in some simple model compounds, such as propenyl-1-trimethylstannane, cycloocta-1,5-diene and the fragment of vitamin D_2.[359] Furthermore, a simple two-dimensional method, which also allows one to measure the vicinal coupling constants between the equivalent protons, has been designed by Radeglia.[360] It involves two classical pulse sequences i.e. ^1H-spin-echo and INEPT techniques.

For the first time, using applied magnetic field gradients in conjunction with 1D ^1H detected ^1H-^{13}C HMQC experiments, Salman et al.[361] have been able to detect, in the ^1H NMR spectrum of ethane-1,2-diol, both the outer and inner ^{13}C satellite resonances together with the ^1H-coupled ^{13}C NMR spectrum. Simulation of these spectra allowed the authors to obtain reliable values for the ^1H-^{13}C and ^1H-^1H coupling constants. Analysis of the vicinal coupling constants measured at various temperatures led to conclusion that, contrary to some recent theoretical and experimental studies, the *trans* rotamer is significantly populated, though the *gauche* rotamer prevails.

It has been shown by Jaffe et al.[362] that the conformer populations determined from *ab initio* electronic structure calculations of the low-energy conformations of dimethoxyethane using a large basis set and including electron correlation effects, allow one to reproduce very well experimental vicinal proton-proton coupling values.

Changes in vicinal ^1H-^1H coupling constants measured in a broad range of solvents have been calibrated against changes of the integrals of peaks corresponding to free (3618 cm^{-1}) and intramolecular hydrogen bonded (3479 cm^{-1}) OH groups observed in IR spectra of hydroxyethers.[363] The temperature dependence of the coupling constants thus obtained allowed to derive the free energy of an intramolecular hydrogen bond as a function of solvation varying between -2.3 to -0.5 kcal mol^{-1}.

An extensive use of the vicinal proton-proton coupling constants has been made by Wache and Taube[364] in order to elucidate the geometry of the C_{12} -allylnickel (II) complexes $[Ni(C_{12}H_{19})]X$ (X = PF_6, $B(C_6H_3(CF_3)_2)_4$, $B(C_6F_5)_3F$). Numerous $^3J(^1H$-$^1H)$ couplings have been obtained and correlated with the torsion angles in the C_{12}-chain *via* the Karplus method. The angles obtained agreed well with those determined by X-ray analysis which was performed for the hexafluorophosphate complex. The vicinal coupling constants have been used to define the structures of some oxomolybdenum (VI) complexes with polyaminocarboxylates,[365] and to assign the solution structures of complexes of the type (bisphosphine)(cycloocta-1,5-diene)iridium(I).[366]

8. Three-Bond Couplings Between Hydrogen and Heteronuclei

Vicinal proton-carbon coupling constants reveal, by analogy to vicinal proton-proton coupling constants, the Karplus-type behaviour. However, one should keep in mind that $^3J(^1H$-$^{13}C)$ values depend also on (i) the pathway involved and (ii) electronegativity of substituents attached to the carbons in question. A careful analysis of both $^2J(^1H$-$^{13}C)$ and $^3J(^1H$-$^{13}C)$ coupling constants has been performed for a series of 6-azabicyclo[3.2.1]octene derivatives by Gács-Baitz et al.[367] The couplings for the HC_{sp3}-C_{sp2}-C_{sp2} pathway, $^3J(H_3C_1)$ = 9 Hz, are considerably larger than those

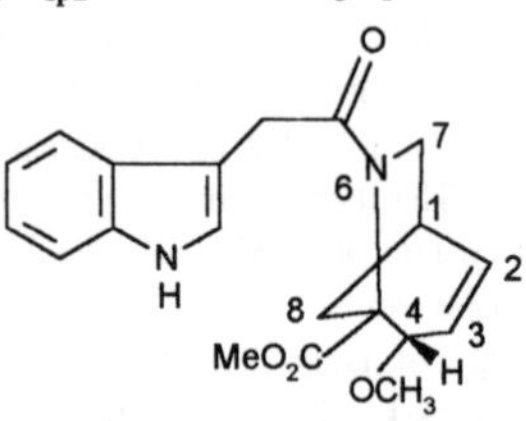

across the $C_{sp3}C_{sp3}C_{sp2}$ path, $^3J(H_2C_8)$ = 3.6 Hz. The influence of electronegative substituents on $^3J(^1H$-$^{13}C)$ and $^2J(^1H$-$^{13}C)$ couplings is also discussed in this paper; a conclusion was drawn that geminal and vicinal proton-carbon couplings accurately reflect the orientation of an electronegative substituent, and therefore their values may serve as suitable guides in stereochemical studies on rigid bicyclic systems.

Three-bond proton-carbon coupling constants play a crucial role in the conformational analysis of saccharides and polysaccharides, since they provide information on the geometry of interglycosidic bond and allow one to assign stereospecifically the exocyclic methylene protons in glycosyl residues. Therefore, it is not surprising that a variety of NMR experiments facilitating measurements of these couplings have been devised, and some new ones have been proposed recently. The ^1H-detected experiments which use selective excitation of ^{13}C have been designed by Adams and Lerner.[19] The use of ^{13}C-filtered 1D total correlation spectroscopy (TOCSY) as a new experiment for the measurement of long-range ^1H-^{13}C coupling constants has been proposed by Poppe et al.[368] The potential of the latter experiment was demonstrated by applying it to the assignment of the C-6-methylene protons of the internal glucosyl residue in panose, D-Glc-α(1→6)- D-Glc-α-(1→4)-D-Glc(α,β) (Glc = glucose).

Conformational analysis by the use of $^3J(^1H-^{13}C)$ couplings has been performed for 4,1′,6′-trichloro-4,1′6′-trideoxy-galacto-sucrose (sucralose),[334] for a cyclic (1→2)-glucan icosamer,[369] for hydrated $6^A,6^C,6^E$-tri-*O*-methylcyclohexaamylose,[370] and for a repeating disaccharide of hyaluronan.[331] An analysis of conformational equilibria has been performed for analogues of maltose in aqueous solutions by the use of $^3J(^1H-^{13}C)$ couplings.[326] They have been also determined for the cardiac glycoside digoxin, an important drug extracted from the leaves of *Digitalis lanata*,[371] and applied in studies on the three-dimensional behaviour of the selectin ligands Lewis^a trisaccharide [Galβ1-3(Fucα1-4)GlcNAc] and its sulphated derivative [$NaSO_3$-3Galβ1-3(Fucα1-4)GlcNAc].[372] It was concluded in the latter case that the introduction of a sulphate group at the 3-position of galactose in Le^a did not cause a detectable change in the conformation or the degree of internal mobility of the trisaccharide, since identical $^3J(^1H-^{13}C)$, NOEs and chemical shift values were observed for both Le^a and SuLe^a.

Peptides and proteins are another group of compounds in the structural studies of which an application of the $^3J(^1H-^{13}C)$ couplings can be especially fruitful. As has been shown by Mierke and Kessler,[222] with more than one coupling about a dihedral angle a great reduction of the allowed conformational space is obtained. This significantly improves molecular dynamics simulation in the determination of peptide structures. The $^3J(^1H-^{13}C)$ couplings have been used in the conformational analysis of the cyclic decapeptide antamanide,[220] and to elucidate the conformation of three cyclic hexapeptides cyclo(-Gly-Pro-Phe-Val-Xaa-Phe), where Xaa = Phe, D-Phe and D-Pro.[230]

A large value of the $^3J(^1H-^{13}C(O))$ constant of ca. 10 Hz estimated for the *cis* (Pro_1) form of H-Trp-Pro-Tyr-OH has allowed to assign the most preferred χ_1 rotamer of Trp (0.8-0.9 molar fraction) as the *trans* one.[373]

Vicinal $^1H-^{13}C$ couplings have been very useful in structural studies on natural products and on the compounds being potential drugs. The values of the main dihedral angles in lisinopril, *N,N*-[(s)-1-carboxy-3-phenylpropyl]-L-lysyl-L-proline, an inhibitor of angiotensin-converting enzyme and a new drug for the treatment of hypertension, have been determined using the experimental values of three-bond $^1H-^{13}C$ and $^1H-^1H$ couplings.[374] Further examples include the macrolide antibiotics roxithromycin, erythromycin A and their metabolites,[352,375] the potent HIV inhibitor, 4′-azidothymidine,[296] and fenamate non-steroidal anti-inflammatory drugs (the anthranilic acid derivatives).[376] The couplings have been used to elucidate the conformation of saturnisporins, the major components of the 20-residue peptaibol mixture isolated from a culture of the fungus *Trichoderma saturnisporum*,[377] and to establish the structure of perylenequinones, biologically active pigments.[378]

Using, among others, vicinal proton-carbon coupling constants, Osborne et al.[379] have been able to assign the correct structure to some Mannich bases, the mono-substituted products derived from 3-hydroxy-2-pyridone. These compounds should be regarded as the 4-isomers rather than, as was previously proposed, the 6-isomers.

The assignment of the configuration on a double CC bond is not always an easy task, especially when the vicinal proton-proton coupling constants are absent. Therefore, any additional piece of information in this respect is very important. Recently, Keah and Rae have demonstrated that the configuration of 3-bromopropenoic acids can be established unequivocally by the use of vicinal proton-carbon coupling constants.[380] It has been shown by them that the ranges for $^3J_{trans}$ and $^3J_{cis}$ do not overlap, the *cis* couplings being always smaller, i.e. 11.3-12.3 Hz and 4.4-4.8 Hz, respectively. This makes unambiguous assignments possible even in such cases when only one of the isomer pair is available. The $^3J(^1H-^{13}C)$ couplings have been also applied in order to assign the configuration on the trisubstituted double CC bond in anilinomethylenemalonic acid derivatives,[381] and to establish the correct configuration on the exocyclic double bond in substituted 5-methylenehydantoins and thiohydantoins.[382]

4-*N*-Methylamino-1,1,1-trifluorobut-3-en-2one is an interesting compound, as it can exists either in *trans* or *cis* configuration and in various *s-trans/s-cis* conformations depending on the solvent and the temperature. Conformational analysis by the use of spectral parameters which included also almost all coupling constants has been performed for it by Wojcik et al.[383]

Three-bond proton-nitrogen coupling constants $^3J(^1H-^{15}N)$ have been used in conformational analysis of the solution structures of the DNA methyl phosphotriester repair domain of *Escherichia coli* Ada,[257] of the periplasmic cyclophilin type *cis-trans* peptidyl-prolyl isomerase from *Escherichia coli*,[267] of calcium-loaded calbidin$_{9k}$,[272] and of the GDP form of a truncated version of the ras p21 protein (residues 1-166).[40]

The relative signs and magnitudes of the five coupling constants, $^3J(^1H-^{19}F)$, $^1J(^{13}C-^{19}F)$, $^2J(^{13}C-^{19}F)$, $^1J(^1H-^{13}C)$ and $^2J(^1H-^{13}C)$, have been measured for several simple aliphatic fluorocarbon molecules containing CF_3C fragments by the use of $^{19}F-^{13}C$, $^1H-^{13}C$ multi and one-bond 2D correlations. The vicinal $^1H-^{19}F$ coupling was assumed to be positive and the remaining couplings were determined relative to it.[384] The purpose of this work was to obtain the patterns which could be further used in analysis of the spectra of metabolic products of drug or pesticide models.

An analysis of the ABMX spectra of the $CH-CH_2F$ fragments in 4-deoxy-4-fluoro-D-glucose and 6-deoxy-6-fluoro-D-galactose has yielded accurate vicinal $^1H-^{19}H$ and $^1H-^{19}F$ coupling constants, which, in turn, *via* iterative analysis of the observed couplings, allowed the determination of both the individual rotamer couplings and the rotamer populations in the compounds under study.[385]

An accurate method for simultaneous estimation of vicinal $^1H-^{31}P$ and $^1H-^1H$ coupling constants along the backbone of DNA fragments has been described by Chary et al.,[386] who designed for this purpose the two-dimensional heteronuclear long-range correlation (2D HELCO) experiment. Vicinal $^1H_{3'}-^{31}P$ coupling constants have been measured for six related octanucleotide duplexes, and a good correlation was observed between the ^{31}P NMR chemical shifts and coupling constants for all these compounds.[387] Conformations of the cyclic phosphate rings have been derived from analysis of the $^1H-^1H$ and $^1H-^{31}P$ vicinal coupling constants performed by Ciesla et al.[388] for the cyclic phosphates of various acyclonucleosides, some with known antiviral activity, including 9-[1,3-dihydroxy-2-propoxy)methyl]guanine and its carbocyclic congener, 9-(3,4-dihydroxybutyl)guanine, 9-[4-hydroxy-2-(hydroxymethyl)-butyl]-guanine, and 9-(2,3-dihydroxypropyl)guanine. Rotamer populations of torsion angle (C3′-C2′-O2′-P) were obtained from $^3J(^1H-^{31}P)$ and $^3J(^{31}P-^{31}P)$ coupling constants for a series of 3′-fluorinated A(2′-5′)A(2′-5′)A fragments,[303] of two related CpG-containing octanucleotide duplexes,[389] of the thrombin-binding DNA aptamer d(GGTTGGTGTGGTTGG),[306] and of a third cyclobutane photodimer of thymidylyl-(3′→5′)-thymidine.[293] The method for measuring $^3J(^1H-^{31}P)$ and $^3J(^{31}P-^{31}P)$ coupling constants based on fitting multiplets containing the heteronuclear couplings to a reference multiplet without couplings has been presented by Schwalbe et al.[390] and applied to Rp-cyclic methylphosphonate.

A strikingly strong dependence of $^3J(^1H-^{31}P)$ couplings on dihedral angle has been observed in transition-metal complexes of di- and tri-phosphines derived from 1,1-bis(diphenylphosphinomethyl)ethene by Bookham et al.[391] The ^3J coupling of ca. 33 Hz was observed between H_A and P_A nuclei, and smaller than 1 Hz for the H_B and P_A pair (see formula below for atom labelling; phenyl groups attached to the phosphorus atoms are omitted for the sake of clarity). The relevant mean dihedral angles determined from X-ray analysis in these rigid polycyclic systems are 193° and 77° . This correlates well with the observed $^3J(^1H-^{31}P)$ values which, according to a Karplus ($\cos^2 \phi$) relationship, should be large for ϕ of ca. 180° , and small for ϕ between 60-90°.

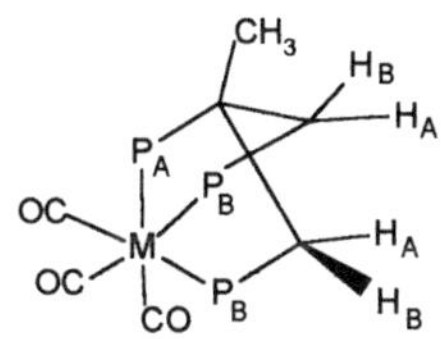

Vicinal proton-phosphorus coupling constants have been applied in a conformational study on bis(5,5-dimethyl-2-X-1,3,2-dioxaphosphorinan-2-yl)sulphanes and polysulphanes performed by Komber and Grossmann:[392] $(CH_3)_2C(CH_2O)_2P(X)$-Y-$(X)P(OCH_2)_2C(CH_3)_2$ (X = O, S; Y = S_n (n = 1-4)).

Studies on vicinal ^{1}H -*cysteine*^{113}Cd coupling constants in Cd-substituted metalloproteins performed by Zerbe et al.[393,394] have shown that also these couplings, like many others, follow a Karplus-type dependence. The $^3J(^1H-^{113}Cd)$ couplings, which varied from 0 to 74.3 Hz, have been determined for cysteine Cβ protons in (^{113}Cd)metallothionein from rat liver and ^{113}Cd-substituted *Desulfovibrio gigas* rubredoxin.[393] The Hβ-Cβ-Sγ-Cd dihedral angles were derived from the respective crystal structures by assuming tetrahedral geometry around Cα.

Evidence for a pentacoordinate structure of some recently synthesized new azastannatranes $ZSn(MeNCH_2CH_2)_3N$ and $ZSn(HNCH_2CH_2)_3N$ (Z = Alk, Ph, Me_3N) has been obtained from vicinal (or geminal) $^1H-^{119}Sn$ coupling constants measured for the protons of the apical groups. The couplings are consistently smaller (by several Hz) than those in the corresponding tetracoordinated compounds. Thus, for example, $^3J(^1H-^{119}Sn)$ of 43.5 Hz was observed for $Me_2NSn(MeNCH_2CH_2)_3N$, and 51.0 Hz for $Sn(NMe_2)_4$.[395] These differences are consistent with the smaller s character of the orbitals forming the apical bond of the five-coordinate compounds and the smaller positive charge on the tin atom.

An indirect ^{199}Hg detection experiment, carried out *via* the $^1H-^{199}Hg$ couplings using the 1D-HMQC sequence, has been performed by Neirinck et al.[396] for 5-acetoxymercuriovanillin and 2-acetoxymercuriosyringaldehyde, two crucial model compounds in the studies on the structure of lignin. The experiment not only provided a reliable information on the aromatic ring substitution pattern, but also gave the chemical shifts of protons coupled to ^{199}Hg nucleus and the precise $J(^1H-^{199}Hg)$ values: $^3J(^1H-^{199}Hg)$ of ca. 220 Hz, $^4J(^1H-^{199}Hg)$ of ca. 68 Hz and $^5J(^1H-^{199}Hg)$ of 22 Hz were observed.

Only one example of the complexes containing N(6)-deprotonated adenine ligands has been reported so far. A new example of this type of the compound has been recently obtained by

Trovo et al.[397] The dinuclear complex *cis*-$[(PMe_3)_2Pt(\mu\text{-9-EtAd(-H)})]_2(NO_3)_2$, where Ad(-H) denotes the N(6)-deprotonated nucleobase, has been characterized by NMR spectroscopy. The H(2) and H(8) resonances, which are indistinguishable in the proton spectrum of the free base, appear to be well differentiated upon platination of the nucleobase, with the H(2) proton coupled to the metal centre, $^3J(^1H\text{-}^{195}Pt)$ = 15.5 Hz.

9. Three-Bond Couplings not Involving Hydrogen

A ^{18}O-labelling has been applied by Mega and van Etten in order to remove the chemical shift degeneracy in succinic acid and dimethylmaleic anhydride.[398] An analysis of the complex ^{13}C NMR spectra arising from the various isotopomers of ^{18}O-labelled $[1,4\text{-}^{13}C_2]$ succinic acid allowed the authors to measure the vicinal $^{13}C\text{-}^{13}C$ coupling constant in this compound and to study the influence of pH on it. These results provide direct access to dihedral angle information in simple succinic acid derivatives giving in turn insight into more complex systems such as DNA-ligand interactions, protein-ligand binding, and a conformation of lipids.

In two papers simple experiments that allow measurement of rather small three-bond $^{13}C\text{-}^{13}C$ coupling constants in 13-C enriched proteins have been described.[399,400] In both cases the stereospecific assignment of γ-methyl groups in valine residues became possible. The $^3J(^{13}C\text{-}^{13}C)$ data for ribonuclease T_1 have been published by Schwalbe et al.,[399] and those for alanine in the 20 kDa complex between the protein calmodulin and a 26-residue synthetic peptide by Grzesiek et al.[400]

A full set of vicinal coupling constants has been derived for valine residues in ribonuclease T_1. This also included carbon-carbon and proton-carbon couplings. These couplings, however, were not considered by the authors in the quantitative evaluation of valine χ_1 angles, but they were rather used for the stereospecific assignment of the methyl resonances.[276]

By analogy to three-bond proton-proton and carbon-carbon couplings, the vicinal coupling constants between the carbon and nitrogen nuclei can be extremely useful in determining backbone and side-chain conformation in peptides and proteins. Recently, an experiment has been described by Vuister et al.[33] which provides quantitative information on the three-bond coupling between methyl carbons and backbone amide ^{15}N nuclei. Combined with $^3J(^{13}C\text{-}^{13}C)$ couplings, this allows one to determine the χ_1 angle in valine, isoleucine and threonine residues. The method applied is essentially a 2D difference experiment which, when applied to uniformly enriched in ^{15}N and ^{13}C the calmodulin-peptide complex, gave $^3J(^{13}C\text{-}^{15}N)$ couplings of 0.6 to 2.26 Hz.

Three-bond $^{13}C\text{-}^{31}P$ coupling constants combined with vicinal $^1H\text{-}^{31}P$ coupling constants play a particularly important role in conformational analysis of DNA and RNA. They have been applied by Agback et al.[301] in their study on the lariat RNA tetramer and pentamer, which mimic the naturally occurring lariat RNA.

Substituent effects on $^3J(^{13}C\text{-}^{31}P)$ in bioactive 4-substituted 2,6,7-trioxa-1-phosphabicyclo[2.2.2.]octane-1-oxides have been studied by Hu et al.[401] One- and three-bond $^{13}C\text{-}^{31}P$ coupling constants in the phosphonate skeleton of dimethyl esters of 2-hydroxy (or 2-methoxy)alkylphosphonic acids have been determined by Belciug et al.[402] from ^{13}C NMR spectra of these compounds recorded in various solvents and in the presence of metal ions. A reasonable correlation between the $^3J(^{13}C\text{-}^{31}P)$ values and the population of the conformer involving the *trans* orientation of the phosphonate group and the C-3 carbon atom was obtained. In this way the $^{13}C\text{-}^{31}P$ couplings can be used in addition to the ^1H NMR spectra for the conformational analysis of the non-rigid, acyclic, phosphonic systems. Vicinal and geminal ^{13}C-^{31}P coupling constants along with five-bond couplings, $^5J(^1H$-$^{31}P)$, have been determined for some novel 6-substituted-2,10-dichloro-12*H*-dibenzo[*d,g*][1,3,2]dioxaphosphocin 6-oxides.[403]

In order to get insight into the solution-phase stereochemistry of the chelate rings and aryl group arrangement of *p*-aryl-substituted chiral ligands in some Rh(I) and Pt(II) complexes, an analysis of $^nJ(^{13}C$-$^{31}P)$ (n = 1-3), $^3J(^1H$-$^{31}P)$ and $^3J(^1H$-$^1H)$ coupling constants has been performed by Toth and Hanson.[404] The complexes studied were: [Rh(diene)L_2]BF_4 and PtL_2Cl_2, where L_2 = DIOP-$(pNMe_2)_4$, BDPP-$(pNMe_2)^4$, Chirophos-$(pNMe_2)_4$, diene = NBD or COD, and [Rh(diene)L_2]$(BF_4)_5$ where L_2=DIOP-$(pNMe_3{}^+)_4$, DIOP-$(pNHMe_2{}^+)_4$, BDPP-$(pNMe_3{}^+)_4$, BDPP-$(pNHMe_2{}^+)_4$, Chiraphos-$(pNMe_3{}^+)_4$, diene=NBD or COD. It was assumed by the authors that, by analogy to one-bond ^{13}C-^{31}P coupling constants, the axial three-bond and two-bond ^{13}C-^{31}P coupling constants are smaller than the equatorial ones. This allowed them to assign the ^{13}C NMR chemical shifts in the compounds studied.

The fact that vicinal ^{13}C-^{199}Hg couplings are strongly dihedral angle dependent showing the Karplus type relationship was recognised a long time ago. The $^3J(^{13}C$-$^{199}Hg)$ values measured for several mercuration products of some aliphatic carboxylic acids, such as 2(chloromercurio)isovaleric acid, 2(chloromercurio)-*n*-valeric acid and 2(chloromercurio)-*n*-butyric acid have been used by Korpar-Colig et al.[405] to estimate the conformation of these compounds. The estimated averaged dihedral angles were 30-50 for *gauche* isomer in 2(chloromercurio)isovaleric acid and 130-150 for *anti* forms in all compounds studied.

Vicinal ^{15}N-^{103}Rh coupling constants of about 0.7 Hz have been observed in the spectra of bond isomeric hexakis(thiocyanato(N)-thiocyanato(S)rhodates $[Rh(NCS)_n(SCN)_{6-n}]^{3-}$, n = 0-4.[406]

^{19}F NMR chemical shifts and $^3J(^{19}F$-$^{19}F)$ coupling constants have been published for a series of aryl (phenyl, naphthyl and phenanthryl) substituted trifluorovinyls and (*E*) and (*Z*)-perfluoropropenyls.[407] Distinctive differences in $^nJ(^{19}F$-$^{19}F)$ values (n = 3-4) made the assignment of all fluorines in the perfluorovinyl system easy: $^3J(^{19}F$-$^{19}F)$ *trans* > $^2J(^{19}F$-$^{19}F)$*gem* > $^3J(^{19}F$-$^{19}F)$ *cis* > $^4J(^{19}F$-$C^{19}F_3)$*cis* > $^4J(^{19}F$-$C^{19}F_3)$*trans* was observed.

The Karplus-like dependence of $^3J(^{31}P$-$^{31}P)$ coupling constants has been used by Hägele et al.[408] in order to elucidate the conformation of *trans*-cyclohexane-1,2-diphosphonic acid and its tetraethyl ester. The $^3J(^{31}P$-$^{31}P)$ coupling constants varying from 81.5 Hz at 198 K to 87.2 Hz at 180 K were obtained for tetraethyl ester providing evidence that for this compound a bisaxial arrangement of the phosphonic groups prevails. The $^3J(^{31}P$-$^{31}P)$ coupling values for the acid are considerably smaller (ca. 20-25 Hz only), and depend on a degree of titration providing evidence that for this compound the bisequatorial conformation prevails.

10. Couplings Across More Than Three Bonds and Through Space

The proton-proton couplings across three and four bonds have been used in the assignment of the relative configuration of 4,8-bis(4-hydroxy-3-metoxyphenyl)-3,7-dioxabicyclo[3.3.0]octan-2-ones, representatives of a major group of the naturally occurring lignans.[409] The coupling constants $^4J(^1H$-$^1H)$ of ca 2.5 Hz and $^3J(^1H$-$^1H)$ of 4.9 to 9.3 Hz have been measured for nitrated aminopyridines.[410] The inter-ring proton-proton coupling constants across four, five, six and seven bonds have been observed in the spectra of some 8-hydroxyquinoline derivatives;[411] it has been suggested that the π -mechanism predominates in the transmission of these couplings. A five-bond proton-proton coupling through the cyclopropane ring has been determined in cyclopropane enolactones.[412] A six-bond coupling constant between the hydroxyl and the formyl protons of ca. 1.2 Hz has been observed in 7-formyl-6-hydroxynaphtho[1,8-*b,c*]pyran and some related compounds.[413] Since a hydrogen bond is formed between the phenolic proton and the carbonyl oxygen, a through-space interaction between the two protons is excluded in this case. It has been therefore suggested by the authors that the coupling observed occurs through the hydrogen bond. It is of interest to note that such interaction was not observed in 1-formyl-8-hydroxynaphthalene.

Some new examples of proton-proton couplings dominated by the through-space mechanism have been found in the H,H COSY spectra of rigid cyclic terpenes by Schröder and Haslinger.[414] Such couplings have been quite often reported for fluorines, but only a few examples have been reported for protons so far.

The temperature dependent coupling constants across six bonds between ^{1}H and ^{19}F nuclei have been observed in 2(4-fluorophenyl)ethene-1,1-dicarbonitrile and 4-fluorobenzylidenemalonaldehyde.[415] The $^{6}J(^{1}H\text{-}^{19}F)$ couplings of 0.217 Hz and 0.568 Hz, respectively, were observed for these compounds at 299 K; at 333 K the values are about 15% larger, which means that the couplings are always positive. The authors indicated that the change was too drastic to arise merely from the intrinsic temperature effects. The large long-range interring H-F and F-F coupling constants of a few hertz were observed in $[Mo(C_6F_6)(C_6H_3F_3\text{-}1,3,5)]$ and $[W(C_6F_6)(C_6H_3F_3\text{-}1,3,5)]$ complexes, which facilitated the interpretation of their ^{19}H NMR spectra.[416]

A through-space coupling of 3 Hz between the $^{1}H_{6,7}$ and ^{31}P nuclei has been found in the 1,4,5,8,9,10-hexahydroanthracene complex with ruthenium, $Ru(\eta^{6}\text{-}C_6H_{14})Cl_2(PPh_3)$; no coupling to protons 5,8 was observed.[417]

In continuation of their studies on the conformational equilibrium in variously substituted aromatic compounds, Schaefer et al. have performed measurements and analysis of $^{4}J(^{1}H\text{-}^{31}P)$ and $^{3}J(^{13}C\text{-}^{31}P)$ coupling constants for 2-(diphenylphosphino)benzaldehyde in polar and nonpolar solutions,[418] and on large perturbations of long-range $^{1}H\text{-}^{1}H$ and $^{1}H\text{-}^{19}F$ coupling constants by the intramolecular hydrogen bonds in 2-mercaptobenzaldehyde, salicylaldehyde and some other derivatives.[419]

A long-range $^{1}H\text{-}^{119}Sn$ coupling across four and five bonds has been observed in the spectrum of six-co-ordinate tris(bidentate ligand tin(IV) complex, $[NH_4]_2[Sn(catechol)_3]$.[420]

A large coupling constant of 125.1 Hz has been observed in the spectrum of 2-(chloromercurio)-*n*-valeric acid between the proton of the methyl group and the Hg nucleus.[405] The large magnitude of this coupling was explained by the authors in terms of through-space interaction between the atoms involved.

Scalar $^{13}C\text{-}^{19}F$ coupling constants between carbonyl carbons and aromatic fluorine, for which the through-space mechanism dominates, have been used by Szczecinski to study the conformation of chelate phosphitodicarbonylchromium and dicarbonyl(phosphine)chromium complexes of fluorobenzenes,[421] and to get insight into the structure of tricarbonylchromium complexes of biphenyl derivatives.[422] The coupling values vary from 0 to 6.9 Hz depending on the distance between the coupled nuclei.

A four-bond coupling constant, $^{4}J(^{13}C\text{-}^{31}P)$ of ca. 2 to 3 Hz, has been invoked as an indication that a close contact exists between a heteroatom Y (Y = O, S, Se) connected with the axial phosphorus P =Y and axial protons H(4,6) in the 1,3-dithiane ring of 1,3-dithianes containing 2-phosphoryl, -(thiophosphoryl), and -(selenophosphoryl) groups.[423]

A study has been continued by Pascal and co-workers[424] on through-space couplings ($J(^{13}C-^{31}P)$ = 7.0-8.2 Hz) between the carbons of the substituted basal aromatic ring and the phosphorus atoms of some new substituted triarylphosphine-containing cyclophanes bearing electron-withdrawing and electron-donating substituents on the basal aromatic ring.

A very carefully planned study on the geometry dependence of through-space couplings between two fluorine nuclei has been published by Ernst et al.[425] $J(^{19}F-^{19}F)$ coupling constants varying from 42.1 to 99.2 Hz were observed in the difluorinated *syn*-[3.3]metacyclophanes and their [2.2]analogues. The fluorine atoms involved are separated formally by eight bonds, and their large values can be explained in terms of a through-space mechanism only. An increase of the $J(^{19}F-^{19}F)$ values correlates well with the decrease of the F,F internuclear distances from 2.78 to 2.48.

The semiempirical calculations performed for bis-(dialkoxythiophosphoryl)-trisulfanes, $(EtO)_2P(S)SSSP(S)(OEt)_2$ have shown that the relatively large four-bond $^{31}P-^{31}P$ coupling constant of 10 to 13 Hz observed in these compounds[426] was caused by conformations with PSSS dihedral angles of about 90°, and not by conformations with W-like or other plane positions of the PSSSP nuclei.

A coupling of 4.8 Hz across seven bonds between two phosphorus atoms has been determined for the azine diphosphine *Z,Z*-PPh_2CH_2C(*t*-Bu)=NN=C(*t*-Bu)CH_2PPH_2 from an analysis of the second order ^{13}C NMR spectra recorded for this compound.[427] A through-space mechanism was invoked to explain its value.

Large through-space coupling constants of ca. 110 Hz have been found between two four-bonded phosphorus nuclei in unsymmetrically bisphosphorylated *N,N'*-dimethyl thiourea derivatives $Ph_2PNMeC(=S)NMePR_2$ (R = F, Me, OMe).[428] A coupling of 124 Hz was earlier reported for $Ph_2PNMeC(=O)NMePMe_2$ by Breker (see ref. 4 in ref. 428).

An unusually large $^{31}P-^{31}P$ coupling constant of 72.8 Hz across eight bonds has been observed by Pastor et al. in the ^{31}P NMR spectrum of 2-{1-{3,5-bis(1,1-dimethylethyl)-2-{[2,4,8,10-tetrakis(1,1-dimethylethyl)dibenzo [*d,f*][1,3,2]dioxaphosphepin-6-yl]oxy}phenyl}ethyl}-4,6-bis(1,1-dimethylethyl)phenylphenyl diphenyl phosphite. The authors suggested that a through-space coupling mechanism must operate in this case, since an intramolecular P,P distance found in this molecule in its X-ray structure is within the sum of the van-der-Waals radii, i.e. 3.67 Å only.[429]

References

1 D.F. Mierke, T. Huber and H. Kessler, *J.Comput.Aid.Molec.Design.*, 1994, **8**, 29.
2 D.M. LeMaster, *Progr.NMR Spectrosc.*, 1994, **26**, 371.
3 G.W. Canters, C.W. Hilbers, M. van de Kamp and S.S. Wijmenga, *Meth.Enzymol.*, 1993, **227**, 243.
4 M.S. Searle, *Progr.NMR Spectrosc.*, 1993, **25**, 403.
5 A.N. Lane, *Progr.NMR Spectrosc.*, 1993, **25**, 481.
6 J.A. Jaeger, J. Santalucia, Jr., and I. Tinoco Jr, *Annu.Rev.Biochem.*, 1993, **62**, 255.
7 K. Bock and J.Φ. Duus, *J.Carbohydr.Chem*, 1994, **13**, 513.
8 R.H. Contreras and J.C. Facelli, *Annual Reports on NMR Spectroscopy*, 1993, **27**, 255.
9 A. Lycka, *Annual Reports on NMR Spectroscopy,* 1993, **26**, 247.
10 G. Barbarella, *Progr.NMR Spectrosc.*, 1993, **25**, 317.

11 A.R. Bassindale and J.N. Iley (1993): *The chem. of sulphur-containing func.groups,* eds. S. Patai and Z. Rappoport, ed. J. Wiley.
12 K. Kamienska-Trela, *Nuclear Magnetic Resonance*, 1993, **23**, 137.
13 R.R. Ernst, *Pulsed Magn.Reson.,NMR,ESR,Opt.*, 1992, 95; eds. Bagguley and M.S Desmond; Oxford University Press.
14 F. del Rio Portilla and R. Freeman, *J.Chem.Soc.Faraday Trans.,* 1993, **89**, 4275.
15 F. del Rio Portilla and R. Freeman, *J.Magn.Reson.*, 1993, **104A**, 358.
16 E. Kupce and R. Freeman, *J.Magn.Reson.*, 1993, **104A**, 234.
17 E. Kupce, J.-M. Nuzillard, V.S. Dimitrov and R. Freeman, *J.Magn.Reson.*, 1994, **107A**, 246.
18 V. Blechta, F. del Rio-Portilla and R. Freeman, *Magn.Reson.Chem.*, 1994, **32**, 134.
19 B. Adams and L. Lerner, *J.Magn.Reson.*, 1993, **103A**, 97.
20 D. Yang and C. Ye, *Magn.Reson.Chem.*, 1993, **31**, 995.
21 H.B. Olsen, S. Ludvigsen and O.W. Soerensen, *J.Magn.Reson.*, 1993, **104A**, 226.
22 T.J. Norwood and K. Jones, *J.Magn.Reson.*, 1993, **104A**, 106.
23 G. Zhu and A. Bax, *J.Magn.Reson.*, 1993, **104A**, 353.
24 P. Huber, C. Zwahlen, S.J.F. Vincent and G. Bodenhausen, *J.Magn.Reson.*, 1993, **103A**, 118.
25 K.E. Koever, O. Prakash and V.J. Hruby, *J.Magn.Reson.*, 1993, **103A**, 92.
26 S. Nicula and G. Bodenhausen, *Chem.Phys.Lett.*, 1994, **219**, 291.
27 J.M. Bernassau and J.M. Nuzillard, *J.Magn.Reson.*, 1993, **104A**, 212.
28 E. Kupce, E. Lukevics and B. Wrackmeyer, *Magn.Reson.Chem.*, 1994, **32**, 326.
29 C. Roumestand, D. Boudot and D. Canet, *J.Magn.Reson.*, 1993, **105A**, 304.
30 P. Huber and G. Bodenhausen, *J.Magn.Reson.*, 1993, **104A**, 96.
31 U. Wollborn, W. Willker and D. Leibfritz, *J.Magn.Reson.*, 1993, **103A**, 86.
32 H.-M. Mons, K. Bergander and H. Guenther, *Helv.Chim.Acta*, 1993, **76**, 1216.
33 G.W. Vuister, A.C. Wang and A. Bax, *J.Am.Chem.Soc.,* 1993, **115**, 5334.
34 G.W. Vuister, T. Yamazaki, D.A. Torchia and A. Bax, *J.Biomol.NMR*, 1993, **3**, 297.
35 G. Vuister and A. Bax, *J.Magn.Reson.*, 1993, **102B**, 228.
36 J.C. Madsen, O.W. Sorensen, P. Sorensen and F.M. Poulsen, *J.Biomol.NMR*, 1993, **3**, 239.
37 H. Oschkinat, T. Mueller and T. Dieckmann, *Angew.Chem.Int.Ed.Engl.*, 1994, **33**, 277.
38 A. Bax and S. Grzesiek, *Acc.Chem.Res.*, 1993, **26**, 131.
39 J. Anglister, S. Grzesiek, A.C. Wang, H. Ren, C.B. Klee and A. Bax, *Biochemistry*, 1994, **33**, 3540.
40 P.J. Kraulis, P.J. Domaille, S.L. Campbell-Burk, T. van Aken and E.D. Laue, *Biochemistry*, 1994, **33**, 3515.
41 J. Chung, J.R. Tolman, K.P. Howard and J.H. Prestegard, *J.Magn.Reson.*, 1993, **102B**, 137.
42 S. Berger and P. Bast, *Magn.Reson.Chem.*, 1993, **31**, 1021.
43 G. Wu and R.E. Wasylishen, *J.Chem.Phys.*, 1994, **100**, 5546.
44 G. Wu and R.E. Wasylishen, *J.Chem.Phys.*, 1994, **100**, 4828.
45 J.R. Tolman and J.H. Prestegard, *J.Magn.Reson.*, 1994, **106A**, 203.
46 S.H. Alarcon, A.C. Olivieri and R.K. Harris, *Solid State Nucl.Magn.Reson.*, 1993, **2**, 325.
47 R.K. Harris, M.S. Jones and A.M. Kenwright, *Magn.Reson.Chem.*, 1993, **31**, 1085.
48 H. Beckmann, G. Grossmann and G. Ohms, *Magn.Reson.Chem.*, 1992, **30**, 860.
49 A. Jarid, M. Moreno, A. Lledos, J.M. Lluch and J. Bertran, *J.Am.Chem.Soc.*, 1993, **115**, 5861.
50 E.M. Hiller and R.A. Harris, *J.Chem.Phys.,* 1994, **100**, 2522.

51 E.M. Hiller and R.A. Harris, *J.Chem.Phys.*, 1993, **98**, 2077.
52 E.M. Hiller and R.H. Harris, *J.Chem.Phys.*, 1993, **99**, 7652.
53 L.D. Field, T.W. Hambley and B.C.K. Yau, *Inorg.Chem.*, 1994, **33**, 2009.
54 G.W. Vuister, F. Delaglio and A. Bax, *J.Biomol.NMR*, 1993, **3**, 67.
55 M.S. Prachand, S. Singh, M.M. Dhingra, U. Singh, S.K. Ghosh, V.R. Mamdapur and M.S. Chadha, *Magn.Reson.Chem.*, 1993, **31**, 944.
56 M. Brakta, R.N. Farr, B. Chaguir, G. Massiot, C. Lavaud, W.R.. Anderson Jr., D. Sinou and G.D. Daves Jr., *J.Org.Chem.*, 1993, **58**, 2992.
57 K.B. Wiberg, B.S. Ross, J.J. Isbell and N. McMurdie, *J.Org.Chem.*, 1993, **58**, 1372.
58 D.P. Kelly, K. Aherne, F. Delgado, J.J. Giansiracusa, W.A. Jensen, K. Karavokiros, R.A. Mantello and M.E. Reum, *J.Am.Chem.Soc.*, 1993, **115**, 12010.
59 G.Y. Ryu and J.H. Shin, *Bull.Korean Chem.Soc*, 1993, **14**, 546.
60 H.J. Reich and R.R. Dykstra, *J.Am.Chem.Soc.*, 1993, **115**, 7041.
61 D. Hoffmann, W. Bauer, F. Hampel, N.J.R. van Eikema Hommes, P. von Rague Schleyer, P. Otto, U. Pieper, D. Stalke, D.S. Wright and R. Snaith, *J.Am.Chem.Soc.*, 1994, **116**, 528.
62 W.T. Dixon, *Theor.Chim.Acta*, 1966, **6**, 359.
63 C.J. Rhodes, J.C. Walton and E.W. Della, *J.Chem.Soc.,Perkin Trans.2*, 1993, 2125.
64 R. Knorr, T. von Roman, J. Freudenreich, T.P. Hoang, J. Mehlstaubl, P. Boehrer, D.S. Stephenson, H. Huber and B. Schubert, *Magn.Reson.Chem.*, 1993, **31**, 557.
65 J.-T. Hung, S.-W. Yang, G.M. Gray and K. Lammertsma, *J.Org.Chem.*, 1993, **58**, 6786.
66 A.L. Esteban, M.P. Galache, E. Diez, R.R. Biekofsky and R.H. Contreras, *Magn.Reson.Chem.*, 1994, **32**, 199.
67 E. Kolehmainen, K. Laihia, R. Kauppinen, R. Gawinecki and D. Rasala, *Magn.Reson.Chem.*, 1993, **31**, 659.
68 H. Ban-Oganowska, *Magn.Reson.Chem.*, 1993, **31**, 1092.
69 R.S. Varma and A. Shukla, *Magn.Reson.Chem.*, 1993, **31**, 612.
70 C. Lopez, R.M. Claramunt, S. Trofimenko and J. Elguero, *Can.J.Chem.*, 1993, **71**, 678.
71 J. Elguero, R.M. Claramunt, C. Lopez and D. Sanz, *Afinidad*, 1993, **50**, 383.
72 A.E. Aliev, A.A. Sinitsina, T.N. Borisova, I.A. Stazharova, N.S. Prostakov and A.V. Varlamov, *Khim.Geterotsikl.Soedin.*, 1993, 75.
73 K. Spirkova and M. Dandarova, *Chem.Papers,* 1993, **47**, 48.
74 J. Imrich, T. Busova, D. Koscik and T. Liptaj, *Chem.Papers*, 1993, **47**, 102.
75 M. Pelecanou, E. Mikros and P. Catsoulacos, *Magn.Reson.Chem.*, 1994, **32**, 178.
76 S. Chimichi, B. Cosimelli, P. Tedeschi and F. Ponticelli, *Magn.Reson.Chem.*, 1994, **32**, 312.
77 S. Chimichi, B. Cosimelli, F. Bruni and A. Constanzo, *J.Chem.Soc.,Perkin Trans.2*, 1993, 209.
78 E. Kolehmainen, K. Laihia, M. Heinanen, K. Rissanen, R. Froehlich, J. Korvola, P. Manttari and R. Kauppinen, *J.Chem.Soc.,Perkin Trans.*2, 1993, 641.
79 A. Wojcicki, *J.Cluster Sci.*, 1993, **4**, 59.
80 E. Kolehmainen, K. Laihia, M.I. Rybinskaya, V.S. Kaganovich and Z.A. Kerzina, *J.Organomet.Chem.*, 1993, **453**, 273.
81 U. Boehme and K.-H. Thiele, *J.Organomet.Chem.*, 1993, **452**, 75.
82 A.E. Aliev, K.D.M. Harris, D.C. Apperley, R.K. Harris and M.M. Sunnetcloglu, *J.Chem.Soc.Faraday Trans.*, 1993, **89**, 3791.
83 J. Jazwinski and L. Stefaniak, Magn.Reson.Chem., 1993, **31**, 447.
84 W. Bocian, J. Jazwinski, W. Kozminski, L. Stefaniak and G. Webb, *J.Chem.Soc.,Perkin Trans.2*, 1994, 1327.

85 R.M. Claramunt, D. Sanz, J. Catalan, F. Fabero, N.A. Garcia, C. Foces-Foces, A.L. Llamas-Saiz and J. Elguero, *J.Chem.Soc.,Perkin Trans.2,* 1993, 1687.

86 P.E. Hansen, A.E. Hansen, A. Lycka and A. Buvari-Barcza, *Acta Chemica Scandinavica,* 1993, **47**, 777.

87 A.R. Bassindale and J. Jiang, *J.Organomet.Chem.*, 1993, **446**, C3.

88 K. Hassler and G. Bauer, *J.Organomet.Chem.*, 1993, **460**, 149.

89 L.J. Procopio, P.J. Carroll and D.H. Berry, *J.Am.Chem.Soc.*, 1994, **116**, 177.

90 R.J. Batchelor, F.W.B. Einstein, I.D. Gay, C.H.W. Jones and R.D. Sharma, Inorg.Chem., 1993, **32**, 4378.

91 K. Bergander, R. He, N. Chandrakumar, O. Eppers and H. Guenther, *Tetrahedron*, 1994, **50**, 5861.

92 R. Taube, S. Wache and J. Sieler, *J.Organomet.Chem.*, 1993, **459**, 335.

93 V.G. Malkin, O.L. Malkina and D.R. Salahub, *Chem.Phys.Lett.*, 1994, **221**, 91.

94 A. Barszczewicz, M. Jaszunski, K. Kamienska-Trela, T. Helgaker, P. Jorgesen and O. Vahtras, *Theor.Chim.Acta,* 1993, **87**, 19.

95 I. Carmichael, D.M. Chipman, C.A. Podlasek and A.S. Serianni, *J.Am.Chem.Soc.*, 1993, **115**, 10863.

96 Z. Hu and C. Zhan, *Jiegou Huaxue,* 1993, **12**, 168.

97 C. Zhan and Z. Hu, *Jiegou Huaxue,* 1993, **12**, 237.

98 T. Nakai and C.A. McDowell, *Mol.Phys.*, 1993, **79**, 965.

99 T. Nakai and C.A. McDowell, *Mol.Phys.*, 1994, **81**, 337.

100 T. Nakai and C.A. McDowell, *J.Magn.Reson.*, 1993, **104A**, 146.

101 T. Nakai and C.A. McDowell, *Bull.Magn.Reson.*, 1993, **15**, 149.

102 J. Kawabata and E. Fukushi, *J.Magn.Reson.*, 1994, **107A**, 173.

103 W. Bocian, J. Jazwinski, L. Stefaniak and G.A. Webb, *J.Chem.Soc.,Perkin Trans.2*, 1994, 1467.

104 A. Dabrowski and K. Kamieńska-Trela, *J.Organomet.Chem.*, 1993, **460**, C1.

105 W. Bauer and C. Griesinger, *J.Am.Chem.Soc.*, 1993, **115**, 10871.

106 K. Kamieńska-Trela, A. Dabrowski and H. Januszewski, *Spectrochim.Acta*, 1993, **49A**, 1613.

107 P. Jankowski, K. Kamieńska-Trela, K. Minksztym and J. Wicha, *J.Organomet.Chem.*, 1993, **460**, 15.

108 M. Witanowski and Z. Biedrzycka, *Magn.Reson.Chem.*, 1994, **32**, 62.

109 N. Krause, R. Wagner and A. Gerold, *J.Am.Chem.Soc.*, 1994, **116**, 381.

110 J. Bluemel, N. Hertkorn, B. Kanellakopoulos, F.H. Koehler, J. Lachmann, G. Mueller and F.E. Wagner, *Organometallics*, 1993, **12**, 3896.

111 M.H. Chisholm, C.M. Cook, J.C. Huffman and J.D. Martin, *Organometallics*, 1993, **12**, 2354.

112 T.R. Forbus Jr and J.C. Martin, *Heteroatom Chem.*, 1993, **4**, 129.

113 M.J. Kochanny, T. Härd and J.A. Katzenellenbogen, *Magn.Reson.Chem.*, 1993, **31**, 977.

114 W. Eisenreich, G. Strauss, U. Werz, G. Fuchs and A. Bacher, *Eur.J.Biochem.*, 1993, **215**, 619.

115 C.T. Evans, B. Sumegi, P.A. Srere, A.D. Sherry and C.R. Malloy, *Biochem.J.*, 1993, **291**, 927.

116 J.M. Duker and A.S. Serianni, *Carbohydr.Res.*, 1993, **249**, 281.

117 S.J. Weiner, S.M. Holl and D.F. Covey, *Magn.Reson.Chem.*, 1994, **32**, 122.

118 J.M. Aramini, P.H. Krygsman and H.J. Vogel, *Biochemistry*, 1994, **33**, 3304.

119 Y. Chu, D. Colclough, J.B. White and W.B. Smith, *Magn.Reson.Chem.*, 1993, **31**, 937.

120 S.G. Lee, *Bull.Korean Chem.Soc.*, 1993, **14**, 416.

121 E. Kolehmainen, K. Rissanen, K. Laihia, P. Malkavaara, J. Korvola and R. Kauppinen, *J.Chem.Soc.,Perkin Trans.2*, 1993,, 437.
122 A.D. Rodriguez, A.L. Acosta and H. Dhasmana, *J.Nat.Prod.*, 1993, **56**, 1843.
123 J.K. Harper, N.K. Dalley, N.L. Owen, S.G. Wood and R.G. Cates, *J.Crystallogr.Spectrosc.Res.*, 1993, **23**, 1005.
124 M.F. Reich, V.J. Lee, J. Ashcroft and G.O. Morton, *J.Org.Chem.*, 1993, **58**, 5288.
125 E.W. Huber, R.J. Barbuch, C.R. Dalton, J.M. Kane and N.L. Velayo, *J.Heterocycl.Chem.*, 1993, **30**, 611.
126 I. Columbus and S.E. Biali, *J.Org.Chem.*, 1993, **58**, 7029.
127 N. Nakayama, A. Aoki and T. Hayashi, *Macromolecules*, 1994, 27.
128 W. Xu, H. Wang and S. Yang, *Fenzi Cuihun*, 1993, **7**, 388.
129 K.K. Laali and P.E. Hansen, *J.Org.Chem.*, 1993, **58**, 4096.
130 I. Schiketanz, A. Racoveanu-Schiketanz, M.D. Gheorghiu and A.T. Balaban, *Rev.Roum.Chim.*, 1992, **37**, 1315.
131 V. Torocheshnikov, D. Rentsch and W. von Philipsborn, *Magn.Reson.Chem.*, 1994, **32**, 348.
132 P. Kofod, N.W. Alcock and P. Moore, *Acta Chem. Scand.*, 1993, **47**, 1083.
133 B. Wrackmeyer, G. Kehr, D. Wettinger and W. Milius, *Main Group Met.Chem.*, 1993, **16**, 445.
134 B. Wrackmeyer and G. Kehr, *Main Group Met.Chem.*, 1993, **16**, 305.
135 M. Herberhold, M. Huebner and B. Wrackmeyer, *Z.Naturforsch.*, 1993, **48B**, 940.
136 F. Huber, M. Vornefeld, G. Ruisi and R. Barbieri, *Appl.Organomet.Chem.*, 1993, **7**, 243.
137 F. Caruso, M. Giomini, A.M. Giuliani and E. Rivarola, *J.Organomet.Chem.*, 1994, **466**, 69.
138 G. Wu and R. Wasylishen, *J.Phys.Chem.*, 1993, **97**, 7863.
139 W.-B. Lee, S.-C. Suen, T.-T. Jong, F.-E. Hong, J.-H. Chen, H.-J. Lin and L.-P. Hwang, *J.Organomet.Chem.*, 1993, **450**, 63.
140 G. Batta, I. Banyai and J. Glaser, *J.Am.Chem.Soc.*, 1993, **115**, 6782.
141 N.W. Mitzel, J. Riede, A. Schier, M. Paul and H. Schmidbaur, *Chem.Ber.*, 1993, **126**, 2027.
142 M. Herberhold, S. Gerstmann, W. Milius and B. Wrackmeyer, *Z.Naturforsch.*, 1993, **48B**, 1041.
143 D. Gudat, *Magn.Reson.Chem.*, 1993, **31**, 925.
144 B. Wrackmeyer and G. Kehr, *J.Magn.Reson.*, 1994, **107A**, 106.
145 B. Wrackmeyer, C. Koehler and E. Kupce, *Magn.Reson.Chem.*, 1993, **31**, 769.
146 B. Wrackmeyer, B. Distler, S. Gerstmann and M. Herberhold, *Z.Naturforsch.*, 1993, **48B**, 1307.
147 A. Fratiello, V. Kubo-Anderson, E.L. Bolanos, D. Haigh, F. Laghaei and R.D. Perrigan, *J.Magn.Reson.*, 1994, **107A**, 56.
148 A. Macchioni, P.S. Pregosin, H. Ruegger, G. van Koten, P.A. van der Schaaf and R.A.T.M. Abbenhuis, *Magn.Reson.Chem.*, 1994, **32**, 235.
149 Y. Qu, N. Farrell, M. Valsecchi, L. de Greco and S. Spinelli, *Magn.Reson.Chem.*, 1993, **31**, 920.
150 J.P. Jokisaari, L.P. Ingman, G.J. Schrobilgen and J.C.P. Sanders, *Magn.Reson.Chem.*, 1994, **32**, 242.
151 E. Parzich, G. Peters and W. Preetz, *Z.Naturforsch.*, 1993, **48B**, 1169.
152 P.K. Jenkner, E. Hengge, R. Czaputa and C. Kratky, *J.Organomet.Chem.*, 1993, **446**, 83.
153 H. Oehme, R. Wustrack, A. Heine, G.M. Sheldrick and D. Stalke, *J.Organomet.Chem.*, 1993, **452**, 33.

154 K. Schenzel and K. Hassler, *Spectrochim.Acta*, 1994, **50A**, 127.
155 T.S. Koloski, D.C. Pestana, P.J. Carroll and D.H. Berry, *Organometallics*, 1994, **13**, 489.
156 G. Szalontai, J. Bakos, S. Aime and R. Gobetto, *Solid State Nucl.Magn.Reson.*, 1993, **2**, 245.
157 J.F. Nixon and G.J.D. Sillet, *J.Organomet.Chem.*, 1993, **461**, 237.
158 K.L. Moran, T.E. Gier, W.T.A. Harrison, G.D. Stucky, H. Eckert, K. Eichele and R.E. Wasylishen, *J.Am.Chem.Soc.*, 1993, **115**, 10553.
159 J.F. Corrigan, M. Dinardo, S. Doherty and A.J. Carty, *J.Cluster Sci.*, 1992, **3**, 313.
160 K. Eichele, R.E. Wasylishen, J.F. Corrigan, S. Doherty, Y. Sun and A.J. Carty, *Inorg.Chem.*, 1993, **32**, 131.
161 P.S. Pregosin (ed.), *Studies in Inorganic Chemistry* 1991, **13**, ed. Elsevier.
162 R. Benn, H. Brenneke, E. Joussen, H. Lehmkuhl and F.L. Ortiz, *Organometallics*, 1990, **9**, 756.
163 D. Steinborn and M. Ludwig, *J.Organomet.Chem.*, 1993, **463**, 65.
164 G.A. Bowmaker, E.J.V. Hanna, P.C. Healy, B.W. Skelton and A.H. White, *J.Chem.Soc.,Dalton Trans.*, 1993, 1387.
165 R.E. Wasylishen, K.C. Wright, K. Eichele and T.S. Cameron, *Inorg.Chem.*, 1994, **33**, 407.
166 J.M. Bevilacqua, J.A. Zuleta and R. Eisenberg, *Inorg.Chem.*, 1994, **33**, 258.
167 P.S. Jarrett, O.M. Ni Dhubhghaill and P.J. Sadler, *J.Chem.Soc.,Dalton Trans.*, 1993, 1863.
168 P.B. Hitchcock, M.F. Lappert, G.A. Lawless and B. Royo, *J.Chem.Soc.,Chem.Commun.*, 1993, 554.
169 N.G. Payne, M.E. Peach and K.J. Schmidt, *J.Fluorine Chem.*, 1993, **64**, 255.
170 M. Krumm, E. Zangrando, L. Randacio, S. Menzer, A. Danzmann, D. Holtenrich and B. Lippert, *Inorg.Chem.*, 1993, **32**, 2183.
171 L. Ernst, V. Boekelheide and H. Hopf, *Magn.Reson.Chem.*, 1993, **31**, 669.
172 S.F. Leonard, P.-J. Jones, D.M. Spero and T.P. Pitner, *Magn.Reson.Chem.*, 1993, **31**, 906.
173 G.J. Moore and L.O. Sillerud, *J.Magn.Reson.*, 1994, **103B**, 87.
174 D.S. Himmelsbach, R.D. Hartley, W.S. Borneman, L. Poppe and H. van Halbeek, *Magn.Reson.Chem.*, 1994, **32**, 158.
175 G. Randazzo, V. Fogliano, A. Ritieni, L. Mannina, E. Rosssi, A. Scarallo and A.L. Segre, *Tetrahedron*, 1993, **49**, 10883.
176 H. Baumann, J.R. Brisson, F. Michon, R. Pon and H.J. Jennings, *Biochemistry*, 1993, **32**, 4007.
177 M. Tata, V.T. John, Y.Y. Waguespack and G.L. McPherson, *J.Phys.Chem.*, 1994,**98**,3809.
178 R.T. Wheelhouse and M.F.G. Stevens, *J.Chem.Soc.,Chem.Commun.*, 1993, 1177.
179 H.E. Mons, H. Guenther and A. Maercker, *Chem.Ber.*, 1993, **126**, 2747.
180 J.V. Hines, G. Varani, S.M. Landry and I. Tinoco Jr., *J.Am.Chem.Soc.*, 1993, **115**, 11002.
181 T. Bandyopadhyay, J. Wu and A.S. Serianni, *J.Org.Chem.*, 1993, **58**, 5513.
182 A.G. Osborne, *Spectrosc.Lett.*, 1994, **27**, 1.
183 A.G. Osborne, J.M. Buley, H. Clarke, R.C.H. Dakin and P.I. Price, *J.Chem.Soc.,Perkin Trans.1*, 1993, 2747.
184 A.G. Osborne and L.A.D. Miller, *J.Chem.Soc.,Perkin Trans.1*, 1993, 181.
185 E. Kolehmainen, I.O.O. Korhonen and R. Kauppinen, *Magn.Reson.Chem.*, 1994, **32**, 186.
186 B.I. Ugrak, Y. Manaev A, V.P. Perevalov and S.A. Shevelev, *Izv.Akad.Nauk,Ser.Khim.*, 1992, 2554.
187 B. Wrackmeyer and C. Koehler, *Magn.Reson.Chem.*, 1993, **31**, 573.
188 V. Baudrillard, D. Davoust and G. Ple, *Magn.Reson.Chem.*, 1994, **32**, 40.
189 N. Bampos, L.D. Field and B.A. Messerle, *Organometallics*, 1993, **12**, 2529.

190 W.J. Boyko, F.J. Duarte and R.M. Giuliano, *Magn.Reson.Chem.*, 1994, **32**, 259.
191 G. Hogarth, M. Arthurs, J.C. Bickerton, L. Daly, C. Piper, D. Ralfe and D.A. Morton-Blake, *J.Organomet.Chem.*, 1994, **467**, 145.
192 C. Kober, J. Kroner and W. Storch, *Angew.Chem.Int.Ed.Engl.*, 1993, **32**, 1608.
193 A. Lycka, I. Tkac, J. Holecek and V. Pejchal, *Magn.Reson.Chem.*, 1994, **32**, 189.
194 Z.-L. Zhou, Y.-Z. Huang, Y. Tang, Z.-H. Chen, L.-P. Shi, X.-L. Jin and Q.-C. Yang, *Organometallics*, 1994, **13**, 1575.
195 S. Berners-Price, U. Frey, J.D. Ranford and P.J. Sadler, *J.Am.Chem.Soc.*, 1993, **115**, 8649.
196 H.J. Reich, J.P. Borst, R.R. Dykstra and D.P. Green, *J.Am.Chem.Soc.*, 1993, **115**, 8728.
197 B. Wrackmeyer and G. Kehr, *Z.Naturforsch.*, 1993, **48B**, 1433.
198 K. Laihia, E. Kolehmainen, P. Mänttäri and R. Kauppinen, *Magn.Reson.Chem.*, 1993, **31**, 512.
199 C.D. Hall, J.H.R. Tucker and S.Y.F. Chu, *J.Organomet.Chem.*, 1993, **448**, 175.
200 B. Wrackmeyer, H.E. Maisel and H. Zhou, *Main Group Met.Chem.*, 1993, **16**, 475.
201 P.A. Morken and D.J. Burton, *J.Org.Chem.*, 1993, **58**, 1167.
202 B. Wrackmeyer, G. Kehr and J. Suess, *Chem.Ber.*, 1993, **126**, 2221.
203 T.R. Prout, T.W. Imiolczyk, F. Barthelemy, S.M. Young, R.C. Haltiwanger and A.D. Norman, *Inorg.Chem.*, 1994, **33**, 1783.
204 H. Beckmann, G. Grossmann, G. Ohms and J. Sieler, *Heteroatom Chem.*, 1994, **5**, 73.
205 B.R. Bender, M. Koller, D. Nanz and W. von Philipsborn, *J.Am.Chem.Soc.*, 1993, **115**, 5889.
206 A.L. Davis and R.J. Goodfellow, *J.Chem.Soc.,Dalton Trans.*, 1993, 2273.
207 M.S. Balakrishna, R. Klein, S. Uhlenbrock, A.A. Pinkerton and R.G. Cavell, *Inorg.Chem.*, 1993, **32**, 5676.
208 P.M. MacDonald, A.D. Hunter, G. Lesley and J. Li, *Solid State Nucl.Magn.Reson.*, 1993, **2**, 47.
209 J. Campora, E. Gutierrez, A. Monge, M.L. Poveda, C. Ruiz and E. Carmona, *Organometallics*, 1993, **12**, 4025.
210 U. Edlund, M. Arshadi and D. Johnels, *J.Organomet.Chem.*, 1993, **456**, 57.
211 G. Matsubayashi and A. Yokozawa, *Inorg.Chim.* Acta, 1993, **208**, 95.
212 P.J. Barrie, R.J.H. Clark, R. Withnall, D.-Y. Chung, K.-W. Kim and M.G. Kanatzidis, *Inorg.Chem.*, 1994, **33**, 1212.
213 M. Björgvinsson, H.P.A. Mercier, K.M. Mitchell, G.J. Schrobilgen and G. Strohe, *Inorg.Chem.*, 1993, **32**, 6046.
214 J. Liu, B. Zhao, C.Y. Rong and M.T. Pope, *Huaxue Xuebao*, 1993, **51**, 368.
215 M.D. Soerensen, J.J. Led and O.W. Soerensen, *J.Biomol.NMR*, 1994, **4**, 135.
216 S.J.F. Vincent, C. Zwahlen and G. Bodenhausen, *J.Am.Chem.Soc.*, 1993, **115**, 9202.
217 G.W. Vuister and A. Bax, *J.Am.Chem.Soc.*, 1993, **115**, 7772.
218 K. Bartik and C. Redfield, *J.Biomol.NMR*, 1993, **3**, 415.
219 J.M. Schmidt, R. Brueschweiler, R.R. Ernst, R.L. Dunbrack Jr, D. Joseph and M. Karplus, *J.Am.Chem.Soc.*, 1993, **115**, 8747.
220 M.J. Blackledge, R. Brueschweiler, C. Griesinger, J.M. Schmidt, P. Xu and R.R. Ernst, *Biochemistry*, 1993, **32**, 10960.
221 G.V. Nikiforovich, O. Prakash, C.A. Gehrig and V.J. Hruby, *J.Am.Chem.Soc.*, 1993, **115**, 3399.
222 D.F. Mierke and H. Kessler, *Biopolymers*, 1993, **33**, 1003.
223 D.F. Mierke, M. Kurz and H. Kessler, *J.Am.Chem.Soc.*, 1994, **116**, 1042.
224 D.F. Mierke, R.M. Scheek and H. Kessler, *Biopolymers*, 1994, **34**, 559.

225 A.E. Torda, R.M. Brunne, T. Huber, H. Kessler and W.F. van Gunsteren, *J.Biomol.NMR*, 1993, **3**, 55.
226 M.S.L. Lim, E.R. Johnston and C.A. Kettner, *J.Med.Chem.*, 1993, **36**, 1831.
227 G.V. Nikiforovich, J.L.F. Kao, K. Plucinska, W.J. Zhang and G.R. Marshall, *Biochemistry*, 1994, **33**, 3591.
228 A.H. van Oijen, S. Behrens, D.F. Mierke, H. Kessler, J.H. van Boom and R.M.J. Liskamp, *J.Org.Chem.*, 1993, **58**, 3722.
229 C. Goerbitz H and J. Krane, *Acta Chem. Scand.*, 1993, **47**, 979.
230 H. Kessler, H. Matter, G. Gemmecker, H.-J. Diehl, C. Isernia and S. Mronga, *Int.J.Peptide Protein Res.*, 1994, **43**, 47.
231 G.V. Nikiforovich, O.M. Prakash, C.A. Gehrig and V.J. Hruby, *Int.J.Peptide Protein Res.*, 1993, **41**, 347.
232 M. Gerz, H. Matter and H. Kessler, *Int.J.Peptide Protein Res.*, 1994, **43**, 248.
233 W. Guo, B. Noszal, I. Schoen and O. Nyeki, *J.Pharmaceut.Biomed.Anal.*, 1993, **11**, 541.
234 X. Liu, J.M. Stewart, L. Gera and G. Kotovych, *Biopolymers*, 1993, **33**, 1237.
235 M.J. Bogusky, A.M. Naylor, M.E. Mertzman, S.M. Pitzenberger, R.F. Nutt, S.F. Brady, C.D. Colton and D.F. Veber, *Biopolymers*, 1993, **33**, 1287.
236 P. Verheyden, I. Van Assche, M.-H. Brichard, T. Demaude, I. Paye, A. Scarso and G. Van Binst, *FEBS Lett.*, 1994, **344**, 55.
237 J.A. Jarvis, S.L.A. Munro and D.J. Craik, *Biochemistry*, 1994, **33**, 33.
238 A. Iida, S. Uesato, T. Shingu, M. Okuda, Y. Nagaoka, Y. Kuroda and T. Fujita, *J.Chem.Soc.,Perkin Trans.1*, 1993, 367.
239 T. Kato, S. Endo, T. Fujiwara and K. Nagayama, *J.Biomol.NMR*, 1993, **3**, 653.
240 D. Frechet, J.D. Guitton, F. Herman, D. Faucher, G. Helynck, B. Monegier du Sorbier, J.P. Ridoux, E. James-Surcouf and M. Vuilhorgne, *Biochemistry*, 1994, **33**, 42.
241 M. Coles, S.L.A. Munro and D.J. Craik, *J.Med.Chem.*, 1994, **37**, 656.
242 G. Mohn, P. Koehl, H. Budzikiewicz and J.-F. Lefevre, *Biochemistry*, 1994, **33**, 2843.
243 A. Berg, A. De Kok and J. Vervoort, *Eur.J.Biochem.*, 1994, **221**, 87.
244 B. Leiting, R. De Francesco, L. Tomei, R. Cortese, G. Otting and K. Wuethrich, *EMBO J.*, 1993, **12**, 1797.
245 B.J. Stockman, A. Euvrard, D.A. Kloosterman, T.A. Scahill and R.P. Swenson, *J.Biomol.NMR*, 1993, **3**, 133.
246 M.J. Lubienski, M. Bycroft, D.N.M. Jones and A.R. Fersht, *FEBS Lett.*, 1993, **332**, 81.
247 S. Golic Grdadolnik, M. Eberstadt, G. Gemmecker, H. Kessler, A. Buhr and B. Erni, *Eur.J.Biochem.*, 1994, **219**, 945.
248 S. Bouaziz, C. van Heijenoort, E. Guittet, J.-C. Huet and J.-C. Pernollet, *Eur.J.Biochem.*, 1994, **220**, 427.
249 M. Shirakawa, W.J. Fairbrother, Y. Serikawa, T. Ohkubo, Y. Kyogoku and P.E. Wright, *Biochemistry*, 1993, **32**, 2144.
250 B.O. Smith, A.K. Downing, T.J. Dudgeon, M. Cunningham, P.C. Driscoll and I.D. I.D. Campbell, *Biochemistry*, 1994, **33**, 2422.
251 C.W.G. Hoitink, P.C. Driscoll, H.A.O. Hill and G.W. Canters, *Biochemistry*, 1994, **33**, 3560.
252 S. Bagby, T.S. Harvey, L.E. Kay, S.G. Eagle, S. Inouye and M. Ikura, *Biochemistry*, 1994, **33**, 2409.
253 P.K. Pallaghy, B.M. Duggan, M.W. Pennington and R.S. Norton, *J.Mol.Biol.*, 1993, **234**, 405.
254 Y.N. Kalia, S.M. Brocklehurst, D.S. Hipps, E. Appella, K. Sakaguchi and R.N. Perham, *J.Mol.Biol.*, 1993, **230**, 323.

255 F.J. Moy, Y.C. Li, P. Rauenbuehler, M.E. Winkler, H.A. Scheraga and G.T. Montelione, *Biochemistry*, 1993, **32**, 7334.
256 C.M. Fletcher, R.A. Harrison, P.J. Lachmann and D. Neuhaus, *Structure*, 1994, **2**, 185.
257 L.C. Myers, G.L. Verdine and G. Wagner, *Biochemistry*, 1993, **32**, 14089.
258 P. Sevilla-Sierra, G. Otting and K. Wuethrich, *J.Mol.Biol.*, 1994, **235**, 1003.
259 X. Li, A.M. Bokman, M. Llinas, R.A.G. Smith and C.M. Dobson, *J.Mol.Biol.*, 1994, **235**, 1548.
260 M.J. Williams, I. Phan, T.S. Harvey, A. Rostagno, L.I. Gold and I.D. Cambell, *J.Mol.Biol.*, 1994, **235**, 1302.
261 W. Lee, M.J. Jablonsky, D.D. Watt and N.R. Krishna, *Biochemistry*, 1994, **33**, 2468.
262 P. Osmark, P. Soerensen and F.M. Poulsen, *Biochemistry*, 1993, **32**, 11007.
263 J.F. O'Connell, P.E. Bougis and K. Wuethrich, *Eur.J.Biochem.*, 1993, **213**, 891.
264 J.H. Bushweller, M. Billeter, A. Holmgren and K. Wuethrich, *J.Mol.Biol.*, 1994, **235**, 1585.
265 J.F. O'Connell, R. Bender, J.W. Engels, K.-P. Koller, M. Scharf and K. Wuethrich, *Eur.J.Biochem.*, 1994, **220**, 763.
266 C. Johansson, M. Ullner and T. Drakenberg, *Biochemistry*, 1993, **32**, 8429.
267 R.T. Clubb, S.B. Ferguson, C.T. Walsh and G. Wagner, *Biochemistry*, 1994, **33**, 2761.
268 H. Baumann, K. Paulsen, H. Kovacs, H. Berglund, A.P.H. Wright, J.-A. Gustafsson and T. Härd, *Biochemistry*, 1993, **32**, 13463.
269 K.V. Andersen and F.M. Poulsen, *J.Biomol.NMR*, 1993, **3**, 271.
270 K.D. Berndt, P. Guentert and K. Wuethrich, *J.Mol.Biol.*, 1993, **234**, 735.
271 J. Santoro, C. Gonzalez, M. Bruix, J.L. Neira, J.L. Nieto, J. Herranz and M. Rico, *J.Mol.Biol.*, 1993, **229**, 722.
272 J. Kordel, N.J. Skelton, M. Akke and W.J. Chazin, *J.Mol.Biol.*, 1993, **231**, 711.
273 R. Bhaskaran, C.C. Huang, D.K. Chang and C. Yu, *J.Mol.Biol.*, 1994, **235**, 1291.
274 K. Bartik, C.M. Dobson and C. Redfield, *Eur.J.Biochem.*, 1993, **215**, 255.
275 J.H. Davis, E.K. Bradley, G.P. Miljanich, L. Nadasdi, J. Ramachandran and V.J. Basus, *Biochemistry*, 1993, **32**, 7396.
276 Y. Karimi-Nejad, J.M. Schmidt, H. Rueterjans, H. Schwalbe and C. Griesinger, *Biochemistry*, 1994, **33**, 5481.
277 L.J. Smith, M.J. Sutcliffe, C. Redfield and C.M. Dobson, *J.Mol.Biol.*, 1993, **229**, 930.
278 D. Morikis, R. Bruschweiler and P.E. Wright, *J.Am.Chem.Soc.*, 1993, **115**, 6238.
279 S.M. Pascal and T.A. Cross, *J.Biomol.NMR*, 1993, **3**, 495.
280 P.J.M. Folkers, M. Nilges, R.H.A. Folmer, R.N.H. Konings and C.W. Hilbers, *J.Mol.Biol.*, 1994, **236**, 229.
281 K.L. Constantine, M.S. Friedrichs, W.J. Metzler, M. Wittekind, P. Hensley and L. Mueller, *J.Mol.Biol.*, 1994, **236**, 310.
282 R.X. Xu, D. Nettesheim, E.T. Olejniczak, R. Meadows, G. Gemmecker and S.W. Fesik, *Biopolymers*, 1993, **33**, 535.
283 Y.Q. Qian, G. Otting, M. Billeter, M. Mueller, W. Gehring and K. Wuethrich, *J.Mol.Biol.*, 1993, **234**, 1070.
284 B.B. Kragelund, K.V. Andersen, J.C. Madsen, J. Knudsen and F.M. Poulsen, *J.Mol.Biol.*, 1993, **230**, 1260.
285 M. Riviere, S. Auge, J. Vercauteren, E. Wisingerova and G. Puzo, *Eur.J.Biochem.*, 1993, **214**, 395.
286 L. Emsley, T.J. Dwyer, H.P. Spielmann and D.E. Wemmer, *J.Am.Chem.Soc.*, 1993, **115**, 7765.

287 V.P. Chuprina, W. Nerdal, E. Sletten, V.I. Poltev and O.Y. Fedoroff, *J.Biomol.Struct.Dyn.*, 1993, **11**, 671.

288 N.B. Ulyanov, U. Schmitz and T.L. James, *J.Biomol.NMR*, 1993, **3**, 547.

289 U. Schmitz, N.B. Ulyanov, A. Kumar and T.L. James, *J.Mol.Biol.*, 1993, **234**, 373.

290 J. Plavec, V. Fabre-Buet, V. Uteza, A. Grouiller and J. Chattopadhyaya, *J.Biochem.Biophys.Meth.*, 1993, **26**, 317.

291 Z. Xi, C. Glemarec and J. Chattopadhyaya, *Tetrahedron*, 1993, **49**, 7525.

292 M.K. Lakshman, W. Xiao, J.M. Sayer, A.M. Cheh and D.M. Jerina, *J.Org.Chem.*, 1994, **59**, 1755.

293 J.L.F. Kao, S. Nadji and J.S. Taylor, *Chem.Res.Toxicol.*, 1993, **6**, 561.

294 W.A. Tabaczynski, D.G.E. Lemaire, B.P. Ruzsicska and J.L. Alderfer, *Biopolymers*, 1993, **33**, 1365.

295 J.K. Kim, S.D. Soni, J.C. Wallace and J.L. Alderfer, *Nucleic Acids Res.*, 1993, **21**, 2755.

296 H. Maag, J.T. Nelson, J.L.R. Steiner and E.J. Prisbe, *J.Med.Chem.*, 1994, **37**, 431.

297 N. Chouini-Lalanne, J.P. Vialaneix, S. Mansouri, M.C. Malet-Martino and R. Martino, *Nucleosides Nucleotides*, 1993, **12**, 331.

298 A. Papchikhin, P. Agback, J. Plavec and J. Chattopadhyaya, *J.Org.Chem.*, 1993, **58**, 2874.

299 R. Chillemi, A. Contino, V. Cucinotta and S. Sciuto, *Nucleosides Nucleotides*, 1994, **13**, 953.

300 C. Glemarec, R.C. Reynolds, P.A. Crooks, J.A. Maddry, M.S. Akhtar, J.A. Montgomery, J.A. Secrist III and J. Chattopadhyaya, *Tetrahedron*, 1993, **49**, 2287.

301 P. Agback, A. Sandstrom, S. Yamakage, C. Sund, C. Glemarec and Chattopadhyaya, *J.Biochem.Biophys.Meth.*, 1993, **27**, 229.

302 N. Boogaard, C. Altona and J. Reedijk, *J.Inorg.Biochem.*, 1993, **49**, 129.

303 J.E. Van Den Boogaart, E.N. Kalinichenko, T.L. Podkopaeva, I.A. Mikhailopulo and C. Altona, *Eur.J.Biochem.*, 1994, **221**, 759.

304 K.L.B. Borden, *Biochemistry*, 1993, **32**, 6506.

305 J.K. James and I. Tinoco, *Nucleic Acids Res.*, 1993, **21**, 3287.

306 P. Schultze, R.F. Macaya and J. Feigon, *J.Mol.Biol.*, 1994, **235**, 1532.

307 A.N. Lane, C.J. Bauer, T.A. Frenkiel and A.J. Birchall, *Eur.Biophys.J.*, 1993, **22**, 135.

308 J.A. Jaeger and I. Tinoco Jr, *Biochemistry*, 1993, **32**, 12522.

309 A.A. Szewczak, P.B. Moore, Y.L. Chan and I.G. Wool, *Proc.Natl.Acad.Sci.USA.*, 1993, **90**, 9581.

310 F.W. Smith and J. Feigon, *Biochemistry*, 1993, **32**, 8682.

311 R.A. Colvin, S.W. White, M.A. Garciablanco and D.W. Hoffman, *Biochemistry*, 1993, **32**, 1105.

312 T. Maltseva, A. Sandstroem, I.M. Ivanova, D.S. Sergeyev, V.F. Zarytova and J. Chattopadhyaya, *J.Biochem.Biophys.Meth.*, 1993, **26**, 173.

313 G. Otting, M. Billeter, K. Wuethrich, H.-J. Roth, C. Leumann and A. Eschenmoser, *Helv.Chim.Acta*, 1993, **76**, 2701.

314 M. Salazar, O.Y. Fedoroff, J.M. Miller, N.S. Ribeiro and B.R. Reid, *Biochemistry*, 1993, **32**, 4207.

315 R. Nibedita, R.A. Kumar, A. Majumdar, R.V. Hosur, G. Govil, K. Majumder and V.S. Chauhan, *Biochemistry*, 1993, **32**, 9053.

316 A.N. Lane, S. Ebel and T. Brown, *Eur.J.Biochem.*, 1993, **215**, 297.

317 J. Santalucia and D.H. Turner, *Biochemistry*, 1993, **32**, 12612.

318 K.D. Bishop, F.J.H. Blocker, W. Egan and T.L. James, *Biochemistry*, 1994, **33**, 427.

319 A. Mujeeb, S.M. Kerwin, G.L. Kenyon and T.L. James, *Biochemistry*, 1993, **32**, 13419.

320 E.M. Meiering, M. Bycroft, M.J. Lubienski and A.R. Fersht, *Biochemistry*, 1993, **32**, 10975.
321 D.R. Brown, M. Kurz, D.R. Kearns and V.L. Hsu, *Biochemistry*, 1994, **33**, 651.
322 C.L. Perrin and K.B. Armstrong, *J.Am.Chem.Soc.*, 1993, **115**, 6825.
323 B. Adams and L.E. Lerner, *Magn.Reson.Chem.*, 1994, **32**, 225.
324 L. Poppe, *J.Am.Chem.Soc.*, 1993, **115**, 8421.
325 S.-K. Chung and Y. Ryu, *Carbohydr.Res.*, 1994, **258**, 145.
326 K. Bock, J.O. Duus and S. Refn, *Carbohydr.Res.*, 1994, **253**, 51.
327 L. Szilagyi and P. Forgo, *Carbohydr.Res.*, 1993, **247**, 129.
328 T. Peters and T. Weimar, *J.Biomol.NMR*, 1994, **4**, 97.
329 G. Pant, M.C. Purohit, G.A. Morris, A.G.W. Halstead and R.I.G. Thompson, *Magn.Reson.Chem.*, 1994, **32**, 213.
330 T. Weimar, B. Meyer and T. Peters, *J.Biomol.NMR*, 1993, **3**, 399.
331 W. Sicinska, B. Adams and L. Lerner, *Carbohydr.Res.*, 1993, **242**, 29.
332 A. Wei and Y. Kishi, *J.Org.Chem.*, 1994, **59**, 88.
333 R.F. Helm and J. Ralph, *Carbohydr.Res.*, 1993, **240**, 23.
334 C. Meyer, S. Perez, C. Herve du Penhoat and V. Michon, *J.Am.Chem.Soc.*, 1993, **115**, 10300.
335 S. Hanessian, H. Hori, Y. Tu and Y. Boulanger, *Tetrahedron*, 1994, **50**, 77.
336 J.W. Timmermans, D. de Wit, H. Tournois, B.R. Leeflang and J.F.G. Vliegenthart, *J.Carbohydr.Chem*, 1993, **12**, 969.
337 A.M. Gil Serrano, G. Franco-Rodriguez, I. Gonzalez-Jimenez, P. Tejero-Mateo, J. Molina Molina, J.A. Dobado, M. Megias and M. J. Romero, *J.Mol.Struct.*, 1993, **301**, 211.
338 V.M. Lo, M.G. Hahn, N.G. Hong, T. Ogawa and H. Vanhalbeek, *Carbohydr.Res.*, 1993, **245**, 333.
339 A.S. Shashkov, E.V. Vinogradov, Y.A. Knirel, N.E. Nifantev, N.K. Kochetkov, J. Dabrowski, E.V. Kholodkova and E.S. Stanislavsky, *Carbohydr.Res.*, 1993, **241**, 177.
340 C.A.G. Haasnoot, *J.Am.Chem.Soc.*, 1993, **115**, 1460.
341 C.A.G. Haasnoot, *J.Am.Chem.Soc.*, 1992, **114**, 882.
342 P. Camps, S. Vazquez, C. Jaime and A. Virgili, *Magn.Reson.Chem.*, 1994, **32**, 210.
343 H. Quast, K. Knoll, E.-M. Peters, K. Peters and H.G. von Schnering, *Liebigs Ann.Chem.*, 1993, 777.
344 Z. Chen and R.A. Sachleben, *J.Chem.Soc.,Perkin Trans.2*, 1994, 537.
345 S.A. Baisalbaeva, B.M. Butin and B. Beketova, *Zh.Obshch.Khim.*, 1992, **62**, 2728.
346 R. Sridharan, U.R. Desai, R.M. Rao and G.K. Trivedi, *Steroids*, 1993, **58**, 170.
347 R.P. Bonar-Law and A.P. Davis, *Tetrahedron*, 1993, **49**, 9845.
348 W.B. Smith, Magn.Reson.Chem., 1994, **32**, 316.
349 T.A. van Beek, G.P. Lelyveld, H.J.M. Gijsen and P.P. Lankhorst, *J.Essent.Oil Res.*, 1993, **5**, 169.
350 Y. Yang, X. Xu, D. Wang and B. Qian, *Magn.Reson.Chem.*, 1994, 32, 71.
351 J. Hoebeke, C. Busatto-Samsoen, D. Davoust and E. Lebrun, *Magn.Reson.Chem.*, 1994, **32**, 220.
352 J. Gharbi-Benarous, P. Ladam, M. Delaforge and J.-P. Girault, J.*Chem.Soc.,Perkin Trans.2*, 1993, 2303.
353 C. Vaccher, P. Berthelot, M. Debaert, G. Vermeersch, R. Guyon, B. Pirard, D.P. Vercauteren, M. Dory, G. Evrard and F. Durant, *J.Mol.Struct.*, 1993, **301**, 199.
354 S.R. Salman, J.C. Lindon, R.D. Farrant and T.A. Carpenter, *Magn.Reson.Chem.*, 1993, **31**, 991.
355 E.S. Lit, F.K. Mallon, H.Y. Tsai and J.D. Roberts, *J.Am.Chem.Soc.*, 1993, **115**, 9563.

356 G. Englert, T. Aakermann and S. Liaaen-Jensen, *Magn.Reson.Chem.*, 1993, **31**, 910.
357 R. Radeglia, H. Poleschner and G. Haufe, *Magn.Reson.Chem.*, 1993, **31**, 639.
358 R. Radeglia, H. Poleschner and G. Haufe, *Magn.Reson.Chem.*, 1993, **31**, 1054.
359 R. Radeglia, H. Poleschner and F. Theil, *J.Prakt.Chem.*, 1993, **335**, 673.
360 R. Radeglia, *J.Prakt.Chem.*, 1993, **335**, 291.
361 S.R. Salman, R.D. Farrant, P.N. Sanderson and J.C. Lindon, *Magn.Reson.Chem.*, 1993, **31**, 585.
362 R.L. Jaffe, G.D. Smith and D.Y. Yoon, *J.Phys.Chem.*, 1993, **97**, 12745.
363 C. Beeson, N. Pham, G. Shipps Jr and T.A. Dix, *J.Am.Chem.Soc.*, 1993, **115**, 6803.
364 S. Wache and R. Taube, *J.Organomet.Chem.*, 1993, **456**, 137.
365 M. Margarida, C.A. Castro, F.G.C. Geraldes and J.A. Peters, *Inorg.Chim. Acta*, 1993, **208**, 123.
366 W.J. Haelg, L.R. Oehrstroem, H. Ruegger and L.M. Venanzi, *Helv.Chim.Acta*, 1993, **76**, 788.
367 E. Gacs-Baitz, H. Bolcskei and C. Szantay, *J.Chem.Soc.,Perkin Trans.2*, 1994, 213.
368 L. Poppe, S. Sheng and H. van Halbeek, *Magn.Reson.Chem.*, 1994, **32**, 97.
369 L. Poppe, W.S. York and H. Vanhalbeek, *J.Biomol.NMR*, 1993, **3**, 81.
370 V. Durier, G. Buisson, E. Duee, H. Driguez and F.-R. Taravel, *J.Chem.Soc.,Perkin Trans.2*, 1993, 1931.
371 A.E. Aulabaugh, R.C. Crouch, G.E. Martin, A. Ragouzeos, J.P. Shockcor, T.D. Spitzer, R.D. Farrant, B.D. Hudson and J.C. Lindon, *Carbohydr.Res.*, 1992, **230**, 201.
372 H. Kogelberg and T.J. Rutherford, *Glycobiology*, 1994, **4**, 49.
373 J. Poznanski, A. Ejchart, K.L. Wierzchowski and M. Ciurak, *Biopolymers*, 1993, **33**, 781.
374 Y. Sakamoto and T. Ishi, *J.Mol.Struct.*, 1993, **298**, 129.
375 P. Ladam, J. Gharbi-Benarous, M. Piotto, M. Delaforge and J.P. Girault, *M agn.Reson.Chem.*, 1994, **32**, 1.
376 S.L.A. Munro and D.J. Craik, *Magn.Reson.Chem.*, 1994, **32**, 335.
377 S. Rebuffat, L. Conraux, M. Massias, C. Auvin-Guette and B. Bodo, *Int.J.Peptide Protein Res.*, 1993, **41**, 74.
378 A. Arnone, L. Merlini, R. Mondelli, G. Nasini, E. Ragg, L. Scaglioni and U. Weiss, *J.Chem.Soc.,Perkin Trans.2*, 1993, 1447.
379 A.G. Osborne, L. Jackson and P.D. Taylor, *Spectrochim.Acta*, 1993, **49A**, 1703.
380 H.H. Keah and I.D. Rae, *Aust.J.Chem.*, 1993, **46**, 1413.
381 C. Buenz, D. Heber, U. Holzgrabe and U. Suerig, *Magn.Reson.Chem.*, 1993, **31**, 371.
382 P. Fresneau, F. Thomasson, J.-M. Morand and M. Cussac, *Spectrosc.Lett.*, 1993, **26**, 1511.
383 J. Wojcik, W. Domalewski, Kamienska-Trela, L. Stefaniak, S.I. Vdovienko, I.I. Gerus and M.G. Gorbunova, *Magn.Reson.Chem.*, 1993, **31**, 808.
384 A. Ribeiro and M.J. Glen, *J.Magn.Reson.*, 1994, **107A**, 158.
385 R.J. Abraham, E.J. Chambers and W.A. Thomas, *Magn.Reson.Chem.*, 1994, **32**, 248.
386 K.V.R. Chary, V.K. Rastogi, G. Govil and H.T. Miles, *J.Chem.Soc.,Chem.Commun.*, 1994, 241.
387 S. El antri, P. Bittoun, O. Mauffret, M. Monnot, O. Convert, E. Lescot and S.Fermandjian, *Biochemistry*, 1993, **32**, 7079.
388 J.M. Ciesla, R. Stolarski and D. Shugar, *Acta Biochim. Polon.*, 1993, **40**, 251.
389 S. El antri, O. Mauffret, M. Monnot, E. Lescot, O. Convert and S. Fermandjian, *J.Mol.Biol.*, 1993, **230**, 373.
390 H. Schwalbe, W. Samstag, J.W. Engels, W. Bermel and C. Griesinger, *J.Biomol.NMR*, 1993, **3**, 479.

391 J.L. Bookham, W. Clegg, W. McFarlane and E.S. Raper, *J.Chem.Soc.,Dalton Trans.*, 1993, 3567.
392 H. Komber and G. Grossmann, *Phosphorus, Sulfur, and Silicon*, 1993, **83**, 233.
393 O. Zerbe, D.L. Pountney, W. von Philipsborn and M. Vasak, *J.Am.Chem.Soc.*, 1994, **116**, 377.
394 C.J. Henehan, D.L. Pountney, O. Zerbe and M. Vasak, *Protein Sci.*, 1993, **2**, 1756.
395 W. Plass and J.G. Verkade, *Inorg.Chem.*, 1993, **32**, 5145.
396 V. Neirinck, D. Robert and R. Nardin, *Magn.Reson.Chem.*, 1993, **31**, 815.
397 G. Trovo, G. Bandoli, M. Nicolini and B. Longato, *Inorg.Chim. Acta*, 1993, **211**, 95.
398 T.L. Mega and R.L. Van Etten, *J.Am.Chem.Soc.*, 1993, **115**, 12056.
399 H. Schwalbe, A. Rexroth, U. Eggenberger, T. Geppert and C. Griesinger, *J.Am.Chem.Soc.*, 1993, **115**, 7878.
400 S. Grzesiek, G.W. Vuister and A. Bax, *J.Biomol.NMR*, 1993, **3**, 487.
401 W. Hu, L. Yun, J. Chen, H. Xu, S. Jia, X. Xu, C. Yuan and S. Li, *Chin.Chem.Lett.*, 1992, **3**, 959.
402 M.P. Belciug, P.L. Wessels and T.A. Modro, *Magn.Reson.Chem.*, 1993, **31**, 1024.
403 C.D. Reddy, K.D. Berlin, R.S. Reddy, C.N. Raju, M. Elmasri and S. Subramanian, *Phosphorus, Sulfur, Silicon,* 1993, **81**, 61.
404 I. Toth and B.E. Hanson, *Organometallics*, 1993, **12**, 1506.
405 B. Korpar-Colig, Z. Popovic, D. Matkovic-Calogvic and D. Vikic-Topic, *Organometallics*, 1993, **12**, 4708.
406 W. Preetz, G. Peters and J.-U. Vogt, *Z.Naturforsch.*, 1993, **48B**, 348.
407 H. Koroniąk, K.W. Palmer, .W.R. Dolbier Jr., and H.-Q. Zhang, *Magn.Reson.Chem.*, 1993, **31**, 748.
408 G. Haegele, R. Peters, K. Kreidler, R. Boese, G. Grossmann, F. Steglich and G. Ohms, *Phosphorus, Sulfur, and Silicon*, 1993, **83**, 77.
409 S. Quideau and J. Ralph, *J.Chem.Soc.,Perkin Trans.1*, 1993, 653.
410 H. Ritter and M. Kaiser, *Magn.Reson.Chem.*, 1993, **31**, 364.
411 N.N. Sveshnikov, A.A. Fomichov, I.V. Vystorop and V.G. Kartsev, *Mendeleev Commun.*, 1993, 107.
412 S. Benayache, J. Courtieu, M. Wanat and R.F. Jullien, *J.Soc.Alger.Chim.*, 1992, **2**, 63.
413 N. Platzer, J.P. Buisson and P. Demerseman, *J.Heterocycl.Chem.*, 1992, **29**, 1149.
414 H. Schroeder and E. Haslinger, *Magn.Reson.Chem.*, 1994, **32**, 12.
415 S. Suntioinen, Laatikainen.R and V. Kral, *Magn.Reson.Chem.*, 1994, **32**, 51.
416 J.J. Barker, A.G. Orpen, A.J. Seeley and P.L. Timens, *J.Chem.Soc., Dalton Trans.*, 1993, 3097.
417 T.J. Beasley, R.D. Brost, C.K. Chu, S.L. Grundy and S.R. Stobart, *Organometallics*, 1993, **12**, 4599.
418 T. Schaefer, R. Sebastian, R.W. Schurko and F.E. Hruska, *Can.J.Chem.*, 1993, **71**, 1384.
419 T. Schaefer, R. Sebastian, D.M. McKinnon, P.W. Spevack, K.J. Cox and C.S. Takeuchi, *Can.J.Chem.*, 1993, **71**, 960.
420 C.I.F. Denekamp, D.F. Evans, J. Parr and J.D. Woolins, *J.Chem.Soc.,Dalton Trans.*, 1993, 1489.
421 P. Szczecinski, *J.Organomet.Chem.*, 1993, **458**, 119.
422 P. Szczecinski and J. Zachara, *J.Organomet.Chem.*, 1993, **447**, 241.
423 M. Mikolajczyk, P.P. Graczyk and M.W. Wieczorek, *J.Org.Chem.*, 1994, **59**, 1672.
424 A.P. West Jr., N. Smyth, C.M. Kraml, D.M. Ho and R.A. Pascal Jr., *J.Org.Chem.*, 1993, **58**, 3502.

425 L. Ernst, K. Ibrom, K. Marat, R.H. Mitchell, G.J. Bodwell and G.W. Bushnell, *Chem.Ber.*, 1994, **127**, 1119.

426 G. Grossmann, B. Schwab, A. John, H. Komber, G. Jeschke and J. Sieler, *Phosphorus, Sulfur, and Silicon*, 1993, **78**, 297.

427 S.D. Perera, B.L. Shaw, M. Thornton-Pett and J.D. Vessey, *Inorg.Chim. Acta*, 1993, **207**, 175.

428 M. Gruber, R. Schmutzler and D. Schomburg, *Phosphorus, Sulfur, and Silicon*, 1993, **80**, 205.

429 S.D. Pastor, S.P. Shum, R.K. Rodebaugh, A.D. Debellis and F.H. Clarke, *Helv.Chim.Acta*, 1993, **76**, 900.

6
Nuclear Spin Relaxation in Liquids and Gases

BY H. WEINGÄRTNER

1. Introduction

In continuation of previous reports of this type,[1] this review covers work on magnetic relaxation and self-diffusion in liquids published mainly between June 1993 and May 1994. A section on magnetic relaxation in gases is added. This topic has last been reviewed in this series four years ago, but even so, little work has come to this reporter's attention. A comprehensive review of gas phase n.m.r. has appeared in 1991.[2]

As in previous periods, this review is limited to work on comparatively simple liquids and solutions of physicochemical and chemical interest, as publications in the field of macromolecular and biological chemistry are covered elsewhere in this volume. Of course, such a distinction is sometimes problematic, as innovative work dealing with solutions of complex molecules may well be of interest for research in the field covered here. Thus, at the risk of duplication, some interesting studies dealing with more complex systems are mentioned briefly.

At the beginning of this chapter it is convenient to quote some authoritative reviews in the subject area. More specialised reviews will be discussed in the corresponding subsections. Also, some important general trends are briefly highlighted here. Details will be discussed later in this chapter.

Among the most important trends is the use of rare gas nuclei for probing molecular structures and local dynamical processes in solution. Since about one decade such studies have dealt with the spin−1/2 nucleus ^{129}Xe, the quadrupolar nucleus ^{131}Xe and to a some extent also with the quadrupolar nucleus ^{83}Kr. Work of this type can now be regarded as being quite routine, and the current state has been reviewed.[3] New interesting applications have however been suggested, and studies of this type have been extended to include the quadrupolar nucleus ^{21}Ne.[4,5] In the past ^{21}Ne has received little attention, owing to its low natural abundance and low n.m.r. sensivity. For the first time, data have also been reported for the β−active quadrupolar nucleus ^{23}Ne in pure liquid neon.[6]

From studies on dynamical processes which continue to be the major topic in the field under review, work on magnetic relaxation and self–diffusion over wide ranges of temperature and pressure may be singled out for special mention. A wealth of information is obtained at temperatures low enough to reach conditions beyond the extreme–narrowing regime, and above all near the glass transition. In fact, magnetic relaxation and self–diffusion studies have contributed much to the current understanding of slow motional processes above and below the glass transition, and it would appear that a concise picture of motional processes near the glass transition can now be developed.[7,8] The study of slow motions near the glass transition requires the application of sophisticated 1D and 2D techniques, often adopted from solid state spectroscopy,[9] and in this context some new multi–dimensional relaxation experiments have been described.[10]

The characterisation of the pressure–dependence of relaxation and self–diffusion processes is another important tool for evaluating a concise picture of the molecular dynamics and structure of fluids. Work in this area has continued, and a comprehensive and authoritative review by Lang and Lüdemann is now available which covers some 600 references on high–pressure n.m.r.[11] Such data form e.g. a major source of information on the unique properties of water and aqueous solutions, above all in the supercooled and metastable state, and a review of this subject is available as well.[12] Sophisticated instrumental set–ups also enable NOE and high–resolution experiments at high pressures,[13,14] which may provide a new tool for studying the pressure–dependence of molecular conformations.

Another interesting development concerns the theoretical and experimental determination of ^{2}H and ^{17}O quadrupole coupling constants (NQCCs) in the liquid phase. In hydrogen–bonded systems, these NQCCs may largely depend on temperature, pressure and on the molecular environment of the probe molecule. Hence, a broad range of values is expected, as is e.g. exemplified by large differences of the NQCCs in gaseous water and ice. A reliable understanding of these effects is an important prerequisite of any theoretical analysis of relaxation processes of quadrupolar nuclei involved in hydrogen–bonding. As suggested about one decade ago by Leyte and coworkers,[15] a combination of ^{1}H, ^{2}H and ^{17}O relaxation time experiments may enable a direct experimental determination of the liquid–state ^{2}H and ^{17}O NQCCs. This experimental approach has now been adopted on a broader basis for studying liquid–phase NQCCs in hydrogen–bonded systems.[16-18] The results are particularly exciting, as accurate quantum–mechanical calculations of electric field gradients and coupling constants can now be carried out with quite complex systems, thus enabling theoretical calculations of NQCCs in molecular clusters.[19,20] Such calculations have been performed with water clusters of various geometries, known from computer simulations to represent snapshots of the liquid structure.[21,22] It would appear to this reviewer that, based on these experimental and theoretical studies, for the first time an understanding of the nature of this coupling in hydrogen-bonded liquids is obtained. A review of this theoretical work[19] is available in an issue of the Zeitschrift für Naturforschung which contains the proceedings of a symposium on nuclear quadrupole resonance. Several other articles in this issue dealing with the theory of quadrupole coupling are also of interest.[23-25]

Another interesting group of studies has dealt with self-diffusion in liquids and solutions. Work in this area is steadily increasing, as obviously, many spectrometers are now routinely equipped with gradient coils for generating magnetic field gradients. While until recently ^{1}H n.m.r. has been the mainstay of the field, the use of less-common nuclei like ^{21}Ne [4] offers promising new aspects for future work. Moreover, new experimental techniques have been developed which offer interesting perspectives. For example, the stray field of cryo-magnets is used to achieve very high magnetic field gradients, thus enabling the observation of very slow diffusion processes. The underlying experimental technique, already employed recently,[26,27] has now been described in detail.[28]

A comparatively new development concerns n.m.r. experiments in the presence of an electric current within the sample. Since 1980 the basis for such techniques has been systematically investigated,[29] and experiments of this type have now been termed DCNMR (direct current n.m.r.).[30] Among the various possible experiments based on DCNMR, the determination of mobilities of charged particles appears to offer the most interesting perspectives, and in the preceding review periods sophisticated 1D and 2D techniques, termed electrophoretic n.m.r. (ENMR) [31] or magnetic resonance electrophoresis (MREP),[32] have been developed. Despite the fact that the potentials of these methods have been impressively demonstrated, they cannot be regarded up to now as routine. The current review period has seen several important publications on this subject, including an authoritative review of the field.[29] A separate section is added in this chapter to describe work in this field.

Finally, it should be noted that the increasing availability of powerful computer facilities is leading to a steadily increasing number of computer-assisted data interpretations. For example, molecular dynamics and Monte-Carlo simulations are now widely used for obtaining information on molecular aggregation and dynamical processes relevant for spin relaxation, or for examining approximations in the underlying analytical theories.[33-38] Classical fields of application are complex relaxation processes caused by intermolecular interactions, e.g. present in the case of intermolecular dipole-dipole relaxation,[33,34] or in the quadrupolar relaxation of mono-atomic ions[35] and rare gases[5,36] in solution. Other interesting applications have dealt with the translational and reorientational dynamics of pure liquids[34,37,38] and electrolyte solutions.[39,40] It has been emphasized [34] that magnetic relaxation provides parameters which can be used to check the quality of the potential functions employed in these simulations. In addition, computer simulations are now also used for guiding the interpretation of relaxation rates due to internal motions in complex molecules of biological interest or simple model compounds for such molecules.[41–46]

In concluding this introductory section, three further reviews may be mentioned: Spiess has given a general review of n.m.r. spectrocopy.[47] Ernst, in his Nobel prize lecture,[48] has reviewed Fourier transform n.m.r. A review of Kowalewski and Laaksonen has dealt with the theoretical parameters of n.m.r. spectroscopy.[49]

2. General, Physical and Experimental Aspects of Nuclear Spin Relaxation

2.1 General Aspects

The Bloch equations are extensively used in the analysis and design of n.m.r. experiments, but to date only approximate, incomplete or numerical solutions to these equations are presented. Morris and Chilvers[50] have now discussed a general analytical solution of the Bloch equations which may speed up calculations in computationally intensive areas such as the description of pulses in the presence of relaxation. Ohara and Slichter[51] have presented a new method for calculating motionally narrowed resonance lines which accounts for the effects of nuclear spin orientation on the environment. The resulting predictions are substantially different from those obtained by conventional methods. Benesi [52] has discussed series expansions of propagators which describe the time–dependence of the density matrix that characterizes relaxation. Van der Maarel [53] has discussed the dynamics of spin $I = 1$ in a radiofrequency (r.f.) field and with both static and fluctuating quadrupolar interaction.

Ernst and coworkers [54] have described an object–orientated programming approach for computer simulation in magnetic resonance. Simulations of magnetic resonance experiments often suffer from the fact that conventional computer languages do not include the data types needed to describe magnetic resonance experiments analytically. The authors have now developed a magnetic resonance "tool kit" termed GAMMA (*g*eneral *a*pproach to *m*agnetic resonance *m*athematical *a*nalysis) which extends the computer language C++ by adding data types typically used in magnetic resonance.

2.2 Experimental Aspects

In the period under review new experimental techniques have been added to the existing plethora of methods for measuring spin–lattice relaxation times T_1, spin–spin relaxation times T_2 and the rotating frame relaxation time $T_{1\rho}$. Several papers have dealt with the measurement of relaxation and magnetisation transfer in the presence of an off–resonance irradiation field. Goldman and Devaux [55,56] have described a method for the measurement of the steady–state magnetisation under strong off–resonance irradiation as a function of the angle between the external field and the effective field in the rotating frame. For purely dipolar relaxation between homonuclear spins, its variation with the angle yields the reorientational correlation time in the liquid state, otherwise deviations are observed from the theoretically predicted shape. The same group has reported on an off–resonance version of the ROESY experiment [57] which deals with the cross relaxation of transverse magnetisations. Bain and Duns [58] have developed an off–resonance saturation experiment for the simultaneous measurement of T_1 and T_2 and have discussed guidelines for its optimization. Kuwata et al.[59] have derived a general relaxation matrix equation for the description of the off–resonance rotating frame spin–lattice experiment.

Kwag et al.[60] have considered experimental conditions under which multiple spin–echos are observed. Such multiple spin–echos are obtained when the repetition time is reduced to be comparable with the relaxation time T_1. Based on a theoretical analysis based on the Bloch equations, the authors propose multiple echos as a means of measuring very long relaxation times. Counsell [61] has demonstrated that the information encoded in multiple echos may enable the simultaneous determination of T_2 and the self–diffusion coefficient.

Mao and coworkers have studied experimentally [62] and theoretically [63] the effects of strong radiation damping on the lineshapes of n.m.r. signals. If a spin system recovers to the equilibrium state under the influence of radiation damping rather than through relaxation, artefacts are obtained in the spectra and spin–lattice and spin–spin relaxation times cannot be measured by conventional techniques. A theory has been developed [63] for describing the limiting case that radiation damping is the dominant mechanism and that relaxation processes can be completely ignored during the detection of the FID.

Other papers have dealt with a new pulse sequence for stimulating a magnetisation mode during coupled relaxation experiments in AX_2 (e.g. $^{13}CH_2$) spin systems,[64] the separation of longitudinal spin–order in cross–correlation experiments may by frequency cycling,[65] and practical aspects of 2D proton–detected ^{15}N relaxation measurements.[66]

2.3 Relaxation in Coupled Spin Systems

As in the preceding periods, cross–correlation effects between dipolar (DD) and chemical shift anisotropy (CSA) interactions have received attention, as have cross–correlations between different dipolar interactions. The work of Farrar et al.[67] on the ^{13}C–^{1}H two–spin system of the acetylene group of phenylacetylene represents a particularly instructive example for the wealth of information obtained from the observation of DD–CSA interferences. As noted in another contribution from the same group,[68] Monte–Carlo simulations can be used for parameter estimation in T_1 measurement of such systems. Another example for this type of investigation is the study of DD–CSA interferences in the ^{13}C–^{1}H spin system of benzene, where some controversy exists about the degree of motional anisotropy in the liquid state.[69] DD–CSA interferences have also been found in triphenyl derivatives of the form $(C_6H_5)_3XH$ with X = ^{13}C, ^{29}Si and ^{119}Sn.[70]

Interferences between different dipolar interactions have been considered as well. Several studies have dealt with relaxation in the $^{13}CH_2$ spin group and a new pulse sequence has been suggested for stimulating magnetisation modes in this group.[64] Gaisin et al.[71] have considered cross–correlation effects in the methylene groups of ethanol (diluted in the fully deuterated analogue), thereby extracting information about rotational motions. A recent study of Grant and coworkers [72] on cross–correlations in the methylene group of ethanol (dissolved in chloroform) has however shown that, when considering $^{13}CH_2$ group relaxation, an appreciable contribution due to DD–CSA interferences must be taken into account. Fuson and Belu[73] have considered the influence of neighbouring intramolecular spins on coupled spin relaxation in the central $^{13}CH_2$ group of nonane.

Other spin groups considered in cross–correlation studies are ^{29}Si–^{1}H in triphenylsilane [70] and ^{119}Sn – ^{1}H in triphenylstannane [70] and tri–n–butyltin fluoride.[74] Hwang et al. [75] have proposed a linear prediction method for the analysis of multi–exponential relaxation decays present in systems exhibiting cross–correlations.

2.4 Dipolar Couplings and Distance Information

NOE experiments have become a routine tool for investigating structures of complex molecules in solution, but this technique may also provide useful information on configurations of molecules of comparatively small or intermediate size[13,76–78] in solution. Distance constraints are now often used together with molecular dynamics simulations to calculate structures of molecules in solution. For non–exchangable protons this can be most accurately done by using a relaxation matrix analysis which accounts for all dipolar $^1H-^1H$ interactions.[79] Exchange can be included by accounting for an appropriate modification of the relaxation matrix. Forster and Mulloy [80] have considered the case that the compounds adopt multiple configurations in solution and have presented an algorithm for accounting for these multiple configurations. The method is based upon representing the effective relaxation matrix as a linear combination of relaxation matrices due to each conformer.

The question how to refine distance determinations by accounting for averaging effects of internal motions has found repeated attention,[81-84] and again, computer simulations may help to judge the magnitude of the relevant effects.[41,42]

2.5 Exchange Spectroscopy

Four papers have dealt with basic theoretical aspects of exchange spectroscopy. McGowan et al.[85] have considered the stability of an N–site magnetic exchange network. If this network is described by the coupled Bloch equations, stable solutions are, in general, obtained. Thereby, the eigenvalues are possessing negative real parts. The authors have demonstrated that this type of analysis is greatly simplified when a detailed balance approach is satisfied. Conditions for the stability of this detailed balance approach have been discussed. In the case of their violation the relaxation takes the form of a damped oscillatory function, with the amount of damping decreasing as the number of sites increases.

Szymanski [86] has reported on a general computer program for simulating 1D multiple quantum spectra in exchanging systems. The program is capable of handling intramolecular exchange in multi–nuclear spin–1/2 systems composed of several interconverting species. Results have also been presented for two model systems containing magnetically equivalent nuclei, where the equivalence is lost in the course of exchange. Seliger [87] has discussed resonance frequencies and lineshapes of a system of nuclei subject to random two–site exchange. London [88] has considered chemical shift and linewidths characteristics of reversibly bound ligands.

Various new or modified experimental techniques for studying exchange phenomena have been proposed. These include a fast inversion recovery sequence in the presence of a saturating field for measuring two–site exchange,[89] a method for optimizing the study of slow exchange by relaxation time measurements,[90] and the use of ^{14}N overtone spectroscopy for studying spin–

exchange.[91] A paper of Macura [92] has dealt with the evaluation of errors in 2D exchange spectroscopy.

Experimental studies in low–molecular weight systems have primarily dealt with proton exchange. Some typical examples may be quoted,[93-99] and the 2D study of Perrin and coworkers [93] on proton exchange between ammonium ions and ammonia may be singled out as a particularly illustrative example. A further notable feature is the increase of n.m.r. studies on molten salts, as is evidenced by work on proton exchange in acidic low–temperature molten salts.[94-96]

Finally, Ernst and coworkers [100] have reported on an interesting new application of exchange studies. 2D techniques have been used in conjunction with the ^{129}Xe resonance to investigate the excange of xenon from the liquid to the gaseous phase in pure xenon.

2.6 Quadrupolar Interactions

Accurate knowledge of coupling constants is a major prerequisite when trying to extract dynamical information from relaxation data of quadrupolar nuclei. If the relaxing nucleus is covalently bound, the NQCC is determined by the electric field gradient (e.f.g.) arising from the intramolecular charge distribution at the nuclear site. Thus, such data provide interesting information on chemical bonds. For example, ^{2}H NQCCs derived from ^{2}H relaxation data have been used to estimate the ionic bond character of the metal–hydrogen bond in transition metal hydride complexes,[101] and ^{11}B and ^{2}H relaxation of cyanoborohydride have been used in conjunction with ab initio quantum mechanical calculations to get information on the e.f.g. at the ^{2}H and ^{11}B sites of this molecule in solution.[102] On a more qualitative level, arguments about the symmetry of the charge distribution at the nuclear site may be applied to estimate the e.f.g. In particular, group III ions in solution may form a variety of ionic complexes with anions which can e.g. be studied ^{27}Al [103] or ^{115}In n.m.r.[104] In this way, it is e.g. possible to distinguish between $AlCl_4^-$ and $Al_2Cl_7^-$ complexes by their characteristic line widths.[103]

Usually, the extraction of NQCCs from liquid state relaxation data relies however on some approximations on correlation times for molecular reorientation entering into the expression for the relaxation rate. The standard procedure is to determine correlation times for molecular reorientation from DD (e.g. ^{13}C–^{1}H) relaxation data, and to use these figures, in turn, for deriving NQCCs for other nuclei in the molecule. This procedure is only correct, if the principal axis of the e.f.g. tensor coincides with the direction of the DD interaction. As noted about one decade ago by Leyte et al.,[15] this is the case for ^{1}H– ^{17}O dipolar and ^{2}H quadrupolar interaction in the OH group of water, but the method can obviously be employed for all OH–groups. This is particularly interesting, the ^{2}H NQCC in hydrogen–bonded OH–groups exhibits large medium effects. Zeidler and coworkers have now performed systematic studies on the ^{2}H (and ^{17}O) NQCC of water and alcohols in the pure liquids and in various binary mixtures over large temperature ranges.[16–18] In conjunction with recent results for other systems of this type[105–109] it is now possible two establish some trends on the temperature– and medium–dependence of ^{2}H QCCs in OH–groups. Of course, changes in hydrogen–bonding with temperature and upon addition of non–electrolytes changes the NQCC strongly, but details are very difficult to interpret without the help of theoretical calculations.

Hitherto, theoretical calculations of the e.f.g. have been limited to isolated molecules,[110] and in this way ab initio calculations have enabled a very accurate prediction of the ^{2}H [111] and ^{17}O [112] NQCCs of water in the gaseous phase in agreement with experimentally determined values. With basis sets of high quality accurate calculations are now also possible in larger molecules. Results have been employed to establish a coorelation which allows the prediction of ^{2}H NQCCs in a few minutes with the help of a desk calculator, if the deuterium bond length is known.[19]

As a logical extension, calculations are also possible with molecular cluster treated as super–molecules. Based on this idea, Huber and coworkers have developed a method for calculating NQCCs of liquid water.[19,20] In detail, bulk water configurations have been produced in molecular dynamics simulations. Clusters of water molecules have been taken randomly from these configurations and used in super–molecule ab initio calculations. The average of many of such calculations yields the NQCC. Data have been presented for ^{2}H [22] as well as ^{17}O NQCCs.[21]

The situation is somewhat different when considering QF–relaxing monoatomic noble gases or ions in solution. In this case the e.f.g. is of intermolecular origin, but again molecular dynamics simulations in conjunction with ab–initio calculations may be employed in such calculations. In the period under review such simulations have been reported for liquid Ne,[113] and for ^{21}Ne, ^{83}Kr and ^{131}Xe in various solvents.[5,36]

In particular, the use of ^{21}Ne ($I = 3/2$) in such studies may be highlighted. In spite of unique possibilities for probing local dynamics and structures by using this nucleus, applications of ^{21}Ne n.m.r. in solution chemistry have been reported only since 1991,[114,115] including the first relaxation study of ^{21}Ne in solution published by Mazitov et al. in 1992 [116] (two decades ago, relaxation data have been reported for ^{21}Ne in the pure liquid [117]). These studies have clearly demonstrated that ^{21}Ne relaxes by QF interaction. In the period under review, in addition first self–diffusion measurements using ^{21}Ne have been reported.[4] ^{21}Ne n.m.r. still turns out to be difficult, owing to the low natural abundance (0.26%), low n.m.r. sensivity and low gas solubilities, and isotopically enriched samples have been used.[4] There are also problems in the preparation of samples containing dissolved gases in liquids, and in the context of n.m.r. studies of dissolved noble gases Mazitov et al.[118] have developed a simple method for sample preparation.

Finally, it may be mentioned that by the aid of β–radiation detected n.m.r. in solid and liquid neon spin–lattice relaxation of ^{23}Ne has been monitored.[6] In this technique, spin–polarized ^{23}Ne atoms are created by capture of polarized cold neutrons and the angular asymmetry of β–radiation is used to monitor the nuclear polarization. Comparing the measured spin–lattice relaxation time (about 37 s) with that of ^{21}Ne the nuclear quadrupole moment of ^{23}Ne has been determined.

The n.m.r. of QF–relaxing noble gas nuclei in solution bears much similarity to that of mono–atomic ionic nuclei like alkali metal, alkaline earth metal, group III or halide nuclei. In contrast to previous review periods, little work on these QF–relaxing nuclei in simple, low–molecular weight systems have however come to this reviewer´s attention. Sacco[119] has reviewed investigations on the structure and dynamics of simple electrolyte solutions and on preferential solvation of alkali and halide ions in binary solvent mixtures. A contribution by the same group [120] has dealt with isotope effects upon the quadrupolar relaxation rates of $^{133}Cs^{+}$ and $^{35}Cl^{-}$ in light and heavy water. These effects are substantially lower than those observed upon other dynamical quantities like the self–diffusion coefficients of the ions and of water, water reorientational correlation times and the viscosity, showing that isotope effects in the direct vicinity of ions may differ from those in the bulk. In the light of potential applications in systems of biological interest,[35,121] an understanding of the basic relaxation processes due to solvent and counter ion effects in simple ionic model systems is needed. Again, molecular dynamics simulations have been employed for this purpose.[35]

Among the alkali metal nuclei, ^{7}Li is an exception, because in addition to the QF interaction ^{7}Li – ^{1}H DD interactions may contribute to the relaxation rate of Li^{+} in H_2O, but this view appears to be controversial. In a brief note, Hasebe et al.[122] have now reported on spin–lattice relaxation times of $^{7}Li^{+}$ in concentrated solutions in light and heavy water at low temperatures beyond the extreme–narrowing regime. Based on these results, it has been concluded that the QF–mechanism is responsible for relaxation. No quantitative estimates are however given for the ^{7}Li – ^{1}H DD interaction. More elaborate work on this subject is available.[123–125] This reporter would like to add that it should not be too difficult to obtain definite figures for the magnitude of the DD interaction by performing ^{7}Li – {^{1}H} 1D or 2D NOE experiments. Also, it may be noted that concentrated aqueous solutions of LiCl represent particularly interesting cases, where the glass transition can be approached. In this situation pulsed neutron diffraction can provide valuable additional information on the structure and dynamics,[126] needed for the interpretation of n.m.r. data.

2.7 Intermolecular Dipolar Interactions in Diamagnetic and Paramagnetic Solutions

A molecular dynamics simulation of a Lennard–Jones system with spin–1/2 nuclei at the molecular centres has been performed [33] to examine the validity of various theories of inter–molecular DD relaxation. The field–dependence of the simulated relaxation rate shows that theoretical models taking account of the liquid structure [127,128] can satisfactorily reproduce the simulation results. Molecular dynamics simulations with two different water models have also been used to get insight upon the intra– and intermolecular contributions to ^1H relaxation.[34]

Dipolar interactions in paramagnetic systems allow an experimental test of such theories, provided that the basic Solomon–Bloembergen–Morgan theory is valid. This is not necessarily so with paramagnetic transition metal ions.[129,130] On the other hand, the use of free radical ions does not suffer from such deficiencies. Beloritzky, Fries and coworkers [131,132] have reported on the relaxation of ^{31}P in the tetramethylphosphonium in the presence of anion radicals at several resonance frequencies. The data have been interpreted by the aid of sophisticated model calculations on the structure and dynamics of ion pairs in solutions.[131] The presence of hyperfine scalar coupling has been shown to have a negligible effect on T_1, but significantly contributes to T_2.[132]

In contrast to the preceding period,[1] little fundamental work has been done on relaxation in the presence of paramagnetic transition metal ions. in the context of an overview of the impact of relaxometry in bio–inorganic chemistry, Luchinat [133] has discussed some basic of the theory of such relaxation processes.

2.8 Slow Motions in Glasses

New developments in the field of molecular reorientation processes have almost exclusively dealt with slow molecular motions in liquids near the glass transition. Here, combined relaxation and self–diffusion studies provide particularly useful information, because obviously, near the glass transition temperature T_c, translational and rotational motions become strongly decoupled. Several papers have dealt with this subject,[7–10,134,135] and it would appear that a concise picture can now be developed which is based on mode coupling theory. It may be mentioned that other papers have dealt with relaxation in highly viscous liquids in the supercooled regime without giving peculiar attention to the peculiar signatures of the glass transition.[122,136] Moreover, the review of Lang and Lüdemann on high-pressure n.m.r. [11] contains many examples for n.m.r. studies of supercooled and solutions near the glass transition.

3. Selected Applications of Nuclear Spin Relaxation

3.1 Pure Liquids and Non-Electrolyte Solutions

Several molecular reorientation studies of simple liquids have been reported. Simcox et al.[137] have investigated anisotropic molecular reorientation of methyl iodide by temperature- and field-dependent ^{13}C relaxation time and NOE measurements. Wakai and Nakahara [138] have determined molecular reorientation times of water, acetonitrile, chloroform and benzene by ^{2}D and ^{14}N relaxation at pressures up to 300 MPa. The data have been represented in terms of rotational friction coefficients, and these have been compared with friction coefficients deduced from self-diffusion and viscosity measurements with respect to their pressure variation.

The reorientational dynamics of liquid benzene has been studied with particular emphasis on the tumbling motion, the time constant of which can be deduced from CSA–DD interferences.[69] 2D techniques for determining CSA–DD interferences in conjunction with conventional relaxation time measurements have been applied for studying molecular reorientation of triphenyl derivatives of the type $(C_6H_5)_3XH$ with X = C, Si and Sn.[70] Fuson and Belu [73] have considered ^{13}C relaxation in the central CH_2 group of nonane. By comparison of data for [5–^{13}C] nonane and its deuterated isotopomer C42H9–13CH2–C42H9, NOESY experiments have served to determine intramolecular cross-relaxation rates. The influence of protons in the neighbouring intramolecular groups upon ^{13}C relaxation in the central CH_2–groups have been determined. Modelling of these interactions as random fields have proved to be sufficient to obtain a reliable evaluation of spectral densities.

Suchanski et al.[139] have investigated molecular motions in 3,5–diemethylpyridine using ^{14}N relaxation. Asknes and Kymtys [140] have reported on ^{1}H relaxation rates of liquid pivalic acid and have compared the results with data for pivalic acid in the plastic phase. NOESY experiments have been performed to study conformations of 2–ethylhexyl benzoate.[13] Gaisin et al.[71] have used cross–correlation effects in the methylene group of ethanol to study its reorientation. Ludwig and Zeidler [16] have performed a comprehensive investigation of spin relaxation in ethanol and propanol over large ranges of temperature. Molecular motions have also been studied in the glass formers 1,3,5–tri–α–naphtylbenzene,[9] m–tricresyl phosphate,[8,135] and ortho–terphenyl.[7]

Several papers have dealt with a variety of aspects of the rotational dynamics of molecules and molecular groups in solution: selected examples are studies of tri–n–butyltin fluoride in n–hexane,[74] cycloalkylidene bisphenols, methyl ethers and benzoate esters in solution,[141] phenylacetylene in toluene and in toluene – alcohol mixtures,[67] and ethanol in chloroform.[72]

Relaxation time measurements of the noble gases ^{21}Ne, ^{83}Kr and ^{131}Xe have been performed in various solvents. Some general aspects of work of this type have been discussed in section 2.6 of this review.

Several papers have dealt with spin relaxation as a means for investigating the composition dependence of molecular motions and molecular aggregation in binary mixtures: tert. butyl iodide – carbon tetrachloride,[142] 1–methylcylohexene – 1–methylchloronaphtene,[143] and carbon dioxide – n–hexadecane in the supercritical regime.[144]

Aggregation effects are particularly interesting, if hydrogen–bonded fluids are diluted in other solvents. In the period under review several systems of this type have found attention: water – dimethyl sulfoxide,[17] methanol – carbon tetrachloride,[145] ethanol – carbon tetrachloride,[16] propanol – carbon tetrachloride,[16] pivalic acid – carbon tetrachloride,[140] and methanol – tetrahydrofuran.[18]

Finally, two studies dealing with the molecular dynamics in the systems aqueous solutions of 2,2–dimethyl–1–propanol [146] and tetrahydrofuran [147] may be mentioned. Both systems exhibit liquid–liquid phase separations. As the relaxation techniques employed in these studies require deuteration of one of the components, the effect of isotopic substitution upon phase separations has been investigated as well. Thereby, Balevicius et al. [147] have observed an disappearance of the closed gap in the water – tetrahydrofuran system upon deuteration of THF, in contrast to what would be expected on the basis of the little work reported hitherto for other systems. Additional data have been reported for ternary systems consisting of water, tetrahydrofuran and various alkali halides. N.m.r data have been used in an attempt to model these phase separations on a microscopic level. To do so, activation energies of molecular reorientation have been determined by ^{2}H relaxation time measurements. Very small differences in activation energies for relaxation in the various systems have then been attributed differences in orientational potentials which, in turn, have been related to parameters in decorated lattice methods for modelling the miscibility gaps. In this reviewer´s opinion, the strong temperature– and medium dependence of ^{2}H QCCs in hydrogen–bonded systems discussed in section 2 may question the interpretation given by these authors.

In another study dealing with the dynamics and structure of aqueous – organic mixtures exhibiting liquid–liquid phase separation, Has and Lüdemann have investigated the pressure dependence of the rotational and translational mobility in aqueous solutions of 2,2–dimethyl–1–propanol as a model system for hydrophobic particles in aqueous solutions.[146]

3.2 Electrolyte Solutions

A series of papers have dealt with spin–relaxation in aqueous electrolyte solutions. An extensive study by Bradl and Lang [136] has dealt with ^{2}H spin–lattice relaxation in aqueous solutions of tetraalkylammonium ions in the supercooled regime down to 180 K and at pressures up to 225 MPa. A comparison with data for aqueous solutions of alkali halides provides insight into the interplay between Coulombic and hydrophobic interactions in such systems. Hasebe et al.[122] have reported on data for $^7Li^+$ relaxation in concentrated solutions of LiCl in light and heavy water. Ion pairing in aqueous solutions has been investigated in a study on ^{31}P relaxation in the tetramethyl phosphonium cation induced by radical anions.[131,132] The effect of hydration and ion pairing upon ^{23}Na relaxation in the model system sodium phosphate – water has been studied by molecular dynamics simulations.[35] Merbach and coworkers [39] have performed Monte–Carlo simulations for tri–valent lanthanide ions in aqueous solutions in order to confirm structural and mechanistic conclusions on water exchange derived from variable–pressure n.m.r. experiments in such systems.

Isotopic ^{1}H – ^{2}H (or ^{3}H) substitution leads to large effects upon the dynamical quantities like the viscosity, self–diffusion coefficient or reorientational correlation time of water. The molecular origin of these effects has been investigated in a molecular dynamics simulation.[37] Sacco et al.[120] have studied the effect of caesium chloride upon these kinetic isotope effects. It has been observed that isotope effects upon the quadrupolar relaxation rates of $^{35}Cl^-$ and $^{133}Cs^+$ are substantially lower than upon those upon the viscosity, ionic self–diffusion coefficients and reorientational correlation time of water in such solutions. It has been concluded that in the direct vicinity of ions such effects may differ considerably from those in the bulk liquid, corresponding to a picture to which the structure–breaking effect extents to the first coordination sphere of the ions. Perrin et al.[93] have used 2D–EXSY techniques for studying proton exchange between ammonium ions and ammonia in mixtures of light and heavy water.

Little work has dealt with spin–relaxation in non–aqueous electrolyte systems and in electrolyte solutions in binary solvent systems. Sacco has reviewed studies dealing with the motional dynamics of solvent molecules in solutions of alkali halides in organic solvents and on the preferential solvation of alkali metal and halide ions in binary solvent mixtures.[119] Mechanistic pathways for Na^+ exchange in complexes with crown–ethers in acetonitrile have been investigated by dynamical ^1H spectroscopy.[148] Hayashi et al.[149] have studied ^7Li relaxation in concentrated solutions of $LiNO_3$ in 1,3–diaminopropane.

3.3 Low–Temperature Molten Salts

The microdynamics of binary and ternary room temperature molten salts containing ethylaluminium dichloride have been examined by ^{13}C and ^{27}Al relaxation as a function of melt composition and temperature.[150] Liquid–state quadrupole coupling constants of ^{27}Al in various complexes have been determined. The room temperature molten salt formed by $AlCl_3$ and 1–ethyl–3–methylimmidazolium chloride has been investigated by ^{27}Al and ^{35}Cl spectroscopy.[103] Proton Exchange processes in these mixtures have been studies by ^1H and ^2D n.m.r. in the acidic [95] and basic [96] regime. The nature of proton species in a room temperature molten salt prepared by reaction of HCl with 1–ethyl–3–methyl–1H–imidazolium chloride has been studied by ^1H n.m.r.[94]

3.4 Transition Metal Complexes

Luchinat has reviewed relaxometry of paramagnetic metal ions in systems of biological interest.[133] Merbach and coworkers [151] have measured ^{17}O longitudinal and transversal relaxation rates of aquo– and PDTA–complexes of Gd^{3+} and of a similar chelate complex serving as contrast agent [152] in magnetic resonance imaging. The measurements have been performed at variable temperature, pressure and at several magnetic fields. The data provide information on water exchange kinetics and reorientation times of complexes in such systems. Gd(III)–induced ^{13}C relaxation has served to characterize inner sphere complexation of lanthanides by propionate, acrylate and propiolate ligands.[153] The QF–relaxing nucleus ^{59}Co n.m.r. has been used to study ethylenediamine complexes of Co in various solvents.[154]

4. Nuclear Spin Relaxation in Gases

The report includes work from 1991 onwards. In 1991 Jameson has given a comprehensive review of gas phase n.m.r.[2]

A paper on ^{14}N spin relaxation of N_2 in buffer gases published in 1991 by Jameson et al.[155] has given rise to a controversy about the correct theory to be used in data evaluation and the applicability of models for molecular motions examined in this context.[156,157] ^1H and ^{13}C n.m.r. has been used to study spin relaxation of methane in various buffer gases, establishing the dominance of spin–rotation interaction.[158] ^{19}F n.m.r. has been applied to study the reorientation of the spherical tops SeF_6 and TeF_6.[159] Etesse et al.[144] have studied ^{13}C spin–lattice relaxation of CO_2 and ^1H spin–lattice relaxation of hexadecane in binary mixtures in both the coexisting liquid and vapour phases. Armstrong et al.[160] have used longitudinal ^2D relaxation in D_2 + He gas mixtures. Full quantum close–coupled scattering calculations have been carried out for an ab initio D_2 – He potential. Excellent agreement between measured and calculated relaxation times has been found.

Other applications of gas phase n.m.r. have dealt with intramolecular kinetics, as exemplified by the ring inversion kinetics in cyclohexane and cyclohexane + bath gas mixtures [161] and N–methylpiperazine.[162]

Ernst and coworkers [100] have used ^{129}Xe n.m.r. for studying xenon exchange between the liquid and gaseous phase. The magnetisation exchange has been monitored by 2D experiments. The technique opens wide applications in the study of liquid–gas equilibrium dynamics.

5. Self–Diffusion in Liquids

5.1 Experimental Aspects

Norwood [163] has reviewed n.m.r. methods for measuring diffusion. Kimmich and Fisher [28] have described diffusion measurements in the fringe field of superconducting magnets. The fringe field provides extremely strong and stable field gradients which can be used for diffusion experiments in the presence of very slow processes. The underlying experimental technique, already used by these [26] and other authors,[27] has now been discussed in detail.[28] 1D and 2D pulse sequences have been described. Test experiments have shown that reliable diffusion coefficients are obtained.

Merrill [164] has discussed the use of a composite pulse sequence in diffusion measurements. Eccles et al.[165] have proposed a semi-automated technique for the determination of the gradient preemphasis parameters. The method is based on a spherical-harmonic expansion of the magnetic field following the gradient pulse.

Van Gelderen et al.[166] have described a single-shot diffusion experiment. The proposed method consists of a simple spin-echo experiment of Stejskal-Tanner type during which a series of gradient echos is created by alternation of the gradient. Because each subsequent echo is attenuated by an additional pair of gradients, the total decay can be acquired in a single experiment. Stepisnik et al.[167] have discussed self-diffusion measurements in the earth´s magnetic field. Self-diffusion measurements require a magnetic field gradient strong enough to achieve a sufficient dephasing of the signal of the migrating spins. The authors have discussed cases in which the strength of the inhomogenous field exceeds that of the main field, as would be fulfilled in measurements in the earth´s magnetic field. In this case the conventional theory is no longer valid, and an extended theory has been developed which deals with this situation. Other experimental work has dealt with the design of a sample spinner for vibration sensitive diffusion experiments [168] and of a computer-controlled bipolar magnetic field gradient driver.[169]

5.2 Selected Examples

Lang and Lüdemann [11] have given a comprehensive review of the work on the pressure-dependence of self-diffusion in liquids. An account of self-diffusion in water and aqueous solutions has also been given.[12] In general, self-diffusion measurements are still based on ^{1}H n.m.r., although a variety of other nuclei may be used. Apart from ^{1}H, ^{13}C is increasingly used and is of prime interest, if molecules do not carry protons. Examples are CO_2 [144] and fullerenes.[170,171] Other interesting nuclei are the quadrupolar alkali metal nuclei, and at least for ^{7}Li, ^{23}Na and ^{133}Cs the spin-spin relaxation times are long enough to provide reasonable conditions for self-diffusion measurements. In the period under review this has been exemplified in a study of $^{133}Cs^{+}$ self-diffusion in aqueous solutions of CsCl.[120] Recently, it has also been noted [172] that rare gas nuclei may provide interesting probes for studying gas diffusion and for obtaining the local dynamics in liquids. Following an earlier study of xenon diffusion in water,[172] ^{21}Ne n.m.r. has been used for the first time for measuring the self-diffusion coefficient of neon in water.[4]

Based on these new as well as on conventional methods, the following systems have been investigated: hydrogen and deuterium fluoride,[173] several glass–forming organic liquids,[8,134,135] the fullerenes C_{60} in toluene [170,171] and in CS_2,[170] and C_{70} in benzene and CS_2,[170] binary mixtures of CO_2 with hexadecane,[144] pivalic acid with carbon tetrachloride,[140] 2,2–dimethylpropanone with water,[146] and CsCl in water.[120]

The more innovative of these studies may be briefly mentioned in more detail: Karger et al.[173] have investigated self–diffusion coefficients of anhydrous liquid ^{1}HF and ^{2}HF as a function of pressure between the melting pressure curve and 373 K at a maximum pressure of 600 MPa. Special equipment had to be developed to overcome corrosion problems encountered with HF. Another investigation from the same group [146] has dealt with 2,2–diemethyl–1–propanol in ^2H_2O used as a model substance for investigating the temperature– and pressure–dependence of the hydrophobic interaction. Etesse et al.[144] have studied the self–diffusion of CO_2 and hexadecane in the coexisting liquid and gaseous phases of their binary mixtures up to the critical point.

Two studies have dealt with the diffusion of fullerenes in solution, and data are now available for C_{60} [170,171] and C_{70}.[170] C_{60} is spherical with a van der Waals radius of about 0.5 nm, while C_{70} is a slightly ellipsoidal molecule, the short semi–axis of which nearly coincides with the radius of C_{60}. This opens interesting possibilities to study the influence of molecular geometry diffusion, as predicted by hydrodynamic theories.[170] Interestingly, despite the large size difference between solute and solvent which would favour the "stick" condition, the "slip" condition of the Stokes–Einstein equation has been experimentally observed. In this context, it is interesting to note that a sophisticated critique of slip and stick hydrodynamics is available for ellipsoidal bodies.[174]

Finally, we note that ^{1}H/^{2}H isotope effect upon self–diffusion have received particular attention in studies of liquid HF,[173] and aqueous solutions of CsCl.[120] Such effects are particularly large for water and aqueous solutions, and a computer simulation on the origin of these effects in pure water is available.[37]

6. Direct Current and Electrophoretic N.M.R.

Holz and coworkers have continued their studies on possible applications of n.m.r. in the presence of an electric current in the sample by studying $^7Li^+$ and $^{133}Cs^+$ mobilities in the ternary system LiCl + CsCl + water.[30] Since 1980 the basis for such techniques has been systematically investigated,[29] and experiments of this type have now been termed DCNMR (direct current n.m.r.).[30] Holz [29] has reviewed this field. Instrumental developments in this field have dealt with the design of a probe[30] and of a computer–controlled field gradient driver.[169]

References

1. H. Weingärtner, in "*Nuclear Magnetic Resonance*", ed. G.A. Webb (Specialist Periodical Reports), The Royal Society of Chemistry, Cambridge, 1994, Vol. **23**, 179.
2. C.J. Jameson, *Chem. Rev.*, 1991, **91**, 1375.
3. J. Jokisaari, *Prog. NMR Spectrosc.*, 1994, **26**, 1.
4. M. Holz, R. Haselmeier, R. K. Mazitov and H. Weingärtner, *J. Amer. Chem. Soc.*, 1994, **116**, 801.
5. J. Vaara, J. Jokisaari, T.T. Rantala and J. Lounila, *Mol. Physics*, 1994, **82**, 13.
6. F. Hilmer, A. Schirmer, H. Ackermann, P. Heitjans and H. J. Stockmann, *Z. Naturforsch. A*, 1994, **49**, 27.
7. E. Bartsch, F. Fujara, B. Geil, M. Kiebel, W. Petry and W. Schnauss, H. Sillescu and J. Wuttke, *Physica A*, 1993, **201**, 223.
8. E. Rössler, A.P. Sokolov, P. Eiermann and U. Warschewski, *Physica A*, 1993, **201**, 237.
9. K. Zemke, K. Schmidtrohr, J. H. Magill, H. Sillescu and H. W. Spiess, *Mol. Phys.*, 1993, **80**, 1317.
10. J. Leisen, K. Schmidtrohr and H.W. Spiess, *Physica A,* 1993, **201**, 79.
11. E.W. Lang and H.D. Lüdemann, *Progr. NMR Spectrosc.*, 1993, **25**, 507.
12. H.D. Lüdemann, *Pol. J. Chem.*, 1994, **68**, 1 (Engl.).
13. S. T. Adamy, S.T. Kerrick and J. Jonas, *Z. Phys. Chem.*, 1994, **184**, 185.
14. J. Jonas, P. Koziol, X. Peng, C. Reiner and D. M. Campbell, *J. Magn. Reson., Ser. B*, 1993, **102**, 299.
15. J. R. C. van der Maarel, D. Lankhorst, J. de Bleijser and J. C. Leyte, *Chem. Phys. Lett.*, 1985, **122**, 541.
16. R. Ludwig and M.D. Zeidler, *Mol. Phys.*, 1994, **82**, 313.
17. R. Ludwig, T.C. Farrar and M.D. Zeidler, *J. Phys. Chem.*, 1994, **98**, 6684.
18. D.S. Gill, J. Singh, R. Ludwig, and M.D. Zeidler, *J. Chem. Soc. Farady Trans.*, 1993, **89**, 3955.
19. H. Huber, *Z. Naturforsch., A,* 1994, **49**, 103.
20. R. Eggenberger, S. Gerber, H. Huber, D. Searles and M. Welker, *J. Comput. Chem.*, 1993, **14**, 1553.
21. R. Eggenberger, S. Gerber, H. Huber, D. Searles and M. Welker, *Mol. Phys.*, 1993, **80**, 1177.
22. R. Eggenberger, S. Gerber, H. Huber, D. Searles and M. Welker, *J. Chem. Phys.*, 1992, **97**, 5898.
23. E.A.C. Lucken, *Z. Naturforsch. A*, 1994, **49**, 133.
24. H. E. Palmer, *Z. Naturforsch. A*, 1994, **49**, 137.
25. H. E. Palmer, *Z. Naturforsch. A,* 1994, **49**, 146.
26. R. Kimmich, W. Unrath, G. Schnur and E. Rommel, *J. Magn. Reson., Ser. A*, 1991, **91**, 136.

27. F. Fujara, B. Geil, H. Sillescu and G. Fleischer, *Z. Phys. B – Condens. Matter*, 1992, **88**, 195.
28. R. Kimmich and E. Fischer, *J. Magn. Reson., Ser. A*, 1994, **106**, 229.
29. M. Holz, *Chem. Soc.Rev.*, 1994, **23**, 3.
30. M. Holz, D. Seiferling and X. Mao, *J. Magn. Reson., Ser. A,* 1993, **105**, 90.
31. C.S. Johnson and Q. He, *Adv. Magn. Reson.,* 1989, **13**, 133.
32. F.M. Coveney, J.H. Strange and E.G. Smith, *Mol. Phys.*, 1992, **75**, 127.
33. M. Odelius, A. Laaksonen, M. H. Levitt and J. Kowalewski, *J. Magn. Reson. A* , 1993, **105**, 289.
34. G. Lippens, D. Vanbelle, S.J. Wodak and J. Jeener, *Mol. Phys*., 1993, **80**, 1469.
35. S.W Chen and P.J. Rossky, *J. Phys. Chem.*, 1993, **97**, 10803.
36. M. Odelius and A. Laaksonen, *Mol. Phys.*, 1994, **82**, 487.
37. I. M. Svitcheff and P.G. Kusalik, *J. Phys. Chem*., 1994, **98**, 728.
38. S.I. Temkin and W.A. Steele, *Chem. Phys. Lett.*, 1993, **215**, 285.
39. S. Galera, J.M. Lluch, A. Oliva, J. Bertran, F. Foglia, L. Helm and A. E. Merbach, *New J. Chem.*, 1993, **17**, 773.
40. Y. Kataoka, *Bull. Chem. Soc. Jpn*, 1993, **66**, 2478.
41. R. Abseher, S. Lüdemann, H. Schreiber and O. Steinhauser, *J. Amer. Chem. Soc.*, 1994, **116**, 4006.
42. S. P. Edmondson, *J. Magn. Reson., Ser. B*, 1994, **103**, 222.
43. L. Troxler and G. Wipff, *J. Amer. Chem. Soc.*, 1994, **116**, 1468.
44. X. Y. Chang and K. F.Freed, *J. Chem. Phys.*, 1993, **99**, 8016.
45. P.J. Hajduk, D.A. Horita and L.E. Lerner, *J. Amer. Chem. Soc.*, 1993, **115**, 9196.
46. S. Balasubramanian, R. Nirmala, D.L. Beveridge and P.H. Bolton, *J. Magn. Reson., Ser. B*, 1994, **104,** 240.
47. H. W. Spiess, *Wiss. Fortschr.*, 1992, **42**, 132.
48. R. R. Ernst, *Angew. Chem.*, 1992, **104**, 817.
49. J. Kowalewski and A. Laaksonen, *Theor. Models Chem. Bonding*, 1991, **3**, 387.
50. G.A. Morris and P.B. Chilvers, *J. Magn. Reson., Ser. A*, 1994, **107**, 236.
51. K.E. Ohara and C.P. Slichter, *J. Magn. Reson., Ser. A*, 1994, **109**, 74.
52. A. J. Benesi, *J. Magn. Reson., Ser. A*, 1993, **103**, 230.
53. J.R.C. Van Der Maarel, *J. Chem. Phys.,* 1993, **99**, 5646.
54. S.A. Smith, T.O. Levante, B.H. Meier and R.R. Ernst, *J. Magn. Reson., Ser. A*, 1994, **106**, 75.
55. H. Desvaux and M. Goldman, *Mol. Phys*, 1994, **81**, 955.
56. H. Desvaux and M. Goldman, *Compt. Rend. Acad. Sci., Ser. II*, 1993, **317**, 749.
57. H. Desvaux, P. Berthault, N. Birlirakis and M. Goldman, *J. Magn. Reson., Ser. A*, 1994, **108**, 219.
58. A.D. Bain and G.J. Duns, *J. Magn. Reson., Ser. A*, 1994, **109**, 56.
59. K. Kuwata, D. Brooks, H. Yang and T. Schleich, *J. Magn. Reson., Ser. B,* 1994, **104,** 11.
60. J.O. Kwag, E.K. Jang and I. Yu, *J. Magn. Reson., Ser. A*, 1993, **103**, 303.
61. C.J.R. Counsell, *J. Magn. Reson., Ser. B*, 1993, **101**, 28.
62. X. Mao, J.X. Guo and C.H. Ye, *Chem. Phys. Lett.*, 1994, **222**, 417.

63. X. Mao and C. Ye, *J. Chem. Phys.*, 1993, **99**, 7455.
64. R.A. Brown and D.M. Grant, *J. Magn. Reson., Ser. A,* 1993, **105**, 284.
65. N. Chadrakumar and S.S. Velan, *J. Magn. Reson., Ser. A,* 1993, **104**, 363.
66. N.J. Skelton, A.G. Palmer, M.Akke, J. Kordel, M. Rance and W.J. Chazin, *J. Magn. Reson., Ser. B*, 1993, **102**, 253.
67. T.C. Farrar, M.J. Jablonski and J.L. Schwartz, *J. Phys. Chem.*, 1994, **98**, 4780
68. J.D. Trudeau, J. Bohmann and T.C. Farrar, *J. Magn. Reson., Ser. A*, 1993, **105**, 151.
69. C. Coupry, M.T. Chenon and L.G. Werbelow, *J. Chem. Phys.*, 1994, **101**, 899.
70. L. Maler, L. Dibari and J. Kowalewski, *J. Phys. Chem.*, 1994, **98**, 6244.
71. N.K. Gaisin, I.R. Manyrov and K.M. Enikeev, *Mol. Phys*., 1993, **80**, 1047.
72. Z. Zheng, C.L. Mayne and D.M. Grant, *J. Magn. Reson., Ser. A*, 1993, **103**, 268.
73. M.M. Fuson and A.M. Belu, *J. Magn. Reson., Ser. A*, 1994, **107**, 1.
74. Y.W. Kim, A. Labouriau, C.M. Taylor, W.L. Earll and L.G. Werbelow, *J. Phys. Chem*., 1994, **98**, 4919.
75. Y.Y.Lin, N.H. Ge and L.P. Hwang, *J. Magn. Reson., Ser. A*, 1993, **105**, 65.
76. C. Rossi, A. Donati and M.R. Sansoni, *Spectrosc. Lett.*, 1993, **26**, 1603.
77. C.E. Keller and W.R. Carper, *Magn. Reson. Chem*., 1993, **31**, 566.
78. Z.H. Chen and R.A. Sachleben, *J. Inclusion Phenom. Mol. Recognition Chem.*, 1993, **16**, 91.
79. H. Liu, A. Kumar, K. Weisz, U: Schmitz, K.D. Bishop, and T.L. James, *J. Amer.Chem. Soc.,* 1993, **115**, 1590.
80. M.J. Forster and B. Mulloy, *J. Comput. Chem.*, 1994, **15**, 155.
81. A.E. Torda, R.M. Scheek and W.F. van Gunsteren, *NATO Adv. Study Institute, Series A*, 1991, **207**, 219.
82. J.H. Prestegard and Y. Kim, *NATO Adv. Study Institute, Series A*, 1991, **207**, 269.
83. F.P. Yip and D.A. Case, *NATO Adv. Study Institute, Series A*, 1991, **207**, 317.
84. R. Kaptein, T.M.G. Koning and R. Boelens, *NATO Adv. Study Institute, Series A*, 1991, **207**, 349.
85. J.C. McGowan, J. Schotland and J.S. Leigh, *J. Magn. Reson., Ser. A*, 1994, **108**, 201.
86. S. Szymanski, *J. Magn. Reson., Ser. A,* 1994, **108**, 151.
87. J. Seliger, *J. Magn. Reson., Ser. A,* 1993, **103**, 175.
88. R.E. London, *J. Magn. Reson., Ser. A*, 1993, **104**, 190.
89. E. Baguet and C. Roby, *J. Magn. Reson., Ser. A,* 1994, **108**, 189.
90. A.D. Bain and J.A. Cramer, *J. Magn. Reson., Ser. A*, 1993, **103**, 217.
91. R. McNamara, C.H. Wu and S.J. Opella, *J. Magn. Reson., Ser. A*, 1993, **103**, 238.
92. S. Macura, *J. Magn. Reson., Ser. B*, 1994, **104**, 168.
93. C.L. Perrin, T.J. Dwyer and P. Baine, *J. Amer. Chem. Soc.*, 1994, **116**, 4044.
94. J.L.E. Campbell, K.E. Johnson and J.R. Torkelson, *Inorg. Chem.*, 1994, **33**, 3340.
95. P.C. Trulove, D.K. Sukumaran and R.A. Osteryoung, *J. Phys. Chem*., 1994, **98**, 141.
96. P.C. Trulove, D.K. Sukumaran and R.A. Osteryoung, *Inorg. Chem*., 1993, **32**, 4396.

97. A Bagno, C. Comuzzi and G. Scorrano, *J. Amer. Chem. Soc.*, 1994, **116**, 916.

98. G.D. Henry and B.D. Sykes, *J. Magn. Reson., Ser. B,* 1993, **102**, 193.

99. H.H. Limbach, *Stud. Phys. Theor. Chem.*, 1992, **78**, 329.

100. M. Tomaselli, B.H. Meier, P. Robyr, U.W. Suter and R.R. Ernst, *Chem. Phys. Lett.*, 1993 **214**, 1.

101. D. Nietlispach, V.I. Bakhmutov and H. Berge, *J. Amer. Chem. Soc.*, 1993, **115**, 9191.

102. G.H. Penner and A.R. Custodio, *Magn. Reson. Chem.,* 1994, **32**, 292.

103. S. Takahashi, M.L. Saboungi, R.J. Klingler, M.J. Chen., and J.W. Rathke, *J. Chem. Soc., Faraday Trans.*, 1993, **89**, 3591.

104. M.A. Malyarick, A.B. Ilyuhin and S.P. Petrosyants, *Polyhedron*, 1993, **12**, 2403.

105. B.C. Gordalla and M.D. Zeidler, *Mol. Phys.*, 1986, **59**, 817.

106. B.C. Gordalla and M.D. Zeidler, *Mol. Phys.*, 1991, **74**, 975.

107. R. Ludwig and M.D. Zeidler, *J. Mol. Liqu.*, 1992, **54**, 181.

108. R. Ludwig, D.S. Gill and M.D. Zeidler, *Z. Naturforsch. A*, 1991, **46**, 89.

109. R. Ludwig, D.S. Gill and M.D. Zeidler, *Z. Naturforsch. A*, 1992, **47**, 857.

110. P. Pyykkö, *Z. Naturforsch. A*, 1992, **47**, 189.

111. S. Gerber and H. Huber, *J. Mol. Spectrosc.*, 1989, **134**, 168.

112. R. Eggenberger, S. Gerber, H. Huber, D. Searles and M. Welker, *J. Mol. Spectrosc.*, 1992, **151**, 474.

113. R. Eggenberger, S. Gerber, H. Huber and M. Welker, *Chem.Phys.*, 1993, **177**, 91.

114. P. Diehl, O. Münster and J. Jokisaari, *Chem. Phys. Lett.*, 1991, **178**, 147.

115. J. Lounila, O. Muenster, J. Jokisaari and P. Diehl, *J. Chem. Phys.*, 1992, **97**, 8977.

116. R.K. Mazitov. H.G. Hertz, R. Haselmeier and M. Holz, *J. Magn. Reson.*, 1992, **96**, 398.

117. R. Henry and R.E. Norberg, *Phys. Rev. B*, 1972, **6**, 1645.

118. R.A. Mazitov, P. Diehl, W.V. Garanin and G. Rotter, *Rev. Sci. Instr.*, 1994, **65**, 2149.

119. A. Sacco, *Chem. Soc. Rev.*, 1994, **23**, 129.

120. A. Sacco, H. Weingärtner, B.M. Braun and M. Holz, *J. Chem. Soc., Faraday Trans.*, 1994, **90**, 849.

121. L.C.A. Groot, J.R.C. van der Maarel and J.C. Leyte, *J. Phys. Chem.*, 1994, **98**, 2699.

122. T. Hasebe, T. Takeda, K. Tanaka and R. Tamamushi, *J. Chem. Soc., Faraday Trans.*, 1993, **89**, 4249.

123. E.W. Lang, W. Fink, H. Radkowitsch and D. Girlich, *Ber. Bunsenges. Phys. Chem.*, 1990, **94**, 342.

124. H. Weingärtner, J. Magn. Reson., 1980, **41**, 74.

125. R. Mazitov, K.J. Müller and H.G. Hertz , *Z. Phys. Chem.*(Munich), 19984, **140**, 55.

126. M. Yamagami, T. Yamaguchi, H. Wakita and M. Misawa, *J. Chem. Phys.*, 1994, **100**, 3122.

127. H. Hwang and J.H. Freed, *J. Chem. Phys.*, 1975, **63**, 4017.

128. P.H. Fries and E. Beloritzky, *J. Phys. (Paris)*, 1978, **39**, 1263.

129. see e.g.: J. Kowalewski, L. Nordenskjöld, N. Benetis and P.O. Westlund, *Prog. NMR Spectrosc.*, 1985, **17**, 141.

130. see e.g.: L. Banci, L. Bertini and C. Luchinat, *"Nuclear and Electron Relaxation"*, VCH, Weinheim, Germany, 1991.

131. M. Jeannin, E. Beloritzky, P.H. Fries and W. Gorecki, *J. Physique II*, 1993, **3**, 1511.

132. E. Beloritzky, P.H. Fries, W. Gorecki, M. Jeannin, and C. Roby, *J. Phys.Chem.*, 1994, **98**, 4517.

133. C. Luchinat, *Magn. Reson. Chem.*, 1993, **31**, Spec. Issue, S145.

134. G. Lahajnar, I. Zupancic, R. Blinc, A. Zidansek, R. Kind and M. Ehrensperger, *Z. Phys. B, – Condens. Matter*, 1994, **95**, 243.

135. E. Rössler and P. Eiermann, *J. Chem. Phys.*, 1994, **100**, 5247.

136. S. Bradl and E.W. Lang, *J. Phys. Chem.*, 1993, **97**, 10463.

137. P.B. Simcox, L.E. Nance and A.A. Rodriguez, *J. Mol. Liqu.*, 1994, **59**, 65.

138. C. Wakai and M. Nakahara, *J. Chem. Phys.*, 1994, **100**, 8347.

139. W. Suchanski, M. Kempka, B. Peplinska and Z. Pajak, *Ber. Bunsenges. Phys. Chem.*, 1994, **98**, 804.

140. D.W. Aksnes and L. Kimtys, *J. Mol Struct.*, 1994, **322**, 187.

141. Y.H. Lee, B.D. Allison and J.B. Grutzner, *J. Phys. Chem.*, 1994, **98**, 1783.

142. D.W. Asknes and L. Kymtis, *Analyst*, 1994, **119**, 641.

143. D.J. Gisser and M.D. Ediger, *J. Phys. Chem.*, 1993, **97**, 10818.

144. P. Etesse, W.G. Chapman and R. Kobayashi, *Mol. Phys.*, 1993, **80**, 1145.

145. U. Wandler and H.G. Hertz, *Z. Phys. Chem.*, 1992, **178**, 217.

146. M. Has and H.–D. Lüdemann, *Z. Naturforsch. A*, 1993, **48**, 793.

147. V. Balevicius, N. Weiden and A. Weiss, *Ber. Bunsenges. Phys. Chem.*, 1994, **98**, 785.

148. Y. Li and l. Echegoyen, *J. Amer. Chem. Soc.*, 1994, **116**, 6832.

149. H. Hayashi, Y. Nakamura and S. Shimikawa, *Ber. Bunsenges. Phys. Chem.*, 1994, **98**, 726.

150. C.E. Keller and W.R. Carper, *J. Phys. Chem.*, 1994, **98**, 6865.

151. K. Micskei, D.H. Powell, L. Helm, E. Brucher and A.E. Merbach, *Magn. Reson. Chem.*, 1993, **31**, 1011.

152. D. Gonzalez, D.H. Powell, V. Tissieres and A.E. Merbach, *J. Phys. Chem.*, 1994, **98**, 53.

153. J. Huskens, J.A. Peters and H. Vanbekkum, *Radiochim. Acta*, 1993, **61**, 195.

154. K.K.W. Ho, W.Y. Choy, C. Yu–Xin and S.C.F. Yeung, *J. Magn. Reson., Ser. A*, 1994, **108**, 196.

155. C.J. Jameson, A.K. Jameson and M.A. ter Horst, *J. Chem. Phys.*, 1991, **95**, 5799.

156. S.I. Temkin and A.I. Burshtein, *J. Chem. Phys.*, 1994, **100**, 1775.

157. C.J. Jameson, A.K. Jameson and M.A. ter Horst, *J. Chem. Phys.*, 1994, **100**, 1777.

158. C.J. Jameson, A.K. Jameson and N.C. Smith, J.K. Hwang and T. Zia, *J. Phys. Chem.*, 1991, **95**, 1092.

159. C.J. Jameson, A.K. Jameson and R.J. Terry, *J. Phys. Chem.*, 1991, **95**, 2982.

160. R.L. Armstrong, M. Bogdan, K.R. Jeffrey, C. Bissonnette and F.R.W. McCourt, *J. Chem. Phys.*, 1993, **99**, 5754.

161. P.O. Moreno and N.S. True, *J. Phys. Chem.*, 1991, **95**, 57.

162. M. Taffazzoli, C. Suarez, N.S. True, C.B. Lemaster and C.L. Lemaster, *J. Mol. Struct.*, 1994, **317**, 137.
163. T.J. Norwood, *J. Magn. Reson., Ser. A,* 1993, **103**, 258.
164. M.R. Merrill, *J. Magn. Reson., Ser. A*, 1993, **103**, 223.
165. C.D. Eccles, S. Crozier, M. Westphal and D.M. Doddrell, *J. Magn. Reson., Ser. A*, 1993, **103**, 135.
166. P. van Gelderen, A. Olson, and C.T.W. Moonen, *J. Magn. Reson., Ser. A*, 1993, **103**, 105.
167. J. Stepisnik, M. Kos, G. Planinsic and V. Erzen, *J. Magn. Reson., Ser. A,* 1994, **107**, 167.
168. D. Wu, W.S. Woodward and C.S. Johnson,jr., *J. Magn. Reson., Ser. A*, 1993, **104**, 231.
169. R.M. Boerner and W.S. Woodward, *J. Magn. Reson., Ser. A*, 1994, **106**, 195.
170. R. Haselmeier, M. Holz, M.M. Kappes, R.H. Michel and D. Fuchs, *Ber. Bunsenges. Phys. Chem.*, 1994, **98**, 878.
171. T. Kato, K. Kikuchi and Y. Achiba, *J. Phys. Chem.*, 1993, **97**, 10251.
172. H. Weingärtner, R. Haselmeier and M. Holz, Chem. Phys. Lett., 1992, **195**, 596.
173. N. Karger, T. Vardag and H.D. Lüdemann, *J. Chem. Phys.*, 1994, **100**, 8271.
174. S.A. Tang and G.T. Evans, *Mol. Phys.*, 1993, **80**, 1443.

7
Solid State NMR

BY M. W. ANDERSON

1 Introduction

The format of this review is, on the whole, consistent with previous years. In total there have been over one thousand references cited in *Chemical Abstracts* this year concerning solid-state NMR. Ninety percent of these papers are concerned with the application of solid-state NMR with the other ten percent concerned with technique or analysis development. Consequently, it is impossible to review comprehensively all areas of application of solid-state NMR within this chapter. The reader is, therefore, directed to the reviews detailed in section 2. Overwhelmingly, the areas which receive greatest attention are polymers, biological materials, zeolites and zeotypes, and minerals. As I have a more detailed knowledge of the field of microporous materials I have tried, where possible, to make a more critical assessment of the papers published on zeolites and zeotypes. In other areas of application only those papers of more general interest have been cited. Papers concerned with theoretical and methodological aspects of solid-state NMR have been reviewed comprehensively.

2 Reviews

During the period covered by this article there have been no comprehensive reviews on methodological areas of solid-state NMR, however, there have been a number of smaller reviews covering applications in a variety of areas. Two very general reviews on state-of-the-art in solid-state NMR[1,2] consider applications in physics and chemistry, molecular motion and magnetic resonance imaging. Reviews covering both high-resolution solution and solid-state NMR[3,4] also cover progress in the instrumentation for NMR spectroscopy. The role of NMR in the study of solids with specific reference to polymers and zeolites is covered in a small review.[5]

Two reviews consider the problem of determining interatomic distances in solids using multidimensional techniques.[6,7] In particular spin-echo double resonance (SEDOR) and rotation-echo double resonance (REDOR) is compared with direct methods using dipolar splittings. Several other new methods are also discussed.

Solid-state NMR of quadrupolar nuclei is routinely used in the study of catalysts and minerals. However, in order to obtain meaningful results the spectrometer conditions must be carefully controlled and spectra interpreted correctly. Problems and pitfalls associated measuring NMR spectra of quadrupolar nuclei such as ^{11}B, ^{17}O, ^{23}Na, ^{27}Al, ^{69}Ga and ^{71}Ga have been reviewed.[8]

The role of rotor-synchronised magic-angle spinning (MAS) coupled with two-dimensional exchange NMR for measuring chemical exchange when coupled with molecular reorientation has been reviewed.[9] Also reviewed are solid-state NMR imaging methods for measuring ultra slow molecular motions with application for the imaging of high-density macromolecular aggregates with very low frequency dynamics.[10]

One single area in chemistry where solid-state NMR continues to make an enormous contribution and where a plethora of papers continue to appear is that of zeolite chemistry. In previous years articles have concentrated on structural aspects, however, this year there have been a large number of papers devoted to NMR techniques applied to catalytic processes. Extensive reviews on catalysis have concentrated on *in situ* NMR methods,[11,12] surface active sites,[13] ^{17}O as an NMR probe[14] and structure and catalytic function of both zeolite and aluminophosphate molecular sieves.[15] A general review of characterisation methods for zeolites and molecular sieves includes a section on the importance of solid-state NMR.[16] From a structural aspect ^{27}Al-double rotation NMR applied to zeolites is reviewed[17] and applications of solid-state NMR to mesoporous materials.[18]

Applications of solid-state NMR to polymer chemistry are too numerous to be covered comprehensively in this article. Five reviews have appeared in this field dealing with structural analysis of synthetic polymers,[19] general polymer characterisation,[20] analysis of thermoplastic polymers, crystalline polymers and blended polymers,[21] and NMR imaging of solid polymers.[22,23]

Other areas of application of solid-state NMR which have been reviewed this year include: applications to material science;[24] studies of host-guest materials with applications in inclusion chemistry,[25,26] stabilisation of pesticides by clays;[27] supported organometallics,[28] polyoxometalate chemistry;[29] glass transition dynamics,[30] silicate melts,[31] pharmaceutical applications;[32,33] applications in natural organic chemistry for soil analysis,[34] forest log analysis,[35] proteins,[36] lignin,[37] carbonaceous solids,[38] coal,[39] and condensed tannins;[40] inorganic chemistry for analysis of powdered alumina and silica gel,[41] chalcogenides and pnictides[42] and xanthates;[43] the importance of the hydrogen bond in inclusion chemistry.[44]

3 Fundamental Aspects: Theoretical and Experimental

This section comprehensively covers the more fundamental aspects of solid-state NMR in terms of theory, new experimental pulse sequences, new methodology and technique, measurement of NMR parameters and fundamental experimental application.

3.1 Theory.

3.1.1 Sidebands. - The treatment of spinning sidebands in both MAS experiments and double angle spinning experiments has received considerable attention. Spinning sidebands contain important information but also often confuse spectral interpretation. Consequently, some papers deal with interpretation of spinning side-bands and others with their suppression.

An approximate method of side-band intensity calculation relies upon expanding the observed free-induction decay as a Taylor series using a basis of irreducible tensors from rank zero to rank infinity.[45] The zero-order irreducible tensors contribute to the sideband intensities and an approximate formula can be set up by truncating the series at the ninth order. This expansion works well as long as the ratio of anisotropy to spinning frequency is less than three. Anisotropic information about the chemical shift anisotropy may be extracted.

Spinning sidebands must be considered in order to render solid-state NMR spectra quantitative. Therefore suppression is not desired in the case of quantitative work. At infinite spinning speeds the sideband intensity will tend to zero. A new approach has been developed to reconstruct NMR spectra at infinite spinning rates using an experimental data set recorded over a range of spinning speeds.[46] The full signal-to-noise advantage of acquiring multiple spectra is reserved in the reconstruction.

Complete averaging of the second order quadrupolar interaction can, in principle, be averaged by spinning a sample at two magic angles as in double orientation rotation (DOR) or dynamic angle spinning (DAS). In DOR the rotation rates are usually about 1kHz for the outer rotor and 7kHz for the inner rotor. As lines are typically broadened by both chemical shift anisotropy (CSA) and quadrupolar interactions a plethora of sidebands results at these low spinning speeds. A complete sideband analysis in DOR NMR spectra at both infinite[47] and finite[48] inner rotor speed have been performed which allows a correct analysis of DOR spectra.

Sideband intensity distributions under cross-polarization (CP) show deviations from patterns recorded without CP. The reason stems from anisotropy in (i) the Hartmann-Hahn match (ii) the CP rate and (iii) the relaxation rate of the I spins in the rotating frame. In order to minimise these deviations a dual approach is proposed whereby a multiple contact pulse sequence partially diminishes deviations and a simple empirical correction reduces the deviations further.[49]

Total suppression of sibebands is considered theoretically by providing analytical solutions to a variety of suppression pulse sequences employing 4, 5, 6 and 9π pulses.[50] Previous treatments have also been critically examined.

3.1.2 Hamiltonian Theory. - Powder NMR line shapes under conditions of magic-angle spinning are calculated using average Hamiltonian theory and shown to be inadequate for treating second order perturbations.[51] This is illustrated by treatment of the centreband in the ^{27}Al MAS NMR spectrum of Y Al Garnet.

A periodic Hamiltonian treatment of the density matrix is derived as an expansion of powers of the inverse modulation frequency.[52] This allows interpretation of the contribution of the dipolar interaction to the sideband intensities.

Average Hamiltonian theory has been applied to the study of J coupling patterns from chemically equivalent nuclear spins[53] and second order two spin (AB) systems under MAS.[54] Expressions for both the positions and intensities of the four allowed transitions have been determined and the theory applied to ^{31}P MAS NMR spectra of phosphole tetramer and cis-1,2-bis(diphenylphosphino)ethylene.

The electron dipolar field in solids containing large numbers of paramagnetic centres has been determined by an exact solution to a multispin Hamiltonian.[55] Measurements of this dipolar field on a number of Argonne Premium coals demonstrated that even brief exposure to air had significant effects on the electron-spin dynamics.

3.1.3 Floquet theory. - Using the Floquet approach expressions for expectation values and observables are derived and analogy made with the spin density approach.[56] Effective Hamiltonians are derived for MAS experiments acquired synchronously. These include DRAMA (dipolar recovery at the magic angle) experiments on a pair of equivalent homonuclear spins, REDOR experiments on heteronuclear spin pairs and the dipolar decay of homonuclear spin pairs.

Second-order dipolar shifts which give rise to field-dependent isotropic shifts are restudied using the Floquet formalism.[57] For ^{13}C-^{1}H systems values of dipolar shifts calculated largely agree with experimental observations. Other second order effects and line broadening are also found to give rise to field dependent frequency shifts.

Floquet's theorem has been used to simulate line shapes for ^{13}C MAS spectra of bullvalene at various temperatures.[58] At *ca.* -10°C two separated centre peaks are observed which coalesce at higher temperatures. This is interpreted in terms of two independent processes, a concerted cope rearrangement reorientation with threefold jumps about the molecular C_3 axes. Analysis of the data yields separate rate constants for the two individual processes.

3.1.4 Quadrupolar effects on spin 1/2 nuclei. - Three papers deal with the effect of quadrupolar coupling on neighbouring spin 1/2 nuclei: in the limit of large quadrupole coupling;[59] concerning the sign of the indirect coupling constant;[60] and in the presence of dipolar and scalar coupling.[61] Equations are derived both for the isotropic shifts and the Pake-like powder patterns for any quadrupolar spin. The treatment is valid also for any orientation of internuclear vector with respect to the unique axis of an axially symmetric quadrupole tensor. Implications regarding ^{13}C,$^{35,37}Cl$, ^{31}P,$^{63,65}Cu$ and ^{119}Sn,$^{35,37}Cl$ cases are discussed and it is shown that for the latter two cases the sign of the indirect coupling constant J can be derived.

3.1.5 Cross-polarization. - Quantitation in CP spectra is always hampered by the possibility of non-uniform cross-polarization and relaxation time constants, T_{cH} and $T_{1\rho H}$ respectively. Therefore each resonance in a spectrum must be corrected individually in order to restore the unrelaxed spectrum. A numerical algorithm has been developed, based upon the concept of a hyperbolic tangent switching function which permits an easy correction to spectral intensities.[62] The technique is tested by fitting experimental spectra having different peak widths, degrees of overlap and relaxation constants.

Under CP conditions different CH_n groupings respond to varying degrees. A method has been developed which allows CP/MAS NMR spectra to be edited into a series of subspectra for CH_n groupings (n = 0, 1, 2 and 3).[63] Special attention is paid to CH and CH_2 groupings which generally behave in a similar fashion under CP but may be differentiated by cross-polarization-depolarization (CPD) techniques. The method is used for spectral editing of a number of test mixtures including complex pharmaceutical products such as calcipotriol monohydrate and vitamin D_3.

3.1.6 Ab initio calculations. - There have been a number of papers which consider ab initio calculations of NMR parameters and most of these are contained within the experimental section. A sample of these is contained in this section

Concentrating on the use of ab initio methods for conformational and structural calculations a study has been performed on peptides and proteins containing L-alanine residues by ^{13}C NMR.[64] Chemical shift calculations have been performed employing coupled-Hartree-Fock methods with gauge-invariant-atomic-orbitals (GIAO). It was found that the chemical shift of the C_β-carbon was related to the main dihedral angles, φ, ψ, but the C_α-carbon was also affected by hydrogen bonding structure.

^{13}C and ^{31}P CP MAS NMR has been combined with ab initio calculations of conformational energies to determine the structure of bis(diethoxythiophosphoryl)trisulfide, $(EtO)_2P(S)SSSP(S)(OEt)_2$.[65] The ^{13}C NMR spectrum resolves four methyl groups and three methylene groups while the ^{31}P NMR displays two signals due to inequivalent phosphorous nuclei within one molecule.

3.1.7 Other systems. - A suite of NMR line shape programs have been written which simulate five common interactions in the solid state.[66] The method is based on a non-linear least squares algorithm and has been used to extract physical parameters such as chemical shift tensor, bond distances and orientation of the chemical shift tensor in the molecular frame.

Another approach to NMR line-shape simulations adopts object oriented programming methodology.[67] Using new definitions such as operators, super operators and tensors an environment is created whereby a wide variety of situations may be simulated both in solution-state and solid-state NMR. A new magnetic resonance tool kit, GAMMA, based on the computer language C++ is introduced.

A theoretical structural analysis is presented for oriented polymers based on analysis of ^{15}N NMR spectra.[68] The analysis, which relies upon transformations of the Principal Axis System to the Molecular Axis System, can be used for the determination of NH and NC bond directions relative to the orientational axis.

3.2 New Pulse Sequences, Development of Old Pulse Sequences and New Methodology

3.2.1 Cross-polarization. - Cross-polarization (CP) is one of the most important techniques for analysis of organic solids, improving both sensitivity and selectivity. For spin 1/2 nuclei at high field strengths spectral widths are usually dominated by chemical shift anisotropy (CSA). In order to eliminate sidebands very high spinning rates are often used, however, this can substantially suppress the dipolar couplings among the abundant protons. This results in a dramatic change in the CP dynamics. At spinning rates of 10kHz only *ca.* 1 - 2kHz of the spectrum is Hartmann-Hahn matched. This match range can be significantly increased (up to *ca.* 8kHz) by simultaneously inverting the two RF channels many times during the contact period.[69] Consequently, intensity distortions are significantly reduced.

An alternative approach is to amplitude modulate the spin-lock field to the S-spins only while maintaining continuous irradiation on the I-spin channel (S-AMCP).[70] A double amplitude modulation scheme (D-AMCP) that modulates amplitudes to both channels is also presented which again significantly broadens the matching conditions. A similar amplitude modulation approach has been used to measure the ^{13}C CP MAS NMR spectrum of an amino acid mixture.[71]

Yet another approach to the problem of CP matching at high spinning rates involves the reintroduction of a heteronuclear coupling interaction by applying short radiofrequency pulses at specific times relative to the sample rotation period.[72] This technique is shown to be applicable up to *ca.* 15kHz sample rotation rate.

CP sequences can be modified by introducing periods for instance for proton dipolar dephasing which allows the experiments to be selective for different chemical groupings. A multipurpose CP pulse sequence is described which can be used to simplify the characterization of π-conjugated compounds such as trans-3,3'-bibenzo[c]thienylidene-1,1'-dione and insoluble poly(isothianaphthene).[73]

A new RF pulse sequence for homonuclear Hartmann-Hahn J-cross polarization (J-HOHAHA) under MAS conditions is described.[74,75] Experiments and numerical simulations of the J-HOHAHA pulse sequence are presented for powdered glycine doubly labelled with ^{13}C nuclei. The magnetization transfers in homonuclear INEPT and J-HOHAHA have also been compared. The dipolar mixing sequence suppresses isotropic and anisotropic chemical shifts and recovers the homonuclear dipolar interactions. The sequence consists of spin-echo and magic-echo sequences which are applied during the first and second halves of 2π sample rotations respectively.

CP usually occurs through magnetization transfer following a Hartmann-Hahn match of the RF fields for the I and S spins. This imposes very stringent conditions upon the strength of these RF fields. A new method is described which allows for efficient polarization transfer without resort to a Hartmann Hahn energy match.[76] The method draws upon concepts known from dynamic nuclear polarization of paramagnetic systems and is especially useful for nuclei with low γ, such as ^{15}N or for I>1/2, e.g. ^{2}H. The method is termed NZPT for non-Zeeman polarization transfer.

A method is described for distinguishing different C resonances by combing CP with polarization inversion (PI) at low to moderate spinning speeds.[77] The method is known as CPPI and relies on the fact that both the cross-relaxation rates and the heat capacity ratios of the I and S spins are different for different carbon groups. ^{13}CH and $^{13}CH_2$ groups can easily be distinguished using this technique.

3.2.2 Multi-dimensional experiments. - Two new 2-dimensional experiments have been described for correlating isotropic chemical shift with chemical shift anisotropy. The first is based on Z. Gan's experiment (1992) and uses slow MAS but instead applies 5 π pulses to the magnetization precessing in the transverse plane.[78] This sequence has the advantage of not sacrificing any magnetization projections out of the transverse plane and an illustrative example of the 5-π 2-D ^{13}C spectrum of 1,2,3-trimethoxybenzene is presented.

Another method achieves the same result by processing a series of 1-D NMR spectra recorded by sample rotation over a range of angle of the spinning axis with respect to the Zeeman field.[79] The advantages of this method are that no sudden mechanical motions are required and no multiple pulse sequences need be used. Applications of the technique are illustrated with an analysis of the CSA for eight distinct ^{13}C sites in tyrosine.

A novel 2-D experiment for monitoring exchange of Zeeman order between homonuclear dipolar-coupled spins is described.[80] Analysis of the magnetization transfer trajectories leads to characteristic powder lineshapes from which information on dipolar coupling and chemical shielding tensors may be extracted. In certain circumstances internuclear distances can be estimated from the dipolar coupling and the technique is applied to ^{13}C-^{13}C spin pairs in zinc acetate and glycine.

2-Dimensional exchange spectroscopy has been extended to spin I=1 nuclei such as deuterium using MAS coupled with synchronization of the mixing time with the rotor period.[81] Theoretical equations are derived for cross-peak intensities and the method is applied to a number of deuterated systems. In particular the method has been applied to dimethylmalonic acid-d_6 where no exchange takes place and to dimethylsulfone-d_6 and thiourea-C_6D_6 inclusion compounds where exchange is observed.

A novel three-dimensional spectroscopy correlates heteronuclear dipolar splitting with separated local field (SLF) and dilute spin exchange (SPEX).[82] The pulse sequence known as 3D SLF/SPEX is diagrammed and illustrated by analysis of ^{15}N-labelled single crystal of glycylglycine.

3.2.3 Quadrupolar nuclei. - One major problem recording NMR spectra of quadrupolar nuclei is that of quantitation. Many papers have dealt with this problem for single pulse excitation and optimal condition are well documented. The problems of quantitation in spin-echo experiments are much less well known and this is the subject of a recent paper.[83] Optimum conditions require the pulse length to be sufficiently short and the interpulse delay to be as short as the duration of the FID. The second pulse phase should also alternate without changing the receiver phase. Using these conditions the relative echo amplitude depends linearly on the first pulse length and quadratically on the second pulse length. Finally the pulse amplitude must be large compared to all dipolar interactions.

Spatial proximity of quadrupolar nuclei with spin 1/2 nuclei may be determined by coherence transfer and dipolar-dephasing double-resonance NMR experiments.[84] Coherence transfer through CP is observed in both directions between ^{27}Al and ^{31}P nuclei in the microporous aluminophosphate systems VPI-5 and $AlPO_4$-8. Dipolar-dephasing difference experiments, rotational-echo double-resonance (REDOR) and transferred-echo double-resonance (TEDOR) experiments with MAS are also demonstrated between quadrupolar and spin 1/2 nuclei.

A new method has been proposed for measuring high-resolution NMR spectra of spin 3/2 and spin 9/2 nuclei which involves exciting multiple quantum transitions.[85] For spin I=3/2 with excitation at $3\nu_o$ and sample spinning at 70.12° 1st and 2nd order quadrupolar interactions are completely removed. Shielding anisotropy is scaled by a factor of 0.98. For I=9/2 with excitation at $7\nu_o$ under MAS conditions resonances are narrower and more clearly separated that at ν_o.

3.2.4 Sideband suppression. - A simple method for removing spinning sidebands relies upon continually varying the spinning rate.[86] This essentially washes out the spinning sidebands by broadening these resonances. An illustration of the method is given for the ^{13}C MAS NMR spectrum of hexamethylbenzene.

A new pulse sequence for the suppression of sidebands involves suppressing the evolution of magnetization during certain timed fractions of the rotor cycle.[87] This is achieved by using pairs of hard RF pulses of equal intensities and opposite phases. The sequences are more robust than previous methods and can be designed to incorporate refocussing pulses.

A short communication details a simple approach to sideband suppression in DOR experiments by application of two inversion pulses.[88] For DAS spectra the sidebands may be eliminated by application of rotor-synchronized π-pulses.[89] This approach retains the full intensity of the isotropic centreband.

3.2.5 Dipolar recoupling. - In solid-state NMR spectroscopy several methods have recently been introduced for the purpose of restoring dipolar couplings into MAS experiments. From such experiments accurate interatomic distances may be sought but more generally they are useful for filtering and correlation experiments. One important example of dipolar recoupling is the REDOR experiment which is a spectrally non-selective reintroduction of dipolar couplings. A triple resonance experiment has been designed which selectively reintroduces heteronuclear dipolar couplings of observed spins with only those non-observed spins lying in exact resonance.[90] Another approach which involves the application of a π-pulse train is effective for reintroducing homonuclear dipolar couplings.[91] The recoupling occurs over a wide range of chemical shift differences and sample rotation speeds.

3.2.6 Imaging. - A proton NMR imaging experiment which uses line narrowing based upon TREV-CRAMPS with a time-dependent magnetic field gradient has been applied to solid adamantane.[92] A spatial resolution of *ca.* 100μm is achieved with a digital resolution of 40μm. This is nearly equivalent to that achievable in solution NMR imaging. High-resolution chemical shift information is retained and can potentially be extracted.

3.2.7 Methodology. - A variety of new methods have been developed which do not involve pulse sequences but which allow novel applications or improve spectral resolution. These methods are unconnected and are therefore listed sequentially.

Magic-angle spinning essentially imposes a cubic symmetry upon a sample by spinning about an axis which represents the body diagonal of a cube with the static field direction along one edge. DOR imposes a higher symmetry and rapid random motion as in solution imposes spherical symmetry. Spherical symmetry removes all broadening interactions. Such spherical symmetry has been imposed upon small solid particles by suspending in a support liquid and irradiating with ultrasound.[93] Evidence is provided for contributions to line narrowing from induced incoherent particle motion and in some cases from induced lattice motions.

In an effort to improve *in situ* NMR studies of catalytic reactions a new design of CAVERN apparatus which facilitates shallow-bed adsorption of reagents is described.[94] The system allows catalysts to be activated under shallow-bed conditions up to 773K followed by adsorption at room temperature. This eliminates the need to handle catalysts in a glove box during activation and adsorption steps which speeds up the sample preparation.

Motional averaging in the solid-state by application of physical manipulation and pulse sequences with particular emphasis on quadrupolar nuclei has been considered in a general article.[95]

An alternative to MAS for motional averaging is magic-angle hopping (MAH). This technique was originally developed by Bax *et al.* (1983) and allows the sample to be in contact with external influences during spectral acquisition. This should in principle allow an extension of various *in situ* solid-state NMR experiments. Some improvements to this technique have been described with modifications to the data acquisition.[96] The new technique is illustrated by measuring the ^{13}C spectrum of 1,2,3-trimethoxybenzene.

A method has been developed to increase the efficiency of coherence transfer in cross-polarization experiments from quadrupolar nuclei which involves dynamic angle spinning (DAS).[97] The CP step is performed while the sample is spinning on an axis oriented parallel to the magnetic field. The angle is then rapidly switched to the magic angle followed by data acquisition. This results in a greatly increased efficiency in the polarization transfer step and the method is illustrated by application for ^{27}Al-^{31}P CP in the large-pore aluminophosphate VPI-5.

A new NMR probe design has been developed for double-resonance spectroscopy.[98] The probe contains no quarter-wavelength elements and yet achieves adequate isolation between the two input channels. The probe contains relatively few components and is both compact and efficient.

A new tune-up procedure has been developed which allows easy alignment of the 4 quadrature channels typical in NMR spectrometers.[99] These methods are independent of the inhomogeneity of the RF coil and allow the amplitudes of the individual channels to be accurately set equal to one another.

3.2.8 Miscellaneous. - ^1H and ^{13}C background signals from stator assemblies can often swamp weak signals. Techniques for suppression of spins originating outside the coil were originally developed by Bax and relied on spatially selective composite 90° pulses. Modified pulse sequences have been developed which essentially completely remove background ^1H and ^{13}C resonances originating from Vespel spinning modules.[100] The methodology may be used under single pulse or CP conditions and is also useful for both T_1 and $T_{1\rho}$ relaxation measurements.

A new method has been developed for obtaining information on homonuclear dipole-dipole couplings which involves spinning the sample on an axis normal to the applied magnetic field.[101] The method is known as normal angle spinning dipolar spectroscopy (NADSY). A series of rotor synchronized pulses are applied during this period of normal angle spinning in order to create non-zero average dipole-dipole couplings. This is followed by a period of MAS during which the signal is acquired. Application of the technique is illustrated by observation of coupled carbonyl and methyl ester carbons and the suppression of uncoupled carbons.

It is often difficult to obtain undistorted ^2H solid-state NMR spectra over wide spectral ranges. Very short RF pulses must be employed which are sometimes impossible to achieve. Spectral excitation width can be dramatically increased by application of multiple pulse sequences. A new method has been devised which allows for simple undistorted spectra to be achieved by simple application of single short pulses.[102] The method works particularly well for large sample volumes which exceed the homogeneity region of the RF coil.

A novel triple resonance experiment has been reported which is useful for probing C-N internuclear distances without ^{15}N labelling by transfer of magnetization from protons.[103] Calculations of the dephasing of ^{13}C magnetization are in good agreement with experimental results obtained at slow spinning speeds for a sample of glycine and an estimate can be made of the dipolar coupling between the nitrogen and directly bonded carbon.

A new technique for detecting spectral spin diffusion in solids that is particularly well suited for accurately measuring cross-relaxation from minor spectral components is presented.[104] The pulse sequence known as SINK (saturation inter-nuclear kinetics) selectively saturates minor spectral components with a rotor synchronized DANTE pulse train and monitors spin diffusion to other peaks with a non-selective 90° pulse. Application of SINK to antimony(III)-doped Ca flouroapatite measures the spin diffusion between a weak peak at 68.6ppm due to fluoride ions associated with Sb^{3+} and other peaks in the spectrum.

3.3 Determination of NMR Parameters and Other Fundamental Information

Chemical shift tensors for the ^{113}Cd CP/MAS NMR spectra of the following seven cadmium complexes have been measured: $[Et_4N]_2[Cd_2(SC_6H_{11})_6]$; $[Me_4N][Cd(S_2CSC_6H_{11})_3]$; $[Cd(SC_5H_9NHMe)_2][ClO_4]_2$; $[Cd\{S(CH_2)_3NHMe_2\}X_2]$ X = Cl; X = Br; $[Cd_4\{S(CH_2)_2NMe_2\}_4Br_4]$; $[Cd_4\{S(CH_2)_2NMe_2\}_6Cl_2]$.[105] The shielding tensor, isotropic chemical shift (σ_{iso}), chemical shielding anisotropy ($\Delta\sigma$) and asymmetry parameter η were determined for all the sites and an orientation of the principal elements of the ^{113}Cd shielding tensor proposed for each cadmium environment.

Chemical shift tensors have been determined for phosphorus, proton and fluorine in monofluorophosphate and phosphite samples as a function of pH, solvent, temperature, concentration and extent of hydration.[106] P-F, P-O and P-H bond distances all decrease as the pH is increased and this affects values of σ_{iso} and $\Delta\sigma$ in the solid state and isotropic J couplings in solution. Chemical shift tensors have also been calculated from first principles using GAUSSIAN and LORG methods and these ab initio methods give excellent agreement with the experimental data.

Spin parameters for homonuclear two-spin systems have been determined using MAS and off-MAS spectra recorded on doubly labelled samples of sodium acetate ($Na^{13}CH_3{}^{13}COO$) and palmitic acid ($CH_3(CH_2)_{13}{}^{13}CH_2{}^{13}COOH$).[107] The method works as long as the isotropic shift difference is large and does not satisfy the rotational resonance condition.

^{15}N NMR measurements of aqueous nitrite NO_2^- under NO atmosphere indicate a significant presence of NO^+ which is confirmed by solid-state NMR measurements of $NOBF_4$.[108] A further broad resonance is assigned to an equilibrium between HNO_2 and $H_2NO_2^+$. CSA parameters in the solid state were obtained on natural abundance ^{15}N for the nitrosonium and nitronium ions in their respective tetraflooroboride salts.

It is found that the lineshapes of NMR spectra of pairs of slightly inequivalent homonuclear spins are dependent upon the spinning frequency even when the spinning rate is much larger than the dipolar coupling.[109] This is demonstrated for ^{15}N-^{15}N dipolar coupled spins in doubly labelled *cis*-azobenzene-$^{15}N_2$ dioxide. The NMR lineshapes have been simulated using average Hamiltonian theory and the comparison with experiment over the spinning range 1.2 to 6.0kHz is excellent. Lineshapes are mainly dependent upon the orientation of the CS tensors with respect to the ^{15}N-^{15}N vector.

Spinning rate dependencies on MAS NMR lineshapes are also observed in the ^{31}P MAS NMR AB spectra of $Cd(NO_3)_2(PPh_3)_2$ although the two phosphorous nuclei are crystallograpically equivalent and have the same isotropic chemical shift.[110] At high spinning rates the spectra are analogous to A_2 solution spectra whilst at low spinning rates they resemble AB spectra. This unusual spinning rate dependency results from weak homonuclear dipolar couplings between nuclei which have different orientation dependencies of the chemical shift tensors. It is predicted that such effects are more readily observable at high magnetic fields.

Solid-state ^{31}P NMR has also been used to study the phosphorus spin pair in the novel compound 3-(dichloromethylene)-trans-1,2-bis(2,4,6-tri-tert-butylphenyl)-1,2-diphosphirane.[111] Average Hamiltonian theory has been used to extract the principle components of the shielding tensors and experiment is compared with theoretical calculations on the model compound $P_2C_2H_4$.

Residual dipolar splittings in the ^{13}C CP MAS NMR spectra of 3-chloroacetanilide isomers resulting from chlorine nuclei have been reported.[112] At medium to high magnetic fields doublets are seen for C-Cl resonances and perturbation theory may be used to explain the appearance of the spectra qualitatively.

Proton Knight shifts in niobium hydride NbH_x have been measured as a function of hydrogen concentration with $0.05 < x \leq 1.05$ using ^1H MAS NMR.[113] The shift variation with $0.05 < x \leq 0.7$ can be correlated with the unit cell volume which is explained by the variation of the density of electronic states at the Fermi level N(0) due to the compression of the crystal lattice. However, with $0.7 < x \leq 1.05$ the Knight shift becomes smaller as the hydrogen concentration increases. This is explained by an increase in the number of electrons in the unit cell resulting in an increase in N(0).

3.4 Other Experimental Aspects

3.4.1 Cross-polarization. - The technique of cross-polarization magic-angle-spinning dipolar decoupling (CP MAS-DD) has been assessed for quantitative analysis of organic materials by measurement of a series of standard samples.[114] It is found that the kinetics of the initial CP process may lead to systematic errors, however, it is believed that this potential source of error can be minimised.

Rotor synchronized two-dimensional ^{13}C CP MAS NMR has been used to evaluate the orientational order in various samples of ultra-high-molecular-weight polyethylene fibres[115] A combination of three morphological components - two crystalline and one amorphous - can be identified and values of $\beta_{1/2}$, the polar angle at which the orientational order decays to half its maximum value can be determined. The limitations of these techniques to polymers in general are analysed.

NMR spectra of aprotic solids, ceramics and glasses often offer poor site resolution owing to large chemical shift distributions. The lack of protons means that cross-polarization to other nuclei becomes favourable as do heteronuclear double-quantum filtered experiments. Such experiments facilitate spectral editing and assignments and have been applied to semiconductors in the chalcopyrite system $CdGeAs_{2-x}P_x$, $CdSiAs_{2-x}P_x$, $Zn_xCd_{1-x}GeP_2$ through the coupling of ^{31}P with ^{113}Cd.[116] The results reveal that compositional dependencies observed in these alloys are dominated by non-local chemical shift contributions.

The CPPI technique, previously described in section 3.2.1 can be manipulated in such a way that $^{13}CH_3$, non-protonated ^{13}C and ^{13}CH signals are suppressed and the $^{13}CH_2$ signals give a methylene only subspectrum.[117] This technique has been demonstrated for monoethyl fumarate and cholesteryl acetate.

3.4.2 Quadrupolar interactions. - A number of lithium and sodium salts complexed with 12-crown-4 and 15-crown-5 have been analysed by ^{7}Li, ^{23}Na and ^{13}C solid-state NMR.[118] The survey focuses on the quadrupolar interaction of the lithium or sodium and highlights temperature-dependent phenomena.

Application of a perturbation approach to the problem of quadrupolar effects on spin-1/2 nuclei is reported which endeavours to explain unexpected field-dependent multiplicities.[119] Particular attention is paid to the ^{13}C, ^{14}N spin pair.

Rotation of a monoclinic single crystal about a single axis is shown to be, in general, sufficient for determining all five components of the quadrupolar coupling tensor.[120] Usually it is necessary to rotate a crystal about two non-orthogonal and non-parallel axes in order to achieve such information. The accuracy of this new single-axis-rotation method depends sensitively upon the choice for the axis of rotation. Application of this method is illustrated by measuring ten independent quadrupolar coupling tensors in bullvalene.

3.4.3 Two-dimensional experiments. - High-resolution solid-state ^{1}H NMR spectra have been recorded by combining MAS with proton dilution using deuterons.[121] The subsequently long T_1 proton relaxation times have been reduced by reverse cross-polarization (RCP) to deuterons. By application of two-dimensional ^{1}H nuclear Overhauser enhancement spectroscopy NOESY spectra proton exchange in solids has been measured and the rates observed correlated with degree of hydrogen bonding. Rate constants of 55 and 58 s^{-1} at 17°C with activation energies of 8.5 and 8.9 kcal/mol over the range -26 to 28°C were found for exchange between the carboxyl groups of the oxalate moiety and the water of hydration in oxalic acid dihydrate.

Main group functionalities have been measured with determination of chemical shift anisotropies using two-dimensional solid-state NMR.[122]

3.4.4 Deuterium NMR. - Static natural abundance deuterium spectra in the solid-state are often precluded owing to the broad spectral width and low sensitivity. It has been demonstrated that in favourable circumstances these problems can be circumvented by application of suitable MAS.[123] This spinning gathers the total spectral intensity into a few sharp spinning sidebands thereby improving the signal/noise ratio. This method is illustrated by application to the methyl groups in hexamethylbenzene and decamethylferrocene.

Intermolecular proton-deuterium cross-polarization has been achieved under MAS conditions on a 20% blend of perdeuterated PMMA in protonated PMMA.[124] A value for the CP time constant of 2.3 ms was determined by monitoring the dependence of the 3rd, 4th and 5th rotational echoes on contact time.

3.4.5 Miscellaneous. - ^{13}C and ^{31}P MAS NMR spectra have been recorded on a number of metal clusters with organic ligands.[125] These clusters include Au_{55}, Ni_8, Pt_{309}, Cu_{36} and Cu_{70}. It is found that the ligands give rise to relatively narrow lines with only the nuclei directly co-ordinated to the metal clusters being significantly broadened. There are no pronounced Knight shifts or indications for metallic-like relaxation.

A number of rare-earth acetates $M(OAc)_3.4H_2O$ (M = Nd, Sm, Eu, Y) and $Pr(OAc)_3.H_2O$ and their deuterated analogues have been studied by ^{13}C MAS NMR.[126] A large range of chemical shifts is observed resulting from the contact interactions with the paramagnetic centres. The spectral spread in proton spectra can be so large that effective decoupling becomes impossible. Ln - ^{13}C distances were estimated from the corresponding spectra which agreed to within 5% for the carboxyl carbons and 15% for the methyl carbons with distances available from single-crystal x-ray diffraction studies.

The temperature dependencies of spin-lattice relaxation times in the laboratory and rotating frames and the second moment of the NMR linewidth for spin 1/2 nuclei has been determined for molecular motions in fully asymmetric triple and double potential wells.[127] A number of examples have been given to emphasize important features of the results.

A set of 15 petroleum cokes produced by delayed coking have been studied by ^{13}C and ^{1}H MAS NMR and CRAMPS (combined rotation and multiple pulse spectroscopy).[128,129] Longitudinal relaxation times have been measured and a relaxation mechanism is discussed. By combining wide-line and high-resolution ^{1}H spectra can be used to derive structural information including aromatic and aliphatic hydrogen to carbon ratios and the average formula per 100 carbon atoms.

A general report covers applications of solid-state NMR to the study of molecular structure.[130]

Two-dimensional NOESY combined with ^{1}H MAS NMR has been used to monitor molecular motions in hydrogels and polymers.[131] The molecular mobility of the super-adsorbing polymer starch-g-poly(acrylonitrile) has been monitored and enhanced mobility is directly revealed by polymer-polymer NOESY connectivities. The presence of an efficient cross-relaxation mechanism between the water of hydration and the polymer is also revealed.

Molecular motion in benzyldimethyl(isopropyl)ammonium bromide $PhCH_2N^+Me_2CHMe_2.Br^-$ has also been measured by solid-state NMR methods.[132] These studies reveal a high activation energy for rotation of one of the N-Me groups (E_a = 28 kJ mol^{-1}). This correlates well with x-ray diffraction results which reveal that rotation of one N-Me group is severely sterically hindered.

High-resolution solid-state NMR relaxometry has been used as a kinetic tool to monitor ultrafast proton transfers in a crystalline powder of dimethyldibenzotetraaza[14]annulene.[133] Rate constants for proton transfer in the milli- to nanosecond timescale can be analysed. This is achieved by monitoring spin-lattice relaxation times T_1 of heteronuclei where the dipolar interaction is modulated by neighbouring mobile protons which are involved in the transfer process. Heteronuclear longitudinal relaxation in the presence of proton transfer is discussed theoretically for the cases of both static powders and rotating powders.

A number of polymers have been imaged by making use of a multiple-pulse coherent averaging method to obtain both high resolution and good sensitivity.[134] The polymers studied include polyether-polysulfone, polycarbonate, polyamide and polystyrene beads.

It has been found that in MAS NMR spectra of heavy atoms such as ^{199}Hg small deviations of the sample rotation axis fro the magic angle gives the false impression of multiple species.[135] This has been calculated theoretically and demonstrated to be the case for mercury(II)acetate. Angle settings should be within $\pm 0.1^{o}$ in order to record accurate spectra. Chemical shift tensors have been derived from spinning sideband analysis and revised values for the anisotropy and asymmetry of mercury(II)acetate are -1732 ppm and 0.08 respectively.

An oriented NMR spectrum may be extracted from an experimental powder spectrum by applying the dePakeing procedure (Bloom *et al.*).[136] This is particularly useful when the powder lineshape is dominated by two or more non-coincident 2nd-rank tensorial interactions. This technique has particular advantages at high magnetic field and is illustrated by application to results obtained on a powder sample of ^{13}C-enriched [3-(5-methyl-2-thienyl)acrylo]imidazole.

Variable-field ^{13}C CP MAS NMR has been applied to several solid arylazonaphthols where the ^{13}C, ^{14}N dipolar interaction is incompletely averaged at low magnetic field for the pure hydrazone form.[137] It is found, however, that in some similar structures which are known to interconvert between keto-hydrazone and enol-azo tautomers do not show significant effects from the nitrogen involved in the proton transfer. This is ascribed to ^{13}C, ^{14}N self-decoupling due to a fast quadrupole relaxation of the nitrogen induced by jumps in its electric field gradient which accompany the proton exchange.

^{13}C MAS NMR spectra have been measured at two fields on a number of urea and thiourea inclusion compounds containing deuterated guest molecules.[138] The spectra for CD, CD_2 and CD_3 groups are triplets, pentets and heptets respectively with the splitting arising from residual dipolar interactions and indirect spin-spin coupling between ^{13}C and ^{2}H. A full analysis of the spectral lineshape is given and the effect of molecular motion is discussed.

4 Chemical Applications

The following section is by no means comprehensive in its coverage. A selection of papers have been made based upon their general relevance particularly in the fields of polymer chemistry, organometallics, biological materials, minerals, coal, soil and inorganic materials. The field of zeolites and zeotypes has been covered more or less comprehensively. Organisation of this section is similar to previous years - that is, first by nucleus and then by subject. Somewhat more subheadings have been included this year in order to facilitate a quick scan of literature within a particular field.

4.1 ^{13}C NMR.

4.1.1 General organic solids. - ^{13}C CP MAS NMR has been used to study inter- and intra-molecular interactions in LDLL mixtures of amino acids with identical chemical structures except for chirality.[139] In another study chemical shielding of protonated and deprotonated carboxylates in amino acids have been measured and, in particular, σ_{11} and σ_{22} exhibit large changes upon protonation.[140] The value of σ_{33} remains constant as this direction is perpendicular to the plane of protonation. A general study records ^{13}C MAS NMR spectra of pharmaceuticals and nucleic acids.[141]

Molecular symmetry in crystalline naphthalene has been studied using ^{13}C MAS NMR by calculation of theoretical shielding values assuming a D_{2h} diffraction structure as observed by diffraction techniques.[142] The results do not fit this symmetry suggesting C_i symmetry as a result of either (i) intermolecular effects or (ii) structural distortion of the molecule. In a separate study it is found that a number of naphthalene derivatives in the solid state have severely distorted naphthalene nuclei.[143]

Molecular motion has been observed in a number of solid-state systems. With a view to studying enzymatic reactions glycine has been studied in frozen solution.[144] The temperature dependence of several relaxation constants has been measured. Upon heating n-octadecanol the ^{13}C chemical shift of internal CH_2 groups moves from 32.8 to 33.3 ppm on going from the crystalline phase to the rotator phase.[145] Such behaviour is similar to that of orthorhombic n-alkanes. Temperature dependent spin-lattice relaxation has also been measured for solid MBBA.[146]

Other high-resolution ^{13}C NMR studies include C,C-linked bipyrazoles and bispyrazolylmethanes,[147] phase changes in adamantanols,[148] structure of 4-substituted 3,4'-diquinolinyl sulfides,[149] a combined x-ray diffraction/NMR/theoretical calculation investigation of benzo-9-crown-3,[150] observation of ^{13}C, $^{35,37}Cl$ residual dipolar coupling in p-chloro benzoylacetones,[151] solution and solid-state tetraphenylchlorins,[152] determination of ionic bonding interactions of N-phenylglycine and N-(2-hydroxy-3-methacryloxypropyl)-N-phenylglycine,[153] location and effects of α-tocopherol, ubiquinone-10 and ubiquinol-10 in sonicated model membranes,[154] degree of deacetylation of chitosan[155] and preparation of low-carbon river sediments.[156]

4.1.2 Fullerenes. - A general study of C_{60} and C_{70} monitors structure and dynamics over a wide temperature range.[157] Small hysteresis in the T_1 curve at 263 K for C_{60} is associated with an orientational phase transition. ^{133}Cs NMR is also used to study $CsRb_2C_{60}$ and Cs_2RbC_{60} and revealed that caesium atoms selectively occupy the larger octahedral site. Other structural studies are concerned with the hydrogenated fullerenes $C_{60}H_{36}$,[158,159] polyfluorinated C_{60} where there is evidence for C-C bond breaking[160] and polymer bound C_{60}.[161]

Molecular dynamics of C_{70} has been measured through both T_1 relaxation measurements and analysis of chemical shift tensors.[162] The latter indicates an axial rotation motion of the C_{70} molecules. Dynamics of C_{60}/n-pentane inclusion compounds indicates a phase transition which occurs over the wide temperature range 210 to 173 K.[163] In the high-temperature phase the two ends of the pentane molecule exchange rapidly. One- and two-dimensional MAS NMR has been used to reveal specific molecular reorientation.[164]

A number of doped fullerenes have been studied by ^{13}C NMR these include organic cations,[165] LaC_{82} and Rb_3C_{60},[166] K_6C_{60}[167] and occupation of fullerene C_{60} interstitial sites by oxygen molecules.[168] This latter study revealed Fermi contact shifts upon adsorption of paramagnetic oxygen and seven evenly spaced resonances were observed corresponding to 0 to 6 of the adjacent interstitial sites being filled with oxygen molecules.

Two studies have revealed that C_{60} can itself be included within the large micropore of the aluminophosphate molecular sieve VPI-5.[169,170]

4.1.3 Polymers. - A large number of papers have been devoted to molecular motion in polymers. These include evidence for local restriction in the segmental chain mobility of two model ionomers (carboxylato-telechelic deuterated polystyrene and protonated polystyrene).[171] Another study measured local motions in mesomorphic main-chain copolyester by changes in the chemical shift tensor with temperature.[172] Other studies of motion in polymers include poly(ethylacrylate) and poly(n-butyl acrylate),[173] side-chain motion of poly(γ-benzylglutamate),[174] dynamics of sub-micron sized droplets of liquid crystals dispersed in a polymer matrix[175] and poly(N-alkylacrylamide) resins.[176]

Interdiffusion measurements have been made of poly(methylmethacrylate), PMMA, and poly(vinylidene fluoride), PVF_2, via a variation of the CP technique.[177] By transfer of magnetization from protons to fluorine the separation between the two polymers may be monitored. Other studies which deal with intermolecular interactions include monitoring interactions in polymer blends via the nuclear Overhauser effect[178] and interdomain communication in a microphase-separated polyurethane elastomer via magnetic spin diffusion.[179]

Other ^{13}C studies of polymer systems include: (i) a magnetic relaxation study of blending in water-soluble polymers such as poly(acrylic acid), poly(methacrylic acid), poly(ethylene oxide) and poly(vinylpyrrolidone);[180] (ii) studies of polymer-blend interfaces using ^{13}C - ^{1}H magnetization transfer such that the ^{13}C signal arises exclusively from chains which are at the polymer interface;[181] (iii) structural studies of monoclinc (α), hexagonal (β) and smectic isotactic polypropylene with different degrees of crystallinity;[182] (iv) two-dimensional chemical exchange experiment on polyaniline which yields the ratio of freely moving rings to rings frozen in place;[183] (v) comparative study of n- and p-doped conducting polyacetylene films;[184] (vi) dynamic-mechanical properties of urethane acrylates via CP time constants to reveal molecular motions in the 60 kHz and near static domains;[185] (vii) analysis of heterogeneity/homogeneity of stabilisers in a polymer matrix by $T_{1\rho}$ measurements;[186] (viii) structural studies of polypeptides by measurement of chemical shift tensors.[187]

4.1.4 Coals. - Effects of high-speed spinning on the study of Argonne Premium coals reveal that CP experiments are prone to errors in quantitation for aromatic fractions and that the method of choice is often a simple Bloch decay.[188] Spin-lattice relaxation times have been measured for a range of coal samples in order to determine the best conditions for measuring ^{13}C single-pulse-excitation NMR spectra. A recycle delay of *ca.* 100 s is found necessary to produce quantitatively reliable spectra.[189]

4.1.5 Biological materials. - ^{13}C CP MAS NMR spectra of lysozyme have been recorded as a function of hydration.[190] As the hydration level is increased line-widths narrow but $T_{1\rho}$ and cross-relaxation times remain constant. This is explained by a reduction in the range of isotropic chemical shifts upon hydration.

Other measurements on biological materials include motional dynamics of protein molecules in bovine, human, rat and chicken ocular lenses where both slow solid-like motions ($\tau_o \geq 10\mu sec$) and mobile rotational reorientational motion (τ_o range of 1-500 nsec) were measured,[191] measurements of the degree of dissociation of carboxylic groups in cellulose membranes by determining the ratio of protonated to deprotonated carboxylic acid groups[192] and spectral editing of ^{13}C spectra of membranes[193] and degradation of wood in old Indian Ocean shipwrecks.[194]

4.1.6 Zeolites and zeotypes. - ^{13}C MAS NMR measurements on zeolites have been dominated by the investigation of catalytic reactions. An *in situ* MAS NMR study of the reactions of acetaldehyde on H-ZSM-5 and the effects of H_2O and O_2 co-adsorption suggests that the strong acid centre in the zeolite catalyses the initial aldol condensation to crotonaldehyde.[195] Co-adsorption of water weakens the acid sites and prevents decomposition of products thereby improving catalytic performance. The conversion of methyl iodides to hydrocarbons on basic zeolites *e.g.* Na-X is preceded by the formation of surface bound methoxy groups.[196] An analogous ethoxy species could also be formed from ethyl iodide. Gaseous products from the reaction of acetone, methanol and ammonia competing for acid sites on H-ZSM-5 and Zn,H-ZSM-5 have been identified by ^{13}C MAS NMR and possible mechanisms for interactions with acid sites are discussed.[197] The formation and reactivity of trialkylonium ions has been studied on zeolites H-ZSM-5 and H-Y.[198] Trimethyloxonium, trimethylsulfonium and trimethylselenonium ions were all formed by adsorption of the corresponding di-methyl chacogenide. Two-dimensional J-resolved spectroscopy has been used to determine the pathway for transfer of the ^{13}C label from the CH_2 group of iso-butyl alcohol into the hydrocarbon skeleton of butene oligomers over H-ZSM-5.[199] The results suggest that the reaction proceeds via a surface tied iso-butyl silyl ether intermediate. Similar experiments have been used to analyse the products from the conversion of ethylene inside zeolite H-ZSM-5.[200] N-hexane cracking on H-ZSM-5 has been performed in a comparative NMR/gas chromatographic (GC) study.[201] Differences between the two sets of data suggests that isomerisation to methylpentanes is the dominant reaction below 300°C. At higher temperatures cracking takes over but formation of dimethylbutanes is precluded by pore-size restrictions. Other reactions studied include cumene-to-n-propylbenzene isomerization over H-

ZSM-11,[202] reactions of ketones on zeolites[203] and propane aromatization over Ga-modified H-ZSM-5.[204]

The formation of surface-bonded methoxy groups by sorption of methanol and methyl iodide on various cation-exchange forms of ZSM-5 has been studied with signals for the surface Me-O groups in the region 53-59 ppm.[205] Interactions of ethanol and ammonia over H-ZSM-5 has also been studied.[206]

Three structural papers monitor the structure-directing effects of crown-ethers in the synthesis of cubic (FAU) and hexagonal (EMT) forms of zeolite Y,[207] the location of Pr_3NH^+ and Et_4N^+ templates clathrated after synthesis of large single crystals of ZSM-5[208] and the characterization of 4,4'-trimethylenebis(1-methylpiperidine) or 1,4-diazabicyclo(2.2.2)octane templates in the novel zeolite CJS-4.[209]

The formation of aminium and bipyridinium cation radicals by electron transfer in H-Y has been observed.[210]

4.1.7 Layered and porous material. - ^{13}C MAS NMR has been used to study the adsorption of 2-phenethylamine on four different aluminium-exchanged clays.[211] Three different types of bound ammonium species were observed. A similar study of benzene chemisorption on copper-montmorillonite which indicated that under anhydrous conditions the benzene polymerised.[212] The reaction of molten sebacic acid has been followed on a layered (Mg/Al) double hydroxide and confirmed the presence of the dianion species $-O_2C(CH_2)_8CO_2-$.[213] ^{13}C CP MAS NMR has been used to monitor the formation of ultramicropores in a carbogenic molecular sieve derived from poly(furfuryl alcohol).[214]

4.1.8 Organometallics. - ^{13}C CP MAS NMR has been used as an indicator of the microstructure of a silicon containing dianhydride of bis(3,4-dicarboxyphenyl)dimethylsilane.[215] These compounds are monomeric models for polyimides and the development of charge-transfer-complexes (CTC) within them supports the belief that CTC also form in polyimides. Metal carbonyl clusters of iron[216] and the cyclopendienyl metal carbonyls of W and Ru[217] have been studied. In the latter study an analysis of the sideband manifold allows one to assess a different semi-bridging character between 2 CO-groups not seen in x-ray diffraction. Other studies on organometallics include structure and dynamics of bis(organothiophosphoryl)disulfides[218] and a variable temperature study of Me_3M propellers (M = Sn, Pb).[219]

4.1.9 Inorganic materials. - ^{13}C enriched CO adsorbed on polycrystalline platinum has been monitored as a function of electrode potential using a spin-echo technique.[220] Narrowing of the resonances at high electrode potential was interpreted as the result of the transformation from a mixed bridge- and linear-bound CO to only linear-bound CO. The quality of diamond films grown by chemical vapour deposition techniques has been studied and indicate that the sp^3 C in such material shows more disorder than that found in high-quality polycrystalline films.[221,222]

4.1.10 Inclusion compounds. - Conformational properties of monosubstituted cyclohexanes in thiourea compounds have been studied by one- and two-dimensional ^{13}C NMR.[223] Relative proportions of axial and equatorial conformers has been ascertained from the NMR spectra under slow exchange conditions. Migration of n-decane molecules into the tunnel structure of urea inclusion compounds has also been studied by a combination of NMR and optical microscopy.[224] Functional group recognition has been monitored by adsorbing molecules of formula $X(CH_2)_nY$ within inclusion compounds of perhydrotriphenylene (PHTP).[225] Direct measurements of different packing arrangements *i.e.* head-to-head, head-to-tail and tail-to-tail can be measured. Molecular motions of γ-valerolactone enantiomers in the chiral cholic acid host have also been monitored.[226]

4.1.11 Amorphous materials. - Amorphous carbon films created using a microwave plasma have been characterised by ^{13}C MAS NMR.[227] Dipolar dephasing experiments indicate a range of T_{dd} time constants making quantifications of quaternary diamond-like carbons difficult. A study of amorphous hydrogenated carbon deposited from acetylene in a d.c. flow discharge measures the fraction of CH_2 carbons.[228] NMR has been used to study molecular motion near the glass transition temperature of 1,3,5-tri-α-naphthylbenzene.[229]

4.1.12 Catalysis. - Ethylene oxidation over a supported silver catalyst has been studied with a view to understanding the surface formed species.[230] These include acetic acid, CO_2, ethane and alkoxy species. The chemisorption of aci-nitromethane on the strongly basic catalysts MgO and CaO as well as zeolite Cs-X shows strong binding.[231] At 473 K decomposition to NH_3 and surface carbonate occurs. Other studies include characterisation of a supported Ziegler-Natta catalyst[232] and a general study of catalytic reactions.[233]

4.2 ^{27}Al and ^{29}Si NMR.

4.2.1 Zeolites and zeotypes. - Double rotation and variable field ^{27}Al NMR has been used in the study of dealuminated zeolite Y which is the most important catalyst for catalytic cracking in the petroleum industry.[234,235] Two types of tetrahedral aluminium are observed framework Al at 59ppm and nonframework at 47ppm. In spectra recorded with MAS the latter signal is observed at 30ppm due to a large second order quadrupolar shift. Chemical shift data from ^{29}Si two-dimensional INADEQUATE experiments have been correlated with various structural parameters obtained from x-ray diffraction for some highly siliceous zeolites.[236] ^{27}Al MAS NMR spectra have been recorded on a large number of ZSM-5 and amorphous silica-alumina samples.[237] The results give the mean quadrupole coupling constants (QCC), linewidths, chemical shift distributions and spin-spin relaxation times. The ^{27}Al chemical shift anisotropy for $AlPO_4$-21 with clathrated pyrrolidine template was determined and the results used to explain coordination in the framework.[238] Site-selective line broadening of ^{29}Si resonances in Cu^{2+}-exchanged zeolites resulted from the close contact of paramagnetic ions with specific silicon sites in the lattice.[239] A ^{29}Si and ^{27}Al MAS NMR study of a single crystal of chabazite revealed inequivalent silicon sites based upon aluminium atoms in the second coordination sphere.[240] A general study monitors both ^{29}Si and ^{27}Al MAS NMR spectra during the chemical modification of a number of zeolites including titanium insertion.[241]

The structure of a novel microporous titanosilicate ETS-10 has been solved in which the titanium adopts octahedral coordination and the silicon tertrahedral coordination.[242] This material has a three-dimensional wide pore and is highly disordered. This disorder makes the structure very difficult to solve and ^{29}Si MAS NMR played a very important role. Two silicon environments are observed Si(1Ti,3Si) and Si(0Ti,4Si) with a further crystallographic splitting of the former. One of the most important new class of materials are the mesoporous MCM-41 type phases with pore sizes in the range 13Å to 100Å. These materials are synthesized with liquid crystal templates and are rapidly finding a myriad of important applications. One ^{29}Si and ^{27}Al MAS NMR study monitors the structure of the essentially amorphous walls in these materials.[243] A study has been made of a variety of framework materials containing three-membered rings such as euclase, phenakite, clinohedrite, VPI-7, ZSM-18 etc. and specific resonances are observed for silicon at the centre of spiro-5 units.[244] Other structural studies of zeolites include ZSM-23,[245] a combined NMR neutron diffraction study of siliceous Y,[246] a combined infrared NMR study of calcined zeolite beta[247] and discrimination between EMT/FAU intergrowths.[248]

Two studies on dealuminated zeolite omega (mazzite) study the silicon aluminium distribution and the influence of acetylacetone treatment.[249,250] Aluminium coordination has been monitored in steamed zeolites and this coordination correlated with Lewis acidity.[251] Evidence is presented which suggest that Al^{V} is responsible for very strong Lewis acidity.

One other study on ZSM-5 investigates the location of naphthalene adsorbed within the pores by a combination of NMR and theoretical calculations.[252]

A number of structural studies have been concerned with aluminophosphate ($AlPO_4$'s) and silicoaluminophosphates (SAPO's). These include: hydration effects in $AlPO_4$-11;[253] the VPI-5 related material $AlPO_4$-H2;[244] framework structure of SAPO-5;[255] reversible modification in template-free SAPO-34 upon hydration-rehydration;[256] thermal stability of SAPO-37.[257,258] One paper deals with hydration effects in MAPO-36 (magnesium aluminophosphate).[259]

A variety of papers have dealt with the synthesis of molecular sieves and these include: synthesis of zeolites from a non-alkaline fluoride media;[260] influences of crystallisation time and H_2O/Al_2O_3 ratio on the synthesis of $AlPO_4$-5;[261] a comparison of crystallisation rates of [aluminium]- and [gallium]-MFI-type zeolites;[262] synthesis of a cobalt substituted aluminophosphate CoAPO-5;[263] synthesis of tetrabutylammonium hydroxide and di-n-propylamine templated VPI-5;[264] synthesis of silicon rich SAPO-44;[265] role of 18-crown-6 and 15-crown-5 ethers in the synthesis of FAU and EMT polytypes of faujasite.[266]

Studies concerned with either catalytic properties of zeolites or the nature of acid sites responsible for catalysis include: selective reduction of nitric oxide by Cu-ZSM-5 as a function of dealumination;[267] the relationship between structural and catalytic properties of ZSM-12;[268] the fate of aluminium during the hydrochlorination of 1-methylcyclohexane by zeolite Y;[269] formation of ketones from lower olefins[270] and disproportionation of ethylbenzene[271] over dealuminated zeolite Y; studies of the acidic properties in rare-earth-exchanged ultrastabilised zeolite Y (REUSY)[272,273] and zeolite β;[274] heterogeneity of the hydroxyl groups in NaH-Y and NaH-X;[275] correlation between infrared and solid-state NMR of acidic hydroxyls in zeolite X and Y;[276]

catalytic studies of tritium exchanged $AlPO_4$-5;[277] improvement of catalytic properties in SAPO-31 synthesised with activated silica.[278]

Isomorphous substitution in zeolites is important for imparting modified chemical properties. Studies in this area include two papers on the modification of mordenite with titanium to make an oxidation catalyst,[279,280] incorporation of germanium in ZSM-5 with characterisation of lattice bond angles,[281] incorporation of silicon or magnesium in the wide pore aluminophosphate VPI-5 in order to generate acidity[282] and the stabilisation of ZSM-5 by incorporation of zinc.[283]

Other ^{29}Si and ^{27}Al studies of zeolites and zeotypes include: cation exchange in zeolite L;[284] the effect of lanthanum incorporation in dealuminated zeolite Y;[285] dealumination of zeolite Y through incorporation of ruthenium trichloride;[286] interactions of silicon and hydrogen upon adsorption of p-xylene in ZSM-4 using CP;[287] interaction of water molecules with the aluminophosphate lattice of VPI-5;[288] combined NMR and Monte Carlo calculations of silicon environments in mordenite;[289] the effects of fluorine, chlorine[290] and ammonium fluoride[291] on H-ZSM-5.

4.2.2 Layered and porous material. - ^{27}Al and ^{29}Si MAS NMR studies of synthesis of clay materials include formation of kaolinite from amorphous silica/alumina mixtures,[292] formation of metakaolinite by flash calcination from kaolinite[293] and the synthesis of gallium substituted sapponite.[294] All these studies attempt to quantify the different silicon and aluminium chemical environments.

The structures of beidellites cross-linked with $NH_2(CH_2)_3Si(OEt)_3$ show that upon calcination silica pillars are formed.[295] Kaolinite, which is important in the ceramics industry, loses crystallinity when subjected to pressure, however, MAS NMR studies reveal very few changes in the local ordering.[296] Other structural studies include a re-expansion of collapsed Li-montmorillonite under high water vapour pressures[297] and a comparative NMR, diffraction study of synthetic leucite.[298]

Studies on pillaring of layered materials include aluminium Keggin ion pillared magadite[299] and saponite,[300,301] and silica-titania pillared saponite.[302]

Other miscellaneous studies include acidity measurements of dealuminated metakaolin and correlation with activity for cumene cracking,[303] a combination of ^{27}Al and ^{25}Mg MAS NMR study of thermal decomposition in hydrotalcite,[304] ^{29}Si NMR studies correlating Q_3 and Q_4 concentrations in expanded clays with catalytic activity,[305] examination of the formation of β-Sialon by carbothermal reduction and nitridation of halloysite clay[306] and measurements of water-softening properties of a layered sodium silicate, SKS-6.[307]

The structure of high-purity porous amorphous silica prepared by various techniques has been studied by a combination of neutron scattering and NMR techniques.[308]

4.2.3 Organometallics. - ^{27}Al MAS NMR has been used to investigate aluminium complexed with acetic, oxalic, lactic and salicylic acids at pH 3.5 - 6.5.[309] At higher pH the aluminium monomers are progressively hydrolysed and tend to condense to form trimers aggregated to Al_{13}. Solid-state ^{29}Si MAS NMR has been used to quantify the relative amounts of silyl species in polysiloxane- bound ether-phosphines and ruthenium complexes.[310]

4.2.4 Inorganic materials. - Double-rotation ^{27}Al NMR has been used in combination with x-ray diffraction for the study of aluminium molybdate $Al_2(MoO_4)_3$.[311] Second order quadrupolar interactions are completely removed and four inequivalent aluminium resonances are observed as expected. Interior hydroxyls in high-surface-area silica gels and their capacity for deuterium exchange have been monitored by ^{29}Si CP MAS NMR.[312] Similar techniques are used to demonstrate that there are strong interactions between methoxysilanes and Mg_2Cl when they are co-milled.[313]

4.2.5 Semiconductors. - Silicon nitride films of 100-300nm thickness have been grown on GaAs wafers by plasma-enhanced chemical vapour deposition (PECVD) and studied by ^{29}Si MAS NMR.[314] This technique discriminates between silicon atoms bonded to one or two hydrogen atoms. Similar PECVD techniques have been used to grow amorphous-SiS_x:H ($x \approx 2$) and ^{29}Si MAS NMR studies show that the material is characterised by corner- and edge-shared $SiS_{4/2}$ tetrahedra.[315]

4.2.6 Minerals. - ^{27}Al and ^{29}Si NMR studies of minerals include: DOR studies of hydrated aluminas bayerite (α-$Al(OH)_3$), boehmite (AlOOH) and gibbsite (γ-$Al(OH)_3$)[316] and aluminium silicate minerals of formula Al_2SiO_5;[317] studies of disorder in alumina hydrates[318] and fibrolitic sillimanite;[319] thermal transformations of hallosite;[320] studies of lead silicates;[321] structural differences in mullite xerogels prepared from different precursors.[322]

4.2.7 Amorphous materials. - ^{27}Al and ^{29}Si MAS NMR studies of glasses and amorphous materials include: sodium oxide-alumina-boron oxide-silica glasses which reveals that Al_2O_3 preferentially associates with Na_2O to form AlO_4 tetrahedra;[323] a structural study of sodium-silicon-oxygen-nitrogen oxynitride glasses;[324] investigation of bond-angle disorder with aluminium content in lithium aluminosilicate glasses;[325] ^{29}Si CP MAS NMR study of surface distribution of silicate species in amorphous alumina.[326]

4.2.8 Catalysis. - The structure of catalysts has been measured by ^{27}Al MAS NMR during the catalytic isomerisation of butene on aluminas.[327] The presence of pentacoordinated aluminium on the surface generates sites which greatly effects product selectivity.

Octahedral aluminium has been observed for the first time in rehydrated calcined Mo/Al_2O_3 hydrotreating catalyst precursors.[328] This resonance is attributed to the presence of hydrated forms of aluminium molybdate such as $[Al(OH)_n(H_2O)_{6-n}]_n(MoO_4)$ ($n = 1$ or 2).

4.3 ^{1}H and ^{2}H NMR.

4.3.1 General organic solids. - Variable temperature natural abundance ^{2}H NMR has been used to study phase transitions in the rotator phase solids adamantane, tetrakis(trimethylsilyl)methane, tetrakis(trimethylsilyl)silane, and 2,2-dimethyl-1,3-

propanediol.[329] Similar experiments over the temperature range 30-120°C have been used to determine cis-trans isomerisation energetics in urea and acetamide.[330]

4.3.2 Polymers. - Two-dimensional combined rotation and multiple-pulse NMR ^1H spectroscopy (CRAMPS) has been used to study the miscibility of poly(para-phenylene vinylene) model compounds in polycarbonate.[331] Miscibility studies have also been performed on blends of polystyrene and poly(2,6-dimethyl-p-phenylene oxide) using 2-dimensional heteronuclear correlation (2D HETCOR) ^1H spectroscopy.[332] Other ^1H NMR studies include measurements of oxygen adsorption on poly(2,6-dimethylphenelene oxide),[333] the use of polyethylene oxide as a chelating agent for mercuric(II) chloride[334] and measurements of cross-relaxation rates for polymer-dispersed liquid crystals.[335]

^2H NMR has been used to study the reorientation of poly(styrene-vinyl-d_3) in a blend of poly(2,6-dimethylphenylene)[336] and for a study of the structure and dynamics of poly(ethylene oxide).[337]

4.3.3 Biological materials. - A novel filtering method is described involving a Hahn echo and spin locking (HAL) for simplifying the complicated ^1H NMR spectra of DNA.[338] Another study monitors ^1H NMR during the ripening process of edible banana tissue.[339] The rigid starch related protons decrease from 20% in a fresh green banana to 2% in a ripe yellow banana.

4.3.4 Zeolites and zeotypes. - ^1H MAS NMR has been used to study the geometry and location of bridging hydroxyl groups in zeolites and SAPO's.[340] From an analysis of the dipolar interaction between ^1H and ^{27}Al an accurate value of the H-Al distance is obtained and there is a correlation between this length and the size of the window containing the acid site. Other ^1H NMR studies on zeolite and zeotypes include highly resolved spectra of dehydrated zeolite H-Y,[341,342] dealuminated H-ZSM-5,[343] weak hydrogen bonding in zeolites and $AlPO_4$'s,[344] sodium aluminogermanate hydroxysodalite hydrate[345] and a general study of acidity in zeolites.[346]

One ^2H NMR study investigates molecular mobility and dehydration of tert-butyl alcohol on H-ZSM-5.[347]

4.3.5 Organometallics. - The mechanism of hydride scrambling in a transition-metal dihydrogen dihydride $Ir(P(CHMe_2)_3)_2(Cl)(\eta\text{-}H_2)(H)_2$ has been measured by single crystal ^1H NMR.[348] Other ^1H NMR studies include C_{12}-allylnickel(II) cation $[Ni(C_{12}H_{19})]^+$,[349] flexidentate behaviour of N-(2-hydroxyethyl)-1,2-ethanediamine (L) in palladium(II) halide complexes,[350] single crystal bis(N,N'-dimethyl-1,2-ethanediamine)copper(II) dinitrate[351] and weak metal-metal bonding interactions in $[Ni_3(\eta^5\text{-}C_5H_5)_3(\mu_3\text{-}S)_2]n$ (n = 1+,0).[352]

One ^2H NMR study investigates the rotational dynamics of substituted phthalocyaninatopolysiloxanes.[353] Uniaxial rotation of the phthalcyanato moiety around the long axis of the polymer is observed.

4.3.6 Inorganic materials. - ^1H MAS NMR has been used to study the nature of Broenstead acidity in $Ag_3PW_{12}O_{40}$ in the presence of hydrogen.[354] Two kinds of protons were observed, one seen at 6.4ppm exists only in the presence of gas phase hydrogen. The radial distribution of hydrogen in a chemical vapour deposited diamond film has been measured by ^1H NMR and found to vary across the sample.[355]

Lateral interactions between ammonia molecules adsorbed on titania (rutile) has been studied by ^2H NMR.[356]

4.3.7 Inclusion compounds. - ^1H NMR has been used to study host-guest molecular association of chiral guests in optically active hosts.[357] ^2H NMR studies have been used to determine the structure and packing of phosphatidylcholines in lamellar and hexagonal liquid-crystalline mixtures[358] and to study the dynamic properties of xylenes in Dianin's inclusion compound.[359]

4.3.8 Catalysis. - One combined ^1H and ^2H study investigates the adsorption of methanol and water on the copper/zinc oxide/alumina methanol synthesis catalyst.[360]

4.4 Multinuclear NMR.

4.4.1 General organic solids. - Multinuclear studies of general organic solids include: conformational distributions of bsclofen analogues;[361] benzene-hexafluorobenzene;[362] 2-D REDOR experiments on Val_1-[1-^{13}C]Gly_2-[^{15}N]Ala_3-gramicidin;[363] curing reactions in bismaleimide;[364] dipolar dephasing in monofluorinated organics;[365] intermolecular proton transfer in 1,1-bis(2,4-dimethylphenyl)but-2-yn-1-ol;[366] 1,8-bis(dimethylamino)naphthalene and its complex with tetrazole;[367] hydrogen bonding in salts of dicarboxylic acids;[368] dynamic properties of p-diiodobenzene.[369]

4.4.2 Fullerenes. - Two multinuclear studies investigate reorientational dynamics of $C_{61}H_2$ and compare this with C_{60} and $C_{60}O$,[370] and synthesis and characterisation of $Na^+[C_{60}]^-(THF)_5$.[371]

4.4.3 Polymers. - Multinuclear studies of polymers include: characterisation of polymers grafted onto solid surfaces;[372] hexafluorophosphate-doped polypyrrole films;[373] curing reactions of cyanate resins;[374] characterisation of nylon 7 crystalline forms;[375] chain motions in polymers;[376] spin diffusion in highly mobile polymers.[377]

4.4.4 Zeolites and zeotypes. - Three multinuclear studies have investigated the 20-ring wide-pore gallophosphate cloverite.[378-380] Quadrupole coupling constants for gallium are found to be so high that MAS techniques are not considered useful. Sodalite as a host for quest species such as organic molecules,[381] carbonates[382] and semiconductors[383] has been investigated. A combination of ^{29}Si, ^{27}Al, ^{13}C and ^1H MAS NMR has been used to investigate the mesoporous MCM-41 materials.[384] Multinuclear studies have also been used to study acid sites in zeolites,[385] surface hydroxyl groups in dealuminated ZSM-5,[386] nitrogen incorporation in $AlPO_4$-5,[387] interaction of bases with acid sites on SAPO-37,[388] characterisation of VAPO-5,[389] interaction of NaY with oxides of vanadium, molybdenum or tungsten[390] and lead exchange into zeolites.[391]

4.4.5 Layered and porous material. - Multinuclear NMR studies on layered materials include investigations on the structure of kaolinites[392] and interlamellar grafting of organics such as DMSO or HCONHMe in kaolinite.[393] The structure of the microporous $((NH_4)_{0.88}(H_3O)_{0.12}AlPO_4(OH)_{0.33}F_{0.67}$ has been studied by ^{27}Al, ^{31}P, ^{19}F, ^{15}N and ^1H NMR.[394]

4.4.6 Organometallics. - Multinuclear NMR studies on organometallics include: dynamic properties of tetrakis(trimethylsilyl)silane;[395] spectroscopic studies of iodoferrocenylruthenocenium$^+$ salts;[396] anisotropic spin interactions in dipotassium tetracyanomercurate(II);[397] architecture of the ion exchanger $[(Me_3Sn^{IV})_4M^{II}(CN)_6]_\infty$ (M = Fe, Ru, Os);[398] spectroscopy of N,N,N',N'-tetramethylethylenediamine-sodium cyclopentadienide;[399] chemical shifts of silatrane;[400] structure of 2,2,4,4,6,6-hexakis(trimethylsilylmethyl)cyclotristannatellurane;[401] Hahn echo extended pulse sequences of sulphur diamides bearing the di-tert-butylphosphinyl substituent;[402] ^{113}Cd NMR of cadmium complexes with glycine, L-alanine and L-cysteine;[403] spectroscopy of cadmium(II) bis(dialkyldithiophosphinates;[404] synthesis of tetraalkylammonium hydrogen telluride and selenide[405] and arylmercury derivatives of α-thiopicolinanilide;[406] structure of chiral rhodium(I) organic complexes;[407] spectroscopy of copper(I) organic complexes;[408] synthesis of tetraselenidotungstate(2-);[409] spectroscopy of cadmium(II) complexes with heterocyclic dithiocarbamate[410] and with 2,3-dihydro-2-thioxo-4(1H)-quinazolinone;[411] spectroscopy and calculations on chiral allyl complexes of palladium(II);[412] studies of organometallic coordination polymers;[413] NMR and diffraction investigations of platinum complexes with tertiary phosphines;[414] a general discussion of NMR on organometallics;[415] spectroscopic study of sodium, lithium-cyclotetraphosphate;[416] multidimensional NMR of di- and tri-nuclear optically active ferrocenylphosphine complexed with mercury(II).[417]

4.4.7 Inorganic materials. - Multinuclear NMR studies on inorganic materials include: ^{19}F - ^{31}P J-resolved MAS NMR spectroscopy of an "inorganic" adamantane KPF_6;[418] phase transitions in tin hexathiohypodiphosphate;[419] structure of aluminium gels formed through base hydrolysis of Al^{3+}aqueous solutions.[420]

4.4.8 Semiconductors. - Multinuclear NMR studies on semiconductor materials include: a quantitative evaluation of gallium phosphide[421] and characterisation of mixed ZnSe/GaP included in sodalites;[422] NMR on a large range of III-V semiconductors prepared by $MX_3 + Na_3Pn \rightarrow MPn + 3NaX$ (M = Al, Ba, In; X = F, Cl, I; Pn = pnictide = P, As, Sb);[423] structure of mercury cadmium telluride[424] and cadmium iron telluride.[425]

4.4.9 Inclusion compounds. - Multinuclear NMR studies on inclusion compounds include: molecular motion of guests trapped in N-methyl-1,3-diaminopropanecadmium(II) tetracyanonickelate(II) host lattices;[426] guest-host interactions in layered intercalation compounds of the type $K_x(H_2O)_y[NbS_2]$;[427] grafting of ethylene oxide onto the interlayer surface of γ-zirconium phosphate.[428]

4.4.10 Catalysis. - Multinuclear NMR studies on catalysis include: measurements of methoxy formation and spillover on Pd/Al_2O_3 by reaction of H_2 with ^{13}CO;[429] an investigation of surface species on NiO-MoO_3-P based hydrotreating catalysts supported on γ-alumina;[430] structure of aluminium phosphate-alumina catalysts;[431] surface complexes formed on V_2O_5-TiO_2-SiO_2 catalysts.[432]

4.5 Other nuclei.

4.5.1 General organic solids. - Pulsed field gradient stimulated spin echoes were measured in order to determine anisotropic self-diffusion coefficients in the smectic liquid crystal CBOOA.[433] ^{15}N NMR is used to measure the chemical shift tensor in p-nitroso-N,N'-dimethylaniline.[434]

4.5.2 Biological materials. - ^{31}P is used to study the dynamic properties of nucleic acids in biosupramolecular systems,[435] and interactions of Cytochrome c in mixed phospholipid membranes.[436,437] ^{15}N CP MAS NMR is used to monitor native soil organic matter on ^{15}N-enriched composts[438] and 2-dimensional separated local dipolar field (2D-SLDF) spectroscopy is used to determine nitrogen-hydrogen bond lengths for simple peptides.[439]

4.5.3 Zeolites and zeotypes and layered materials. - ^{71}Ga MAS NMR has been used to investigate aromatization over gallium containing H-ZSM-5,[440] to quantify gallium in zeolites,[441] in a study of the gallium analogue of Nu-23/ferrierite[442] and for characterisation of the gallosilicate analogue of mordenite.[443] ^{31}P MAS NMR has been used to study pillaring of rectorite catalysts,[444] for monitoring the interlayer collapse during dehydration of synthetic beidellite,[445] for characterisation of a novel layered titanium phosphate with formula $TiO(OH)(H_2PO_4).2H_2O$[446] and to study the phase transition between VPI-5 and $AlPO_4$-8.[447] The structure of the borosilicates with MFI[448] and ANA[449] structure have been studied with ^{11}B MAS NMR. ^{23}Na NMR reveals the existence of secondary anchoring sites for Na^+ cations between the supercage site II and the nucleophilic methyl groups in MeM (M = Zn, Cd) occluded zeolite Y.[450] Cation sites for sodium are also investigated in lanthanum-exchanged Na-Y.[451] $^{47,49}Ti$ and ^{17}O MAS NMR is used to study the incorporation of titanium in silicalite.[452] Titanium is found in both framework and extra-framework sites. ^{77}Se and ^{125}Te MAS NMR has been used to study zinc chalcogenides contained in sodalite hosts.[453]

4.5.4 Organometallics. - ^{31}P MAS NMR has been used in the following investigations: correlation of shielding parameters to molecular structure in bis(organothiophosphoryl) disulfides[454] and phosphoroorganic disulfides;[455] study of distortions in linear mono-, bi- and trimetallic bis(tri-n-butylphosphine) nickel and palladium complexes with 1,4-tetrafluorophenylene bridges;[456] structures of mercury(II) halides with phosphine ligands;[457] copper(I) complexes of tetradentate-N_2P_2 ligands;[458] solvation effects on tertiary phosphine derivatives and platinum complexes.[459]

Studies using ^{119}Sn MAS NMR include: six-coordinate complexes R_2SnCl_22L and R_2SnCl_2L' (R = Me or Ph; L = py; L' = bipy or phen);[460] reaction of $(n\text{-}C_4H_9)_3SnH$ with the surface of a silica-alumina;[461] correlations of solution and solid-state structures of poly(pyrazolyl)borate complexes of tin(II);[462] synthesis and structure of crystalline $[K\{Sn(CH_2Bu\text{-}tert)_3\}(\eta\text{-}C_6H_5Me)_3]$ which includes an observation of ^{119}Sn - ^{39}K coupling;[463] study of the hydrolysis of monoorganotin halides.[464]

^{7}Li MAS NMR has been used for the study of N,N,N',N'-tetramethylethylenediamine-trimethylsilylcyclopentadienyllithium yielding very large quadrupolar couplings,[465] for investigations of alkali-metal iminoaminosulfinates[466] and phenyllithium aggregates.[467]

Other miscellaneous nuclei observed include: ^{15}N studies of selenium diimides[468] and cyclic sulphur diimides of silicon;[469] ^{113}Cd NMR of the three structural isomers of hexadecakis(benzenethiolato)tetrasulfidodecacadmate(4-);[470] ^{59}Co NMR studies of metal-ligand interaction in axially symmetric diamagnetic cobalt(III) complexes;[471] ^{201}Hg NMR investigation of solid potassium tetracyanomercurate(II);[472] ^{195}Pt NMR studies of $Pt(en)Cl_x$ (en = 1,2-diaminoethane; x = 2, 3 or 4).[473]

4.5.5 Inorganic materials. - MAS NMR has been used in the study of inorganic materials using a wide variety of nuclear probes which include: ^{63}Cu for a MAS NMR and diffraction study of copper(I) bromide;[474] ^{19}F NMR investigation of fluorinated phyllosilicates and phyllogermanates;[475] ^{15}N study of phase transitions in ammonium nitrate;[476] ^{17}O NMR for detection of atomic scale phase separation in titania-silica gels;[477] ^{31}P NMR relaxation study of silica supported polyoxoanions of the Keggin and Dawson structures;[478] ^{119}Sn NMR temperature dependent study of neodymium, samarium and yttrium samarium tin oxides.[479]

4.5.6 Semiconductors. - ^{31}P MAS NMR has been used to study the organometallic synthesis of crystalline and glassy ternary semiconductors $M^{II}M^{IV}P_2$ (M^{II} = Zn and Cd, M^{IV} = Ge and Sn).[480] ^{31}P and ^{77}Se MAS NMR have been used to study phosphine chalcogenide species coordinating the surface of CdSe nanocrystals.[481]

4.5.7 Inclusion compounds. - Intercalation of cobalt phthalocyanine into the layered host material antimony hydrogen phosphate $HSb(PO_4)_2$ has been monitored by ^{31}P MAS NMR.[482] ^{7}Li solid-state NMR has been used to investigate the metal-like behaviour of lithium intercalated in molybdenum cluster chalgogenides.[483]

4.5.8 Glasses. - Structural equilibria in the glass system phosphorus-arsenic-selenium[484] has been examined using ^{31}P MAS NMR and also calcium phosphate glasses.[485] ^{11}B MAS NMR has been applied to the study of borate and borosilicate glasses[486] and ^{23}Na MAS NMR to the study of local coordination environments in sodium in silicate crystals, melts and glasses.[487]

4.5.9 Catalysis. - ^{31}P MAS NMR has been applied to the study of the synthesis[488] and n-butane oxidation to maleic anhydride[489,490] over a vanadium phosphate (VPO) oxidation catalyst. Base catalysed nucleophilic displacement reactions of pentacoordinate phosphorous in cyclic oxyphosphoranes has also been studied by ^{31}P MAS NMR.[491]

References

1 R.K. Harris, *Chem. Br.*, 1993, **29**, 601-604.
2 G.E. Pake, *Phys. Today*, 1993, **46**, 46-50.
3 T. Erata, *Bunseki*, 1993, **4**, 264-268.
4 S. Kuroki and I. Ando, *Sen'i Gakkaishi*, 1993, **49**, 135-139.
5 I. Ando and E Kato, *Kagaku*, 1994, **49**, 175-179.
6 S.O. Smith, *Curr. Opin. Struct. Biol.*, 1993, **3**, 755-759.
7 H. Saito and A. Naito, *Kagaku*, 1993, **48**, 220-221.
8 L.B. Alemany, *Appl. Magn. Reson.*, 1993, **4**, 179-201.
9 Z. Liz, H.W. Spiess and J.J. Titman, *Isr. J. Chem.*, 1992, **32**, 145-160.
10 F. De Luca, N. Lugeri, B.C. De Simone and B. Maraviglia, *Colloids Surf. A.*, 1993, **72**, 63-70.
11 M.W. Anderson and J. Klinowski, *NATO ASI Ser., Ser. C*, 1993, **386**, 473-494.
12 J. Klinowski, *Top. Inclusion Sci.*, 1992, **4**, 115-136.
13 V.M. Mastikhin, *Colloids Surf., A*, 1993, **78**, 143-166.
14 A. Maijanen, A. Annila, P. Qvintus-Leino and A. Hase, *VTT Tied.*, 1993, **1448**, 26.
15 J. Klinowski, *Anal. Chim. Acta.*, 1993, **283**, 929-965.
16 M.E. Davis and J.B. Higgins, *Charact. Catal. Mater.*, 1992, 129-148.
17 S. Nakata, *Zeoraito*, 1993, **10**, 184-188.
18 M.J. Hudson and A.D. Workman, *NATO ASI Ser. Ser. C*, 1993, **400**, 347-371.
19 N.J. Clayden, *NMR*, 1993, **29**, 91-126.
20 R.L. Silvestri and J.L. Koenig, *Anal. Chim. Acta*, 1993, **283**, 997-1005.
21 A. Matsumoto, *Kagaku to kogyo*, 1993, **67**, 507-513.
22 D.G. Cory and W.E. Maas, *Bull. Magn. Reson.*, 1993, **15**, 160-164.
23 J. Liu, *Kobunja Kwahak Kwa Kisul.*, 1993, **4**, 287-304.
24 T. Akai, *Osaka Kogyo, Gijutsu Shikensho Kiho*, 1993, **44**, 66-83.
25 J.A. Ratcliffe and I. Christopher, *Top. Inclusion Sci.*, 1992, **4**, 1-27.
26 S. Hayashi, *Zeoraito*, 1993, **10**, 140-147.
27 L. Marguiles, H. Rozen, T. Stern, G. Rytwo, B. Rubin, L.O. Ruzo, S. Nir and E. Cohen, *Arch. Inset Biochem. Physiol.*, 1993, **22**, 467-486.
28 L. Reven, *J. Mol. Catal.*, 1994, **86**, 447-477.
29 T. Okuhara, *Kikan Kagaku Sosetsu*, 1993, **20**, 149-159.
30 J.L Koenig, *Chemtracts. Macromel. Chem.*, 1992, **3**, 233-236.
31 H. Maekawa, *Nippon Kessho Gakkaishi*, 1992, **34**, 347-353.
32 D.E. Bugay, *Pharm. Res.*, 1993, **10**, 317-327.
33 A. Makriyannis, D.P. Yang and T. Mavro-moustakos, *NIDA Res. Monorg.*, 1991, **112**, 106-128.
34 K. Yonebayashi, *Nippon Dojo Hiryogaku Zasshi*, 1993, **64**, 206-211.
35 C.M. Preston, *Can. J. Appl. Spectrosc.*, 1993, **38**, 61-69.
36 H. Saito, A. Naito and S. Tuzi, *Seibutsu Bursuri*, 1993, **33**, 274-280.
37 G.J. Leary and R.H. Newman, *Methods Lignin. Chem.*, 1992, 146-161.
38 R.A. Wind, G.E. Maciel and R.E. Botto, *Adv. Chem. Ser.*, 1993, **229**, 3-26.
39 K.L. Smith, L.D. Smoot and T.H. Fletcher, *Coal Sci. Technol.*, 1993, **20**, 131-298.
40 R.H. Newman and L.J. Porter, *Basic Life Sci.*, 1992, **59**, 399-347.
41 S. Hayashi, *Funtai Kogaku Kaishi*, 1994, **31**, 207-213.
42 H. Eckert, K. Moran, D. Franke and C. Hudalla, *NATO ASI Ser., Ser. C*, 1993, 49-73.
43 E.R.T. Tiekini and G. Winter, *Rev. Inorg. Chem.*, 1992, **12**, 183-302.

44 J. Elguero, *An. Quim.*, 1993, **89**, 43-50.
45 B.Q. Sun and A. Pines, *Appl. Magn. Reson.*, 1993, **5**, 43-61.
46 N.C. Elbaum and J.F. Haw, *Gov. Rep. Announce. Index*, 1992, **92**.
47 E. Cochon and J.P. Amoureux, *Solid State Nucl. Magn. Reson.*, 1993, **2**, 205-222.
48 J.P. Amoureux and E. Cochon, *Solid State Nucl. Magn. Reson.*, 1993, **2**, 223-234.
49 G. Jeschke and G. Grossmann, *J. Magn. Reson. Ser. A*, 1993, **103**, 323-328.
50 O.N. Antzutkin, Z. Song, X. Feng and M.H. Levitt, *J. Chem. Phys.*, 1994, **100**, 130-140.
51 S.N. Stuart, *J. Magn. Reson. Ser. A,* 1993 **101**, 327-328.
52 M. Goldman, *J. Magn. Reson. Ser. A*, 1993, **102**, 173-179.
53 G. Wu and R.E. Wasylishen, *J. Chem .Phys.*, 1993, **98**, 6138-6149.
54 G. Wu and R.E. Wasylishen, *J. Chem .Phys.*, 1993, **99**, 6321-6329.
55 M.K. Bowman, *Adv. Chem. Ser.*, 1993, **229**, 605-626.
56 O. Weintraub and S. Vega, *J. Magn. Reson. Ser. A*, 1993, **105**, 245-267.
57 M.P. Augustine, K.W. Zilm and D.B. Zax, *J. Chem. Phys.*, 1993, **98**, 9432-9443.
58 Z. Luz, R. Poupko and S. Alexander, *J. Chem. Phys.*, 1993, **99**, 7544-7553.
59 A.C. Olivieri, *Solid State Nucl. Magn. Reson.*, 1993, **1**, 345-353.
60 S.H. Alarcon, A.C. Olivieri and R.K. Harris, *Solid State Nucl. Magn. Reson.*, 1993, **2**, 325-334.
61 A.C. Olivieri, *J. Magn. Reson. Ser. A*, 1993, **101**, 313-316.
62 H. Collazo-Lopez, M. Koch and W.T. Cooper, *Comput. Chem.*, 1993, **17**, 73-79.
63 R. Sangill, N. Rastrup-Andersen, H. Bildsoe, H.J. Jakobsen and N.C. Nielsen, *J. Magn. Reson., Ser. A*, 1994, **107**, 67-78.
64 N. Asakawa, H. Kurosu, I. Ando, A. Shoji and T. Ozaki, *J. Mol. Struct.*, 1994, **317**, 119-129.
65 G. Grossmann, B. Schwab, A. John, H. Komber, G. Jeschke and J. Sieler, *Phosphorus, Sulfur Silicon Relat. Elem.*, 1993, **78**, 297-308.
66 J.M Koons, E. Hughes and P.D. Ellis, *Anal. Chim. Acta*, 1993, **283**, 1045-1058.
67 S.A. Smith, T.O. Levante, B.H. Meier and R.R. Ernst, *J. Magn. Reson., Ser. A*, 1994, **106**, 75-105.
68 T. Asakura and J.H. Yeo, *Kobunshi Ronbunshu*, 1994, **51**, 43-46.
69 X. Wu and K.W. Zilm, *J. Magn. Reson. Ser. A*, 1993, **104**, 154-165.
70 S. Hedifer, B.H. Meier and R.R. Ernst, *Chem. Phys. Lett.*, 1993, **213**, 627-635.
71 O.B. Peersen, X. Wu, I. Kustanovich and S.O. Smith, *J. Magn. Reson. Ser. A*, 1993, **104**, 334-339.
72 H. Green, J.J. Titman and H.W. Spiess, *Chem. Phys. Lett.*, 1993, **213**, 145-152.
73 I. Hoogmartens, P. Adriaensens, D. Vanderzance and J. Gelan, *Anal. Chim. Acta*, 1993, **283**, 1025-1031.
74 A. Ramamoorthy, T. Fujiwara and K. Nagayama, *J. Magn. Reson., Ser. A*, 1993, **104**, 366-368.
75 T. Fujiwara, A. Ramamoorthy, K. Nagayama, K. Hioka and T. Fujito. *Chem. Phys. Lett.*, 1993, **212**, 81-84.
76 H.M. Vieth and C.S. Yannoni, *Chem. Phys. Lett.*, 1993, **205**, 153-156.
77 X. Wu and K.W. Zilm, *J. Magn. Reson., Ser. A*, 1993, **102**, 205-213.
78 J.Z. Hu, D.W. Alderman, C. Ye, R.J. Pugmire and D.M. Grant, *J. Magn. Reson., Ser. A*, 1993, **105**, 82-87.
79 L. Frydman, G.C. Chingas, Y.K. Lee, P.J. Grandinetti, M.A. Eastman, G.A. Barrall and A. Pines, *Isr. J. Chem.*, 1992, **32**, 161-164.
80 N.C. Nielsen, F. Creuzet and R.G. Griffin, *J. Magn. Reson., Ser. A*, 1993, **103**, 245-252.
81 D. Reichert, Z. Olender, R. Poupko, H. Zimmermann and Z. Luz, *J. Chem. Phys.*, 1993, **98**, 7699-7710.
82 B.S.A Kumar and S.J. Opella, *J. Magn. Reson., Ser A*, 1993, **101**, 333-336
83 P.P. Man, *Appl. Magn. Reson.*, 1993, **4**, 65-87.
84 C.A. Fyfe, K.T. Mueller, H. Grondey and K.C. Wong-Moon, *J. Phys. Chem.*, 1993, **97**, 13484-13495.
85 J.P. Amoureux, *Solid State Nucl. Magn. Reson.*, 1993, **2**, 83-88.
86 T.J. Bongamba, H. Panepucci, *J. Magn. Reson., Ser. A*, 1993, **103**, 103-104.
87 J. Hong and G.S. Harbison, *J. Magn. Reson., Ser. A*, 1993, **105**, 128-136.
88 A. Samoson, *Chem. Phys. Lett.*, 1993, **214**, 456-458.
89 S.L. Gann, J.H. Baltisberger and A. Pines, *Chem Phys. Lett.*, 1993, **210**, 405-410.

90 A.E. Bennett, L.R. Becerra and R.G. Griffin, *J. Chem. Phys.*, 1994, **100**, 812-814.
91 D.K. Sodickson, M.H. Levitt, S. Vega and R.G. Griffin, *J. Chem. Phys.*, 1993, **98**, 6742-6748.
92 M.L. Buszko and G.E. Maciel, *J. Magn. Reson., Ser. A*, 1993, **104**, 172-176.
93 J. Homer and M.J. Howard, *J. Chem. Soc. Faraday Trans.*, 1993, **89**, 3029-3038.
94 E.J. Munson, D.K. Murray and J.F. Haw, *J. Catal.*, 1993, **141**, 733-736.
95 A. Somoson, B.Q. Sun and A. Pines, *Pulsed Magn. Reson. NMR, ESR, Opt.*, 1992, 80-94.
96 J. Hu, A.M. Orendt, D.W. Alderman, C. Ye, R.J. Pugmire and D.M. Grant, *Solid State Nucl. Magn. Reson.*, 1993, **2**, 235-243.
97 C.A. Fyfe, K.C. Wong-Moon, H. Grondey and K.T. Mueller, *J. Phys. Chem.*, 1994, **98**, 2139-2142.
98 P. Tang, W.J. Wei and G.S. Harbison, *Solid State Nucl. Magn. Reson.*, 1993, **2**, 343-348.
99 D.G. Cory and W.E.J.R. Maas, *J. Magn. Reson., Ser. A*, 1993, **102**, 190-194.
100 J.L White, L.W. Beck, D.B. Ferguson and J.F. Haw, *Gov. Rep. Announce. Index (U.S.)*, 1992, **92**.
101 R. Tycko, *J. Am. Chem. Soc.*, 1994, **116**, 2217-2218.
102 D. Reichert and H. Schneider, *Meas. Sci. Technol.*, 1993, **4**, 1376-1379.
103 C.P. Grey, W.S. Veeman and A.J. Vega, *J. Chem. Phys.*, 1993, **98**, 7711-7724.
104 L.B. Moran, J.K. Berkowitz and J.P. Yesinowski, *Solid State Nucl. Magn. Reson.*, 1993, **1**, 307-311.
105 J. Sola, P. Gonzalez-Duarte, J. Sanz, I. Casals, T. Alsina, I. Sobrados, A. Alvarez-Larena, J.F. Piniella and X. Solans, *J. Am. Chem. Soc.*, 1993, **115**, 10018-10028.
106 T.C. Farrar and J.D. Trudeau, *NATO ASI Ser., Ser. C*, 1993, **386**, 27-48.
107 T. Nakai and C.A. McDowell, *Bull. Magn. Reson.*, 1993, **15**, 149-153.
108 F.T. Bonner, G.B. Crull and G.S. Harbison, *Inorg. Chem.*, 1993, **32**, 3316-3319.
109 G. Wu and R.E. Wasylishen, *J. Magn. Reson., Ser. A*, 1993, **102**, 183-189.
110 K. Eichele, G. Wu and R.E. Wasylishen, *J. Magn. Reson., Ser. A*, 1993, **101**, 157-161.
111 R. Challoner, C.A. McDowell, M. Yoshifuji, K. Toyota and J.A. Tossell, *J. Magn. Reson., Ser. A*, 1993, **104**, 258-267.
112 R.K. Harris, M.M. Sunnetciogly, K.S. Cameron and F.G. Riddell, *Magn. Reson. Chem.*, 1993, **31**, 963-965.
113 T. Ueda, S. Hayashi and H. Kikuko, *Solid State Commun.*, 1993, **87**, 429-434.
114 D.G. Rethwisch, M.A. Jacintha and C.R. Dybowski, *Anal. Chim. Acta*, 1993, **283**, 1033-1043.
115 D.L. Rzou, T.H. Huang, P. Desai and A.S. Abhiraman, *J. Polym. Sci., Part B: Polym. Phys.*, 1993, **31**, 1005-1012.
116 D. Franke, C. Hudalla and H. Eckert, *Solid State Nucl. Magn. Reson.*, 1993, **1**, 297-306.
117 X. Wu and K.W. Zilm, *J. Magn. Reson., Ser. A*, 1993, **104**, 119-122.
118 T. Petrass and P.K. Burkert, *Magn. Reson. Chem.*, 1993, **31**, 709-713.
119 P. Grondona and A.C. Oliveiri, *Concepts Magn. Reson.*, 1993, **5**, 319-339.
120 B. Tesche, H. Zimmermann, R. Poupko and U. Haeberlen, *J. Magn. Reson., Ser. A*, 1993, **104**, 68-77.
121 L. Zheng, K.W. Fishbein, R.G. Griffin and J. Herzfeld, *J. Am. Chem. Soc.*, 1993, **115**, 6254-6261.
122 J.C. Duchamp, *Diss. Abstr. Int. B*, 1993, **53**, 5735-5736.
123 R. Poupko, Z. Olender, D. Reichert and Z. Luz, *J. Magn. Reson., Ser. A*, 1994, **106**, 113-115.
124 N. Zumbulyadis and J.M. O'Reilly, *J. Am. Chem. Soc.*, 1993, **115**, 4407-4408.
125 A.C. Kolbert, H.J.M. de Groot, D. van der Putten, H.B. Brom, L.J. de Jongh, G. Shmid, H. Krautscheid and D. Fenske, *Z. Phys. D: At., Mol Clusters*, 1993, **26**, 24-26.
126 A.R. Brough, C.P. Grey and C.M. Dobson, *J. Am. Chem. Soc.*, 1993, **115**, 7318-7327.
127 E. Knop, L. Latanowicz and E.C. Reynhardt, *Ber. Bunsen-Ges. Phys. Chem.*, 1993, **97**, 1457-1466.
128 D. Michel, M. Pruski and B.C. Gerstein, *Carbon*, 1994, **32**, 32-40.
129 M. Pruski and B.C. Gerstein, *Carbon*, 1994, **32**, 41-49.
130 R.D. Curtis, *Diss. Abstr. Int. B*, 1992, **53**, 351.
131 S. Ganapathy, P.R. Rajamohanan, P.M. Ramanujulu, A.B. Mandhare and R.A.

Mashelkar, *Polymer*, 1994, **35**, 888-891.
132 F.G. Riddel, P.G. Bruce, P. Lightfoot and M. Rogerson, *J. Chem. Soc., Chem. Commun.*, 1994, **2**, 209-211.
133 C.G. Hoelger, B. Wehrle, H. Benedict and H.H. Limbach, *J. Phys. Chem.*, 1994, **98**, 843-851.
134 B.J. Miller, D.G. Cory and A.N. Garroway, *Polym. Prepr. (Am. Chem. Soc., Div. Polym. Chem.)*, 1992, **33**, 765-766.
135 J.C. Groombridge, *Magn. Reson. Chem.*, 1993, **31**, 380-387.
136 R.D. Curtis, P.J. Tonge, I.C.P. Smith and H.C. Jarrell, *J. Magn. Reson., Ser. A*, 1993, **102**, 110-113.
137 S.H. Alacron, A.C. Olivieri and P. Jonsen, *J. Chem. Soc., Perkin Trans. 2*, 1993, **10**, 1783-1786.
138 A.E. Aliev, K.D.M. Harris, D.C. Apperley, R.K. Harris and M.M. Sunnetcioglu, *J. Chem. Soc., Faraday, Trans.*, 1993, **89**, 3791-3796.
139 W.E. Schmidt, A.D. Mitchell, M.J. Line and J.B. Reeves, *Solid State Nucl. Magn. Reson.*, 1993, **2**, 11-20.
140 Z. Gu and A. McDermott, *J. Am. Chem. Soc.*, 1993, **115**, 4282-4285.
141 N.S. Cauchon, *Diss. Abstr. Int. B*, 1993, **53**, 4677-4678.
142 J.C. Facelli and D.M. Grant, *Nature*, 1993, **365**, 325-327.
143 P.R. Ashton, G.R. Brown, A.J. Foubister, D.R. Smith, N. Spencer, J.F. Stoddard and D.J. Williams, *Tetrahedron Lett.*, 1993, **34**, 8333-8336.
144 R.J. Applyard and J.N.S. Evans, *J. Magn. Reson., Ser. B*, 1993, **102**, 245-252.
145 S. Ishikawa and I. Ando, *J. Mol. Struct.*, 1993, **29**, 183-190.
146 V.K. Dolganov, I. Pocsik and L. Rosta, *Liq. Cryst.*, 1993, **14**, 1895-1900.
147 M. Angeles Monge, E. Gutierrez Puebla, J. Elguero, C. Toiron, W. Meutermans and I. Sobrados, *Spectrochem. Acta, Part A*, 1994, **50A**, 727-734.
148 S.R. Salman, J.C. Lindon and T.A. Carpenter, *Thermochim Acta*, 1994, **232**, 171-176.
149 A. Maslankiewicz, M. Wyszomirski, S. Boryczka, A. Gogoll and T. Glowiak, *Phosphorus, Sulfur Silicon Relat. Elem.*, 1993, **74**, 429-430.
150 G.W. Buchanan, A.B. Driega, A. Moghimi, C. Bensimon, R.A. Kirby and T.D. Bouman, *Can. J. Chem.*, 1993, **71**, 1983-1989.
151 R.M. Cravero, M. Gonzalez-Sierra, C. Fernandez and A.C. Olivieri, *J. Chem. Soc., Chem. Commun.*, 1993, **16**, 1253-1254.
152 T. Nozawa, M. Nishimura, M. Hatano and M. Sato, *Bull. Chem. Soc. Jpn.*, 1993, **66**, 692-697.
153 L.E. Wolinsky, R.W. Armstrong and R.R. Seghi, *J. Dent. Res.*, 1993, **72**, 72-77.
154 J. Salgado, J. Villalain and J.C. Gomez-Fernandez, *Eur. Biophys. J.*, 1993, **22**, 151-155.
155 L. Raymond, F.G. Morin and R.H. Marchessault, *Carbohydr. Res.*, 1993, **246**, 331-336.
156 J.A. Leenhee, G.E. Maciel and T.I. Noyes, *Am. Chem. Soc. Div. Environ. Chem.*, 1993, **33**, 1740177.
157 K. Mizoguchi, Y. Maniwa and K. Kume, *Mater. Sci. Eng., B*, 1993, **B19**, 146-153.
158 L.E. Hall, D.R. McKenzie, M.I. Attalla, A.M. Vassallo, R.L. Davis, J.B. Dunlop and D.J.H. Cockayne, *J. Phys. Chem.*, 1993, **97**, 5741-5744.
159 M.L. Attalla, A.M. Vassallo, B.N. Tattam and J.V. Hanna, *J. Phys. Chem.*, 1993, **97**, 6329-6331.
160 D.M. Cox, S.D. Cameron, A. Tuinman, A. Gakh, J.L. Adcock, R.N. Crompton, E.W. Hagaman, K. Kniaz and J.E. Fischer, *J. Am. Chem. Soc.*, 1994, **116**, 1115-1120.
161 K.E. Geckeler, A. Hirsch, *J. Am. Chem. Soc.*, 1993, **115**, 3850-3851.
162 L. Firlej, Z. Belahmer, P. Bernier, A. Zahab, M. Ribet, N. Coustel and R. Aznar, *Solid State Commun.*, 1993, **87**, 669-673.
163 A.E. Aliev, K.D.M. Harris, M. Tegze and S. Pekker, *J. Mater. Chem.*, 1993, **3**, 1091-1094.
164 R Tycko, G. Dabbagh, G.B.M. Vaughan, P.A. Heiney, R.M. Strongin, M.A. Cichy and A.B. Smith, *J. Chem. Phys.*, 1993, **99**, 7554-7564.
165 R. Zuzok, P. Wzietek, M. Fourmigue, P. Batail, P. Auban-Senzier, S. Roth and D. Jerome, *Synth. Met.*, 1993, **56**, 3235-3239.
166 C.S. Yannoni, H.R. Wendt, M.S. de Vries, R.L. Siemens, J.R. Salem, J. Lyerla, R.D. Johnson, M. Hoinkis and M.S. Crowder, *Synth Met.*, 1993, **59**, 279-295.
167 F. Rachdi, J. Reichenback, L. Firlej, P. Bernier, M. Ribet, R. Aznar, G. Zimmer, M.

Helmle and M. Mehring, *Solid State Commun.*, 1993, **87**, 547-550.
168 R.A. Assink, D.A. Loy, J.E. Schirber and B. Morosin, *Mater. Res. Soc. Symp. Proc.*, 1992, **270** 255-260.
169 M.W. Anderson, J. Shi, D.A. Leigh, A.E. Moody, F.A. Wade, B. Hamilton and S.W. Carr, *J. Chem. Soc., Chem. Commun.*, 1993, 533.
170 A. Guegel, K. Muellen, H. Reichert, W. Schmidt, G. Schoen, F. Schueth, J. Spickermann, J. Titman and K. Unger, *Angew. Chem.*, 1993, **105**, 618-619.
171 P. Vanhoorne, R. Jerome, P. Teyssie and F. Laupretre, *Macromolecules*, 1994, **27**, 2548-2552.
172 A. Gerard, F. Laupretre and L. Monnerie, *Macromolecules*, 1993, **26**, 3313-3318.
173 S.K. Hideyuki, *Macromolecules*, 1993, **26**, 7326-7332.
174 M. Yamaguchi and A. Tsutsumi, *Polym. J. (Tokyo)*, 1993, **25**, 427-433.
175 K.L. Buchert, J.L. Koeing, S.Q. Wang and J.L. West, *Appl. Spectrosc.*, 1993, **47**, 933-941.
176 Y. Yu and G.R. Brown, *Macromolecules*, 1993, **26**, 4872-4877.
177 W.E.J.R. Maas, C.H.M. Papavoine, W.S. Veeman, G.H. Wermeus Buning and J.M.J. Vankan, *J. Polym. Sci., Part B: Polym. Phys.*, 1994, **32**, 785-789.
178 J.L White and P. Mirau, *Macromolecules*, 1993, **26**, 3049-3054.
179 L.A. Belfiore, *Prog. Pac. Polym. Sci. 2, Proc. Pac Polym. Conf., 2nd*, 1992, **ISS/PTTL**, 113-129.
180 S.L. Maunu, J. Kinnuen, K. Soljamo and F. Sundholm, *Polymer*, 1993, **34**, 1141-1145.
181 M. Afeworki, R.A. McKay and J. Schaefer, *Mater. Sci. Eng. A*, 1993, **A162**, 221-228.
182 H. Tanaka, Y. Fujiwara, T. Asano and S. Polizzi, *Macromol. Chem.*, 1993, **194**, 2279-2286.
183 P.C. Stein, W.L. Earl and A. Ray, *Synth. Met.*, 1993, **55**, 702-707.
184 A. Vainrub, I. Heinmaa, E. Lippmaa and G.I. Kozub, *Synth. Met.*, 1993, **55**, 660-665.
185 P.E.M. Allen, A.B. Clayton and D.R.G. Williams, *Eur. Polym. J.*, 1994, **30**, 427-432.
186 W. Barendswaard, J. Moonen and M. Neilen, *Anal. Chim. Acta*, 1993, **283**, 1007-1024.
187 A. Shoji, *Kenkyu Hokoku - Asahi Garasu Zaidan*, 1992, **61**, 35-43.
188 J.A. Franz and J.C. Linehan, *Advn. Chem. Ser.*, 1993, **229**, 377-400.
189 P.J. Barrie, E.S. Madrali and R. Kankiyoti, *Energy Fuels*, 1993, **7**, 479-481.
190 R.B. Gregory, M. Gangoda, R.K. Gilpin and W. Su, *Biopolymers*, 1993, **33**, 513-519.
191 J.M. Rydzewski, S.X. Wang, A. Stevens, C. Serdahl and T. Schleich, *Exp. Eye Res.*, 1993, **56**, 305-316.
192 R. Weber, I. Nydegger and P. Skrabal, *Recents Prog. Genie Procedes*, 1992, **6**, 149-153.
193 B. Montez, E. Oldfield, J.A. Urbina, S. Pekerar, C. Husted and J. Patterson, *Biochim. Biophys. Acta*, 1993, **115**, 314-318.
194 M.A. Wilson, I.M. Godfrey, J.V. Hanna, R.A. Quezada and K.S. Finnie, *Org. Geochem.*, 1993, **20**, 599-610.
195 E.J. Munson and J.F. Haw, *Angew. Chem.*, 1993, **105**, 643-644.
196 D.K. Murray, J.W. Chang and J.F. Haw, *J. Am. Chem. Soc.*, 1993, **115**, 4732-4741.
197 J. Novakova, L. Bosacek, Z. Dolejsek and L. Kubelkova, *Stud. Surf. Sci. Catal*, 1993, **75**, 1732-1736.
198 E.J. Munson, A.A. Kheir and J.F. Haw, *J. Phys. Chem.*, 1993, **97**, 7321-7327.
199 A.G. Stepanov and K.I. Zamaraev, *Catal. Lett.*, 1993, **19**, 153-158.
200 A.G. Stepanov, V.N. Zudin and K.I. Zamaraev, *Solid State Nucl. Magn. Reson.*, 1993, **2**, 89-93.
201 A.K. Nowak, A.E. Wilson, K. Roberts and K.P. Datema, *J. Catal.*, 1993, **144**, 495-505.
202 I.I. Ivanova, D. Brunel, J.B. Nagy, G. Daelen and E.G. Derouane, *Stud. Surf. Sci. Catal.*, 1993, **78**, 587-594.
203 T. Xu, E.J. Munson and J.F. Haw, *J. Am. Chem. Soc.*, 1994, **116**, 1962-1972.
204 E.G. Derouane, S.B. Abdul Hamid, I.I. Ivanova, N. Blom and P.E. Hoejlund-Nielsen, *J. Mol. Catal.*, 1994, **86**, 371-400.
205 V. Bosacek, *J. Phys. Chem.*, 1993, **97**, 10732-10737.
206 J. Novakova, V. Bosacek, Z. Dolejsek and L. Kubelkova, *J. Mol. Catal.*, 1993, **78**, 181-187.
207 S.L. Burkett and M.E. Davis, *Microporous Mater.*, 1993, **1**, 265-282.
208 D. Zhao, S. Qiu and W. Pang, *Zeolites*, 1993, **13**, 478-480.
209 D. Zhao, S. Qiu and W. Pang, *Chem. Res. Chin. Univ.*, 1992, **8**, 185-187.

210 D. Brunel, J.B. Nagy, G. Daelen, E.G. Derouane, P. Geneste and D.J. Vanderveken, *Appl. Catal., A*, 1993, **99**, 9-20.
211 J.F. Bank, G. Ofori-Okai and S. Bank, *Clays Clay Miner.*, 1993, **41**, 95-102.
212 Z.R. Hinedi, C.T. Johnson and C. Erickson, *Clays Clay Miner.*, 1993, **41**, 87-94.
213 S. Carolino and M.J. Hudson, *J. Mater. Chem.*, 1994, **4**, 99-104.
214 R.K. Mariwala and H.C. Foley, *Ind. Eng. Chem. Res.*, 1994, **33**, 607-615.
215 F.J. Dinan, R.A. Wolfe, D.S. Hojinicki, W.T. Schwartz, T. St. Clair and J.R. Pratt, *High Perform. Polym.*, 1992, **4**, 131-139.
216 S. Aime and R. Gobetto, *J. Cluster Sci.*, 1993, **4**, 1-8.
217 S. Aime, L. Cordero, R. Gobetto and G. Szalontai, *Spectrochim. Acta, Part A*, 1993, **49A**, 1307-1314.
218 P. Knopi,, L. Luczak, M.J. Potrzebowski, J. Michalski, J. Blaszczyk and M.W. Wieczorek, *J. Chem. Soc., Dalton Trans.*, 1993, **18**, 2749-2757.
219 J. Kuemmerlen, A. Sebald and E. Sendermann, *Organometallics*, 1994, **13**, 802-807.
220 P.J. Slezak and A. Wieckowski, *J. Magn. Reson., Ser. A*, 1993, **102**, 166-172.
221 K.M. McNamara, K.K. Gleason, D.J. Vestyck and J.E. Butler, *Diamond, Relat. Mater.*, 1992, **1**, 1145-155.
222 L.H. Merwin, D.E. Johnson and W.A. Weimer, *J. Mater. Res.*, 1994, **9**, 631-635.
223 A.E. Aliev and K.D.M. Harris, *J. Am. Chem. Soc.*, 1993, **115**, 6369-6377.
224 A. Mahdyarfar and K.D.M. Harris, *J. Chem. Soc., Chem. Commun.*, 1993, **1**, 51-53.
225 M.D. Hollingworth and A.R. Palmer, *J. Am. Chem. Soc.*, 1993, **115**, 5881-5882.
226 F. Imashiro, D. Kuwahara and T. Terao, *J. Chem. Soc., Perkin Trans. 2*, 1993, **10**, 1759-1763.
227 G. Beamson, W.J. Brennan, N.J. Clayden and R.C.K. Jennings, *J. Polym Sci., Part B: Polym Phys.*, 1993, **31**, 1205-1211.
228 P.B. Lukins, D.R. McKenzie, A.M. Vassallo and J.V. Hanna, *Carbon*, 1993, **31**, 569-575.
229 K. Zemke, K. Schmidt-Rohr, J.H. Magill, H. Sillescu and H.W. Spiess, *Mol. Phys.*, 1993, **80**, 1317-1330.
230 S. Hosseini, *Energy Res. Abstr.*, 1992, **17**.
231 A.A. Kheir and J.F. Haw, *J. Am. Chem. Soc.*, 1994, **116**, 817-181.
232 J.C. Vizzini, J.F. Shi and J.C.W. Chien, *J. Appl. Polym. Sci.*, 1993, **48**, 2173-2179.
233 J.F. Haw, *Spec. Publ. - R. Soc. Chem.*, 1992, **114**, 1-31.
234 G.J. Ray and A. Samoson, *Zeolites*, 1993, **13**, 410-413.
235 G.W. Haddix, M. Narayan, W.D. Gillespie, M.B. Georgellis and Y. Wu, *J. Am. Chem. Soc.*, 1994, **116**, 672-674.
236 C.A. Fyfe, Y. Feng and H. Grondey, *Microporous. Mater.*, 1993, **1**, 393-400.
237 E. Oldfield, J. Haase, K.D. Schmitt and S.E. Schramm, *Zeolites*, 1994, **14**
238 A. Somoson, P. Sarv, J.P. van Braam Houckgeest and B. Kraushaar-Czarnetzki, *Appl. Magn. Reson.*, 1993, **4**, 171-178.
239 J.H. Kwak and R. Ryoo, *J. Phys. chem.*, 1993, **97**, 11154-11156.
240 M. Stocker, *Acta. Chem. Scand.*, 1993, **47**, 935-936.
241 A. Diaz, J. Lujano and J. Tejada, *Rev. Soc. Venez. Catal.*, 1992, **6**, 107-118.
242 M.W. Anderson, O. Terasaki, T. Ohsuna, A. Philippou, S.P. MacKay, A. Ferreira, J. Rocha and S. Lidin, *Nature*, 1994, **367**, 347-351.
243 C.Y. Chen, H.X. Li and M.E. Davis, *Microporous Mater.*, 1993, **2**, 17-26.
244 M.J. Annen and M.E. Davis, *Microporous Mater.*, 1993, **1**, 57-65.
245 B. Marler, C. Deroche, H. Gies, C.A. Fyfe, H. Groundey, G.T. Kokotailo, Y Feng, S. Ernst, J. Weitkamp and D.E. Cox, *J. Appl. Crystallogr.*, 1993, **26**, 636-644
246 J.A. Hriljac, M.M. Eddy, A.K. Cheetham, J.A. Donohue and G.J. Ray, *J. Solid State Chem.*, 1993, **106**, 66-72.
247 C. Jia, P. Massiani and D. Barthomeuf, *J. Chem. Soc., Faraday Trans.*, 1993, **89**, 3659-3665.
248 J.A. Martens, Y.L. Xiong, E.J.P. Feijen, P.J. Grobet and P.A. Jacobs, *J. Phys. Chem.*, 1993, **97**, 5132-5135.
249 P. Sun, B. Li, Q. Jin, J. Wang, D. Ding and C. Hu, *Huaxue Wuli Xuebao*, 1993, **6**, 39-46.
250 W.A. Buckermann, C.B. Huong, F. Fajula and C. Gueguen, *Zeolites*, 1993, **13**, 448-453.
251 H. Yong, D. Coster, F.R. Chen, J.G. Davis and J.J. Fripiat, *Stud. Surf. Sci. Catal.*, 1993, **75**, 1159-1170.

252 B.F. Mentzen, M. Sacerdote-Peronnet, J.F. Berar and F. Lefebvre, *Zeolites*, 1993, **13**, 485-492.
253 M.P.J. Peeters, J.W. de Haan, L.J.M. van de Ven and J.H.C. van Hooff, *J. Phys. Chem.*, 1993, **97**, 5363-5369.
254 H.X. Li and M.E. Davis, *J. Chem. Soc., Faraday, Trans.*, 1993, **89**, 951-956.
255 D. Zhao, J. Zhang, Y. Wu, F. Pei, S. Qiu and W. Pang, *Wuli Huaxue Xuebao*, 1993, **9**, 706-708.
256 R. Vomscheid, M. Briend, M.J. Peltre, P. Massiani, P.P. Man and D. Barthomeuf, *J. Chem. Soc., Chem. Commun.*, 1993, **6**, 544-546.
257 M. Briend, A. Lamy, M.J. Peltre, P.P. Man and D. Barthomeuf, *Zeolites*, 1993, **13**, 201-211.
258 M. Derewinski, M.J. Peltre, M. Briend, D. Barthomeuf and P.P. Man, *J. Chem. Soc., Faraday Trans.*, 1993, **89**, 1823-1828.
259 K. Nakashiro, U. ono, N. Shinichi and Y. Mormura, *Zeolites*, 1993, **13**, 561-564.
260 D. Zhao, J. Zhang, Y. Wu, F. Pei, S. Qiu and W. Pang, *Yingyong Huaxue*, 1993, **10**, 17-19.
261 S.H. Jhung, S.B. Hong, Y.S. Uh and H. Chon, *J. Korean Chem. Soc.*, 1993, **37**, 867-873.
262 G. Giannetto, R. Monque, J. Perez, J. Papa and L. Garcia, *Zeolites*, 1993, **13**, 557-560.
263 A.M. Prakash, K.V. Pao and S. Unnikrishnan, *Indian J. Chem., Sect. A: Inorg., Bio-inorg., Phys., Theor. Anal. Chem.*, 1993, **32A**, 947-951.
264. H. He and J. Klinowski, *J. Phys. Chem.*, 1994, **98**, 1192-1197.
265 A.M. Prakash, S. Unnikrishnan and K.V. Rao, *Appl. Catal., A*, 1994, **110**, 1-10.
266 E.J.P. Feijen, K. De Vadder, M.H. Bosscharets, J.L. Lievens, J.A. Martens, P.J. Grobet and P.A. Jacobs, *J. Am. Chen. Soc.*, 1994, **116**, 2950-2957.
267 R.A. Grinsted, H.W. Jen, C. N. Montreuil, M.J. Rokosz and M. Shelef, *Zeolites*, 1993, **13**, 602-606.
268 Z. Suo, C. Zhu, Q. Li, L. Zhou and H. Xu, *Shiyou Huagong*, 1993, **22**, 20-26.
269 W. Kolodziejski, L. Delaude, P. Laszlo, M.T. Montaufier and J. Klinowski, *Appl. Catal., A*, 1993, **98**, 71-79.
270 N. Mizuno, H. Mori, M. Tajima, S. Kagawa, H. Ueno, H. Yahiro and M. Iwamoto, *J. Mol. Catal.*, 1993, **80**, 229-242.
271 N.P. Rhodes and R. Rudham, *J. Chem. Soc., Faraday Trans.*, 1993, **89**, 2551-2557.
272 S. Bao, J. Fu, Q. Xu, G. Tong and C. Ruan, *Ranliao, Huaxue Xuebao*, 1993, **21**, 154-158.
273 G. Niu, C. Zhu, Q. Li, Z. Xiu, Y Zhao, Y. Hou and M. Li, *Wuli Huaxue Xuebao*, 1993, **9**, 374-381.
274 I. Kiricsi, C. Flego, G. Pazzuconi, W.O. Parker Jnr., R. Millini, C. Perego and G. Bellussi, *J. Phys. Chem.*, 1994, **98**, 4627-4634.
275 J. Datka and B. Gil, *J. Catal.*, 1994, **145**, 372-376.
276 B. Gil, E. Broclawik, J. Datka and J. Klinowski, *J. Phys. Chem.*, 1994, **98**, 930-933.
277 M.A. Long, S.J.X. He and C. Than, *J. Labelled Compd. Radiopharm*, 1993, **33**, 381-390.
278 H.L. Zubowa, G. Lischke, B. Parlitz, E. Schreier, R. Eckelt, G. Schulz and R. Fricke, *Appl. Catal., A*, 1994, **110**, 27-38.
279 G.J. Kim, B.R. Cho and J.H. Kim, *Catal. Lett.*, 1993, **22**, 259-270.
280 G.J. Kim and B.R. Cho, *Hwahak Konghak*, 1993, **31**, 658-666.
281 H. Kosslick, V.A. Tuan, R. Fricke, C. Peuker, W. Pilz and W. Storek, *J. Phys. Chem.*, 1993, **97**, 5678-5684.
282 B. Krauxhaar-Czarnetzki, R.J. Dogterom, W.H.J. Stork, K.A. Emeis and J.P. van Braam Houckgeest, *J. Catal.*, 1993, **141**, 140-147.
283 Y. Chen, Z. Xue and L. Dai, *Huaxue Tongbao*, 1993, **11**, 32-34.
284 A. Dyer, S. Amini, H. Enamy, H.A. El-Naggar and M.W. Anderson, *Zeolites*, 1993, **13**, 281-290.
285 J. Fu, S. Bao, Q. Xu and C. Hu, *Gaodeng Xuexiao Huaxue Xuebao*, 1993, **14**, 907-910.
286 C.L. Tway, S.J. Bonafede, A.M. Ghazi, C.P. Reed, R.J. DeAngelis and T.M. Apple, *ACS Symp. Ser.*, 1993, **517**, 372-382.
287 F. Lefebvre, M. Sacerdot-Peronnet and B.F. Mentzen, *C.R. Acad. Sci., Ser. II*, 1993, **316**, 1549-1555.
288 S. Prasad and R. Vetrivel, *J. Phys. Chem.*, 1994, **98**, 1579-1583.

289 D. Ding, P. Sun, Q. Jin, B. Li and J. Wang, *Zeolites*, 1994, **14**, 65-73.
290 P. Qiu, Y. Zang, K. Tao, F. Gao and H. Li, *Cuihua Xuebao*, 1993, **14**, 466-472.
291 P. Qiu, Y. Zang, K. Tao, F. Gao and H. Li, *Ranliao Huaxue Xuebao*, 1992, **20**, 355-360.
292 R. Miyawaki, S. Tomura, K. Inukai, Y. Shibasaki, M. Okazaki, S. Samejima and S. Satokawa, *Clay Sci.*, 1992, **8**, 273-284.
293 R.H. Meinhold, R.C.T. Slade and T.W. Davies, *Appl. Magn. Reson.*, 1993, **4**, 141-155.
294 Q. Zhou, C. Wang, D. Huang, D. Jiang and E. Min, *Cuihua Xuebao*, 1993, **14**, 478-483.
295 L. Li, Y. Zhang, Y. Ge and L. Li, *Shiyou Xuebao, Shiyou Jiagong*, 1992, **8**, 59-65.
296 A. La Iglesia, *Clay Miner.*, 1993, **28**, 311-319.
297 J.M. Trillo, M.D. Alba, R. Alvero and M.A. Castro, *J. Chem. Soc., Chem. Commun.*, 1993, **24**, 1809-1811.
298 A.M.T. Bell, C.M.B. Henderson, S.A.T. Redfern, R.J. Cernik, P.E. Champness, A.N. Fitch and S.C. Kohn, *Acta Crystallogr., Sect. B: Struct. Sci.*, 1994, **B50**, 31-41.
299 S.T. Wong and S. Cheng, *chem. Mater.*, 1993, **5**, 770-777.
300 Z. Li, D. Huang, C. Wang and D. Jiang, *Jilin Daxue Ziran Kexue Xuebao*, 1992, **3**, 90-94.
301 J.F. Lamber, S. Chevalier, R. Franck, H. Suquet and D. Barthomeuf, *J. Chem. Soc., Faraday Trans.*, 1994, **90**, 675-682.
302 S. Komarneni, *Clays Clay Miner.*, 1993, **41**, 472-483.
303 J.C.D. Macedo, C.J.A. Mota, S.M.C. de Menezes and V. Camorim, *Appl. Clay Sci.*, 1994, **8**, 321-330.
304 K.J.D. MacKenzi, R.H. Meinhil, B.L. Sherriff and Z. Xu, *J. Mater. Chem.*, 1993, **3**, 1263-1269.
305 M.L. Occelli, K. Takahama, M. Yokoyama and S. Hirao, *Synth. Microporous Mater.*, 1992, 57-80.
306 G.S. Neal, M.E. Smith, M.B. Trigg and J. Drennan, *J. Mater. Chem.*, 1994, **4**, 245-251.
307 E.N. Coker and L.V.C. Rees, *J. Mater. Chem.*, 1993, **3**, 523-529.
308 L.F. Gladden, P. Chiaranussati, M. Vignaux, R.W. Griffiths, S.D. Jackson, J.R. Jones, A.P. Sharratt, F.J. Robertson and G. Webb, *Eur. Mater. Res. Soc. Monogr.*, 1992, **5**, 85-92.
309 F. Thomas, A. Masion, J.Y. Bottero and D. Canet, *Natl. Meet. - Am. Chem. Soc., Div. Environ. Chem.*, 1993, **33**, 152-154.
310 E. Lindner, M. Kemmler, H.A. Mayer and P. Wegner, *J. Am. Chem. Soc.*, 1994, **116**, 348-361.
311 G.W. Haddix, M. Narayana, S.C. Tang and T. Wu, *J. Phys. Chem.*, 1993, **97**, 4624-4627.
312 I.S. Chuang, D.R. Kinney and G.E. Maciel, *J. Am. Chem. Soc.*, 1993, **115**, 8695-8705.
313 N.J. Clayden and P.J.V. Jones, *J. Polym. Sci., Part A: Polym. Chem.*, 1993, **31**, 2835-2838.
314 J.W. Lee, K.I. Cho, R. Ryoo and M.S. Jhon, *Proc. - Electrochem. Soc.*, 1993, **93-27**, 245-255.
315 R.K. Shibao, V.I. Srdanov, M. Hay and H. Eckert, *Chem. Mater.*, 1994, **6**, 306-310.
316 T.J. Bastow, J.S. Hall, M.E. Smith and S. Steuernagel, *Mater. Lett.*, 1994, **18**, 197-200.
317 Z. Xu and B.L. Sherriff, *Appl. Magn. Reson.*, 1993, **4**, 203-211.
318 R.H. Meinhold, R.C.T. Slade and R.H. Newman, *Appl. Magn. Reson.*, 1993, **4**, 121-140.
319 J.F. Stebbins, C.W. Burnham and D.L. Bish, *Am. Mineral*, 1993, **78**, 461-464.
320 M.E. Smith, G. Neal, M.B. Trigg and J. Drennan, *Appl. Magn. Reson.*, 1993, **4**, 157-170.
321 C. Bessada, D. Massiot, J. Coutures, A. Douy, J-P Coutures and F. Taulelle, *J. Non-Cryst. Solids*, 1994, **168**, 76-85.
322 U. Selvaraj, S. Komarneni and R. Rustum, *J. Solid State Chem.*, 1993, **106**, 73-82.
323 G. El-Damrawi, W. Mueller-Warmuth, H. Doweidar and I.A. Gohar, *Phys. Chem. Glasses*, 1993, **34**, 52-57.
324 Government Industrial Development Laboratory, Hokkadio, *Hokkaido Kogyo Kaihatsu Shikensho Hokoku*, 1992, **57**, 8-13.

325 T. Parot-Rajaona, B. Cote, C. Bessada, D. Massiot and F. Gervais, *J. Non-Crystl. Solids*, 1994, **169**, 1-14.
326 A.P. Legrand, H. Taibi, H. Hommel, P. Tougne and S. Leonardelli, *J. Non-Crystl. Solids*, 1993, **155**, 122-130.
327 Y. Hong, F.R. Chen and J.J. Fripiat, *Catal. Lett.*, 1993, **17**, 187-195.
328 J.C. Edwards and E.C. Decanio, *Catal. Lett.*, 1993, **19**, 121-130.
329 A.E. Aliev and K.D.M. Harris, *Mendeleev Commun.*, 1993, **4**, 153-155.
330 J.C. Williams and A.E. McDermott, *J. Phys. Chem.*, 1993, **97**, 12393-12398.
331 J. Nouwen, P. Adriaensens, D. Vanderzande, J. Gelan, H. Paulussen, G. Verreyt, Z. Yang and H.J. Geise, *Synth. Metl.*, 1993, **59**, 171-179.
332 S. Li, D.M. Rice and F.E. Karasz, *Macromolecules*, 1994, **27**, 2211-2218.
333 D. Capitani, M. Clericuzio, P. Fiordiponti, F. Lillo and A.L. Serge, *Eur. Polym. J.*, 1993, **29**, 1451-1456.
334 Y. Frere and P. Gramain, *Ion Exch. Adv. Proc. IEX `92*, 1992, 9-16.
335 C.W. Cross and B.M. Fung, *J. Chem. Phys.*, 1993, **99**, 1425-1428.
336 Y.H. Chin, P.T. Inglefield and A.A. Jones, *Macromolecules*, 1993, **26**, 5372-5378.
337 P.A. Mirau, G. Koegler and G.E. Johnson, *Polym. Prepr. (Am. Chem. Soc., Div. Polym Chem.)*, 1993, **33**, 84-85.
338 C. Kohima and Y. Kyogoku, *J. Magn. Reson., Ser. B*, 1993, **102**, 214-217.
339 X. Qing and T.M. Eads, *J. Agric. Food Chem.*, 1993, **41**, 1035-1040.
340 M. Hunger, M.W. Anderson, A. Ojo and H. Pfeifer, *Microporous Mater.*, 1993, **1**, 17-32.
341 P.G. Clarke, K. Gosling, R.K. Harris and E. Smith, *Zeolites*, 1993, **13**, 388-393.
342 F. Deng, Y. Du, K. Wang, D. Ding, J. Wang and H. Li, *Chin. Chem. Lett.*, 1993, **4**, 453-456.
343 E. Brunner, *Microporous Mater.*, 1993, **1**, 431-435.
344 M.A. Makarova, A.F. Ojo, K. Karim, M. Hunger and J. Dwyer, *J. Phys. Chem.*, 1994, **98**, 3619-3623.
345 M. Wiebcke, P. Sieger, J. Felsche, G. Engelhardt, P. Behrens and J. Schefer, *Z. Anorg. Allg. Chem.*, 1993, **619**, 1321-1329.
346 J.L. White, *Diss. Abstr. Int. B*, 1992, **53**, 2898-2899.
347 A.G. Stepanov, A.G. Maryasov, V.N. Romannikov and K.I. Zamaraev, *Magn. Reson. Chem.*, 1994, **32**, 16-23.
348 L.L. Wisniewski, M. Mediati, C.M. Jensen and K.W. Zilm, *J. Am. Chem. Soc.*, 1993, **115**, 7533-7534.
349 S. Wache and R. Taube, *J. Organomet. Chem.*, 1993, **456**, 137-145.
350 S. Koner, A. Ghosh, N.R. Chaudhuri, M. Mukherjee and A.L. Mukherjee, *Polyhedron*, 1993, **12**, 2551-2556.
351 S. Koner, A. Ghosh, N.R. Chaudhuri, A.K. Mukherjee, M. Mukherjee and R. Ikeda, *Polyhedron*, 1993, **12**, 1311-1317.
352 T.E. North, J.B. Thoden, B. Spencer and L.F. Dahl, *Organometallics*, 1993, **12**, 1299-1313.
353 S. Schwiegk, M Werth, J. Leisen, G. Wegner and H.W. Spiess, *Acta. Polym.*, 1993, **44**, 31-38.
354 T. Baba, M. Nomura, Y. Ohno, Y. Hiroyama and Y. Ono, *Stud. Surf. Sci. Catal.*, 1993, **77**, 281-284.
355 K.M. McNamara and K.K. Gleason, *Chem. Mater.*, 1994, **6**, 39-43.
356 B. Boddenberg and R. Niepmann, *Mol. Phys.*, 1993, **79**, 405-412.
357 F. Toda, K. Tanaka, M. Ootani, A. Hayashi, I. Miyahara and K. Hirotsu, *J. Chem. Soc., Chem. Commun.*, 1993, **18**, 1413-1415.
358 R.L. Thurmond, D. Otten, M.F. Brown and K. Beyer, *J. Phys. Chem.*, 1994, **98**, 972-983.
359 E. Zaborowski and S. Vega, *Mol. Phys.*, 1993, **78**, 703-732.
360 A. Bendada, J.B.C. Cobb, B.T. Heaton and J.A. Iggo, *Spec. Publ. - R. Soc. Chem.*, 1992, **114**, 155-164.
361 C. Vaccher, P. Berthelot, M. Debaert, G. Vermeersch, R. Guyon, B. Picard, D.P. Vercauteren, M. Dory and G. Evard, *J. Mol. Struct.*, 1993, **301**, 199-210.
362 M. Duer, *J. Chem. Soc., Faraday Trans.*, 1993, **89**, 823-826.
363 A.W. Hing and J. Schaefer, *Biochemistry*, 1993, **32**, 7593-7604.
364 S. Shibahara, T. Yamamoto, S. Tatsumiya and K. Miki, *Netsu Kokasei Jushi*, 1992, **13**, 156-165.

365 E.W. Hagaman, *J. Magn. Reson., Ser. A*, 1993, **104**, 125-131.
366 F. Toda and K. Tanaka, *J. Chem. Soc., Chem Commun.*, 1993, **14**, 1139-1142.
367 E. Grech, L. Stefaniak, I. Ando, H. Yoshimizu, G.A. Webb and L. Sobczyk, *Bull. Chem. Soc. Jpn.*, 1990, **63**, 2716-2717.
368 N. Kalsbeek, K. Schaumburg and S. Larsen, *J. Mol. Struct.*, 1993, **299**, 155-170.
369 A.E. Aliev, K.D.M. Harris, X. Alcobe and E. Estop, *J. Chem. Soc., Faraday Trans.*, 1993, **89**, 3797-3800.
370 M. Ricco, L. Cristofolini, G. Viola and E. Dalcanale, *J. Phys. Chem. Solids*, 1993, **54**, 1487-1490.
371 R.E. Douthwiate, A.R. Brough and M.L.H. Green, *J. Chem. Soc., Chem. Commun.*, 1994, **3**, 267-268.
372 T. Tajouri, *J. Soc. Chem. Tunis.*, 1990, **2**, 39-60.
373 M. Forsyth and M.E. Smith, *synth. Met.*, 1993, **55**, 714-719.
374 C.A. Fyfe, J. Niu, S. Rettig, M.E. Burlinson, C. Reidsema, D. Wang and M. Poliks, *Bull. Magn. Reson.*, 1993, **15**, 179-182.
375 C.G. Johnson and L.J. Mathias, *Polym. Prepr. (Am. chem. Soc., Div. Polym. Chem.)*, 1992, **33**, 148-149.
376 H.W. Spiess and K. Schmidt-Roht, *Polym Prepr. (Am. Chem. Soc., Div. Polym. Chem.)*, 1992, **33**, 68-69.
377 S. Spiegel, K. Schmidt-Rohr, C. Boeffel and H.W. Spiess, *Polymer*, 1993, **34**, 4566-4569.
378 B. Zibrowius, M.W. Anderson, W. Schmidt, F.F. Schueth, A.E. Aliev and K.D.M. Harris, *Zeolites*, 1993, **13**, 607-610.
379 H. Meyer zu Altenschildesche, H.J. Muhr and R. Nesper, *Microporous Mater.*, 1993, **1**, 257-264.
380 S.M. Bradley, R.F. Howe and J.V. Hanna, *Solid State Nucl. Magn. Reson.*, 1993, **2**, 37-46.
381 G. Engelhardt, P. Sieger and J. Felsche, *Anal. Chim. Acta*, 1993, **283**, 967-985.
382 J.-Ch. Buhl, *Thermochim. Acta*, 1993, **219**, 205-214.
383 R. Jelinek, B.F. Chmelka, A. Stein and G.A. Ozin, *Mater. Res. Soc. Symp. Proc.*, 1992, **272**, 235-241.
384 W. Kolodziejski, A. Corma, M.T. Navarro and J. Perez-Pariente, *Solid State Nucl. Magn. Reson.*, 1993, **2**, 253-259.
385 H. Ernst, D. Freud and I. Wolf, *Chem. Phys. Lett.*, 1993, **212**, 588-596.
386 A. Maijanen, E.G. Derouane and J.B. Nagy, *Appl. Surf. Sci.*, 1994, **75**, 204-212.
387 A. Stein, B. Wehrle and M. Jansen, *Zeolites*, 1993, **13**, 291-298.
388 M. Derewinski, M. Briend, M.J. Peltre, P.P. Man and D. Barthomeuf, *J. Phys. Chem.*, 1993, **97**, 13730-13735.
389 M.S. Rigutto and H. van Bekkum, *J. Mol. Catal.*, 1993, **81**, 77-98.
390 J. Thoret, C. Marchal, C. Doremieux-Morin, P.P. Man, M. Gruia and J. Fraissard, *Zeolites*, 1993, **13**, 269-275.
391 J.J. Liang and B.L. Sherriff, *Geochim. Cosmochim. Acta*, 1993, **57**, 3885-3894.
392 S. Hayashi, T. Ueda, K. Hayamizu and E. Akiba, *Bull. Magn. Reson.*, 1992, **14**, 282-284.
393 J.J. Tunney and C. Detellier, *Chem. Mater.*, 1993, **5**, 747-748.
394 F. Taulelle, T. Loiseau, J. Maquet, J. Livage and G. Ferey, *J. Solid State Chem.*, 1993, **105**, 191-196.
395 A.E. Aliev, K.D.M. Harris and D.C. Apperley, *J. Chem. Soc., Chem. Commun.*, 1993, **3**, 251-253.
396 M. Watanabe, T. Iwamoto, H. Sano and I. Motoyama, *J. Organomet. Chem.*, 1993, **446**, 177-181.
397 G. Wu and R.E. Wasylishen, *J. Phys. Chem.*, 1993, **97**, 7863-7869.
398 S. Eller, P. Schwarz, A.K. Brimah, R.D. Fischer, D.C. Apperley, D.A. Davies and R.K. Harris, *Organometallics*, 1993, **12**, 3232-3240.
399 T. Pietrass and P.K. Burkert, *Inorg. Chim. Acta*, 1993, **207**, 253-254.
400 J.H. Iwamiya and G.E. Maciel, *J. Am. Chem. Soc.*, 1993, **115**, 6835-6842.
401 F.W.B. Einstein, I.D.G. Colin, H.W. Jones, A. Riesen and R.D. Sharma, *Acta Crystallogr., Sect. C: Cryst. Struct. Commun.*, 1993, **C49**, 470-472.
402 B. Wrackmeyer, S. Gerstmann and M. Herberhold, *Magn. Reson. Chem.*, 1993, **31**, 499-504.
403 P.J. Barrie, A. Gyani, M. Motevalli and P. O'Brien, *Inorg. Chem.*, 1993, **32**, 3862-

3867.
404 J.S. Casa, M.S. Garcia-Tasende, A. Sanchez and J. Sordo, *Inorg. Chim. Acta*, 1993, **209**, 137-142.
405 R.J. Batchelo, F.W.B. Einstein, I.D. Gay, C.H.W. Jones and R.D. Sharma, *Inorg. Chem.*, 1993, **32**, 4378-4383.
406 M. Song, Y. Wu, L. Yang and J. Chen, *Chin. J. Chem.*, 1993, **11**, 45-52.
407 H. Berger, R. Nesper, P.S. Pregosin, H. Ruegger and M. Woerle, *Helv. Chim. Acta*, 1993, **76**, 1520-1538.
408 N. Vijayashree, A.G. Samuelson and M. Nethaji, *Curr. Sci.*, 1993, **65**, 57-67.
409 C.G. Christuk and J.A. Ibers, *Inorg. Chem.*, 1993, **32**, 5105-5107.
410 S. Garcia-Fontan, P. Rodriguez-Seoane, J.S. Casas, J. Sordo and M.M. Jones, *Inorg. Chim. Acta*, 1993, **211**, 211-215.
411 F.J. Martinez-Martinez, A. Ariza-Castolo, V. Ramos-Nava, N. Barba-Behrens and R. Contreras, *Magn. Reson. Chem.*, 1993, **31**, 832-835.
412 P.S. Pregosin, H. Ruegger, R. Salzmann, A. Albinati, F. Lianza and R.W. Kunz, *Organometallics*, 1994, **13**, 83-90.
413 R.K. Harris, D.G. Apperley, N.A. Davies, and R.D. Fischer, *Bull. Magn. Reson.*, 1993, **15**, 22-28.
414 S.G. Dutremez, *Diss. Abstr. Int. B*, 1992, **53**, 269.
415 A.J. Kim, *Diss. Abstr. Int. B*, 1993, **53**, 4098.
416 M. Bdiri, A. Jouini, G. Villeneuve and G. Le Flem, *J. Solid State Chem.*, 1994, **108**, 134-140.
417 R.E. Blumer, F. Lianza, P.S. Pregosin, H. Ruegger and A. Togni, *Inorg. Chem.*, 1993, **32**, 2663-2669.
418 R. Challoner, T. Schaller and A. Sebald, *J. Magn. Reson., Ser. A*, 1993, **101**, 106-108.
419 D.C. Apperley, R.K. Harris and D.A. Cleary, *Chem. Mater.*, 1993, **5**, 1772-1774.
420 S.M. Bradley, R.A. Kydd and R.F. Howe, *J. Collid Interface Sci.*, 1993, **159**, 405-412.
421 D. Franke, H. Eckert, R.B. Kaner and R.E. Treece, *Anal. Chim. Acta*, 1993, **203**, 987-995.
422 K.L. Moran, T.E. Gier, W.T.A. Harrison, G.D. Stucky, H. Eckert, K. Eichele and R.E. Wasylishen, *J. Am. Chem. Soc.*, 1993, **115**, 10553-10558.
423 R.E. Treece, G.S. Macala, L. Rao, D. Franke, H. Eckert and R.B. Kaner, *Inorg. Chem.*, 1993, **32**, 2745-2752.
424 D.B. Zax, D. Zamir and S. Vega, *Phys. Rev. B: Condens. Matter*, 1993, **47**, 6304-6311.
425 M. Gavish, S. Vega and D. Zamir, *Phys. Rev. B: Condens. Matter*, 1993, **48**, 2191-2199.
426 S. Nishikiori, C.I. Ratcliffe and J.A. Ripmeester, *Can. J. Chem.*, 1993, **71**, 1810-1815.
427 M. Lobert, W. Mueller-Warmuth, W. Paulus and R. Schoelhorn, *Colloids Surf., A*, 1993, **72**, 345-351.
428 H. Nakano, T. Ohno and S. Yamanaka, *Chem. Lett.*, 1994, **1**, 9-12.
429 O.H. Han, G. Larsen, K.W. Zilm and G.L. Haller, *Stud. Surf. Sci. Catal.*, 1993, **77**, 223-228.
430 O.H. Han, C.Y. Lin, N. Sustache, M. McMillan, J. Donald Carruthers, K.W. Zilm and G.L Haller, *Appl. Catal. A*, 1993, **98**, 195-210.
431 F.M. Bautista, J.M. Campelo, A. Garcia, D. Luna, J.M. Marinas and A.A. Romero, *Appl. Catal., A*, 1993, **96**, 175-199.
432 O.B. Lapina, A.V. Nosov, V.M. Mastikhin, K.A. Dubkov and V.V. Mokrinski, *J. Mol. Catal.*, 1994, **87**, 57-66.
433 S. Miyajima, A.F. McDowell and R.M. Cotts, *Chem. Phys. Lett.*, 1993, **212**, 277-282.
434 M.D. Lumsden, G. Wu, R.E. Wasylishen and R.D. Curtis, *J. Am. Chem. Soc.*, 1993, **115**, 2825-2832.
435 T. Odahara, S. Nishimoto, N. Katsutani, Y. Kyogoku, Y. Morimoto, A. Matsushiro and H. Akutsu, *J. Biochem.*, 1994, **115**, 270-278.
436 T.J.T. Pinheiro and A. Watts, *Biochemistry*, 1994, **33**, 2459-2467.
437 T.J.T. Pinheiro and A. Watts, *Biochemistry*, 1994, **33**, 2451-2458.
438 H. Knicker, R. Freund and H.D. Luedemann, *Naturwissenschaften*, 1993, **80**, 219-221.
439 A. Naito, S. Tuzi and H. Saito, *Bull. Chem. Soc. Jpn.*, 1993, **66**, 3309-3313.
440 C.R. Baijense, *Sci. Tech Aerosp. Rep.*, 1992, **30**.
441 A. Diaz, R. Monque and M. Bussolo, *Bull. Magn. Reson.*, 1993, **15**, 112-116.
442 N.E. Jacob, P.N. Joshi, A.A. Shaikh and V.P. Shiralkar, *Zeolites*, 1993, **13**, 430-434.
443 A.J. Chandwadkar, R.A. Abdulla, S.G. Hedge and J. B-Nagy, *Zeolites*, 1993, **13**, 470-

477.
444 M.L. Occelli, J.M. Dominguez and H. Eckert, *J. Catal.*, 1993, **141**, 510-523.
445 J.T. Kloprogge, J.B.H. Jansen, R.D. Schuiling and J.W. Geus, *Clays. Clay Miner.*, 1992, **40**, 561-566.
446 Y.J. Li and M.S. Whittingham, *Solid State Ionics*, 1993, **63-65**, 391-395.
447 D. Akporiaye and M. Stocker, *Microporous Mater.*, 1993, **1**, 423-430.
448 S.A. Axon and J. Klinowski, *J. Phys. Chem.*, 1994, **98**, 1929-1932.
449 R. Millini, L. Montanari and G. Bellussi, *Microporous Mater.*, 1993, **1**, 9-15.
450 M.R. Steele, P.M. Macdonald and G.A. Ozin, *J. Am. Chem. Soc.*, 1993, **115**, 7285-7292.
451 H.K. Beyer, G. Pal-Borbely and H.G. Karge, *Microporous Mater.*, 1993, **1**, 67-71.
452 A. Lopez, M.H. Tuilier, J.L. Guth, L. Delmotte and J.M. Popa, *J. Solid State Chem.*, 1993, **102**, 480-491.
453 K.L. Moran, A.W. Ott, T.E. Gier, W.T.A. Harrison, H. Eckert and G.D. Stucky, *Mater. Res. Soc. Symp. Proc.*, 1992, **242**, 249-254.
454 M.J. Potrzebowski, *J. Chem. Soc., Perkin Trans 2*, 1993, **1**, 63-66.
455 M.J. Potrzebowski, *Phosphorus, Sulfur Silicon Relat. Elem.*, 1993, **74**, 435-436.
456 P.M. Macdonald, A.D. Hunter, G. Lesley and J. Li, *Solid State Nucl. Magn. Reson*, 1993, **2**, 47-55.
457 G.A. Bowmaker, H.J. Clase, N.W. Alcock, J.M. Kessler, J.H. Nelson and J.S. Frye, *Inorg. Chim. Acta*, 1993, **210**, 107-124.
458 B. Mohr, J. Schmidt, E.E. Brooks, M.J. Heeg, D.M. Ho and E. Deutsch, *Inorg. Chem.*, 1993, **32**, 3236-3240.
459 J.A. Davies and S.G. Dutremez, *Magn. Reson. Chem.*, 1993, **31**, 439-443.
460 N.A. Davies, K.B. Dillon, R.K. Harris, G.F. Hewitson and L. Toms, *Polyhedron*, 1994, **13**, 19-22.
461 C. Nedez, A. Choplin, F. Lefebvre and J.M. Basset, *Inorg. Chem.*, 1994, **33**, 1575-1578.
462 D.L. Reger, M.F. Huff, S.J. Knox, R.J. Adams, D.C. Apperley and R.K. Harris, *Inorg. Chem.*, 1993, **32**, 4472-4473.
463 P.B. Hitchcock, M.F. Lappert, G.A. Lawless and B. Royo, *J. Chem. Soc., Chem. Commun.*, 1993, **6**, 554-555.
464 H. Reuter and A. Sebald, *Z. Naturforsch, B: Chem. Sci.*, 1993, **48**, 195-198.
465 T. Pietrass and P.K. Burkert, *Z. Naturforsch, B: Chem. Sci.*. 1993, **48**, 1555-1557.
466 S. Freitag, W. Kolozdiejski, F. Pauer and D. Stalke, *J. Chem. Soc., Dalton Trans.*, 1993, **23**, 3479-3488.
467 D. Johnels, *J. Organomet. Chem.*, 1993, **445**, 1-5.
468 B. Wrackmeyer, B. Distler, S. Gerstmann and M. Herberhold, *Z. Naturforsch, B: Chem. Sci.*, 1993, **48**, 1307-1314.
469 M. Herberhold, S. Gerstmann, W. Milius and B. Wrackmeyer, *Z. Naturforsch., B: Chem. Sci.*, 1993, **48**, 1041-1050.
470 G.S.H. Lee, K.J. Fisher, A.M. Vassallo, J.V. Hanna and I.c. Dance, *Inorg. Chem.*, 1993, **32**, 66-72.
471 S.C. Chung, J.C.C. Chan, S.C.F. Au-Yeung and X. Xu, *J. Phys. Chem.*, 1993, **97**, 12685-12690.
472 G. Wu and R.E. Wasylishen, *Magn. Reson. Chem.*, 1993, **31**, 537-539.
473 E.J.W. Austin, P.J. Barrie and R.J.H. Clark, *J. Chem. Soc., Chem Commun.*, 1993, **18**, 1404-1405.
474 K. Endo, K. Yamamoto and K. Deguchi, *J. Phys. Chem. Solids*, 1993, **54**, 357-364.
475 L. Huve, D. Saehr, L. Delmotte, J. Baron and R. Le Dred, *C.R. Acad. Sci., Ser. II*, 1992, **315**, 545-549.
476 K.L. Anderson-Altmann and D.M. Grant, *J. Phys. Chem.*, 1993, **97**, 11096-11102.
477 M.E. Smith and H.J. Whitfield, *J. Chem. Soc., Chem. Commun.*, 1994, **6**, 723-724.
478 R. Contant, C. Rocchiccioli-Deltcheff, M. Fournier and R. Thouvenot, *Colloids Surf., A*, 1993, **72**, 301-306.
479 C.P. Grey, A.K. Cheetham and C.M. Dobson, *J. Magn. Reson., Ser. A*, 1993, **101**, 299-306.
480 S.C. Goel, W.E. Buhro, N.L. Adolphi and M.S. Conradi, *J. Organomet. Chem.*, 1993, **449**, 9-18.
481 L.R. Becerra, C.B. Murray, R.G. Griffin and M.G. Bawendi, *J. Chem. Phys.*, 1994, **100**, 3297-3300.

482 M.J. Hudson, W.J. Locke, P.C.H. Mitchell and X. Hu, *Solid State Ionics*, 1993, **61**, 131-137.
483 C. Prigge, W. Mueller-Warmuth, E. Gocke and R. Schoellhorn, *Chem. Mater.*, 1993, **5**, 1493-1498.
484 D. Lathrop and H. Eckert, *J. Non-Cryst. Solids*, 1993, **160**, 111-115.
485 J.P. Fletcher, R.J. Kirkpartic, D. Howell and S.H. Risbud, *J. Chem. Soc., Faraday Trans.*, 1993, **89**, 3297-3299.
486 L. van Wuellen and W. Mueller-Warmuth, *Solid State Nucl. Magn. Reson.*, 1993, **2**, 279-284.
487 X. Xue and J.F. Stebbins, *Phys. Chem. Miner.*, 1993, **20**, 297-307.
488 N.H. Batis, H. Batis, A. Ghorbel, J.C. Vedrine and J.C. Volta, *Colloq. Fr.-Maghrebin Catal., 1st*, 1990 (pub. 1993), **2**, 15-29.
489 Y. Zhang-Lin, M. Forissier, J.C. Vedrine and J.C. Volta, *J. Catal.*, 1994, **145**, 267-275.
490 J.C. Volta, K. Bere, Y.J. Zhang and R. Olier, *ACS Symp. Ser.*, 1993, **523**, 217-230.
491 R.R. Holmes, *Report*, 1991, 16.

8
Multiple Pulse NMR

BY L. Y. LIAN

1 Introduction

The aim of this report is to cover the progress of work in the field of multiple pulse NMR over a period of twelve months from June 1993 to May 1994 and is a continuation of the report from last year.[1]

Two substantial reviews on spectral assignments and structure analysis of proteins contain within them many of the three and four-dimensional triple-resonance experiments reported in the literature up to the end of 1993.[2,3] The review by Oschkinat *et al.*[3] describes some of the basics of multidimensional NMR and give examples of how these experiments are applied to protein spectral assignments and structure determination. LeMaster[2] describes in great detail the isotopic-labelling strategies that are adopted for spectral assignments of proteins. Another review which deals with multidimensional NMR methodology by Geronthanassis[4] concentrates on the detection of water molecules when bound to peptide and proteins in solution. The determination of the structures of protein and nucleic acid in solution using NMR is now well-established, although there is still some uncertainty about the interpretation of the details of these structures; in particular when conformational mobility or disorder affect the spectra. In order to understand the resulting dynamics data thus obtained, Bruschweiler and Case[5] give a detailed description of how homonuclear cross-relaxation can be used to characterise biomolecular structure and dynamics, concentrating on the fundamentals of spin-relaxation; including the estimation of cross-relaxation rates and the solving of the Soloman equations. Artifacts in multipulse experiments are problematic and Bain *et al.*[6] give a comprehensive discussion of artifacts, their origins and ways to remove or reduce these artifacts. With increasing popularity with the use of magnetic field gradients, Norwood's short review on this subject is timely; the article deals with some fundamentals of field gradients including problems that can occur when using them.[7]

Much development work on selective excitation pulses have been reported. Considerable efforts have also been devoted to obtaining coupling constants since the success in determining the structure of a molecule depends on being able to reliably measure these scalar coupling values.

The following abbreviations will be used frequently:
1D, 2D, 3D, 4D: One-, Two-, Three and Four-dimensional; FT: Fourier Transformation; S/N: Signal-to-Noise; RF: Radiofrequency; TOCSY: Total Correlation Spectroscopy; HOHAHA: Homonuclear Hartmann Hahn; NOESY: Nuclear Overhauser Enhancement Spectroscopy; ROESY: Rotating-frame Nuclear Overhauser Spectroscopy; COSY: Correlation Spectroscopy; DEPT: Distortionless Enhancement by Polarization Transfer; INEPT: Insensitive Nucleus Enhancement by Polarization Transfer; HMQC: Heteronuclear Multiple Quantum Coherence; HSQC: Heteronuclear Single Quantum Coherence; ZQ, SQ, DQ, TQ, MQ: zero-, single-, double-, triple-, and multiple-quantum coherence; CPMG: Carl-Purcell Meiboom Gill; BURP: Band-selective Uniform-Response Pure-Phase.

With the current advances in techniques, particularly due to the widespread use in proton detected spectroscopy, it is difficult to write a review of this type under separate headings. For example, pulse field gradient spectroscopy, decoupling sequences and selective excitation methods are used in both homonuclear and heteronuclear 1D, 2D, 3D and 4D experiments. Both homonuclear and heteronuclear strategies are being used to obtain coupling constant values. Attempts are therefore made to keep the categories the same as in previous reports although there will inevitably be some degree of overlap between the different sections.

2 Variation of the Radiofrequency Pulse

2.1 Composite and Decoupling Pulses - Starcuk *et al.*[8] describe the application of adiabatic pulses for heteronuclear broadband decoupling. The cyclicity is superior even in the presence of time delays between successive pulses. Reduced sample heating and low sensitivity to RF inhomogeneity can be acheived using this method.

Composite pulse sequences are extensively used; however, one of their main drawbacks is that many of them are very elaborate and long. In addressing this problem, Yang *et al.*[9] have proposed short two- and three-pulse composite sequences that are as effective as the longer ones. Instead of using phase-alternating composite pulses, arbitrary-phase composite pulses are designed which are also insensitive to the frequency offset.

The TOCSY experiment is now routinely used for obtaining scalar connectivities in an NMR spectrum. This experiment is characterised by a mixing period often consisting of a windowless multiple-pulse sequence. A variety of pulse sequences have been proposed for obtaining the desired mixing Hamiltonian and the actual pulse sequence employed essentially consists of a basic pulse cycle of total duration τ_{cyc} that is repeated n times so as to obtain the desired mixing Hamiltoninan for the duration of $n\tau_{cyc}$. The effective isotropic-mixing Hamiltonian created during the mixing sequence depends on factors including the actual pulse sequence, the chemical-shift range δ_{kl} of the spins involved, the scalar coupling constant J_{kl}, and the RF amplitude ω_1. Unwanted linear and bilinear cross-terms can be produced that can degrade the effectiveness of the mixing and hence a reduction in the transfer frequency and/or amplitude. Depending upon the range of chemical shifts over which the coherence transfer is to be achieved, the maximum available RF power for the mixing sequence, the range of the RF pulse flip and phase errors to be expected for the different sequences will be optimal under different experimental conditions. Bai and Ramachandran[10] have attempted to numerically design homonuclear isotropic-mixing sequences with good performance characteristics, as assessed by the 'global offset quality factor Q_a; a high Q_a ensures that coherence transfer can be efficiently achieved over the chemical-shift range of interest. The commonly-used isotropic mixing sequences IICT-1, IICT-2, IICT-3, IICT-4, MLEV-16, DIPSI-2 and FLOPSY-2 are analysed using this numerical approach.

One of the main problems in TOCSY experiments is that cross-relaxation can contribute to magnetization transfer during the TOCSY mixing sequence, causing noticeable artifacts in the TOCSY spectrum. This incoherent magnetization transfer results from a superposition of longitudinal (s1) and transverse relaxation (s2) rates. For cancellation of the cross-relaxation effects to occur, the ratio of the cross-relaxation rates s1/s2 approaches -1/2 for a rigid pair of isolated protons in the limit of slow isotropic molecular tumbling. By arranging a trajectory such that the spins spend twice the time along the z axis as in the tranverse plane, the two relaxation processes cancel each other. Several papers have been published suggesting modified mixing sequences in order to null the unwanted relaxation pathways to give 'clean' TOCSY spectra. Kadkhodaei *et al.*[11] propose a simple new windowless mixing sequence, TOWNY (TOCSY without NOESY), which achieves suppression of cross-relaxation peaks successfully.

With increase in static magnetic fields, broader carbon decoupling ranges are required in proton-detected inverse experiments. Although an increase in decoupling RF power can meet this demand this is not necessarily desirable because of the increased heating effect on the sample. Fujiwara *et al.*[12] examine several frequency-switched decoupling pulses, MPFn, for their effective bandwidth, and propose extensions to the original sequences.

2.2 Selective Excitation Pulses - The development of phase- and amplitude-modulated RF pulses which achieve nearly ideal frequency profiles has been quite successful and this has led to widespread use of these pulses in one- and multi-dimensional experiments. Apart from a few publications, a comprehensive analysis of the effects of relaxation on the many shaped pulses currently available has not been reported. Hajduk *et al.*[13] presents an analysis of the effects of short T_1 and T_2, using numerical integration of the Bloch equations, taking into account T_1 and T_2 to demonstrate these relaxation effects on the frequency-domain profiles of a variety of pulse patterns, which include rectangular, Gaussian 180° and Gaussian 270°, Hermitian 180°, hyperbolic secant, the BURP series, G3 and G4 Gaussian cascades and Q3 and Q5 quaternion pulses. The effects of cross relaxation during these shape pulses are also analysed by the same authors.[13] The authors conclude that all the pulses are significantly affected by short relaxation times, although most pulses behave in an irregular fashion and it is not possible to conclude which is the best pulse. The choice of pulses is dependent on the specific application and sample characteristics. In general, if only a single resonance is to be excited or inverted, then the shape of the near-resonance profile is unimportant, and only the magnitude of the on-resonance magnetization need to be considered; the choice of pulse type is therefore dependent on the selectivity requirements, with the EBURP or Q5 pulses giving the best 90° and the IBURP or Q3 for the best 180°. If a multiplet is to be excited or inverted, then the shape of the near-resonance profile is of paramount importance. The best 90° pulses are either the EBURP and the UBURP pulses, and the best 180° either the IBURP2 or the REBURP pulses. For the effects of cross-relaxation, a description of the buildup of NOE during the selective pulse, and the impact of this relaxation on the final quantitative results is given. The authors also conclude that for large macromolecules, where spectral crowding requires high selectivity, that is, long pulses, there is significant NOE even at zero mixing time, and the initial buildup curve after the selective pulse may no longer give a true estimation of the cross-relaxation rate.

Nuzillard and Freeman [14] also consider the implications of the T_1 and T_2 relaxation. The process of generating band-selective pulses to accomodate relaxation, starting with the BURP pulses, are discussed.

Selective shaped pulse, such as the Gaussian and sinc pulses have drawbacks in that they generate a large frequency-dependent phase shift, attributable to the mandatory delay between the centre of the pulse and the start of data acquisition. Many "pure-phase" pulses have been reported, all having the property that off-resonance magnetization trajectories are brought to a focus at the time of signal acquisition so that a phase gradient in the spectrum is removed. Many schemes involve iterative numerical optimization. However, it has been difficult to devise direct methods where the desired frequency-domain excitation profile is translated into a RF pulse shape. Kupce and Freeman [15] describe a new approach of pulse design (polychromatic pulse) in which a "comb" of discrete frequency components is arrayed in a suitable pattern in frequency space. In the simplest form, these components would be regularly spaced but have different intensities. The required excitation profile is obtained by adjusting their relative intensities. In another paper, the authors [16] describe a polychromatic selective pulse (Janus pulse) which is used for excitation of antiphase magnetization in a selective coherence-transfer experiment.

Lui *et al.*[17] discussed tailored notch excitation pulses which are designed to produce no excitation over a selected frequency region and uniform 90° excitation over the remainder of the frequency range of interest. Design optimization and the practical implementation of tailored $\pi/2$ pulses are discussed.

Of the many selective excitation and inversion pulses that have been reported, the pure amplitude-modulated pulses are the most often used. Many criteria have to be taken into consideration when designing and using these pulses, such as an interrelationship between selectivity, RF power and pulse lengths. Starcuk *et al.*[18] describe a new approach for designing asymmetric, highly selective spin-inversion pulses and show how the new pulses can be used for the construction of band-selective decoupling pulses.

Konrat *et al.*[19] show, by simulation and by experiment, that amplitude-modulated 'soft' pulses (RF pulses with low amplitude) can be designed in order to induce well-defined transformations of spin states in either a symmetrical or antisymmetrical manner. A detail theoretical description of how to design selective excitation is given.

Bernassau and Nuzillard [20] report on the use of soft pulses when they are applied to heteronuclei in order to record F_1-selective 2D correlation spectra using proton detection. One of the main problems with selective pulse experiments is the calibration of the power or length of soft pulses. Apart from relaxation effects, the most common interaction present in an IS spin system are I,S offset effects, their mutual coupling and the I and S radiofrequency pulses. The remedy for the coupling effects is decoupling. The authors propose several heteronuclear experiments which makes use of selective S pulses with simultaneous spin-locking of I spins for decoupling.

2.3 Gradient Enhanced Spectroscopy - The use of field gradients as an alternative to phase cycling in high resolution experiments has now been widely reported. By incorporating pulsed field gradients at specific times in a pulse sequence, it is possible to selectively rephase coherences, while dephasing those that do not follow the desired coherence pathway. Applications of pulsed gradients are described in the following sections on heteronuclear 2D and 3D spectroscopy (Sections 5.4 and 7.3).

Maas *et al.*[21] report on the use of RF, rather than field, gradients, for coherence selection. Many of the difficulties associated with field gradients, such as eddy currents and B_0 shifts, lock channel interference, long swtiching times, lineshape distortion and the requirement for gradient selectivity, can be circumvented in using this RF gradient.

2.4 Other Variations of RF pulses -There is increasing interest in being able to distinguish and separate different kinds of motional and magnetic interactions among the spins in two-dimensional NMR spectra of large molecules. The basic phenomena of spin-spin coupling, cross-relaxation and chemical exchange can occur simultaneously and it can be difficult to tell which interaction or combination of interactions is responsible for a particular 2D peak.Specialized pulse sequences that null unwanted mechanisms for magnetization transfer can be very useful. Norton *et al.*[22] have described new amplitude-modulated windowless pulse shapes for obtaining 2D spectra showing only peaks due to chemical exchange. The pulse sequences that were proposed remove cross-relaxation responses by continuous rapid averaging of the transverse and longitudinal cross-relaxation rates, these rates having opposite signs for large molecules.

2.5 Solvent Suppression - Solvent suppression techniques can be divided into two broad categories: those that achieve water presaturation and those that aim to suppress water by achieving a net non-excitation of the water while at the same time exciting as much of the rest of the spectrum as possible. Many of these non-excitation pulses suffer from large frequency-dependent phase shifts and rolling baselines due to precession during the longer excitation periods. Smallcombe [23] describe in detail the use of symmetrically-shifted (SS) shaped pulses, which are non-excitation pulses. When used in combination with linear prediction coupled with data shifting, these SS pulses provide efficient water suppression.

Sodano and Delepierre [24] describe a post-acquisition method for clean and distortionless water suppression. The method is simple and is based on convolution difference of the time domain data. Convolution difference of the time-domain data uses the fact that low-frequency components resulting from the residual water signal of the FID can be approximated by an average of neighbouring time-domain data points. The authors improve on the original idea of data convolution by discarding first points in the averaged FID. The same authors have also described the combined use of short hard and spin-lock pulses to afford water suppression in very weak samples.[25]

3 Homonuclear Correlation Spectroscopy

Incomplete relaxation is inevitable during data acquisition because it is impractical to allow for the theoretical relaxation time delay for optimal sensitivity. Artifacts produced as a result of high repetition rates appear at integral multiples of the resonance offset in the F_1 dimension. Turner and Hutton [26] have investigated three different data-acquisition schemes - interleaved hypercomplex data acquisition, transposed acquisition, and conventional data acquisition - to obtain comparison of whether these artifacts can be minimized by using schemes other than the conventional ones. In the transposed scheme, instead of acquiring all the steps in the phase-cycle before incrementing the evolution time, all the increments of the evolution time are acquired before the phase relationship between the pulses is changed. The quality of the partially relaxed 2D data can be improved by rearranging the order in which the data is acquired. In very short phase cycles, where dummy scans are undesirable, the transposed data-acquisition scheme produces the best quality data. In longer phase cycles, conventional data acquisition produces superior data provided that the phase cycle is designed to suppress repetition-rate artifacts.

The sensitivity of COSY is severely reduced when linewidths exceed coupling constants. It is commonly believed that to overcome cancellation of antiphase cross-peaks in a absorption-mode spectrum under these circumstances, magnitude mode COSY should be used; Turner [27] has analysed the signal to noise ratio in the various methods of obtaining COSY spectra and report that a pure-phase dispersion-mode spectrum gives the best sensitivity.

3.1 Determination of Scalar Coupling Constants - Spin-spin coupling constants can sometimes be measured directly from one- or two-dimensional spectra, with the help of software routines. Where the coupling constants are comparable with the linewidth or where there are several overlapping resonances, more sophisticated methods are required. Rio Portillo and Freeman [28] suggest a new method which is more direct and which can be used both for homonuclear and heteronuclear spin-spin coupling; what is required is a well-digitized multiplet as the raw data set. This data can be extracted from a 1D spectrum or from a suitable section through a 2D cross-peak. The splitting is then decoupled by standard coherent double-irradiation methods, giving a decoupled spectrum. Reconvolution can now be used to artificially introduce a trial splitting to give a multiplet with as good a match as possible to the original spectrum, this match being determined by a least squares procedure.

As mentioned above, obtaining scalar coupling constants directly from a spectrum is not straightforward or always feasible. In many cases, the multiplet fine structure is affected by the lineshapes, by couplings to other spins, and by partial cancellation of signals of opposite signs. Huber *et al.*[29] adapted an earlier reported method (which makes use of the convolution of the COSY and TOCSY peaks) to compare the multiplets from the soft-COSY and PICSY (pure in-phase correlation spectroscopy) in order to get the required scalar coupling constant.

Chandrakumar and Velan [30] describe the frequency-cycled longitudinal multispin order method to determine the relative signs of couplings, for spectral editing and for measurement of relaxation rates of the longitudinal multispin orders. The procedure uses frequency-selective inversion, followed by a hard read-pulse of fixed phase and flip angle ϕ.

Norwood and Jones [31] examine the effects of relaxation in the E.COSY experiments when these experiments are used for the measurement of scalar coupling constants. The authors report, based on simulation studies, that when a scalar coupling is of the same order as the transverse relaxation rate the value of the coupling constant obtained from the E.COSY spectra can be erroneous.

3.2 Zero and Multiple Quantum Experiments - It has been known for some time that two types of MQ artifacts can be present in the 2D COSY spectra if repetition time between the COSY pulse sequences is not long enough to allow the nuclei to relax to thermal equilibrium. For repetition time less than T_1, longitudinal MQ artifacts can be seen in the 2D COSY spectra. For repetition times less than the spin-spin relaxation time T_2, tranverse MQ artifacts can also occur. Bowden and Prandolini [32] describe the theoretical aspects of how a previously reported pulse sequence can help to eliminate the longitudinal MQ artifacts in the 2D COSY spectra.

Zero-quantum coherence has the property of being independent of magnetic field inhomogeneity and hence was expected to be widely used if not for the fact that it has two major drawbacks: it is not possible to obtain absorptive phase-sensitive cross-peaks and the resultant spectra may contain intense autocorrelation peaks arising from both coupled and uncoupled spins. Norwood[33] describe a method for acquiring absorptive phase-sensitive zero-quantum correlation, discuss the source of autocorrelation peaks, and propose several methods for their elimination.

Bowden *et al.*[34] describe the zeeman-incremented multiple-quantum experiment in which instead of the evolution t_1 period being incremented as in the conventional experiment, it is kept fixed. In its place, a uniform magnetic field $\Delta\omega_s = \gamma\Delta B_z$ is increased. The theoretical basis of the experiment are given in detail. The advantage of this approach is that both the separation of the MQ signals and their linewidths can be experimentally controlled and the experiment is suitable for samples with short T_2 values.

3.3 Variations of Correlation Spectroscopy - It is generally accepted that the sensitivity of two-dimensional spectroscopy is comparable to that of the 1D analogue performed in the same total accumulation time. However, when the inherent sensitivity is sufficient, the 1D analogues are adequate, provided the same amount of information is gathered as in the 2D experiments. For example, often suitable resonance frequencies which can be selectively excited can be obtained from the normal 1D spectrum. This N-separate 1D correlations could be carried out with soft-pulse excitation in a time much shorter than that of the corresponding 2D experiment. Kupce and Freeman [35] examine the different practical approaches to multisite excitation. The authors considered three methods of generating simultaneous pulses, namely, amplitude modulation, interleaved, phase-ramped DANTE sequence, and a new concept which is a combination of three soft pulses, each being broken down into eight segments. In the latter case the different excitation frequencies are achieved by phase incrementation. The same authors [36] suggest using an experimental NMR FID to define the soft-pulse excitation, generating a large array of simultaneous soft pulses whose Fourier spectrum exactly matches the NMR spectrum in frequency space. This new method for multiple irradiation uses a 'template' which is designed to excite the desirable features of a high-resolution spectrum while suppressing all others. In this approach, inverse Fourier transformation is applied to the experimental spectrum; this 'regenerated' FID is used to calculate the shape of the pulse envelope. In another example, the authors describe a correlation experiment, based on Hadamard spectroscopy.[37] The experiment involves using with N simultaneous soft pulses, and if necessary, 2 x N simultaneous spin-lock fields, thereby gathering all the useful information in each scan. BURP-2 pulses are used to generate the selective pulses. The authors summarise these multi-site experiments and show how the BURP type pulse trains can be used.[38] The resultant spectrum is a 1D spectrum containing only signals from the two coupled sites, with the active splitting in an antiphase mode.

The DQF-COSY experiment has proven to be reliable and useful for determining scalar connectivities. However, the method can be somewhat insensitive, particularly when working with weak solutions. Girvin [39] describe a constant-evolution-time version of the experiment which may increase the sensitivity by as much as 20-fold over the DQF-COSY experiment. Because in the constant time period the cross-peaks are decoupled in t_1, there is no loss of intensity since the cross-peaks are in-phase in t_1.

Konrat *et al.* [40] propose novel methods for generating various type of coherence transfer, such as longitudinal two-spin order, antiphase magnetization, and multistage transfer between selected spins. The authors use semi-selective RF pulses, based on 270° self-refocusing Gaussian pulses, modulated by sine and cosine functions.

In the conventional multidimensional experiments, spectral resolution is essentially determined by the length of the sampling period. Discrete Fourier transform requires that the data is 'sampled' at uniform intervals. With the availability of other forms of data processing, such as maximum entropy, this restriction needs no longer be true, and comparable resolution may be achieved by collecting fewer points spanning the same data acquisition time. Nonlinear sampling in the t_1 dimension is applied to a P.E.COSY spectrum by Schmieder *et al.* [41]. The sine dependence of the time domain signals for the cross peaks is matched by a nonlinear sampling scheme that samples most densely around the maximum of the sine function.

3.4 TOCSY-type experiments - Briand and Ernst [42] make a quantitative comparison of the achievable sensitivities, as measured by the cross-peak intensity, in the TOCSY and COSY experiments. The sensitivity is govern by a combination of three factors: coupling constant, relaxation and spin-diffusion effects. Bearing in mind the differences between the different spectral characteristics of the two experiments (single vs multiple-step transfer, transfer frequency as determined by the J coupling constants and relaxation, and cross-peak structures), the authors have performed theoretical analysis for a two-spin model with dipolar and external random-field relaxation and examined the balance between the different factors that affect sensitivity. Extension to multiple-spin systems are also discussed. The data presented showed that for cross-peaks multiplets with well-resolved splitting, the TOCSY and COSY intensities are very similar. However, when the active couplings disappear into the linewidths due to relaxation or passive couplings, the TOCSY intensities become more favorable than COSY, although only a sensitivity gain of 3 rather than the theoretical sentivity gain of between 5 and 9, is normally achieved.

Many multiple pulse NMR experiments include a period during which magnetization is rotated temporarily to the z axis. Such a period may be a fundamental part of the mixing process or is used to generate a spectrum in which all the peaks have the same phase, a process referred to as z-filtration. NOESY and z-filtered TOCSY are examples of these two types of experiments. Either phase cycling or a field gradient pulse is used to select z magnetization during the z-filter delay but these selection procedures are not capable of separating zero-quantum coherence from the desired z magnetization. This ZQ coherence is transferred into observable antiphase magnetization, thereby making the z filter imperfect. The purge pulse, which is a long spin-lock pulse, is also used for generating spectra with all the peaks in the same phase. During such a pulse all the magnetization perpendicular to the spin-lock axis is dephased, although there will be some that are not dephased completely. Davis *et al.*[43] show that the deficiencies in the z filters and the purging pulses have the same origin. The authors also demonstrated that a period of spin locking in an inhomogeneous B_0 or B_1 field can be used to form near perfect purging pulses and z filters. Relaxation and cross-relaxation during the spin-locking pulses are the main drawbacks of this approach.

Glaser [44] investigates the efficiency, in terms of transfer amplitude and transfer time, of selective and non-selective cross-polarization in homo- and heteronuclear TOCSY and TACSY experiments. Even in the absence of relaxation, polarization transfer depends on a large number of parameters: pulse sequence parameters such as multiple pulse scheme, irradiation frequency, mean RF power, etc.) as well as spin-system parameters (chemical shifts, J coupling constants). For most multiple-pulse sequences, these parameters may be condensed into effective coupling tensors; a theoretical description of the TOCSY and TACSY experiments is given in terms of these coupling tensors. The analysis is quantified and presented in the form of maps as a function of the relative coupling constants in systems consisting of three coupled spins. Characteristic differences in the polarisation efficiency of TOCSY and TACSY variants are shown.

In high resolution experiments, coherence can be transferred in a stepwise manner along an extended chain of coupled protons. The mechanism of coherence transfer along a typical chain A $\rightarrow$ B $\rightarrow$ C $\rightarrow$ D involves Hartmann Hahn mixing by application of matched RF at sites A and B for a time $1/(J_{AB})$, followed by selective irradiation at sites B and C for a time $1/(J_{BC})$ and then by irradiation at sites C and D for a time $1/(J_{CD})$. These multistage transfer experiments would be more conveniently implemented if a single mixing time which cover a range of experimental coupling constants could be used. In systems of several coupled spins further complications can be caused by passive couplings, thereby giving rise to mismatch of the Hartmann-Hahn condition. In order to address these two questions of variation in J values and mismatch of Hartmann-Hahn condition, Blechta and Freeman [45] use J-compensated pulse sequence.

4 Dipolar Coupling, Chemical Exchange and Relaxation Time Experiments

4.1 Dipolar Coupling and Chemical Exchange -Many routine experiments involve the continuous application of radio frequency (rf) fields to a homonuclear ensemble of spins; examples of these are the TOCSY or HOHAHA, CAMELSPIN or ROESY, and $T_{1\rho}$ experiments. In such experiments, several processes occur during the application of RF: relaxation between eigenstates, including coherent magnetization transfer, ZQ,SQ and MQ evolution, etc. It has been the aim, when designing and performing these experiments, to find conditions or experimental procedures in which one of the processes is isolated for observation while the others are suppressed or removed. Ravikimar and Bothner-By [46] propose the acquisition of a 2D data-set where, in the second dimension, a Fourier transformation is performed with respect to the time of RF excitation. The following pulse sequence is used: [RD-spin lock(t_1)-FID(t_2)]. In the resulting 2D spectrum, peaks are present at frequencies which represent the frequencies of precession about the effective field in the presence of the RF.

Zwahlen *et al.*[47] also address the question of interference from unwanted pathways when making quantitative measurements using the transient NOE experiment. Spin-diffusion effects are problematic in this case and the authors make use of selective pulses to restrict the NOE transfer to the proton pairs in question, inhibiting spin-diffusion via other pathways. The Q^3 Gaussian cascades are used as inversion pulses.

The use of rate matrix analysis for refinement of solution structures of biopolymers on the basis of NMR-derived structural constraints has greatly enhanced the self-consistency of the general approach. However, present implementations of this approach appears not to consider important issues which can prevent the quantitative interpretation of the NOESY spectra of proteins; these factors include the effects of chemical and physical exchange, internal motion, the presence of paramagnetic centers, and the presence of heteronuclei, the direct and indirect consequences of solvent suppression by presatuarion, incomplete recovery of equilibrium during the course of the experiment, and randomization of relaxation pathways by deuterium introduced as a frequency lock. Dellwo *et al.*[48] show how the rate matrix should be modified to accomodate many of these effects; additionally, simulations are presented to illustrate the relative importance of each of these factors. From the simulations, the effects of nonequilibrium populations generated by overpulsing and solvent presaturation are generally large enough to cause problems.

Eads [49] describes the use of RF (B_1) inhomogeneity for coherence selection in 2D exchange or Overhauser effect spectroscopy. The method allows phase-sensitive NOESY spectra to be acquired using a single scan per t_1 increment. The main advantage of this approach is that, given that sufficient material is available, a full phase-sensitive correlation map can be obtained with a single scan per t_1 increment. With long, inhomogeneous B_1 pulses, it is possible to directly select Cartesian product-operator components of the density matrix via inhomogeneous rotations in the rotating frame.

The NOE is modulated by fluctuations in the orientation of the internuclear vector which includes contributions from global motions due to molecular tumbling and localized internal motions due to dynamic fluctuations in molecular structure. Because of the quantitative usage of the NOE data, it is important to be able to evaluate the internal and anisotropic motions. Edmondson [50] describes a molecular dynamics simulation of the effects of methyl rotation and other protein motions on the NOE. From these simulations, it can be concluded that the relaxation rates of the methyl group (and hence NOEs at short mixing times) are considerably reduced relative to static methyl groups by proton jumps between equivalent proton positions.

Scalar and dipolar connectivities within protons experiencing large hyperfine shifts and fast relaxation because they are close to the paramagnetic centre are difficult to detect. One strategy, termed the NOE-NOESY experiment, described by Bertini *et al.*[51], which makes use of the known three-dimensional structure, leads to the recognition of connectivities between the hyperfine shift signals and the diamagnetic ones, the latter being assignable specifically. The method used is based on the NOESY sequence, preceded by selective saturation of the hyperfine-shifted signal for a time which can vary from a few milliseconds to hundreds of milliseconds. This time will allow steady-state NOEs to develop from all the diamagnetic signals that are dipole-dipole coupled to the saturated signal.The experiments are alternated with off-resonance experiments where full NOESY intensities are recovered; taking the algebraic difference between the on- and off-resonance experiment, a NOESY difference spectrum is obtained in which only the dipolar connectivities involving the signals that experience NOE from the irradiated signal are detected.

Wang *et al.*[52] considered the effects of chemical exchange on heteronuclear ^{31}P-$\{^1H\}$ NOE, as applied during the analysis of the steady-state transferred NOE experiments. Numerical simulations which characterise the behaviour of the steady-state transferred NOE formalism, assuming broadband proton saturation, are performed for different distribution of the free and bound states, ligand exchange rates, bound ligand correlation times and varying order parameters under the extreme narrowing condition for bound ligand internal motion. The calculations show a pronounced effect on the transferred NOE when chemical exchange is present.

Grzesiek and Bax [53] describe a rapid and sensitive 2D experiment for measuring amide proton exchange rates and the NOE interaction between amide protons and water. The approach is applicable to doubly $^{13}C/^{15}N$-enriched proteins and can measure magnetization exchange rates between 0.02 to > 20s^{-1} range. The first part of the pulse scheme selectively inverts the H_2O resonance. Signals from the ^{13}C-attached protons are eliminated by a 90° ^{13}C purge pulse. This is followed by a NOESY- or ROESY-type transfer to amide protons, which are dispersed by the amide ^{15}N frequencies in an HSQC-type experiment.

Another method described for the measurement of fast proton exchange rates, which makes use of isotopically labelled proteins, is proposed by Gemmecker *et al.* [54] Here, the fast exchange rate is measured by an evaluation of the buildup rates of signals that are caused by the exchange process itself. The crucial feature in the proposed method is the use of a pulse sequence that is designed to selectively detect only the resonances whose amide protons are in exchange with water. This is achieved by using a double ^{15}N and ^{13}C filter to suppress all protons which are attached to nitrogen or carbon; the water protons are at the same time flipped into the z-axis. Following a variable mixing time, exchange of the z magnetization from water to the labile amide protons occurs and these protons are subsequently detected using a HMQC or HSQC experiment.

4.2 Relaxation Time Measurements - With the wide spread use of heteronuclear relaxation parameters to elucidate dynamics properties in biomolecues, it is important to be able to perform experiments that will yield reliable results with minimal systematic errors and maximum sensitivity. Skelton *et al.* [55] have described in detail many aspects of the acquisition and processing of the NMR data. Pulse sequences are described that will reduce the time required to for data acquisition, without compromising precision or accuracy of the final results. Additionally, a number of factors that influence the accuracy and precision of measured relaxation parameters, including uncertainties in peak intensities, overlap between resonances, and amide-exchange with solvent, are examined theoretically and experimentally.Sensitivity-enhanced pulse schemes were described to achieve shorter data acquisition times. Various post-acquisition methods, such as baseplane correction, and linear prediction, were used to improve the quality of the final data from which quantitative values are derived.

Of the various relaxation measurements in protein, the heteronuclear $^{15}N\{H\}$ NOE is perhaps one that can be difficult because of low sensitivity (due to the small γ value of nitrogen), exacerbated by the fact that the water resonance has to be suppressed without saturating the amide protons. Up to now, a compromise must be found by using very low power presaturation of water. Ross *et al.*[56] describe how the problem can be alleviated by using pulse field gradients for water supression. No water presaturation is required.

5 Heteronuclear Experiments

5.1 Direct Detection Correlation Spectroscopy - The INADEQUATE experiment has played an important role as a method for establishing connectivities of ^{13}C nuclei and for determining ^{13}C-^{13}C J-coupling constants. Nakai and McDowell [57] describe in detail the refocussed INADEQUATE method, which gives in-phase J doublets from ^{13}C-^{13}C spin pairs in natural abundance samples. Methods for obtaining a pure absorptive mode INADEQUATE spectrum, using the TPPI method, is also discussed.

5.2 Inverse Proton-Detected Correlation Spectroscopy -
The 2D inverse experiment can be converted to analogous 1D experiments with the aid of selective pulses. Because of the high dispersion of ^{13}C frequencies, selective ^{13}C excitation is convenient. Willker *et al.* [58] present a 2D version of the HSQC experiment with frequency-selective carbon excitation, which records those protons directly bound to a selected carbon. The HSQC sequence is more desirable than the HMQC sequence since the selective ^{13}C pulse can be introduced without having to compromise on the proton-decoupling or the length of the presaturation time.

One-dimensional versions of many 2D inverse detection experiments are popular as a time-saving device; however, difficulties arise when the magnitude of the heteronuclear coupling constant J_{XH} becomes comparable to that of the homonuclear proton coupling constants J_{HH}, mainly due to phase modulation which occurs during the large evolution delays which are dictated by the weak heteronuclear coupling constants. Several methods have been proposed to overcome this phase modulation and this is summarised by Roumestand *et al.*[59]. The authors also propose using an alternative solution which incorporates a selective DANTE pulse instead of a hard π refocussing pulse on the proton channel in a 1D HMQC experiment.

Krishnamurthy [60] address the question of water suppression in experiments where the signals of interest are not displaced symmetrically around the water resonance, such as in the HMQC experiments. The author describe experiments using shifted laminar pulses which is a train of τ pulses, each of which is a frequency-shifted laminar pulse. The frequency shift is afforded by a phase change on successive pulses. This method of water suppression means that only data in the region of interest needs to be collected, thereby improving digital resolution and saving expensive spectrometer time.

Chary *et al.*[61,62]describe a method, based on the heteronuclear multiple quantum experiment, to observe long-range ^{31}P-^{1}H correlation. In this experiment, both the direct and long-range connectivities are detected.

Pratum and Moore [63] report on how ^{13}C multiple quantum coherence can be observed using proton detection. The final spectrum contains the proton frequency in one dimension, and the multiple quantum frequency in the ^{13}C dimension. This experiment has a better sensitivity when compared with the carbon-detected INADEQUATE experiment.

When HMQC experiments are performed on non-isotopically enriched samples, one of the main problems is the removal of the proton signals bound to ^{12}C or ^{14}N. Solutions have been proposed to remove these unwanted signals before acquisition by suppression of the origins of these signals, using sequences such as bilinear rotation pulses and TANGO-sequences. Nuzillard *et al.*[64] have reported the simultaneous use of the BIRD-relaxation and spin-locking methods for enhanced suppression of unwanted signals.

5.3 Scalar Coupling Constants Using Heteronuclear Proton-Detection Experiments - Many of the method available for measurement of J coupling rely on the E.COSY principle, in which the J coupling is measured from the frequency difference between two resonances in a 2D or 3D spectrum. However, a different approach can be taken which relies on quantifying coherence transfer or measuring the magnetization loss due to dephasing caused by unresolved J coupling.

5.3.1 Coupling Constant from Resolved Peak Multiplet - The measurement of heteronuclear long-range coupling constants is of great importance for the determination of the three-dimensional structure of biomolecules. One way of determining heteronuclear coupling constants is by using a two-dimensional heteronucleus-coupled ω_1 hetero-half-filtered proton-proton correlation (HETLOC) experiment. This method is used to determine heteronuclear long-range couplings ($^{LR}J_{HX}$) at natural abundance. The $^{LR}J_{HX}$ values are measured from signal displacement in two F_1 slices through each signal of the cross-peak doublet split by $^1J_{HX}$ in the F_1 dimension. The technique is advantageous in that, firstly, coupling constants smaller than the linewidth can still be determined accurately, and secondly, the pulse sequence contains no delays to evolve $^{LR}J_{HX}$, that is the spectra are free from J_{HH} modulation and can be phased to pure absorption. Fukushi and Kawabata [65] describes a selective 1D variant of the HETLOC experiment which gives two 1D subspectra corresponding to the two slices used in the 2D version of the experiment. They propose the paired satellite selection technique that separately excites each signal of the satellite peak. In this method, the combination of an appropriate delay and ^1H pulse offset, in addition to a selective X pulse, makes it possible to get the requisite pair of 1D spectra. The selective X pulse excites one isotopomer that contains one X nucleus at the single position. Nuzillard and Bernassau [66] also propose a 1D version of the previously reported 2D X-filtered TOCSY experiment.

Adams and Lerner [67] give a brief summary of the many reported methods for obtaining long-range ^1H-^{13}C coupling constants and describes yet another method for using selective excitation of carbon-13. Their method allows optimization of the measurement to a given carbon. One of the main advantage of the reported experiment is that it does not require selective excitation of the proton, and hence can be used for a molecule where the proton spectrum is crowded but the carbon signals are well-separated. A low-power rectangular pulse is used to obtain carbon-13 selectivity.

Kover *et al.*[68] use the z-filtered heteronuclear coupled HSQC-TOSCY experiment, which gives in-phase pure absorption cross-peaks, to measure long-range heteronuclear coupling constants. This experiment is essentially the HSQC-TOSCY experiment extended by a z-filter just prior to detection. The extraction of the coupling constants relies on the comparison of linewidths of multiplets in the ^{13}C -coupled SQC-TOCSY and in the z-filtered homonuclear TOCSY spectra. Subsequently, the same authors [69] describe the 1D variants of the z-filtered homonuclear TOCSY and the z-filtered HSQC/HMQC-TOCSY experiments for measurement of homo- and heteronuclear coupling constants. The selective filtration is achieved using chemical-shift filters.The frequency selection is achieved by setting the transmitter or decoupler carrier frequency to the frequency of the selected carbon or proton resonance, respectively, and by systematic incrementation of the chemical-shift evolution period.

Another method that can be used for obtaining long-range ^1H-^{13}C coupling constant is the 2D HMQC experiment. Zhu and Bax [70] describe a method where the signal intensities of an HMQC spectrum and a 'reference' spectrum are compared. The reference spectrum is acquired with the same pulse sequence as the HMQC spectrum with the exception that no carbon pulses are used; instead these pulses are replaced with time delays equivalent to the length of the carbon pulses.

Kawabata and Fukushi [71] suggest using symmetric phase-alternating composite pulses in DEPT C-C relay experiments in order to alleviate phase distortion (due to resonance-offset effects) and sensitivity loss. By improving this experiment, more reliable ^{13}C-^{13}C coupling constants can be obtained.

The E.COSY-type experiments are considered one of the most reliable for measuring J coupling constants. These experiments work on the basis that coherence is transferred from either spins 1 or 2 to a third spin 3, leaving the spin which is passive in the coherence transfer (1 or 2) unperturbed. The most important condition to fulfill is that the passive spins are unperturbed; this can be achieved when the passive spin is a heteronucleus to both active spins, such as the measurement of the homonuclear coupling constant $J_{NH\alpha H}$, where the αH is attached to a carbon centre and NH to nitrogen. In situations where the desired coupling are between two protons attached to the same heteronuclei, selective nonperturbation of the passive spin may be a problem. Olesen *et al.*[72] describe pulse sequences that are designed to limit the perturbation of the passive spin. They have introduced small-angle heteronuclear bilinear rotation pulses for coherence transfer at the beginning of the pulse sequence, thereby obtaining a significant gain in sensitivity.

The E.COSY class of experiments are aimed at providing J_{HH} coupling constants and recent HCCH E.COSY experiments, using ^{13}C-labelled samples are based on the original proton homonuclear ones. Olsen *et al.*[73] propose a simplification to the previously published sequence in order to reduce the number of pulses required, thereby reducing sensitivity losses as a result of pulse imperfections.

In order to overcome resonance overlaps in the 2D H,C TOSCY spectrum (obtained by using an X half-filter), Wollborn *et al.*[74] use multiplicity editing, in combination with a 2D H,C TOSCY sequence, for determining long-range heteronuclear coupling constants. These sequences record TOCSY spectra of only one multiplicity, CH or CH_2 thereby considerably simplifying the spectrum.

Two classes of long-range couplings, namely, J_{HH} and J_{XH}, are frequently measured. In order to generate E.COSY-type multiplet patterns measurement of J_{HH} relies on X $\Leftrightarrow$H coherence transfer typically over two bonds, while measurement of J_{XH} relies on H $\Leftrightarrow$ H coherence transfer over four bonds. The H $\Leftrightarrow$ H transfer can be effected via NOESY or TOCSY and requires only one X-spin filter as part of the preparation period. The most sensitive way to effect X $\Leftrightarrow$ H transfer in fully labelled molecules employ one-bond couplings exclusively and requires two directly bound nonproton spins. At low levels of labelling or at natural abundance, this approach is difficult. Sorensen *et al.*[75] suggest that is may be more advantageous to use heteronuclear long-range couplings to obtain the X $\Leftrightarrow$ H transfer in one step as only one X spin is required. The pulse sequence is similar to the HMBC experiment and the constant delay τ is tuned to the heteronuclear J_{XH} ling-range coupling. A selective version of this experiment is described by the same authors.[76]

Schwalbe *et al.*[77] describe a method for measuring J(C,P) and J(H,P) coupling constants, based on fitting a target multiplet containing the heteronuclear coupling to a reference multiplet that does not contain the heteronuclear coupling. The experimental multiplets are obtained from ^{31}P-coupled and ^{31}P-decoupled ^{1}H,^{13}C HSQC spectra.

Grezesiek *et al.*[78] show how the three bond ^{13}C-^{13}C J coupling value can be obtained from a 2D ^{13}C-{^{13}CO} spin-echo difference constant-time heteronuclear single-quantum (CT-HSQC) experiment from which the J_{CC} coupling between the methyl groups and the ^{13}CO carbons can be obtained. The experiment works on the basis that two datasets are obtained : one with the ^{13}C-^{13}CO J coupling eliminated and the other with this coupling still active, thereby attenuating the signal by $\cos(\pi J_{CCO}T)$. Schwalbe *et al.* [79] describe another method for obtaining the three-bond $J_{C'C\gamma}$ coupling constant in valine residues, hence affording strereospecific assignments of valine methyl groups. The HCCC-COSY experiment will give E.COSY type multiplets.

5.3.2 Coupling Constants from Peak Intensities - Constantine *et al.*[80] suggest estimating the vicinal ^{1}H-^{1}H coupling constants from intensity ratios of the cross-peak to the diagonal-peak derived from 3D ^{13}C-edited TOCSY-HMQC spectra.

Vuister *et al.*[81] describe a {^{15}N} spin-echo difference CT-HSQC experiment that will give quantitative information on the J coupling between methyl carbons and backbone amide ^{15}N nuclei. Together with the measurement of $^{3}J_{CC}$ this new method allows the determination of χ_1 angle in valine, isoleucine and threonine.

Vuister and Bax [81] propose a new method for measuring homonuclear H^N-H^α scalar coupling in ^{15}N-enriched proteins. The methods relies om measurement of the diagonal-peak to cross-peak intensity ratio in a 3D ^{15}N-separated quantitative J-correlation HNHA spectrum which correlates $^{1}H^N$,^{15}N,and $^{1}H^\alpha$ resonances.

5.4 Gradient Enhanced Inverse Detection Heteronuclear Experiments

5.4.1 B_0 Gradients

The use of pulsed B_0 gradients is again well-reported, their applications being mainly in simplifying complicated phase-cycling schemes, for reducing the experimental time, provided the intrinsic sensitivity of the particular multi-dimensional experiment permits a single transient per incremental delay.

For optimization of multi-dimensional phase-sensitive heteronuclear NMR spectra Van Zijl [82] describe a new method which combines frequency-independent water suppression using diffusion filters with minimal phase cycling and the use of dephasing -rephasing additional gradient pairs. The diffusion filters are applied in constant-time periods, allowing phase-sensitive data accumulation with optimum sensitivity. Example is given for the implementation of these gradient pulses in an HMQC experiment. Gradients in x,y, and z axis are necessary to achieve enough gradient strength with the constant-time period.

The exchange between the NH protons of proteins and solvent water can lead to a loss of sensitivity in HSQC spectra when using pulsed field gradients either for coherence-pathway selection or for purging unwanted magnetization. Stonehouse *et al.*[83] describe how sensitivity loss can be recovered by ensuring that the magnetization from the water protons is returned to the +z axis by simple modifications to the pulse sequences, prior to data acquisition.

Kriwacki *et al.*[84] show how pulsed field gradients can be used, together with isotope filtering methods and selective excitation, to characterise the rates of exchange of the labile amide protons in a protein. A range of amide proton-water exchange rate, ranging 100us to 10ms, can be obtained by inserting pulse field gradients into the NOESY-HMQC, ROESY-HMQC and TOCSY-HMQC pulse sequences.

Muhandiram *et al.*[85] describe a gradient ^{13}C NOESY-HSQC experiment for recording NOESY spectra of ^{13}C-labelled proteins in water. The pulsed field gradients are incorporated into the NOESY-HSQC sequence, with gradients being used for water suppression.

Lee *et al.*[86] use pulse field gradients for coherence selection (as well as water suppression) in ω_2 half-filtered NOESY experiments.

Medvedeva *et al.*[87] report the application of gradient-enhanced experiments in proton-detected ^{13}C-^{1}H experiments using natural abundance proteins, and demonstrate that it is possible to obtain data of good S/N.

5.4.2 B_1 Gradients

Despite their wide-spread use currently, B_0 gradients are rather demanding on the hardware, with their attendant inconveniences. There is hence a growing interest in using B_1 gradients for coherence pathway selection . B_1 gradient methods are potentially frequency selective, have short switching times, do not interfere with the lock channel, and do not introduce line broadening. Cory *et al.*[88] have discuss several B_1-gradient DQF-COSY experiments with particular focus on those that may be implemented with a radial-gradient field. Another example of the use of B_1 gradients is given by Mutzenhardt *et al.*[89] who use B_1 gradients to generate a correlation spectrum with a single scan per t_1 value, to show that satisfactory results regarding coherence transfer to yield magnitude and absorption mode spectra can be obtained, and for solvent suppression.

B_1 gradients have also been reported for use in solvent suppression by Canet *et al.*[90]; the authors propose using a simple DANTE train of B_1-gradient pulses.

Over the past few years, detection of hydration water has attracted alot of interest.Because of the rapid exchange between hydration water and bulk water, the ^{1}H chemical shifts of the hydration water protons are averaged with the dominant bulk water signal leading the the lineshape of the NOE and chemical exchange cross-peaks being that of the bulk water signal. Otting [91] address the question of radiation damping which can lead ot a very broad water lineshape in NOESY and ROESY experiments, and present a generally applicable method for the suppression of radiation damping using B_1 gradients. Radiation damping originates from coupling of the precessing water magnetization with the RF coil of the probehead, which rotates any excited water magnetization back to its equilibrium position along the z-axis of the main magnetic field at a rate proportional to the transverse component of the water magnetization.Since in typical NOESY and ROESY experiments the evolution times are longer than 50 ms, radiation damping contributes noticeably to the apparent relaxation rate of the water magnetization during the evolution period, and therefore leads to a broadened water signal. The most substantial line-narrowing effect is achieved when radiation damping is suppressed during the evolution period t_1. Otting show how the can be done by using a pair of short spin-lock pulses, which serve as B_1 field gradients to defocus the magnetization before the evolution time and to refocus it afterward.

6 Methods for the Improvement of Data Quality

Instead of using selective excitation to acquire data on the region of interest, and hence to improve digital resolution, Rosen [92] propose a selective detection approach, using digital filtering. Digital filtering simply limits the bandwidth of the signals in a FID but does not reduce the number of data points. To reduce the number of data points in a FID, the digital-filtering process must be followed by downsampling to limit the spectral width of the final time-domain data set.

For successful analysis of multidimensional data, in particular where quantitative results are desirable, flat baselines are necessary. Many methods have been proposed, using either improved data-acquisition methods or post-acquisition data improvement methods. Froystein [93] analyse briefly the factors that affect spectral baselines and present solutions to the problem. The author investigates the combined use of the spin-echo detection and synchronised sampling. The synchronised sampling is delayed past the initial part of the filter response ringing pattern; the spin-echo takes care of the rephasing prior to detection.

Waltho and Cavanagh [94] also deal with the problem of recording multidimensional spectra in water which contain flat baselines and highlight applications of some of the reported methods to inverse detection heteronuclear experiments. The most common source of baseline distortion in the indirectly detected dimensions of multidimensional spectra results from the necessity for linear phase correction. In the ω_1 dimension, several proposals have previously been reported which involves modifying the starting point for the acquisition of the first FID. In the present paper the authors describe preferences for starting the acquisition at the second t_1 data point. Baseline distortions in the acquisition dimension, apart from those created by the water resonance, normally result from a compromise between allowing for transient response of the spectrometer on commencement of data acquisition and the undesirable phase evolution of the required data. Again the use of spin-echoes to remove this phase distortion is examined. The residual water resonance can make a significant contribution to baseline distortion, and the use of a convolution filter is discussed. Each data point that defines the water resonance is replaced by a weighted average of its neighbouring points and the time-domain equivalent of a frequency-selective spectrum centered about the water frequency is generated. The generated data is subtracted from the original time-domain data to simulate the action of a frequency-dependent filter.

The frequency resolution obtained from discrete Fourier transformation of data is limited when acquisition times are short, as in the case for 3D and 4D NMR experiments. Extrapolation of the FID by linear prediction prior to FT has been one method used for increasing the resolution of highly truncated datasets although this approach can lead to distortions in lineshapes and generation of artifacts. As an alternative, Chylla and Markley [95] have developed the Bayesian approach, which has already been applied to 1D NMR, to analyse N-dimensional datasets. The

NMR data acquired using the constant-time approach is particularly amenable to the Bayesian analysis as the mathematical models describing such data need not include decay terms.

Application of RF power (in the form of composite pulses for decoupling and isotropic mixing) in multidimensional NMR experiments can significantly increase the sample temperature, this effect being particularly adverse in high-ionic strength samples. Because of the current probe designs and position of the thermocouple, when different experiments are recorded with the same 'set' temperature, the actual sample temperature can be quite different. The resulting changes in chemical shifts can be problematic. Another even more serious problem is when the amount of RF heating changes with each increment of one dimension. This can lead to distortions in the lineshapes. One remedy for this is to maintain the same level of RF heating throughout the course of an experiment, and Wang and Bax [96]describe such an approach.

7 Triple Resonance NMR

7.1 Heteronuclear Triple Resonance Three-Dimensional Experiments - Djikstra *et al.*[97] describe an alternative to the HCACO experiment; instead of starting the coherence transfer from protons, the pulse sequence suggested, known as the COCAH experiment, begins with a carbonyl $\pi/2$ pulse. The sensitivity of the COCAH experiment is good, despite the fact that the first $\pi/2$ pulse is applied to the carbonyl ^{13}C spins with their low gyromagnetic ratio. The loss in sensitivity is partially compensated by the fact that relaxation losses are minimized during the shorter pulse sequence. Since no proton polarization is needed at the start of the sequence, a broadband presaturation of all proton spins is possible. Saturation of the HDO signal during the relaxation delay does not appear to weaken the Hα spins that resonate near the HDO frequency, whereas these are weakened significantly in the HCACO spectrum.

Grzesiek and Bax [98] describe the origin and removal of artifacts in their previously proposed constant-time 3D HCACO experiment. These artifacts arise from from the fact that, although the experiment is designed for detection of magnetization transfer from C^{α} centers, it is inevitable that the C^{β} resonances also experience the effects of hard pulse applied to the nearby C^{α} spins. The resulting undesired coherence pathway results in a multiplet in the F_2 dimension of the 3D spectrum. The overcome this problem the authors propose a different phase cycling routine (compared with their originally proposed sequence) for some of the 90° carbon pulses. This change in the phase cycling give rise to a spin order state that cannot be transformed into observable magnetization, thereby removing the spurious signals.

With uniformly ^{13}C labelled molecules, it is possible to use heteronuclear cross-polarization , instead of pulsed-based sequences (such as INEPT and reverse INEPT) to transfer magnetization between carbon and protons. Majumdar *et al.* [98] use this heteronuclear CP method for obtaining a 3D HCCH spectrum. Initial analysis of the coherence transfer equations appear to suggest that fine-tuning of mixing period, extensive phase cycling, excessive use of trim pulses are all required for this experiment to work properly; however, in practice many of the undesirable pathways are removed or attenuated by the anisotropy of the heteronuclear cross-polarization.

Four-dimensional experiments which establish intra- and inter-residue heteronuclear scalar connectivities are attractive methods since only a few of these experiments need to be performed to obtain full sequential assignments of large biomolecules. However, these experiments often suffer from low sensitivity, particularly because of the short T_2 relaxation times of the $^{13}C\alpha$ in large proteins. In an effort to overcome some of these drawbacks, Szyperski *et al.* [99] describe methods for reducing the dimensionality of these experiments, improving sensitivity whilst retaining the information content. The HACANNH experiment is used as an example.

The HCN experiments for correlating base and sugar resonances in $^{15}N,^{13}C$-labelled RNA oligonucleotides have been described by Sklenar *et al.*[100]. Selective ^{13}C-refocusing and ^{15}N-inversion pulses are applied to optimise the magnetization transfer along a desired pathway.

Further experiments which allow assignments of nucleic acid spectra, in particular to establish connectvities between the sugar and the base moiety within a purine nucleotide, using scalar coupling connectivities, have been described by Farmer II *et al.* [101]. Similar experiments which make scalar connectivities within a pyrimidine nucleotide have also been proposed by the same authors. [102] The experiments described have increased complexities, compared with their counterparts as applied to protein spectra, these complexities arising from the more intricate network of scalar coupling patterns which are present in the spectra of nucleic acids. The same group has also described a 3D (^1H, ^{13}C, 1H) HCCH-TOCSY experiment which establishes correlations between adenine H2 to H8 protons in isotopically-enriched ribonucleic acids.[103] The ^{13}C-^{13}C magnetization transfer is achieved using a FLOPSY-8 spin-lock pulse technique; Marino *et al.*[104] report a 2D version of the same experiment.

7.2 Triple Resonance Two-Dimensional Experiments - Presaturation of the water resonance , while necessary, is undesirable in many types of NMR experiments on proteins and nucleic acids, since it obliterates the resonances of rapidly exchanging protons. In addition, through spin diffusion or direct NOE with water, attenuation throughout the spectrum can also occur. In an attempt to address these problems, Grzesiek and Bax [105] propose the water flip-back experiment, in which the the water is left unperturbed along the B_o axis during most of the NMR pulse sequence, thereby avoiding water saturation; this is done by using a combination of long low power and gradient pulses which are sited at the appropriate points in the pulse sequence.

In view of the amount of spectrometer time required to collect a 3D and its complexities, it is advantageous to devise 2D versions of the 3D experiments. Simorre *et al.*[106] describe such experiments, using the HNCO and HNCA experiments as examples. The experiments reported rely on the sampling of multiple-quantum coherences and are given the names MQ-HNCO and MQHNCA experiments. The original HNCO and HNCA sequences contain a ^{15}N 180° pulse which is applied to suppress the ^{15}N frequency; removal of the pulse will enable ^{13}C-^{15}N multiple quantum (ZQ and 2Q) coherence to be subsequently created.

Burgering *et al.* [107] describe a time-shared [$^{15}N,^{13}C$] half-filter NOE experiment which yields, in a single experiment, all NOE interactions between the protons attached to the labelled and unlabelled centres.

Sklenar *et al.* [108] describe a two-dimensional triple-resonance HCNCH experiment in order to obtain direct correlation of ribose H1' and the base protons four bonds away. The pulse scheme includes semiselective pulses to direct the transfer along the desired coherence pathway.

7.3 Gradient Enhanced 3D and 4D Heteronuclear Experiments - The use of pulsed field gradients allows the selection of NMR signals which have certain desired properties and the elimination of artifacts that are difficult to suppress with nongradient methods. Despite the advantages gained in using gradient pulses, the main disadvantages of many of the experiments proposed so far is the loss in sensitivity; this is due to the fact that N- or P-type coherent pathways, but not both, are selected during a single scan. Several approaches have been previously described to obtain pure-absorption spectra with coherence-transfer selection via gradient pulses, which offers substantial improvements in sensitivity over other experiments that were recorded with gradients. These improvements are achieved by accumulating both N- and P-type signals and using post-acquisition processing to generate pure-absorption data. Muhandiram and Kay [109] discuss in detail several enhanced-sensitivity $^{15}N,^{13}C$ gradient 3D triple resonance experiments with NH detection where the gradients are used to select the coherence-transfer pathway which involves magnetization passing through ^{15}N (rather than selecting the carbon magnetization). The experiments described are the HNCO, CBCA(CO)NH and HNCACB experiments. The same group also describes an enhanced-sensitivity gradient 4D $^{15}N,^{13}C$-edited NOESY experiment.[110] The gradients are used to suppress artifacts and water signals as well as to select coherence-transfer pathway involving ^{15}N magnetization during t_3.

7.4 Selective 3D Experiments - All 3D sequences suffer from lengthy experimental time, because of the necessity to evolve in two dimensions. Berger [111] describe a simple 2D version of a 3D experiment by using selective pulses on the third nucleus. In the triple resonnace ^1H, ^{13}C, ^{31}P experiment described, a double INEPT polarization transfer between ^1H and ^{13}C is used; the selective 90° half Gaussian ^{31}P pulse is applied in the middle of a d_3 period which replaces an incrementable t_2 period.

One serious problem with using selective excitation techniques for multidimensional expeimrnts is that frequency-dependent phase shifts are introduced in the excited spectrum due to the relatively long duration of the excitation sequence. Holmbeck *et al.* [112] describe the use of self-refocusing shaped pulses such as the BURP pulse to obtain pseudo-3D TOCSY-NOESY and TOCSY-ROESY spectra.

8 Computation

8.1 Data Processing -Fourier transformation is the most common method for converting time-domain NMR signals into frequency-domain data. A delay in sampling the first data point of the FID can result in distortion of the baseline of the requency domain spectrum. However, when using discrete Fourier transformation, baseline offset and curvature are minimized if sampling of the FID is delayed by half a dwell time. Zhu *et al.* [113] have discussed the mathematics for this relationship between sampling delay and spectral baseline.

The digital resolution of 2D NMR spectra in the ω_1 dimension is usually rather low since the available spectrometer time limits the number of time increments t_1 for which signals can be recorded. Beneicke and Kalbitzer [114] describe a data-filling approach to improve the digital resolution in the t_1 dimension. Instead of zero-filling by addition of FIDs to the data matrix which contains zeroes only, the method that is proposed replaces the zeroes in the zero-filled time-domain data by the symmetry-related data that already exists within the data-set.

The quality of the structural and dynamical information derived from 2D NOE spectra depends on the accuracy with which the intensities of the signals can be determined. Gesmar *et al.*[115] describe how a least-squares estimate of a 2D NMR signal intensity can be calculated from two 1D intensities, as estimated from a row and column.

Jeong *et al.*[116] analyse the maximum-likelihood method for spectral deconvolution and assessed its performance when used to obtain quantitative NOE measurements. The authors conclude that more accurate quantification of cross-peaks using this method of data processing can be obtained as opposed to conventional fast Fourier transformation.

Levy *et al.*[117] have described a totally automated 2D baseplane correction routine. The protocol they report consists of three steps: a) the initial global baseplane information is extracted from the spectral region, and a new baseplane spectrum is constructed with interpolated baseplane information, (b) a refined 2D baseplane is constructed, following inverse FT, using a 2D window function on the time domain, and (c) subtraction of the refined baseplane spectrum from the original NMR spectrum yields a distortionless 2D spectrum. The method depends on the fact that typical 2D NMR spectra contain more than 70% non-peak areas and less than 30% resonance peak areas.

8.2 Computer-Aided Assignments and Procedures - All computer-aided assignment strategies require a search through several multidimensional NMR spectra for patterns of peaks. One definition of a peak pattern is a set of peaks in which each member shares one or more coordinates with at least one other member of the set. It is the successful identification of the peak patterns and the determination of the chemical shifts of signals within those patterns that proved the information that can be used to assign the resonances. Eads and Shaffer [118] have written a computer programme which interprets a simple langauge that can specify and identify patterns in peak lists extracted from multidimensional spectra.

Hare and Prestegrad [119] report the use of neural networks for automated assignment of protein spectra. The use of neural networks offer the advantage that it can be trained on examples of the type of data, rather than on sets of fixed rules; it is also very flexible.

Meadows *et al.* [120] describe a semiautomated protocol for the spectral assignment of heteronuclear multidimensional data of proteins. The programme is also integrated with the identification of distance restraints from the NOE data and with the determination of 3D structures.

Olson and Markley [121] evaluate a package designed for connectivity tracing in the assignment process. The algorithm produces sequential assignments from appropriate peak lists of multidimensional NMR data.

Fuzzy mathematics was used by Xu *et al.*[122] for complete automated assignment of peptides. The first part of the procedure involves firstly, creation and identification of spin-coupling topological systems, and secondly, with a fuzzy-graph pattern recognition algorithm, each residue is assigned a set of possible candidates, each of which is a spin soupling network. Up to this point, only DQF-COSY and TOCSY peak sets are used. The second part creates a forest of spin coupling networks and an algorithm then searches this 'forest' and finds the optimum sequence-specific assignment based on a NOESY dataset. Cross-peak lists of DQF-COSY, TOCSY, and NOESY spectra are the input data sets.

Craig and Kuntz [123] describe a methodology for computerized analysis of 2D correlation spectra using discrete matrix representations.

8.3 Computer Simulations - Computer simulations are becoming increasingly necessary as NMR experiments become more complex. In addition, at magnetic fields greater than 11.75T, the neglect of certain relaxation phenomenon, such as cross-correlation effects between dipole-dipole interactions, and relaxation by the chemical shift anisotropy (CSA) mechanism must be considered in such treatments. Many of the existing computer simulation programmes are prohibitively time-consuming; Smith *et al.* [124], however, have develop an efficient formalism using Liouville space superoperators, which leads to simplification of such simulations programs. The implementation of the formalism developed in the paper is accomplished by using the GAMMA magnetic resonance simulation platform, which adopts an object-orientated programming approach.[125]

References

(1) L.Y.Lian, in "*Nuclear Magnetic Resonance*", ed. G.Webb, (Specialist Periodical Reports). The Royal Society of Chemistry, **1993**, *23*, 243
(2) D.M.LeMaster *Prog. NMR Spectroc.* **1994**, *26*, 371.
(3) H.Oshkinat; T.Muller; T.Dieckmann *Angew. Chem. Int. Ed. Engl.* **1994**, *33*, 277.
(4) I.P.Gerothanassis *Prog. in NMR Spectr.* **1994**, *26*, 171.
(5) R.Bruschweiler; D.A.Case *Prog. in NMR Spectr.* **1994**, *26*, 27.
(6) A.D.Bain; I.W.Burton; W.F.Reynolds *Prog. in NMR Spectr.* **1994**, *26*, 59.
(7) T.J.Norwood *Chem. Soc. Rev.* **1994**, *23*, 59.
(8) Z.Starcuk, J.; K.Bartusek; Z.Starcuk *J.Mag. Reson.* **1994**, *Series A* ***107***, 24.
(9) X.Yang; M.A.Keniry; B.C.Sanctuary *J. Magn. Reson.* **1993**, *Series A* ***105***, 295.
(10) N.S.Bai; Ramachandran, R. *J. Magn. Reson.* **1993**, *Series A* ***105***, 298.
(11) M.Kadkhodaei; T-L. Hwang; J.Tang; A.J.Shaka *J. Mag. Reson.* **1993**, *Series A* ***105***, 104.
(12) T.Fujiwara; T.Anai; N.Kurihara; K.Nagayama *J. Mag. Reson.* **1993**, *Series A* ***104***, 103.
(13) P.J.Hajduk; D.A.Horita; L.E.Lerner *J. Magn. Reson.* **1993**, *Series A* ***103***, 40.
(14) J.M.Nuzillard; R.Freeman *J. Mag. Reson.* **1994**, *Series A* ***107***, 113.
(15) E.Kupce; R.Freeman *J. Magn. Reson.* **1993**, *Series A* ***103***, 358.
(16) E.Kupce; R.Freeman *J. Mag. Reson.* **1993**, *Series A* ***104***, 234.
(17) H.Liu; K.Weisz; T.L.James *J. Mag. Reson.* **1993**, *Series A* ***105***, 184.
(18) Z.Starcuk, J.; K.Bartusek; Z.Starcuk *J. Mag. Reson.* **1993**, *Series A* ***104***, 340.
(19) R.Konrat; G.Zieger; H.Sterk; W.Bermel *J. Mag. Reson.* **1994**, *Series A* ***106***, 23.
(20) J-M.Bernassau; J-M.Nuzillard *J. Mag. Reson.* **1993**, *Series A* ***104***, 212.
(21) W.E.Maas; F.Laukien; D.G.Cory *J. Magn. Reson.* **1993**, *Series A* ***103***, 115.
(22) A.Norton; D.Galambos; T.L.Hwang; M.Stimson, a.; A.J.Shaka *J. Magn. Reson.* **1994**, *Series A* *108*, 51.
(23) S.H.Smallcombe *J. Amer. Chem. Soc.* **1993**, *115*, 4776.

(24) P.Sodano; M.Delepierre *J. Mag. Reson.* **1993**, *Series A **104***, 88.
(25) P.Sodano; M.Delepierre *J. Biom. NMR* **1993**, *3*, 471.
(26) C.J.Turner; W.C.Hutton *J. Mag. Reson.* **1993**, *Series A **105***, 72.
(27) D.Turner *J. Mag. Reson.* **1993**, *Series A **104***, 197.
(28) F.DelRio-Portilla; R.Freeman *J. Mag. Reson.* **1993**, *Series A **104***, 358.
(29) P.Huber; C.Zwahlen; S.J.F.Vincent; G.Bodenhausen *J. Magn. Reson.* **1993**, *Series A **103***, 118.
(30) N. Chandrakumar; S.S.Velan *J. Magn. Reson.* **1993**, *Series A **104***, 363.
(31) T.J.Norwood; K.Jones *J. Mag. Reson.* **1993**, *Series A **104***, 106.
(32) G.J.Bowden; M.J.Prandolini *J. Mag. Reson.* **1994**, *Series A **107***, 32.
(33) T.J.Norwood *J. Mag. Reson.* **1993**, *Series A **105***, 193.
(34) G.J.Bowden; M.J.Prandolini; J.M.Pope; F.Separovic *J. Magn. Reson.* **1993**, *Series A **103***, 307.
(35) E.Kupce; R.Freeman *J. Mag. Reson.* **1993**, *Series A **105***, 234.
(36) E.Kupce; R.Freeman *J. Mag. Reson.* **1994**, *Series A **106***, 135.
(37) E.Kupce; R.Freeman *J. Mag. Reson.* **1993**, *Series A **105***, 310.
(38) E.Kupce; J.M.Nuzillard; V.S.Dimitrov; R.Freeman *J. Magn. Reson.* **1994**, *Series A **107***, 246.
(39) M.E.Girvin *J. Magn. Reson.* **1994**, *Series A **104***, 99.
(40) R.Konrat; G.Zieger; H.Sterk *J. Mag. Reson.* **1993**, *Series A **104***, 63.
(41) P.Schmieder; A.S.Stern; G.Wagner; J.C.Hoch *J. Biom. NMR.* **1993**, *3*, 569.
(42) J.Briand; R.R.Ernst *J.Magn. Reson.* **1993**, *Series A **104***, 54.
(43) A.L.Davis; G.Estcourt; J.Keeler; E.D.Laue; J.Titman *J. Mag. Reson.* **1993**, *Series A **105***, 167.
(44) S.Glaser *J. Mag. Reson.* **1993**, *Series A **104***, 283.
(45) V.Blechta; R.Freeman *J. Magn. Reson.* **1993**, *Series A **103***, 349.
(46) M.Ravikumar; A.A.Bothner-By *J. Amer. Chem. Soc.* **115**, *115*, 7537.
(47) C.Zwahlen; S.J.F.Vincent; L.Di Bari; M.H.Levitt; G.Bodenhausen *J. Amer. Chem. Soc.* **1994**, *116*, 362.
(48) M.J.Dellwo; D.M.Schneider; A.J.Wand *J. Magn. Reson.* **1994**, *Series B **103***, 1.
(49) C.D.Eads *J. Mag. Reson.* **1994**, *Series A **107***, 109.
(50) S.P.Edmondson *J. Mag. Reson.* **1994**, *Series B **103***, 222.
(51) I.Bertini; A.Dikiy; C.Luchinat; M.Piccioli, a.; D.Tarchi *J. Mag. Reson.* **1994**, *Series B **103***, 278.
(52) S.X.Wang; G.H.Caines; T.Schleich *J. Mag. Reson.* **1993**, *Series B **102***, 47.
(53) S.Grzesiek; A.Bax *J. Biom. NMR.* **1993**, *3*, 627.
(54) G.Gemmecker; W.Jahnke; H.Kessler *J. Amer. Chem. Soc.* **1993**, *115*, 11620.
(55) N.J.Skelton; A.G.Palmer III; M.Akke; J.Kordel; M.Rance; W.J.Chazin *J. Magn. Reson.* **1994**, *102*, 253.
(56) A.Ross; M.Czisch; T.Zink; T.A.Holak *J. Magn. Reson.* **1993**, *Series B **102***, 314.
(57) T.Nakai; C.A.McDowell *J. Mag. Reson.* **1993**, *Series A 104*, 146.
(58) W.Willker; J.Stelten; D.Leibfritz *J. Mag. Reson.* **1994**, *Series A **107***, 94.
(59) C.Roumestand; D.Boudot; Canet, D. *J. Mag. Reson.* **1993**, *Series A **105***, 304.
(60) V.V.Krishnamurthy *J. Mag. Reson.* **1993**, *Series B **102***, 77.
(61) K.V.R.Chary; V.K.Rastogi; G.Govil *J. Mag. Reson.* **1993**, *Series B **102***, 81.
(62) K.V.R.Chary; V.K.Rastogi; G.Govil; M.H.Todd *J.C.S. Chem Commun.*, **1994**, *(3)*, 241.

(63) T.K.Pratum; B.S.Moore *J. Mag. Reson.* **1993**, *Series B* ***102***, 91.
(64) J-M.Nuzillard; G.Gasmi; J-M.Bernassau *J. Mag. Reson.* **1993**, *Series A* ***104***, 83.
(65) E.Fukishi, a.; J.Kawabata *J.Mag. Reson.* **1994**, *Series A 108*, 103.
(66) J.M.Nuzillard; J.M.Bernassau *J. Mag. Reson.* **1994**, *Series B* ***103***, 284.
(67) B.Adams; L.Lerner *J. Magn. Reson.* **1993**, *Series A* ***103***, 97.
(68) K.E.Kover; O.Prakash; V.J.Hruby *J. Magn. Reson.* **1993**, *Series A* ***103***, 92.
(69) K.Kover; D.Jiao; D.Uhrin; P.Forgo; V.J.Hruby *J. Mag. Reson.* **1994**, *Series A* ***106***, 119.
(70) G.Zhu; A.Bax *J. Mag. Reson.* **1993**, *Series A* ***104***, 353.
(71) J. Kawabata; E.Fukushi *J. Magn. Reson.* **1994**, *Series A* ***107***, 173.
(72) H.E.Olsen; S.Ludvigsen; O.W.Sorensen *J. Mag. Reson.* **1993**, *Series A* ***104***, 226.
(73) H.E.Olsen; S.Ludvigsen; O.W.Sorensen *J. Mag. Reson.* **1993**, *Series A* ***105***, 321.
(74) U.Wollborn; W.Willker; D.Leibfritz *J. Magn. Reson.* **1993**, *Series A* ***103***, 86.
(75) M.D.Sorensen; S.M.Kristensen; J.J.Led; O.W.Sorensen *J. Magn. Reson.* **1993**, *Series A* ***103***, 364.
(76) M.D.Sorensen; J.J.Led; O.W.Sorensen *J. Biom. NMR.* **1994**, *4*, 135.
(77) H.Schwalbe; W.Samstag; J.W.Engels; W.Bermel; C.Griesinger *J. Biom. NMR* **1993**, *3*, 479.
(78) S.Grzesiek; G.W.Vuister; A.Bax *J. Biom. NMR* **1993**, *3*, 487.
(79) H.Schwalbe; A.Rexroth; U.Eggenberger; T.Geppert; C.Griesinger *J. Amer. Chem. Soc.* **1993**, *115*, 7878.
(80) K.L.Constantine; M.S.Friedrichs; L.Mueller *J. Magn. Reson.* **1994**, *Series B* ***104*** ,62.
(81) G.W.Vuister; A.C.Wang; A.Bax *J. Amer. Chem. Soc.* **1993**, *115*, 5334.
(82) P.C.M. Van Zijl; M.O'Neil Johnsson, a.; C.Abeygunawardana *J. Mag. Reson.* **1994**, *Series A 108*, 116.
(83) J.Stonehouse; G.L.Shaw; J.Keeler; E.D.Laue *J. Magn. Reson.* **1994**, *Series A* ***107***, 178.
(84) R.W.Kriwacki; R.B.Hill; J.M.Flanagan; J.P.Caradonna; J.H.Prestegard *J. Amer. Chem. Soc.* **1993**, *115*, 8907.
(85) D.R.Muhandiram; N.A.Farrow; G-Y. Xu; S.H.Smallcombe; L.E.Kay *J. Magn. Reson.* **1993**, *Series B* ***102***, 317.
(86) J.Lee; J.Fejzo; G.Wagner *J. Magn. Reson.* **1993**, *Series B* ***102***, 322.
(87) S.Medvedeva; J.P.Simmore; B.Brutscher; F.Guerlesquin; D.Marion *FEBS Letts.* **1993**, *333*, 251.
(88) D.G.Cory; F.H.Laukien; W.E.Maas *J. Mag. Reson.* **1993**, *Series A* ***105***, 223.
(89) P.Mutzenhardt; J. Brondeau, a.; D.Canet *J. Mag. Reson.* **1994**, *Series A !08*, 110.
(90) D.Canet; J.Brondeau; E.Mischler; F.Humbert *J. Mag. Reson.* **1993**, *Series A* ***105***, 239.
(91) G.Otting *J. Mag. Reson.* **1994**, *Series B* ***103***, 288.
(92) M.E.Rosen *J. Mag. Reson.* **1994**, *Series A* ***107***, 119.
(93) N.A.Froystein *J. Magn. Reson.* **1993**, *Series A* ***103***, 332.
(94) J.P.Waltho; J.Cavanagh *J. Magn. Reson.* **1993**, *Series A* ***103***, 338.
(95) R.Chylla; J.L.Markley *J. Biom. NMR* **1993**, *3*, 515.
(96) A.C.Wang; A.Bax *J. Biom. NMR.* **1993**, *3*, 715.
(97) K.Djikstra; G.J.A.Kroon; N.A.J.Van Nuland; R.M.Scheek *J. Mag. Reson.* **1994**, *Series A* ***107***, 102.
(98) S.Grzesiek; A.Bax *J. Mag. Reson.* **1993**, *Series B* ***102***, 103.
(99) T.Szyperski; G.Wider; J.H.Bushweller; K.Wuthrich *J. Amer. Chem. Soc.* **1993**, *115*, 9307.
(100) V.Sklenar; R.D.Peterson; M.R.Rejante; J.Feigon *J. Biom. NMR.* **1993**, *3*, 721.
(101) B.T.Farmer II; L.Muller; E.P.Nikonowicz; A.Pardi *J. Biom. NMR* **1994**, *4*, 129.

(102) B.T.Farmer II; L.Muller; E.P.Nikonowicz; A.Pardi *J. Amer. Chem. Soc.* **1993**, *115*, 11040.
(103) P.Legault; B.T.Farmer II; L.Muller; A.Pardi *J. Amer. Chem. Soc.* **1994**, *116*, 2203.
(104) J.P.Marino; J.H.Prestegard; D.M.Crothers *J. Amer. Chem. Soc.* **1994**, *116*, 2205.
(105) S.Grzesiek; A.Bax *J.Amer. Chem. Soc.* **1993**, *115*, 12593.
(106) J.P.Simorre; B.Brutscher; M.S.Caffrey; D.Marion *J. Biom. NMR.* **1994**, *4*, 325.
(107) M.Burgering; R.Boelens; R.Kaptein *J.Biom. NMR* **1993**, *3*, 709.
(108) V.Sklenar; R.D.Peterson; M.R.Rejante; E.Wang; J.Feigon *J. Amer. Chem. Soc.* **1993**, *115*, 12181.
(109) D.R.Muhandiram; L.E.Kay *J. Mag. Reson.* **1994**, *Series B* ***103***, 203.
(110) D.R.Muhandiram; G.Y.Xu; L.E.Kay *J. Biom. NMR* **1993**, *3*, 463.
(111) S.Berger *J. Mag. Reson.* **1993**, *Series A* ***105***, 95.
(112) S.Holmbeck; P.Hajduk; L.Lerner *J. Mag. Reson.* **1993**, *Series B* ***102***, 107.
(113) G.Zhu; D.A.Torchia; A.Bax *J. Mag. Reson.* **1993**, *Series A* ***105***, 219.
(114) W.Beneicke; H.R.Kalbitzer *J. Magn. Reson.* **1994**, *Series A* ***107***, 134.
(115) H.Gesmar; P.F.Nielsen; J.J.Led *J. Mag. Reson.* **1994**, *Series B* ***103***, 10.
(116) G.W.Jeong;P.N.Borer; S.S.Wang; G.C.Levy *J.Magn.Reson.* **1993**, *Series A* ***103***, 123.
(117) G.C.Levy; I.Pelczer; G.Jeong; K.Wang *J. Mag. Reson.* **1993**, *Series A* ***105***, 316.
(118) C.D.Eads; J.D.Shaffer *J. Magn. Reson.* **1994**, *Series B* ***102***, 310.
(119) B.J.Hare; J.H.Prestegard *J. Biom. NMR* **1994**, *4*, 35.
(120) R.P.Meadows; E.T.Olejniczak; S.W.Fesik. *J. Biom. NMR* **1994**, *4*, 79.
(121) J.B.Olson Jr.; J.L.Markley *J. Biom. NMR.* **1994**, *4*, 385.
(122) J.Xu; S.K.Straus; B.C.Sanctuary; L.Trimble *J. Magn. Reson.* **1994**, *Series B* ***103***, 53.
(123) E.C.Craig; I.D.Kuntz *J. Magn. Reson.* **1993**, *Series B* ***102***, 265.
(124) S.A.Smith *J. Magn. Reson.* **1994**, *Series A* ***106***, 57.
(125) S.A.Smith; T.O.Levante; B.H.Meier; R.R.Ernst *J. Magn. Reson.* **1994**, *Series A* ***106***, 75.

9
Natural Macromolecules

BY J. R. P. ARNOLD AND J. FISHER

1. INTRODUCTION

Over the last decade, the application of NMR to the study of natural macromolecules has been a spectacular area of progress. Thus now, in the mid-1990's, reports of NMR studies of natural macromolecules are appearing at a great rate, covering an ever-widening range of biological systems. Here we present a fairly comprehensive collection of the last year's reports, subdivided into sections representing the main categories for natural macromolecules. We consider high resolution NMR studies in the solution state, referring to original reports published in English in the last year or so, and excluding conference proceedings, reviews and books. We use the nomenclature employed in the original papers; this can lead to different entries for the same biological system having slightly different names, depending on authors and/or journal.

2. PEPTIDES

Peptides span a wide range of systems, but can generally be classified either as fragments of protein sequences, or as intact biomolecules such as hormones or toxins. Much NMR work involves the effect of solvent environment on the peptide conformation. Many peptides are apparently unstructured in aqueous solvents, but show indications of secondary structure structure in the presence of organic solvents and detergents. There can be a mixture of conformations, but usually only one set of ^{1}H signals is observed. NMR-derived parameters are thus averaged, each on its appropriate timescale. The influence of dynamics means that the term 'structure' has a somewhat different meaning than in, say, a small globular protein. The dynamic behaviour of proteins is now receiving significant attention (section 3.3); reports of the application of such methods to peptides are now appearing.

The distinction between some peptides and proteins can be semantic. Some entries in this section, (such as some snake toxins which form well-defined structures and are relatively large) could well be classified as proteins. There is thus some overlap between this section and the one concerning proteins.

There has been increasing use of SDS micelles as a mimic of membrane environment. Alamethicin, a 20 amino acid peptide which produces voltage-dependent conductance in membranes, has been incorporated into SDS micelles and its structure determined in the manner of many peptide sytems. In this case there is a helical N-terminal region and a more flexible C-terminal segment. That the peptide was effectively isolated from surrounding water was demonstrated by the line-broadening effects of aqueous paramagnetic reagents [1].

Peptides related to HIV are being increasingly studied, for example fragments of envelope glycoprotein gp120 [2,3] and transmembrane protein [4], for which there is interest in any relationship between structure and immunogenicity. This is an important general theme for many other peptide NMR studies, but often non-aqueous solvents have to be used to see any peptide structure at all. Peptide-related compounds with anti-HIV activity are also being increasingly reported (there are several examples in Table 1); these are derived from extremely diverse sources, reflecting the enormous world-wide search for HIV treatments. Another increasingly important medical condition in Western societies is Alzheimer's disease. This condition is accompanied by deposits of β/A4 amyloid protein, and this irreversible aggregation has been subject to much study. Peptide fragments of this protein are generally very difficult to solubilise, so harsh chemical treatment has often been required. Formic acid yields a soluble form, which turns out to be due to formylation of a serine residue [5]. A soluble fragment of the protein has been studied and its structure (actually, a mixture of conformations) compared to a mutant which arises in a related clinical condition [6]. Table 1 lists other NMR studies of peptides reported this year.

Table 1. Peptides studied by NMR

System	Remarks	Reference
47aa seminal plasmin	random coil in water; detergent promotes 2° structure	7
19, 22aa of Pro-rich salivary prot.	extended coil in 10% DMSO;90% *trans* Pro	8
19, 22aa of Pro-rich salivary prot.	binding to a polyphenol: peptides remain extended	9
ω-conotoxin GVIA	smallest reported peptide to contain triple-stranded β-sheet	10
ω-conotoxin GVIA	solution structure; triple stranded β-sheet	11
ω-conotoxin GVIA	solution structure	12
Neurotoxin III (27aa)	pH-dependent self-association; struct contains a series of reverse turns	13
Neurotoxin III (sea anemone)	full ^{1}H assignments, chem. shifts compared to related toxins	14
G- & H-helices from myoglobin	G-helix is unstructured in water, H- is significantly ordered	15
G-H turn region from myoglobin	series of peptides containing turn but no helical content in water	16
G-H helical hairpin of myoglobin	helical & turn regions: turn retained but only limited helicity	17
$(Pro\text{-}Hyp\text{-}Gly)_{10}$ and derivative	collagen models;form trimers, parallel strands in triple helix	18
$(Pro\text{-}Hyp\text{-}Gly)_{10}$ and derivative	H-D exchange, ^{15}N used to probe backbone dynamics, confirm model	19
$(Pro\text{-}Hyp\text{-}Gly)_{10}$ and derivatives	effects of various sequence interruptions in middle region;	20
16aa triple-helix-forming peptide	Aggregation effects on structure and dynamics	21
Cartilage protein fragments	nOe suggests no stable 2° struct., relevance of binding of hyaluronan	22
Collagen α-chain C-telopeptides	type II and type III examples; both essentially extended structures	23
Collagen α-chain C-telopeptides	in MeOH/water; 19aa type II has β-turn, 14aa type III + some helicity	24
N-terminal region of p21^{ras}	Gly-rich nucleotide binding site; interaction with GTP	25
17aa from caldesmon	Calmodulin-binding domain; α-helical in TFE and when bound	26
Oxytocin	5aa with a disulphide; structural changes on protonation in DMSO	27
Oxytocin	bound to neurophysin I in fast exchange with free form; tr-nOe shows free and bound structures are quite different	28
8aa cyclic Angiotensin analogues	structures from NMR in DMSO and energy calcs. compared	29
Angiotensin II and analogue	function of terminal NH_2 probed-important for struct & activity	30
Angiotensin II analogues	methylation of His ring effects structure; relate to function	31
30aa glucagon-like peptide	in micelles; 2 helical regions with hydrophobic side chains on one side	32
acetylcholine receptor analogues	main immunogenic region; correlate structure with biological activity	33
acetylcholine receptor analogues	10aa, in DMSO; rigid segment, flexible C-terminal region	34
18aa model amphipathic α-helix	helicity in water; not dependent on concentration	35
10aa analogues of bradykinin	agonists contain type I β-turns while antagonists are extended	36
Snake Curaremimetic toxin α	two epitopes; H-D exchange mapped in free and bound to antibodies	37
Tertiapin (21aa bee venom peptide)	solution struct. is compact with helical region and type-I reverse turn	38
6aa substance P analogues	two peptides, both extended in TFE/water, reverse turn in $CDCl_3$	39
Tachykinin peptides	substance P and neurokinin analogues; structs in various solvents	40
Senktide (neurokinin B analogue)	modified Phe causes slow *cis/trans* peptide bond isomerisation; *trans* isomer binds more strongly to neg. charged vesicles	41
Substance P analogues	correlating structure with activity	42
Snake neurotoxin II (61aa)	solution structure, much β-sheet	43
Snake cardiotoxin V	solution structure similar to related toxins; dynamics may explain unique biological activity	44
Snake cardiotoxin CTX IIb	solution struct.; relationship with function & specificity	45

Cyclic 6aa NK-2 antagonist	structure contains two β1 (or βV') turns	46
8aa analogues of CCK-8	use of z-filtered RELAY to define side chain conformations	47
Scorpion neurotoxin variant 1	19 non-Pro residues, obtain similar structure to wild-type	48
Scorpion toxin analogue P05-NH_2	solution structure; local electric charges may control recognition	49
Monocyclic endothelin analogue	solution struct. has 3 defined regions, similar to parent compound	50
9aa related to MUC1 core	carcinoma related; substitution alters β-turn characteristics	51
human muc-1 core fragments	1,2 & 3 tandem repeats of 20aa unit; structural changes with length	52
Praying mantis neuropeptide	Emp-AKH; structure contains β-turn and 5-residue β-sheet	53
Sea anemone neurotoxin ShI	refined NMR structure; type I β-turn at functionally important region	54
Urotensen II (12aa)	solution struct.; core well defined, N-terminal region disordered	55
Eggshell protein analogues	5-residue repeating unit; left-handed helix formed with D-amino acids	56
Cyclic 6aa somatostatin analogue	In 1:1 DMSO:water, structure includes type II' β-turn	57
Dermenkephalin (7aa δ-opioid)	quantitative NMR structure as starting model for modelling etc.	58
Deltorphin-I (7aa δ-opioid)	in DMSO; dynamic structure study	59
δ-opioid peptides	cyclic and acyclic enkephalin-related, membrane-bound and free conformations compared	60
Uteroglobin fragments	11 & 17 aa; in SDS micelles, a helical region in each peptide	61
N-terminal 25aa of β-casein	highly flexible; neither fully extended nor random coil; Ca^{2+}-effects	62
Nisin	structure in various non-aqueous systems, differs from that in water	63
ω-Aga-IVB (a spider toxin)	48aa, four disulphides present; NMR to identify Cys-Cys pairs	64
ω-Aga-IVB	solution structure and disulphide bridge configuration	65
25aa M32 peptide	in 30% hexafluoro-2-propanol; structure may be loop-like	66
29aa signal peptide	membrane-bound conformation deduced from tr-nOe	67
19aa signal peptide	in micelles, single α-helical region over most of peptide	68
E.coli OmpA signal peptides	helical in TFE/water & SDS micelles;goodcorrelation between *in vivo* activity and SDS-induced helicity but not TFE-induced helicity	69
17aa sorting signal	helical region and β-turn, turn is destabilised by some mutations	70
PACAP38 & PACAP27	neuropeptides; ordered structs. in 50% TFE; helical regions located	71
Neuropeptide K	in 28% TFE/water; amphipathic α-helical region; cf. to neorokinin A	72
Polymer epitopes	long peptide chains of differing immunlogical activity; compared	73
Ubiquitin fragments (25 & 35aa)	structures in methanol/water has similarities with corresponding regions in partially unfolded A-state of parent protein	74
Bis-headed inhibitor of Trypsin	engineered relative of squash trypsin inhibitor; compare structs.	75
15aa of BPTI	N-terminal region; evidence for amide-aromatic interaction	76
5aa related to elastin	conformational properties strongly concentration-dependent	77
Snake cardiotoxin γ (60aa)	refined solution structure; conserved residues have ordered side chains	78
Albolabrin (snake toxin)	73aa, 2° struct has RGD-motif in a mobile segment	79
Type-α transforming growth factor	Human, 50aa, ^{15}N-labelled, structure det., temperature effects	80
growth hormone releasing factor	Human, 18aa, random structure in water, helical in SDS micelles	81
32aa fragment of annexin I	helix-loop-helix region; structure strongly environment-dependent	82
α-Bungarotoxin	bound to 12aa fragment of receptor, which is essentially extended in complex, while α-bungarotoxin is distorted	83
Lophyrotomin (8aa)	toxic peptide; no persistent 2° struct.; flexible	84
Calcitonin	*cis/trans* isomerisation at one of 2 Pro's catalysed by cyclophilin	85
Calcitonin analogue	same structure as calitonin, but different biological activity	86
Human parathyroid hormone	N-terminal 34aa; structure contains α-helical region	87
Human parathyroid hormone	4 fragments which cover all regions of intact hormone; all require TFE to display structure; variations in helical propensity	88
Human parathyroid hormone	N-terminal 34aa; in micelles and various solvents; helicity observed	89
Human parathyroid hormone	N-terminal 34aa in 10% TFE/water; 2 helical regions linked by 'unordered' segment	90
Thymosin β_4 (43aa)	no unique struct. in water, but some helicity present	91
Oostatic factor (10aa)	Trypsin-modulator; left-handed helix; no β-turn apparent	92
Scorpion venom neurotoxin	variant-3; solution struct. determined, compared to crystal struct.	93
SPKK-containing peptides	8 & 9aa in DMSO; β-turns apparent	94
Saturnisporins SAII & SAIV	fungal antibiotics, 20aa, α-helical with local distortions	95
Mouse γ-interferon	N-terminal 39aa; 40% TFE/water; flexible loop links2 helical regions	96
5aa Fe(II)-binding peptides	Cys-containing, fluorinated; ^{19}F relaxation & temperature dependence	97
28aa from Met-tRNA synthetase	Zn-binding domain; well-defined solution struct. compared to others differs from corresponding region in crystal of intact protein	98
15aa of fibronectin	heparin binding domain; C-terminal structured	99
Pilin-derived 17aa antigen	disulphide bridge; 2 β-turns; relevance to immunogenic response	100
43aa of a pyruvate dehydrogenase multi-enzyme complex	from peripheral subunit-binding domain of dihydrolipoamide acetyl-transferase; has some features typical of larger structures	101
Tyrosine kinase insulin receptor	12aa, substrate; sequential phosphorylation of the 3 Tyr observed	102
Interleukin-6 19aa c-terminal	human & murine; unstructured in water, pH-dependent chem.shift dispersion; murine helical in TFE	103

Interleukin 1β fragments	epitope mapping; NMR to establish structural effects of a mutation	104
Ribonuclease A fragments	8 & 14aa; disulphide loop; stabilised by β-turns	105
Protozoan pheremone Er-10	38aa ("protein"); well-defined solution struct. with distinct features	106
RP 71955 (21aa, tricyclic)	active against HIV-1; characterisation and solution struct.	107
Tachypesin I	isolated from a crab; structure compared to T22, a derivative which has anti-HIV activity	108
Salinomycin (Na^+ complex)	first nOe-derived structure of ionophore-metal complex	109
Melittin (a hexacosa apeptide)	variation of $^{13}C_{\alpha}$ chem. shifts with solvent and solvent-induced confs.	110
HIV gp120-related peptide (40aa)	envelope protein; immunogenic induces HIV-neutralising antibodies, in water largely disordered but critical regions show hints of structure	111
17aa model helical peptide	evidence for multiple conformations	112
Cytochrome c fragments	separate C-terminal & heme-containing regions form tight complex which mimics folding intermediate of intact protein	113
Farnesyl transferase substrate	7aa, protein-bound conf. deduced from tr-nOe, type I β-turn present	114
Nitric oxide synthetase fragment	23aa of putative calmodulin-binding domain; TFE-induced structure	115
Major histocompatability complex-derived peptide	Class I, murine, 17aa, micelle-bound conf. has helical region and type I β-turn	116
Bradykinin analogues	three 9aa antagonists; presence or absence of turns in structures	117
Gonadotropin-releasing hormone antagonist	10aa, potent; dicyclic though hormone is linear; 'β-hairpin', tail region not strongly structured	118
Neuropeptide FMRFamide	7aa analogues; preliminary struct. studies in DMSO	119
Barnase modules	29 & 17aa; similar structures to those in intact protein	120
antigenic-derived peptide	10aa, no structure in water, TFE induces helical region and turn	121
Anatamanide (cyclic, 10aa)	struct. & dynamics study; multiconfomational search algorithm used	122
23aa of transthyretin	plasma protein which can form amyloids; solvent-dependence of structure.	123
9aa delta sleep-inducing peptide	mixture of ordered and disordered in water; helix and β-turn more evident as TFE is introduced	124

3. PROTEINS

3.1 NMR Parameters and Protein Structure: One indicator of the march of progress in biomolecular NMR is the increasing molecular weight of proteins for which detailed structural information can be obtained. The effectiveness of pulse sequences is generally limited by the relaxation characteristics of the protein resonances. Thus as the size of the protein rises NMR linewidths generally increase, and multipulse sequences (especially those containing long fixed refocussing delays) become less effective. A theoretical analysis of relaxation arising from molecular tumbling [125] suggests that this reduction in pulse sequence effectiveness should not really matter until molecular weights are two to three times larger than what is actually found. Another vital factor is, of course, the tendency of proteins to associate and aggregate. If this can be overcome, larger proteins become accessible to detailed study by NMR. Inceasing the tumbling rate by raising the sample temperature will generally reduce linewidths. However, many proteins will not tolerate this; they just irreversibly denature. Indeed, some fragile proteins do this at quite moderate temperatures. To overcome this, stabilisation of (folded) proteins in solution has been reported using osmolytes, such as 2M predeuterated glycine [126].

The advent of uniformly doubly (^{13}C,^{15}N) labelled proteins has led to a variety of triple resonance experiments and associated assignment strategies. Manually analysing a selection of higher dimensional spectra can be unwieldy; computer programs to assist in assignment have been reported [127,128]. Several strategies concentrating on the backbone nuclei [129,130,131], and a procedure which also determines the side chain type [132] have been presented. A 3D triple resonance experiment which relates backbone and side chain signals, yielding six appropriate frequencies to aid sequential assignment [133] has been devised, as has a procedure which can potentially obtain complete assignments from only two 4D experiments [134]. A variety of other new higher dimensional experiments and associated assignment strategies have also appeared [135-139]. Prolines inevitably interrupt sequential backbone assignments; to overcome this a triple resonance method (but using only 2 dimensions) which picks out residues preceeding prolines [140] has been reported. The structural consequences of the presence of proline, i.e. *cis*/*trans* configuration of the appropriate peptide bond, can be probed using selective labelling [141]. Selective labelling has also been used for assignment purposes in larger proteins [142]. This is labelling in an 'active' sense; an alternative is deuterium labelling which effectively removes NMR (^1H) signals, rather than introducing additional resonances. Selective D-labelling reveals spectral details by comparision with spectra of unlabelled material. This has been demonstrated by dihydrofolate reductase in which the leucines were stereospecifically deuterated, from which stereospecific ^1H assignments could be derived [143]. The advantages from monodeuteration of methylene positions have been demonstrated [144], such as improved performance of certain heteronuclear experiments. Selective deuteration of aromatic amino acids has been shown to be valuable in identifying a folding intermediate of a thioredoxin, which was otherwise undetectable [145]. Extensive D-labelling can be used to simplify the ^1H spectrum, and also reduces spin diffusion amonst the remaining ^1H signals, which makes long range nOe contacts more reliable. These advantages have been well illustrated in a study of a phage transcription factor [146].

An alternative to labelling is to exploit natural abundance ^{13}C, and this has been demonstrated using a 79-residue ferrocyctochrome [147]; this was greatly assisted by a sample concentration of 5.2 mM and the use of pulsed field gradients.

Another approach is to use ^{19}F as a label. This can be done by biosynthetically incorporating fluorinated amino acids, usually with modified aromatic side chains. An example of this is the CheY (I) protein containing 4-fluorophenylalanines, in which the ^{19}F signals were indicators of local conformation [148,149]. 3-Fluorotyrosines have been incorporated into elongation factor Tu, and the ^{19}F chemical shifts, and their changes on ligand binding, were used as indicators of the differing environments of the 10 tyrosines [150]. There have been several reports of the use of fluorine-containing ligands bound to proteins. A fluorinated borane bound to α-chymotrypsin has been used to probe the local environment of the ligand and the dynamics of binding [151]. Fluorinated analogues of folate and dUMP have been used to study ligand interactions with thymidylate synthase [152], and fluoro-dUMP has also been used in studies of inactivated [153] and mutant [154] versions of the enzyme. ^{19}F has been used in a study of the binding of a fluorinated ketone to phospholipase A_2 [155].

Full and selective ^{13}C and ^{15}N-labelling of proteins means that an increasing number of scalar coupling constants are potentially accessible, many of which can provide structural information. New experiments have been devised to extract accurate values of various coupling constants [156-162]. Accurately-determined one-bond ^{13}C-^1H coupling constants have been shown to be useful [163]. Relaxation effects may distort the splittings in cross peaks, and this should be taken into account in spectra of larger proteins [164]. Despite this progress, important coupling constants can still be extracted from homonuclear 2D spectra. The shape and fine structure of COSY NH-$C_\alpha H$ 'fingerprint' cross peaks has been investigated, and rules correlating these features with $C_\alpha H$-$C_\beta H$

coupling constants established [165]. To match the advances in the determination of coupling constants, more sophisticated use of these values in structural calculations is being made. The direct input of experimental scalar coupling constants as time-averaged, rather than instantaneous values, in a molecular dynamics treatment [166] has been reported.

The interpretation of the chemical shifts in proteins promises rich reward, and progress is continuing in this area. Theoretical studies using *ab initio* calculations have indicated the relative contributions of such factors as torsion angles, electrostatics and H-bonding to the shifts of ^{1}H, ^{13}C, ^{15}N and ^{19}F [167,168]. For ^{19}F, a significant contribution may be made by van der Waals interactions [169]. The ^{19}F chemical shifts in a protein containing 5-fluorotryptophan residues [170] have been used to probe the effects of charge fields in the protein and on its surface. Where a structure is known, comparision of calculated and observed C_{α} shifts can be used to refine that structure [171].

Despite the increasing use of 3 and 4D spectra, analysis of homonuclear 2D spectra is still very important and potentially laborious, and is an inviting test for computer methods. A neural network has been used to classify cross peaks [172]. Treatments based on 'fuzzy' mathematics and graph theory [173,174], spin system topology analysis [175], artificial intelligence techniques [176], and pattern-matching [177] have also been reported. Compared to 2D, homonuclear 3D spectra are significantly more complicated to analyse; semi-automatic assignment algorithms [178] have now been developed to assist in this.

The determination of solution structure of proteins by NMR has become well established, but the levels of resolution varies considerably. At the coarse level, it has been demonstrated that the overall fold can be deduced with NMR data from as few as 2 spins per residue [179].

Relaxation matrix analysis has been established in recent years, and has now been refined to take into account incomplete relaxation, solvent presaturation, and the presence of heteronuclei [180]. Structural determinations are generally strongly reliant on NOESY correlations, but these are often ambiguous. An automated approach to this problem has been devised, which uses all the possible nOe assignments, and sorts out the lowest energy conformation from these [181]. NMR-derived structures are generally refined, often with molecular dynamics. The use of nOe data for refinement is potentially advantageous as it makes the most direct use of experimental data. This has been demonstrated with crambin [182,183], and another nOe-driven refinement procedure has been applied to a zinc finger [184]. The practical use of homonuclear 3D nOe-nOe spectra in structure determinations has been addressed [185]. The sensitivity of the final structure on the calibration of NOESY cross peak intensities has also been noted [186].

A number of proteins exist as symmetric dimers, for which structure determination is hindered by the ambiguity between inter- and intra-monomer nOe's. A molecular dynamics-based approach to this problem has been presented [187], and a related method applied to leucine zipper proteins [188].

A study of molecular dynamics simulations of a thioredoxin showed the importance of including solvent water, and also demonstrated how ten 100 ps simulations spanned conformational space more efficiently than a single 1 ns run [189].

An NMR structure determination often yields a family of structures, in which each individual member satisfies the NMR constraints within the appropriate error spans. This ensemble may be

represented by a single structure which can be chosen in various ways, such as an average or by the lowest energy; the merits of how this choice is made, if at all, has been discussed [190].
There has been continuing discussion of the reliability of NMR-derived protein structures [191]. The way certain details of the apparent structure of the lac repressor headpiece vary with the method used to calculate that structure has been demonstrated [192].

3.2 Protein Structure Determination: NMR investigations of a huge range of biological macromolecules have been reported this year, a significant proportion of these are described in Table 2. There is diversity in the nature of the systems studied, in their molecular weights and in the level of structural detail obtained. The general layout of Table 2 is such that reports of secondary structure and/or assignments appear first, followed by those for folding topology or gross structural information. Finally reports of determinations of tertiary structure are included in the lower part of Table 2.
It is noticeable that a majority of proteins structures reported in the last year have been uniformly labelled with ^{15}N, and a significant number have been doubly labelled with both ^{15}N and ^{13}C. In one study it was shown how extensive assignment and subsequently structural information could be obtained for a 130aa protein from only 3mg of ^{15}N-labelled protein and 2mg of $^{15}N/^{13}C$ doubly labelled protein [193]. However, there are several examples in which ^1H properties alone have been used to generate molecular structures. The largest molecular weight encountered was around 26 kDa for full structural determination [194]; and several 20 kDa systems have been investigated in similar detail [195,196]. There have been examples of refinements of previous NMR structures by the acquisition of more experimental constraints [197,198]. The structure of Ribonuclease A was improved by acquiring further data sets at 600 MHz [197]. The rms deviation over all heavy atoms was reduced from 1.8 Å to 1.5 Å, an improvement from data obtained at 360 MHz. This was further improved to 1.2 Å when dihedral angle data were included.

Table 2 Protein structure determinations by NMR

System	Remarks	Reference
Ribonuclease H(*E.coli*) 155aa	^1H, ^{15}N, ^{13}C, 95% assigned. Ten types of 3D expt. used. chemical shift/structure analysis.	199
Ribonuclease B	Comparison of NH-D exchange rates between glycosylated and non-glycosylted form at binding site.	200
Src SH3 domain	^1H, ^{15}N, six β-strands forming two anti-parallel β-sheets.	201
PI3K SH3 domain	2 β-turns, two anti-parallel β-sheets, 4 short α-helices; 3D struct.	202
Bacteriorhodopsin	^{13}C of labelled methionines, ^{15}N co-labelled amino acids Met environment determined.	203
Fibronectin I and EGF modules from tissue type plasminogen activator.	Units linked without struct. distortion	204
Fibronectin I modules human 94aa)	Linked modules same struct. as seperate. modules. pH and temp. varied, sequential assign., 2°-struct,	205
Fibronectin I modules (human, N-term.)	Linked modules Tertiary structure.	206
Soyabean legHb(15.9kDa)	2D, 3D-^1H, 2 temps, 2 pH values, 94% backbone assign., 70% sidechain.	207
Barstar (89aa)	Backbone ^1H, ^{15}N, 2°-struct.	208
Enzyme-II^{mtl}(*E.coli*)	Backbone assign., 2°-struct, 3D-het. NMR.	209

Azurin (*Alcaligenes denitrificous*) (129aa)	Complete ^1H, ^{15}N assign., 2°-struct., solvent exposure	210
Glucose transporter protein (*E.coli*).	^1H, ^{15}N, ^{13}C, assigned, 2°-struct.	211
apo-Calmodulin	Ca^{2+} free. 2°-struct.	212
Calmodulin	Sidechain pKa of Lys by ^{13}C (enriched)	213
Apolipoprotein E	Micro-environments of ^{13}C labelled Lys.	214
Protein S (173aa)	^1H, ^{15}N, ^{13}C, 2°-struct., unusual helix, Greek keys	215
Adenylate kinase (21.6 kDa)	With Mg-AP_5A bound. ^1H, ^{15}N, ^{13}C, 2°-struct. pH 7.1 monomer.	216
Histone H1(globular domain, 75aa)	Sequential assign., 2°-struct., cf. X-ray struct.	217
Calcineurin B (170aa, human)	^1H, ^{15}N, ^{13}C, 2°-struct. in detergent CHAPS.	218,219
Che Y (*E.coli*, 129aa)	^1H, ^{15}N, assignments different to previous reports. preliminary tertiary structure.	220
Oct-1 (POU specific DNA binding domain)	^1H, ^{15}N, 2°-struct., 2D, 3D NMR. 4 α helices all amphipathic	221
Oct-3 (POU homeodomain, 66aa, mouse)	^1H, ^{15}N, 2°-struct., helix-turn-helix motif. NH exchange	222
Prolyl isomerase (*E.coli*, 167aa)	^1H, ^{15}N, ^{13}C, 165aa assigned at least in part.; 8-stranded anti-parallel β-sheet, 2 helices. Topology similar to T-cell.	223
Human melanoma growth stimulatin activity.	^1H, 2°-struct., dimer interface similar to IL8.	224
c-Myb protein (DNA-binding domain)	^1H, ^{15}N, ^{13}C, almost compl. assign., 2°-struct.	225
Phosphatidylinositol 3'-kinase SH3 domain (human)	^1H, ^{15}N, 2°-struct.; 2 anti-parallel β-sheets, 6 β-strands, 2 β-turns, 4 short helices	226
Human C-Ha-ras protein	^1H, ^{15}N(selective) 2°-struct., GDP bound, backbone assign..	227
c-Ab1, Src homology 2 domain	^1H, ^{15}N, 2°-struct., 2 α-helices, 8 β-strands.	228
γ-Interferon (mouse, 39aa)	40% TFE, distance geom., restrained MD/MM, ^1H, 2°-struct., 2 α-helices, no inter-helix nOes.	229
Mu T enzyme (129aa)	^1H, ^{15}N, ^{13}C. 2°-struct.; chemical shift changes on cofactor binding	230,231
EIAV-Tat protein (75aa)	^1H, pH 3.0, near compl. assign., some α-helix., no overall stable struct.	232
EIAV-Tat protein	40% TFE/water mix., mainly helical.	233
α-Herpes virus (fragment)	2°-struct.; novel amongst Zn-containing proteins.	234
Ribosomal protein S17	^1H, ^{15}N, assignments almost complete, anti-parallel β-sheet, 'Greek key' topology.	235
F_1F_O ATP synthase (subunit C, 79aa)	In chloroform/methanol/water mixture; 2°-struct. is α-helical, folding topology.	236
Acanthamoeba profilin I	triple resonance; 5 stranded anti-parallel β-sheet, 3 α-helices, folding topology.	237,238
Human profilin (15 kDa)	^1H, ^{15}N, ^{13}C, 1186 constraints; 7 stranded anti-parallel β-sheet, 3 α-helices; global fold similar to A.Profilin (above)	239
Mammalian hnRNP A1 (92aa, RNA bind. dom.)	2°-struct., $\beta\alpha\beta\beta\alpha\beta$ folding topology.	240
Human sterol carrier protein 2 (123aa)	^1H, ^{15}N, topology from distance geometry treatment of 625 nOes and 95 dihedral angle constraints	241
Thyroid TF 1 homeo-domain(rat, 68aa)	2°-struct., helix-turn-helix topology.	242
Human CD2 (glycosylated adhesion domain, 105aa)	98% backbone, 80% sidechain ^1H assign., ^{15}N, topology.	243,244,245
Human stromelysin 1 (catalytic domain, 19.5 kDa)	^1H, ^{15}N, ^{13}C 2°-struct., folding topology.	246
Human stromelysin (short form)	^1H, ^{15}N, ^{13}C 2°-struct.; Zn ligation effects	247
CD59 (human, 77aa)	Almost compl. assign., also resonances of N-glycan. Similar topology to snake venom equivalent protein.	248
CD59 (70aa, extracell. region)	2D ^1H. 2°-struct. and toplogy.	249
Bacterioopsin (1-71)	Methanol/chloroform mixture and in SDS micelles. ^1H, ^{15}N, 651 nOes (org. solvent), 520 (micelles). Topology det.	250
p6 protein (fromHIV-1 gag)	Possibly a single α-helix, overall no stable single structure	251
Fd coat protein	α-helices and their arrangement determined in solution and solid state	252
γS resolvase (20.5 kDa)	dimerisation supported by linewidth measurements.	253
Ada protein (N-term frags. 14, 16, & 20 kDa, *E.coli*)	^1H, ^{15}N. 14 kDa protein has 2 α-helices and β-sheet with mixed parallel and anti-parallel strands; folding topology	254

Ada protein (*E.coli*, 10 kDa frag. DNA repair domain.)	Zn-binding motif unlike that in other transcription factors.	255
Kringle 4 (horse plasminogen, ε-aminocaproic acid bound)	3D structure determination	256
Urokinase-type plasminogen activator.(Kringle domain)	946 nOes, 48 dihedral angles; simulated annealing, global fold similar to tissue type plas. act. Kringle 2.	257
Urokinase-type plasminogen activator.(N-terminal frag. and Kringle domain)	1299 nOes, ^{1}H, ^{15}N, ^{13}C; compared with homologous proteins.	258
Dendrotoxin K (57aa)	Struct. similar to BPTI, 38 diastereospecific assignments; 818 nOes, 123 dihedral angles.	259
Arc repressor (*Salmonella* phage P22)	Structure determination using ensemble relaxation matrix calcs.	260
Glucocorticoid receptor (DNA-binding domain)	906 nOes, 43 dihedral angles. RMS deviation for backbone atoms 0.7Å, 1.44 Å for all heavy atoms.	261
Mot A protein (from T4)	Combined x-ray/NMR approach	262
Kunitz-type proteinase inhibitor (sea anemone, 55aa)	pH 4.6, 36°C, 666 nOes, 122 dihedral angles; DIANA. High quality struct., compared with 4 homologus proteins.	263
Lysozyme (turkey egg white, 129aa)	All backboneassigned, 75% side chain assign., nOe, H-D exchange dihedral angles. Structure identical to hen egg white lysozyme	264
IGF-I (human)	320 nOe, of which 12 related to disulphide bridges. Dist. geo., RMS deviation 1.9 Å for backbone atoms.	265
Cold-shock protein, CspB (*B.subtilis*, 67aa)	3D structure very similar to *Staph*. nuclease.	266
Flavoridin (snake venom, 70aa)	2 short regions of anti-parallel β-sheet, no other classical 2° struct. motif.	267
HMG-1 (hamster, DNA bind. domain)	α-helices fold to 'V'.shape; mode of interaction with 4-way DNA junction suggested.	268
Interleukin-4 (human, ca. 15kDa.)	2973 constraints, hybrid dist. geo., simulated annealing high resolution 3D structure.	269
Human TGFα (50aa)	pH 6.5 & 3.5, 10° & 30°C, 357 nOe distances, 26 dihedral angles.	270
LFB1/HNF1 transcrip. factor (rat liver, 92aa)	^{15}N, ^{13}C, ^{1}H, non-classical helix-turn-helix	271
FABP (bovine heart, 15 kDa)	1027 nOe distances; deuterated fatty acid ligand; DIANA, β-barrel of 10 anti-parallel β-strands, short helix-turn-helix.	272
HPr (*Staph.* aureus)	2D ^{1}H, restrained mol. dyn., twisted anti-parallel β-sheet, similar to other HPr's	273
Calbindin D_{9k}	1176 constraints;16 per residue overall, 32 per residue in core. Dist. geometry., restr.mol.dyn. to generate high resolution. struct.	274
Calbindin D_{9k} (disulphides in EF-hand subdomain dimers)	^{1}H, ^{113}Cd, Ca^{2+}. ^{1}H assign. confirm S-S bond.	275
Dihydrolipoamide acetyl transferase (43aa)	429 nOes, 13 dihedral angles; 3D structure.	276
Jun oncoprotein dimer (Leu zipper, 9.8 kDa)	Almost compl. ^{1}H assign., H-D exch., coiled-coil arrange. of parallel α-helical strands.	277
Antibody F_V fragment (^{15}N-labelled in V_L domain)	HSQC, 3D NOESY-HSQC; 105 of 106 NH's assigned	278
Spasmolytic protein (porcine growth factor, 106aa)	Almost compl. assign., 2D, 3D homonuclear; trefoil motif, unusual polypeptide fold. dist. geometry.	279
Antibody V_L domain (isolated)	930 nOes, 113 dihedral angles. Rmsd 0.85 Å for backbone atoms bone atoms, 1.29 Å for all heavy atoms.	280
Antibody F_V fragment (26 kDa)	Assign. achieved by ^{13}C,^{15}N labelling domains separately	194
Mutant α-amylase inhibitor(R19L)	845 nOes, 154 dihedral angles;DG,rMM; localised differences to WT	281
P22 c2 repressor (DNA-binding domain)	^{1}H, ^{15}N, ^{13}C. similar to 434 repressor.	282
Chicken crystatin (phospho- & unphosphorylated varients)	2D, 3D and 4D NMR; structures similar; 5-stranded β-sheet wrapped around 5-turn helix; 2nd helix absent but seen in x-ray structure	283
M13 coat protein (gVIIIp)	In SDS;.no aggregation at high SDS conc.; 2 α-helices present	284
Acyl co A-binding protein (bovine liver)	Structure refinement using extra nOes and dihedrals; auto.assign using 'PRONTO'; 1096 nOes, 124 dihedral angles, DIANA, XPLOR.	198
Ras p21.GDP (166aa)	3D, 4D expts. almost compl. ^{1}H, ^{15}N, ^{13}C. 3369 constraints Rmsd backbone atoms 0.55 Å, all heavy atoms 1.0 Å.	195
Periplasmic Cyclophilin (*E.coli*. 167aa)	3D structure. 1720 constraints. Dist. geo.; 8-stranded β-barrels capped by α-helix. signif. differences with. x-ray structure near binding site	196

3.3 Protein Dynamics: In recent years NMR has made a great contribution to the understanding of proteins in solution as dynamic species, rather than inert, static structures. This is changing the terms in which proteins are generally considered.

NMR is an excellent probe of dynamics, as it can be used to investigate an extremely wide range of timescales. Various NMR relaxation parameters are available for the analysis of dynamics, but these are not always accessible in practice. Methods have been devised for accessing the relaxation of protons in crowded spectra [285]. There are an increasing number of reports which exploit the possibilities offered by $^{15}N/^{13}C$-labelled proteins. Certain dynamic data, such as amide exchange rates, are often included in reports of structure determinations. In this section we highlight reports which have dynamics as the principle consideration.

The last year has seen the addition of ^{15}N $T_{1\rho}$ [286] to the range of ^{15}N-related parameters which can be measured in labelled proteins. ^{15}N-based methods have been used to characterise backbone dynamics in human profilin [287], (1-71)bacterioopsin in different media [288], FK506-binding protein (free [289] and in a complex [290]), a peptide bound to calmodulin [291], and apo- and Cd^{2+}-calbindin D_{9k} [292]. ^{15}N-relaxation measurements have been compared to molecular dynamics simulations in the DNA-binding domain of the glucocorticoid receptor [293]. A fluctuating system can be expressed in terms of a 'generalised order parameter', S^2; values of this from the simulations and NMR data generally agree except where motions are inadequately sampled by the simulation. This occurs for large-scale, slow, motions for which the simulation had not been run long enough. NMR-derived general order parameters have been related to thermodynamic quantities [294] so the contribution of backbone fluctuations to the free energy can be assessed. It is differences in free energy, rather than absolute values, that are generally accessible, so Ca^{2+}-binding to calbindin D_{9k} was analysed in this way.

The rates of exchange of amide protons with solvent water are important structural and dynamic indicators. The rates can span a wide range of timescale. A method has been demonstrated for determining relatively fast exchange rates ranging from 0.5 to 50 Hz in labelled proteins [295]. For much slower timescales, another demonstration of linear prediction for determining rates as slow as 0.1 h^{-1} has appeared [296]. Amide exchange kinetics, and the activation energies for exchange, have been used to probe amide environments in hen egg white lysozyme [297]. Amide exchange kinetics in free *Staphylcoccal* nuclease and in a ternary complex with Ca^{2+} and thymidine-bis-phosphate indicated how the complexation alters global and local fluctuations [298]; the opening of the structure necessary to expose certain interior amides can be explained in terms of global unfolding. This study is a good illustration of the dynamic nature of protein structures in solution which is increasingly being realised.

Tritium NMR is fairly infrequent in studies of natural macromolecules, but it has been used in a study of chymotypsin, labelled by tosylation [299]. The tosyl group appeared highly mobile in solution, spending only about half the time in the location observed in a crystal structure.

3.4 Protein Folding: NMR is continuing to make strong contributions to the understanding of how proteins fold into their native states. Such studies are often supported by data from other techniques, such as calorimetry. Folding intermediates have been observed and characterised in a number of cases, most notably the so-called molten globules. Studies of this sort have continued to appear during the last year. Most proteins are not amenable to reversible unfolding in the NMR tube. Those which can be studied in this way may not necessarily constitute a representative subset of Nature's proteins, but they are enabling an understanding of unfolded states and folding transitions.

Once a protein has been completely denatured, there is obviously an absence of tertiary interactions. Resonance assignments cannot then rely on nOes. Therefore a general method for assigning denatured proteins has been devised, which uses triple resonance and does not involve nOe steps, instead relying only on scalar coupling. This has been demonstrated with FK506-binding protein [300]. When unfolded in urea and guanidinium chloride, this protein shows extensive conformational averaging, but there are defined areas of secondary structure which differ slightly between the two solvents. For the most part this residual secondary structure corresponds to that in natured state except at C terminus which is different [301].

Molten globule states have been characterised in guinea pig α-lactalbumin. At low pH, amide exchange indicates protected amides in 2 hydrophobic regions, stable sub-domains [302]. Progress is also being made in the unfolding of these globule states. For human, bovine and guinea-pig α-lactalbumins, the molten globule 'A-state' existing at pH 2.0; can be further unfolded by addition of TFE. The resulting further partially-folded state has some resemblances to the A-state [303]. The unfolding of bovine α-lactalbumin A-state molten globule by urea has been shown not to proceed as a 2-state cooperative process [304]. The pH-induced unfolding of bovine serum albumin does display structural transitions [305].

There have been several reports of folding studies of the well-characterised protein *Staphylcoccal* nuclease. The refolding of the thermally-denatured protein has two distinct phases, due to *cis-trans* isomerisation at a Lys-Pro peptide bond [306]. Mutation of this Pro to Gly simplifies matters somewhat, and a folding intermediate of this mutant has been characterised by amide hydrogen exchange [307]. Another mutation, a Gly-to-Val substitution, destabilses the native structure but also reduces sensitivity to a denaturant. It does this by alteration to the denatured state; the mutation perturbs the cooperative breakdown of the folded conformation so that the denatured state is more ordered and more stable than the wild type [308]. Deletion of parts of both termini gives rise to a partially folded species, which has some simarlities with the native form [309].

Alcohol-induced unfolding of platelet factor 4 leads to 'O-states' which retain considerable secondary structure [310]. The urea-induced folding intermediate of tryptophan synthase α-subunit has molten globule characteristics [311], and a different study [312] has shown other related intermediates as the urea concentration is varied. The protease domain of a urokinase-type plasminogen activator has been observed in partially folded states with stable subdomains [313]. The partially denatured A-state of human ubiquitin has been characterised in greater detail than previously by using ^{15}N-labelled protein [314]. An apo-plastocyanin has been denatured to a molten globule-like state; the transition to this involving successive *trans-cis* isomerisation at two Pro residues [315]. Subtilisin BPN' has been observed as a folding intermediate with molten globule-like characteristics; a 77-residue pro- sequence is required for this to fold to the native state; presumably this stabilises an otherwise prohibitivly high-energy transition state [316]. The pro-region required for the folding of carboxypeptidase Y is, when isolated, a partially-folded system

[317]. Equine milk lysozyme has been observed to undergo a two-stage unfolding process, with the first step being sensitive to the presence of Ca^{2+}. The intermediate state has resemblances to the molten globules of α-lactalbumins [318].

BPTI has served as a model in numerous NMR studies, and this is so in folding studies. The BPTI folding pathway involves non-native disulphide bridges. These intermediates have been characterised and the non-native disulphides found to be in non-folded regions, and not disrupting folded regions [319]. A BPTI analogue, with all the Cys residues replaced by Ala, has no stable secondary structure but there is some evidence that it contains a hydrophobic cluster [320]. Mutants of BPTI have been used to correlate amide exchange with transient unfolding [321], the core of the protein exchanging very slowly. Another trypsin inhibitor, EETI II, contains three disulphide bridges; reduction of these and subsequent reoxidation produces a stable two-disulphide species, which is also formed by gentle reduction of the native form [322]. Structurally, this intermediate is very similar to the native, with major differences confined to regions around the disrupted disulphide.

The kinetics of folding is also being studied, and again such work often combines NMR data with information provided by other techniques. H-D exchange, observed by 2D NMR, has been applied to the folding of ribonuclease T_1 [323]. This uses rapid mixing techniques and relies on the differences in solvent exchange rates between H-bonded and non-H-bonded amides, monitored as the protein folds. This study shows at least two parallel pathways, each with at least one intermediate, though these intermediates have different characteristics. The kinetics of the folding of the all β-sheet protein interleukin-1β has been studied, and demonstrates some very slow late steps [324].

3.5 Mutagenesis: Site specific mutagenesis may be applied to the probing of structure-function relations for individual amino acids in proteins. The structural and dynamic consequences of a mutation can be assessed in great detail by comparison of NMR data with that from the wild type, this is often supported by activity/binding data. Table 3 shows the range of such reports in the last year, which are not covered in the other sections of this review. In general, the design of a mutant protein is a compromise; the change should be interesting enough to warrant study, but not so disruptive that the protein is no longer viable.

To a less general extent chemical modifications can also be used in this sort of study.

Table 3 NMR Studies of mutant proteins

Mutated Protein	Remarks	Reference
Dihydrofolate reductase (*E.coli*)	loop of 3 aa's replaced by Gly: effects dynamics	325
Staphylcoccal nuclease	mutations of residue preceeding a Pro; normally *cis* peptide bond can become *trans*	326
Calbindin D_{9k}	mutatations to give disuplhide bridges between monomers are stabilised, Ca^{2+} uneffected	327
Calbindin D_{9k}	mutations at Ca^{2+}-binding site, localised structural effects	328
Phospholipase A_2	mutation of catalytic His shows structural and catalytic roles	329
Antifreeze protein type III	mutate Pro to remove confusing *cis-trans* isomerisation	330
Phosphotyrosyl protein phosphatase	2 conserved His mutated: one is important for catalysis, other has structural role	331
Human ubiquitin	Phe mutated to Trp; only effects chem.shifts of neighbouring residues.	332
T4 lysozyme	mutations to probe enivonment changes at a Trp residue	333

Yeast chorismate mutase	mutant locked in activated state compared to wild type	334
Kringle 2 domain of plasminogen activator	global fold of mutants assessed from high field methyl signals	335
Kringle 2 domain of plasminogen activator	conserved Trp mutated; important for structural stability but not binding	336
Kringle 2 domain of plasminogen activator	mutants causing only localised struct.changes, effects on binding constants of various ligands	337
Human c-Ha-Ras protein	mutations of 2 Tyr residues show differing functional roles	338
Murine interleukin-6	replace last 5 residues with human sequence, also mutant with these 5 deleted: region has little struct. or funct.role	339
Adenylate kinase	Thr involved in catalysis, indicated by proximity of side chain with bound substrate analogue	340
Calmodulin	effects of mutation of unique trimethyl Lys	341
Epidermal growth factor	different mutations of conserved Asn cause variable changes in ^{1}H spectrum.	342
Adhesion domain of CD2	mutations used to locate binding site of CD58	343
M13 gene V protein	binds single-stranded DNA; full structure determination with Tyr mutated to His, consistent with binding data	344
Cytochrome c-550	mutation of conserved Lys effects structure and function	345
HPr	mutants used to locate sites of deamidation	346
Ribonuclease & lysozyme	^{15}N-tetranitromethane to modify Tyr's; new reporter group	347

3.6 Ligand binding: In this section we are concerned with NMR studies of protein function; for the most part the interactions between proteins and ligands, the latter ranging from metal ions to large receptors (but not including oligonucleotides, these complexes are considered in section 3.7). For NMR studies of enzymes, the complexes characterised by NMR are usually catalytically inactive ('dead-end') since NMR is not generally sensitive enough to be able to monitor the progress of a reaction and provide detailed structural information at the same time. Complexes may be tightly bound, or looser so that exchange effects are observed. There are examples of these in Table 4, which lists reports of protein-ligand interactions.

Table 4 NMR Studies of Protein-Ligand Interactions

System	Remarks	Reference
Acanthamoeba prolifin I	effect on 2°-struct and site of binding of poly-L-Pro determined	348
Human prolifin	binding of poly-L-Pro has highly localised effects	349
Dihydrofolate reductase (*L.casei*)	tautomerism and ionisation states of bound ^{13}C-labelled folate	350
Dihydrofolate reductase (human)	complex with labelled folate; bound tautomer identified,relevance of pK_a of active site carboxyl to hydride transfer questioned,	351
Thymidylate synthase (*L.casei*)	inactivated by carboxypeptidase A; interaction with 5-F-dUMP	352
Thymidylate synthase (*E.coli*)	mutation effect on ligand binding; 5-F-dUMP and tetrahydrofolate	353
N^{10}-formyltetrahydrofolate synthase	conformations of bound ADP and ATP	354
FK506-binding protein	secondary struct. in complex with ascomycin; compared to uncomplexed protein and crystal structure	355
FK506-binding protein	dynamics of bound FK506; solvent exposure and mobilities	356
Calbindin D_{9k}	mutant binding only 1 Ca^{2+} ion; able to characterise this binding	357
Interleukin-8	complex with 40aa of N-terminal of receptor; binding surface defined	358
Class II MHC	binding of ^{13}C-labelled 16aa peptide; preliminary model obtained	359
Rhodopsin	bound conf. of 11aa peptide from G-protein related to rhodopsin state	360
Myosin subfragment I	^{3}H-ATP; hydrolysed to ADP by protein, ^{3}H NMR of bound ADP	361
Osteocalcin (from dog bone)	Ca^{2+}, Lu^{3+} binding studied by 1D ^{1}H NMR; binding constants	362
Barnase	indentification of binding site of the inhibitor barstar	363
Barnase	3'-GMP binding; no major structural changes but does effect dynamics	364
IgG interaction with FcγRII	mobility change of a Leu in hinge peptide from Fc of IgG on binding	365
IgGbinding of *Staphylcoccal* protein A	^{13}C-labelled Fc fragments of IgG; chem. shift changes on binding to define contact regions	366
Calmodulin	^{13}C-labelled dimethyl Leu's, intact and fragments; tentative assignments; Leu's not involved in Ca^{2+} binding	367
Calmodulin	interaction with phospholambin cytoplasmic domain characterised	368
Calmodulin	Mg^{2+} interference with Ca^{2+} and Cd^{2+} binding	369
Phosphotriesterase	2 different metal ion binding sites characterised by ^{113}Cd	370

Bovine & human serum albumin	Ni^{2+}, Cu^{2+}, Co^{2+}, Cd^{2+}, Al^{3+} interactions with N-terminal regions	371
Ribonuclease Pb_1	complex with phosphonate analogue of GpU; mode of binding	372
Ribonuclease H	^{25}Mg NMR; variation with pH to characterise metal binding site	373
Human N-ras p21	protein conf. changes between GDP & GTPγS complexes	374
Human c-H-ras p21	timecourse using lowfield-shifted protons of bound GTP; to GDP	375
Human p21-ras	wild type and mutant binding to GTP analogues; protein confs.	376
Ada protein	^{113}Cd used to show metal ion may act as conf. switch	377
LIM-domains	^{113}Cd to characterise metal ion coordination	378
Human tyrosine hydroxylase	bound phenylalanine conf. and interaction with metal ion binding site	379
Acyl-coenzyme A binding protein	solution structure of palmitoyl-coA complex; specific interactions	380
Cyclophilin	peptidyl prolyl isomerase activity; conf. of bound model substrate	381
Cardiac troponin C	structural effects of Ca^{2+} binding at various sites; structural roles in N- & C-terminal regions	382
α-thrombin & prothrombin	interaction with hirudin analogue; binding constants	383
Mouse monoclonal antibodies	anti-4-hydroxy-3-nitrophenylcetyl; hydrophobicity of binding site	384
α-lytic protease & chymotrypsin	^{11}B with boron-containing peptide analogue; tetrahedral boron in both His- and Ser- adduct complexes	385
Retinol-binding proteins	binding of retinol analogues in 2 proteins compared	386
Mouse ribonucleotide reductase	mobile C-terminal of one subunit is restricted on binding other subunit	387
UDP-galactose 4-epimerase	signals from bound labelled NAD^+ perturbed by binding of UDP; consequences for catalytic mechanism	388

3.7 Protein-Nucleic Acid Interactions: Protein-nucleic acid interactions are of enormous biological importance, and this topic is being increasingly studied by NMR. Thus here is a section devoted to such studies. Whilst the number of reports of this category is not large, the importance of these systems justifies this section, especially in the light of recent developments. Prior to the recent advances in methodology involving protein NMR, complexes of proteins and DNA duplexes were rather too large for detailed NMR studies, as the homonuclear 2D spectra were generally too crowded for extensive analysis. This has now been overcome with the advent of $^{15}N/^{13}C$-labelled proteins, which allow the distinction between resonances of the protein and DNA. This has enabled the solution structure determinations of several protein-DNA complexes, with details of the recognition interfaces becoming apparent. NOe's between the protein and DNA can be unambiguously identified. This is the case in the structures of a complex of the lac repressor headpiece and an 11 base pair DNA duplex (representing a half operator) [389], the complex of an Antennapedia homeodomain and 14 base pair DNA duplex [390,391], and a DNA complex of a zinc-containing DNA-binding domain of chicken erythroid transcription factor [392]. In each of these cases a number of specific side chain-DNA interactions are observed, but also a significant degree of hydrophobic interactions. Interactions between ^{15}N-labelled ribonuclease H1 and 12 base pair duplexes of DNA, RNA and a DNA-RNA hydrid have been studied [393]; distortion of the protein was only observed in the case of the hybrid binding.

The interaction of the 65-residue N-terminal region of GAL4 and a 10 base pair DNA duplex has been studied by 2D homonuclear NMR [394], showing that this monomeric system has the same recognition features as the natural dimeric arrangement. Preliminary NMR details of the interaction of histone-like HU_α with DNA have been reported, which apparently involve binding to the DNA minor groove [395].

When confined to 1 dimension, ^1H spectra of protein-nucleic acid complexes are very crowded in most regions. However, progress can be made by observing NMR signals which are unique to the nucleic acid, and hence can be studied without confusing overlap from protein signals. Generally this can be done using ^{31}P and H-bonded imino ^1H resonances. Changes in the ^{31}P spectra of a range of duplexes on interaction with the lac repressor headpiece [396] show that in tight binding complexes the inherent phosphate ester flexibility is retained, but is quenched in weaker binding complexes. Localised backbone perturbations in various DNA duplexes have been suggested in ^{31}P studies of binding with the *trp* repressor [397]. Imino ^{1}H signals have been used to study T4 endonuclease interactions with different DNA duplexes [398].

Another approach is to introduce a selective label, such as fluorine, into the nucleic acid. ^{19}F signals from 5-fluorodeoxyuridine incorporated into RNA and DNA in a structural study of transcription by T7 RNA polymerase have been used as indicators of changes in local environment [399]. ^{19}F signals from fluorine-containing DNA duplexes have been used to study base sequence symmetry effects in interactions with the lac repressor headpiece [4] .

Alternatively, rather than introducing new resonances, signals can be removed (quenched) instead. This can be done using spin-labelled DNA which broadens signals of nearby protein protons, whose correlations are then absent from homonuclear 2D spectra. This approach has been applied to the DNA-binding domains of M13 and IKe gene V proteins and their interactions with a trinucleotide [401,402].

3.8 Zinc Fingers: Zinc fingers are peptide domains of around 30 residues which are able to bind Zn^{2+} ions, the coordination of which is via the side chains of His and Cys residues, in various combinations. Zinc fingers occur in a wide range of proteins, and can act as DNA-binding motifs. Interest in these systems is increasing. There have been a number of reports of studies of zinc-finger domains; and these range across any notional division between proteins and peptides. Therefore this year we collect these reports in one section; they are listed in Table 5.

Table 5 NMR Studies of Zinc-Finger Domains

System	Remarks	Reference
HIV-2 nucleocapsid protein 1st Zn-finger	compact folded struct., 2 slowly interconverting forms	403
HIV-1 nucleocapsid protein 1st Zn finger	NC_p7, His-Cis mutation alters structure, loss of activity	404
HIV-1 nucleocapsid protein 1st Zn finger	complexes with DNA; d(AGCC) binds in hydrophobic cleft	405
HIV-1 nucleocapsid protein; 2 Zn fingers	2 fingers with basic linker sequence; modification here removes activity and effects spatial proximity of the 2 fingers	406
Eukaryotic transcriptional elongation factor TFIIS	solution structure; new motif described as a 'zinc-ribbon', contains 3-stranded β-sheet	407,408
Human retanoic acid receptor-β DNA binding domain	2 Zn-fingers, structure resembles glutocorticoid receptor but differs in C-terminal region	409
Human retanoic acid receptor-β DNA binding domain	Use of ^{113}Cd; heteronuclear experiments show that 1st eight Cys coordinate metal ion, ninth Cys does not	410
Human retanoic acid receptor-α DNA binding domain	2 Zn-fingers; again structural similarities to other receptors, function of additional helix in this molecule discussed	411
ADRI-UAS1	C-terminal Zn-finger solution structure, 2 Pro residues mutated to Ala removes conformational hetergeneity	412
ADRI-UAS1	N-terminal Zn-finger; mutations that reduce DNA-binding do not cause significant structural changes to peptide; side chains make direct contact with DNA	413

3.9 Metalloproteins: Metalloproteins are those in which metal ions are integral components of the molecular structure. Often the metal ion is paramagnetic, which causes wide chemical shift ranges and distinctive relaxation behaviour. Generally rapid relaxation of ^{1}H does not preclude the use of multidimensional experiments. The application and adaptation of homonuclear 2D experiments in such circumstances has been demonstrated [414]. An nOe-nOe spectrum of a myoglobin derivative provides the first example of a homonuclear 3D experiment on a paramagnetic system [415]. The metal ion itself may provide an NMR signal, or can be substituted for an NMR-active species such as Cd^{113}. Use of Cd^{2+} is fairly common, as it can mimic several different ions. For example, it can replace zinc with coordination by cysteine sulphurs. In such a case, a Karplus relation has been found to apply to three-bond ^{1}H-^{113}Cd scalar coupling constants in a Cd-substituted rubredoxin [416]. Here the dihedral angles were extracted from a crystal structure of the native protein. Reports using other NMR-active metal ions in proteins are also starting to appear. ^{205}Tl, a spin $I=^{1}/_{2}$ nucleus, has been used in a study of ovatransferrins [417]. Here the Tl^{3+} was found to have a highly anisotropic chemical shift, which dominated its relaxation behaviour. The binding of Al^{3+} to transferrins has been investigated using ^{27}Al NMR at different field strengths [418]. Only the central $m=^{1}/_{2} \rightarrow -^{1}/_{2}$ quadrupolar transition was observed. This was also the case for ^{45}Sc NMR with Sc^{3+} bound to ovatransferrin [419], and is due to the far from extreme narrowing conditions that operate in proteins. The quadrupole coupling constants and ionic rotational correlation times were derived, demonstrating that quadrupolar nuclei can provide useful physical information. More applications of this sort can be expected in future. Metalloproteins can provide excellent demonstrations of aspects of NMR physics which are not generally evident in other classes of natural macromolecules. Table 6 lists the reports of metalloprotein studies.

Table 6 NMR Studies of Metalloproteins

System	Remarks	Reference
IP	NMR consistent with fast exchange between 2 different orientations of Fe cluster in protein frame	420
HiPIP I	characterisation of electronic structure and electron self-exchange process	421
HiPIP, HiPIP II	modified; electronic structures of $[Fe_4Se_4]^{4+}$ clusters deduced	422
HiPIP	assignment of aromatic residues	423
Glutaredoxin (*E.coli*)	mutant in mixed disuphide with glutathione, full structure det.	424
Glutaredoxin (*E.coli*)	^{13}C,^{15}N labelling, assignments of mutant mixed disulphide with glutathione	425
Hemoglobin	modified porphyrin moiety, effect on structure	426
Hemoglobin	myoglobin-like monomer component IV, from marine annelid. Hyperfine resonance assignments of heme protons	427
Hemoglobin	α-globin from human adult, with Zn-porphyrin; assignments and changes with CO binding determined, compared to x-ray crystal of normal subunits	428
Myoglobin	from mollusc, cyanide-ligated ferric complex: internal dynamics	429
Myoglobin	from sperm whale, CO-form, His pH titrations	430
Myoglobin	from horse heart & sperm whale, bound to trimethylphosphine, assignments of heme pockets and other residues	431
Myoglobin	equine; thiocyanate complexation slows transition between acid & alkali forms	432
Myoglobin	mollusc & shark, acid-alkaline transitions compared; different distal residues	433
Myoglobin	equine, mollusc, shark: transfer of assignments in high spin forms to low spin using saturation transfer	434
Myoglobin	human; NMR relaxation shows water ligand of Fe in mutants	435
Myoglobin	Mn(III)-form, study of binding of ^{15}N-labelled CN^- and SCN^-	436
Myoglobin	sperm whale apo-protein with iron porphyrin, so heme side chains absent	437
Myoglobin	human; mutants to probe role of His in ligand binding	438
Myoglobin	sperm whale; mutants effecting ligand orientation	439
Myoglobin	from two sources, mutants to probe distal H-bonding in ligation	440
Myoglobin	*Aplysia limacina*, active site structure determined	441
Myoglobin	sperm whale; active site study in cyano-Met form of a mutant	442
Myoglobin	sulph-myoglobin from horse, with deuterated heme; characterised	443
Myoglobin	sulf-myoglobin with deuterated heme; 2 prosthetic groups probed, 2D ^{1}H spectra of paramagnetic species using viscous solvent	444
Myoglobin	sperm whale; mutants to probe ligand stabilisation	445
Myoglobin	elephant; Met-cyano complex; heme cavity study	446
Myoglobin	sperm whale; combined use of metaquo- and carbonmonoy- complexes to derive zero-field splitting constant	447
Myoglobin	more assignments of hyperfine-shifted proton signals, T_1 relaxation mechanisms	448

Leghemoglobin	new procedure for ^{1}H assignments and conf for heme propionates by 2 & 3D	449
Cytochrome p450	reduced form, broad signals; relation of this with electronic structure	448
Peroxidase	horseradish, lignin & manganese; binding to their substrates: characterisation of interactions with heme active sites	450
Peroxidase	lignin and manganese; ^{15}N-labelled cyanide complexes	451
Peroxidase	horseradish; esterified heme propionate group; binding of SCN$^-$ demonstrates salt bridge between distal amino group & propionate	452
Peroxidase	horseradish, interaction with EDTA monitored primarily by relaxation rates binding only observed at low pH (below 5)	453
Cytochrome p450	newly isolated; free and substrate-bound forms characterised	454
Cytochrome c553	protein expressed in different organism; isolated oxidized form exists in 2 non-interconverting conformations	455
Cytochrome c551	comparision of observed and calculated shifts	456
Cytochrome c6	ferro- & ferri- forms studied; unusual heme axial coordination apparent	457
Cytochrome c3	assignment of heme & aromatic resonances	458
Cytochrome c3	resonance assignments, compared to calculated chem. shifts	459
Cytochrome c3	resonance assignments	460
Cytochrome c	mutations to probe interior of protein	461
Cytochrome c	double mutation; structural effects confined to small region, side chain interactions observed	462
Cytochrome c	from horse, pyridine binding: rates and binding constants	463
Cytochrome c	horse; oxidized; structure & dynamics with THF as solvent	464
Cytochrome b5	bovine ribosomal; recombinant wild type and truncated versions characterised	465
Cytochrome b5	rat; assignments from diamagnetic protein transferred to paramagnetic	466
Cytochrome f	heme Me linewidths in free protein and cross-linked to plastocyanin to determine electron transfer rate in complex	467
Ferricytochrome c551	characterisation of axial-Met conformation, pK_a's determined	468
Ferricytochrome c"	assignment of heme cavity resonances	469
Iso-1-Ferricytochrome c	identification of a Lys as an Fe ligand in one alkaline form of the protein	470
Iso-1-Ferricytochrome c	exchange between free and non-covalent complexes with 2 forms of cytochrome c peroxidase; electron transfer rate, complex lifetime	471
Cytochrome c peroxidase	cyanide form complexed with different ferricytochromes c; structural effects	472
Ferricytochrome b5	various sources, hyperfine shift analysis to probe protein-heme interactions	473
Ferricytochrome c551	2D ^{1}H methods to assign fully heme resonances	474
Superoxide dismutase	Co-substitution, now studied by 2D ^{1}H experiments	475
Superoxide dismutase	Fe(II), from *E.coli*, find same active site arrangement as in Fe(III)form	476
Ferredoxin	diamagnetic, ^{15}N-labelling, comprehensive resonance assignments & 2°struct.	477
Ferredoxin	4 Cys systematically mutated to Ser; decrease in stability	478
Ferrodoxin	mutations of 2 Pro's has little effect on structure, but effect redox potentials	479
Ferrodoxin	from 2 sources, sequential assignments and comparisions	480
Ferrodoxin	Ru-modification of a specific His confirmed by NMR	481
Flavodoxin	desulfovibrio vulgaris, oxidized, ^{1}H, ^{15}N; assignments & secondary structure	482
Thioredoxin	2 active site Cys replaced by Ser; resembles wild type	483
Thioredoxin Ch2	from chloroplasts of an algea; first example from photosynthetic system oxidized and reduced forms assigned, secondary structure and global fold	484
Ribonucleotide reductase R2	Co(II)-form; 87 kDa but remarkably high resolution paramag. shifted ^{1}H signals, some preliminary resonance assignments	485
Sulphite reductase	hemoprotein subunit; assignments in various ligated states	486
Pyoverdin GM-II	Ga^{3+}-analogue used, solution structure determined	487
Metallothionein	Cu(I) and Ag(I)-substituted, full sequential ^{1}H assignments, 2 forms similar in metal coord.; C-terminii conformational heterogeneity , use of ^{1}H-^{109}Ag HMQC.	488
Lactoperoxidase	goat & bovine; NO_2^- & SCN$^-$ binding	489
Pseudoazurin	a type I blue-Cu protein; partial ^{1}H assignments; self exchange rate constant for 1:1 mixture of oxidised and reduced forms	490

4. DNA

NMR studies of deoxyribonucleic acids have been reported in reasonable numbers. There is continued interest in using NMR to elucidate sequence-dependent structural variations of DNA but the size of the sequences being investigated has still not increased much beyond 20 base-pairs, and indeed the majority are 12 base-pairs or less. Researchers have used ^{1}H and ^{31}P [499,502,511,512,520,543,545,547,568] NMR in the main but several groups have used ^{13}C and ^{15}N nuclei, both in natural abundance [510,511,583,589] and enriched [570], to probe the conformation and dynamics of DNA. Irrespective of the precise approach, the most appropriate way to classify the publications is by virtue of the secondary structure of the sequence under consideration. There have been studies of 'natural' duplexes [491-515], where natural' refers to naturally occurring nucleic acids but not necessarily natural (Watson-Crick) base-pairing, of non-natural duplexes [514-523] (i.e. containing modified nucleic acids), of duplexes that are hybrids of DNA and RNA [541,542] and of single DNA strands that fold to form unimolecular secondary structure elements such as hairpins [543-549]. More complicated structures have been investigated; the majority of these being of the DNA junction [550-562] rather than triplex-type [563-566]. The next largest area of interest after natural duplexes appears to be that of DNA-drug or small molecule interactions.

It remains the case that the majority of the NMR studies reported involve a qualitative interpretation of NMR-derived structural parameters, but there are some exceptions [534,535]. There has been a very thorough investigation of, in sequence terms, a rather unusual non-self-complementary 13-mer duplex [491]. This sequence forms part of a repeating unit found in the genome of the Hepatitus B virus. Imino proton exchange rates were monitored, 501 distance constraints were extracted from NOESY data sets and torsion angles were determined for 22 of the 26 nucleotide residues. Distance geometry followed by restrained molecular dynamics treatments of the data indicated a disruption of B-type helical parameters at the T.A base-pair at position 9 in the sequence. It was also noted that the imino proton of T_9 exchanged at twice the rate of other non-terminal thymine iminos. In a similar manner a conserved 13-mer sequence from the genome of HIV-1 has been studied in some detail [492]. NOe distance constraints, and torsional angles from the measurement of scalar coupling constants, have been used as the restraints in molecular dynamics calculations. The starting structures for the spectral simulation and dynamics runs were A-type and B-type helices, and these converged to essentially the same structure. Although the 13-mer is B-type there appears to be some longitudinal compression of the helix. A 14-mer duplex with a 5'-CCAA-3' sequence that is recognised by several transcription factors has been investigated [495]. A set of 260 constraints which included 73 torsion angles, 79 nOes and 108 hydrogen bonds, were derived and used to produce a distance-geometry model that includes a bend and base-stacking that differs from that found in standard B- or A-type DNA. In another study of a dodecamer, although 250 distance contraints were derived from NOESY data sets this was deemed an insufficient number of constraints to determine the structure of the duplex following the usual molecular dynamics, molecular mechanics and spectral back-calculation protocols [504]. In general the quality of structures depends on the number of constraints available and efforts continue to be made to maximise this number. ^{31}P [511,512], ^{13}C [511,513], ^{15}N [510] heteronuclear NMR and ^{1}H-only studies; specific ^{2}H labelling to simplify spectral interpretation [509], have been used to obtain structural or dynamic constraints, over and above the normal set.

Attention has been paid as to how better to deal with NMR data relating to DNA structure [507,508]. A novel method, based on a restrained Metropolis Monte Carlo calculation, has been developed which uses a starting 'model' of generalised double helical parameters as opposed to cartesian co-ordinates [507]. This method is used to simulate an annealing process and is more efficient than other approaches because a simplified energy profile is used.

Several investigations have concentrated on analysing local structural features of DNA sequences [493,494,496-500]. Measurement of ^1H scalar couplings for a sequence in the activating transcription factor region of the E2A promoter suggested that sugar puckers were in the C3'-*endo*-C4'-*endo* range typical of an A-type helix [494]. Different sequences from the tryptophan repressor DNA operator have been investigated [497,498]. In one study two sequences, differing in having an AT step at opposite ends of the sequence, were investigated in some detail. Interpretation of coupling contants indicated typical B-type helical parameters [497]. In another $T_{1\rho}$ and T_2 measurements were made and a flattening of the propeller twist was invoked at an A.T base-pair [498].

In the cell DNA is stored in a highly compact fashion which induces some stress in the double helix structure. In an attempt to elucidate the effect of stress on helical parameters, DNAs have been prepared with disulphide cross-linkers, of various lengths [500] and these have been the subject of NMR investigations.

There is continued interest in the structural features of A-tracts [501-503], T-tracts [504,505], and TA or AT steps [503,506]. A 9-mer duplex containing a central A_3 tract was shown to have a narrow minor groove at this site, from spectra recorded at low temperature. This feature was found to extend to a G.C rich region downstream of the A-tract [501]. Three self-complementary duplexes, each containing AhaIII and PmeI restriction sequences, were investigated to establish the reason for the susceptibility to cleavage at TpA steps of these sequences [505]. It was suggested that the DNAs possessed extra mobility at these sites due to the observation that adenine H2 protons were broad at TpA steps and sharp at all other sites.

There have been several investigations of DNA duplexes stabilised by base-pairings involving protonated and non-protonated cytidine [514,515], thymine-thymine interactions [514], gaunine-adenine [519-522], and guanine-guanine interactions [523] or mismatches, and also parallel stranded duplexes [516,517]; in one instance one of the strands had the β-ribose form [518].

Many DNA sequences containing modified nucleic acids have been studied. Modifications in some cases have been quite subtle; such as methylation of cytidine resulting in a stabilised cytidine-cytidine base-pair at low pH [524,525]. The NMR properties of several DNA octamers have been compared, the sequences varying in the presence or otherwise of a methoxy-cytidine, mimicing a transition mutant [526]. Although wobble base-pairs were detected near the substitution site the overall structure could be described as B-type helical. Other simple modifications have included glycol [529], hydroxymethyl [527], and propanodeoxy [528,530] derivatives of the natural bases many of these inducing 'kinks'. More dramatic modifications have involved the introduction of amino-fluorene derivatives [531,532], benzo [c] phenanthrene [533], phenazinium [534] bizelesin [536], and methylhypoxanthine moieties [535]. Many of these additional groups have been found to intercalate [533,536,586], in some instances resulting in Hoogsteen base-pairing interactions between base-pairs flanking the intercalation site [536,537]. There have also been studies of DNA duplexes that have been modified by photochemically induced cross-linking reactions [539,540].

In recent years there have been several studies of RNA hairpin structures as these have been implicated in many protein recognition processes. Similar secondary structure motifs have been found in DNA molecules, and so interest has grown in establishing the structural requirements for DNA hairpin (stem-loop) formation [543-549]. A fifteen (or sixteen) base sequence containing a

central T_3 (or T_4) tract has been shown to fold to form a hairpin [543]. A dodecamer containing a central 5'-'TTCG-3' sequence has been investigated and compared with the analogous RNA system, previously shown to be 'unusually stable' [546]. Longer sequences have also been shown to fold in this fashion [547,548]. In one instance a 20-mer included a T_8A_8 base-pair stem and a C_4 loop; the 'C' and 'T' bases connected to each other *via* a 5'-5' switchback arrangement [547].
Higher order structures such as triplexes [563-569], 3-way junctions [550,551,553] and quadruplexes [552,554-562] have been studied by a number of groups. In three of the five triplex papers reported here non-natural nucleotides or ligated nucleotides were involved and found to intercalate into the duplex [563,564,567]. Following 2D- and 3D-homonuclear NMR experiments [550,553] on three-stranded DNA systems it was concluded that the base-pair stacking potential of nucleotides near the three-way junction was important in determining the stability of such structures [550,551]. Quadruplexes that are G and T rich have been studied [552,554,555,557,559,560,561], the majority of these have involved G_nT_n tracts, where n=3 or more, but in one instance n=1 with a G_3 tract at either end of the sequence [559]. Several reports of quantitative analyses of data have appeared involving complete relaxation matrix treatment of NOESY data [557], and molecular mechanics and restrained molecular dynamics calculations [556,557]. The effects of monovalent [552,559,560] and divalent [559] cations on the formation of these higher order DNA structures have been investigated.
The effects of metal [589-593] or metal-complex [580,581] binding on duplex DNA has been assessed. ^{25}Mg NMR was used to show that Mg^{2+} binds 40-100 times more strongly at G.C base-pairs than at A.T base-pairs [593]. ^{1}H NMR investigations of 10 dodecamers revealed that Mg^{2+} preferentially binds to the 5'-side of guanosine residues [592]. Circular dichroism studies of 'GC' containing octamer duplexes have suggested a B- to A-helical transition when Co^{III} is introduced to the sample [590,591], this transition has been reflected in ^{59}Co chemical shifts, and longitudinal and transverse relaxation times. Phenathrenequinone-, and derivatives thereof, rhodium [581] and ruthenium [580] complexes have been shown to intercalate with duplex DNA in the major [581] and minor [580] grooves. Both enantiomers of a chiral ruthenium complex, labelled **Δ** and **V**, have been shown to bind in the minor groove of a dodecamer; $d(CGCGATCGCG)_2$ [580]. Enantiomer **Δ** binds and distorts the duplex, **V** binds without inducing distortions.
Antibiotic binding to duplex DNA continues to be of interest [568-579,582-588,596]. There have been several reported studies of distamycin and distamycin-analogue binding to DNA [570,575,578,588,596]. Similarly for mithramycin [577,579], bleomycin [569] and anti-tumour drugs such as calicheamicin [594,595] and (+)CC-1065; produced by *Streptomyces zelensis* [585]. ^{1}H NMR coupled with hydroxyl radical footprinting and non-denaturaing gel electrophoresis studies of a dodecamer duplex has shown that drug binding shifts the locus of an A-tract bend by one base-pair [585]. A detailed study of Hoechst 33258 bound to $d(GTGGAATTCCAC)_2$ showed this to be a dynamic system; this could explain the differences in the various crystal structures reported for this complex [571].

5. RNA

NMR is becoming established as the principle technique for high resolution structural studies of RNA. Preparation of adequate material is now much easier than in the past, this can be labelled in various ways, and a range of multidimensional experiments can be applied. The ground is thus set for significant increase in the understanding of RNA structure and function. The number of reports of NMR studies of RNA during the last year is still relatively small, but this is an important topic.

Following last year's reports of the preparation of $^{13}C/^{15}N$-labelled RNA, further strategies exploiting the various coherence transfer possibilities have been presented. A triple resonance technique [597] has been presented which only uses only scalar coupling correlations, and hence provides unambiguous matching of intra-nucleotide sugar and base resonances. Tested on a 12-mer duplex containing 8 Watson-Crick base pairs, the technique worked for pyrimidines, but only one guanosine out of five yielded a detectable signal, there were no adenines, and for the pyrimidine signals there was significant variation in the intensities of the correlations. This is an indication that dynamic effects within RNA can compromise the performance of sophisticated pulse sequences. There have been several reports of the use of various heteronuclear (double resonance) experiments to assign uniformly labelled RNA [598-600]. Alternatively, labelling can be confined to residue type; this has been demonstrated in a study of a 29-mer derived from the HIV-1 TAR system [601]. This indicated that a mismatched A had an unusual environment.

There have been further reports of structural studies of unlabelled RNAs. Structural features of a 14-mer stem-loop representing a truncated version of the HIV-1 TAR system have been investigated [602]. The loop region demonstrates considerable flexibility, with some evidence that the A-helix is continued on the 3'-side of the loop. An imino proton study of a 21-mer RNA indicated a stem-loop secondary structure [603], this was an mRNA sequence and the formation of this motif had been proposed as a transcription termination signal.

Self-cleaving RNA sequences, or ribozymes, are important targets for structural studies, since the structural features which lead to RNA catalysis are currently obscure. To this end, a small lariat RNA which self-cleaves has been charactersided by NMR [604,605]. Although this system cleaves far slower than the naturally-occurring ribozymes and does not have share their strong metal ion dependences, it is a very useful ribozyme model. The full solution structure proved impossible to define because of the lack of internucleotide nOe's, but some structural and dynamic features were apparent. Preliminary NMR studies of a 74-mer RNA derived from another self-cleaving RNA, a domain of the human hepatitus delta virus, have been reported [606]. The thermally stable UUCG loop sequence was incorporated, to introduce extra rigidity as well as to act as an NMR marker. The NMR data suggested multiple conformations in some regions, based on an overall pseudoknot secondary structure. Various secondary structures for this domain have been proposed, but other evidence also supports this pseudoknot form. Last year saw the NMR identification of a 'nucleotide triple' in a model core of a self-cleaving group I intron, formed at the junction of two helices where a base of a single-stranded portion interacts with the 2'-OH group of a base pair. This has now been pursued further, by testing this interaction amonst different bases [607]. Triples were observed with conserved bases, though the structures appeared to vary somewhat. Triples were not observed with non-conserved bases. Nucleic acid ^{31}P chemical shifts have long been discussed as potential structural flags. The GAAA tetraloop sequence has been shown previously to show distinct ^{31}P chemical shifts, which have now been seen to be retained when this tetraloop has been used to cap the ends of helices in certain self-cleaving RNA motifs [608]. The structural features of the tetraloop are thus essentially conserved, though this would be expected in these cases as the loops are relatively remote from the catalytically active centres. Some unusually shifted phosphorus resonances were also noted, though how these relate to structural features of the catalytic core is not yet known.

The solution structure of the duplex r(GGCGAGCC)$_2$ has been determined [609]; the tandem G·A mismatches have strong interstrand stacking interactions (A-A, G-G). As a model of the 3'-terminal portion of the aminoacyl stem of a tRNA, the stability of a defined 7 base pair duplex with various 4-nucleotide single stranded tails attached has been studied [610]. The tail stabilised the duplex in a sequence-dependent manner. Interactions of these RNAs with Mn^{2+} has also been studied [611]. Imino ^1H NMR has been used to confirm the secondary structures in a study of a series of 17-mers representing the anticodon arm of yeast $tRNA^{Phe}$ [612]. ^1H NMR has assisted in the identification of a modified cytosine found in $tRNA^{Met}$ from bovine liver mitochondria [613], and a modified uridine in a minor $tRNA^{Arg}$ from *E.coli* [614]. The codon-anticodon interactions in two $tRNA^{Gly}$ have been studied using 3- and 4-mer codon-mimics, one of which contained an ^{15}N-labelled G; with most attention on the imino region of the ^1H spectrum [615]. A tRNA molecule is a large RNA in terms of NMR; but is dwarfed in a report of the 69 kDa complex of Phe-$tRNA^{Phe}$ with elongation factor Tu and GTP [616]. The imino proton signals indicated that the complexation causes changes in the tRNA structure. Where there is significant overlap of imino ^1H signals, this can be alleviated by using ^{15}N-correlating heteronuclear experiments. A technique of this type, efficient enough to work at natural abundance levels of ^{15}N, has been presented [617] which is a version of the HMQC experiment and includes pulsed field gradients. This technique, together with ^{31}P and natural abundance ^{13}C heteronuclear, and 3D heteronuclear (^{31}P-^1H-^1H) experiments have been applied to a 29-mer based on the sarcin/ricin loop of 28S rRNA [618], to probe the solution structure. Overall this was a stem-loop, with a compact tetraloop and unusual interactions in the upper stem region.

6. BOUND WATER MOLECULES

The interaction with solvent water is an important aspect of biomolecular structure and function. The number of reports of NMR studies of biomolecular interactions with solvent water is relatively small, but we consider the topic to warrant attention here.

The exchange lifetimes of bound waters can be determined by a new technique which combines pulsed field gradients, isotopic editing and selective excitation, which has been demonstrated using ^{15}N-labelled DnaJ protein [619]. The residence lifetimes of waters on the surface of BPTI have been investigated by molecular dynamics calculations, with simulations run for the long time of 1.4 ns [620]. This is long enough to cover the residence lifetimes of many of the waters; the results are in encouragingly good agreement with experimental NMR data.

A new 2D experiment using $^{13}C/^{15}N$-labelled protein has been applied to HIV protease-inhibitor complexes to detect an interior water molecule, which had been observed in x-ray crystal structures [621]. This water is present in one complex, but absent in another in which the inhibitor had been purposely modified to eliminate it. This inhibitor thus binds more tightly to the protein.

Following last year's initial reports, there have been further studies of DNA hydration. Spines of hydration have been observed along the full length of both grooves of a non-selfcomplementary 12 base pair duplex [503]. A correlation between increasing duplex stability and decreased bound water in the minor grooves has been claimed [622], but this is placing strong emphasis on the absence of nOe contacts in a series of spectra.

In principle, bound waters can also be detected by using ^{17}O-enriched solvent. Bound waters will then effect the relaxation rates of nearby protons. This has been tested using the ascomycin complex of FK506-binding protein [623], in which internal waters have previously been detected by NMR. There were in fact no pronounced effects on the relaxation parameters of resonances of the complex; evidently any ^{17}O-induced effects are smothered by other factors.

7. CARBOHYDRATES

As in previous years there have been (literally) hundreds of reports on oligo- and polysaccharides that have involved the use of NMR in 'structure' analysis. More often than not compositional analysis has actually been achieved; saccharides that are isolated are generally degraded using well-established methods and the lowest molecular weight carbohydrate components are analysed by NMR and a variety of other methods. In this fashion the sequence of the oligo- or polysaccharide is determined, in some instances with additional information as to branching sites. Hence, the vast majority of carbohydrate work utilising NMR is not appropriate for inclusion in a review of macromolecular NMR. Consequently here we consider examples of systems of trisaccharides or larger that have been analysed using NMR as the principle technique and for which extensive assignment data is available or detailed three-dimensional structures have been determined. We have also considered novel approaches towards NMR studies of oligo- or polysaccharides.

Most of the NMR studies reported have employed both ^{13}C and ^{1}H resonances, the chemical shifts of which have been used, for example, to determine the positions of acetyl groups in O-specific polysaccharides of Klebsiella pneumoniae serotype O1 and O8 liposaccharides [624]. Similarly ^{1}H and ^{13}C data have been acquired for a series of chondroitin sulphate epitopes that may be useful when analysing larger chondroitin sulphate oligosaccharides [625].

Extensive chemical shift assignment is required for branch points or linkage sites to be determined. Investigations of Kaempferol-3-O-glucosyl (1-2) rhamnoside from Ginkgo biloba have shown that NMR can be used to distinguish 1-2, 1-3, and 1-4 linked glucopyrosylrhamnopyranosides [626]. A novel C4-branched 3,6-dideoxy-α-xylo-hexopyranose has been investigated and complete ^{13}C and ^{1}H assignment has been used to analyse penta- and hexa-mannose derivatives [631] and the effects of introducing an additional mannose with an α-1-2-link.

Transverse and longitudinal relaxation times of ^{1}H [631,632] and ^{13}C [633] nuclei have been measured in order to determine local correlation times in a range of saccharides. Cross-relaxation rates among protons of an oyster glycogen have been examined by steady-state and transient nOe experiments [632]. NOes measured directly were compared with those derived from relaxation times and hence correlation times. NMR studies of a pentasaccharide, linked on one side to a 7-amino acid chain and on the other side to a 3-amino acid chain, have revealed that although the presence of the peptide chains increases local correlation times they do not perturb the O-glycoside conformation [634]. Correlation times from T_1 measurements for ^{13}C nuclei and simulations of ^1H-^1H nOes, using a program called NOEMOL, have suggested that the solution conformation of heparin is similar to that determined by X-ray crystallography [633].

Several systems have been the subject of detailed structural analysis. A tetrasaccharide receptor for E-selectin has been shown to have rigid and flexible regions *via* complete relaxation matrix analysis and molecular mechanics and molecular dynamics simulations [635]. Overall two conformations have been proposed that are in fast exchange. Another tetrasaccharide, a sialyl-Lewis X antigen, has been the subject of ROESY experiments. Spectra have been back-calculated from molecular dynamics simulations of the tetramer, run for 5 ns [636]. The structure of the backbone of the core lipid-A region of lipopolysaccharides from vibrio cholerae strain H11 (non-O1) has been determined following Monte Carlo simulations and the identification of several long range nOes [637].

The solution structure of a trisaccharide bound to an antibody has been investigated [638]. Utilising transfered nOes, two conformations for the 'bound' saccharide have been determined, one of which has been observed in the X-ray crystallographic structure of a related complex. A β-glucan octa- and an undecasaccharide of Candida albicans uniformly ^{13}C-labelled has been the subject of HCCH-COSY, HCCH-TOCSY and HMBC experiments [639], the results have enabled a determination as to where cell wall glucanase cleaves the saccharides. has been achieved utilising a variety of experiments including HOHAHA, HMQC, and ROESY [627]. A series of Keratan sulphates, tetrasaccharides and fucosylated pentasaccharides, have been investigated and complete and partial ^{13}C assignments, respectively, have been determined from spectra recorded at 600 MHz (^1H frequency) and 60° C [628]. Two octa-, two dodeca-, and a tridecasaccharide from human milk have been shown, through extensive resonance assignment, to have a 'new core structure' [629]. The anomeric configuration and the position of glycosyl linkages have been determined by the measurement of vicinal coupling constants and extensive resonance assignment of an oligosaccharide from a clinical isolate of vibrio vulnificus strain B062316 [630]. ^1H and ^{13}C chemical shifts assigned through the use of HOHAHA and HMBC experiments have

NMR investigations of saccharides are complicated by the conformational freedom of six membered rings and the considerable resonance overlap that occurs in ^1H NMR spectra. There is considerable interest therefore in approaches to simplifying the spectral analysis process. This year has seen some novel approaches to this problem. Natural abundance ^{13}C ω_1 half-filtered ^1H-^1H ROESY spectra have been recorded for a cyclic β-(1->2) glucan [640] demonstrating a way in which chemically equivalent H_2 protons in adjacent glycosyl residues can be distinguished. A series of pseudo-3D experiments has been utilised to alleviate resonance overlap problems [641]. Selective pulses and chemical shift selective filters have been used for double selection of magnetization during a range of pulse sequences; COSY-ROESY, COSY-TOCSY, COSY-NOESY, and COSY-ROESY. These sequences and several variations have been applied to oligo- and poly-saccharides. Finally, in recent years neural networks have been employed in attempts to peak-pick and then peak assign in 2D spectra of proteins; now neural networks have also been applied to the analysis of ^{13}C spectra of polysaccharides [642]. Three approaches to the analysis were utilised, the quasi-Newton method yielded the best results and required fewer training cycles.

Acknowledgement: JRPA would like to acknowledge personal support from The Wellcome Trust.

References

1. J.C.Franklin, J.F.Ellena, S.Jayasinghe, L.P.Kelsh, and D.S.Cafiso, *Biochemistry*, 1994, **33**, 4036-45.
2. G.Gupta, G.M.Anantharamaiah, D.R.Scott, J.H.Eldridge, and G.Myers, *J.Biomol.Struct.Dyn.*, 1993, **11**, 345-66.
3. M.Dettin, A.De Rossi, M.Autiero, J.Guardiola, L.Checo-Bianchi, and C.Di Bello, *Biochem.Biophys.Res.Commun.*, 1993, **191**, 364-70.
4. P.J.Klasse, R.Pipkorn, J.Blomberg, K-H.Han, B.Hilton, and J.A.Ferretti, *FEBS Lett.*, 1993, **323**, 68-72.
5. W.E.Klunk, C.J.Xu, and J.W.Pettegrew, *J.Neurochem.*, 1994, **62**, 349-54.
6. K.Sorimachi and D.J.Craik, *Eur.J.Biochem.*, 1994, **219**, 237-51.
7. T.Maurer and H.Rueterjans, *Eur.J.Biochem.*, 1994, **220**, 193-200.
8. N.J.Murray and M.P.Williamson, *Eur.J.Biochem.*, 1994, **219**, 915-21.
9. N.J.Murray, M.P.Williamson, T.H.Lilley, and E.Haslam, *Eir.J.Biochem.*, 1994, **219**, 923-35.
10. J.H.Davis, E.K.Bradley, G.P.Miljanich, L.Nadasdi, J.Ramachandran, and V.J.Basus, *Biochemistry*, 1993, **32**, 7396-405.
11. J.J.Skalicky, W.J.Metzler, D.J.Ciesla, A.Galdes, and A.Pardi, *Protein Sci.*, 1993, **2**, 1591-603.
12. P.Sevilla, M.Bruix, J.Santaro, F.Gago, A.G.Garcia, and M.Rico, *Biochem.Biophys.Res.Commun.*, 1993, **192**, 1238-44.
13. R.S.Norton, K.Cross, V.Braach-Maksvvytis, and E.Wachter, *Biochem.J.*, 1993, **293**, 545-51.
14. M.G.Hinds and R.S.Norton, *J.Protein Chem.*, 1993, **12**, 371-8.
15. J.P.Waltho, V.A.Feher, G.Merutka, J.H.Dyson, and P.E.Wright, *Biochemistry*, 1993, **32**, 6337-47.
16. H.C.Shin, G.Merutka, J.P.Waltho, P.E.Wright, and J.H.Dyson, *Biochemistry*, 1993, **32**, 6348-55.
17. H.C.Shin, G.Merutka, J.P.Waltho, L.L.Tennant, J.H.Dyson, and P.E.Wright, *Biochemistry*, 1993, **32**, 6356-64.
18. M.H.Li, P.Fan, B.Brodsky, and J.Baum, *Biochemistry*, 1993, **32**, 7377-87.
19. P.Fan, M.H.Li, B.Brodsky, and J.Baum, *Biochemistry*, 1993, **32**, 13299-309.
20. C.G.Long, E.Braswell, D.Zhu, J.Apigo, J.Baum, and B.Brodsky, *Biochemistry*, 1993, **32**, 11688-95.
21. V.A.Daragan, E.Ilyina, and K.H.Mayo, *Biopolymers*, 1993, **33**, 521-33.
22. D.A.Horita, P.J.Hajduk, P.F.Goetinck, and L.Lerner, *J.Biol.Chem.*, 1994, **269**, 1699-704.
23. X.Liu, A.Otter, P.G.Scott, J.R.Cann, and G.Kotovych, *J.Biomol.Struct.Dyn.*, 1993, **11**, 541-55.
24. A.Otter, P.G.Scott, and G.Kotovych, *Biopolymers*, 1993, **33**, 1443-59.
25. N.E.Keane, D.P.Molloy, R.J.A.Grand, D.Owen, B.A.Levine, and L.Ellis, *Biochem.Soc.Trans.*, 1993, **21**, 267S.
26. M.Zhang and H.J.Vogel, *Biochemistry*, 1994, **33**, 1163-71.
27. T.Kato, S.Endo, T.Fujiwara, and K.Nagayama, *J.Biomol.NMR*, 1993, **3**, 653-73.
28. G.Lippens, D.Van Belle, S.J.Wodak, K.Hallenga, N.R.Nirmala, P.Hill, K.C.Russell, D.D.Smith, and V.J.Hruby, *Biochemistry*, 1993, **32**, 9423-34.
29. G.V.Nikiforovich, J.L-F.Kao, K.Plucinska, W.J.Zhang, and G.R.Marshall, *Biochemistry*, 1994, **33**, 3591-8.
30. J.M.Matsoukas, J.Hondrelis, M.Keramida, T.Mavromoustakos, A.Makriyannis, R.Yamdagni, Q.Wu, and G.J.Moore, *J.Biol.Chem.*, 1994, **269**, 5303-12.
31. G.J.Moore, M.H.Goghari, and K.J.Franklin, *Int.J.Pept.Prot.Res.*, 1993, **42**, 445-9.
32. K.Thornton and D.G.Gorenstein, *Biochemistry*, 1994, **33**, 3532-9.
33. E.Detsikas, V.Tsikaris, M.Sakarellos-Daitsiotis, C.Sakarellos, M.T.Cung, M.Marraud, E.Vataki, and S.J.Tzartos, *Pept.Res.*, 1993, **6**, 17-23.
34. V.Tsikaris, E.Detsikas, M.Sakarellos-Daitsiotis, C.Sakarellos, E.Vatzaki, S.J.Tartos, M.Marraud, and M.T.Cung, *Biopolymers*, 1993, **33**, 1123-34.
35. N.E.Zhou, C.M.Kay, B.D.Sykes, and R.S.Hodges, *Biochemistry*, 1993, **32**, 6190-7.
36. A.Otter, P.Bigler, J.M.Stewart, and G.Kotovych, *Biopolymers*, 1993, **33**, 769-80.
37. S.Zinn-Justin, C.Roumetand, P.Drevet, A.Mene, and F.Toma, *Biochemistry*, 1993, **32**, 6884-91.
38. X.Xu and J.W.Nelson, *Proteins:Struct.,Funct.,Genet.*, 1993, **17**, 124-37.
39. T.C.Wong, C.M.Lee, W.Guo, and D.K.Chang, *Int.J.Pept.Prot.Res.*, 1993, **41**, 185-95.
40. M.Zhang and T.C.Wong, *Biopolymers*, 1993, **33**, 1901-8.
41. B.Bersch, P.Koehl, Y.Nakatani, G.Ourisson, and A.Milon, *J.Biomol.NMR*, 1993, **3**, 443-61.
42. S.Lavielle, G.Chassaing, A.Brunissen, M.Rodriguez, J.Martinez, O.Convert, A.Carruette, C.Garret, and F.Petitet, *Int.J.Pept.Prot.Res.*, 1993, **42**, 270-7.
43. A.P.Golovanov, A.L.Lomize, A.S.Arseniev, Y.N.Utkin, and V.I.Tsetlin, *Eur.J.Biochem.*, 1993, **213**, 1213-23.
44. A.K.Singhal, K.Y.Chien, W.G.Wu, and G.S.Rule, *Biochemistry*, 1993, **32**, 8036-44.
45. J.F.O'Connell, P.E.Bougis, and K.Wüthrich, *Eur.J.Biochem.*, 1993, **213**, 891-900.
46. T.J.Siahaan and K.Lutz, *J.Pharm.Biomed.Anal.*, 1994, **12**, 65-71.
47. K.E.Kover, D.Jiang, S.Fang, and V.J.Hruby, *Magn.Reson.Chem.*, 1993, **31**, 1072-6.
48. W.Lee, M.J.Jablonsky, D.D.Watt, and N.R.Krihna, *Biochemistry*, 1994, **33**, 2468-75.

49. S.Meunier, J.M.Bernassau, J.M.Sabatier, M.F.Martin-Eauclaire, J.Van Ritschoten, C.Cambillau, and H.Darbon, *Biochemistry*, 1993, **32**, 11969-76.
50. M.Coles, S.L.A.Munro, and D.J.Craik, *J.Med.Chem.*, 1994, **37**, 656-64.
51. M.J.Scanlon, M.R.Price, and S.J.B.Tendler, *Int.J.Oncol.*, 1993, **2**, 271-3.
52. J.D.Fontenot, N.Tjandra, D.Bu, C.Ho, R.C.Montelaro, and O.J.Finn, *Cancer Res.*, 1993, **53**, 5386-94.
53. I.Z.Zubrycki and G.Gaede, *Biochem.Biophys.Res.Commun.*, 1994, **198**, 228-35.
54. G.R.Wilcox, R.H.Fogh, and R.S.Norton, *J.Biol.Chem.*, 1993, **268**, 24707-19.
55. R.Bhaskaran, A.I.Arunkumar, and C.Yu, *Biochim.Biophys.Acta*, 1994, **1199**, 115-22.
56. C.R.Middaugh, J.A.Thomson, C.J.Burke, H.Mach, A.M.Naylor, M.J.Bogusky, J.A.Ryan, S.M.Pitzberger, H.Ji, and J.S.Cordingley, *Protein Sci.*, 1993, **2**, 900-14.
57. N.G.J.Delaet, P.Verheyden, B.Velkeniers, E.L.Hooghe-Peters, C.Bruns, D.Tourwe, and G.Van Binst, *Pept.Res.*, 1993, **6**, 24-30.
58. M.Naim, S.Charpentier, P.Nicolas, and D.Baron, *Biopolymers*, 1993, **33**, 1889-900.
59. D.Duchesne, M.Naim, P.Nicolas, and D.Baron, *Biochim.Biophys.Acta*, 1993, **195**, 630-6.
60. T.O.Matsunaga, N.Collins, V.Ramaswami, S.H.Yamamura, D.F.O'Brien, and V.J.Hruby, *Biochemistry*, 1993, **32**, 13180-9.
61. M.Tassari, M.T.Foffani, S.Mammi, and E.Peggion, *Biopolymers*, 1993, **33**, 1877-87.
62. N.M.Wahlgren, J.Leonil, P.Dejmek, and T.Drakenberg, *Biochim.Biophys.Acta*, 1993, **1202**, 121-8.
63. H.W.van den Hooven, F.Fogolari, H.S.Rollema, R.N.H.Konings, C.W.Hilbers, and F.J.M.van de Ven, *FEBS Lett.*, 1993, **319**, 189-94.
64. H.Yu, M.K.Rosen, N.A.Saccomano, and D.Phillips, *Biochemistry*, 1993, **32**, 13123-9.
65. M.E.Adams, I.M.Mintz, M.D.Reilly, V.Thanabal, and B.P.Bean, *Mol.Pharmacol.*, 1993, **44**, 681-8.
66. K.Arvidsson, T.Land, U.Langel, T.Bartfai, and A.Ehrenberg, *Biochemistry*, 1993, **32**, 7787-98.
67. Z.Wang, J.D.Jones, J.Rizo, and L.M.Gierasch, *Biochemistry*, 1993, **32**, 13991-9.
68. K.Thornton, Y.Wang, H.Weiner, and D.G.Gorenstein, *J.Biol.Chem.*, 1993, **268**, 19906-14.
69. J.Rizo, F.J.Blanco, B.Kobe, M.D.Bruch, and L.M.Gierrasch, *Biochemistry*, 1993, **32**, 4881-94.
70. B.Aroeti, P.A.Kosen, I.D.Kuntz, F.E.Cohen, and K.E.Mostov, *J.Cell Biol.*, 1993, **123**, 1149-60.
71. V.Wray, C.Kakoschke, K.Nokihara, and S.Naruse, *Biochemistry*, 1993, **32**, 5832-41.
72. J.Horne, M.Sadek, and D.J.Craik, *Biochemistry*, 1993, **32**, 7406-12.
73. K.A.Muszkat, B.Schechter, and M.Sela, *Int.Immunol.*, 1993, **5**, 591-7.
74. J.P.L.Cox, P.A.Evans, L.C.Packman, D.H.Williams, and D.N.Woolfson, *J.Mol.Biol.*, 1993, **234**, 483-92.
75. L.Chiche, A.Heinz, A.Padilla, D.Le-Nguyen, and B.Cistro, *Protein Eng.*, 1993, **6**, 675-82.
76. J.Kemmink, C.P.M.Van Mierlo, R.M.Scheek, and T.E.Creighton, *J.Mol.Biol.*, 1993, **230**, 312-22.
77. M.A.Catiglione-Morelli, M.DeBiasi, A.DeStradis, and A.M.Tamburro, *J.Biomol.Struct.Dyn.*, 1993, **11**, 181-90.
78. B.Gilquin, C.Roumestand, S.Zinn-Justin, A.Menez, and F.Toma, *Biopolymers*, 1993, **33**, 1659-7.
79. M.Jaseya, K.J.Smith, X.Lu, J.A.Williams, H.Trayer, I.Trayer, and E.I.Hyde, *Eur.J.Biochem.*, 1993, **218**, 853-60.
80. F.J.Moy, Y.C.Li, P.Rauenbuehler, M.E.Winkler, H.A.Sheraga, and G.T.Montelione, *Biochemistry*, 1993, **32**, 7334-53.
81. D.A.Kloosterman, T.A.Scahill, and A.R.Friedman, *Pept.Res.*, 1993, **6**, 211-8.
82. F.Macquaire, F.Baleux, T.Huynh-Dinh, D.Rouge, J.M.Neumann, and A.Sanson, *Biochemistry*, 1993, **32**, 7244-54.
83. V.J.Basus, G.Song, and E.Hawrot, *Biochemistry*, 1993, **32**, 12290-8.
84. N.L.Daly, A.R.Atkins, and R.Smith, *Int.J.Pept.Prot.Res.*, 1993, **42**, 366-71.
85. D.Kern, T.Drakenberg, M.Wikstoem, S.Forsen, H.Bang, and G.Fischer, *FEBS Lett.*, 1993, **323**, 198-202.
86. J.P.O'Connell, S.M.Kelly, D.P.Raleigh, J.A.M.Hubbard, N.C.Price, C.M.Dobson, and B.J.Smith, *Biochem.J.*, 1993, **291**, 205-10.
87. J.A.Barden and B.E.Kemp, *Biochemistry*, 1993, **32**, 7126-32.
88. V.Wray, T.Federau, W.Gronwald, H.Mayer, D.Schomburg, W.Terner, and E.Wingender, *Biochemistry*, 1994, **33**, 1684-93.
89. L.A.Strickland, R.P.Bozzato, and K.A.Kronis, *Biochemistry*, 1993, **32**, 6050-7.
90. J.A.Barden and R.M.Cuthbertson, *Eur.J.Biochem.*, 1993, **215**, 315-21.
91. M.S.Czisch, M.Scleicher, S.Hoerger, W.Voelter, and T.A.Holak, *Eur.J.Biochem.*, 1993, **218**, 335-44.
92. E.V.Curto, M.A.Jarpe, J.E.Blalock, D.Borovsky, and N.R.Krishna, *Biochem.Biophys.Res.Commun.*, 1993, **193**, 688-93.
93. W.Lee, C.H.Moore, D.D.Watt, and N.R.Krishna, *Eur.J.Biochem.*, 1994, **219**, 89-95.
94. M.Suzuki, M.Gerstein, and T.Johnson, *Protein Eng.*, 1993, **6**, 565-74.
95. S.Rebuffat, L.Conraux, M.Massias, C.Auvin-Guette, and B.Bodo, *Int.J.Pept.Prot.Res.*, 1993, **41**, 74-84.
96. T.T.Sakai, M.J.Jablonsky, P.A.DeMuth, N.R.Krishna, M.A.Jarpe, and H.M.Johnson, *Biochemistry*, 1993, **32**, 5650-5.
97. W.Y.Sun, N.Uleyama, and A.Nakamura, *J.Chem.Soc.,Dalton Trans.*, 1993, 1871-4.
98. D.Fourmy, F.Dardel, and S.Blanquet, *J.Mol.Biol.*, 1993, **231**, 1078-89.
99. S.L.Drake, J.Varnum, K.H.Mayo, P.C.Letourneau, L.T.Furcht, and J.B.McCarthy, *J.Biol.Chem.*., 1993, **268**, 15859-67.
100. C.McInnes, F.D.Sonnichsen, C.M.Kay, R.S.Hodges, and B.D.Sykes, *Biochemistry*, 1993, **32**, 13432-40.
101. Y.N.Kalia, S.M.Brocklehurst, D.S.Hipps, E.Appella, K.Sakaguchi, and R.N.Percham, *J.Mol.Biol.*, 1993, **230**, 323-41.

102. N.E.Keane, B.A.Levine, B.Calas, A.Chavanieu, F.Grigorescu, and L.Ellis, *Biochem.Soc.Trans.*, 1993, **21**, 266S.
103. C.J.Morton, R.J.Simpson, and R.S.Norton, *Eur.J.Biochem.*, 1994, **219**, 97-107.
104. P.L.Simon, V.Kumar, J.S.Lillquist, P.Bhatnagar, R.Einstein, T.Porter, D.Green, G.Sathe, and P.R.Young, *J.Biol.Chem.*, 1993, **268**, 9771-9.
105. S.Talluri, C.M.Falcomer, and H.A.Scheraga, *J.Am.Chem.Soc.*, 1993, **115**, 3041-7.
106. L.B.Brown, S.Mronga, R.A.Bradshaw, C.Ortenzi, P.Luporini, and K.Wüthrich, *J.Mol.Biol.*, 1993, **231**, 800-16.
107. D.Frechet, J.D.Guitton, F.Herman, D.Faucher, G.Helynck, B.Monegier du Sorbier, J.P.Ridoux, E.James-Surcouf, and M.Vuilhorgne, *Biochemistry*, 1994, **33**, 42-50.
108. H.Tamamura, M.Kuroda, M.Masuda, A.Otaka, S.Funakoshi, H.Nakashima, N.Yamamoto, M.Waki, and A.Matsumoto, *Biochim.Biophys.Acta*, 1993, **1163**, 209-16.
109. S.Mronga, G.Mueller, J.Fischer, and F.Riddell, *J.Am.Chem.Soc.*, 1993, **115**, 8414-20.
110. P.Buckley, A.S.Edison, M.D.Kemple, and F.G.Prendergast, *J.Biomol.NMR*, 1993, **3**, 639-52.
111. R.de Lorimier, M.A.Moody, B.F.Haynes, and L.D.Spicer, *Biochemistry*, 1994, **33**, 2055-62.
112. G.Merutka, D.Morikis, R.Bushweiler, and P.E.Wright, *Biochemistry*, 1993, **32**, 13089-97.
113. L.C.Wu, P.B.Laub, G.A.Elove, J.Carey, and H.Roder, *Biochemistry*, 1993, **32**, 10271-6.
114. S.J.Stradley, J.Rizo, and L.M.Gierasch, *Biochemistry*, 1993, **32**, 12586-90.
115. M.Zhang and H.J.Vogel, *J.Biol.Chem.*, 1994, **269**, 981-5.
116. K.L.Constantine, C.Mapelli, C.A.Meyers, M.S.Friedrichs, S.Krystek, and L.Mueller, *J.Biol.Chem.*, 1993, **268**, 22830-7.
117. X.Liu, J.M.Stewart, L.Gera, and G.Kotovych, *Biopolymers*, 1993, **33**, 1237-47.
118. R.J.Bienstock, J.Rizo, S.C.Koerber, J.E.Rivier, A.T.Hagler, and L.M.Gierasch, *J.Med.Chem.*, 1993, **36**, 3265-73.
119. R.Geraghty, D.J.S.Guthrie, G.B.Irvine, and C.H.Williams, *Biochem.Soc.Trans.*, 1993, **21**, 261S.
120. T.Ikura, N.Go, D.Kohda, F.Inagaki, H.Yanagawa, M.Kawabata, S.Kawabata, S.Iwanaga, T.Noguti, and M.Go, *Proteins:Struct.,Funct.,Genet.*, 1993, **16**, 341-56.
121. G.Esposito, F.Fogolari, P.Viglino, S.Cattarinussi, M.T.De Magistris, L.Chiappinelli, and A.Pessi, *Eur.J.Biochem.*, 1993, **217**, 171-87.
122. M.J.Blackledge, R.Brueschweiler, C.Griesinger, J.M.Scmidt, P.Xu, and R.R.Ernst, *Biochemistry*, 1993, **32**, 10960-74.
123. J.A.Jarvis, S.L.A.Munro, and D.J.Craik, *Biochemistry*, 1994, **33**, 33-41.
124. R.A.Gray, D.G.Vander Velde, C.J.Burke, M.C.Manning, C.R.Middaugh, and R.T.Borchardt, *Biochemistry*, 1994, **33**, 1323-31.
125. G.Wagner, *J.Biomol.NMR*, 1993, **3**, 375-85.
126. S.J.Matthews and R.J.Leatherbarrow, *J.Biomol.NMR*, 1993, **3**, 597-600.
127. C.D.Eads and J.D.Shaffer, *J.Magn.Reson.,Ser.B*, 1993, **103**, 310-313.
128. R.Bernstein, C.Cieslar, A.Ross, H.Oschkinat, J.Freund, and T.A.Holak, *J.Biomol.NMR*, 1993, **3**, 245-51.
129. T.Szyperski, G.Wider, J.H.Bushweller, and K.Wüthrich, *J.Biomol.NMR*, 1993, **3**, 127-32.
130. R.Weismann, H.Reuterjans, and W.Bernel, *J.Biomol.NMR*, 1993, **3**, 113-20.
131. S.Seip, J.Balhach, and H.Kessler, *J.Biomol.NMR*, 1993, **3**, 233-7.
132. S.Grzesiek and A.Bax, *J.Biomol.NMR*, 1993, **3**, 185-204.
133. M.Wittekind and L.Mueller, *J.Magn.Reson.,Ser.B*, 1993, **101**, 201-5.
134. R.T.Clowes, W.Boucher, C.H.Hardman, P.J.Domaille, and E.D.Laue, *J.Biomol.NMR*, 1993, **3**, 349-54.
135. B.A.Lyons, M.Tashiro, L.Cedergren, B.Bilsson, and G.T.Montelione, *Biochemistry*, 1993, **32**, 7839-45.
136. T.Yamazaki, M.Yoshida, and K.Nagayama, *Biochemistry*, 1993, **32**, 5656-69.
137. B.A.Lyons and G.T.Montellone, *J.Magn.Reson.,Ser.B*, 1993, **101**, 206-9.
138. J.M.Richardson, R.T.Clowes, W.Boucher, P.J.Domaille, C.H.Hardman, J.Keeler, and E.D.Laue, *J.Magn.Reson.,Ser.B*, 1993, **101**, 223-7.
139. M.Wittekind, W.J.Metzler, and L.Mueller, *J.Magn.Reson.,Ser.B*, 1993, **101**, 214-7.
140. E.T.Olejniczak and S.W.Fesik, *J.Am.Chem.Soc.*, 1994, **116**, 2215-6.
141. A.P.Hinck, E.S.Eberhardt, and J.L.Markley, *Biochemistry*, 1993, **32**, 11810-18.
142. G.S.Rule, N.Tjandra, V.Simplaceanu, and C.Ho, *J.Magn.Reson.,Ser.B*, 1993, **102**, 126-8.
143. G.Ostler, A.Soteriou, C.M.Moody, J.A.Khan, B.Birdsall, M.D.Carr, D.W.Young, and J.Feeney, *FEBS Lett.*, 1993, **318**, 177-80.
144. D.M.Kushlan and D.M.LeMaster, *J.Biomol.NMR*, 1993, **3**, 701-8.
145. D.S.Wishart, B.D.Sykes, and F.M.Richards, *Biochim.Biophys.Acta*, 1993, **1164**, 36-46.
146. J.M.Reisman, V.L.Hsu, I.Jariel-Encontre, C.Lecou, M.H.Sayre, D.R.Kearns, and J.Parello, *Eur.J.Biochem.*, 1993, **213**, 865-73.
147. S.Madvedeva, J-P.Simorre, B.Brutscher, F.Guerlesquin, and D.Marion, *FEBS Lett.*, 1993, **333**, 251-6.
148. S.K.Drake, R.B.Bourret, L.A.Luck, M.I.Simon, and J.J.Falke, *J.Biol.Chem.*, 1993, **268**, 13081-8.
149. R.B.Bourret, S.K.Drake, S.A.Chervitz, M.I.Simon, and J.J.Falke, *J.Biol.Chem.*, 1993, **268**, 13089-96.
150. J.F.Eccleston, D.P.Molloy, M.G.Hinds, R.W.King, and J.Feeney, *Eur.J.Biochem.*, 1993, **218**, 1041-7.
151. L.A.Sylvia and J.T.Gerig, *Biochim.Biophys.Acta*, 1993, **32**, 3418-24.
152. T.J.Connick, R.T.Reilly, R.B.Dunlap, and P.D.Ellis, *Biochemistry*, 1993, **32**, 9888-95.
153. R.J.Cisneros, J.W.Zapf, and R.B.Dunlap, *J.Biol.Chem.*, 1993, **268**, 10102-8.
154. J.Zapf, M.S.Weir, V.Emerick, J.E.Villafranca, and R.B.Dunlap,*Biochemistry*, 1993, **32**, 9274-81.

155. L.A.Trimble, I.P.Street, H.Perrier, N.M.Trimble, P.K.Weech, and M.A.Bernstein, *Biochemistry*, 1993, **32**, 12560-5.
156. G.W.Vuister, A.C.Wang, and A.Bax, *J.Am.Chem.Soc.*, 1993, **115**, 5334-5.
157. G.W.Vuister and A.Bax, *J.Magn.Reson.,Ser.B*, 1993, **102**, 228-31.
158. S.Grzesiek, G.W.Vuister, and A.Bax, *J.Biomol.NMR*, 1993, **3**, 487-93.
159. G.W.Vuister, T.Yamazaki, D.A.Torchia, and A.Bax, *J.Biomol.NMR*, 1993, **3**, 297-306.
160. H.Schwalbe, A.Rexroth, U.Eggenberger, T.Geppert, and C.Griesinger, *J.Am.Chem.Soc.*, 1993, **115**, 7878-9.
161. G.W.Vuister and A.Bax, *J.Am.Chem.Soc.*, 1993, **115**, 7772-7.
162. J.C.Madsen, O.W.Soerensen, and F.M.Poulsen, *J.Biomol.NMR*, 1993, **3**, 239-44.
163. G.W.Vuister, F.Delaglio, and A.Bax, *J.Biomol.NMR*, 1993, **3**, 67-80.
164. M.Gorlach, M.Wittekind, B.T.Farmer, L.E.Kay, and L.Mueller, *J.Magn.Reson.,Ser.B*, 1993, **101**, 194-7.
165. K.Bartik and C.Redfield, *J.Biomol.NMR*, 1993, **3**, 415-28.
166. A.E.Torda, R.M.Brunne, T.Huber, H.Kessler, and W.F.Van Gunsteren, *J.Biomol.NMR*, 1993, **3**, 55-66.
167. A.C.de Dios, J.G.Pearson, and E.Oldfield, *Science*, 1993, **260**, 1491-6.
168. A.C.de Dios, J.G.Pearson, and E.Oldfield, *J.Am.Chem.Soc.*, 1993, **115**, 9768-73.
169. S.E.Chambers, E.Y.Lau, and J.T.Gerig, *J.Am.Chem.Soc.*, 1994, **116**, 3603-4.
170. J.G.Pearson, E.Oldfield, F.S.Lee, and O.Warshel, *J.Am.Chem.Soc.*, 1993, **115**, 6851-62.
171. D.D.Laws, A.C.de Dios, and E.Oldfield, *J.Biomol.NMR*, 1993, **3**, 607-12.
172. S.A.Corne, J.Fisher, and A.P.Johnson, *Anal.Chim.Acta*, 1993, **278**, 149-58.
173. J.Xu, S.K.Straus, B.C.Sanctuary, and L.Trimble, *J.Chem.Inf.Comput.Sci.*, 1993, **33**, 668-82.
174. J.Xu, S.K.Straus, B.C.Sanctuary, and L.Trimble, *J.Magn.Reson.,Ser.B*, 1994, **103**, 53-8.
175. J.Xu, L.R.Brown, W.Cai, and M.Zhang, *Sci.China,Ser.B*, 1993, **36**, 526-39.
176. R.Wehrens, C.Lucasius, L.Buydens, and G.Kateman, *Anal.Chim.Acta*, 1993, **277**, 313-24.
177. V.W.Soo, J-F.Hwang, T-B.Chen, and C.Yu, *J.Comput.Chem.*, 1993, **14**, 1164-71.
178. G.J.Kleywegt, G.W.Vuister, A.Padilla, R.M.A.Knegtel, R.Boelens, and R.Kaptein, *J.Magn.Reson.,Ser.B*, 1993, **102**, 166-76.
179. P.J.Connolly, A.S.Stern, and J.C.Hoch, *J.Am.Chem.Soc.*, 1994, **116**, 2675-6.
180. M.Dellwo, D.M.Schneider, and A.J.Wand, *J.Magn.Reson.,Ser.B*, 1994, **104**, 1-9.
181. R.Bernstein, A.Schnuchel, and C.Cieslar, *J.Magn.Reson.,Ser.B*, 1993, **102**, 116-9.
182. A.M.J.J.Bonvin, R.Boelens, and R.Kaptein, *Biopolymers*, 1994, **34**, 39-50.
183. A.M.J.J.Bonvin, J.A.C.Rullmann, R.M.J.N.Lamerichs, R.Boelens, and R.Kaptein, *Proteins:Struct.Funct.Genet.*, 1993, **15**, 385-400.
184. R.C.Hoffman, R.X.Xu, R.E.Klevit, and J.R.Herriott, *J.Magn.Reson.,Ser.B*, 1993, **102**, 61-72.
185. B.Bernstein, A.Ross, C.Cieslar, and T.A.Holak, *J.Magn.Reson.,Ser.B*, 1993, **101**, 185-8.
186. H.Widmar, A.Widmar, and W.Braun, *J.Biomol.NMR*, 1993, **3**, 307-24.
187. M.Nilges, *Proteins:Struct.,Funct.,Genet.*, 1993, **17**, 297-309.
188. S.I.O'Donoghue, F.K.Junius, and G.F.King, *Protein Eng.*, 1993, **6**, 557-64.
189. A.Elofsson and L.Nilsson, *J.Mol.Biol.*, 1993, **233**, 766-80.
190. M.J.Sutcliffe, *Protein Sci.*, 1993, **2**, 936-44.
191. G.M.Clore, M.A.Robien, and A.M.Gronenborn, *J.Mol.Biol.*, 1993, **231**, 82-102.
192. R.B.Altman, R.Pachter, and O.Jardetzky, *Appl.Magn.Reson.*, 1993, **4**, 441-60.
193. G.W.Vuister, S.J.Kim, C.Wu, and A.Bax, *Biochemistry*, 1994, **33**, 10-16.
194. C.Freund, A.Ross, A.Pluekthun, and T.A.Holak, *Biochemistry*, 1994, **33**, 3296-303.
195. P.J.Kraulis, P.J.Domaille, S.L.Campbell-Burk, T.van Aken, and E.D.Laue, *Biochemistry*, 1994, **33**, 3515-31.
196. R.T.Clubb, S.B.Ferguson, C.T.Walsh, and G.Wagner, *Biochemistry*, 1994, **33**, 2761-72.
197. M.Rico, J.Santoro, C.Gonzalez, M.Bruix, J.L.Neira, and J.F.Nieto, *Appl.Mag.Res.*, 1993, **4**, 385-415.
198. K.V.Andersen and F.M.Poulsen, *J.Biomol. NMR*, 1993, **3**, 271-84.
199. T.Yamazaki, M.Yoshida, and K.Nagayama, *Biochemistry*, 1993, **32**, 5656-69.
200. H.C.Joas and R.A.Dwek, *Eur.J.Biochem.*, 1993, **218**, 239-44.
201. H.Yu, M.K.Rosen, and S.L.Schreiber, *FEBS Lett.*, 1993, **324**, 87-92.
202. S.Koyama, H.Yu, D.C.Dalgarno, T.B.Shin, L.D.Zydowsky, and S.L.Schreiber, *Cell*, 1993, **72**, 945-52.
203. M.Seigneuret and M.Kainosho, *FEBS Lett.*, 1993, **327**, 7-12.
204. B.O.Smith, A.K.Downing, T.J.Dudgeon, M.Cunningham, P.C.Driscoll, and I.D.Campbell, *Biochemistry*, 1994, **33**, 2422-9.
205. M.J.Williams, I.Phan, M.Baron, P.C.Driscoll, and I.D.Campbell, *Biochemistry*, 1993, **32**, 7388-95.
206. M.J.Williams, I.Phan, T.S.Harvey, A.Rostagno, L.I.Gold, and I.D.Campbell, *J.Mol.Biol.*, 1994, **235**, 1302-11.
207. D.Morikis, C.A.Lepre, and P.E.Wright, *Eur.J.Biochem.*, 1994, **219**, 611-26.
208. M.J.Labienski, M.Bycroft, D.N.M.Jones, and A.R.Fersht, *FEBS Lett.*, 1993, **332**, 81-7.
209. G.J.A.Kroon, J.Groetzinger, K.Dijkstra, M.R.Scheek, and G.T.Robillard, *Prot.Sci.*, 1993, **2**, 11331-41.
210. C.W.G.Hoitink, P.C.Driscoll, H.A.O.Hill, and G.W.Canters, *Biochemistry*, 1994, **33**, 3060-71.
211. S.Golic Grdadolnik, M.Eberstadt, G.Gemmecker, H.Kessler, A.Buhr, and B.Erni, *Eur.J.Biochem.*, 1994, **219**, 945-54.
212. B.E.Finn, T.Drakenberg, and S.Forsen, *FEBS Lett.*, 1993, **336**, 368-74.
213. M.Zhang and H.J.Vogel, *J.Biol.Chem.*, 1993, **268**, 22420-8.
214. S.Lund-Katz and K.H.Weisgraber, *J.Biol.Chem.*, 1993, **268**, 23008-15.

215. S.Bagby, T.S.Harvey, L.E.Kay, S.G.Eagle, S.Inouye, and M.Ikura, *Biochemistry*, 1994, **33**, 2409-21.
216. I.J.L.Byeon, H.Yan, A.S.Edison, E.S.Mooberry, F.Abildgaard, J.L.Markley, and M.D.Tsai, *Biochemistry*, 1993, **32**, 12508-21.
217. C.Cerf, G.Lippens, S.Muyldermans, A.Segers, V.Ramakrishnan, S.J.Wodak, K.Hallenga, and L.Wyns, *Biochemistry*, 1993, **32**, 11345-51.
218. J.Anglister, S.Grzesiek, A.C.Wang, H.Ren, C.B.Klee, and A.Bax, *Biochemistry*, 1994, **33**, 3540-7.
219. J.Anglister, S.Grzesiek, A.C.Wang, H.Ren, C.B.Klee, and A.Bax, *J.Biomol.NMR*, 1993, **3**, 121-6.
220. M.Bruix, J.Pascual, J.Santoro, J.Prieto, L.Serrano, and M.Rico, *Eur.J.Biochem.*, 1993, **215**, 573-85.
221. M.Cox, N.Dekker, R.Boelens, C.P.Verrijzer, P.C. van der Vliet, and R.Kaptein, *Biochemistry*, 1993, **32**, 6032-40.
222. E.H.Morita, M.Shirakawa, F.Hayashi, M.Imagawa, and Y.Kyogoku, *FEBS Lett.*, 1993, **321**, 107-10.
223. R.T.Clubb, V.Thanabal, J.Fejzo, S.B.Ferguson, L.Zydowsky, H.C.Baker, C.T.Walsh, and G.Wagner, *Biochemistry*, 1993, **32**, 6391-401.
224. W.J.Fairbrother, D.Reilly, T.Colby, and R.Hruk, *FEBS Lett.*, 1993, **320**, 302-6.
225. N.Jamin, O.S.Gabrielsen, N.Gilles, P.N.Lirsac, and F.Toma, *Eur.J.Biochem.*, 1993, **216**, 147-54.
226. S.Koyama, H.Yu, D.C.Dalgarno, T.B.Shin, L.D.Zydowsky, and S.L.Schreiber, *FEBS Lett.*, 1993, **324**, 93-8.
227. Y.Muto, K.Yamasaki, Y.Ito, S.Yajima, H.Masaki, T.Uozumi, M.Walchli, S.Nishimura, T.Miyazawa, and S.Yokoama, *J.Biomol.NMR*, 1993, **3**, 165-84.
228. M.Overduin, B.Mayer, C.B.Rias, D.Baltimore, and D.Cowburn, *Proc.Natl.Acad.Sci.*, 1992, **89**, 11673-7.
229. T.T.Sakai, M.A.Jablonsky, A.P.DeMuth, N.R.Krishna, M.A.Jarpe, and H.M. Johnson, *Biochemistry*, 1993, **32**, 5650-5.
230. C.Abeygunawardana, D.J.Weber, D.N.Frick, M.J.Bossman, and A.S.Mildvan, *Biochemistry*, 1993, **32**, 13071-80.
231. D.J.Weber, C.Abeygunawardana, M.J.Bessman, A.S.Mildvan, *Biochemistry*, 1993, **32**, 13081-8.
232. D.Willbold, U.Krueger, R.Frank, R.Rosin-Arbesfeld, A.Gazit, A.Yaniv, and P.Roesch, *Biochemistry*, 1993, **32**, 8439-45.
233. H.Sticht, D.Willbold, P.Bayer, A.Ejchart, F.Herrmann, R.Rosin-Arbesfeld, A.Gazit, A.Yaniv, R.Frank, and P.Roesch, *Eur.J.Biochem.*, 1993, **218**, 973-6.
234. R.D.Everett, P.Barlow, A.Milner, B.Luisi, A.Orr, G.Hope, and D.Lyon, *J.Mol.Biol.*, 1993, **234**, 1038-47.
235. B.L.Golden, D.W.Hoffman, V.Ramakrishnan, and S.W.White, *Biochemistry*, 1993, **32**, 12812-20.
236. M.E.Girvin and R.H.Fillingame, *Biochemistry*, 1993, **32**, 12167-77.
237. S.J.Archer, V.K.Vinson, T.D.Pollard, and D.A.Torchia, *Biochemistry*, 1993, **32**, 66807.
238. V.K.Vinson, S.J.Archer, E.E.Lattman, T.D.Pollard, and D.M.Torchia, *J.Cell.Biol.*, 1993, **122**, 1277-83.
239. W.Metzler, K.L.Constantine, M.S.Friedrichs, A.J.Bell, E.G.Ernst, T.B.Lavoie, and L.Mueller, *Biochemistry*, 1993, **32**, 13818-29.
240. D.S.Garrett, P.J.Lodi, Y.Shamoo, K.R.Williams, G.M.Clore, and A.M.Gronenborn, *Biochemistry*, 1994, **33**, 2852-8.
241. T.Szyperski, S.Scheek, J.Johansson, G.Assmann, U.Serdorp, and K.Wüthrich, *FEBS Lett.*, 1993, **335**, 18-26.
242. P.Viglino, F.Fogolari, S.Formisano, N.Bortolotti, G.Damante, R.DiLauro, and G.Esposito, *FEBS Lett.*, 1993, **336**, 397-402.
243. D.F.Wyss, J.M.Withka, H.M.Knappers, K.A.Sterne, M.A.Rency, and G.Wagner, *Biochemistry*, 1993, **32**, 10995-1006.
244. P.C.Driscoll, J.G.Cyster, C.Somoza, A.D.Crawford, P.Howe, T.S.Harvey, B.Kieffer, I.D.Campbell, and A.F.Williams, *Biochem.Soc.Trans.*, 1993, **21**, 947-52.
245. J.M.Withka, D.F.Wyss, G.Wagner, A.R.N.Arulanandam, E.L.Reinherz, and M.A.Recny, *Structure*, 1993, **1**, 69-81.
246. S.R.van Doren, A.V.Kurochkin, Q.Z.Ye, L.L.Johnson, D.J.Hupe, and E.R.P.Zuiderweg, *Biochemistry*, 1993, **32**, 13109-22.
247. P.R.Gooley, B.A.Johnson, A.I.Marcy, G.C.Cuca, S.P.Salowe, W.K.Hagmann, C.K.Esser, and J.P.Springer, *Biochemistry*, 1993, **32**, 13098-108.
248. C.M.Fletcher, R.A.Harrison, P.J.Lachmann, and D.Neuhaus, *Prot. Sci.*, 1993, **2**, 2015-27.
249. B.Kieffer, P.C.Driscoll, I.D.Campbell, A.C.Willis, P.A. van der Merwe, and S.J.Davis, *Biochemistry*, 1994, **33**, 4471-82.
250. K.V.Pervushin, V.Y.Orekhov, A.I.Popov, L.Y.Musina, and A.S.Arseniev, *Eur.J.Biochem.*, 1994, **219**, 571-83.
251. D.Stys, I.Blaha, and P.Strop, *Biophys.Biochem. Acta.*, 1993, **1182**, 157-61.
252. P.A.McDonnell, K.Shon, Y.Kim, S.J.Opella, *J.Mol.Biol.*, 1993, **233**, 447-63.
253. T.Liu, D.Liu, E.F.DeRose, and G.R.Mullen, *J.Biol.Chem.*, 1993, **268**, 16309-15.
254. H.Sakashita, T.Sakuma, T.Ohkubo, M.Kainosho, K.Sakumi, M.Sekiguchi, and K.Morikawa, *FEBS Lett.*, 1993, **323**, 252-6.
255. L.C.Myers, G.L.Verdine, and G.Wagner, *Biochemistry*, 1993, **32**, 14089-94.
256. M.Cox, J.Schaller, R.Boelens, R.Kaptein, E.Rickli, and M.Llinas, *Chem.Phys.Lipids*, 1994, **67-68**, 43-58.
257. X.Li, A.M.Bokman, M.Llinas, R.A.G.Smith, and C.M.Dobson, *J.Mol.Biol.*, 1994, **235**, 1548-59.
258. A.P.Hansen, A.M.Petros, R.P.Meadows, D.G.Nettesheim, A.P.Mazar, E.T.Olejniczak, R.X.Xu, T.M.Pederson, J.Henkin, and S.W.Fesik, *Biochemistry*, 1994, **33**, 4847-64.
259. K.Berndt, P.Guentert, and K.Wüthrich, *J.Mol.Biol.*, 1993, **234**, 735-50.

260. A.M.J.J.Bonvin, H.Vis, J.N.Berg, M.J.M.Burgering, R.Boelens, and R.Kaptein, *J.Mol.Biol.*, 1994, **236**, 328-41.
261. H.Baumann, K.Paulsen, H.Kovacs, H.Berglund, A.P.H.Wright, J.A.Gustafsson, and T.Haerd, *Biochemistry*, 1993, **32**, 13463-71.
262. M.S.Finnin, D.W.Hoffman, K.N.Kreuzer, S.J.Porter, R.P.Schmidt, and S.W.White, *J.Mol.Biol.*, 1993, **232**, 301-4.
263. W.Antuch, K.D.Berndt, M.A.Chavez, J.Delfin, K.Wüthrich, *Eur.J.Biochem.*, 1993, **212**, 675-84.
264. K.Bartik, C.M.Dobson, and C.Redfield, *Eur.J.Biochem.*, 1993, **215**, 255-66.
265. A.Sato, S.Nishlmura, T.Ohkubo, Y.Kyogoku, S.Koyama, M.Kobayashi, T.Yasuda, and Y.Kobayashi, *Int.J.Pept.Protein Res.*, 1993, **41**, 433-40.
266. A.Schnuchel, R.Wiltscheck, M.Czisch, M.Herrler, G.Willimsky, P.Graumann, M.A.Marahiel, and T.A.Holak, *Nature*, 1993, **364**, 169-71.
267. H.Senn and W.Klaus, *J.Mol.Biol.*, 1993, **232**, 907-25.
268. C.M.Read, P.D.Cary, C.Cran-Robinson, P.C.Driscoll, and D.G.Norman, *Nucl.Acids.Res.*, 1993, **21**, 3427-36.
269. R.Powers, D.S.Garrett, C.J.March, E.A.Frieden, A.M.Gronenborn, G.M.Clore, *Biochemistry*, 1993, **32**, 6744-62.
270. F.J.Moy, Y.C.Li, P.Rauenbuehler, E.M.Winkler, H.A.Scheraga, and G.T.Montelione, *Biochemistry*, 1993, **32**, 7334-53.
271. B.Leiting, R.DeFrancesco, L.Tomei, R.Cortese, G.Otting, and K.Wuthrich, *EMBO J.*, 1993, **12**, 1797-803.
272. D.Lassen, C.Luecke, A.Kromminga, A.Lezius, F.Spener, and H.Rueterjans, *Mol.Cell.Biochem.*, 1993, **123**, 15-22.
273. H.R.Kalbitzer and W.Hengstenberg, *Eur.J.Biochem.*, 1993, **216**, 205-14.
274. J.Kordel, N.J.Skelton, M.Akke, and W.J.Chazin, *J.Mol.Biol.*, 1993, **231**, 711-34.
275. S.Linse, E.Thulin, and P.Sellers, *Protein Sci.*, 1993, **2**, 985-1000.
276. Y.N.Kalia, S.M.Brocklehurst, D.S.Hipps, E.Appella, K.Sakayuchi, and R.N.Perham, *J.Mol.Biol.*, 1993, **230**, 323-41.
277. F.K.Junius, A.S.Weiss, and G.F.King, *Eur.J.Biochem.*, 1993, **214**, 415-24.
278. V.Goldfarb, M.Wittekind, P.D.Jeffrey, L.Mueller, and K.L.Constantine, *J.Mol.Biol.*, 1993, **232**, 15-22.
279. M.D.Carr, C.J.Bauer, M.J.Gradwell, and J.Feeney, *Proc.Natl.Acad.Sci.*, 1994, **91**, 2206-10.
280. K.LConstantine, M.S.Friedrichs, W.J.Metzler, M.Wittekin, P.Hensley, and L.Mueller, *J.Mol.Biol.*, 1994, **236**, 310-27.
281. J.F.O'Connell, R.Bender, J.W.Engels, K.P.Koller, M.Scharf, and K.Wuthrich, *Eur.J.Biochem.*, 1994, **220**, 763-70.
282. P.Sevilla-Sierra, G.Otting, and K.Wüthrich, *J.Mol.Biol.*, 1994, **235**, 1003-20.
283. T.Dieckmann, L.Mitschang, M.Hofmann, J.Kos, V.Turk, E.A.Auerswald, R.Jaenicke, and H.Oschkinat, *J.Mol.Biol.*, 1993, **234**, 1048-59.
284. F.J.M.van der Ven, J.W.M. van Os, J.M.A.Aelen, S.S.Wymenga, M.L.Remerowski, R.N.H.Konings, and C.W.Hilbers, *Biochemistry*, 1993, **32**, 8322-8.
285. B.Boulat and G.Bodenhausen, *J.Biomol.NMR*, 1993, **3**, 335-48.
286. T.Szyperski, P.Luginbuehl, G.Otting, P.Guentert, and K.Wüthrich, *J.Biomol.NMR*, 1993, **3**, 151-64.
287. K.L.Constantine, M.S.Friedrichs, A.J.Bell, T.B.Lavoie, L.Mueller, and W.J.Metzler, *FEBS Lett.*, 1993, **336**, 457-61.
288. V.Y.Orekhov, K.V.Pervushin, and A.S.Arseniev, *Eur.J.Biochem.*, 1994, **219**, 887-96.
289. J-W.Cheng, C.A.Lepre, S.P.Chambers, J.R.Fulghum, J.A.Thomson, and J.M.Moore, *Biochemistry*, 1993, **32**, 9000-10.
290. J-W.Cheng, C.A.Lepre, and J.M.Moore, *Biochemistry*, 1994, **33**, 4093-100.
291. C.Chen, Y.Feng, J.H.Short, and J.A.Wand, *Arch.Biochem.Biophys.*, 1993, **306**, 510-14.
292. M.Akke, N.J.Skelton, J.Kordel, A.G.Palmer, and W.J.Chazin, *Biochemistry*, 1993, **32**, 9832-44.
293. M.A.L.Eriksson, H.Berglund, T.Haerd, and L.Nilsson, *Proteins:Struct.,Funct.,Genet.*, 1993, **17**, 375-90.
294. M.Akke, R.Bueschweiler, and A.G.Palmer, *J.Am.Chem.Soc.*, 1993, **115**, 9832-3.
295. G.Gemmecker, W.Jahne, and H.Kessler, *J.Am.Chem.Soc.*, 1993, **115**, 11620-1.
296. R.Moss, H.Gesmar, and J.J.Led, *J.Am.Chem.Soc.*, 1994, **116**, 747-9.
297. N.K.Thomsen and F.M.Poulsen, *J.Mol.Biol.*, 1993, **234**, 234-41.
298. S.N.Loh, K.E.Prehoda, J.Wang, and J.L.Markley, *Biochemistry*, 1993, **32**, 11022-8.
299. T.M.O'Connell, P.G.Williams, and J.T.Gerig, *J.Labelled Compd.Radiopharm.*, 1993, **33**, 371-80.
300. T.M.Logan, E.T.Olejiniczak, R.X.Xu, and S.W.Fesik, *J.Biomol.NMR*, 1993, **3**, 225-31.
301. T.M.Logan, Y.Theriault, and S.W.Fesik, *J.Mol.Biol.*, 1994, **236**, 637-48.
302. C.L.Chyan, C.Wormald, C.M.Dobson, P.A.Evans, and J.Baum, *Biochemistry*, 1993, **32**, 5681-91.
303. A.T.Alexandrescu, Y.L.Ng, and C.M.Dobson, *J.Mol.Biol.*, 1994, **235**, 587-99.
304. A.Shimizu, M.Ikelguchi, and S.Sugai, *Biochemistry*, 1993, **32**, 13198-203.
305. P.J.Sadler and A.Tucker, *Eur.J.Biochem..*, 1993, **212**, 811-7.
306. R.A.Krautz and R.O.Fox, *Protein Sci.*, 1993, **2**, 851-8.
307. M.D.Jacob and R.O.Fox, *Proc.Natl.Acad.Sci.*, 1994, **91**, 449-53.
308. D.Shortle and C.Abeygunawardana, *Structure*, 1993, **1**, 121-34.
309. A.T.Alexandrescu, C.Abeygunawardana, and D.Shortle, *Biochemistry*, 1994, **33**, 1063-72.
310. Y.Yang and K.H.Mayo, *Biochemistry*, 1993, **32**, 8661-71.
311. K.Ogasahara, E.Matsushita, and K.Yutani, *J.Mol.Biol.*, 1993, **234**, 1197-206.

312. G.Saab-Rincon, C.L.Froebe, and C.R.Matthews, *Biochemistry*, 1993, **32**, 13981-90.
313. U.K.Novak, A.Cooper, D.Saunders, R.A.G.Smith, and C.M.Dobson, *Biochemistry*, 1994, **33**, 2951-60.
314. B.J.Stockman, A.Euvrard, and T.A.Scahill, *J.Biomol.NMR*, 1993, **3**, 285-96.
315. S.Koide, J.H.Dyson, and P.E.Wright, *Biochemistry*, 1993, **32**, 12299-310.
316. J.Eder, M.Rheinnecker, and A.R.Fersht, *Biochemistry*, 1993, **32**, 18-26.
317. P.Soerensen, J.R.Winther, N.C.Kaarholm and F.M.Poulsen, *Biochemistry*, 1993, **32**, 12160-6.
318. H.Van Dael, P.Haezebrouk, L.Morozova, C.Arico-Muendel, and C.M.Dobson, *Biochemistry*, 1993, **32**, 11886-94.
319. C.P.M.van Mierlo, J.Kemmink, D.Neuhaus, N.J.Darby, and T.E.Creighton, *J.Mol.Biol.*, 1994, **235**, 1044-61.
320. K.J.Lumb and P.S.Kim, *J.Mol.Biol.*, 1994, **236**, 412-20.
321. S.K.Kim, J.A.Fuchs, and C.K.Woodward, *Biochemistry*, 1993, **32**, 9600-8.
322. D.LeNguyen, A.Heinz, L.Chiche, M.El Hajji, and B.Castro, *Protein Sci.*, 1993, **2**, 165-74.
323. L.S.Mullins, C.N.Pace, and F.M.Raushel, *Biochemistry*, 1993, **32**, 6152-6.
324. P.Varley, A.M.Gronenborn, H.Christensen, P.T.Wingfield, R.H.Pain, and G.M.Clore, *Science*, 1993, **260**, 1110-13.
325. C.J.Falzone, P.E.Wright, and S.J.Benkovic, *Biochemistry*, 1994, **33**, 439-42.
326. A.Hodel, R.A.Kautz, M.D.Jacobs, and R.O.Fox, *Protein Sci.*, 1993, **2**, 838-50.
327. S.Linse, E.Thulin, and P.Sellers, *Protein Sci.*, 1993, **2**, 985-1000.
328. C.Johansson, M.Ullner, and T.Drakenberg, *Biochemistry*, 1993, **32**, 8429-38.
329. Y.Li and M-D.Tsai, *J.Am.Chem.Soc.*, 1993, **115**, 8523-6.
330. H.Chao, P.L.Davis, B.D.Sykes, and F.D.Sonnichsen, *Protein Sci.*, 1993, **2**, 1411-28.
331. J.P.Davis, M.M.Zhou, and R.L.Van Etten, *Biochemistry*, 1994, **33**, 1278-86.
332. S.Khorasanizadeh, I.D.Peters, T.R.Butt, and H.Roder, *Biochemistry*, 1993, **32**, 7054-63.
333. P.Pjura, L.P.McIntosh, J.A.Wozniak, and B.W.Matthews, *Proteins:Struct.,Funct.,Genet.*, 1993, **15**, 401-12.
334. C.Ramilo, G.Braus, and J.N.S.Evans, *Biochim.Biophys.Acta*, 1993, **1203**, 71-6.
335. V.S.De Serrano and F.J.Castellino, *Biochemistry*, 1994, **33**, 3509-14.
336. V.S.De Serrano and F.J.Castellino, *Biochemistry*, 1994, **33**, 1340-4.
337. V.S.De Serrano and F.J.Castellino, *Biochemistry*, 1993, **32**, 3540-8.
338. K.Yamasaki, M.Shirouzu, Y.Muto, J.Fujita-Yoshigaki, H.Koide, Y.Ito, G.Kawai, S.Hattori, and S.Yokoyama, *Biochemistry*, 1994, **33**, 65-73.
339. L.D.Ward, A.Hammacher, J-G.Zhang, J.Weinstock, K.Yasukawa, C.J.Morton, R.S.Norton, and R.J.Simpson, *Protien Sci.*, 1993, **2**, 1472-81.
340. Z.Shi, I.J.L.Byeon, T.Ru, and M-D.Tsai, *Biochemistry*, 1993, **32**, 6450-8.
341. M.Zhang, E.Huque, and H.J.Vogel, *J.Biol.Chem.*, 1994, **269**, 5099-105.
342. S.R.Campion, C.Biamonti, G.T.Montelione, and S.K.Niyogi, *Protein Eng.*, 1993, **6**, 651-9.
343. A.R.N.Arulanandam, J.M.Withka, D.F.Wyss, G.Wagner, A.Kister, P.Pallai, M.A.Recny, and E.L.Reinherz, *Proc.Natl.Acad.Sci.*, 1993, **90**, 11613-7.
344. P.J.M.Folkers, M.Nilges, R.H.A.Folmer, R.N.H.Konings, and C.W.Hilbers, *J.Mol.Biol.*, 1994, **236**, 229-46.
345. M.Ubbink and G.W.Canters, *Biochemistry*, 1993, **32**, 13893-901.
346. S.Sharma, P.K.Hammen, J.W.Anderson, A.Lueng, F.Georges, W.Hengstenberg, R.E.Klevit, and B.E.Waygood, *J.Biol.Chem.*, 1993, **268**, 17695-704.
347. W.J.Skawinski, F.Adebodun, J.T.Cheng, F.Jordan, and R.Mendelsohn, *Biochim.Biophys.Acta*, 1993, **1162**, 297-308.
348. S.J.Archer, V.K.Vinson, T.D.Pollard, and D.A.Torchia, *FEBS Lett.*, 1994, **337**, 145-51.
349. W.J.Metzler, A.J.Bell, E.Ernst, T.B.Lavoie, and L.Mueller, *J.Biol.Chem.*, 1994, **269**, 4620-5.
350. H.T.A.Cheung, B.Birdsall, T.A.Frenkiel, D.D.Chau, and J.Feeney, *Biochemistry*, 1993, **32**, 6846-54.
351. R.L.Blakley, J.R.Appleman, J.H.Freisheim, and M.J.Jablonsky, *Arch.Biochem.Biophys.*, 1993, **306**, 501-9.
352. R.J.Cisneros, J.W.Zapf, and R.B.Dunlap, *J.Biol.Chem.*, 1993, **268**, 10102-8.
353. J.W.Zapf, M.S.Weir, V.Emerick, J.E.Villafranca, and R.D.Dunlap, *Biochemistry*, 1993, **32**, 9274-81.
354. S.Song, D.Vander Velde, C.W.Gunn, and R.H.Himes, *Biochemistry*, 1994, **33**, 693-8.
355. R.X.Xu, D.Nettesheim, E.T.Olejniczak, R.Meadows, G.Gemmecker, and S.W.Fesik, *Biopolymers*, 1993, **33**, 535-50.
356. C.A.Lepre, J.W.Cheng, and J.M.Moore, *J.Am.Chem.Soc.*, 1993, **115**, 4929-30.
357. G.Carlstrom and W.J.Chazin, *J.Mol.Biol.*, 1993, **231**, 415-30.
358. R.T.Clubb, J.G.Omichinski, G.M.Clore, and A.M.Gronenborn, *FEBS Lett.*, 1994, **338**, 93-7.
359. P.C.Driscoll, J.D.Altman, J.J.Boniface, K.Sakaguchi, P.A.Reay, J.G.Ominchinski, E.Appella, and M.M.Davis, *J.Mol.Biol.*, 1993, **232**, 342-50.
360. E.A.Dratz, J.E.Furstenau, C.G.Lambert, D.L.Thireault, H.Rarick, T.Schepers, S.Pakhevaniants, and H.E.Hamm, *Nature*, 1993, **363**, 276-81.
361. S.Highsmith, M.Kubinec, D.K.Jaiswal, H.Morimoto, P.G.Williams, and D.E.Wemmer, *J.Biomol.NMR*, 1993, **3**, 325-34.
362. D.T.Isbell, S.Du, A.G.Schroering, G.Colombo and J.G.Shelling, *Biochemistry*, 1993, **32**, 11352-62.
363. D.N.M.Jones, M.Bycroft, M.J.Lubienski, and A.R.Fersht, *FEBS Lett.*, 1993, **331**, 165-72.
364. E.M.Meiering, M.Bycroft, M.J.Lubienski, and A.R.Fersht, *Biochemistry*, 1993, **32**, 10975-87.
365. S.A.Hindley, Y.Gao, P.H.Nash, C.Sautes, J.Lund, M.Goodall, and R.Jefferis, *Biochem.Soc.Trans.*, 1993, **21**, 337S.
366. K.Kato, H.Gouda, W.Takaha, A.Yoshino, C.Matsunaga, and Y.Arata, *FEBS Lett.*, 1993, **328**, 49-54.

367 M.E.Huque and H.J.Vogel, *J.Protein Chem.*, 1993, **12**, 695-707
368 D.A.Slatter, B.A.Levine, V.B.Patchall, A.C.Wang, M.Reedy, and G.M.Strasburg, *Biochem.Soc.Trans.*, 1993, **21**, 265S.
369 S.Ohki, U.Iwamoto, S.Aimoto, M.Yazawa, and K.Hikichi, *J.Biol.Chem.*, 1993, **268**, 12388- 92.
370 G.A.Omburo, L.S.Mullins, and F.M.Raushel, *Biochemistry*, 1993, **32**, 9148-55.
371. P.J.Sadler, A.Tucker, and J.H.Viles, *Eur.J.Biochem.*, 1994, **220**, 193-200.
372. G.I.Yakovlev and G.P.Moiseyev, *Biochim.Biophys.Acta*, 1993, **1202**, 143-8.
373. H-W.Huang and J.A.Cowan, *Eur.J.Biochem.*, 1994, **219**, 253-60.
374. J.S.Hu and A.G.Redfield, *Biochemistry*, 1993, **32**, 6763-72.
375. A.Loew, M.Sprinzl and S.Limmer, *Eur.J.Biochem.*, 1993, **213**, 781-8.
376. A.F.Miller, C.J.Halkides and A.G.Redfield, *Biochemistry*, 1993, **32**, 7367-76.
377. L.C.Myers, M.P.Terranova, A.E.Ferentz, G.Wagner, and G.L.Verdine, *Science*, 1993, **261**, 1164-7.
378. J.L.Kosa, J.W.Michelsen, H.A.Louis, J.I.Olsen, D.R.Davis, and M.C.Beckerle, *Biochemistry*, 1994, **33**, 468-77.
379. A.Martinez, C.Abeygunawardana, J.Haavik, T.Flatmark, and A.S.Mildvan, *Biochemistry*, 1993, **32**, 6381-90.
380. B.B.Kragelund, K.V.Andersen, J.C.Madsen, J.Knudsen, and P.M.Flemming, *J.Mol.Biol.*, 1993, **230**, 1260-77.
381. L.T.Kakalis and I.M.Armitage, *Biochemistry*, 1994, **34**, 1495-501.
382. R.M.M.Brito, G.A.Krudy, J.C.Negele, J.A.Putkey, and P.R.Rosevear, *J.Biol.Chem.*, 1993, **268**, 20966-73.
383. F.Ni, Q.Ning, C.M.Jackson, and J.W.Fenton, *J.Biol.Chem.*, 1993, **268**, 16899-902.
384. T.Nakayama, Y.Arata, and I.Shimada, *Biochemistry*, 1993, **32**, 13961-8.
385. E.Tsilikounas, C.A.Kettner, and W.W.Bachovchin, *Biochemistry*, 1993, **32**, 12651-5.
386. D.Rong, A.J.Lovey, M.Rosenberger, A.d'Avignon, J.Ponder, and E.Li, *J.Biol.Chem.*, 1993, **268**, 7929-34.
387. P.O.Lycksell, R.Ingemarson, R.Davis, A.Graeslund, and L.Thelander, *Biochemistry*, 1994, **33**, 2838-42.
388. J.R.Burke and P.A.Frey, *Biochemistry*, 1993, **32**, 13220-30.
389. V.P.Chuprina, J.A.C.Rullman, R.M.J.N.Lamerichs, J.H.van Boom, R.Boelens, and R.Kaptein, *J.Mol.Biol.*, 1993, **234**, 446-62.
390. M.Billeter, Y.Q.Qian, G.Otting, W.Gehring, and K.Wüthrich, *J.Mol.Biol.*, 1993, **234**, 1084-93.
391. Y.Q.Qian, G.Otting, M.Billeter, M.Mueller, W.Gehring, and K.Wüthrich, *J.Mol.Biol.*, 1993, **234**, 1070-83.
392. J.G.Omichinski, G.M.Clore, O.Schaad, G.Felsenfeld, C.Trainor, E.Appella, S.J.Stahl, and A.M.Gronenborn, *Science*, 1993, **261**, 438-46.
393. Y.Oda, S.Iwai, E.Ohtsuka, M.Ishikawa, M.Ikehara, and N.Hakamura, *Nucl.Acids Res.*, 1993, **21**, 4690-5.
394. J.D.Baleja, T.Mau, and G.Wagner, *Biochemistry*, 1994, **33**, 3071-8.
395. H.Shindo, H.Kurumizaka, C.Sakuma, U.Matsumoto, A.Yanagida, N.Goshima, Y.Kano, and F.Imamoto, *Biol.Pharm.Bull.*, 1993, **16**, 437-43.
396. M.V.Botuyan, D.A.Keire, C.Kroen, and D.G.Gorenstein, *Biochemistry*, 1993, **32**, 6863-74.
397. P.Beckmann, S.R.Martin, and A.N.Lane, *Eur.Biophys.J.*, 1993, **21**, 417-24.
398. B.J.Lee, H.Sakashita, T.Ohkubo, M.Ikehara, T.Doi, K.Morikawa, Y.Kyogoku, T.Osafune, S.Iwai, and E.Ohtsuka, *Biochemistry*, 1994, **33**, 57-64.
399. F.Rastinejad and P.Lu, *J.Mol.Biol.*, 1993, **232**, 105-22.
400. F.Rastinejad, P.Artz, and P.Lu, *J.Mol.Biol.*, 1993, **233**, 389-99.
401. P.J.M.Folkers, J.P.M.van Duynhoven, H.T.M.van Lieshout, B.J.M.Harmsen, J.H.van Boom, G.I.Tesser, R.N.H.Konings, and C.W.Hilbers, *Biochemistry*, 1993, **32**, 9407-16.
402. J.P.M.van Duynhoven, I.M.A.Nooren, D.W.Swinkels, P.J.M.Folkers, B.J.M.Harmsen, R.N.H.Konings, G.I.Tesser, and C.W.Hilbers, *Eur.J.Biochem.*, 1993, **216**, 507-17.
403. J.P.Laussac, G.Peyrou, H.Marzarguil, M.Erard, M.Bourdonneau, and T.C.Manh, *New J.Chem.*, 1993, **17**, 607-12.
404. N.Julian, H.Demene, N.Morellet, B.Maigret, and B.P.Roques, *FEBS Lett.*, 1993, **331**, 43-8.
405. T.L.South and M.F.Summers, *Protein Sci.*, 1993, **2**, 3-19.
406. N.Morellet, H.de Rocquigny, Y.Mely, N.Jullian, H.Demene, M.Ottmann, D.Gerard, J.L.Darlix, M.C.Fourne-Zaluski, and B.P.Roques, *J.Mol.Biol.*, 1994, **235**, 287-301.
407. X.Qian, C.Jeon, H.Yoon, K.Agarwal, and M.A.Weiss, *Nature*, 1993, **365**, 277-9.
408. X.Qian, S.N.Gozani, H.Yoon, C.Jeon, K.Agarwal, and M.A.Weiss, *Biochemistry*, 1993, **32**, 9944-59.
409. R.M.A.Knegtel, M.Katahira, J.G.Schilthuis, A.M.J.J.Bonvin, R.Boelens, D.Eib, P.T.Van der Saag, and R.Kaptein, *J.Biomol.NMR*, 1993, **3**, 1-17.
410. R.M.A.Knegtel, R.Boelens, M.L.Ganadu, A.V.E.George, P.T.Van der Saag, and R.Kaptein, *Biochem.Biophys.res.Commun.*, 1993, **192**, 492-8.
411. M.S.Lee, S.A.Kliewer, J.Provencal, P.E.Wright, and R.M.Evans, *Science*, 1993, **260**, 1117-21.
412. B.E.Bernstein, R.C.Hoffman, S.Horvath, J.R.Herriott, and R.E.Klevit, *Biochemistry*, 1994, **33**, 4460-70.
413. R.C.Hoffman, S.J.Horvath, and R.E.Klevit, *Protein Sci.*, 1993, **2**, 951-65.
414. L.Banci, I.Bertini, C.Fuchinat, L.Messori, and P.Turano, *Appl.Magn.Reson.*, 1993, **4**, 461-76.
415. L.Banci, W.Bermel, C.Luchinat, R.Pierattelli, and D.Tarchi, *Magn.Reson.Chem.*, 1993, **31**, S3-S7.
416. C.J.Henehan, D.L.Pountney, O.Zerbe, and M.Vasak, *Protein Sci*, 1993, **2**, 1756-64.
417. J.M.Aramini, P.H.Krygsman, and H.J.Vogel, *Biochemistry*, 1994, **33**, 3304-11.
418. J.M.Aramini, M.W.Germann, and H.J.Vogel, *J.Am.Chem.Soc.*, 1993, **115**, 9750-3.
419. J.M.Aramini and H.J.Vogel, *J.Am.Chem.Soc.*, 1994, **116**, 1988-93.

420. L.Banci, I.Bertini, S.Ciurli, S.Ferretti, C.Luchinat, and M.Piccioli, *Biochemistry*, 1993, **32**, 9387-97.
421. I.Bertini, A.Gaudemer, C.Luchinat, and M.Piccioli, *Biochemistry*, 1993, **32**, 12887-93.
422. I.Bertini, S.Ciurli, A.Dikiy, and C.Luchinat, *J.Am.Chem.Soc.*, 1993, **115**, 12020-8.
423. H.Rueterjans, L.Messori, O.Ohlenschlaeger, F.Briganti, and I.Bertini, *Appl.Magn.Reson.*, 1993, **4**, 477-89.
424. J.H.Bushweller, M.Billeter, A.Holmgren, and K.Wüthrich, *J.Mol.Biol.*, 1994, **235**, 1585-97.
425. J.H.Bushweller, A.Holmgren, and K.Wüthrich, *Eur.J.Biochem.*, 1993, **218**, 327-34.
426. M.Fujii, H.Hori, G.Miyazaki, H.Morimoto, and T.Yonetani, *J.Biol.Chem.*., 1993, **268**, 15386-93.
427. S.L.Alam and J.D.Satterlee, *Biochemistry*, 1994, **33**, 4008-18.
428. L.Martineau and C.T.Craescu, *Eur.J.Biochem.*, 1993, **214**, 383-93.
429. Y.Yamamoto and T.Suzuki, *Biochim.Biophys.Acta*, 1993, **1163**, 287-96.
430. D.Bashford, D.A.Case, C.Dalvit, L.Tennant, and P.E.Wright, *Biochemistry*, 1993, **32**, 8045-56.
431. C.Brunel, A.Bondon, and G.Simonneaux, *Eur.J.Biochem.*, 1993, **214**, 405-14.
432. Y.Yamamoto, *Biochem.Biophys.Res.Commun.*, 1993, **196**, 348-54,
433. Y.Yamamoto, T.Suzuki, and H.Hori, *Biochim.Biophys.Acta*, 1993, **1203**, 267-75.
434. Y.Yamamoto, Y.Inoue, and T.Suzuki, *Magn.Reson.Chem.*., 1993, **31**, S8-16.
435. T.E.Zewert, H.B.Gray, and I.Bertini, *Biochem.Biophys.Res.Commun.*, 1994, **198**, 228-35.
436. M.S.Mondal, S.Mazumdar, and S.Mitra, *Inorg.Chem.*, 1993, **32**, 5362-7.
437. S.Neya, N.Funasaki, T.Sato, N.Igarashi, and N.Tanaka, *J.Biol.Chem.*, 1993, **268**, 8935-42.
438. J.Qin, G.N.La Mar, Y.Dou, S.J.Admiraal, and M.Ikedo-Saito, *J.Biol.Chem.*, 1994, **269**, 1083-90.
439. J.Qin, G.N.La Mar, F.Cutruzzola, C.T.Allocatelli, A.Brancaccio, and M.Brunori, *Biophys.J.*, 1993, **65**, 2178-90.
440. J.Qin, U.Pande, G.N.La Mar, F.Ascoli, P.Ascenzi, F.Cutruzzola, C.Travaglini-Allocatelli, and M.Brunori, *J.Biol.Chem.*, 1993, **268**, 24012-21.
441. J.Qin, G.N.La Mar, F.Ascoli, and M.Brunori, *J.Mol.Biol.*, 1993, **231**, 1009-23.
442. K.Rajarathnam, J.Qin, G.N.La Mar, M.L.Chiu, and S.G.Sligar, *Biochemistry*, 1993, **32**, 5670-80.
443. M.A.Scharberg and G.N.La Mar, *J.Am.Chem.Soc.*, 1993, **115**, 6513-21.
444. M.A.Scharberg and G.N.La Mar, *J.Am.Chem.Soc.*, 1993, **115**, 6522-8.
445. C.Travaglini-Allocatelli, F.Cutruzzola, A.Brancaccio, M.Brunori, J.Qin, and G.N.La Mar, *Biochemistry*, 1993, **32**, 6041-9.
446. K.Vyas, K.Rajarathnam, L.P.Yu, S.D.Emerson, G.N.La Mar, R.Krishnamoorthi, and H.Mizukami, *J.Biol.Chem.*, 1993, **268**, 14826-35.
447. Y.H.Kao and J.T.J.Lecomte, *J.Am.Chem.Soc.*, 1993, **115**, 9754-62.
448. L.Banci, I.Bertini, S.Marconni, and R.Pierattelli, *Eur.J.Biochem.*, 1993, **215**, 431-7.
449. D.Morikis, R.Brushweiler, and P.E.Wright, *J.Am.Chem.Soc.*, 1993, **115**, 6238-46.
450. L.Banci, I.Bertini, T.Bini, M.Tien, and P.Turano, *Biochemistry*, 1993, **32**, 5825-31.
451. L.Banci, I.Bertini, I.C.Kuan, M.Tien, P.Turano, and A.J.Vila, *Biochemistry*, 1993, **32**, 13483-9.
452. S.Modi, D.V.Behere, and S.Mitra, *Biochim.Biophys.Acta*, 1994, **1204**, 14-8.
453. S.Modi, *Biochim.Biophys.Acta*, 1993, **1162**, 121-6.
454. L.Banci, I.Bertini, L.D.Eltis, and R.Pierattelli, *Biophys.J.*, 1993, **65**, 806-13.
455. L.Blanchard, D.Marion, B.Pollock, G.Voordouw, J.Wall, M.Bruschi, and F.Guerlesquin, *Eur.J.Biochem.*, 1993, **218**, 293-301.
456. R.Timkovich and M.Cai, *Biochemistry*, 1993, **32**, 11516-23.
457. A.P.Campos, A.P.Aguiar, M.Hervas, M.Regalla, J.A.Navarro, J.M.Ortega, A.V.Xavier, M.A.De la Rosa, and M.Teixeira, *Eur.J.Biochem.*, 1993, **216**, 329-41.
458. I.B.Coutinho, D.L.Turner, J.LeGall, and A.V.Xavier, *Biochem.J.*, 1993, **294**, 899-908.
459. M.A.Picarra-Pereira, D.L.Turner, J.LeGall, and A.V.Xavier, *Biochem.J.*, 1993, **294**, 909-15.
460. J.S.Park and S.W.Kang, *Bull.Korean Chem.Soc.*, 1993, **14**, 588-92.
461. A.M.Davies, J.G.Guillemette, M.Smith, C.Greenwood, A.G.P.Thurgood, A.G.Mauk, and G.R.Moore, *Biochemistry*, 1993, **32**, 5431-5.
462. R.M.Greene, S.F.Betz, S.Hilgen-Willis, D.S.Auld, J.B.Fencl, and G.J.Pielak, *Inorg.Biochem.*, 1993, **51**, 663-76.
463. W.Shao, Y.Yao, G.Liu and W.Tang, *Inorg.Chem.*, 1993, **32**, 6112-4.
464. J.Wu and D.G.Gorenstein, *J.Am.Chem.Soc.*, 1993, **115**, 6843-50.
465. R.Hewson, R.J.Newbold, and D.Whitford, *Protein Eng.*, 1993, **6**, 953-64.
466. R.D.Guiles, V.J.Basus, S.Sarma, S.Malpure, K.M.Fox, I.D.Kuntz, and L.Waskell, *Biochemistry*, 1993, **32**, 8329-40.
467. L.Qin and N.M.Kostic, *Biochemistry*, 1993, **32**, 6073-80.
468. L.Banci, I.Bertini, M.T.Cambria, F.Capozzi, and A.Dikiy, *Eur.J.Biochem.*, 1994, **219**, 663-9.
469. H.S.Costa, H.Santos, and D.L.Turner, *Eur.J.Biochem.*, 1993, **215**, 817-24.
470. J.C.Ferrer, J.G.Guillemette, R.Bogumil, S.C.Inglis, M.Smith, and A.G.Mauk, *J.Am.Chem.Soc.*, 1993, **115**, 7505-8.
471. Q.Yi, J.E.Erman, and J.D.Satterlee, *J.Am.Chem.Soc.*, 1994, **116**, 1981-7.
472. Q.Yi, J.E.Erman, and J.D.Satterlee, *Biochemistry*, 1993, **32**, 10988-94.
473. K-B.Lee, G.N.La Mar, K.E.Mansfield, K.M.Smith, T.C.Pochapsky, and S.G.Sligar, *Biochim.Biophys.Acta*, 1993, **1202**, 189-99.
474. J.M.Moratal, A.Donaire, J.Salgado,H.R.Jimenez, J.Castells, and M.Piccioli, *FEBS Lett.*, 1993, **324**, 305-8.
475. L.Banci, I.Bertini, C.Luchinat, M.Piccioli, and A.Scozzafava, *Gazz.Chim.Ital.*, 1993, **123**, 95-100.

476. L.J.Ming, J.B.Lynch, R.C.Holz and L.Que, *Inorg.Chem.*, 1994, **33**, 83-7.
477. Y.K.Chae, F.Abildgaard, E.S.Mooberry and J.L.Markley, *Biochemistry*, 1994, **33**, 3287-95.
478. H.Cheng, B.Xia, G.H.Reed, and J.L.Markley, *Biochemistry*, 1994, **33**, 3155-64.
479. J.Gaillard, I.Quinkal, and J.M.Moulis, *Biochemistry*, 1993, **32**, 9881-7.
480. J.Gaillard, J.M.Moulis, R.Kummerle, and J.Meyer, *Magn.Reson.Chem.*, 1993, **31**, S27-33.
481. E.Lloyd, N.P.Tomkinson, G.A.Salmon, and A.G.Sykes, *Biochim.Biophys.Acta*, 1993, **1202**, 113-20.
482. B.J.Stockman, A.Euvrard, D.A.Kloosterman, T.A.Scahill, and R.P.Swenson, *J.Biomol.NMR*, 1993, **3**, 133-49.
483. J.H.Dyson, M-F.Jeng, P.Model, and A.Holmgren, *FEBS Lett.*, 1994, **339**, 11-7.
484. J.M.Lancelin, M.Stein, and J.P.Jacquot, *J.Biochem (Tokyo)*, 1993, **114**, 421-31.
485. T.E.Elgren, L.J.Ming, and L.Que, *Inorg.Chem.*, 1994, **33**, 891-4.
486. J.Kaufman, L.M.Siegel, and L.D.Spicer, *Biochemistry*, 1993, **32**, 8782-91.
487. G.Mohn, P.Koehl, H.Budzikiewicz, and J-F.Lefevre, *Biochemistry*, 1994, **33**, 2843-51.
488. S.S.Narula, D.R.Winge, and I.M.Armitage, *Biochemistry*, 1993, **32**, 6773-87.
489. G.S.Lukat, M.B.Doran, L.M.Utschig, and H.M.Goff, *J.Inorg.Biochem.*, 1993, **50**, 157-71.
490. C.Dennison, T.Kohzuma, W.McFarlane, S.Suzuki, and A.G.Sykes, *J.Chem.Soc.,Dalton Trans.*, 1994, 437-43.
491. K.D.Bishop, F.J.H.Blocker, W. Egan, and T.L. James, *Biochemistry*, 1994, **33**, 427-438.
492. A.Mujeeb, S.M.Kerwin, G.L.Kenyon, and T.L.James, *Biochemistry*, 1993, **32**, 13419-31.
493. R.Michalczyk and I.M.Russu, *FEBS Lett*, 1993, **331**, 217-22.
494. K.L.B.Borden, *Biochemistry*, 1993, **32**, 6506-14.
495. R.Nibedita, R.A.Kumar, A.Mujumdar, R.V.Hosur, G.Govil, K.Majumder, and V.S.Chauhan, *Biochemistry*, 1993, **32**, 9053-64.
496. T.V.Maltseva, S.I.Yamakage, P.Agback, and J.Chattopadhyaya, *Nucl.Acids Res.*, 1993, **21**, 4288-95.
497. A.N.Lane, C.J.Bauer, T.A.Frenkiel, and A.J.Birchall, *Eur.Biophys. J.*, 1993, **22**, 135-43.
498. A.N.Lane, C.J.Bauer, and T.A.Frenkiel, *Eur.Biophys. J.*, 1993, **21**, 425-431.
499. S.El Antri, P.Bittoun, O.Mauffret, M.Monnot, O.Convert, E.Lescot, and S.Fermandjian, *Biochemistry*, 1993, **32**, 7079-88.
500. S.A.Wolfe and G.I.Verdine, *J.Am.Chem.Soc.*, 1993, **115**, 12585-6.
501. N.B. Ulyanov, M.H.Sarma, V.B.Zhurkin, and R.H.Sarma, *Biochemistry*, 1993, **32**, 6875-83.
502. Z.Reich, P.Friedman, S.Levin-Zaidman, and A.Minsky, *J.Biol.Chem.*, 1993, **268**, 8261-6.
503. S.A.Fawthrop, J-C.Yang, and J.Fisher, *Nucl.Acids Res.*, 1993, **21**, 4860-6.
504. V.P.Chuprina, E.Sletten, and O.U.Fedoroff, *J.Biomol.Struct.Dyn.*, 1993, **10**, 693-707.
505. M.A.Kennedy, S.T.Nuutero, J.T.Davis, G.P.Drobny, and B.R.Reid, *Biochemistry*, 1993, **32**, 8022-35.
506. K.J.Walters and I.M.Russu, *Biopolymers*, 1993, **33**, 943-51.
507. N.B. Ulyanov, U.Schmitz, and T.L.James, *J.Biomol. NMR*, 1993, **3**, 547-68.
508. U.Schmitz, N.B. Ulyanov, A.Kumar, and T.L.James, *J.Mol.Biol., 1993*, **243**, 373-89.
509. S.Yamakage, T.V.Maltseva, F.P.Nilson, A.Foldesi, and J.Chattopadhyaya, *Nucl.Acids Res.*, 1993, **21**, 5005-11.
510. A.A.Szewczak, G.W.Kellogg, and P.B.Moore, *FEBS Lett.*, 1993, **327**, 261-4.
511. S.R.LaPlante, N.Zanatta, A.Hakkinen, A.H-J, Wang, and P.N.Borer, *Biochemistry*, **33**, 2430-40.
512. G.W.Kellogg and B.I.Schweitzer, *J.Biomol.NMR.*, 1993, **3**, 577-95.
513. P.N.Borer, S.R.LaPlante, A.Kumar,N.Zanatta, A.Martin, A.Hakkinen, and G.C.Levy, *Biochemistry*, 1994, **33**, 2441-50.
514. T.N.Jaishree and A.H-J.Wang, *Nucl.Acids Res.*, 1993, **21**, 3839-44.
515. D.S.Pilch and R.H.Shafer*J.Am.Chem.Soc.*, 1993, **115**, 2565-77.
516. H.Robinson, J.H.van Boom, and A.H.J.Wang,*J.Am.Chem.Soc.*, 1994, **116**, 1565-6.
517. H.Robinson and A.H.J.Wang, *Proc.Natl.Acad.Sci.*, 1993, **90**, 5224-8.
518. W.H.Gmeiner, B.Rayner, F.Morvan, J.L.Imbach, and J.W.Lown,*J.Biomol.NMR*, 1992, **2**, 275-88.
519. K.Maskos, B.M.Gunn, D.A.LeBlanc, and K.M.Morden, *Biochemistry*, 1993, **32**, 3583-95.
520. K.L.Greene, R.L.Jones, H.Robinson, A.H.J.Wang, G.Zon, and W.D.Wilson, *Biochemistry*, 1994, **33**, 1053-62.
521. M.Katahira, H.Sato, K.Mishima, S.Uesugi, and S.Fujii, *Nucl.Acids Res.*, 1993, **21**, 5418-24.
522. J.W.Cheng, S.H.Chou, and B.R.Reid, *J.Mol.Biol.*, 1992, **228**, 1037-41.
523. P.K.Radha, A.Madan, R.V.Hosur, *J.Chem.Soc.,Chem.Commun*, 1993, 1491-2.
524. C.C.Hardin, M.Corregan, B.A.Brown, and L.N.Frederick, *Biochemistry*, 1993, **32**, 5870-80.
525. T.N.Jaishree and A.H.J.Wang, *FEBS Lett.*, 1994, **337**, 139-44.
526. A.N.R.Nedderman, M.J.Stone, D.H.Williams, P.K.T.Lin, and D.M.Brown, *J.Mol.Biol.*, 1993, **230**, 1068-76.
527. S.Mellac, G.V.Fazakerley, L.C.Sowers, *Biochemistry*, 1993, **32**, 7779-86.
528. G.E.Plum, A.P.Grollman, F.Johnson, and K.J.Breslauer, *Biochemistry*, 1992, **31**, 12096-102.
529. J.Y.Kao, I.Goljer, T.A.Phan, and P.H.Bolton, *J.Biol.Chem.*, 1993, **268**, 17787-93.
530. P.Huang, D.J.Patel, and M.Eisenberg, *Biochemistry*, 1993, **32**, 3852-66.
531. B.P.Cho, F.A.Beland, and M.M.Marquis, *Biochemistry*, 1994, **33**, 1374-84.
532. S.F.O'Handley, D.G.Sanford, R.Xu, C.C.Lester, B.E.Hingerty, S.Broyde, and T.R.Krugh, *Biochemistry*, 1993, **32**, 2481-97.
533. M.Cosman, R.Fiala, B.E.Hingerty, A.Laryea, H.Lee, R.G.Harvey, S.Amin, N.E.Geacintov, S.Broyde, and D.Patel, *Biochemistry*, 1993, **32**, 12488-97.

534. T.Maltseva, A.Sanstroem, I.M.Ivanova, S.D.Sergeyev, V.F.Zarytova, and J.Chattopahyaya, *J.Biochem.Biophys.Methods*, 1993, **26**, 173-236.
535. D.Yang, Y.Gao, H.Robinson, G.A.van der Marel, J.H. van Boom, and A.H.J.Wang, *Biochemistry*, 1993, **32**, 8672-81.
536. F.C.Seaman and L.Hurley, *Biochemistry*, 1993, **32**, 12577-85.
537. U.S.Singh, J.G.Moe, G.R.Reddy, J.P.Weisenseel, L.J.Marnett, and M.P.Stone, *Chem.Res.Toxicol.*, 1993, **6**, 825-36.
538. V.Gervais, A.Guy, R.Teoule, and G.V.Fazakerley, *Nucl.Acids Res.*, 1992, **20**, 6455-60.
539. J.K.Kim, S.D.Soni, J.C.Wallace, and J.L.Alderfer, *Nucl.Acids Res.*, 1993, **21**, 2755-9.
540. J.L.F.Kao, S.Nadji, and J.S.Taylor, *Chem.Res.Toxicol.*, 1993, **6**, 561-7.
541. T.N.Jaishree, G.A.van der Marel, J.H. van Boom, and A.H.J.Wang, *Biochemistry*, 1993, **32**, 4903-11.
542. A.N.Lane, S.Abel, and T.Brown, *Eur.J.Biochem.*, 1993, **215**, 297-306.
543. S.M.Baxter, M.B.Greizerstein, D.M.Kushlan, and G.W.Ashley, *Biochemistry*, 1993, **32**, 8702-11.
544. N.B.Blatt, S.E.Osborne, R.J.Cain, and G.D.Glick, *Biochimie*, 1993, **75**, 433-41.
545. Y.Cho, F.C.Zhu, B.A.Luxon, and D.G.Gorenstein, *J.Biomol.Strct.Dyn.*, 1993, **11**, 685-702.
546. J.K.James and I.Tinoco Jr., *Nucl.Acids Res.*, 1993, **21**, 3287-93.
547. N.Zhou, M.W.Germann, J.H.van de Sande, N.Pattabiraman, and H.J.Vogel, *Biochemistry*, 1993, **32**, 646-56.
548. N.Zhou and H.J.Vogel, *Bull.Mag.Res.*, 1993, **15**, 80-3.
549. C.Spiro, J.P.Richards, S.Chandrasekaran, R.G.Brennan, and C.T.McMurray, *Proc.Natl.Acad.Sci.*, 1993, **90**, 4606-10.
550. M.A.Rosen and D.J.Patel, *Biochemistry*, 1993, **32**, 6563-75.
551. M.A.Rosen and D.J.Patel, *Biochemistry*, 1993, **32**, 6576-87.
552. F.W.Smith and J.Feigon, *Biochemistry*, 1993, **32**, 8682-92.
553. N.B.Leontis, M.T.Hills, M.Piotto, A.Malhotra, J.Nussbaum, and D.G.Gorenstein, *J.Biomol.Struct.Dyn.*, 1993, **11**, 215-23.
554. Y.Wang and D.J.Patel, *J.Mol.Biol.*, 1993, **234**, 1171-83.
555. D.Ding, H.Sun, Y.Yao, W.Tang, C.Wang, and L.Zhang, *Spectrosc.Lett.*, 1993, **26**, 1437-46.
556. K.Gehring, J.L.Leroy, and M.Gueron, *Nature*, 1993, **363**, 561-4.
557. G.Gupta, A.E.Garcia, Q.Guo, M.Lu, and N.R.Kallenbach, *Biochemistry*, 1993, **32**, 7098-103.
558. J.L.Leroy, K.Gehring, A.Kettani, and M.Gueron, *Biochemistry*, 1993, **32**, 6019-31.
559. E.A.Venczel and D.Sen, *Biochemistry*, 1993, **32**, 6220-8.
560. Q.Xu, H.Deng, and W.H.Braunlin, *Biochemistry*, 1993, **32**, 13130-7.
561. R.F.Macaya, P.Schultze, F.W.Smith, J.A.Roe, and J.Feigon, *Proc.Natl.Acad.Sci.*, 1993, **90**, 3745-9.
562. S.M.Chen, F.Heffron, and W.J.Chazin, *Biochemistry*, 1993, **32**, 319-26.
563. K.M.Koshlap, P.Gillespie, P.B.Dervan, and J.Feigon, *J.Am.Chem.Soc.*, 1993, **115**, 7908-9.
564. I.Radhakrishnan, D.J.Patel, E.S.Priestley, H.M.Nash, and P.B.Dervan, *Biochemistry*, 1993, **32**, 11228-34.
565. I.Radhakrishnan, C. de los Santos, and D.J.Patel, *J.Mol.Biol.*, 1993, **234**, 188-97.
566. H.P.Hopkins, D.D.Hamilton, W.D.Wilson, and G.Zon, *J.Phys.Chem.*, 1993, **97**, 6555-63.
567. X.Ye, K.Kimura, and D.J.Patel, *J.Am.Chem.Soc.*, 1993, **115**, 9325-6.
568. D.R.Brown, M.Kurz, V.L.Hsu, and D.R.Kearns, *Biochemistry*, 1994, **33**, 651-64.
569. H.Urata, Y.Ueda, Y.Usami, and M.Akagi, *J.Am.Chem.Soc.*, 1993, **115**, 7135-8.
570. Y.Rhee, C.Wang, B.L.Gaffney, and R.A.Jones, *J.Am.Chem.Soc.*, 1993, **115**, 8742-6.
571. A.Fede, M.Billeter, W.Leupin, and K.Wüthrich, *Structure*, 1993, **1**, 177-86.
572. S.Gopalakrishnan and D.J.Patel, *J.Am.Chem.Soc.*, 1993, **115**, 9321-2.
573. K.E.Barr, R.A.Russell, R.N.Warrener, and J.G.Collins, *Febs Lett.*, 1993, **322**, 173-6.
574. M.Cosman, C.de los Santos, R.Fiala, B.E.Hingerty, V.Ibanex, E.Luna, R.Harvey, N.E.Geacintov, S.Broyde, and D.J.Patel, *Biochemistry*, 1993, **32**, 4145-55.
575. A.Blasko and T.C.Bruice, *Proc.Natl.Acad.Sci.*, 1993, **90**, 10018-22.
576. W.Nerdal, O.M.Andersen, and E.Sletten, *Acta.Chem.Scand.*, 1993, **47**, 658-62.
577. M.A.Keniry, D.L.Banville, P.M.Dimmonds, and R.Shafer, *J.Mol.Biol.*, 1993, **231**, 753-67.
578. B.H.Geierstanger, P.J.Jacobsen, M.Mrsich, P.B.Dervan, and D.E.Wemmer, *Biochemistry*, 1994, **33**, 3005-62.
579. M.Sastry and D.J.Patel, *Biochemistry*, 1993, **32**, 6588-604.
580. M.Eriksson, M.Leijon, C.Hiort, B.Norden, and A.Magdalena, *Biochemistry*, 1994, **33**, 5031-40.
581. S.S.David and J.K.Barton, *J.Am.Chem.Soc.*, 1993, **115**, 2984-5.
582. T.C.Jenkins, A.N.Lane, S.Neidle, and D.G.Brown, *Eur.J.Biochem.*, 1993, **213**, 1175-84.
583. D.Sun, M.Hansen, J.J.Clement, and L.H.Hurley, *Biochemistry*, 1993, **32**, 8068-74.
584. M.Mrksich, W.S.Wade, T.J.Dwyer, B.H.Geierstanger, D.E.Wemmer, and P.B.Dervan, *Proc.Natl.Acad.Sci.*, 1992, **89**, 7586-90.
585. D.Sun, C.H.Lin, and L.H.Hurley, *Biochemistry*, 1993, **32**, 4487-95.
586. U.AsselineT.Rozelle, G.Lancelot, and T.T.Nguyen, *Nucl.Acids Res.*, 1992, **20**, 4705-9.
587. E.Trotta, E.D'Ambrosio, N. Del Grosso, G.Ravagnan, M.Cirilli, and M.Paci, *J.Biol.Chem.*, 1993, **268**, 3944-51.
588. B.H.Geierstanger, T.J.Dwyer, Y.Bathini, J.W.Lown, and D.E.Wemmer, *J.Am.Chem.Soc.*, 1993, **115**, 4474-82.
589. N.A.Froeystein and E.Sletten, *J.Am.Chem.Soc.*, 1994, **116**, 3240-50.
590. Q.Xu, R.K.Shoemaker, and W.H.Braunlin, *Biophys.J.*, 1993, **65**, 1039-49.

591. Q.Xu, S.R.B.Jampani, and W.H.Braunlin, *Biochemistry*, 1993, **32**, 11754-60.
592. N.A.Froeystein, J.T.Davis, B.R.Reid, and E.Sletten, *Acta.Chem.Scand.*, 1993, **47**, 649-59.
593. C.J.Cowan, H.W.Huang, and L.Y.Hsu, *J.Inorg.Chem.*, 1993, **52**, 121-9.
594. S.Walker, J.Murnick, and D.Kahne, *J.Am.Chem.Soc.*, 1993, **115**, 7954-61.
595. S.Walker, A.H.Andreotti, and D.Kahne, *Tetrahedron*, 1994, **50**, 1351-60.
596. M.Rosaria Conte, E.Fattorusso, L.Gomez Paloma, and L.Mayol, *Bioorg.Med.Chem.Lett.*, 1992, **2**, 1299-304.
597. B.T.Bennett, L.Mueller, E.P.Nikonowicz, and A.Pardi, *J.Am.Chem.Soc.*, 1993, **115**, 11040-1.
598. E.P.Nikonowicz and A.Pardi, *J.Mol.Biol.*, 1993, **232**, 1141-56.
599. P.Legault, B.T.Farmer, L.Mueller, and A.Pardi, *J.Am.Chem.Soc.*, 1994, **116**, 2203-4.
600. J.P.Marino, J.H.Prestegard, and D.M.Crothers, *J.Am.Chem.Soc.*, 1994, **116**, 2205-6.
601. M.J.Michnicka, J.W.Harper, and G.C.King, *Biochemistry*, 1993, **32**, 395-400.
602. J.A.Jaeger and I.Tinoco, *Biochemistry*, 1993, **32**, 12522-30.
603. V.Ramesh, *Nucl.Acids Res.*, 1993, **21**, 5485-88.
604. P.Agback, C.Glemarec, L.Yin, A.Sandstrom, J.Plavec, C.Sund, S.Yamakage, G.Viswanadham, B.Rousse, N.Puri, and J.Chattapadhyaya, *Tetrahedron Lett.*, 1993, **34**, 3929-32.
605. B.Rousse, N.Puri, G.Viswanadham, P.Agback, C.Glemarec, A.Sandstrom, C.Sund and J.Chattopadhyaya, *Tetrahedron*, 1994, **50**, 1777-810.
606. B-S.Lee, H-N.Wu, and T-H.Huang, *FEBS Lett*, 1993, **324**, 296-300.
607. M.Chastain and I.Tinoco, *Biochemistry*, 1993, **32**, 14220-8.
608. P.Legault and A.Pardi, *J.Magn.Reson.,Ser.B*, 1994, **103**, 82-6.
609. J.Santa-Lucia and D.H.Turner, *Biochemistry*, 1993, **32**, 12612-23.
610. S.Limmer, H-P.Hofman, G.Ott, and M.Sprinzl, *Proc.Natl.Acad.Sci.*, 1993, **90**, 6199-202.
611. G.Ott, L.Arnold, and S.Limmer, *Nucl.Acids Res.*, 1993, **21**, 5859-64.
612. Y.Chen, H.Sierzputowska, R.Guenther, K.Everett, and P.F.Agris, *Biochemistry*, 1993, **32**, 10249-53.
613. J.Moriya, T.Yokogawa, K.Wakita, T.Ueda, K.Nishikawa, P.F.Crain, T.Hashizume, S.C.Pomerantz, and J.A.McCloskey, *Biochemistry*, 1994, **33**, 2234-9.
614. K.Sakamato, G.Kawai, T.Niimi, T.Satoh, M.Sekine, Z.Yamaizumi, S.Nishimura, T.Miyazawa, and S.Yokoyama, *Eur.J.Biochem.*, 1993, **216**, 369-75.
615. M.Amano and Y.Kyogoku, *Eur.J.Biochem.*, 1993, **217**, 131-6.
616. J.P.Abrahams, M.J.van Raaij, L.Bosch, B.Kraak, A.L.I.Johannes, and S.Wijmenga, *Spectroscopy (Amersterdam)*, 1993, **11**, 37-44.
617. A.A.Szewczak, G.W.Kellog, and P.B.Moore, *FEBS Lett.*, 1993, **327**, 261-4.
618. A.A.Szewczak, P.B.Moore, Y-L.Chan, and I.G.Wool, *Proc.Natl.Acad.Sci.*, 1993, **90**, 9581-5.
619. R.W.Kriwacki, R.B.Hill, J.M.Caradonna, and J.H.Prestegard, *J.Am.Chem.Soc.*, 1993, **115**, 8907-11.
620. R.M.Brunne, E.Liepinsh, G.Otting, K.Wüthrich, and W.F.van Gunsteren, *J.Mol.Biol.*, 1993, **231**, 1040-8.
621. S.Grzesiak, A.Bax, L.K.Nicholson, T.Yamazaki, P.Wingfield, S.J.Stahl, C.J.Eyermann, D.A.Torchia, and C.N.Hodge, *J.Am.Chem.Soc.*, 1994, **116**, 1581-2.
622. T.V.Maltseva, P.Agback, and J.Chattopadhyaya, *Nucl.Acids Res.*, 1993, **21**, 4246-4252.
623. L.Yu, E.T.Olejniczak, and S.W.Fesik, *J.Magn.Reson.,Ser.B*, 1993, **102**, 218-21.
624. R.F.Kelly, W.B.Severn, J.C.Richards, M.B.Perry, L.L.MacLean, J.M.Tomas, S.Merino, and C.Whitfield, *Mol.Microbiol.*, 1993, **10**, 615-25.
625. S.M.T.D'Arcy, S.L.Carney, and T.J.Howe, *Carbo.Hyd.Res.*, 1993, **255**, 41-59.
626. K.R.Markham, H.Geiger, and H.Jaggy, *Phytochemistry*, 1992, **31**, 1009-11.
627. M.Gilleron, J.Verauteren, and G.Puzo, *Biochemistry*, 1994, **33**, 1930-7.
628. T.N.Huckerby, J.M.Dickenson, G.H.Tai, R.M.Lauder, G.M.Brown, and I.A.Nieduszynski, *Magn.Reson.Chem.*, 1993, **31**, 394-8.
629. S.Haeuw-Fievre, T.M.Wieruszeski, Y.Plancke, J.C.Michalski, J.Montreuil, and G.Strecker, *Eur.J.Biochem.*, 1993, **215**, 361-71.
630. G.P.Reddy, U.Hayat, C.A.Bush, and J.G.Morris Jr., *Anal.Biochem.*, 1993, **214**, 106-15.
631. N.Shibata, C.Kejima, Y.Satah, R.Satah, A.Suzuki, H.Kobayashi, and S.Suzuki, *Eur.J.Biochem.*, 1993, **217**, 1-12.
632. W.Chen, M.J.Avison, X.H.Zhu, and R.G.Shulman, *Biochemistry*, 1993, **32**, 11483-7.
633. B.Mulloy, M.J.Forster, C.Jones, and D.B.Davies, *Biochem.J.*, 1993, **293**, 849-58.
634. J.T.Davis, S.Hirani, C.Bartlett, and B.R.Reid, *J.Biol.Chem.*, 1994, **269**, 3331-8.
635. C.Mukhopadhyay, K.E.Miller, and C.A.Bush, *Biopolymers*, 1994, **34**, 21-9.
636. T.J.Rutherford, D.G.Spackman, P.J.Simpson, and S.W.Homans, *Glycobiol.*, 1994, **4**, 59-68.
637. G.V.Vinogradov, R.Stuike-Prill, K.Bock, O.Holst, and H.Brade, *Eur.J.Biochem.*, 1993, **218**, 543-54.
638. D.R.Bundle, H.Baumann, J-R.Brisson, S.M.Gagne, A.Zdanov, and M.Cygler, *Biochemistry*, 1994, **33**, 5183-92.
639. L.Yu, R.Goldman, P.Sullivan, G.F.Walker, and S.W.Fesik, *J.Biomol.NMR*, 1993, **3**, 429-41.
640. L.Poppe, W.S.York, and H.van Halbeek, *J.Biomol.NMR*, 1993, **3**, 81-9.
641. D.Uhrin, J.R.Brisson, and D.R.Bundle, *J.Biomol.NMR*, 1993, **3**, 367-73.
642. J.W.Ball and P.C.Jurs, *Anal.Chem.*, 1993, **65**, 3615-21.

10
Synthetic Macromolecules

BY A. H. FAWCETT

1 Introduction: Reviews

The insight into polymer microstructure that is almost uniquely provided by NMR spectroscopy, when experiments are performed upon solutions to obtain high resolution spectra, has been reviewed by Japanese, English and American workers during the current period[1-4]. The particular structures that derive from the chemistry of initiation, and depend upon the manner in which initiators function, have also been reviewed[5]. The chemistry of polymerization has best been elucidated by such studies: high field NMR instruments were developed principally for this reason. The time domain behaviour that is accessible, particularly through Fourier Transform spectrometers, has made possible the study of the kinetic behaviour of macromolecules[6,7] in solution and in melts. Such dynamic behaviour is also monitored in a complementary manner by dielectric means, if electric dipoles are present[8]. The interesting ideas of Mandelbrot which have implications for the fractal nature of certain polymer systems are covered in a review of the NMR response that reflects the dynamic behaviour of polymeric liquids and gels[9]. The liquid crystalline state, a state that depends upon the complex interplay of mesogenic or ordering and of fluid-contributing components, may be created by a variety of polymer molecular architectures, and lends itself to exploration by multidimensional NMR methods, as Spiess describes[10].

While such ideas in the traditional areas of work are still developing, as these reviews and many of the papers below demonstrate, a second field of NMR spectroscopy is now flourishing within the physics of polymers - that field which, moreover, provides their commercially interesting behaviour - as a number of reviews discuss[11-14]. A most interesting recent development is the discovery and demonstration of two-, three- and now four-dimensional methods of NMR spectroscopy, as is now described in a book by Schmidt-Rohr and Speiss[15]. The formation of blends promoted by hydrogen bonding and the similarly-facilitated formation of interpenetrating polymer networks are covered in one of these[12]. We have noted the introduction of Xenon NMR spectroscopy in earlier reports to the study of polymer morphology: the field is now covered by a review[16].

Several reviews have appeared on the topic of NMR imaging of polymeric materials[17-20]; they cover respectively the development of the methods themselves, their applications and potential impact, the charactertization of melts, liquids and elastomers, and finally for solid polymers the line-narrowing approach to imaging, gradient decoupling and time sequence second averaging.

2 Polymer Microstructure from Solution Studies[1-4]

Initiation[5] and other Minor Events, Propagating Species

Despite the peaks from the minor components being so small that they are almost negligible when compared to those from the free radical aqueous slurry-polymerised acrylonitrile residues, it has been found that comparing the high field NMR spectra of polymers made with distinctly different molecular weight allows the shifts from initiator and chain transfer fragments to be identified in samples prepared under conditions corresponding to those in a commercial reactor[21]. Other minor structures from end group hydrolysis, but not those structures derived from a termination reaction, were identified. A similar comparison of polystyrenes with different molecular weights has allowed direct evidence of the geminal methines of the head-to-head type of structure provided by a combination reaction[22]. The 2-D INADEQUATE method, employing multiple quantum coherence, revealed correlations in the 2D plot that identified sites with adjacent 13-C labels. The shifts of initiator fragments on polyisobutylenes were readily recognised in the proton and ^{13}C spectrum because the polymer residues had been prepared from perdeuterated monomer, a DEPT spectrum being used for the carbon signals[23]. Direct evidence for the links between blocks of polymethylmethacrylate and polyethylmethacrylate in polymers prepared by group transfer polymerization has been obtained by observing the shift of the methylene group that links the blocks by means of DS INEPT and H-C-C RELAY pulse sequences. The spectra of the block copolymers also lack the signals from the ends that are present in homopolymers prepared by the same means[24]. Polyolefins produced by metalocene catalysts have one saturated and one vinyl end group. Randal has now characterised both types of end for the four sequence dyads of an ethylene-propylene copolymer[25]; it was thus possible to deduce that ethylene ends suffer chain transfer much less frequently than propylene ends, and that initiating units may be exclusively propylene. The residues near propylene oxide end groups have been distinguished: their shifts are sensitive to the DP_n[26].

Besides these studies on conventional systems, new chemistry has been explored with the aid of NMR spectroscopy; for example new end-functionalised polystyrenes were obtained from styrene with living anionic systems based upon $SnCl_4$ - activated adducts of a series of vinyl ethers[27], a new method of stabilising polyoxymethylene with acetate end groups has been introduced[28], perfluoroalkyl groups have been placed at the ends of anionically-formed polymers[29], an iniferter system that introduces a phenyl group to poly(methylmethacrylate) has been examined[30] and allyl end group has been placed upon an acrylic chain[31]. By standard procedures followed with care polystyrene telechelics have been synthesised in which the terminal styrene residue was deuterated and the converse case of a perdeuterated polystyrene has been terminated with a normal styrene residue, the anionic functions then being converted to carboxylate groups[32]. The modes of function of an azo initiator have been followed, both in the effectiveness of the free radical initiation process and in the manner in which the two furan groups placed upon it, and therefore becoming placed at each end of the telechelic styryl or acrylic polymers, performed Diels-Alder additions with dienophiles such as maleic anhydride and similarly produced networks with bis dienophiles[33,34].

The propagating species have been the subject of an investigation that employed several NMR active nuclei: ^{1}H, ^{13}C, ^{19}F, ^{31}P, ^{75}As and ^{121}Sb resonances were observed in one study of the cationic polymerization of N-benzoyl-8-octanelactam, the Sb and the As nuclei being present in the counterions $SbCl_6$ or AsF_6 and the P within an endcapping triphenyl phosphine[35]. It appears that the counterion and the nearest benzoylamide group co-ordinate with the acylium ion. A quasi classical "living" cationic polymerization of polystyrene promoted by a Lewis acid system based upon $SnCl^4$ and Bu_4NCl displays, in the ^{119}Sn NMR spectrum, evidence for a number of different anions including an oligomeric form, the proportions of which and their rates of exchange depended upon the salt/Sn ratio[36]. During the isotactic polymerization of polymethylmethacrylate promoted by t-butylMgBr in toluene at -78°C the penultimate and antepenultimate ester groups appear to be co-ordinated to the metal counterion[37]. ^{7}Li NMR spectroscopy has been used to study the complexes that form about a model for living PMMA at various temperatures: tetrameric and dimeric forms of the living ends have been identified, and other species appear to form in the presence of LiCl[38]. Ion pair equilibria were also identified within a homogeneous Titanium-^{27}Al Zeigler catalyst system that is effective in ethylene polymerization[39]. The solvent-separated ion pair was found to be most active, and a novel arene solvated cation ion pair (SCAP) to be least active of the complexes recognised. The insertion of saturated nitriles into the Ti-N bonds of tetraaziridotitanium(IV) $[Ti(N(CH_3)_2)_4]$ initiator may be monitored with proton NMR[40]: atactic polyacrylonitriles were obtained.

Li-7 spectroscopy of dilithium initiators in hydrocarbon solvents indicated aggregations, and new species were formed when isoprene became initiated[41]. It was possible to see that propagation was less facile that initiation. C-13 spectroscopy was sensitive enough to monitor the manner in which Titanium-Aluminium catalytic systems behaved at different temperatures, in different solvents, and how they evolved once olefins were added. A complex, probably a solvent separated ion pair, was promoted by the addition of CD_2Cl_2 to the solution in toluene[42]. By terminating the group-transfer formed poly(methyl methacrylate with ^{13}C-labelled methyliodide a signal from the chain ends was obtained with sufficient intensity that the two diastereooptic chain ends of the PMMA[43]. A rather complex E/Z model of the manner of the end reacting, that might be second order Markov, is considered but is not required for this reaction, nor for the propagation reaction, which are shown to be adequately given by Bernoullian statistics involving a bias towards syndiotacticity.

Chain Microstructure[1-4]

Acrylonitrile has been copolymerized with styrene[44], with butyl acrylate[45], with methyl acrylate in the presence of zinc chloride[46], and with vinyl propionate or vinyl butyrate[47], in each case by free radical means. 13-C NMR spectroscopy permitted the recognition of sequence triads, whose proportions were also predicted well by models that predicted the microstructure on the basis of comonomer reactivity ratios[44,45]. For the styrene-acrylonitrile case the behaviour in non-polar and polar solvents differed from that in the bulk, the recognition of which lead to the penultimate effect model being modified to allow for specific solvent effects in a manner termed a modified bootstrap model. The $ZnCl_2$ functioned as a retarder for the copolymers of methylacrylate in the basic solvent DMF: the AN - MA copolymers were recognised as simply alternating[46]. For the copolymers of acrylonitrile and vinyl acetate formed in an aqueous slurry process a penultimate model gave the best account to the acrylonitrile-centred triad populations of several copolymers[48]. Methacrylonitrile homo[48] and copolymers[49-52] have also received attention. Many of the previously-made assignments have been corrected by a current study of the homopolymer that uses double-quantum-filtered phase-sensitive COSY and proton-detected ^{1}H-^{13}C shift correlated NMR spectroscopies[48], two-dimensional methods that resolve peaks that overlap in the one dimensional mode. The methylene group was resolved at the hexad level and the tetrad levels in the proton and carbon spectra respectively. Again a penultimate model has been found appropriate to describe a sequence distribution, in this case within copolymers of methacrylonitrile and styrene[50] made at low conversions; in contrast, copolymers of methacrylonitrile and alkyl methacrylates grafted onto amylomaize with high conversion had proportionately higher than random runs of of the alkyl methacrylates on account of solubility effects. Methacrylonitrile copolymers with alkyl methacrylates have been studied, the alkyl groups being ethyl, isobutyl and methyl[52]. Copolymers of the latter have a lower coisotacticity parameter (0.87) than the others (0.98), which was interpreted in terms of a specific nitrile-ester interaction.

The electro-negativity of the fluorine in alpha-fluoroacrylonitrile promotes the formation of alternating copolymers with 1,3-cyclohexadiene[53] and ethyl-vinyl ether[54], in the first case alternation being complete and the diene residues being both 1,4- and 1,2-addition structures. Poly(alpha-fluoroacrylonitrile) itself is predominantly atactic[55], as is poly(ethyl alpha-fluoroacrylate)[55]. The sequence distributions and compositions of copolymers of maleic anhydride with p-methoxy styrene were such that the participation of the charge-transfer complex was indicated: for the less-strongly alternating copolymers with p-chlorostyrene the penultimate effect model was equally appropriate to interpret the same type of data[56]. Vinylidene cyanide and methylvinylidene cyanide copolymers with substituted styrenes have been made by free radical means: the former are entirely alternating, the latter were found to contain runs of styrenes[57]. Styrene - CO copolymers made in the presence of $Pd(OAc)_2$, as with the similarly made ethylene - CO copolymers, are alternating, according to their NMR spectra[58]. Polyethylene formed upon a Titanium catalyst, in contrast, incorporates a small proportion of CO residues as ketones when ^{13}CO is admitted[59]. Carbon dioxide and cyclohexene oxide or propylene oxide form regularly alternating copolymers - essentially polycarbonates[60] - in the presence of catalysts such as phenol/$ZnEt_2$. These epoxide copolymers have now been joined by a new set of materials based upon CO_2 and diacetylene molecules, a nickel catalyst being the required agent[61]. The structure thus formed is a 2-pyrone link. Alternation in copolymerizations of myrcene (an octadiene) decreases in the order methylmethacrylate > p-fluorostyrene > styrene according to the analysis of polymer composition / feed data. The reactivity ratios were obtained, as is proper from such information, by a nonlinear least squares error-in-variables method[62].

Two-dimensional spectroscopy methods have been used to examine two homopolymers, polyvinylchloride[63] and polyacrylic acid[64]. For PVC a 2D-INADEQUATE spectrum has shown that there are two **mmr** peaks, one of which lies close to the **mrm** peak and was previously unresolved in earlier 2D experiments. The relative areas of the methylene peaks now agree with those predicted by Bernoullian propagation statistics. The 2D HSQC-TOCSY method was pseudo three dimensional, as ^{13}C-^{1}H as well as ^{1}H-^{1}H correlations were found by the single experiment[64]. An **mmr** tetrad resonance of the methylene carbon was correlated to the **mm** and **mr** triad resonances of the methine carbon. This self-consistent technique for assignment is independent of the conventional methods based upon peak areas and assumptions about sequence populations, and was applied to the assignment of all tetrads and several pentad resonances. Features from the ends of chains were also assigned.

Finely resolved ^{13}C NMR spectra of poly(methyl methacrylate) have been obtained from atactic and tactic polymers, and the assignments made, generally at the heptad level with the aid of a first-order Markov model, for the quaternary carbon and those carbons in the side chain[65]. Three different poly(1-naphthylalkyl methacrylate)s have also been examined and found to be Bernoullian in their tacticity[66]. For diphenyl-o-methoxyphenylmethylacrylate polymers made with optically active catalysts, the polymers themselves were optically active.

The effect of the presence of lithium chloride during the anionic polymerization at -78°C of methyl methacrylate and t-butyl methacrylate depends upon the alkyl group: in the former case syndiotactic placements were favoured[67], and in the second place isotacticity was enhanced[68]. The effects are attributed in each case to complex formation at the living end in the particular solvent system employed. Oligomers of t-butyl acrylate have been obtained by an anionic method and their resonances assigned[69]. For the tetramer and the pentamer an f^1-decoupled COSY 2D-NMR method was employed: for the tetramer meso and racemic forms were found in proportions of 0.78:0.22 , for the pentamer the proportions were **mm:mr:rm** = 0.635:0.185:0.180, **rr** isomers being negligible.

The standard rotational isomeric state (RIS) models of polyethylene and polyvinyl acetate have been combined to describe the microstructurally-sensitive shifts of the various sequences encountered in copolymers and polyvinyacetate[70]; a good account of the shift effects was thus obtained. Following interesting force field calculations of the potential energy surfaces of meso and racemic 2,4-diphenyl pentane, models for the corresponding dyads of polystyrene chain, the conformational energies of the polymer and of a trimeric model have been obtained. Using standard RIS methods the conformational populations were predicted for a two-state model of the polymer, trimers and the dimers, and mean values of the coupling constants were thus found[71]. The values corresponded quite well with experimental observations. The same ideas have been employed to assist the replication of the widely-dispersed 13-C shift patterns of the methylene and methyl carbons of polyisobutylene in frozen conformations within the solid state below T_g. Chemical shifts were predicted by <u>ab initio</u> IGLO methods for each conformation, and weighted according to the relative energies[72]. This most interesting innovation confirms that fine structure has principally a conformational origin, may ultimately allow a check on RIS models, the development of better force fields, the production of theoretical chemical shift anisotropies, and a method for conformational statistics of amorphous chains.

Computer-based models of a second type have been developed to simulate the distribution of tacticities found in polymers such as polypropylene and polymethylmethacrylate, the study being designed to show how microstructure depends upon the sequences of propagation steps that are governed by probalistic parameters governing the mode of propagation at a site in a consecutive and concurrent manner, as well as multiple initiator or catalyst sites. The simulation approach[73] presents the consequences for microstructure as might be analysed by NMR spectroscopy. Similar models have been introduced to describe propylene polymerization by Ziegler-Natta catalysts[74] and the copolymerization of ethylene and oct-1-ene[75]. They ranged from Bernoullian to second order Markov, each site on the heterogeneous surface having its propagation and termination probabilities. NMR tacticity and GPC data together indicate the need for a three site model in the polypropylene case[74] and a two site model was required to explain the sequence information in the copolymerization[75]. This approach may also be used for NMR data alone, if various catalyst treatments vary the relative proportions of the different sites[76] for the polymerization of polypropylene. In a parallel study of fractionated products a two site model was considered appropriate[77]. A maximum likelihood method has been introduced to deal with discrepancies between the pentad fine structure of polypropylene and the populations expected from conservation laws[78]. Conservation laws for hexads and heptads are also presented.

The effects of adding hydrogen to one Zeigler-Natta catalyst system and of adding an electron donor to a second have been monitored. In the first case a chain transfer reaction was identified[79], in the second case atactic and low isotactic polymer diminished in proportion[80]. The relationship between Zeigler-Natta catalysts and ring-opening metathesis (ROMP) catalysts has been of some interest. Grubs has found that a most effective Titanium-based ROMP catalyst, containing a titanacyclobutane, can be converted into a sigma-alkyltitanocene complex which functions as a catalyst for polyethylene formation. Block copolymers are envisaged[81]. Ethylene has been copolymerized with 5-vinyl-2-norbornene using the dicyclopentadienyl Zirconium dichloride/methylaluminiumoxane catalyst system; as the cyclic double bond entered the chain, the vinyl groups were pendent[82], and methods of converting them to other functions were described.

Oxiranes such as propylene oxide are polymerised by anionic, cationic and free radical means: each unit is truly chiral, and may be entrained in the normal head-to-tail sense as well as with head-to-head and tail-to-tail linkages (regio regular and regioirregular), as has been found when organotin-alkyl phosphate condensates are used as catalysts. The carbon spectra of a benzene/hexane soluble fraction show regiosequence effects at the triad level[83,84], which have been assigned using an optically active polymer obtained from (S)-(-)-propylene oxide[84]; the polymers prepared from racemic monomers display the extra features of dyad ($-CH_2-$) and triad (-CH-) sensitivity[83]. The fine structure is inconsistent with earlier RIS treatments. The benzene/hexane insoluble components are predominantly isotactic and regio regular. Isotactic and less tactic components were also obtained with an alkyzinc/alcohol catalyst[85]. Amorphous poly(epichlorohydrin) has also been shown to have the same features: once a 2-dimensional heteronuclear correlation NMR spectrum had been obtained and analysed, 1st order Markov statistics were found to apply and it was recognised that head to tail sequences came in blocks[86]. The cyclohexeneoxide residue within the polymer have the threo-enchainment produced by inversion upon cleavage of the C--O bond and are synditactic (P_s = 0.58 to 0.80). The cationic polymerization mechanism of glycidol seems to involve two simultaneous mechanisms, an attack of the hydroxyl group of the polymer upon the proton-activated monomer and the attack of the monomer upon the activated oxonium ion at the end of the chain. Up to 50% of the residues form from the first process, as ^{13}C NMR and ^{29}Si NMR (on the silylated secondary hydroxyl groups) demonstrated[87].

Other ring opening polymerizations have been effected: a living process of caprolactone promoted by a Titanium complex[88], a cleavage-insertion process of the same monomer by a series of rare earth co-ordination catalysts[89], an anionic process involving a biclycic bis(gamma-lactone) and a second essential monomer, an epoxide, the radical polymerization of hybrid monomers of vinylcyclopropane and cyclic ketene acetates[91], a free radical polymerization of alpha-cycloproplstyrenes[92], a thermal ring-opening polymerization of [2]ferrocenophanes with a hydrocarbon bridge[93], a Zirconium complex catalysed ring-opening process involving a beta-alkyl shift for methylenecyclobutane[94], and the cationic and partially ring opening polymerization of 2-methylene-7-oxabicyclo[2.2.1]heptane[95]. In each case the structure or structures within the polymer were established by NMR spectroscopy so that the mechanism of the polymerization of the difunctional monomers was indicated.

Ring opening polymerization by metathesis catalyst systems (ROMP) yield cis and trans isomers for the double bond within the main chain[96-99], structures to which ^{13}C shifts are sensitive. With norborn-2-ene derivatives there is also sensitivity to orientation (regio) sense, and the manner in which a particular catalyst functions may depend upon whether a substituent is in the exo- or endo-position, and whether the catalyst attacks the double bond from the exo or endo orientation. Such considerations have lead to the suggestion that, in 7-t-butoxy and similar derivatives the lone pairs of the oxygen atoms may themselves feature in a [3 + 2] transition state, which is thought to be distinguishable from a chelating process[99]. Hetero atoms within such monomers are not only tolerated by the present generation of carbenes, but may even be required for a particular functioning of the catalyst, when, for example, oxygen atoms are in the 7-position, as is obtained in a bicyclic diene produced by the Diels Alder reaction of furan with an acetylene[100] or of maleimide[101]. In order to prevent the catalyst being destroyed by a ketone group it was flanked sterically by alkyl groups[102]: the first use of a particular Molybdenum catalyst effective in such cases has been reported[102]. In a similar acyclic diene metathesis polymerization (ADMET) a poly(siloxalkenylene-co-biphenylene) has been obtained and characterised[103]. A standard carbene catalyst has caused polymers to form from 1,5-cyclooctadiene and [2.2]paracyclophane-1,9-diene: the latter furnished p-phenylenevinylene units units, and these became randomly distributed within the chain[104].

Dendrimers, Fractals and Stars

The microstructure of linear molecules has been regularly studies by NMR spectroscopy, the microstructure of regularly-branched molecules such as starburst dendrimers[105-107], arborols [108] and cascade polymers have residues progressively distanced from the centre are in slightly different environments. High field NMR can detect these minor effects, and so is useful in characterisation as well as checking the effectiveness of the reactions that develop the successive generations in structure, whether they are built up by convergent or divergent approaches. Thus the formation of a 94-mer dendrimer from three 31-mer wedges of a phenyl acetylenic monomer has been described[106]. The production of polynitrile and polyamine functionalised poly(trimethylenimine) dendrimers from acrylonitrile at the third generation level has similarly been monitored[107]. Arborols with spirane or biphenyl cores have no tendency to aggregate in water[108]. Reactions linking the centres of such molecules with a telomeric unit have similarly been performed, to yield an unusual triblock copolymer[109].

The much less regular fractal structures that form when trifunctional and difunctional monomers are linked have also been elucidated. The strategy that was adopted, using the succinic acid in different proportions and its anhydride, allowed some control of the manner in which the oligomers formed with glycerol; its shifts were found to be sensitive not only to how the other alcohol functions had reacted within that residue but to whether the second group of the acid had reacted, in some cases even detecting whether the other glycerol residue had formed the ester link through its primary or secondary site[110]. The chemistry of the formation of 4-micto arm stars has been developed and demonstrated: living polymers of polystyrene, polybutadiene, polyisoprene and poly(4-methyl styrene) were individually attached in turn to a $SiCl_4$ core[111]. In another development cyclotriphosphazene was used as the core for a star-shaped polymer derived from 2-methyl-2-oxazolines[112].

Reactions on Polymers and Rotoxanes

The chemistry of polymer reactions is readily followed by NMR spectroscopy, proton NMR being most appropriate on account of its quantitative nature, and ^{13}C NMR on account of its sensitivity to the details of structural changes. The Diels-Alder reactions of furans pendent to acrylamide backbones with an excess of the dienophile dimethylbutynedioate (acetylene dimethylcarboxylate) have been followed by proton NMR spectroscopy and found to deviate from pseudo-first order kinetics through neighbouring group effects and the influence of the neighbouring backbone, when the spacer is short[113]. When polystyrene is chlorinated not only does the aromatic ring become substituted, but the backbone methylene groups also react[114]. Chlorination of cis-polybutadiene in the presence of aryl nucleophiles such as phenol not only produced dichlorinated residues but also O-linked structures[115]. The same polymer has been partially reduced towards polyethylene, the remaining internal double bonds then being converted to terminal olefins by a metathesis reaction with ethylene to yield telomeric products[116]. A further study has been reported of the addition of chlorofluorocarbene to a polymer with a double bond in the backbone; the mode of addition was characterised by the ^{29}Si NMR of the methyl-phenyl silane residues also present, as well as through the other NMR-sensitive nuclei[117]. The microstructure of butadiene oligomers and the changes brought about by converting the double bonds to epoxy groups have been characterised by ^{13}C NMR spectroscopy[118]. The extent of the process was reliably measured by ^{13}C and ^{1}H NMR, but chemical methods for this purpose were less useful.

The conversion of vinyl groups pendent to a predominately polyethylene chain to epoxy and hydroxy groups has been accomplished[82] and hydroxyl groups have been silylated[87], as demonstrated by NMR analyses. NMR spectroscopy has assisted the radiation chemistry study of cis-polybutadiene and ethylene-propylene copolymer, among others, that was facilitated by proton and carbon NMR spectroscopy, and allowed the identification of the units formed by scission and crosslinking events. These reactions were randomly distributed[119]. The crosslinking reactions, except for that of polychloroprene, were found to be chain reactions[120].

Rotoxanes, complexes of ring molecules about a chain, have been prepared in a number of ways, NMR spectroscopy being an essential aid in this process. Thus cyclic ethers have been strung upon a polyurethane backbone[121], the threading efficiency rising as ring size increased, and cyclodextrins have been strung upon a polyethylene glycol chain[122-124], the system being stabilised by placing a large unit at the end of the chain. The stoichiometry of the complexes was 2 residues : 1 cyclodextrin[123].

Dynamic Studies[6-9]

The old problem of the need for a function to describe the wide distribution of correlation times suitable for a polymeric substance in solution when performing time domain measurements upon, for example carbon T_1 and NOE values, was tackled by Schaeffer in the early days with a log chi^2 function. It has been shown that the mean value of the correlation times behaves in a manner that is intuitively satisfying for a number of applications, in which respect the tau_0 parameter of the function itself is less satisfactory[125]. The log normal distribution was preferred for interpreting T_1 data[125]. The spectral density functions themselves, for deuterium nuclei in perdeuterated polyethylene oxide, have been obtained at six applied field strengths by magnetic relaxation measurements. At low and moderate concentrations the spectral density function corresponds to an anisotropic relaxation of a hydrodynamic unit, and at higher concentrations a new regime is found in which the activation energy and the spectral density curve become dependent upon concentration, and an intramolecular effect between proton-containing polymers develops[126]. In a separate study of solutions of polyethylene oxide and poly(tetramethylene oxide) the geometries as well as the time scales of local motions have been obtained from the ^{13}C spectrum obtained without proton decoupling[127]. For each $^{13}CH_2$ spin system three magnetization modes were measured at various time delays, and from them three second-rank Cartesian correlation times were found. Differences between the two polymers were attributed to differences in barrier-hopping rates of C--C and C--O bonds. A molecular dynamics simulation with an explicit atom model of an oligomethylene chain has C-H vector reorientational correlation times that are in good agreement with values deduced from T_1 measurements[128]. The pattern of movement is of unconcerted conformational transitions. For a copolymer of vinyl chloride and vinylidene chloride T_1 and NOE measurements have been interpreted in terms of a local libration at high frequencies and a slower diffusion of a motion along the chain[129]. T_1 studies on isotactic and syndiotactic PMMA in a series of n-alkyl solvents have been performed to investigate the manner in which motions were influenced by polymer-solvent interactions[130], thus shedding light upon the polymer dissolution process: it was deduced that polymer dissolution is limited by transport of solvent through the local motions of polymer segments.

The presence of helices in solution has been demonstrated for isotactic chloral oligomeric isomers[131]: well-resolved peaks merge as raising the temperature promotes a helix reversal. The two-dimensional NOESY spectrum has negative features, not from the NOE effect, as the sites are remote, but from the site exchange process, the energy barrier to which was found to increase with oligomer size.

3 The Solid State[11-14]

Homopolymer Systems

Polyethylene samples crystallised from the melt at high pressures have much greater extents of crystallinity and a much thicker amorphous-crystalline interface, indeed the amorphous content is negligible if chain lengths are low. The normal three-phase model is does fit the NMR data, but an unusually long ^{13}C T_1 has been obtained for a high molecular weight sample, a value which nevertheless is shorter than predicted by a model that depends upon perfect chains within the thick lamellae. The relaxation is associated with defects in the crystals[132]. Isotactic polypropylene has at least three crystal forms, as well as an amorphous domain, phases whose proportions affect cross-polarization between proton and ^{13}C nuclei. Amorphous regions are ineffective in general, though a sample that was predominantly smectic had contributions to cross-polarization from the amorphous region where some chains are somewhat constrained[133]. Cross polarization/dipolar dephasing studies of mixtures of normal and deuterated polystyrenes obtained by freeze drying solutions have shown that the segments of the two polymers were intimately mixed before and after they were annealed[134].

A high resolution solid state study of syndiotactic polypropylene has discovered a sensitivity to a variety of packing effects of the methyl group signals upon the chiral helical chains. Below the glass transition conformations were distinguished within the amorphous phase[135], for, as with polyisobutylene[72], these became frozen out and could be identified using the gamma-gauche parameter. Such conformational effects and also packing effects have been invoked to explain the multiplicities of the carbonyl carbons of poly(lactides) in spectra obtained from partially crystalline polymer of the L form, from a crystalline stereocomplex of the L and D forms: a urea L-lactic acid polymer canal complex lacks the fine structure[136]. Within such canal complexes polymers are confined to a narrow tube, a condition that constrains the number of conformations that the main chain may adopt: modelling studies have shown how some polymers may not only have conformational freedom within this type of complex but also are capable of conformational changes[137], much like a single chain within a crystal. Most of the bonds in the eleven carbon spacer between the mesogenic blocks of a main chain liquid crystalline polymer have been recognised as having trans conformations while certain bonds were either trans or gauche in a conformationally-disordered manner within a low temperature ordered phase[138]. Extra bonds (but not all) became conformationally disordered at a higher temperature within a new phase. Packing effects within the chain of poly(gamma-benzyl-DL-glutamate) in the solid state differ from those in the (L) homopolymer, the former showing evidence for side chain stacking, especially, and curiously, within the spacer linking the benzyl group to the backbone[139]. Polyphosphazenes may show a liquid crystalline as well as crystalline and amorphous phases, in a particular case the first two being recognisable for well resolved ^{31}P shifts may be seen by MAS NMR spectroscopy[140].

Layered Structures

The recognition and production of polymeric systems that naturally have layered structures on the nanometer scale has been recorded[141-144]. A useful feature encouraging such structure is a stiff backbone, as is encountered with main chain mesogenic units such as polyaramids, where hydrogen bonds between chains promote the layers[141]. Aliphatic (non-polar) side chains, that may be regarded as tied solvent, migrate to the interlaminar space, and, depending upon the temperature, may be in a quasi liquid or a crystalline state, as the carbon shifts indicate. This feature has also been found with poly(n-alkyl-L-glutamate)s, where the alpha-helix is present in the backbone[142] and also, perhaps with more surprise to some, in poly(olefin sulphone)s with long side chains[143-144], whose backbones despite being atactic do adopt helical conformations, and probably associate in layers through their polarity. Frozen conformations in backbones[143] and side chains[141,143,144] have been recognised in the ^{13}C CP-DD-MAS and WISE[4] spectra: their dynamics and organization, as measured by proton T_2 measurements depend upon the length of the hairs about the hairy rods[142]. The layered structures self-assemble when the melts are cooled. Similar assemblies of layers are obtained by the Langmuir-Blodgett technique with maleic anhydride alkene copolymers[145]. Their signals have to be distinguished from those of the substrate, but it seems that the assemblies are less mobile than the bulk homopolymer. Gauche conformations within the aliphatic spacers of the backbone have been indicated in another type of system when the aromatic mesogenic groups form layers normal to the general direction of the backbone[146].

Block Copolymers

The microphases that are found within block copolymer systems have a great variety of morphologies and may be observed directly using the electron microscope, and have profound consequences for the material properties. Of the repertoire of solid state NMR techniques, proton spin diffusion measurements are particularly sensitive to the details when a mobile and a glassy or crystalline phase are present. Spin diffusion coefficient measurements, through the carbon spectrum, have been performed with a diblock copolymer of known phase structure to calibrate the technique[147]. Such segmented copolymers based upon poly(dimethylsiloxane), poly(p-hydroxy styrene) and/or polysulphone have been studied with the extended Goldman-Shen pulse sequence for spin-spin diffusion rates, as well as by ^{13}C and ^{29}Si spectroscopy. A four phase model, reflecting partial miscibility of one component in another phase as well as separate pure phases of each type was inferred, a possibility made through the different mobilities within each phase[148]. MDI-based polyurethanes have a variety of microphases - several crystalline polymorphs, amorphous, interfacial - according to an NMR study of such systems and linear poly(ether urethane) block copolymers[149]. The ^{13}C NMR line width of styrene-isoprene block copolymers has been measured at temperatures above and below the temperature at which the lamellar ordering occurs[150], to provide information on the homogeneity of the material. Proton logitudinal and transverse relaxation measurements on styrene-(ethylene/propylene) block copolymers dispersed in octane-d^1 revealed that the micelles had slightly swollen polystyrene cores with about half the residues near to the boundary[151]. Within other block copolymers the diffusion behaviour of homopolymer chains have been studied as a function of molecular weight by the pulsed-field gradient NMR method[152]. The diffusion depends upon the spatial distribution within the microphase domains.

Deuterium NMR Studies

Deuteration at the first methylene site of the side chains in poly(hexyl isocyanate) provides a means, through Deuterium NMR spectroscopy, of measuring the order within solutions of this lyotropic main chain polymer at different concentrations[153]. (As the chains are chiral helices, two splittings are seen from the single CD_2 group.) A new treatment for interpreting the details of the line shape has been developed to allow for the inherent flexibility of the backbone. For a siloxane side chain nematic liquid crystalline polymer the deuterium NMR spectrum has been recorded at intervals during the magnetic field-driven realignment of the director[154] through 90°. The response of the quadrupolar splitting patterns was modelled to yield the evolving director distribution functions. During this splay Freedricks transition optical microscopy simultaneously probed the azimuthal orientation of the director. The technique of using a deuterated solvent, benzene-d_6, to probe order within a side chain liquid crystalline polymer has been discussed[155]. Flexible deuterated molecules, oligomers of ethylene oxide, have been used as probes within a standard liquid crystal[156]: the preponderance of gauche conformations seems to have suppressed the tendency for the medium to induce order in the probe, though the chains are somewhat extended by the medium.

By varying the temperature the deuterium spectrum patterns change considerably, higher temperatures leading to motional narrowing of Pake-type spectra as has been demonstrated for polymers with partially-deuterated mesogenic side chains in proportions sufficient and insufficient to generate liquid crystallinity[157]. Two dimensional exchange spectra were obtained and simulated. As well as discrete 180° flips about the long axis of the mesogen, continuous rotations were also disclosed in a fairly permanently ordered liquid state. Phenyl group ring flips in poly(p-phenylene terephthalamide) and in oligo(p-phenylene)s have been examined in other studies. The process in the polyaramid is eased by inhomogeneities and by proximity to the fibre surface[158]. Torsional librations (± 20°) and 180° flips were simulated in the p-phenylenes, as experiments had found[159]. Within the columnar hexatic phase of the hairy rod polymers based upon phthalocyanines the discs have been shown to rotate independently in steps of 90°, rather than as a consequence of whole molecule rotation according to a solid-echo deuterium NMR study[160]. Deuterated components have also been used in some of the studies described below in the sections on blends and on motional studies.

Blends

Blends of polymers rarely form in a completely miscible manner unless a specific attraction is present between the residues of the two different polymers. Solid state NMR is able to probe such interactions, and to measure the size of heterogeneous structures present as a result of immiscibility. NMR may also probe motions within the material that may contribute to impact resistance. Thus an understanding may be obtained of any synergistic effects that are discovered in properties. In the present period the following polymer systems have been examined: polystyrene-poly(phenylene oxide)[161-163], polystyrene-polycarbonate[165-165], polystyrene-poly(methyl methacrylate)-benzene-d_6[166], polystyrene-poly(vinyl methyl ether)[167], polystyrene-'polysilastyrene'[168], normal and deuterated polystyrene[134], normal and deuterated poly(methylmethacrylate)[169], poly(phenylene oxide)-poly(3-octyl thio-phene)[170,171], polycarbonate-poly(p-phenylene vinylene)[172], polycarbonate-poly(methyl methacrylate)[173], polycarbonate-poly(epsilon caprolactone)[174], poly(ethylene terephthalate)-poly(p-hydroxybenzoic acid-co-p-hydroxynaphthoic acid) (Vectra-A)[175], poly(ethylene terephthalate)-poly(ethylene naphthalenedicarboxylate)[176], poly-(butylene terephthalate)-polyarylate[177], polyvinyl alcohol-poly(N-vinyl-2-pyrrolidone)[178], poly(methyl methacrylate)-poly(vinylidene fluoride)[179,180], polyacrylic acid-polyacrylamide[181], polyvinyl chloride-poly(ethyleneoxide)[182], polyvinylchloride-poly(methylymethacrylate-co-methyl-acrylate)[183], poly(vinyl phenol)-poly(methacrylate)[184], poly-(vinyl phenol)-poly(methylmethacrylate)[184], poly(vinyl phenol)-poly(ethylene oxide)[185], polypropylene-polysilastyrene[168], poly(1,2-butadiene)-poly(cis-1,4-isoprene)[186], polybutadiene-polyisoprene[187-189] and high impact polypropylene (PP + EPR rubber)[190]. This list simply provides the reader with the individual systems studied, so that he may locate those of interest to him.

The techniques that were used and the questions that were investigated were very varied. The shifts of different carbons, in the absence of specific interactions between polymers, are not sensitive to the composition of the blend[175], though their relative intensities may not be proportional to their proportions as cross polarization efficiencies can vary. Conformational effects within a molecule may give rise to splittings in blends, as in pure polymers, and be sensitive to blend structure[175]. The CP/MAS spectral shifts themselves did show that each polymer of the polyvinyl alcohol-poly(N-vinyl-2-pyrrolidone) pair were in different environments in blends and in the pure polymers[178]: the polymers became intimately mixed through hydrogen bond formation. In contrast the shifts of poly(butylene terephthalate) and a polyarylate were insensitive to composition[177]. Mirau's two dimensional heteronuclear correlation NMR spectroscopy study in the solid state found composition-sensitive shifts for poly(vinyl phenol) in two acrylic ester systems that again were H-bond acceptors[184]. Similar considerations apply to the blends of poly(vinylphenol) and poly(ethylene oxide)[185]: within the miscible phase backbone motions were monitored through the ^{13}C PEO methylene line width. The complexes that formed between polyacrylic acid and polyacrylamide did so in a manner that is sensitive to the extent of neutralization of the acid groups[181]: as that increased the formation of carboxylic acid dimers was suppressed, a network structure was broken and both types of chain became more mobile, as T_1 measurements indicated. Proton spin relaxations of the two polymers, poly(ethylene terephthalate)-poly(p-hydroxy-benzoic acid-co-p-hydroxynaphthoic acid) (Vectra-A), were found to be different in blends of the two polymers by measurements of proton inversion-recovery ^{13}C-MAS spectra[175]. Upon annealing a 50/50 mixture a minor slow process developed in the signal from the aliphatic protons, with a rate only slightly slower than that of the protons in pure Vectra: a miscible phase has been identified. Spin diffusion experiments depend strongly upon the distance between nuclei; when performed in a rotor-driven manner[162], they are particularly sensitive[175].

Such an investigation has found that in the 2D plots there develops cross peaks upon annealing that indicate again that residues of the two polymers are intimately mixed. Rather than a cross esterification reaction, the effect was attributed to the introduction of a proportion of the stiff Vectra molecules within an amorphous PET phase facilitated by gauche conformations of the aliphatic links.

Rotor-driven spin diffusion, requiring a simple integer for the quotient of the rotor frequency and the difference between resonating line shifts[163], has also been demonstrated for the polystyrene-PPO system[163] using a single methyl ^{13}C-labelled PPO. It is anticipated that intermolecular distances may be obtained from such measurements upon glasses. Two different studies of the polystyrene-PPO system have employed partial deuteration of the polystyrenes[161,163]. In the first[161] the phenyl group was deuterated and solid state 2D HETCOR spectra were obtained by a new method that provided the proportion of contacts between the two different polymers. In the second[163] the styrene backbones were deuterated, and well resolved two dimensional deuterium exchange spectra were obtained at temperatures near T_g for different mixing times that reflected the molecular motions. In contrast to the case with pure polystyrene the blend displayed at temperatures below the T_g evidence for rapid motions. A motional model linked with a statistical lattice model was devised to account for the bimodal range of correlation times. A similar model has been employed for the polycarbonate-poly(methyl methacrylate) studied (the second polymer being perdeuterated) when proton lineshape and proton T_{1rho} measurements were performed[173]. The blend of 1,4-polyisoprene and 1,2-polybutadiene displayed a broad glass transition which, after two-dimensional deuterium exchange NMR experiments was attributed to differences in the dynamics of the two chains that were emphasised by variations in blend compositions[188]. For this system, with deuterated 1,2-polybutadiene, cross-polarization studies revealed that the different polymers in the blend had closer Van der Vaals contacts between the chains than in the pure form[186]. The glass transition temperature phenomenon is yielding to NMR studies. The rate of return to a homogeneous structure of this type of blend after cooling from above a lower critical solution temperature has been monitored using ^{29}Xe NMR[189], a technique that has a higher resolution of blend morphology than DSC. Intermolecular proton-deuterium polarization transfer rates in a blend of normal and perdeuterated PMMA have been obtained by monitoring the rotational echoes at different contact times[169].

Triple resonance NMR experiments involving protons, fluorine and 13-C nuclei have been performed upon a 60/40 poly(methyl methacrylate)-poly(vinylidene fluoride) blend to examine the phase separations induced by annealing[179]. The proton-fluorine cross depolarization study was sensitive to small separations on the molecular scale of the order of 2-3 nm and to the fractions of the two polymers that were intimately mixed. Diffusion of chains from separate phases of these two polymers into each other has been measured; it seems that the diffusion coefficients may be concentration-dependent[180]. Measurements were made of the proportion of the two polymers that were within a few Å of each other by a cross polarisation method. In this an initial proton magnetization transfers to adjacent protons by CP, so that only the remote protons provided a signal.

Signals from the polycarbonate at the interface with polystyrene in a blend have been exclusively promoted by means of a technique involving free radicals. Their polarization was promoted by irradiation at the difference between the radical and proton Larmour frequencies, and was transferred via the protons to the carbons within the polycarbonate[164-165]. One consequence of direct transfer of polarization to carbons from the radical over a range of 30 to 60 Å was an asymmetry in the sideband pattern in a related experiment[191].

Immiscibility at the molecular level has been recognised in the case of poly(phenylene oxide)-poly(3-octyl thiophene)[170,171], mixtures, for the proton T_{1rho} values of the polymers in mixtures were the same as for the separate polymers. The second polymer suffers thermal stress in the mixtures, for the T^1 values of the side chains are increased on account of increased flexibility[170]. In the case of polyvinylchloride-poly(methylymethacrylate-co-methylacrylate)[183] the T_{1rho} values indicated that for a 20% PVC mixture the two phases had different proportion of the two polymers, consistent with it being visually opaque.

Xenon[15], other Gasses and Ions

The mechanism by which the shift of dissolved ^{129}Xe responds to the polymeric matrix has been modelled[192]. The shape and size of the cavities within the polymer have been derived from a 6-12 Lennard-Jones potential, and the manner in which these change as the temperature rises has been shown to lead to a good account of xenon shift and its temperature coefficient in a number of polymeric substances. These shift effects are large and gas residence times are sufficiently long to have allowed the recognition within blends[189,190] of separate domains both while cooling, when a rubber mixture became homogeneous[189], and upon, heating when a precipitated mixture of polypropylene and EP rubber developed a second phase[19]. 13-C labelled CO_2 within polyisoprene relaxes by a number of mechanisms, the relative importance of which have been established by T_1 and T_2 measurements at different frequencies and temperatures[193] and from which the correlation times of the different types of motion of the molecule were obtained. Oxygen is particularly well absorbed by polyphenylene oxide[194], the mechanism and the motions within the material below its T_g being determined by a number of NMR techniques.

Salts such as $LiClO_4$, BaI_2 and $Ba(ClO_4)_2$ form complexes, when dispersed in polyethylene oxide, that restrict the polymer chain mobility, according to proton and carbon T_{1rho} measurements[195]. A single relaxation process was not found. The demonstration of a postulated ionomer morphology - that residues of ionic residues associate and have restricted mobility - has been achieved with telechelic polystyrenes whose carboxylate end groups were on residues distinguishable from the remainder of the chains by being either deuterated on normally protonated[32]. The solvent for polyacrylonitrile, propylene carbonate, permits the formation of a gel on cooling, the crosslinks between otherwise highly mobile chains being microcrystals, involving several chains and possibly some solvent, as X-ray and NMR measurements suggest[196]. Systems of PVC with ketone solvents are homogeneous according to proton T_1 measurements but have more than one domain, according to T_{1rho} results[197].

Motional Studies[6-9]

Motions within liquid and solid polymers continue to be probed by NMR methods[128,198-203], the approach generally being through measurements of proton and carbon T_1 and T_{1rho} values over a wide range of temperatures, but deuterium NMR is also of value[204-206]. Recent studies have been performed upon poly(di-n-alkyl siloxanes)[198], poly(ethylacrylate) and poly(n-butyl acrylate)[199], poly(dialkylfumarates)[201] and poly(4-methyl-pent-1-ene)[202]. In general two relaxation regimes may be encountered, as when chain modes and order director fluctuations were invoked[198], or when conventional[128] slow main chain and faster side chain conformational transitions were used[199]. The alpha-relaxation dynamics of polystyrene has been probed by a novel method, utilising the second harmonic generation characteristics from molecules of disperse red, thus extending our understanding previously acquired by dynamic NMR and dielectric means[200]. Power law relationships between mean square displacements and time or frequency, as are found within a Rouse description of a chain within constraints are not consistent with experimental behaviour[203], but may be reconciled by assuming a correlation between displacement direction and segmental orientation. For the studies utilising Deuterium NMR the extra dimensions of pressure was also varied[204-206]. An unexpectedly weak influence of pressure upon the main chain alpha process was encountered[204]. The phenyl groups, from work on the phenyl-deuterated sample, move independently of the backbones, apparently by ring flips as well as by a slower process[205]. The effectiveness of both process for spin relaxation has also been examined[206].

A three-dimensional NMR experiment has been introduced to characterise the very slow motions, such as the screw translation of isotactic polypropylene within the crystal, has been introduced[207]. It provides three dimensional correlations of position. A second new technique, through the refocussing of multiple quantum NMR coherences, has been introduced and applied to the ^{19}F NMR of poly(tetrafluoroethylene)[208], and has been shown chain motions to be sensitive to the thermal history.

Oriented Samples

The oriental distribution within a bundle of polyaramid fibres[209], within gel-spun and melt-spun fibres of polyethylene[210], stretched poly(methylmethacrylate)[211] and for the ethylene oxide residues[212] within the homopolymer and within polyethyleneterephthalate have been measured. The techniques included two new solid state ^{13}C methods[209] - ORDER (orientation distributions with enhanced resolution) that employed rotor-synchronised three-dimensional magic angle spinning and DECODER (direction exchange with correlation of orientation distribution evaluation and reconstruction) - and a rotor-synchronised two-dimensional method[210]: they provide order parameters as accurately as do X-ray measurements. The aramid fibres were well ordered, according to the second, forth, sixth and eighth moments, but the polyethylene samples displayed much greater inconsistencies between and within the different phases. The accuracies of the ^{13}C shift tensors in polyethylene and polyoxymethylene have been improved[212].

Poly(ethyl methacrylate) and to a lesser extent poly(methyl methacrylate) display, at temperatures above the T_g, an unusual anisotropy of motion, according to ^{13}C spectra obtained from the partially ^{13}C enriched carbonyl carbon[213]. The anisotropy has two components, and is characteristic of motion about a fairly persistent backbone axis: below the T_g three components are seen, and at elevated temperatures only one. Upon stretching the glassy polymer the three-component spectrum distorts in the directions along and normal to the draw direction, as the chains become partially aligned[211]. The order parameter agrees with a conformational dissorder model rather than a pseudo afine model.

Imaging[17-20]

The T_2 flash sequence has been used to image the layered structure that formed within a low density polyethylene sample upon extrusion into a rod-shaped mould after loading with a mobile solvent[214]. Mobility similarly provided the contrast in a polysiloxane rubber within which a filler had been dispersed[215]. The gradient-echo images were also useful when obtained in three dimensions. The technique of NMR imaging has been compared with that of microscopy[216].

References

1) Hatada, K., Kitayama, T., Ute, K., Annu. Rep., NMR. Spectrosc., 1993, **26**, 100.

2) Bunn, A., NMR, 1993, **29**, 127.

3) Mirau, P.A., p. 37, Polym. Charact., Ed. Hunt, B.J., James, M.I., Blackie, Glasgow, 1993.

4) Smith, C.G., Smith, P.B., Paztor, A.J., McKelvey, M.L., Meunier, D.M., Froelicher, S.W., Ellaboudy, A.S., Anal. Chem. 1993, **65**, 217.

5) Bevington, J., Trends in Polymer Science, 1993, **1**, 68.

6) Cohen-Addad, Lecture Notes in Physics, 1993, 415 (Complex Fluids), 113.

7) Jones, A.A., Polym. Prepr., (Am. Chem. Soc., Div. Polym. Chem.) 1992, **33(1)**, 63.

8) Pratt, G.J., Smith, M.J.A., Elect., Opt. Acoustic. Prop. Polym., Int. Conf., 2nd, 1990, 13/1, Plast. Rubber Inst. London.

9) Addad, J.P. C., Prog. Nucl. Magn. Reson. Spectrosc., 1993, **25**, 1.

10) Spiess, H.W., Ber. Bunsen-Ges. Phys. Chem., 1993, **97**, 1294.

11) Silvestri, R.L., Koenig, J.L., Anal. Chem. Acta., 1993, **283**, 997.

12) Matsumoto, A., Kagaku to Kogyo (Osaka), 1993, **67**, 507.

13) Pearce, E.M., Kwei, T.K., Stud. Polym. Sci., 1992, **11** (Polymer Solutions, Blends and Interfaces), 133.

14) Segre, A.L., Capitani, D., Trends Polymer Sci., 1993, **1**, 280.

15) Schmidt-Rohr, K., Spiess, H.W., Multidimensional NMR and Polymers, Academic Press, New York, 1994.

16) Miller, J.B., Rubber Chem. Technol. 1993, **66**, 455.

17) Bluemich, B., Bluemich, P., Guenther, E., Jansen, J., Schauss, G., Spiess, H.W., p 167 in Magn. Reson. Microsc., 1992, Ed Bluemich, B., Kuhn, W., VCH: Weinheim, Germany.

18) Koenig, J.L., p 187 in Magn. Reson. Microsc., 1992, Ed Bluemich, B., Kuhn, W., VCH, Weinheim, Germany.

19) Bluemich, B., Bluemler, P., Makromol. Chem., 1993, **194**, 2133.

20) Cory, D.G., Maas, W.E., Bull. Magn. Reson., 1993, **15**, 160.

21) Ebdon, J.R., Huckerby, T.N., Hunter, T.C., Polymer, 1994, **35**, 250.

22) Hensley, D.R., Goodrich, S.D., Harwood, H.J., Rinaldi, P.L., Macromolecules, 1994, **27**, 2351.

23) Lubnin, A.V., Kennedy, J.P., J. Macromol. Sci., Pure Appl. Chem., 1994, **A31**, 655.

24) Kriz, J., Masar, B., Vlek, P., Makromol. Chem., 1993, **194**, 1435.

25) Randal, J.C., Rucker, S.P., Macromolecules, 1994, **27**, 2120.

26) Heatley, F., Ding, J., Yu. G., Booth, C., Blease, T.G., Makromol. Chem., Rapid. Commun., 1993, **14**, 819.

27) Miyashita, K., Kamigaito, M., Sawamoto, M., Higashimura, T., Macromolecules, 1994, **27**, 1093.

28) Masamoto, J., Matsuzaki, K., Iwaisako, T., Yoshida, K., Kagawa, K., Nagahara, H., J. Appl. Polym. Sci., 1993, **50**, 1317.

29) Hunt, M.O., Belu, A.M., Linton, R.W., DeSimone, J.M., Macromolecules, 1993, **26**, 4854.

30) Haque, S.A., Clouet, G., Macromol. Chem. Phys., 1994, **195**, 315.

31) Yamada, B., Kobatake, S., Satake, M., Otsu, T., J. Polym. Sci., A, Polym. Chem., 1993, **31**, 1551.

32) Vanhoorne, P., Jerome, R., Teyssie, P., Laupretre, F., Macromolecules, 1994, **27**, 2548.

33) Edelmann, D., Ritter, H., Makromol. Chem., 1993, **194**, 1183.

34) Edelmann, D., Ritter, H., Makromol. Chem., 1993, **194**, 2375.

35) Kriz, J., Puffr, R., J. Polym. Sci., B, Polym. Phys., 1994, **32**, 1129.

36) Lin, C.H., Xiang, J.S., Matyjaszewski, K., Macromolecules, 1993, **26**, 2785.

37) Kitayama, T., Fujimoto, N., Hatada, K., Makromol. Chem., Makromol. Symp., 1993, **67** (EPF Workshop on Anionic Polymerization) 137.

38) Wang, J.S., Warin, R., Jerome, R., Teyssie, P., Macromolecules, 1993, **26**, 6776.

39) Eisch, J.J., Pombrik, S.L., Zheng, G.X., Organometalics, 1993, **12**, 3856.

40) Billingham, N.C., Lees, P.D., Makromol. Chem., 1993, **194**, 1445.

41) Gatzke, A.L., Green, D. P., Macromoleecules, 1994, **27**, 2249.

42) Tritto, I., Li, S., Sacchi, M.C., Zannoni, G., Macromolecules, 1993, **26**, 7111.

43) Banerjee, K.G., Hogen-Esh, T.E., Macromolecules, 1994, **27**, 2187.

44) Klumperman, B., Kraeger, I.R., Macromolecules, 1994, **27**, 1529.

45) Brar, A.S., Sunita, Polymer, 1993, **34**, 3391.

46) Rai, J.S.P., J.Macromol. Sci., Pure Appl. Chem., 1993, **A30**, 541.

47) Bixamaiah, B., Yoganandam, P., Rao, P.R., Sundaram, E.V., Macromol. Rep., 1993, **A30**, 67.

48) Cheetham, P.F., Huckerby, T..N., Tabner, B.J., Europ. Polym. J., 1994, **30**, 581.

49) Dong, L., Hill, D.J.T., O'Donnell, J.H., Whittaker, A.K., Macromolecules, 1994, **27**, 1830.

50) Hill, D.J.T., Dong, L., O'Donnell, J.H., J. Polym. Sci., A, Polym. Chem., 1993, **31**, 2951.

51) Vasquez, B., Gurruchaga, M., Goni, I., Valero, M., Martin Guzman, G., Macromolecules, 1993, **26**, 4298.

52) Vasques, B., Valero, M., Guzman, G.M., San Roman, J., Polymer, 1993, **34**, 1755.

53) Panchalingam, V., Reynolds, J.R., Report, 1991, TR-34, Order no. AD-A242257, NTIS (US).

54) Victor, M.W., Saffariannour, M., Reynolds, J.R., Polym. Prepr. (A.C.S. Div. Polym. Chem.), 1992, **33(1)**, 1162.

55) Victor, M.W., Saffariannour, M., Reynolds, J.R., J. Macromol. Sci. Pure Appl. Chem., 1994, **A31**, 721.

56) Brown, P.G., Fujimori, K., Brown, A.S., Tucker, D.J., Makromol. Chem., 1993, **194**, 1357.

57) Montheard, J.P., Boinon, B., Belfkira, A., Raihane, M., Pham, Q.T., Makromol. Chem., 1993, **194**, 2839.

58) Busico, V., Corradini, P., Landriani, L., Trifuoggi, M., Makromol. Chem., Rapid Commun., 1993, **14**, 261.

59) Shiono, T., Ohgizawa, M., Soga, K., Makromol. Chem., 1993, **194**, 2075.

60) Kuran, W., Listos, T., Macromol. Chem. Phys., 1994, **195**, 977.

61) Tsuda, T., Kitaike, Y., Ooi, O., Macromolecules, 1993, **26**, 4956.

62) Trumbo, D.L., Polym. Bull (Berlin), 1993, **31**, 629.

63) Nakayama, N., Aoki, A., Hayashi, T., Macromolecules, 1994, **27**, 63.

64) Beshah, K., Makromol. Chem., 1993, **194**, 3311.

65) Kawamura, T., Toshima, N., Matsuzaki, K., Makromol. Chem., Rapid Commun., 1993, **14**, 719.

66) Radiotis, T., Spyros, A., Dais, P. Polymer, 1993, **34**, 1846.

67) Wang, J.S., Jerome, R., Warin, R., Teysie, P., Macromolecules, 1993, **26**, 5984.

68) Varshney, S.K., Gao, Z., Zhong, X.F., Eisenberg, A., Macromolecules, 1994, **27**, 1076.

69) Suchoparek, M., Spevacek, J., Polym. J. (Tokyo), 1993, **25**, 789.

70) Tart, E., Wood, G., Wernsman, D., Sangwatanaroj, U., Howe, C., Zhou, Q., Zhang, S., Tonelli, A.E., Macromolecules, 1993, **26**, 4283.

71) Rapold, R.F., Suter, U.W., Macromol. Theory Simul., 1994, **3**, 1.

72) Born, H., Speiss, H.W., Kutzelnigg, W., Fleischer, U., Schindler, M., Macromolecules, 1994, **27**, 1500.

73) Cheng, H.N., Kasehagen, L.J., Macromolecules, 1993, **26**, 4774.

74) van der Burg, M.W., Chadwick, J.C., Sudmeijer, O., Tulleken, H.J.A.F., Makromol. Chem., Theory Simul., 1993, **2**, 399.

75) Bailey, A.L., Kale, L.T., Tchir, W.J., J. Appl. Polym. Sci., 1994, **51**, 547.

76) Cheng, H.N., Makromol. Chem., Theory Simul., 1993, **2**, 901.

77) Paukkeri, R., Vaananen, T., Lehtinen, A., Polymer, 1993, **34**, 2488.

78) van der Burg, M.W., Chadwick, J.C., Sudmeijer, O., Tulleken, H.J.A.F., Makromol. Chem., Theory Simul., 1993, **2**, 385.

79) Chadwick, J.C., Miedema, A., Sudmeijer, O., Macromol. Chem. Phys., 1994, **195**, 167.

80) Kioka, M., Makio, H., Mizuno, A., Kashiwa, N, Polymer, 1994, **35**, 580.

81) Tritto, I., Sacchi, M.C., Grubbs, R.H., J. Mol. Catal., 1993, **82**, 103.

82) Marthe, S., Sivaram, S., Macromolecules, 1994, **27**, 1083.

83) Miura, Kitayama, T., Hatada, K., Nakata, T., Polymer J., 1993, **25**, 685.

84) Miura, Kitayama, T., Hatada, K., Nakata, T., Polymer J., 1993, **25**, 697.

85) Kuran, W., Listos, T., Macromol. Chem. Phys., 1994, **195**, 401.

86) Lindfors, K.R., Pan, Sheng, Dreyfuss, P., Macromolecules, 1993, **26**, 2919.

87) Tokar, R., Kubisa, P., Penczek, S. Dworak, A., Macromolecules, 1994, **27**, 320.

88) Okuda, J., Rushkin, I., Macromolecules, 1993, **26**, 5530.

89) Shen, Z., Chen, X., Shen, Y., Zhang, Y., J. Polym. Sci., A,: Polym. Chem., 1994, **32**, 597.

90) Tadokoro, A., Takata, T., Endo, T., Macromolecules, 1993, **26**, 4400.

91) Sanda, F., Takata, T., Endo, T., Macromolecules, 1994, **27**, 1099.

92) Sanda, F., Takata, T., Endo, T., Macromolecules, 1993, **26**, 5748.

93) Nelson, J.M., Rengel, Heiko, Manners, I., J. Amer. Chem. Soc., 1993, **115**, 7035.

94) Yang, Xinmin, J.L., Marks, T.J., J. Amer. Chem. Soc., 1993, **115**, 3392.

95) Bai, R., Li, Shihua, Zou, T., Pan, C., Uryu, T., Macromol. Chem. Phys., 1994, **195**, 119.

96) Ivin, K.J., Lam, L.M., Rooney, J.J., Makromol. Chem., 1993, **194**, 4393.

97) Ivin, K.J., Lam, L.M., Rooney, J.J., Macromol. Chem. Phys., 1994, **195**, 1189.

98) Lu, S-Y., Amass, J.M., Majid, N., Glennon, D., Booth, C., Yeates, S.G., Padget, J.C., Macromol. Chem. Phys., 1994, **195**, 1273.

99) Ivin, K.J., Lam, L.M., Rooney, J.J., Macromol. Chem., 1993, **194**, 2907.

100) France, M.B., Grubbs, R.H., McGrath, D.V., Paciello, R.A., Macromolecules, 1993, **26**, 4742.

101) Viswanath, T., Jethmalani, J., J. Appl.Polym. Sci., 1993, **48**, 1289.

102) Wagener, K.B., Patton, J.T., Forbes, M.D.E., Myers, T.L., Maynard, H.D., Polym. Int., 1993, **32**, 411.

103) Smith, D.,W., Wagener, K.B., Polym. Prep. (ACS Div. Polym. Chem.) 1992, **33(2)**, 112.

104) Thorn-Csanyi, E., Pfug, K.P., Macromol. Chem., 1993, **194**, 2287.

105) Xu, Z., Kahr, M., Walker, K., Wilkins, C.L., Moore, J.S., J. Amer. Chem. Soc., 1994, **116**, 4537

106) Xu, Z., Moore, J.S., Angew. Chem., 1993, 105, 261; Angew. Chem.Int. Eng. Edn., 1993, **105**, 246.

107) Woerner, C., Muelhaupt, R., Angew. Chem., 1993, **105**, 1367;

108) Newkome, G.R., Lin, X., Yaxiong, C., Escamilla, G.H., J. Org. Chem., 1993, **58**, 3123.

109) Gitsov, I., Wooley, K.L., Hawker, C.J., Ivanova, P.T., Frechet, J.M., Macromolecules, 1993, **26**, 5621.

110) Fawcett, A.H., Hania, M., Lo., K.W., Patty, A., J. Polym. Sci., A: Polym. Chem., 1994, **32**, 815.

111) Iatrou, H., Hadjichristidis, N., Macromolecules, 1993, **26**, 2479.

112) Chang, J.Y., Ji, H.J.,Han, M.J., Rhee, S.B., Cheong, S., Yoon, M,, Macromolecules, 1994, **27**, 1376.

113) Ritter, H., Sperber, R., Weissbuhn, C.M., Macromol. Chem., 1993, **194**, 1721.

114) Camps, M., Jebri, A., European Polymer J., 1993, **29**, 957.

115) Hahn, S.F., Dreyfuss, M.P., J. Polym. Sci., A, Polym. Chem., 1993, **31**, 3039.

116) Shiono, T., Naga, N., Soga, K., Makromol. Chem., Rapid Commun., 1993, **14**, 323.

117) Wang, L., Liao, C.X., Weber, W.P., J. Polym. Sci., A, Polym. Chem., 1993, **31**, 1789.

118) Maenz, K., Stadermann, European Polym. J., 1993, **29**, 855.

119) Hill, D.J., Whittaker, A.K., J. Macromol. Sci., Pure and Applied Chem., 1993, **A30**, 633.

120) Hill, D.J.T., O'Donnell, J.H., Senake, P.M.C., Pomery, P.J., ACS Symp. Ser., 1993, **527** (Irradiation of Polymeric Materials) 74.

121) Shen, Y.X., Xie, D., Gibson, H.W., J. Amer. Chem. Soc., 1994, **116**, 537.

122) Harada, A., Li, J., Kamachi, M., J. Amer. Chem. Soc., 1994, **116**, 3192.3

123) Harada, A., Li, J., Kamachi, M., Macromolecules, 1993, **26**, 5698.

124) Harada, A., Li, J., Kamachi, M., J. Org. Chem. Soc., 1994, **58**, 7524.

125) O'Connor, R.D., Blum, F.D., Macromolecules, 1994, **27**, 1654.

126) Bieze, T.W.N., van der Maarel, J.R.C., Eisenbach, C.D., Leyte, J.C., Macromolecules, 1993, **27**, 3218.

127) Fuson, F.M., Miller, J.B., Macromolecules, 1993, **27**, 3218.

128) Smith, G.D., Yoon, D.Y., J. Chem. Phys., 1994, **100**, 649.

129) Menge, H., Schneider, H., Polymer, 1993, **34**, 4208.

130) Ryoo, B.K., Kim, D.W., Lee, J.H., Park, J.K., Polym. Eng. Sci., 1993, **33**, 622.

131) Ute, K., Hirose, K., Kashimoto, H., Nakayama, H., Hatada, K., Vogl, O., Polymer J., 1993, **25**, 1175.

132) Kitamaru, R., Horii, F., Zhu, Q., Bassett, D.C., Olley, R.H., Polymer, 1994, **35**, 1171.

133) Tanaka, H., Fujiwara, Y., Asano, T., Polizzi, S., Makromol. Chem., 1993, **194**, 2279.

134) McGrath, K.J., Roland, C.M., Weiss, R.G., Macromolecules, 1993, **26**, 6127.

135) Sozani, P., Simonutti, R., Galimberti, M., Macromolecules, 1993, **26**, 5782.

136) Howe, C., Sankar, S., Tonelli, A.E., Polymer, 1993, **34**, 3674.

137) Tonelli, A.E., Makromol. Chem., Macromol. Symp., 1993, **65**, 133.

138) Cheng, J., Jin, Y., Chen, W., Wunderlich, B., Jonsson, H., Hult, A., Geddge, U.W., J. Polym. Sci., B; Polym. Phys., 1994, **32**, 721.

139) Yamaguchi, M., Tsutsumi, A., Polym. J., 1993, **25**, 427.

140) Taylor, S.A., White, J.L., Elbaum, N.C., Crosby, R.C., Campbell, G.C., Report, 1992, No. AD-A250883, NTIS.

141) Kricheldorf, H.R., Domschke, A., Macromolecules, 1994, **27**, 1509.

142) Yamanobe, T., Tsukamoto, H., Uematsu, Y., Ando, I., Uematsu, I., J. Mol. Struct., 1993, **295**, 25.

143) Dass, N.N., Date, R.W., Fawcett, A.H., McLaughlin, J.D., Sosanwo, O.A., Macromolecules, 1993, **26**, 4192.

144) Dass, N.N., Date, R.W., Fawcett, A.H., McLaughlin, J.D., Sosanwo, O.A., Polym. Prepr., ACS. Div. Poly. Chem., 1994, **34.1**, 156.

145) Davis, F., Heatley, F., Hodge, P., Towns, C., Polymer, 1994, **35**, 885.

146) Kricheldorf, H.R., Schwarz, G., Berghahn, M., de Abajo, J. de la Campa, J., Macromolecules, 1994, **27**, 2540.

147) Spiegel, S., Schmidt-Rohr, K., Boeffel, C., Spiess, H.W., Polymer, 1993, **34**, 4566.

148) Feng, H., Shi, L., Feng, Z., Macromol. Chem., 1993, **194**, 2257.

149) Okamoto, D.t., O'Connell, E.M., Cooper, S.L., Root, T.W., J. Polym. Sci., B; Polym. Phys., 1993, **31**, 1163.

150) Prud'homme, J., Perrier, M., Denault, J., Macromolecules, 1994, **27**, 1493.

151) Godward, J., Heatley, F., Price, C., J. Chem. Soc., Faraday Trans., 1993, **89**, 4371.

152) Baba, J., Kubo, T., Takano, A., Nose, T., Polymer, 1994, **35**, 145.

153) Wang, H., Wittebort, R.J., DuPre, D.B., J. Chem. Phys., 1993, **99**, 7449.

154) Schwenk, N., Spiess, H.W., J. Phys. II France, 1993, **3**, 865.

155) Zachmann, H.G., Chemtracts; Macromol. Chem., 1992, **3**, 229.

156) Abe, A., Iizumi, E., Sasanuma, Y., Polym J., 1993, **25**, 1087.

157) Leisen, J., Ohlemacher, A., Boeffel, C., Spiess, H.W., Ber. Bunsen-Ges. Phys., 1993, **97**, 1306.

158) Schadt, R.J., Cain, E.J., Gardner, K.H., Gabara, V., Allen, S.R., English, A.D., Macromolecules, 1993, **26**, 6503.

159) Socci, E.P., Farmer, B.L., Adams, W.W., J. Polym. Sci, B; Polym. Phys., 1993, **31**, 1975.

160) Schwiegk, S., Werth, M., Leisen, J., Wegner, G., Spiess, H.W., Acta. Polym., 1993, **44**, 31.

161) Li, S., Rice, D.M., Karascz, F.E., Macromolecules, 1994, **27**, 2211.

162) Pavlovskaya, G., Hansen, M., Jones, A.A., Inglefield, P.T., Macromolecules, 1993, **26**, 6310.

163) Chin, Y.H., Inglefield, P.T., Jones, A.A., Macromolecules, 1993, **26**, 5372.

164) Afewoki, M., McKay, R.A., Schaefer, J., Report, 1991, TR-1; Order No. AD-A242 574.

165) Afewoki, M., McKay, R.A., Schaefer, J., Mater. Sci. Eng. A., 1993, **A162**, 221.

166) Miyashita,, N., Nose, T., J. Chem. Phys., 1994, **100**, 6028.

167) White, J.L., Mirau, P., Macromolecules, 1993, **26**, 3049.

168) Asuke, T., Chien-Hua, Y., West, R., Macromolecules, 1994, **27**, 3023.

134) Mcgrath, K.J., Roland, C.M., Weiss, R.G., Macromolecules, 1993, **26**, 6127.

169) Zumbulyadis, N., O'Reilly, J.M., J. Amer. Chem. Soc., 1993, **115**, 4407.

170) Schantz, S., Ljungqvist, N., Macromolecules, 1993, **26**, 6517.

171) Schantz, S., Ljungqvist, N., Synth. Met., 1993, **57**, 4383.

172) Nouwen, J., Adriaensens, P., Vanderzande, D., Gelan, J., Paulussen, H., Verreyt, G., Yang, Z., Geise, H.J., Synth. Meth., 1993, **59**, 171.

173) de los Santos Jones, H., Liu. Y., Inglefield, P.T., Jones, A.A., Kim, C.K., Paul, D.R., Polymer, 1994, **35**, 57.

174) Cheung, Y.W., Stein, R.S., Macromolecules, 1994, **27**, 2512.

175) Tang, P., Reimer, J.A., Denn, M.M., Macromolecules, 1993, **26**, 4269.

176) Guo, M., Zachmann, H.G., Polymer, 1993, **34**, 2503.

177) Huo, P.P., Cebe, P., Macromolecules, 1993, **26**, 5561.

178) Feng, H., Feng, Z., Shen, L., Polymer, 1993, **34**, 2516.

179) Papavione, C.H.M., Maas, W.E.J.R., Veeman, W.S., Werumeus, B, G.H., Vankan, J.M.J., Macromolecules, 1993, **26**, 6611.

180) Maas, W.E.J.R., Papavoine, C.H.M., Veeman, W.S., Weremus, B.G.H., Vankan, J.M.J., J. Polym. Sci., B; Polym. Phys., 1994, **32**, 785.

181) Garces, F.O., Sivadsan, K., Somasundaran, P., Turro, N.J., Macromolecules, 1994, **27**, 272.

182) Marco, C., Gomez, M.A., Fatou, J.G., Exteberria, A., Elorza, M.M., Iruin, J.J., European Polymer J., 1993, **29**, 1477.

183) Sofue, S., Tamura, N., J. Appl. Polym. Sci., Appl. Polym. Symp., 1993, **52**, 261

184) White, J.L., Mirau, P.A., Macromolecules, 1994, **27**, 1648.

185) Zhang, X., Takegoshi, K., Hikichi, K., Macromolecules, 1993, **26**, 2198.

186) Roland, C.M., Miller, J.B., Macromolecules, 1993, **26**, 4967.

187) Kawahara, S., Sato, K., Akiyama, S., J. Polym. Sci., B, Polym. Phys., 1994, **32**, 15.

188) Chung, C., Korngield, J.A., Smith, S.D., Macromolecules, 1994, **27**, 964.

189) Walton, J.H., Miller, J.B., Roland, C.M., Nagode, J.B., Macromolecules, 1993, **26**, 4052.

190) Mirabella, F.M, J. Polym. Sci., B: Polym. Phys., 1994, **32**, 1205.

191) Afeworki, M., Vega, S., Schaefer, J., Report, 1991, TR-4; Order no. AD-A242 572, NTIS.

192) Miller, J.B., Walton, J.H., Roland, C.M., Macromolecules, 1993, **26**, 5602.

193) Dong, Z.P., Cauley, B.J., Bandis, A., Mou, C.W., Inglefield, C.E., Jones, A.A., Inglefield, P.T., Wen, W.Y., J. Polym. Sci., B; Polym. Phys, 1993, **31**, 1213.

194) Capitani, D., Clericuzo, M., Fiordiponti, P., Lillo, F., Segre, A.L., Europ. Polym. J., 1993, **29**, 1451.

195) Maunu, S.L., Soljamo, K., Laantera, M., Sundholm, F., Macromol. Chem. Phys., 1994, 195, 723.

196) Herbert, I.R., Tipping, A., Bashir, Z., J. Polym. Sci., B; Polym. Phys., 1993, **31**, 1459.

197) Tavares, M.I.B., Monteiro, E.E.C., Kenwright, A., Haris, R., Polym. Bull., 1993, **30**, 677.

198) Kimmich, R., Stapf, S., Moeller, M., Out, R., Seitter, R.O., Macromolecules, 1994, **27**, 1505.

199) Kikuchi, H., Tokumitsu, H., Seki, K., Macromolecules, 1993, **26**, 7362.

200) Dhinojwala, A., Wong, G.K., Torkelson, J.M., J. Chem. Phys., 1994, **100**, 6046.

201) Joruso, H. Yamada, T., Ando, I., Sato, K., Otsu, T., J. Mol. Struct., 1993, **300**, 303.

202) Ma, S., Chen, M.C., Wang, S.p., J. Appl. Polym. Sci., 1994, **51**, 1861.

203) Kimmich, R., Weber, H.W., J. Chem. Phys., 1993, **98**, 5847.

204) Kulik, A.S., Prins, K.O., Polymer, 1993, **34**, 4629.

205) Kulik, A.S., Prins, K.O., Polymer, 1993, **34**, 4635.

206) Kulik, A.S., Prins, K.O., Polymer, 1993, **34**, 4642.

207) Frydman, L., Lee, Y.K. Emsley, L. Chingas, G.C., Pines, A., J. Amer. Chem. Soc., 1993, **115**, 4825.

208) Lathrop, D.A., Gleason, K.K., Macromolecules, 1993, **26**, 4652.

209) Wilhelm, M., Feaux de Lacroix, S., Titman, J.J., Schmidt-Rohr, K., Spiess, H.W., Acta. Polym., 1993, **44**, 279.

210) Tzou, D.L., Huang, T.H., Desai, P., Abhiraman, A.S., J. Polym. Sci., B: Polym. Phys., 1993, **31**, 1005.

211) Kulik, A.S., Spiess, H.W., Macromol. Chem. Phys, 1994, **195**, 1755.

212) Schmidt-Rohr, K., Wilhelm, M., Johansson, A., Spiess, H.W., Magn. Reson. Chem., 1933, **31**, 352.

213) Kulik, A.S., Radloff, D., Spiess, H.W., Macromolecules, 1994, **27**, 3111.

214) Okazaki, M., Toriyama, K., Bull. Chem. Soc. Japan., 1993, **66**, 2515.

215) Bluemler, P., Bluemich, B., Acta. Polym., 1993, **44**, 125.

216) Garrido, L., Polym. Preprints, ACS Div. Polym. Chem., 1992, **33(1)**, 761.

11
Conformational Analysis

BY JILL BARBER AND BRIAN WOOD

1. Introduction

The format of this review resembles that of previous years. The volume of reports concerning biological macromolecules continues to expand and protein secondary (as opposed to tertiary) structure determinations have been omitted.

2. Methods

New methods for the accurate measurement of J_{HH}[1-3], J_{CH}[4,5], J_{CC}[4], and long range heteronuclear couplings[5-9] have been described. A method for evaluating J coupling constants from peak amplitudes of TOCSY spectra has been applied to model compounds.[10] Various numerical techniques to alleviate crowding in 2D NMR spectra have been evaluated with a view to facilitating measurement of J coupling constants.[11] Structure refinement using time-averaged J-coupling constant restraints has been reported.[12] A new method for the quantitative analysis of the conformation of 6-membered ring compounds in solution has been presented.[13] Endocyclic vicinal HH coupling constants are translated into conformational parameters using a combination of a generalised Karplus equation and the recently developed truncated Fourier formalism.

Streamlined pseudo-3D TOCSY-NOESY and TOCSY-ROESY techniques employing minimal number of delays and phase-cycle steps have been developed with the use of self-refocussing shaped pulses. Their application has been described in conformational analysis of model systems.[14] Bimolecular structures have been calculated from homonuclear 3D NOE-NOE spectra using an iterative procedure to obtain the relaxation matrix.[15] The 3D HMQC-NOESY heteroMQC NOE experiment was used to illustrate the SLE (set of linear equations) approach to calculating intensities in multidimensional NMR spectroscopy.[16]

Computational methods for analysing NMR and x-ray diffraction data of macromolecules were reviewed[17] and a review containing discussion of the use of NMR data in different levels of molecular modelling was published.[18] A further series of reviews focussed on NMR of biological systems.[19] Distance geometry calculations for proteins (DIANA) were applied to Sandostatin and bovine pancreatic trypsin inhibitor[20] and a least-squares approach to obtaining protein structures from NMR data has been discussed.[21] An algorithm rationalising NOE data for compounds adopting multiple conformations in solution has been developed. This involves representing the effective relaxation matrix for a set of proposed conformers as linear combinations of relaxation matrices due to each conformer.[22] A new approach to conformational analysis of highly flexible molecules involves lanthanide-induced proton NMR shifts in combination with a steepest-descent search algorithm to find the minimum characterised by the 'R' factor.[23] Chemical shift data have been used to provide orientational constraints leading to very high resolution structures from samples aligned with respect to the magnetic field. Gramicidin channel structures were used as an example.[24]

3. Small Organic Molecules

A large volume of literature concerns conformational analysis of natural products and other biologically active molecules in attempts to correlate structure with activity. Conformational analysis of natural products has been reviewed.[25] Table 1 summarises studies carried out on

Table 1: Small Molecules with Biological Activity

Compound	Comments	Ref.
Abscisic acid	Empirical force field calculations and NMR	26
Amphotericin-B in DMSO	Head to tail dimer suggested	27
4′-Azidothymidine	Potent HIV inhibitor	28
AZT	Selective deuteriation allowed detection of the pseudo-rotational parameters of the sugar ring. Equal amounts of C-2′-endo and C-3′-endo species exist in equilibrium.	29
Bacitracin A in D_2O		30
Baclofen analogues	NMR and molecular modelling	31
Bilirubin analogues	Conformational information from ^{13}C T_1 values in DMSO	32
	Effect of C(10) gem dimethyl substitution	33
	Diamides stabilised in ridge-tile shape by H-bonding	34
Caffeine and isocaffeine	Conformational barriers in 7-membered ring assessed	35
Cascarosides A,B,C and D	Conformations and absolute configurations from ^1H NOEs	36
Chromomycin A_3	2:1 complex with Mg^{2+} in methanol is octahedral. Inplications for mode of action are discussed.	37
Cyclazocine	Mixed agonist-antagonist opioid narcotic; ^{13}C NMR relaxation times to determine dynamics and internal mobility	38
Cytochalasin D in DMSO	^1H and ^{13}C relaxation rates and NOESY spectra	39
Digoxin in DMSO	NOE and $^3J_{CH}$ data used to derive conformation	40
Elsinochromes B_1, B_2, G and C_2	Perylenequinone pigments	41
cis-Enantholactam	^{13}C NMR shows 2 conformers interconverting above -110°	42
Fenpropimorph	Fungicide with several conformation in aqueous soln	43
FK506 analogue	Water soluble	44
Folic acid	Folded glutamic acid stabilised by H-bonding in soln	45
Hopane and isohopane	Sidechain conformations	46
Imipramine and amitriptyline	Antidepressants with folded conformations in vesicles	47
Lisinopril	New drug against hypertension	48
Nefopam	Non narcotic analgesic with many conformers in soln.	49
(±)-N-demethylnefopam	Averaged twist-chair - flattened chair in solution	50
Nervosin	Chinese herbal medicine	51
Nigericin (ionophore)	Complexes with Na^+ and K^+ in $CDCl_3$	52
Penicillin V	Solid state ^{13}C NMR indicates 4 distinct orientations of phenoxy ring	53
Phenylmorphan + analogue	Opioid	54
Pilocarpine	Evidence for a puckered furanone ring in aqueous soln	55
Rapamycin in DMSO	Well defined conformation found	56
Scopalamine	Reassessed stereochemistry and NMR parameters	57
Selegiline and analogues	Rotamer populations detected	58
Sparteine	CYP266 model substrate, pH delendence of conformation	59
Streptonigrin in THF	Antitumour drug	60
Timolol	Non-selective β-blocker	61
Timolol derivatives	Analysis based on temperature dependence of vicinal coupling constants.	62
Tofisopam	Two separate conformers detected	63
Vitamin D	Conformational flexibility indicated by lanthanide shift measurements and force field calculations	64

named drugs and natural products. The effect of water on the conformations of the antimitotic agent taxol and the related taxotere and analogues has been investigated,[65-67] the results indicating that hydrophobic interactions contribute to a change in conformation on contact with water. Other natural products analysed by NMR include a humulenedione from *Lippia integrifolia*[68] and tetrahydropyrimidines from *Streptomyces parvulus*.[69] The conformations of the major metabolites of the antibiotics roxithromycin and erythromycin A have been elucidated by molecular dynamics and NMR in $CDCl_3$ solution.[70] The preferred conformations of sympathomimetic catecholamines were investigated using a range of synthetic conformationally restrained analogues.[71] C-nucleoside analogues of triazolo-triazolo-indoles have been shown to exist mainly in a planar zigzag arrangement of carbon atoms.[72] NMR analysis of 2′,3′-*cis*-fused pyrrolidino-β-D-nucleosides has revealed that the glycosyl conformation is *anti* in these compounds.[73] The equilibrium between E and A-conformers of proanthrocyandins is controlled by A-strain and 1,3-diaxial interactions.[74] ^{1}H NMR studies have been used to determine the conformations of the sidechains of 12-hydroxylated neoclerodanes.[75] Conformational analysis of colchicinoids with electron deficient aromatic rings on the B-ring,[76]bromosubstituted 4-en-3-one steroids,[77] cyclohexane-derived glutamic acid analogues[78] and morphin-6-one alkaloids[79] has been carried out using NMR methods.

Antiviral cyclic phosphates of various acylnucleosides were investigated as substrates and inhibitors of phosphodiesterases and related enzymes. Activity was correlated with conformation.[80] The stereochemistry and conformation of decahydroisoquinoline-3(S)-carboxylic acid and its stereoisomers were established using NMR. A derivative of the (S,S,S) isomer is a potent HIV proteinase inhibitor, and its conformation was compared with the parent compound.[81] Antileukemic tropoloisoquinoline alkaloids exhibited puckering of the tropolone ring.[82] The solid and solution conformations of isotiazofurin, a thiazole nucleoside analogue of the antitumour agent tiazofurin were determined.[83] Immunosuppressive activity of boronic acid dipeptides was correlated with conformation.[84] The solution conformations of meropenem and its metabolites were determined using ^{1}H and ^{13}C chemical shifts and NOE measurements.[85] Variable temperature ^{1}H NMR and MacroModel calculations revealed conformational flexibility in 4,5-dihydro-1-phenyl-1H-2,4-benzodiazepines.[86] Enzyme inhibitors examined include indolactam V

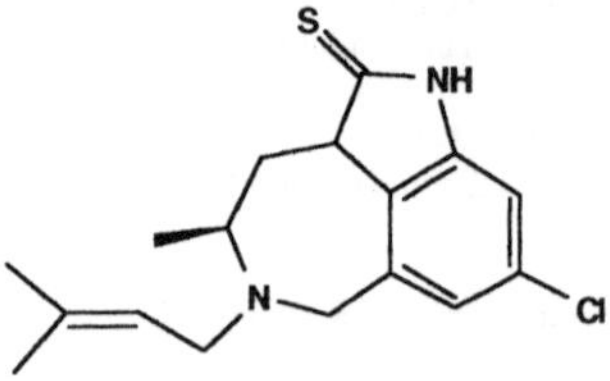

1

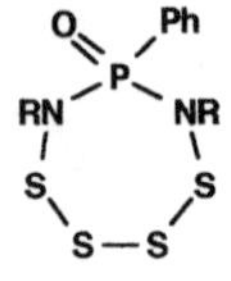

2

which inhibits protein kinase C[87], the reverse transcriptase inhibitor 1[88] and 3-β-hydroxy-3-α-alkoxycarbonyl tropanes which inhibit GABA binding to synaptosomal brain membranes.[89]

The conformations of a number of calixarenes and cyclophanes have been elucidated by NMR.[90-102] Several conformational studies of organosulphur compounds have been conducted.[103-113] These include a dynamic solid state study of hindered naphthyl sulphoxides[103] and the observation that the novel heterocycle 2 exists in a pseudo-chair conformation.[104] Time-dependent changes in high resolution solid state ^{13}C NMR spectra of thiourea inclusion compounds containing a monosubstituted cyclohexane guest molecule show that chair-chair inversion takes place in the guest molecules within the thiourea tunnel.[105] Organophosphorus compounds other than nucleosides and derivatives have attracted less attention, but the conformations of phosphoranes with 5-coordinate phosphorus,[114] 2-Z-3-phenyl-1,3,2-oxaphosphorinones,[115] 2-

(diphenylphosphino)benzaldehyde[116] and some 1,3, alkadiene-2-phosphonates[117] have been established. Methylene splitting patterns and variable temperature NMR have been used to ascertain conformational preferences of β-functional acyclic organosilicon compounds.[118]

Many conformational studies of nitrogen heterocycles have been carried out.[119-139] 2,6-Substituted N-isopropylpyridones show conformational isomers at -30° because of restricted rotation of the N-isopropyl group.[119] NMR isotope shifts and $^3J_{CNCD}$ in the ^{13}C spectra of N-neopentyl-4-tert-butylpiperidine-1′-d provide evidence for an unusual, eclipsed, *syn* geometry for the sidechain.[120] In protonated 3-fluoropiperidines in water the axial orientation of fluorine was preferred.[121] Substituted oxazines have attracted some attention[140,141] as have 1,3-dioxanes.[142]

In conformationally restricted O-, C-, S- and N-furanosides, O- and S- substituents were shown to take up a pseudoaxial conformation whereas C- substituents took up a pseudoequatorial position.[143] Substituted crown ethers have been shown to behave similarly.[144,145] The first hypho-carbaborane, $C_3B_4H_{12}$ was found to exist in an open 7-vortex carbaborane framework.[146] Conformational analysis of nitropolycyclic fluoranthenes,[147] *cis*-cyclohexane-1,3-diol,[148] 1,1′-ethano-9,9′-bifluorenyl derivatives,[149] 6-β-eudesmanes,[150] *cis*-fused eudesm-11-en-4-ols[151] and dihydrotetraphenylmethanes[152] have been carried out. *Ab initio* IGLO studies revealed the conformational dependence of the γ-effect in the ^{13}C NMR spectra of cyclic hydrocarbons.[153] Bent aromatic rings were found in the solid and solution states in highly substituted naphthalene derivatives.[154] *Trans*-diastereomers of substituted methan-3-ones, 3, were shown to exist in a twist conformation with pseudoaxial alkyl groups.[155] Correlated, uncorrelated and localised disrotatory rotations in propeller chains have been investigated.[156]

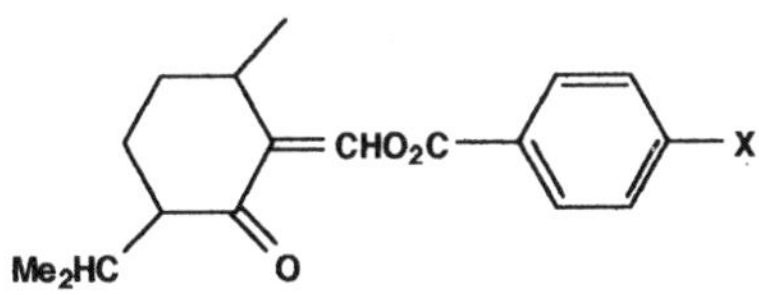

3

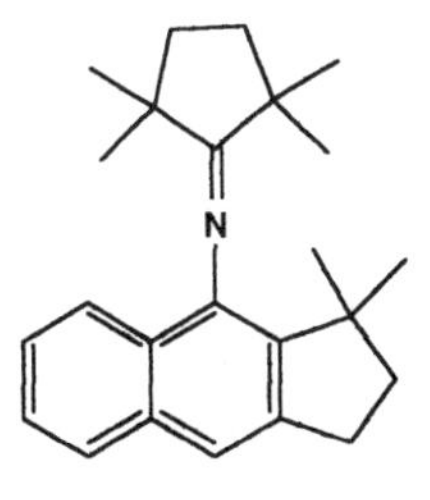

4

Barriers to internal rotation in many compounds were studied. Variable temperature NMR was used to distinguish between C:N double bond rotation and N inversion in compound 4.[157] Variable temperature NMR was also used to determine the barriers to rotation about the C(O)-N bond for 13 α-alkyl and α-halo-substituted N-ethyl-N-methyl amides. The steric effect of the α-carbonyl group was shown to be critical in determining isomer preference and barrier to rotation.[158] The barrier to inversion in 1,2-ethanocorannulene was found to be at least twice that in the parent compound corannulene.[159] ^{13}C lineshape analysis was used to study the conformational dynamics of cyclic *cis* vicinal tertiary diamines.[160] Rotation about the central N-C bond in R_2NCOOR (R=Bu, hexyl, octyl, dodecyl, octadecyl) was investigated in various solvents. Even solvents such as hexadecane and mineral oil did not substantially impede rotation, revealing a surprising absence of solvent friction induced by chain entanglement.[161] Conformational preferences and barriers to internal rotation in fluorothioanisoles were explored using long-range spin-spin couplings, as well as non-NMR methods.[162] The influence of protective groups on the relative stability of ground-state rotational isomers in chiral alkenes has been studied.[163]

[2]- and [3]-rotaxanes with porphyrin stoppers have been synthesised using transition metals to direct threading. NMR showed profound conformational changes on demetallation.[164] The conformations adopted by the alkoxy chains in $(PhO(CH_2)_nO)_3P$ (n=2,3,4) in acetone and 3 nematic solvents were investigated and the energy differences between *gauche* and *trans* conformers derived. These were found to depend upon chain length.[165] Succinic acid in fully protonated, monoprotonated and fully deprotonated states was analysed in liquid crystal mesophases. Direct dipolar couplings were used to determine the conformational equilibria of the molecule in each protonation state.[166] The chain conformations of ionic surfactants absorbed on solid surfaces were shown to resemble the micellar state.[167] Gradient-enhanced inverse detection (HMQC method) was used to show that the *trans* rotamer of ethane-1,2-diol is significantly populated.[168]

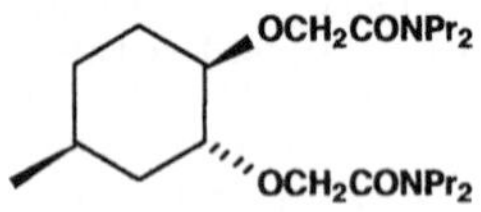

5

Other conformational studies included 1,2:3,4-diepoxides,[169] the ketone and epoxide precursors to β-chimigrene,[170] a novel calcium ionophore (5),[171] kessane,[172] spiperone,[173] benzene-1,4-diones,[174] tricyclohexylmethane,[175] a number of oxamide derivatives,[176] 3-phenylpentane,[177] methylenemalonaldehyde,[178] Me_3CO_2C-Val-NRCONHR (R=cyclohexyl),[179] chlorinated, oxygenated chiral alkenes,[180] C3-substituted C1 chiral alkenes,[181] $Me_2NCOCH_2CONHCH_2CONHMe$,[182] benzyl, 1-phenylethyl and cumyl formates,[183] butanechoic acid[184] and n-alkanes enclathrated in Dianin's compound.[185]

4 Nucleic Acids

The conformations of DNA oligonucleotides continue to be investigated. Unusual structures such as quadruplexes and bulges continue to attract especial interest. A review of oligonucleotide NMR has been published,[186] and guanine quartets have also been reviewed specifically.[187] Drug-DNA interactions have been reviewed.[188] There has been an increase in the volume of work on RNA and reviews have appeared on RNA structure and thermodynamics,[189] tertiary structure of RNA,[190] the structure of catalytic RNA[191] and on DNA/RNA hybrids.[192] A general review of the structure and dynamics of nucleic acids and their complexes has also appeared.[193] Table 2 summarises NMR investigations of oligonucleotide structure, and table 3 those of ligand-nucleic acid interactions not mentioned specifically in the text.

Studies of unusual DNAs include analysis of the first purine-purine-pyrimidine DNA triplex. It was found to be stabilised by G.GC and T.AT base triples with the third strand in an *anti* orientation.[259] The geometry of base pairing in GA mismatches varies enormously and has been shown to depend entirely on the flanking sequences.[260] Conformational differences between bulged pyrimidines (C-C) and purines (A-A, I-I) (I =5-(hydroxymethyl)-2′-deoxyuridine) at the branch point of 3-stranded DNA junctions were observed in oligonucleotide models. The data are consistent with the bulges stacking among themselves.[261] In another study I-A base pairs were found to adopt a Watson-Crick conformation whereas I-G were unstable at low pH, but adopted a Watson-Crick conformation at high pH.[262] The modified oligonucleotide d(CGC[m^2I]AATTCGCG) where m^2I is a methylhypoxanthine guanine analogue was studied by NMR and x-ray crystallography.[263] $d(C\text{-}T)_n$ sequences showed pH dependent polymorphism.[264] Model systems including those chemically modified to lift equivalence suggested that antiparallel-stranded duplexes are formed at pH7 and pH3, but that there is a different conformation at pH5.[265]

The solution structure of the parallel stranded G-quadruplex DNA d(TTGGGT) has been determined. Four GGGG tetrad planes stack with minimal overlap of adjacent base planes within individual strands.[266] The E2A promoter region of the adenovirus-activating transcrition factor binding site has been shown to have a novel conformation with characteristics normally associated with both B- and A-DNA.[267] The wild type sequence of the -35 region of the *trp* promoter has been compared by NMR with the same sequence containing an AT transversion. Only local changes were found in the spectra, supporting the hypothesis that flexible oligonucleotides can readily accommodate small changes of this type.[268] Conformational transition rates in the *trp* promoter have been measured using rotating frame T_1 and cross-relaxation rate measurements. Flattening of the propeller twist in AT base pairs of the YAAY sections of both the -35 and -10 regions were observed.[269]

Chemically modified DNAs have been widely studied in efforts to understand why DNA has evolved into its present structure, and also to understand the bases of mutagenesis. A *cis-anti* - [benzo[a]pyrenyl]dG adduct in a DNA duplex shows intercalation of the benzo[a]pyrenyl ring with displacement of the modified dG.[270] With the (+)-*trans-anti*-[benzo[c]phenanthrene]dA adduct opposite dT, however, intercalation occurs without disruption of the modified base pair.[271] Conformations and intercalation of 3′-5′-cyclic oligothymidylic acids have been examined.[272] N^4-methoxycytosine, a mutagenic base analogue, is able to form Watson-Crick base pairs with A, but a different conformation with G. This may be the basis of mutagenesis.[273] By contrast, in methoxyamine-induced mutagenesis, products can form Watson-Crick base pairs but a wobble conformation is also possible.[274] Watson-Crick base pairing is observed for O-silyl derivatives of 8-oxoguanosine and the mutagenic 8-oxo-2′-deoxyguanosine in oligonucleotides, but Hoogsteen base-pairing is also present.[275] In a propane deoxyguanosine adduct in a $(CG)_3$ frameshift hotspot of *Salmonella typhimurium* hisD3052 Watson-Crick base-pairing becomes disfavoured and Hoogsteen-like base-pairing is observed.[276] Thymidine hydroperoxides are the primary products of DNA oxidation by free radicals. The pyrimidine ring was shown to be puckered in these adducts with the hydroxy or hydroperoxy substituents predominantly axial.[277] Adducts of *cis*- and *trans*-DDP with d(GpTpG) have been studied.[278] In efforts to understand why nature uses deoxyribose, hexose nucleic acids have been synthesised. In self-complementary oligonucleotides Watson-Crick base-pairing is observed, but the oligonucleotide cores are partially molten.[279] $d(CGCGAATTCGCG)_2$ with an unnatural L-deoxyguanosine at G4 forms B-DNA with S-type sugar conformation.[280,281] In 3′,5′-O-(di-tertbutylsilanediyl)nucleosides the furanose ring is forced into the C-3′-endo conformation[282] but in 4′-C-methylnucleosides the 3′-exo furanose ring conformation is favoured over 3′-endo.[283] Gauche and anomeric effects on the pseudorotational equilibrium of the pentofuranose moiety of nucleosides have been evaluated.[284] Conformational studies of some C1′-branched β-D-nucleosides[285] and of d(TpA) and r(UpA) conjugated with histamine and ethylenediamine[286] have been conducted.

Differences between the substituent effects on the puckering mode of the cyclobutane ring and the glycosyl bond of *cis-syn* DNA and RNA photodimers have been observed.[287] The conformation of the riboside analogue of cyanothymidine was found to be solvent dependent.[288] When RNA was incorporated into DNA oligonucleotides partial A-DNA character was discerned.[289] The non-natural ribonucleoside D_3 was incorporated into an intramolecular DNA triplex specifically by intercalation. The nucleoside recognised T.A and C.G pairs when incorporated into py.pur.py triplexes.[290] Nonameric and decameric branched RNAs have been investigated as models for splicing in mRNAs.[291] The active site of lariat-RNA, the smallest RNA to catalyse ribozyme-like cleavage, was characterised.[292] A synthetic 21-mer was used to provide NMR evidence for RNA stem-loop structures in transcription attenuation of the *E. coli* trp operon.[293]

Various NMR parameters have been used and evaluated as tools for determining nucleic acid conformations. It has been suggested that ^{13}C chemical shifts can be used to predict conformational parameters in DNA.[294] The effects of phosphate backbone conformation on ^{31}P

Table 2: Conformational Studies of Oligonucleotides

Oligonucleotide	Comments	Ref
A•AT base triple in intramolecular purine•purine•pyrimidine DNA triplexes		194
d(AGCTTGCCTTGAG).d(CTCAAGGCAAGCT)	Conserved sequence of HIV-1 genome	195
2-Aminofluorene-modified 15-mer DNA duplex from ras proto-oncogene	Conformational heterogeneity	196
$d(ATGAGCGAATA)_2$	Adjacent G.A mismatches stabilised by cross-strand base-stacking and B_{II} phosphate groups	197
C-C bulge in 3-stranded DNA junction		198
ColE1 cruciform		199
$d(CCATGAATGG)_2$	GA mismatch	200
d(CCTAAATTTGCC)..d(GGCAAATTTAGG)	Essentially B-DNA	201
d(CCTCAAACTCC).d(GGAGTTTGAGG)	At low temp. AT region has narrow minor groove.	202
d(CGACGAC)	Forms unusual homobase-paired parallel-stranded double helix at low pH.	203
α-d(CGCAATTCGC).β-d(GCGTTAAGCG)	Parallel stranded. B DNA geometry.	204
$d(CGCTTpS_2\text{-}AAGCG)_2$	Distorted A-DNA like duplex	205
DNA three way junction		206
G:A mismatches in oligoribonucleotides	Duplexes modelled after hammerhead ribozyme and small metallo-ribozyme	207
$d(GAATTTAAATTC)_2$		208
d(GCGAACAAGCG).d(CGCTTTTCGC)	Extrahelical cytosine (C bulge) examined	209
d(GCGAAGC)	Very stable hairpin-turn	210
$d(GCGAATAAGCG)_2$	G.A and A.A mismatches	211

Table 2: cont.

$d(GGATTGGCCAATCC)_2$	Contains GCCAAT, recognition site for several transcription factors. Has a bend and is not a simple B or A DNA structure	212
d(GGAC(TTCG)GTCC)	DNA analogue of unusually stable RNA hairpin	213
d(GGCAGAGGTGAAA).d(TTTCACCTCTGCC)	Trisdecamer duplex	214
d(GGTTGGTGTGGTTGG), thrombin-binding aptamer	Forms a unimolecular quadruplex in solution	215-216
$d(G_4T_4G_4)$, containing Oxytricha telomere repeat	Quadruplex containing wide, narrow and 2 medium grooves	217
d(GTGAACTT).r(AAGUUCAC)	Helical structure of A series	218
d(GTTAATNATTAAC) with G-G mismatch at N site	Pair to loop equilibrium observed at mismatch site	219
Holliday junctions	Base pairing scheme of 32-base pair synthetic immobile junctions	220
Deoxy-m^5HO^6-uridylyl(5-5(3′-5′)deoxyuridine	Photoadduct	221
Lariat RNA	Tetramer and pentamer which mimic natural lariat RNA	222
Nucleoside triples	From self-splicing group I intron	223
dT_4G_4	Parallel stranded tetramer	224
d(TCCCCC)	Tetramer with protonated cytosine-cytosine base pairs	225
TTT and TTTT loops in DNA hairpins	Contained in d(CGATCG-T_n-CGATCG)	226
$d(T_8C_4A_8)$	Parallel stranded hairpin	227
r(CGCUUUCGC)	Hairpin stem-loop, distorted A form, flexible loop	228
$r(GGCGAGCC)_2$	Contains tandem G-A mismatches	229
Sarcin / ricin loop from 28S ribosomal RNA	Compact with several purine-purine base pairs	230
TAR element hairpin	From HIV-1	231
Thymine dimers	Conformational properties of *trans-syn*-I photodimer in DNA	232
	Identification of a *trans-syn*-II product	233
Thymine-glycol containing oligonucleotide	NMR suggests thymine glycol is extrahelical	234
Urea-deoxyribose containing oligonucleotide	2 isomers. Globally B DNA with distortions.	235

Table 3: Conformational Studies of Ligand-Nucleic Acid Complexes

Complex	Comments	Ref
Actinomycin D / DNA hairpin		236
Altromycin B (N7-guanine)-DNA adduct	Proposed DNA adduct structure for the potent antitumour pluramycin antibiotics	237
Anthracycline SN-07 / Y.RY DNA triplex	Site-specific intercalation with covalent adduct formation	238
Bizelesin / d(CGTAATTACG)$_2$	Interstrand cross-linking produces a Watson-Crick to Hoogsteen base-pairing transition region	239
Berenil and pentamidine/ d(CGCAAATTTGCG)$_2$	Both antitrypanosomal drugs	240
Calicheamicin / ACCT-containing non-selfcomplementary duplex	Claicheamicin cleaves DNA double strands at ACCT	241-2
(+)-CG-1065 / A-tracts and related sequences	Bending of DNA is induced or enhanced.	243
4´,6-Diamidino-2-phenylindole / [d(GCGATCGC)$_2$]	Ligand binds in the minor groove	244
Distamycin / d(CGCAAGTTGGC)d(GCCAACTTGCG)	Also a distamycin analogue	245
Distamycin analogues / d(CGCGAATTCGCG)$_2$	Ligands span AATT in minor groove	246
Distamycin / d(GCCTAACAAGG)d(CCTTGTTAGGC)	Also an analogue	247
Echinomycin / d(ACGT)$_2$ and d(TCGA)$_2$		248
Hg(II) / d(CGCGAATTCGCG)$_2$	B-DNA with changes in the AT region	249
HMG-I/Y peptide fragments / d(CGCAAAAAAGGC). d(GCCTTTTTTGCG)	Peptides bind to minor groove in AT region	250
Meta-Hoechst 33258 / d(CGCGAATTCGCG)$_2$	Analogue of Hoechst 33258 with slightly modified DNA binding	251
1-Methylimidiazole-2-carboxamide netropsin /d(TGACT)	Ligand binds 2:1 with oligonucleotides containing this sequence	252
Mithramycin and Chromomycin / d(ACCCGGGT)$_2$	Oligonucleotide binds two dimers of mithramycin but only one of chromomycin	253
Kaempferol 7-neohesperidoside/ oligonucleotide	Plant flavonoid which binds in the minor groove	254
Δ-Rh(phen)$_2$phi^{3+} / d(GTCGAC)$_2$	Phi (=phenanthrenequinone diamine) found in major groove	255
Δ-[Ru(phen)$_3$]$^{2+}$ and Λ-[Ru(phen)$_3$]$^{2+}$ /d(CGCGATCGCG)$_2$	Both ligands bind in the minor groove of DNA	256
SN6999 / alkylated DNA dodecamer	Minor groove binding; SN6999 has antiviral and antitumour activity	257
Sterigmatocystin-formamidopyrimidine adduct in duplex DNA	Aromatic chromophore intercalates into helix on 5´ side of adduct	258

spectra has been investigated for DNA duplexes.[295] Selective 2D experiments were applied to the determination of DNA conformation and 3-bond coupling constants found to be especially useful parameters for this purpose.[296] Exchangeable protons distance constraints have to been used to enhance the structure determinations for two oligonucleotides, in efforts to obtain more precise DNA structures from NMR data.[297] ^{1}H coupling constants were used to obtain the deoxyribose ring conformations of *trans-syn* cyclobutane photodimers of dTpdU.[298]

An attempt has been made to reconcile various data concerning the binding of Hoechst 33258 in the minor groove of DNA. NMR and crystallography studies using $d(GTGGAATTCCAC)_2$ suggest multiple binding modes.[299] The conformations of the cyclohexane rings of the isomers produced by reaction of (1R,3S-cyclohexanediamine)platinum(II) with d(GpG) have been assessed.[300] Conformational studies of thymidine dimers containing sulphonate and sulponamide linkages[301] and of nucleic acid analogues with conformational flexibility in the sugar-phosphate backbone[302] have been carried out. The conformations of two thymidine-containing dinucleoside monophosphates were studied before and after γ-irradiation; all sugar moieties were found to be predominantly in the 2′-endo conformation.[303] Barriers to internal rotation about the C-N bond of synthetic dimethylaminomethylene nucleosides were found to depend on the oxo function in the nucleobase.[304]

5. Proteins and Peptides

Despite the instrument time required for such analyses hundreds more protein 3D structures have been analysed. General reviews on protein structure determination by NMR,[305-308] and a review concentrating on 3D and 4D NMR methods[309] have been published. Reviews concerned with NMR of 3D structures around metal centres,[310] paramagnetic proteins,[311] iron-sulphur clusters,[312] phosphoglycerate kinase,[313] enzymes in general,[314] the use of transferred NOEs in determining conformations of bound peptides,[315] interactions of peptides with receptors generally,[316] the cell adhesion RGD motif[317] and proteins and peptides of relevance to drug design[318] have appeared. In addition, high resolution solid state NMR of proteins has been reviewed,[319] and methods of modification of peptides to assist conformational evaluation have been evaluated.[320] Table 4 summarises individual protein and peptide three dimensional structure determinations.

Many further attempts have been made to improve the quality of the 3D structures obtained from NMR data. Several groups have tried to make better use of individual NMR parameters. Precise vicinal coupling constants $^{3}J_{HN\alpha}$ have been obtained from non-linear fits of J-modulated [^{15}N,^{1}H]COSY experiments.[556] $^{1}J_{C\alpha H\alpha}$ coupling constants were used to probe backbone conformations of proteins.[557] A discussion of the application of chemical shift to protein structure determination has appeared,[558] and the relationship between conformation and theoretical chemical shift was explored for glycine and glycine amide.[559] There has been further work on relaxation matrix analysis,[560] including the development of an iterative "ensemble" approach.[561] Modifications of the rate matrix for the quantitative analysis of NOESY spectra of proteins in order to limit problems of solvent exchange and nonequilibrium populations have been described.[562] A range of computational methods for analysing NMR data were published,[563-571] including strategies for dealing with incomplete NOESY data.[572,573] The limits of precision and accuracy of NMR-derived protein structures has been explored using the IgG binding domain of streptococcal protein G as an example,[574] and has been compared with crystal structures.[575] A calculation strategy for symmetrical dimers was described.[576] A ^{13}C labelling strategy for the determination of the configuration of XaaPro bonds has been developed.[577]

There is now abundant evidence that the 3D structures of domains and modules are very often preserved in the intact protein, so a number of structural studies of such domains have been

Table 4: Secondary and Tertiary Structure of Proteins and Peptides

Protein or Peptide	Comments	Reference
Acyl-coenzyme A binding protein	Complex with palmitoyl-coenzyme A	321
	Structural refinement	322
Ada protein, N-terminal fragment	Folding and DNA binding	323
Ada protein, repair domain	Repairs DNA methyl phosphotriesters	324
Adenylate cyclase activating polypeptide		325
Adenylate kinase	Substrate binding sites	326
ω-Aga-IVB	Spider toxin	327-8
Cyclo(β-Ala-L-Pro-β-Ala-L-Val)	Presence of γ-turns investigated	329
Albolabrin	RGD-containing snake toxin	330
α/β alterable peptides	KVLDA, LQKVLDAL, LMVKVLDAV	331
Amb t V and Amb a V allergens	From ragweed	332
Amyloid protein extracellular fragments		333
Angiotensin	Role of N terminal domain	334
	Cyclic analogues	335
Antennapedia homeodomain mutant	Complexed with DNA	336
Antibody V_L domain	Isolated domain	337
Antigen representing receptor binding domain of *Pseudomonas aeruginosa*		338
Annexin I fragment	Influence of phospholipid micelles	339
Arc repressor		340
$(Asn\text{-}Pro\text{-}Asn\text{-}Ala)_n$	Also gly analogues	341
Aureobasidin E	Cyclic depsipeptide with potent antifungal activity	342
Bacterioopsin (1-36) and (1-71)	In organic solution and micelles	343
Bacteriorhodopsin	^{13}C solid state study	344
Barstar	Secondary structure	345
Bovine pancreatic trypsin inhibitor	Partially folded non-native state	346
Bovine serum albumin	Solid state ^{13}C study: influence of hydration on conformation	347
Bradykinin		348
Bradykinin antagonists	Turn like structure may be important	349,350
C_3HC_4 motif	Of alpha herpes virus protein family	351
CalbindinD_{9k}	Calcium loaded	352
CalbindinD_{9k} mutants	Calcium binding site can adopt different folds	353
Calcium binding cystinyl cyclic heptapeptide		354
apo-Calmodulin	Carboxy-terminal domain	355
Carboxypeptidase Y	pro region	356
CCK-B agonists		357
CD2	Structure-function studies	358
CD2 glycosylated adhesion domain	From human T lymphocyte	359
	CD58 binding site	360
hnCD59	Cell surface glycoprotein	361
CD59	Extracellular region	362
Cetrorelix	LH-RH antagonist	363
Che Y	^{19}F studies on native and mutant	364
	Solution structure	365
Cholecystokinin analogue, CCK9		366
Cold shock protein	From *Bacillus subtilis*	367

Table 4 cont.

Collagen α-1 chain C-telopeptides	Type II and type III collagen	368
Collagen α-1 chain N-telopeptides	Type II and type III collagen	369
Collagen-like peptides	GXY repeating sequence interrupted	370
Complement modules	15th and 16th modules of human complement factor H	371
ω-Conotoxin GVIA		372-3 374
Conotoxin G1	Two possible conformations found	375
Crambin	Refinement using DINOSAUR	376
βB2-Crystallin	From bovine eye lens	377-8
CsE-v3 neurotoxin	From scorpion *C. sculpturatus*	379
Cyclophilin	From *E. coli* periplasm	380
Cyclosporin analogue in water	Nearly identical to cyclophilin-bound cyclosporin conformation	381
Cystatin	Variation of x-ray and NMR structures	382
Cyclo[(R)-cysteinyl-(R)-cysteine]	2 conformers with *cis*-amide bond	383
Cytochrome c_3	Argues in favour of conservation of haem core architecture	384
Delta-sleep-inducing-peptide		385
Deltorphin-I	δ-selective opioid heptapeptide	386
Dendrotoxin K	From black mamba	387
Dermenkephalin	Potent δ-opiod peptide	388
DPDPE	Tyr-cyclo[D-Pen-Gly-Phe-D-Pen]OH	389
Duramycin B and C	Lantibiotics related to cinnamycin	390
Dynophin A (1-17)	Bound to perdeuteriated lipid micelle	391
Elongation factor Tu complexes	^{19}F studies of conformational changes	392
Empusa adipokinetic hormone	Peptide of AKH/RPCH family	393
Endothelin analogues		394-5
Endothelin antagonists	Cyclic pentapeptide	396-7
Dpp-Met^5-enkephalin	Type 1′ β-bend between Gly^2-Gly^3	398
Er-10	Pheromone from ciliated protozoan *Euplotes raikovi*	399
F_1F_0 ATP synthase	Subunit c - helical structure and folding	400
Fatty acid binding protein	From bovine heart	401
Fibronectin (human plasma)	Fragment containing disulphide bonds	402
	4th & 5th type 1 module pair	403
FK506 binding protein	Unfolded in urea and guanidine hydrochloride	404
Flavoridin	Antagonist of platelet GP IIb-IIIa	405
Flavodoxin from *D. vulgaris*	Secondary structure	406
FMRFamide analogues	Invertebrate neuropeptide	407
Gene V single stranded DNA binding protein mutant	Of bacteriophage M13	408
cyclo(Gln-Trp-Phe-Leu-Met)	NK-2 antagonist	409
Boc-D-Glu-Ala-Gly-Lys-Ala-Leu-OMe	Deduced to be 3_{10} helix	410
Glucocorticoid receptor	DNA-binding domain	411
Glucagon-like peptide(7-36)	In dodecylphosphocholine micelle	412
Glutaredoxin	Mixed disulphide with glutathione	413
Oxidised glutathione	No single conformation	414
Gonadotropin-releasing hormone antagonist	Dicyclic peptide	415

Table 4: cont.

Gramicidin A	In multilamellar dimyristoylphosphatidylcholine dispersions. REDOR methods used.	416
	Solid state NMR in lipid bilayers	417
Gramicidin S analogue	β-turn found	418
	Cyclic peptides incorporating benzodiazepine β-turn mimetics	419
Heat shock factor from Drosophila	Similarity with transciption factors	420
Histidine containing protein (HPr)	From *Staphylococcus aureus*	421
Histidine-X_4-histidine zinc finger	Reference to protein-DNA recognition	422
HIV-1 protease	Complexed to a novel inhibitor	423
HIV-1 transmembrane proteins	Peptides and mutants	424
HIV-2 nucleocapsid protein	Fragment	425
HMG1 DNA-binding domain	Box 2 of the hamster protein	426
Human growth hormone releasing factor analogue	Significant α-helical character	427
HUα (histone like protein)	And its complex with oligoDNAs	428
Hymenistatin 1	Solvent effects	429
Icaria chemotactic peptide	And analogues and complexes	430
γ-Interferon from mouse	Receptor binding domain	431
Interleukin 2 fragment 66-72	Complexed with monoclonal antibody	432
Interleukin-4		433
Interleukin-6	Peptides of C-terminal helix	434
Interleukin-8	Binding surface for type 1 human interleukin-8 receptor	435
Insulin-like growth factor-1		436
Jun oncoprotein homodimer	Leucine zipper motif	437
Kunitz-type proteinase inhibitor	From sea anemone	438
α-Lactalbumin	Molten globule state	439-440
	Partially folded state	441
LamB signal peptide analogue	Membrane bound using trNOEs	442
Lophyroromin	Toxic octapeptide	443
Luciferase (bacterial)	Each subunit a molten globule	444
Lysozyme	Partially folded equine	445
M32	Chimeric peptide recognised by galanin and neuropeptide Y receptors	446
Magainin	Structure and orientation in membranes	447
Melanoma growth stimulating activity	Dimer homologous to interleukin-8	448
N-formyl-Met-Pro-Phe-OMe	γ-turn found	449
Methionyl-tRNA synthetase	Zinc-binding domain, *E. coli* enzyme	450
MotA protein from bacteriophage T4	Preliminary analysis shows 2 domains	451
MutT enzyme	Secondary structure	452
c-Myb protooncogene product	Recognition of DNA sequences	453
	Conformational changes	454
Myoglobin (elephant, met-cyano)	Haem cavity	455
Myoglobin (sperm whale) mutants	Role of arginine (E10)	456
Myoglobin	Effect of trimethylphosphine binding	457
	Peptide models of folding initiation sites	458
	Cyanomet complexes of mutants	459
NC_p7 from HIV-1	Mutant with reduced activity	460
	Conformations of active and inactive forms	461
Neurotoxin Sh1	From sea anemone	462
Neurokinin A analogue		463

Table 4: cont.

Neurokinin-2 antagonistic cyclic peptides	No single conformer in DMSO	464
Neuropeptide K		465
Neurotoxin II	From Asian cobra Naja naja oxiana	466
Nisin	Non-aqueous environments	467
OCT-1-POU specific domain	Shows remarkable similarity to bacteriophage λ repressor DNA binding domain	468
OmpA signal peptides	Membrane-mimic environments	469
Opiod peptide analogue		470
Oxytocin	Structural changes on protonation	471
P05-NH_2 (scorpion toxin analogue)	Binds apamin-sensitive K^+ channel	472
$p21^{ras}$	Effect of GTP and GDP	473
	Solution structure and dynamics	474
$p21^{H\text{-}ras}$	Transforming and non-transforming glycine-12 mutants	475
P22 c2 repressor (1-76)	DNA binding domain	476
Parathyroid hormone (human)	In presence of solvents and micelles	477
	Residues 1-34	478
	Several fragments covering the whole structure	479
Pardaxin peptides	Ordered structures observed	480
Ac-Pen-Arg-Gly-Asp-Cys-OH	Fibrinogen receptor antagonist	481
Peptide T	Inhibitor of HIV reproduction	482
Pertussis toxin T cell epitope	Structure-activity relationships	483
D-Phe-Pro-boroArg-OH	Potent thrombin inhibitor	484
Cyclo(L-Phe-L-Pro-Gly-L-Pro)$_2$		485
Phosphatidylinositol 3-kinase	SH3 domain	486
Plasmin (seminal)	In presence of micelles	487
Plasminogen kringle 4 domain	From horse	488
Polymeric immunoglobulin receptor	Basolateral sorting signal	489
Porcine spasmolytic protein	Containing trefoil motif	490
Cyclo(Pro-Arg-Gly-Asp-Gly-D-Pro) + Cyclo(Pro-Arg-Gly-Asp-D-Pro-Gly)	Isomeric peptides show quite different conformational mobilities	491
Pre-Pro-GnRH/GAP protein	Initial study of conformation	492
cyclo(Pro-Pro-Gly)	Evidence for novel β-bend structure	493
Profilin I	From Acanthamoeba	494-6
Protein S, calcium loaded	Helix-containing Greek keys found	497
Pyoverdin GM-II		498
L-Pyroglutamyl-L-histidylglycine	2 stable conformers in DMSO	499
Pyruvate dehydrogenase multienzyme complex	Peripheral subunit-binding domain of dihydrolipoamide acetyltransferase	500
Retinoic acid receptor-β	DNA binding domain	501-2
Retinoid X receptor α	DNA binding domain	503
Ribonuclease A	Sidechain dynamics	504
Ribosomal S17	Potential RNA-binding sites identified	505
hnRNP A1 protein	RNA binding domain	506
Sarcotoxin 1A	2 helical regions joined by hinge	507
Saturnisporins SA II and SA IV		508
Senkide	Membrane bound conformation using transferred NOESY	509
Somatostatin analogues	Hexapeptide	510
	With medicinal activity	511
	With bridge modifications	512
	Cyclic hexapeptides	513
SPKK (DNA-binding motif)	Interaction with DNA modelled	514

Table 4: cont.

Staphylococcal nuclease	Thermal quench refolding	515
	Denatured fragment	516
	Denatured mutant	517
Sterol carrier protein 2	Human	518
Stromelysin		519-20
Substance P hexapeptides		521-2
Substance P analogues	Importance of leucine side-chain	523
Substance P methyl ester	Also neurokinin A fragment	524
[R19L]Tendamistat	Mutant amylase inhibitor	525
Tertiapin	Compact tertiary structure found	526
TGN38/41 cytoplasmic domain	Tyr-containing internalisation motif	527
Thymosin β_4	Multiple conformations	528
δ-Toxin and derivatives	Many have relatively stable helix	529
Toxin γ from snake	Possible phospholipid binding site	530
Transcription factor 1	Selective deuterium labelling	531
	Homeodomain (TTF-1 HD)	532
Transcription factor LFB1/HNF1	Fragment 195-286	533
Transcriptional elongation factor TFIIS	Nucleic acid binding domain	534
Transforming growth factor (Type-α)		535
Transthyretin	Fragment 71-93	536
Triple-helical peptides	$(Pro\text{-}Hyp\text{-}Gly)_{10}$ and a collagen related peptide	537
H-Trp-$(Pro)_n$-Tyr-OH	n=0-5	538
Trypsin carboxypeptidase peptide inhibitor (TCPI)	Comparison with EETI-II and PCI	539
Trypsin inhibitor EETI II	[9-21, 15-27] disulphide folding intermediate	540
Trypsin modulating oostatic factor	A left handed helix	541
Boc-Tyr-Gly-Gly-Pro-Leu-OMe	2 conformers in solution	542
H-Tyr-Gln-Asn-Pro-Asp-Gly-Ser-Gln-Ala-OH	Adopts β-hairpins in aqueous solution	543
Tyrosine hydroxylase	Phenylalanine and its interaction with divalent cation at active site	544
Tyrosine kinase substrate	Free and Ca^{2+} bound conformations	545
Ubiquitin A state	Only 2 β sheet segments (of 5) remain	546-7
Uteroglobin fragments	In presence of micelles	548
Urokinase-type plasminogen activator	Amino terminal fragment	549
Urotensin II		550
Boc-Val-Val-Ile-OH	Forms micelles in chloroform	551
Variant-1 neurotoxin (CsE-v1)	From New World scorpion	552
Triglycyl-lysyl-vasorepressin		553
V3 loop of HIV	Principal neutralising determinant of Thailand isolate	554
Zinc finger domain	DNA-binding mutant	555

completed. A 4-helix bundle is the most likely structure of the dimerisation domain of LFB1/HNF1 related transcription factors.[578] The interactions between the zinc-binding domain of GATA-1 and the major groove of DNA are mainly hydrophobic.[579] The urokinase kringle molecule relates structurally but not functionally to the K2 domain of tissue-type plasminogen activator.[580] The structures and interactions of fibronectin type 1 and EGF modules from tissue type plasminogen activator have also been studied.[581] The 3MDa protein titin seems to be modular with type I and type II domains. Three type II domains have β-sheet conformations but different stabilities.[582]

Other structural studies of proteins in solution include a determination of the structure of the HMG box motif of HMG1. The L-shaped molecule is structurally distinct from other DNA binding motifs.[583] Two breast carcinoma-associated MUC1 core related proteins were shown to contain modified β-turns.[584] NMR and differential scanning calorimetry were used to identify partly folded states and stable subdomains in urokinase-type plasminogen activator.[585] NOESY studies on peptides containing tandem repeats from cartilage link protein suggest the possibility of a long lived extended chain conformation.[586] A myoglobin double mutant was used to assess the coordination of His^{64}(E11) at the sixth position in both the high and low spin states.[587] Proline mutants were of value in solving the solution structure of type III antifreeze protein.[588] The F_V fragment and single-chain F_V fragment of the phosphorylcholine antibody McPC603 in the presence of hapten were shown to have limited flexibility in the backbone, but a very mobile linker.[589] A salivary protein proline-rich repeat sequence has been investigated.[590]

A conformational change has been observed in the *ras* product p21 when GDP is replaced by GTP. In solution this change appears not to be identical with that seen in the crystal.[591] The conformational requirements for the recognition of the main immunogenic region of acetylcholine receptor have been clarified using monoclonal antibodies raised against the receptorst the receptor.[592] The interactions of several DNAs including thymine photodimers with T_4 endonuclease V have been investigated by NMR.[593] Solid state ^{13}C NMR has been used to determine the influence of hydration on the conformation of lysozyme. Hydration appears to limit the range of conformations available, as shown by line-sharpening.[594] ^{15}N solid state NMR has been used in structure determination of uniaxially aligned silk fibroin fibres,[595] and poly(L-proline) form II was examined by high resolution solid state ^{13}C NMR.[596]

Many studies on the conformations of biologically important peptides have also been conducted. A synthetic peptide corresponding to the N terminus of the CCHC zinc finger of HIV shows sequence-dependent binding to ribonucleotide fragments. The bound nucleic acid mainly adopts a single-stranded A conformation.[597] The novel anti-HIV peptide T22 was shown to have a similar conformation to the membrane protein tachyplesin I.[598] The solution conformations of a hybrid synthetic peptide containing multiple epitopes of the envelope protein gp120 from the RF strain of HIV were determined.[599] A synthetic peptide analogue of the eggshell protein of *Schistosoma mansoni* $(Gly\text{-}Tyr\text{-}Asp\text{-}Lys\text{-}Tyr)_6$ was reported to be the first example of a natural left-handed helix.[600] Two pairs of agonist/antagonist bradykinin analogues were synthesised. The agonists but not the antagonist peptides contained β-turns.[601] A type II β-turn was found in NCO-Met-Δ^zLeu-Phe-OMe, an active analogue of chemotactic N-formyl-tripeptides.[602] The role of proline in defining β-turn conformation in cyclic peptides was investigated using a range of L- and D-proline-containing peptides.[603] NMR and molecular mechanics revealed the presence of a γ-turn in a methanomethionine-containing antiopiate peptide.[604] Two conformers of the δ-selective cyclic opioid Tyr-cyclo[D-Cys-Phe-D-Pen]OH were found both in solution and in the solid state.[605] Z-filtered relay NMR was used to study four β-methylphenylalanine stereoisomers in a bioactive peptide.[606] The GlcNAc-1 C2-N2 bond of an Asn-linked glycopeptide was found to be rigid in aqueous solution.[607] A solid state NMR study of the indole ring conformations in gramicidin channels gave results consistent with studies in micelles.[608]

Model peptide studies include observations of a C-terminal asparagine cap stabilising an α-helix in a synthetic peptide.[609] Helix-helix transitions were detected in the lyotropic liquid crystalline state in poly(β-phenethyl-L-aspartate) using ^{2}H NMR.[610] Cyclic hexapeptides with *cis* peptide bonds have been used to identify a pseudo-β1-turn.[611] Formylmethionyl tripeptides containing α,β-dehydrophenylalanine and dehydroleucine were constructed; these are conformationally constrained with type II β-turns.[612] Variable temperature NMR and molecular mechanics were used to examine hydrogen-bond mediated folding of depsipeptide models of β-turns and α-helical turns.[613] Tripeptides and pentapeptides have also been used as β-turn models.[614,615] Three α-hydroxyisobutyric acid-containing depsipeptides have been shown to contain β-bends of various types.[616] The conformational prperties of β-methylphenylalanine in a linear octapeptide,[617] pseudopeptides with modified amide bonds,[618] four carbocyclic muramyl dipeptide analogues,[619] model dipeptides containing dehydroaminobutyric acid residues,[620] ψ[CH_2NH] pseudopeptides,[621] Δ^{z}Phe-containing peptides,[622] and a cyclic phophopeptide containing a phosphodiester linkage[623] have been investigated by NMR. Oxazepines have been evaluated as conformationally constrained gauche(-)-dipeptide mimics.[624] The effect of N-amination and N-hydroxylation on the conformation of peptides has been investigated.[625] NMR and other techniques have been used to investigate the conformations of basic amphiphilic peptides in small unilamellar vesicles.[626] Nine-membered cyclodepsitripeptides have been prepared in order to investigate the factors influencing cyclodepsitripeptide formation and stability,[627] and the competing effects of a disulphide bridge and an α-aminoisobutyryl residue in determining peptide conformation have also been examined using appropriate synthetic peptides.[628]

6. Ligand Binding

Conformational studies of ligands bound to macromolecules continue to arouse interest because of their importance in fundamental biochemistry and their huge potential importance in drug design. NMR-structure based drug design, including computational aspects has been reviewed.[629] These investigations, especially of slow exchange systems, remain difficult, typically requiring developments in molecular biology as well as in NMR. The transferred NOE experiment allows conformational information to be obtained for bound ligands in fast exchange with the free state, and although the experimenter must be wary of spin diffusion, this type of experiment is generally more likely to be successful than investigations involving slow exchange systems. Consequently, transferred NOE studies represent a high proportion of the published work on ligand binding. The use of transferred NOEs in the study of receptor-G protein interactions has been reviewed,[630] and practical aspects of the study of peptides bound to chaperones by the same technique have been discussed.[631]

The transferred NOE experiment has been used to show that oxytocin bound to bovine neurophysin I at pH 2.1 differs in conformation from the crystal structures of both free and bound oxytocin.[632] When the heptapeptide substrate KTKCVFM is bound to farnesyl transferase the sequence CVFM is intimately involved in binding.[633] The conformations of ATP and ADP when bound to N^{10}-formyltetrahydrofolate synthetase have been determined.[634] The calmodulin-binding domain of caldesmon, a 17-residue peptide, was shown to bind to calmodulin in an α-helical conformation.[635] The conformation of cyclophillin when bound to proline isomerase was elucidated.[636]

Conventional NOESY experiments were used to show that barnase undergoes little conformational change on binding to 3′-GMP,[637] and to show that penicillin-derived inhibitors of the HIV aspartate protease bind symmetrically with an S shape.[638] The conformation of FK506 bound to FK binding protein 12 has been determined by 2D NMR experiments and the surfaces of the drug identified by studying the effect of paramagnetic probes on T_1 measurements.[639] Three dimensional TOCSY-NOESY and NOESY-NOESY experiments were used to determine the conformations of haem propionates in haemoglobin, an example of local ligand conformations in slowly tumbling proteins.[640] The interaction of the cytoplasmic domain of phospholamban with calmodulin[641] and of α-bungarotoxin with the functionally important region of nicotinic actylcholine receptor have also been examined.[642]

The introduction of selective pulses was shown to give increased resolution in the very crowded NMR spectrum of a dothiepine-cyclodextrin inclusion complex.[643] ^{13}C NMR of retinols bound to rhodopsin and bacteriorhodopsin led to theoretical estimations of the ring-chain conformations.[644] The preferred orientations of 4′-hydroxyacetanilide (acetaminophen) bound to cytochromes P450 1A1 and 2B1 were determined by ^{13}C and ^{15}N relaxation studies.[645]

7. Carbohydrates

The determination of the 3D structure of carbohydrates using NMR and x-ray crystallography has been reviewed.[646] The correlation between $^1J_{CC}$ and C-C dihedral angle in OH-C-C-OH fragments, as revealed by NMR studies of ^{13}C enriched carbohydrates, has been explored in detail by measuring and computing $^1J_{CC}$ values in simple model compounds.[647] It was found that in addition to the relative C-C torsion, it was also important to consider the conformational behaviour of the C-O bonds. In order to determine the effect of the choice of potential function used in the conformational analysis of a carbohydrate, the NMR spectrum of methyl β-xylobioside was analysed using calculated $J_{C\text{-}H}$ coupling constants and NOEs for protons across the anomeric linkage.[648] A conformational analysis of methyl β-cellobiose in aqueous solution by ROESY NMR and MD simulations, in combination with the CROSREL method, revealed a similar extended conformation to that found in the solid state.[649] Two major conformations about the C_5-C_6 bond in β-gentibiose have been found.[650] The conformational dependence of the hydroxymethyl group on substituent at the anomeric carbon in carbohydrates has been investigated.[651]

There have been a number of reports of conformational analyses using NMR and molecular mechanics calculations.[652-659] N,N-diacyl derivatives of mannopyranosyl amides have a conformation lying between a $B_{2,5}$ boat and a $^1S_5$ twist boat.[652] Only one dominant conformer population was found for a series of 8 N-glycopeptides in three different solvents.[653] The conformations of the tetrasaccharide, Man(α1-3)Man(α1-6)Man(β1-4)GlcNAcβ1-OCD_3,[654] 1,2,-anhydro-3,4,6-tri-*O*-benzyl-β-D-talopyranose,[655] a hydroxymethyl furanoside,[656] the hydrochlorides of α- and β-D-glucosamines,[657] mono methyl ethers of methyl β-lactoside,[658] as well sisomicin and penta-N-di-*O*-acetylsisomicin[659] have also been studied. Molecular modelling and NMR experiments have been used to study the conformation of a trichlorinated derivative of sucralose, a sugar which is 650 times sweeter than sucrose.[660] ^{13}C enriched sucrose was used to assess the furanose ring and glycosidic bond conformations in aqueous solution.[661] 2,5 Anhydro-3,4-*O*-(1,2-ethanediyl)-D-mannitol has been shown, by VT NMR, to have a fixed $^4T_3$ conformation between 25-80^{0}C in D_2O.[662] The proportions of axial anomers of various glucopyranosylamines were shown to be small, thus proving the reverse anomeric effect does not exist.[663]

A conformational analysis of muramic acid δ-lactam structures and their 4-O-(2-acetamido-2-deoxy-β-D-glucopyranosyl) derivatives, which are characteristic of bacterial spore peptidoglycan, has been performed.[664] The conformations of di- to pentasaccharides related to O-specific polsaccharide of *Shigella dysenteriae* type 1 have been shown to partially mimic those of the parent polsaccharide.[665] Glycosylated derivatives of helper-T-cell immunogenic peptides from hen-egg lysozyme show random solution conformations.[666] The carbohydrate portions of T and T_N haptens have been investigated. For the T hapten, the two sugar moieties are in a $^4C_1$ conformation with quasi-parallel orientation.[667] The trisaccharide-F_{ab} complex α-D-Galp(1-2)[α-D-Abep(1-3)]-α-D-Manp]-OMe was shown, using TRNOE build up rates, to undergoe a conformational shift about the Gal-Man glycosidic linkage on binding to monoclonal antibody.[668] Two equilibrating conformations were found for a tetrasaccharide which was taken as a model of the oligosaccharide receptor for E-selection.[669] An extended helical conformation was observed for the epitope of α(2-8)polysialic acid. Although this conformation is not stabilised in the reduced form of the epitope, it does survive following substitution with bulky N-acyl groups.[670]

The solution conformation of de-N-sulphated, re-N-acetylated heparin appears to be similar to that observed in the fibrous state.[671] The influence of sulphate and carboxylate groups on the conformation of chondroitin sulphate related dissacharides has been investigated.[672] A conformational analysis of D-fructans, nystose and kestopentaose has shown that an equivalent inulin enviroment is not attained in nystose, and only partially attained in kestopentaose.[673] Other carbohydrates studied include: model sugars for the linkage region in certain D-linked glycoproteins,[674] inhibitors of endo-α-mannosidase,[675] a number of C-glycosides,[676] some glycosylamines,[677] $3^A,6^A$-anhydro-cyclomatohexaoese,[678] 6A,6C,6E-tri-*O*-methylcyclohexa-amylose,[679] and α-D-Fuc-(1-4)-β-D-GlcNAc-OMe.[680]

8. Membranes

NMR work on the conformations of membranes and membrane components continues to involve, in large measure, the use of ^{2}H NMR of oriented samples. This general area has been reviewed.[681] Backbone conformations of glycosphingolipids and phospholipids in cholesterol-containing bilayers have been determined by this approach. Both were found to be similar to the x-ray derived structures.[682] Transmembrane potential was shown to have little effect on phospholipid bilayer headgroup conformation.[683] The conformational order of amphiphilic chain molecules in lyotropic liquid-crystalline phases was investigated by the same method.[684] ^{31}P NMR in conjunction with ^{2}H studies was used to determine acyl chain orientation in large unilamellar vesicles,[685] and to show that glycerol backbones of phosphatidylcholines in lamellar and hexagonal liquid crystalline mixtures have almost identical conformations in the two phases.[686] ^{13}C NMR in addition to ^{31}P and ^{2}H NMR was used to show that the antiviral peptide carbobenzoxy-D-phenylalanyl-L-phenylalanylglycine changes the average conformation of phospholipids in membranes.[687] Aromatic detergents also change the orientations of molecules in lipid bilayers; this was determined by ^{31}P NMR.[688] The orientation of vitamin E in liposomal membranes was established using ^{19}F studies.[689] The binding of δ-selective opioid peptides to membranes was investigated using the transferred NOE experiment. A cyclic derivatised enkephalin, DPDPE, was shown to undergo little conformational change on binding, whereas binding substantially affected the conformation of the corresponding acyclic peptide.[690]

9. Inorganic and organometallic compounds

^{19}F NMR was used to show that, at a sufficiently low temperature, the CF_3 groups within complexes of the type $(\mu\text{-H})Os_3(CO)_{10}L[ON(CF_3)_2]$ become magnetically inequivalent due to restricted rotation about the N-O bond.[691] Variable temperature ^{19}F NMR was also used to reveal unexpectedly low barriers to propeller rotation of the C_2F_4 ligand about the Ru-olefin axis in a series of ruthenium metallocyclopropanes.[692] The dynamics of the Me_3Sn groups in solid bis(trimethyltin)-dimesylamide-hydroxide was studied using VT ^{13}C NMR.[693] The free energy of activation for C_3 rotation about the metal-C_3 axis of η^3-cyclopropenyl complexes of cobalt, rhodium and iridium has been shown to increase down the group in order Co<Rh<Ir.[694] For bis(alkyne)tunsten(0) complexes it has been shown that only a conrotatory motion of the two alkyne ligands is consistent with the NMR data.[695] The dynamics of p-aryl-substituted chiral ligands in Rh(I) and Pt(II) complexes,[696] trimethylcyclopropenyl and cyclopentadienyl ligands in ruthenium complexes,[697] trimethylene methane(TMM) in compounds of the type η^4-TMMFe$(CO)_2$L (L=tert-BuNC, $AsPh_3$, tertiary phosphines),[698] and imide ligands in $Mo(CO)_2$(N-methylmaleimido)$_2$(bipyridine)[699] have also been investigated.

^{73}Ge and ^{13}C NMR, as well as molecular mechanics calculations, were used to show that 3-germabicyclo[3.1.0]hexanes assume a boat conformation.[700] ^{13}C-^{19}F couplings were used to determine the structure and conformation of tricarbonylchromium complexes of fluorinated biphenyl derivatives.[701] A biphenyl pendant-capped porphyrin model of catalase and its Fe(III) complex has also been studied.[702] The conformational properties of nonbridged bis(indenyl) metal type complexes, which are used as catalysts in propene polymerisation, appear to be important in determining the stereoselectivity of the polymerisation.[703] Conformational studies of some octahedral bis(maleate)molybdenum(0) complexes,[704] Eu(III) complexes of polymethylene bridged derivatives of terpyridine,[705] and Ni(II) and Co(II) complexes of potentially quinquedentate macrobicyclic ligands[706] have also been reported. Methyl and ammonia group reorientation in the bilayered compound bis(decylammonium)tetrachlorocadmate[707] and in methylammonium hexabromoselenate[708] have been studied.

The solution and solid states of 1,5,6-trimethylbenzimidazole Re(V) and Ru(II) complexes were found to adopt the rare head-to-head conformation.[709] The $[Ni(C_{12}H_{19})]^+$ ion was shown (using J couplings) to have the same conformation in solution as in the crystalline state.[710] ^{31}P NMR data were used to suggest that $[(\eta^5\text{-}C_5H_5)_2ZrCl(P_4H)]$,$[(\eta^5\text{-}C_5Me_5)_2ZrCl(P_4H)]$ and $[(\eta^5\text{-}C_5H_5)_3Zr(P_4H)]$ exist in the exo, endo isomer as well as in lower abundance isomers.[711] ^1H NMR data were used to show that Me_8[16]aneS_4 exists in the *syn* conformation in $[Mo_2S_2(\mu\text{-}S)(syn\text{-}Me_8[16]aneS_4)_2]BF_4$ and related new molybdenum complexes.[712] Conformational analyses have been carried out on nickel(II) Schiff base complexes with benzoyl groups and long alkyl chains,[713] and on chelate phosphitodicarbonylchromium and dicarbonyl(phosphine)chromium complexes of fluorobenzenes.[714] Conformational changes in simple silica immobilised systems have been reviewed.[715]

1. F. Del Rio Portilla and R. Freeman, J. Magn. Reson. Ser. A 1993, 104, 358.
2. H.B. Olsen, S. Ludvigsen and O.W. Soerensen, J. Magn. Reson. Ser. A 1993, 104, 226.
3. M.D. Soerensen, S.M. Kristensen, J. Led and O.W. Soerensen, J. Magn. Reson. Ser. A 1993, 103, 364.
4. E. Kupce and R. Freeman, J. Magn. Reson. Ser. A 1993, 104, 234.
5. P. Huber, C. Zwahlen, S.J.F. Vincent and G. Bodenhausen, J. Magn. Reson. Ser. A 1993, 103, 118.
6. U. Wollborn, W. Willker and D. Liebfritz, J. Magn. Reson. Ser. A 1993, 103, 86.
7. K. Kover, O. Prakash and V.R. Hruby, J. Magn. Reson. Ser. A 1993, 103, 92.
8. B. Adams and L. Lerner, J. Magn. Reson. Ser. A 1993, 103, 97.
9. W. Willker, U. Wollborn and D. Liebfritz, J. Magn. Reson. Ser. B 1993, 101, 83.
10. F. Fogolati, G. Esposito, S. Cauci and P. Viglino, J. Magn. Reson. Ser. A 1993, 102, 49.
11. J.A. Jones, D.S. Grainger, P.J. Hore and G.J. Daniell, J. Magn. Reson. Ser. A 1993, 101, 162.
12. A.E. Tords, R.M. Brunne, T. Huber, H. Kessler and W.F. Van Gunsteren, J. Biomol. NMR 1993, 3, 127.
13. C.A.G. Haasnoot, J. Amer. Chem. Soc. 1993, 115, 1460.

14. S. Holmbeck, P. Hajduk and L. Lerner, J. Magn. Reson. Ser. B 1993, 102, 107.
15. R. Bernstein, A. Ross, C. Cieslar and T.A. Holak, J. Magn. Reson. Ser. B 1993, 101, 185.
16. C. Cieslar, A. Ross, T. Zink and T.A. Holak, J. Magn. Reson. Ser. B 1993, 101, 97.
17. A.T. Brunger and M. Nilges, Q. Rev. Biophys. 1993, 26, 49.
18. A.E. Torda and W.F. van Gunsteren, Rev. Comput. Chem. 1992, 3, 143.
19. C. Jones, B. Molloy *et al* Methods Mol. Biol. 1993, 17, several chapters.
20. H. Widmer, A. Widmer and W. Braun, J. Biomol. NMR 1993, 3, 307.
21. J.X. Yang and T.F. Havel, J. Biomol. NMR 1993, 3, 355.
22. M.J. Forster and B. Molloy, J. Comput. Chem. 1994, 15, 155.
23. J. Lehmann and E. Kleinpeter, Magn. Reson. Chem. 1993, 31, 367.
24. W. Mai, W. Hu, C. Wang and T.A. Cross, Protein Sci. 1993, 2, 532.
25. B. Zhou, Adv. Nat. Prod. Chem., Proc. Int. Symp. Pak.-U.S. Binatl. Workshop 5th, 1992, 413.
26. R.D. Willows and B.V. Milborrow, Phytochemistry 1993, 34, 233.
27. A. Balakrishnan and K.R.K. Easwaran, Biochim. Biophys. Acta 1993, 1148, 269.
28. H. Maag, J.T. Nelson, J.L.R. Steiner and E.J. Prisbe, J. Med. Chem. 1994, 37, 431.
29. M.K. Gurjar, A.C. Kunwar, D.V. Reddy, A. Islam, S.V.S. Lalitha, B. Jagannadh and R.V.A. Rao, Tetrahedron 1993, 49, 4373.
30. X. Han, J. Ni, X. Miao, J. Hu, J. Zhuang, G. Wong, X. Zhong, J. Yang, X. Dai *et al* Chinn. Chem. Lett. 1993, 4, 427.
31. C. Vaccher, P. Berthelot, M. Debaert, G. Vermeersch, R. Guyon, B. Pirard, D.P. Vercauteren, M. Dory, G. Evrard *et al.*, J. Mol. Struct. 1993, 301, 199.
32. D.P. Shrout and D.A. Lightner, Spectrosc. Lett. 1993, 26, 461.
33. M. Xie, D.L. Holmes and D.A. Lightner, Tetrahedron 1993, 49, 9235.
34. S.E. Boiadjiev, R.V. Person and D.A. Lightner, Tetrahedron: Asymmetry 1993, 4, 491.
35. Y.E. Shapiro, E.I. Ivanov, E.G. Pykhteeva, I.M. Yaroshchenko, V.Y. Gorbatyuk and G.D. Kalayanov, Zh. Strukt. Khim. 1992, 33, 102.
36 P. Manitto, D. Monti, G. Speranza, N. Mulinacci, F.F. Vincieri, A. Griffini and G. Pifferi, J. Chem. Soc. Perkin Trans. 1 1993, 1577.
37. D.J. Silva and D.E. Kahne J. Am. Chem. Soc. 1993, 115, 7962.
38. D. Carbone, A. Grassi, G.C. Pappatardo and L. Radics, Magn. Reson. Chem. 1993, 31, 823.
39. A. Maccotta, G. Valensin, N. Gaggelli and E. Gaggelli, J. Chem. Soc. Perkin Trans. 2 1993, 729.
40. A.E. Aulabaugh, R.C. Crouch, G.E. Martin, A. Ragouzeos, J.P. Shockcor, T.D. Spitzer, R.D. Farrant, B.D. Hudson and J.C. Lindon, Carbohydr. Res. 1992, 230, 201.
41. A. Arnone, L. Merlini, R. Mondelli, G. Nasini, E. Ragg and L. Scaglioni, Gazz. Chim. Ital. 1993, 123, 131.
42. G. Borgen, J. Dale and F. Rise, Magn. Reson. Chem. 1993, 31, 855.
43. J. Sandholm Jensen, F.S. Joergensen, P.D. Klemmensen, U. Hacksell and I. Pettersson, Pestic. Sci. 1992, 36, 309.
44. A.M. Petros, J.R. Luly, H. Liang and S.W. Fesik, J. Amer. Chem. Soc. 1993, 115, 9920.
45. C. Rossi, A. Donati and M.R. Sansoni, Spectrosc. Lett. 1993, 26, 1603.
46. K. Shiojima, H. Suzuki and S. Nakamura, Chem. Pharm. Bull. 1993, 41, 1939.
47. M.G. Casarotto and D.J. Craik, J. Colloid Interface Sci. 1993, 158, 326.
48. Y. Sakamoto and T. Ishi, J. Mol. Struct. 1993, 298, 129.
49. R. Glaser, J. Blumenfeld and S. Geresh, Magn. Reson. Chem. 1993, 31, 845.
50. R. Glaser, S. Geresh, J. Blumenfeld, D. Donnell, N. Sugisaka, M. Dronin and A. Michel, J. Pharm. Sci. 1993, 82, 276.
51. Y.Yang, X. Xu, D. Wang and B. Qian, Magn. Reson. Chem. 1994, 32, 71.
52. J.C. Beloeil, V. Biou, G. Dauphin, J. Garnier, N. Morellet and F. Vaufrey, Magn. Reson.Chem. 1994, 33, 1003.
53. J. Fattah, J.M. Twyman, S.J. Heyes, D.J. Watkin, A.J. Edwards, K. Prout and C.M. Dobson, J. Amer. Chem. Soc. 1993, 115, 5636.
54. C.M. DiMeglio, M. Froimewitz and A. Makriyannis, Pharm. Res. 1993, 10 1200.
55. E. Gaggelli, N. Gaggelli, G. Valensin and A. Vivi, Can. J. Chem. 1993, 71, 738.
56. H. Kessler, R. Haessner and W. Schueler, Helv. Chim. Acta 1993, 76, 117.
57. R. Glaser, Magn. Reson. Chem. 1993, 31, 335.
58. B. Podanyi, Acta Pharm. Hung. 1992, 62, 218.
59. F. Schneider, P. Fischer, T. Ebner and C.O. Meese, Bioorg. Med. Chem. Lett. 1993, 3, 1667.
60. M.M. Harding, G.V. Long and C.L. Brown, J. Med. Chem. 1993, 36, 3056.
61. A. Saran, S. Srivastava, V.M. Kulkarni and E. Coutinho, THEOCHEM, 1993, 98, 281.
62. J. Kivikoski, J. Vepsalainen, R. Suontamo, E. Pohjala and R. Laatikainen, Tetrahedron; Asymmetry 1993, 4, 709.
63. I. Kovesdi and K. Ujszaszy, Acta Pharm. Hung. 1993, 63, 53.
64. O. Hofer, H. Kachlig and W. Reischl, Monatsh. Chem. 1993, 124, 185.

65. D.G. Vander Velde, G.I. Georg, G.L. Grunewald, C.W. Gunn and L.A. Mitscher, J. Amer. Chem. Soc. 1993, 115, 11650.
66. H.J. Williams, A.I. Scott, R.A. Dieden, C.S. Swindell, L.E. Chirlian, M.M. Frend, J.M. Heerding and N.E. Krauss, Tetrahedron 1993, 49, 6545.
67. J. Dubois, D. Guenard, F. Gueritte-Voegelein, N. Guedira, P. Potier, B. Gillet and J.C. Beloeil, Tetrahedron 1993, 49, 6533.
68. C.A.N. Catalan, M.E.P. De Lampasona and I.J.S. De Fenik, J. Nat. Prod. 1993, 56, 381.
69. L. Inbar, F. Frolow and A. Lapidot, Eur. J. Biochem. 1993, 214, 897.
70. J. Gharbi-Benarous, P. Ladam, M. Delaforge and J.P. Girault, J. Chem. Soc. Perkin Trans. 2 1993, 2303.
71. B. Macchia, A. Balsamo, M.C. Breschi, G. Chiellini, A. Lapucci, M. Macchia, C. Manera, A. Martinelli, C. Martini *et al.* J. Med. Chem. 1993, 36, 3077.
72. N. Rashed, A. El Nemr and E.S.H. El Ashry, Spectrosc. Lett. 1993, 26, 1817.
73. Z. Xi, C. Glemarec and J. Chattopadhyaya, Tetrahedron 1993, 49, 7525.
74. J.P. Steynberg, E.V. Brandt, M.J.H. Hoffman, R.W. Hemingway and D. Ferriera, Basic Life Sci. 1992, 59, 501.
75. J. Jimenez-Barbero, Tetrahedron 1993, 49, 6921.
76. A.E. Pyles and S.B. Hastie, J. Org. Chem. 1993, 58, 2751.
77. R. Sridharan, U.R. Desai, M.R. Rao and G.K. Trivedi, Steroids 1993, 58, 170.
78. N. Morelle, J. Gharbi-Benarous, F. Archer, G. Valle, M. Crisma, C. Toniolo, R. Azerad and J.P. Girault, J. Chem. Soc. Perkin Trans. 2 1993, 525.
79. G.W. Caldwell, A.D. Gauthier, J.E. Mills and M.N. Greco, Magn. Reson. Chem. 1993, 31, 309.
80. J.M. Ciesla, R. Stolarski and D. Shugar, Acta Biochim. Pol. 1993, 40 251.
81. J.C. Gilbert, S. Redshaw, H.S. Simmonite, W.A. Thomas and I.W.A. Whitcombe, J. Chem. Soc. Perkin Trans. 2 1993, 475.
82. H. Morita, K. Matsumoto, K. Takeya and H. Itokawa, Chem. Pharm. Bull. 1993, 41, 1478.
83. H. Li, S.D. Kennedy and B.M. Goldstein, Acta Crystallogr. Sect. B: Struct. Sci. 1993, B49, 729.
84. T.A. Kelly, J. Adams, W.W. Bachovchin, R.W. Randall, S.J. Campbell, S.J. Coutts, C.A. Kennedy and R.J. Snow, J. Am. Chem. Soc. 1993, 115, 12637.
85. Y. Takeuchi, T. Inoue and M. Sunagawa, J. Antibiot. 1993, 46, 827.
86. M.A. Sanner, K. Josef, R.E. Johnson, T. D'Ambra, P. Kowalczyk, R.K. Kullnig and F. Michaels, J. Org. Chem. 1993, 58, 6417.
87. A.P. Kozikowski, D. Ma, P.Y. Pang, P. Shum, V. Likic, P.K. Mishra, S. Macura, A. Basu, J.S. Lazo and R.G. Ball, J. Amer. Chem. Soc. 1993, 115, 3957.
88. G.W. Caldwell, A.D. Ganthier, G.C. Leo and M.J. Kukla, Tetrahedron Lett. 1993, 34, 2063.
89. C. Burgos, M.L. Izquierdo, M.S. Arias, E. Galvez, E. Matesanz, M. Martinez-Ripoll, J. Sanz-Aparicio, I. Fonseca, J. Bellanato et al. J. Mol. Struct. 1993, 301, 95.
90. K. Araki, K. Inada, H. Otsuka and S. Shinkai, Tetrahedron 1993, 49, 9465.
91. S. Kanamathareddy and C.D. Gutsche, J. Am. Chem. Soc. 1993, 115, 6572.
92. Y.H. Lai and A.H.T. Yap, J. Chem. Soc. Perkin Trans. 2 1993, 1373.
93. T. Yamoto, Y. Saruwatari, L.K. Doamekpor, K. Hasegawa and M. Koike, Chem. Ber. 1993, 126, 2501.
94. A. Casnati, E. Cornelli, M. Fabbi, V. Bocchi, G. Mori, F. Ugozzoli, A.M. Lanotti Lanfredi, A. Pochini and R. Ungaro, Recl. Trav. Chim. Phys.-Bas 1993, 112, 384.
95. P. Neri, C. Rocco, G.M.L. Consoli and M. Piattelli, J. Org. Chem. 1993, 58, 6535.
96. T. Yamoto, K. Hasegawa, Y. Saruwatari and L.K. Doamekpor, Chem. Ber. 1993, 126, 1435.
97. C.G. Gibbs and C.D. Gutsche, J. Amer. Chem. Soc. 1993, 115, 5338.
98. E. Nomura, H. Taniguchi, K. Kawaguchi and Y. Otsuji, J. Org. Chem. 1993, 58, 4709.
99. G.D. Andretti, V. Boehmer, J.G. Jordan, M. Tabatabai, F. Ugozzoli, W. Vogt and A. Wolff, J. Org. Chem. 1993, 58, 4023.
100. P. Neri, G. Ferguson, J.F. Gallagher and S. Pappalardo, Tetrahedron Lett. 1992, 33, 7403.
101. C. Krieger, M. Dernbach, G. Voit, T. Carell and H.A. Staab, Chem. Ber. 1993, 126, 811.
102. T. Wong, M.S.M. Yuen, T.C.W. Mak and H.N.C. Wong, J. Org. Chem. 1993, 58, 3118.
103. D. Casarini, E. Foresti, F. Gasparrini, L. Lunazzi, D. Misiti, D. Macciantelli and C. Villani, J. Org. Chem. 1993, 58, 5674.
104. S.E. Bottle, R.C. Bott, I.D. Jenkins, C.H.L. Kennar, G. Smith and A.P. Wells, J. Chem. Soc. Chem. Commun. 1993, 1684.
105. A.E. Aliev and K.D.M. Harris, J. Amer. Chem. Soc. 1993, 115, 6369.
106. S.F. Leonard, P.J. Jones, D.M. Spero and T.P. Pitner, Magn. Reson. Chem. 1993, 31, 906.
107. R. Kimmelma, Acta Chem. Scand 1993, 47, 1201.
108. J. Froehlich, R. Ilk and K. Mereiter, J. Mol. Struct. 1993, 293, 5.
109. A. Pandya and H.W. Gibson, J. Org. Chem. 1993, 58, 2851.

110. A. Bozzoli, E. Marotta, S. Piani and G. Rosili, Tetrahedron Lett. 1993, 34, 3759.
111. K. Pihlaja, R. Sillanpaa, M. Dahlqvist, G. Stajer and M. Ahlgren, Struct. Chem. 1993, 4, 203.
112. G. Toth, A. Levai, A. Szollosy and H. Duddeck, Tetrahedron 1993, 49, 863.
113. K. Pandiarajan, R. Sekar, T. Rangarajan, R. Sarumathi and D. Marko, Indian J. Chem. Sect. B 1993, 32B, 535.
114. Y. Huang, A.E. Sopchik, A.M. Arif and W.G. Bentrude, J. Amer. Chem. Soc. 1993, 115, 4031.
115. Y. Huang, A.M. Arif and W.G. Bentrude, J. Org. Chem. 1993, 58, 6235.
116. T. Schaefer, R. Sebastian, R.W. Schurko and F.E. Hruska, Can. J. Chem. 1993, 71, 1384.
117. I.E. Gurevich, A.V. Dogadina, S.S. Ligai, B.I. Ionin and A.A. Petrov, Zh. Obshch. Khim. 1993, 63, 311.
118. M.S. Beevers, S. Bratt, A.W.P. Jarvie and M.C. Perry, J. Chem. Soc. Perkin Trans. 2 1993, 1569.
119. G. Bartoli, M. Bosco, L. Lunazzi and D. Macciantelli, Tetrahedron 1994, 50, 2561.
120. D.A. Forsyth and S.M. johnson, J. Amer. Chem. Soc. 1993, 115, 3364.
121. D.C. Lankin, N.S. Chandrakumar, S.N. Rao, D.P. Spangler and J.P. Snyder, J. Amer. Chem. Soc. 1993, 115, 3356.
122. J.W. Lygo, R.N. Henrie II, G.A. Meier, R.W. Creekmore and R.M. Patera, Magn. Reson. Chem. 1993, 31, 323.
123. M. Gelbcke and D. Tytgat, Bull. Soc. Chim. Belg. 1993, 102, 67.
124. A.E. Aliev, A.A. Sinitsiria, L.M. Kirillova and N.S. Prostakov, Khim. Geterotsikl. Soedin. 1993, 71.
125. M. Czugler, A. Kalman, M. Kajtar-Peredy, E. Kollat, J. Kajtar, Z. Majer, O. Farkas and M. Hollosi, Tetrahedron 1993, 49, 6661.
126. G.W. Caldwell, D.A. Gauthier and M.J. Kukla, Spectrosc. Lett. 1993, 26, 1005.
127. M.I.M. Wazeer and S.A. Ali, Can. J. Appl. Spectrosc. 1993, 38, 22.
128. A. Capretta, H.N. Hunter, C.S. Frampton and R.A. Bell, Can. J. Chem. 1993, 71, 96.
129. M.L. Izquierdo, E. Galvez, C. Burgos, M.S. Arias, J. Sang-Aparicio, I. Fonseca and J. Bellanato, J. Mol. Struct. 1993, 291, 1.
130. J.E. Anderson, D. Casarini, J.E.T. Corrie and L. Lunazzi, J. Chem. Soc. Perkin Trans. 2 1993, 1299.
131. A.L. Cholli and M.L. Lau, Appl. Spectrosc. 1993, 47, 357.
132. S.R. Salman, R.D. Farrant, R.C. Glen and J.C. Lindon, Spectrosc. Lett. 1993, 26, 745.
133. M. Okamura, Y. Mikata, N. Yamazaki, A. Tsutsumi and A. Ohno, Bull. Chem. Soc. Jpn. 1993, 66, 1197.
134. A. Hassner, R. Maurya, O. Friedman, H.E. Gottlieb, A. Padwa and D. Austin, J. Org. Chem. 1993, 58, 4539.
135. R. Jokela and M. Lounasmaa, Heterocycles 1993, 36, 2373.
136. H.S.P. Rao and B. Bharathi, J. Chem. Res. Synop. 1994, 87.
137. M.S. Morales-Rios, M.A. Bucio and P. Joseph-Nathan, Magn. Reson.Chem. 1993, 31, 960.
138. I. Iriepa, B. Gil-Alberdi, E. Galvez, J. Sang-Aparicio, I. Fonseca and J. Bellanato, J. Mol. Struct. 1993, 293, 49.
139. C. Burgos, E. Galvez, M.L. Izquierdo, M.S. Arias, E. Matesanz and M. Martinez-Ripoll, J. Mol. Struct. 1993, 293, 37.
140. H.A. Al-Muallem and S.A. Ali, J. Phys. Org. Chem. 1993, 6, 326.
141. A.R. Katritzky, I.V. Shcherbakova, B. Mancheno and R.D. Tack, Magn. Reson. Chem. 1993, 31, 615.
142. S.D. Rychnovsky, G. Yang and J.P. Powers, J. Org. Chem. 1993, 58, 5251.
143. U. Ellervik and G. Magnusson, J. Am. Chem. Soc. 1994, 116, 2340.
144. Z. Chen and R.A. Sachleben, J. Chem. Soc. Perkin Trans. 2 1994, 537.
145. Z. Chen and R.A. Sachleben, J. Inclusion Phenom. Mol. Recognit. Chem. 1993, 16, 91.
146. M.A. Fox, R. Greatrex, N.N. Greenwood and M. Kirk, Polyhedron 1993, 12, 1849.
147. B.P. Cho, M. Kim and R.G. Harvey, J. Org. Chem. 1993, 58, 5788.
148. R.J. Abraham, E.J. Chambers and W.A. Thomas, J. Chem. Soc. Perkin Trans. 2 1993, 1061.
149. Y.H. Lai, S.M. Lee, S.Y. Lee and M. Yi, Tetrahedron 1993, 49, 2437.
150. J. Molina Molina and A. Parra Sanchez, J. Mol. Struct. 1993, 296, 153.
151. R.P.W. Kesselmans, J.B.P.A. Wynberg, R. Schrijvers, E.J.R. Sudholter and A. de Groot, Red. Trav. Chim. Pays-Bas 1993, 112, 226.
152. M.C. Grossel, A.K. Cheetham, D.A.O. Hope and S.C. Westron, J. Org. Chem. 1993, 58, 6654.
153. M. Barfield, NATO ASI Ser. Ser. C 1993, 386, 523.
154. P.R. Ashton, G.R. Brown, A.J. Foubister, D.R. Smith, N. Spencer, J.F. Stoddart and D.J. Williams, Tetrahedron Lett. 1993, 34, 8333.
155. Y.E. Shapiro, L.A. Kutulya, A.R. Kalyuskii, N.S. Pivnenko, S.N. Yarmolenko and V.V. Vashchenko, Zh. Obshch. Khim. 1993, 63, 1381.

156. A.B. Lindner, F. Grynszpan and S.E. Biali, J. Org. Chem. 1993, 58, 6662.
157. R. Knorr, J. Ruhdorfer, J. Mehlstaeubl, P. Boehrer and D.S. Stephenson, Chem. Ber. 1993, 126 747.
158. H.G. Bonacorso, M.S.B. Caro, N. Zanatta and M.A.P. Martins, Magn. Reson. Chem. 1993, 31, 451.
159. A.H. Abdourazak, A. Sygula and P.W. Rabioleau, J. Amer. Chem. Soc. 1993, 115, 3010.
160 G. Fraenkel, V. Balasubramanian, H.L. Chang and J. Gallucci, J. Amer. Chem. Soc. 1993, 115, 6795.
161. F.M. Menger and C.E. Mounier, J. Org. Chem. 1993, 58, 1655.
162. D. Chmielewski, N.H. Werstiuk and T.A. Wildman, Can. J. Chem. 1993, 71, 1741.
163. B.W. Gung and M.A. Wolf, J. Org. Chem. 1993, 58, 7038.
164. J.C. Chambron, V. Heitz and J.P. Sauvage, J. Am. Chem. Soc. 1993, 115, 12378.
165. S.T. Cheung and J.W. Emsley, Liq. Cryst. 1993, 13, 265.
166. G. Chidichimo, P. Formoso, A. Golemme and D. Imbardelli, Mol. Phys. 1993, 79, 25.
167. E. Soederlind and P. Stilbs, Langmuir 1993, 9, 1678.
168. S.R. Salman, R.D. Farrant, P. Sanderson and J.C. Lindon, Magn. Reson. Chem. 1993, 31, 585.
169. D. Bur, M. Nikles, U. Sequin and M. Neuburger, Helv. Chim. Acta 1993, 76, 1863.
170. S.F. Leonard, P.J. Jones, D.M. Spero and T.P. Pitner, Magn. Reson. Chem. 1993, 31, 906.
171. M. Raban, D.L. Burch, E.R. Hortelano, D. Durocher and D. Kost, J. Org. Chem. 1994, 59, 1283.
172. P.P. Lankhorst, T.A. van Beek and C.A.G. Haasnoot, Magn. Reson. Chem. 1993, 31, 590.
173. C.E. Keller and W.R. Carper, Magn. Reson. Chem. 1993, 31, 566.
174. I. Takahashi, J. Fukuhara, O. Muramatsu and H. Kitajima, Chem. Express. 1993, 8, 725.
175. I. Columbus and S.E. Biaki, J. Org. Chem. 1993, 58, 7029.
176. F.J. Martinez-Martinez, A. Ariza-Castolo, H. Tlahuextl, M. Tlahuextl and R. Contreras, J. Chem. Soc. Perkin Trans. 2 1993, 1481.
177. T. Schaefer and L.B.L. Lee, Can. J. Chem. 1993, 71, 520.
178. K. Tuppurainen and R. Laatikainen, THEOCHEM 1993, 100, 15.
179. R. Jayakmar and V. Pattabhi, Bioorg. Med. Chem. Lett. 1993, 3, 153.
180. B.W. Gung, M.A. Wolf and Z.Zhu, J. Org. Chem. 1993, 58, 3350.
181. B.W. Gung, M.A. Wolf, K. Ohm and A.J. Peat, Tetrahedron Lett. 1993, 34, 1417.
182. G.P. Dado and S.H. Gellmann, J. Amer. Chem. Soc. 1993, 115, 4228.
183. H. Suezawa, A. Mori, M. Sato, R. Ehama, I. Akai, K. Sakakibara, M. Hirota, M. Nishio and Y. Kodama, J. Phys. Org. Chem. 1993, 6, 399.
184. E.S. Lit, F.K. Mallon, H.Y. Tsai and J.D. Roberts, J. Amer. Chem. Soc. 1993, 115, 9563.
185. F. Isashiro, J. Amer. Chem. Soc. 1993, 115, 2231.
186. J.W. Jaroszewski, S. Roy and J.S. Cohen, Methods Mol. Biol. 1994, 26, 301.
187. J.R. Williamson, Curr. Opin. Struct. Biol. 1993, 3, 357.
188. M.S. Searle, Prog. Nucl. Magn. Reson. Spectrosc. 1993, 25, 403.
189. J.A. Jaeger, J. SantaLucia Jr and I. Tinoco Jr, Annu. Rev. Biochem. 1993, 62, 255.
190. G. Kawai, Seikagaku 1994, 66, 60.
191. A. Pardi, Nucleic Acids Symp. Ser. 1993, 29, 117.
192. K.B. Hall, Curr. Opin. Struct. Biol. 1993, 3, 336.
193. J.R. Everett and S.A. Redshaw, Annu. Rev. Prog. Chem. Sect. B 1991, 88, 3.
194. I. Radhakrishnan, C. de los Santos and D.J. Patel, J. Mol. Biol. 1993, 234, 188.
195. A. Mujeeb, S.M. Kerwin, G.L. Kenyon and T.L. James, Biochemistry 1993, 32, 13419.
196. S.H. Chou, J.W. Cheng and B.R. Reid, J. Mol. Biol. 1992, 228, 138.
197. B.P. Cho, F.A. Beland and M.M. Marques, Biochemistry 1994, 33, 1373.
198. M.A. Rosen and D.J. Patel, Biochemistry 1993, 32, 6563.
199. N.B. Blatt, S.E. Osborne, R.J. Cain and G.D. Glick, Biochimie 1993, 75, 433.
200. K.L. Greene, R.L. Jones, Y. Li, H. Robinson, A.H.-J. Wang, G. Zon and W.D. Wilson, Biochemistry 1994, 33, 1053.
201. S.A. Fawthrop, J.C. Yang and J. Fisher, Nucl. Acids Res. 1993, 21 4860.
202. N.B. Ulyanov, M.H. Sarma, V.B. Zhurkin and R.H. Sarma, Biochemistry 1993, 32, 6875.
203. H. Robinson and A.H.J. Wang, Proc. Natl. Sci. USA 1993, 90, 5224.
204. W.H. Gmeiner, B. Rayner, F. Morvan, J.L. Imbach and W.J. Lown, J. Biomol. NMR 1992, 2, 275.
205. Y. Cho, F.C. Zhu, B.A. Luxon and D.G. Gorenstein, Biomol. Struct. Dyn. 1993, 11, 685.
206. N.B. Leontis, M.T. Hills, M. Piotto, A. Malhotra, J. Nussbaum and D.G. Gorenstein, J. Biomol. Struct. Dyn. 1993, 11, 215.
207. M. Katahira, H. Sato, K. Mishima, S. Uesugi and S. Fujii, Nucl. Acids Res. 1993, 21 5418.
208. V.P. Chuprina, E. Sletten and O.Y. Fedroff, J. Struct. Dyn. 1993, 10, 693.
209. K.M. Morden and K. Maskos, Biopolymers 1993, 33, 27.
210. I. Hirao, G. Kawai, S. Yoshizawa, Y. Nishimura, Y. Ishido, K. Watanabe and K. Miura, Nucleic

Acids Symp. Ser. 1993, 29, 205.
211. K. Maskos, B.M. Gunn, D.A. LeBlanc and K.M. Morden, Biochemistry 1993, 32, 3583.
212. R. Nibedita, R.A. Kumar, A. Majumdar, R.V. Hosur, G. Govil, K. Majumder and V.S. Chauhan, Biochemistry 1993, 32, 9053.
213. J.K. James and I. Tinoco Jr. Nucl. Acids Res. 1993, 21 3287.
214. K.D. Bishop, F.J.H. Blocker, W. Egan and T.L. James, Biochemistry 1994, 33, 427.
215. R.F. Macaya, P. Schultze, F.W. Smith, J.A. Roe and J. Feigon, Proc. Natl. Sci. USA 1993, 90, 3745.
216. K.Y. Wang, S.H. Krawczyk, N. Bischofberger, S. Swaminathan and P.H. Bolton, Biochemistry 1993, 32, 11285.
217. F.W. Smith and J. Feigon, Biochemistry 1993, 32, 8682.
218. A.N. Lane, S. Ebel and T. Brown, Eur. J. Biochem. 1993, 215, 297.
219. P.K. Radha, A. Madan and R.V. Hosur, J. Chem. Soc. Chem. Commun. 1993, 1491.
220. S.M. Chen, F. Heffron and W.J. Chazin, Biochemistry 1993, 32, 319.
221. J.K. Kim, S.D. Soni, J.C. Wallace and J.L. Alderfer, Nucl. Acids Res. 1993, 21 2755.
222. P. Agback, A. Sandstroem, S. Yamakage, C. Sund, C. Glemarec and J. Chattopadhyaya, J. Biochem. Biophys. Methods 1993, 27, 229.
223. M. Chastain and I. Tinoco Jr., Biochemistry 1993, 32, 14220.
224. G. Gupta, A.E. Garcia, Q. Guo, M. Lu and N.R. Kallenbach, Biochemistry 1993, 32, 7098.
225. K. Gehring, J.L. Leroy and M. Gueron, Nature 1993, 363, 561.
226. S.M. Baxter, M.B. Greizerstein, D.M. Kushlan and G.W. Ashley, Biochemistry 1993, 32, 8702.
227. N. Zhou, M.W. Germann, J.H. van de Sande, N. Pattabiraman and H.J. Vogel, Biochemistry 1993, 32, 646. N. Zhou and H.J. Vogel, Bull. Magn. Reson. 1993, 15, 80.
228. P.W. Davis, W. Thurmes and I. Tinoco Jr. Nucl. Acids Res. 1993, 21, 537.
229. J. Santa Lucia Jr. and D.H. Turner, Biochemistry 1993, 32, 12612.
230. A.A. Szewczak, P.B. Moore, Y.L. Chan and I.G. Wool, Proc. Natl. Sci. USA 1993, 90, 9581.
231. J.A. Jaeger and I. Tinoco Jr. Biochemistry 1993, 32, 12522.
232. C.I. Wang and J.S. Taylor, Chem. Res. Toxicol. 1993, 6, 519.
233. J.L.F. Kao, S. Nadji and J.S. Taylor, Chem. Res. Toxicol. 1993, 6, 561.
234. J.Y. Kao, I. Goljer, T.A. Phan and P.H. Bolton, J. Biol. Chem. 1993, 268, 17787.
235. V. Gervais, A. Guy, R. Teoule and G.V. Fazakerley, Nucl. Acids Res. 1992, 20, 6455.
236. D.R. Brown, M. Kurz, V.L. Hsu and D.R. Kearns, Biochemistry 1994, 33, 651.
237. D. Sun, M. Hansen, J.J. Clement and L.H. Hurley, Biochemistry 1993, 32, 8068.
238. X. Ye, K. Kimura and D.J. Patel, J. Am. Chem. Soc. 1993, 115, 9325.
239. F.C. Seaman and L. Hurley, Biochemistry 1993, 32, 11285.
240. T.C. Jenkins, A.N. Lane, S. Neidle and D.G. Brown, Eur. J. Biochem. 1993, 213 1175.
241. S. Walker, J. Murnik and D. Kahne, J. Am. Chem. Soc. 1993, 115, 7954.
242. S.L. Walker, A.H. Andreotti and D.E. Kahne, Tetrahedron 1994, 50, 1351.
243. D. Sun, C.H. Lin and L.H. Hurley, Biochemistry 1993, 32, 4487.
244. E. Trotta, E. DAmbrosio, N. Del Grosso, G. Ravagnan, M. Cirilli and M. Paci, J. Biol. Chem. 1993, 268, 3944.
245. B.H. Geierstanger, T.J. Dwyer, Y. Bathini, W.J. Lown and D.E. Wemmer, J. Am. Chem. Soc. 1993, 115, 4474.
246. M. Rosaria Conte, E. Fattorusso, L. Gomez Paloma and L. Mayol, Bioorg. Med. Chem. Lett. 1992, 2 1299.
247. B.H. Geierstanger, J.P. Jacobsen, M. Mrksich, P.B. Dervan and D.E. Wemmer, Biochemistry 1994, 33, 3055.
248. J. Gallego, A.R. Ortiz and F. Gago, J. Med. Chem. 1993, 36, 1548.
249. N.A. Froeystein and E. Sletten, J. Am. Chem. Soc. 1994, 116, 3240.
250. B.H. Geierstanger, B.F. Volkman, W. Kremer and D.E. Wemmer, Biochemistry 1994, 33, 5347.
251. S.E. Ebrahimi, J.A. Parkinson, K.R. Fox, J.H. McKie, J. Barber and K.T. Douglas, J. Chem. Soc. Chem. Commun. 1992, 1398.
252. M. Mrksich, W.S. Wade, T.J. Dwyer, B.H. Geierstanger, D.E. Wemmer and P.B. Dervan, Proc. Natl. Acad. Sci. USA 1992, 89, 7586.
253. M.A. Keniry, D.L. Banville, P.M. Simmonds and R. Scafer, J. Mol. Biol. 1993, 231, 753.
254. W. Nerdal, O.M. Anderson and E. Sletten, Acta Chem. Scand. 1993, 47, 658.
255. S.S. David and J.K. Barton, J. Am. Chem. Soc. 1993, 115, 2984.
256. M. Eriksson, M. Leijon, C. Hiort, B. Norden and A. Graeslund, Biochemistry 1994, 33, 5031.
257. Y.G. Gao, M. Sriram, W.A. Denny and A.H.J. Wang, Biochemistry 1993, 32, 9639.
258. S. Gopalakrishnan and D.J. Patel, J. Am. Chem. Soc. 1993, 115, 9321.
259. I. Radhakrishnan and D.J. Patel, J. Am. Chem. Soc. 1993, 115, 1615.
260. J.W. Cheng, S.H. Chou and B.R. Reid, J. Mol. Biol. 1992, 228, 1037.

261. M.A. Rosen and D.J. Patel, Biochemistry 1993, 32, 6563.
262. S. Mellac, G.V. Fazakerley and L.C. Sowers, Biochemistry 1993, 32, 7779.
263. D. Yang, Y. Gao, H. Robinson, G.A. van der Marel, J.H. van Boom and A.H.J. Wang, Biochemistry 1993, 32, 8672.
264. T.N. Jaishree and A.W.J. Wang, Nucl. Acids Res. 1993, 21 3839.
265. T.N. Jaishree and A.H.-J. Wang, FEBS Lett. 1994, 337, 139.
266. Y. Wang and D.J. Patel, J. Mol. Biol. 1993, 234, 1171.
267. K.L.B. Borden, Biochemistry 1993, 32, 6506.
268. A.N. Lane, C.J. Bauer, T.A. Frenkiel and A.J. Birchall, Biophys. J. 1993, 22, 135.
269. A.N. Lane, C.J. Bauer and T.A. Frenkiel, Eur. Biophys. J. 1993, 21 , 425.
270. M. Cosman, C. de los Santos, R. Fiala, B.E. Hingerty, V. Ibanez, E. Luna, R. Harvey, N.E. Geacintov, S. Broyde and D.J. Patel, Biochemistry 1993, 32, 4145.
271. M. Cosman, R. Fiala, B.E. Hingerty, A. Laryea, H. Lee, R.G. Harvey, S. Amin, N.E. Gieacintov, S. Broyde and D.J. Patel, Biochemistry 1993, 32, 12488.
272. G. Hamoir and E. Sonveaux, Bull. Soc. Chim.Belg. 1993, 102, 335.
273. G.V. Fazakerley, Z. Gdaniec and L.C. Sowers, J. Mol. Biol. 1993, 230, 6.
274. A.N.R. Nedderman, M.J. Stone, D.H. Williams, P.K.T. Lin and D.M. Brown, J. Mol. Biol. 1993, 230, 1068.
275. P.M. Gannett and P.T. Sura, Chem. Res. Toxicol. 1993, 6, 690.
276. U.S. Singh, J.G. Moe, G.R. Reddy, J.P. Weisenseel, L.J. Marnett and M.P. Stone, Chem. Res. Toxicol. 1993, 6, 825.
277. J.R. Wagner, J.E. van Lier, M. Berger and J. Cadet, J. Am. Chem. Soc. 1994, 116, 2235.
278. N. Boogaard, C. Altona and J. Reedijk, J. Inorg. Biochem. 1993, 49, 129.
279. G. Otting, M. Billeter, K. Wuethrich, H.J. Roth, C. Leumann and A. Eschenmoser, Helv. Chim. Acta 1993, 76, 2701.
280. H. Urata, Y. Ueda, H. Suhara, E. Nishioka and M. Akagi, J. Am. Chem. Soc. 1993, 115, 9852.
281. H. Urata, Y. Ueda, H. Suhara, E. Nishioka and M. Akagi, Nucleic Acids Symp. Ser. 1993, 29, 69.
282. T. Katsura, K. Ueno and K. Furusawa, Magn. Reson. Chem. 1993, 31, 1039.
283. T. Waga, T. Nishizaki, I. Miyakawa, H. Ohrui and H. Meguro, Biosci. Biotechnol. Biochem. 1993, 57, 1433.
284. J. Plavec, W. Tong and J. Chattopadhyaya, J. Am. Chem. Soc. 1993, 115, 9734.
285. J. Plavec, V. Fabre-Buet, V. Uteza, A. Grouiller and J. Chattopadhyaya, J. Biochem. Biophys. Methods 1993, 26, 317.
286. T.P. Prakash, R.K. Kumar and K.N. Ganesh, Tetrahedron 1993, 49, 4035.
287. J.K. Kim, J.C. Wallace and J.L. Alderfer, Biopolymers 1993, 33, 713.
288. M.J. Camarasa, M.A. Calvo and M.L. Jimeno, An. Quim. 1993, 89, 258.
289. T.N. Jaishree, G.A. van der Marel, J.H. van Boom and A.H.J. Wang, Biochemistry 1993, 32, 4903.
290. K.M. Koshlap, P. Gillespie, P.B. Dervan and J. Feigon, J. Am. Chem. Soc. 1993, 115, 7908.
291. P. Agback, C. Glemarec, C. Sund and J. Chattopadhyaya, Tetrahedron 1992, 48, 6537.
292. P. Agback, C. Glemarec, L. Yin, A. Sandstroem, J. Plavec, C. Sund, S. Yamakage, G. Viswanadham, B. Rousse *et al.* Tetrahedron Lett 1993, 34, 3929.
293. V. Ramesh, Nucleic Acids Res. 1993, 21, 5485.
294. Y. Zhang, K. Wang, P.N. Borer and G.C. Levy, Sci. China Ser. B 1992, 35, 1416.
295. S. El antri, P. Bittoun, O. Mauffret, M. Monnot, O. Convert, E. Lescot and S. Fermandijian, Biochemistry 1993, 32, 7079.
296. L. Emsley, T.J. Dwyer, H.P. Spielmann and D.E. Wemmer, J. Am. Chem. Soc. 1993, 115, 7765.
297. R. Michalczyk and I.M. Russu, FEBS Lett. 1993, 331, 217.
298. W.A. Tabaczynski, A. Walter, D.G.E. Lemaire, B.P. Ruzsicska and J.L. Alderfer, Biopolymers 1993, 33, 1365.
299. A. Fede, M. Billeter, W. Leupin and K. Wuethrich, Structure 1993, 1, 177.
300. K. Inagaki and K. Sawaki, Bull. Chem. Soc. Jpn 1993, 66, 1822.
301. C. Glemarec, R.C. Reynolds, P.A. Crooks, J.A. Maddry, M.S. Akhtar, J.A. Montgomery, J.A. Secrist III and J. Chattopadhyaya, Tetrahedron 1993, 49, 2287.
302. M. Tartov, M. Bolli, B. Schweizer and C. Leumann, Helv. Chim. Acta 1993, 76, 481.
303. J.D. Baleja, G.W. Buchko, M. Weinfeld and B.D. Sykes, J. Biomol. Struct. Dyn. 1993, 10, 747.
304. E.Y. Osei-Twum, Can. J. Chem. 1993, 71, 1006.
305. T. Tanaka and M. Ikura, Yuki Gosei Kagaku Kyokaishi 1993, 51, 491.
306. H. Tamaoki and K. Haruhiko, Saibo Koggaku 1993, 12, 15.
307. J. Xu, L.R. Brown, W. Cai and M. Zhang, Sci. China, Ser. B 1993, 36, 526.
308. S.A. Sherman and M.E. Johnson, Prog. Biophys. Mol. Biol. 1993, 59, 285.

309. H. Oschkinat, T. Mueller and T. Dieckmann, Angew. Chem. 1994, 106, 284.
310. A.V. Xavier, D.L. Turner and H. Santos, Methods Enzymol. 1993, 227, 1. H. Thomann and M. Bernardo, *ibid*, 118. G.W. Canters, C.W. Hilbers, M. Van De Mart and S.S. Wijmenga, *ibid*, 244.
311. L. Banci, I. Bertini, C. Luchinat, L. Messori and P. Turano, Appl. Magn. Reson. 1993, 4, 461.
312. L. Banci, I. Bertini, S. Forretti, C. Luchinat and M. Piccioli, J. Mol. Struct. 1993, 292, 207.
313. H. Joao and R.J.P. Williams, Eur. J. Biochem. 1993, 216, 1.
314. J.N.S. Evans, Pulsed Magn. Reson.: NMR, ESR, Opt. 1992, 123.
315. B.D. Sykes, Curr. Opin. Biotechnol. 1993, 4, 392.
316. M.P. Williamson, Biochem. Soc. Trans. 1994, 22, 140.
317. T. Yamada and M. Kikuchi, Baiosaiensu to Indasutori 1993, 51 635.
318. E.R.P. Zuiderweg, S.R. van Doren, A.V. Kurochkin, R.R. Neubig and A. Majumdar, Perspect. Drug Discovery Des. 1993, 1, 391.
319. H. Saito, A. Naito and S. Tuzi, Seibutsu Butsuri 1993, 33, 274.
320. D.F. Meirke and H. Kessler, Biopolymers 1993, 33, 1003.
321. B.B. Kragelund, K.V. Andersen, J.C. Madsen, J. Knudsen and F.M. Poulsen, J. Mol. Biol. 1993, 230, 1260.
322. K.V. Andersen and F.M. Poulsen, J. Biomol. NMR 1993, 3, 271.
323. H. Sakashita, T. Sakuma, T. Ohkubo, M. Kainosho, K. Sakumi and M. Sekiguchi, FEBS Lett. 1993, 323, 252.
324. L.C. Myers, G.L. Verdine and G. Wagner, Biochemistry 1993, 32, 14089.
325. V. Wray, C. Kakoschke, K. Nokihara and S. Naruse, Biochemistry 1993, 32, 5832.
326. I.J.L. Byeon, H. Yan, A.S. Edison, E.S. Mooberry, F. Abildgaard, J.L. Markley and M.D. Tsai, Biochemistry 1993, 32, 12508.
327. H. Yu, M.K. Rosen, N.A. Saccomano, D. Phillips, R.A. Volkmann and S.L. Schreiber, Biochemistry 1993, 32, 13123.
328. M.E. Adams, I.M. Mintz, M.D. Reily, V. Thanabal and B.P. Bean, Mol. Pharmacol. 1993, 44, 681.
329. B. Di Blasio, A. Lombardi, G. DAuria, M. Saviano, C. Isernia, O. Maglio, L. Paolillo, C. Pedone and V. Pavone, Biopolymers 1993, 33, 621.
330. M. Jaseja, K.J. Smith, X. Lu, J.A. Williams, H. Trayer, I.P. Trayer and E.I. Hyde, Eur. J. Biochem. 1993, 218, 853.
331. M. Xu, L. Xu and Z. Lu, Shengwu Huaxue Yu Shengwu Wuli Xuebao 1993, 25 1.
332. W.J. Metzler, L. Mueller, S.K. Huang and D.G. Marsh, Mol. Biol. Immunol. Allergens 1993, 53.
333. K. Sorimachi and D.J. Craik, Eur. J. Biochem. 1994, 219 237.
334. J.M. Matsoukas, J. Hondrelis, M. Keramida, T. Mavromoustakos, A. Makriyannis, R. Yamdagni, Q. Wu and G.J. Moore, J. Biol. Chem. 1994, 269, 5303.
335. G.V. Nikiforovich, J.L.-F. Kao, K. Plucinska, W.J. Zhang and G.R. Marshall, Biochemistry 1994, 33, 3591.
336. M. Billeter, Y.Q. Qian, G. Otting, M. Mueller, W. Gehring and K. Wuethrich, J. Mol. Biol. 1993, 234, 1084.
337. K.L. Constantine, M.S. Friedrichs, W.J. Metzler, M. Wittekind, P. Hensley and L. Mueller, J. Mol. Biol. 1994, 236, 310.
338. C. McInnes, F.D. Sonnichsen, C.M. Kay, R.S. Hodges and B.D. Sykes, Biochemistry 1993, 32, 13432.
339. F. Macquaire, F. Baleux, T. Huynh Dinh, D. Rouge, J.M. Neumann and A. Sanson, Biochemistry 1993, 32, 7244.
340. A.M.J.J. Bonvin, H. Vis, J.N. Breg, M.J.M. Burgering, R. Boelens and R. Kaptein, J. Mol. Biol. 1994, 236, 328.
341. S. Isokawa, M. Narita and K. Umemoto, Makromol. Chem. 1993, 194, 3247.
342. A. Fujikawa, Y. In, M. Inoue, T. Ishida, N. Nemoto, Y. Kobayashi, R. Kataoka, K. Ikai, K. Takesako and I. Kato, J. Org. Chem. 1994, 59, 570.
343. K.V. Pervushin, A.G. Sobol, L.Y. Musina, G.V. Abdulaeva and A.S. Arseniev, Mol. Biol. (Moscow) 1992, 26, 1397. K.V. Pervushin, V.Y. Orekhov, A.I. Popov, L.Y. Musina and A.S. Arseniev, Eur. J. Biochem. 1994, 219, 571.
344. S. Tuzi, A. Naito and H. Saito, Eur. J. Biochem. 1993, 218, 837.
345. M.J. Lubienski, M. Bycroft, D.N.M. Jones and A.R. Ferscht, FEBS Lett. 1993, 332, 81.
346. C.P.M. van Mierlo, J. Kemmink, D. Neuhaus, N.J. Darby and T.E. Creighton, J. Mol. Biol. 1994, 235, 1044.
347. R.B. Gregory, M. Gangoda, R.K. Gilpin and W. Su, Biopolymers 1993, 33, 1871.
348. W. Yang, S. Wang, L. Xu and Z. Lu, Shengwu Huaxue Yu Shengwu Wuli Xuebao 1993, 25, 11.
349. X. Liu, J.M. Stewart, L. Gera and G. Kotovych, Biopolymers 1993, 33, 1237.

350. D.J. Kyle, P.R. Blake, D. Smithwick, L.M. Green, J.A. Martin, J.A. Sinsko and M.F. Summers, J. Med. Chem. 1993, 36, 1450.
351. R.D. Everett, P. Barlow, A. Milner, B. Luisi, A. Orr, G. Hope and D. Lyon, J. Mol. Biol. 1993, 234, 1038.
352. J. Kordel, N.J. Skelton, M. Akke and W.J. Chazin, J. Mol. Biol. 1993, 231, 711.
353. C. Johannson, M. Ullner and T. Drakenberg, Biochemistry 1993, 32, 8429.
354. G. Zanotti, A. Maione, F. Rossi, M. Saviano, C. Pedone and T. Tancredi, Biopolymers 1993, 33, 1083.
355. B.E. Finn, T. Drakenberg and S. Forsen, FEBS Lett. 1993, 336, 368.
356. P. Soerensen, J.R. Winther, N.C. Kaarsholm and F.M. Poulsen, Biochemistry 1993, 32, 12160.
357. N. Goudreau, J.H. Weng and B.P. Roques, Biopolymers 1994, 34, 155.
358. P.C. Driscoll, J.G. Cyster, C. Somoza, D.A. Crawford, P. Howe, T.S. Harvey, B. Kieffer, I.D. Campbell and A.F. Williams, Biochem. Soc. Trans. 1993, 21, 947.
359. J.M. Withka, D.F. Wyss, G. Wagner, A.R.N. Arulanandam, E.L. Reinherz and M.A. Recny, Structure 1993, 1, 69.
360. A.R.N. Arulanandam, J.M. Withka, D.F. Wyss, G. Wagner, A. Kister, P. Pallai, M.A. Recney and E.L. Reinherz, Proc. Natl. Sci. USA 1993, 90, 11613.
361. C.M. Fletcher, R.A. Harrison, P.J. Lachmann and D. Neuhaus, Protein Sci. 1993, 2, 2015.
362. B. Kieffer, P.C. Driscoll, I.D. Campbell, A.C. Willis, P.A. van der Merwe and S.J. Davis, Biochemistry 1994, 33, 4471.
363. A. Mueller, E. Busker, J. Engel, B. Kutscher, M. Bernd and A.V. Schally, Int. J. Pept. Protein Res. 1994, 43, 264.
364. S.K. Drake, R.B. Bourret, L.A. Luck, M.I. Simon and J.J. Falke, J. Biol. Chem. 1993, 268, 13081 and following paper.
365. M. Bruix, J. Pascual, J. Santoro, J. Prieto, L. Serrano and M. Rico, Eur. J. Biochem. 1993, 215 573.
366. L. Moroder, A. dUrsi, D. Picone, P. Amodeo and P.A. Temussi, Biochem. Biophys. Res. Commun. 1993, 190, 741.
367. A. Schnuchel, R. Wiltscheck, M. Czisch, M. Herrler, G. Willimsky, P. Graumann, M.A. Marahiel and T.A. Holak, Nature 1993, 364, 169.
368. X. Liu, A. Otter, P.G. Scott, J.R. Cann and G. Kotovych, J. Biomol. Struct. Dyn. 1993, 11, 541.
369. A. Otter, P.G. Scott and G. Kotovych, Biopolymers 1993, 33, 1443.
370. C.G. Long, E. Braswell, D. Zhu, J. Apigo, J. Baum and B. Brodsky, Biochemistry 1993, 32, 11688.
371. P.N. Barlow, A. Steinkasserer, D.G. Norman, B. Kieffer, A.P. Wiles, R.B. Sim and I.D. Campbell, J. Mol. Biol. 1993, 232, 268.
372. P. Sevilla, M. Bruix, J. Santoro, F. Gago, A.G. Garcia and M. Rico, Biochem. Biophys. Res. Commun. 1993, 192, 1238.
373. J.J. Skalicky, W.J. Metzler, D.J. Ciesla, A. Galdes and A. Pardi, Protein Sci. 1993, 2, 1591.
374. J.H. Davis, E.K. Bradle, G.P. Milijanich, L. Nadsadi, J. Ramachandran and V.J. Basus, Biochemistry 1993, 32, 7396.
375. K.S. Kim, Bull. Korean Chem. Soc. 1993, 14, 18.
376. A.M.J.J. Bonvin, R. Boelens and R. Kaptein, Biopolymers 1994, 34, 39.
377. J.A. Carver, P.G. Cooper and R.J.W. Truscott, Eur. J. Biochem. 1993, 213, 313.
378. P.G. Cooper, J.A. Carver and R.J.W. Truscott, Eur. J. Biochem. 1993, 213, 321.
379. W. Lee, C.H. Moore, D.D. Watt and N.R. Krishna, Eur. J. Biochem. 1994, 219, 89.
380. R.T. Clubb, S.B. Ferguson, C.T. Walsh and G. Wagner, Biochemistry 1994, 33, 2761.
381. R.M. Wenger, J. France, G. Bovermann, L. Walliser, A. Widmer and H. Widmer, FEBS Lett. 1994, 340, 255.
382. R.A. Engh, T. Dieckmann, W. Bode, E.A. Auerswald, V. Turk, R. Huber and H. Oschkinst, J. Mol. Biol. 1993, 234, 1060.
383. A. Horne, M. North, J.A. Parkinson and I.H. Sadler, Tetrahedron 1993, 49, 5891.
384. I.B. Coutinho, D.L. Turner, J. LeGall and A.V. Xavier, Biochem. J. 1993, 294, 899. M.A. Picarra-Periera, D.L. Turner, J. LeGall and A.V. Xavier, *ibid*, 909.
385. R.A. Gray, D.G. Vander Velde, C.J. Burke, M.C. Manning, C.R. Middaugh and R.T. Borchardt, Biochemistry 1994, 33, 1323.
386. D. Duchesne, M. Naim, P. Nicholas and D. Baron, Biochem. Biophys. Res. Commun. 1993, 195, 630.
387. K. Berndt, P. Guentert and K. Wuethrich, J. Mol. Biol. 1993, 234, 735.
388. M. Naim, S. Charpentier, P. Nicolas and D. Baron, Biopolymers 1993, 33, 1889.
389. G.V. Nikiforovich, O.M. Prakash, C.A. Gehrig and V.J. Hruby, Int. J. Pept. Protein Res. 1993, 41, 347.
390. N. Zimmermann, S. Freund, A. Fredenhagen and G. Jung, Eur. J. Biochem. 1993, 216, 419.

391. D.A. Kallick, J. Amer. Chem. Soc. 1993, 115, 9317.
392. J.F. Eccleston, D.P. Molloy, M.G. Hinds, R.W. King and J. Feeney, Eur. J. Biochem. 1993, 218, 1041.
393. I.Z. Zubrzycki and G. Gaede, Biochem. Biophys. Res. Commun. 1994, 198, 228.
394. R.C. Miller, J.T. Pelton and R.C. Hiley, Methods Neurosci. 1993, 13, 139.
395. M. Coles, S.L.A. Munro and D.J. Craik, J. Med. Chem. 1994, 37, 656.
396. M. Coles, V. Sowemimo, D. Scanlon, S.L.A. Munro and D.J. Craik, J. Med. Chem. 1993, 36, 2658.
397. M.J. Bogusky, S.F. Brady, J.T. Sisko, R.F. Nutt and G.M. Smith, Int. J. Pept. Protein Res. 1993, 42, 194.
398. D.D. Smith, K.G. Boyd, D. Hopton, R.L. Baxter and R. Ramage, J. Chem. Soc. Perkin Trans. 1 1993, 551.
399. L.R. Brown, S. Mronga, R.A. Bradshaw, C. Ortenzi, P. Luporini and K. Wuethrich, J. Mol. Biol. 1993, 231, 800.
400. M.E. Girvin and R.H. Fillingame, Biochemistry 1993, 32, 12167.
401. D. Lassen, C. Luecke, A. Kromminga, A. Lezius, F. Spener and H. Rueterjans, Mol. Cell Biochem. 1993, 123, 15.
402. L. Kar, C.S. Lai, C.E. Wolff, D. Nettesheim, S. Sherman and M.E. Johnson, J. Biol. Chem. 1993, 268, 8580.
403. M. Williams, I. Phan, T.S. Harvey, A. Rostagno, L.I. Gold and I.D. Campbell, J. Mol. Biol. 1994, 235, 1302.
404. T.M. Logan, Y. Theriault and S.W. Fesik, J. Mol. Biol. 1994, 236, 637.
405. H. Senn and W. Klaus, J. Mol. Biol. 1993, 232, 907.
406. M.A. Knauf, F. Loehr, P.G. Curley, P. OFarrell, S.G. Mayhew, F. Mueller and H. Rueterjans, Eur. J. Biochem. 1993, 213, 167.
407. R. Geraghty, D.J.S. Guthrie, G.B. Irvine, C.H. Williams, Biochem. Soc. Trans. 1993, 21, 261S.
408. P.J.M. Folkers, M. Nilges, F.H.A. Folmer, R.N.H. Konings and C.W. Hilbers, J. Mol. Biol. 1994, 236, 229.
409. T.J. Siahan and K. Lutz, J. Pharm. Biomed. Anal. 1994, 12, 65.
410. V. Bobde, S. Beri and S. Durani, Tetrahedron 1993, 49, 5397.
411. H. Baumann, K. Paulsen, H. Kovacs, H. Berglund, A.P.H. Wright, J.A. Gustafsson and T. Haerd, Biochemistry 1993, 32, 13463.
412. K. Thornton and D.G. Gorenstein, Biochemistry 1994, 33, 3532.
413. J.H. Bushweller, M. Billeter, A. Holmgren and K. Wuethrich, J. Mol. Biol. 1994, 235, 1585.
414. C. Rossi, A. Donati, S. Ulgiati and M.R. Sansoni, Can. J. Chem. 1993, 71, 506.
415. R.J. Beinstock, J. Rizo, S.C. Koerber, J.E. Rivier, A.T. Hagler and L.M. Gierasch, J. Med. Chem. 1993, 36, 3265.
416. A.W. Hing and J. Schaefer, Biochemistry 1993, 32, 7593.
417. R.R. Ketchem, W. Hu and T.A. Cross, Science 1993, 261, 1457.
418. W.C. Ripka, G.V. De Lucca, A.C. Bach II, R.S. Pottorf and J.M. Blaney, Tetrahedron 1993, 49, 3593.
419. W.C. Ripka, G.V. De Lucca, A.C. Bach II, R.S. Pottorf and J.M. Blaney, Tetrahedron 1993, 49, 3609.
420. G.W. Vuister, S.J. Kim, C. Wu and A. Bax, Biochemistry 1994, 33, 10.
421. H.R. Kalbitzer and W. Hengstenberg, Eur. J. Biochem. 1993, 216, 205.
422. B.E. Bernstein, R.C. Hoffman, S. Horvath, J.R. Herriott and R.E. Klevit, Biochemistry 1994, 33, 4460.
423. T. Yamazaki, L.K. Nicholson, D.A. Torchia, S.J. Stahl, J.D. Kaufman, P.T. Wingfield, P.J. Domaille and S. Campbell-Burk, Eur. J. Biochem. 1994, 219, 707.
424. P.J. Klasse, R. Pipkorn, J. Blomberg, K-H. Han, B. Hilton and J.A. Ferretti, FEBS Lett. 1993, 323, 68.
425. J.P. Laussac, G. Peyrou, H. Mazarguil, M. Erard, M. Bourdonneau and T,C. Manh, New J. Chem. 1993, 17, 607.
426. C.M. Read, P.D. Cary, C. Crane-Robinson, P.C. Driscoll and D.G. Norman, Nucleic Acids Res. 1993, 21, 3427.
427. C.L. Stevenson, M.E. Donlan, A.R. Friedman and R.T. Borchardt, Int. J. Pept. Protein Res. 1993, 42, 24.
428. H. Kurumizaka, A. Furubayashi, C. Sakuma, U. Matsumoto, A. Yanagida, N. Goshima, Y. Kano and F. Imamoto, Biol. Pharm. Bull. 1993, 16, 437.
429. R.K. Konat, D.F. Mierke, H. Kessler, B. Kutscher, M. Bernd and R. Voegli, Helv. Chim. Acta 1993, 76, 1649.
430. I. Shimada, Kenkyu HoHoku - Asahi Garasu Zaidan 1992, 61 177.
431. T.T. Sakai, M.J. Jablonsky, P.A. DeMuth, N.R. Krishna, M.A. Jarpe and H.M. Johnson, Biochemistry 1993, 32, 5650.

432. T.A. Balashova, V.S. Pashkov, D.E. Nolde, L.V. Onoprienko, I.I. Mikhaleva, L.V. Samokhvalova, G.V. Malakhova and A.S. Arseniev, Bioorg. Khim. 1993, 19 21.
433. R. Powers, D.S. Garrett, C.J. March, E.A. Frieden, A.M. Gronenborn and G.M. Clore, Biochemistry 1993, 32, 6744.
434. C.J. Morton, R.J. Simpson and R.S. Norton, Eur. J. Biochem. 1994, 219, 97.
435. R.T. Clubb, J.G. Omichinski, G.M. Clore and A.M. Gronenborn, FEBS Lett. 1994, 338, 93.
436. A. Sato, S. Nishimura, T. Ohkubo, Y. Kyogoku, S. Koyama, M. Kobayashi, T. Yasuda and Y. Kobayashi, Int. J. Pept. Protein Res. 1993, 41, 433.
437. F.K. Junius, A.S. Weiss and G.F. King, Eur. J. Biochem. 1993, 214, 415.
438. W. Antuch, K.D. Berndt, M.A. Chavez, J. Delfin and K. Wuethrich, Eur. J. Biochem. 1993, 212, 675.
439. C.L. Chyan, C. Wormald, C.M. Dobson, P.A. Evans and J. Baum, Biochemistry 1993, 32, 5681.
440. A. Shimizu, M. Ikeguchi and S. Sugai, Biochemistry 1993, 32, 13198.
441. A.T. Alexandrescu, Y.L. Ng and C.M. Dobson, J. Mol. Biol. 1994, 235, 587.
442. Z. Wang, J.D. Jones, J. Rizo and L.M. Gierasch, Biochemistry 1993, 32, 13991.
443. N.L. Daly, A.R. Atkins and R. Smith, Int. J. Pept. Protein Res. 1993, 42, 366.
444. G.C. Flynn, C.J.M. Beckers, W.A. Baase and F.W. Dahlquist, Proc. Natl. Sci. USA 1993, 90, 10826.
445. H. Van Dael, P. Haezebrouck, L. Morozova, C. Arico-Muendel and C.M. Dobson, Biochemistry 1993, 32, 11886.
446. K. Ardvisson, T. Land, U. Langel, T. Bartfai and A. Ehrenberg, Biochemistry 1993, 32, 7787.
447. B. Bechinger, M. Zasloff and S.J. Opella, Protein Sci. 1993, 2 2077.
448. W.J. Fairbrother, D. Reilly, T. Colby and R. Horuk, FEBS Lett. 1993, 330 302.
449. H. Dugas, M. Laroche, M. Ptak and H. Labbe, Int. J. Pept. Protein Res. 1993, 41, 595.
450. D. Fourmy, F. Dardel and S. Blanquet, J. Mol. Biol. 1993, 231, 1078.
451. M.S. Finnin, D.W. Hoffman, K.N. Kreuzer, S.J. Porter, R.P. Schmidt and S.W. White, J. Mol. Biol. 1993, 232, 301.
452. D.J. Weber, C. Abeygunawardana, M.J. Bessman and A.S. Mildvan, Biochemistry 1993, 32, 13081.
453. J. Tanikawa, T. Yasukawa, M. Enari, K. Ogata, Y. Nishimura, S. Ishii and A. Sarai, Proc. Natl. Sci. USA 1993, 90, 9320.
454. A.H. Myrset, A. Bostad, N. Jamin, P-N. Lirsac, F. Toma and O.S. Gabrielsen, EMBO J. 1993, 12 4625.
455. K. Vyas, K. Rajarathnam, L.P. Yu, S.D. Emerson, G.N. La Mar, R. Krishnamoorthi and H. Mizukami, J. Biol. Chem. 1993, 268, 14826.
456. C.A. Travaglini, F. Cutruzzola, A. Brancaccio, M. Brunori, J. Qin and G.N. La Mar, Biochemistry 1993, 32, 6041.
457. C. Brunel, A. Bondon and G. Simonneaux, Eur. J. Biochem. 1993, 214, 405.
458. J.P. Waltho, V.A. Feher, G. Merutka, H.J. Dyson and P.E. Wright, Biochemistry 1993, 32, 6337 and following papers.
459. J. Qin, G.N. La Mar, F. Cutruzzola, C.T. Allocatelli, A. Brancaccio and M. Brunori, Biophys. J. 1993, 65, 2178.
460. N. Julian, H. Demene, N. Morellet, B. Maigret and B.P. Roques, FEBS Lett. 1993, 331, 43.
461. N. Morllet, H. de Rocquigny, Y. Mely, N. Julian, H. Demene, M. Ottoman, D. Gerard, J.L. Darlix, M.C. Fournie-Zaluski and B.P. Roques, J. Mol. Biol. 1994, 235, 287.
462. G.R. Wilcox, R.H. Fogh and R.S. Norton, J. Biol. Chem. 1993, 268, 24707.
463. R. Husain, A.A. Abu Shanab, D.J.S. Guthrie, G.B. Irvine, P.J. Stevenson and J.M. Allen, Biochem. Soc. Trans. 1992, 20, 628.
464. U. Wollborn, R.M. Brunne, J. Harting, G. Hoelzemann and D. Liebfritz, Int. J. Pept. Protein Res. 1993, 41, 376.
465. J. Horne, M. Sadek and D.J. Craik, Biochemistry 1993, 32, 7406.
466. A.P. Golovanov, A.L. Lomize, A.S. Arseniev, Y.N. Utkin and V.I. Tsetlin, Eur. J. Biochem. 1993, 213, 1213.
467. H.W. van den Hooven, F. Fogolari, H.S. Rollema, R.N.H. Konings, C.W. Hilbers and F.J.M. van de Ven, FEBS Lett. 1993, 319, 189.
468. N. Assa-Munt, R.J. Mortshire-Smith, R. Aurora, W. Herr and P.E. Wright, Cell 1993, 73 193.
469. J. Rizo, F.J. Blanco, B. Kobe, M.D. Bruch and L.M. Gierasch, Biochemistry 1993, 32, 4881.
470. L-C. Chuang, S-T. Chen and C. Ya, Biochim. Biophys. Acta 1993, 1158, 209.
471. T. Kato, S. Endo, T. Fujiwara and K. Nagayama, J. Biomol. NMR 1993, 3, 653.
472. S. Meunier, J.M. Bernassau, J.M. Sabatier, M.F. Martin-Eauclaire, J. Van Rietschoten, C. Cambillau and H. Darbon, Biochemistry 1993, 32, 11969.
473. A.F. Miller, C.J. Halkides and A.G. Redfield, Biochemistry 1993, 32, 7367.

474. P.J. Kraulis, P.J. Domaille, S.L. Campbell-Burk, T. Van Aken and E.D. Laue, Biochemistry 1994, 33, 3515.
475. S.M. Franken, A.J. Scheidig, U. Krengel, H. Rensland, A. Lautwein, M. Geyer, K. Scheffzek, R.S. Goody, H.R. Kalbitzer et.al. Biochemistry 1993, 32, 8411.
476. P. Sevilla-Sierra, G. Otting and K. Wuethrich, J. Mol. Biol. 1994, 235, 1003.
477. L.A. Strickland, R.P. Bozzato and K.A. Kronis, Biochemistry 1993, 32, 6050.
478. J.A. Barden and B.E. Kemp, Biochemistry 1993, 32, 7126; J.A. Barden and R.M. Cuthbertson, Eur. J. Biochem. 1993, 215, 315.
479. V. Wray, T. Federau, W. Gronwald, H. Mayer, D. Schomburg, W. Tegge and E. Wingender, Biochemistry 1994, 33, 1684.
480. G. Saberwal and R. Nagaraj, J. Biol. Chem. 1993, 268, 14081.
481. M.J. Bogusky, A.M. Naylor, M.E. Mertzman, S.M. Pitzenberger, R.F. Nutt, S.F. Brady, C.D. Colton and D.F. Veber, Biopolymers 1993, 33, 1287.
482. A.M. Andrianov and A.A. Akirem, Mol. Biol. (Moscow) 1993, 27, 934.
483. G. Esposito, F. Fogolari, P. Viglino, S. Cattarinussi, M.T. De Magistris, L. Chiappinelli, A. Pessi, Eur. J. Biochem. 1993, 217, 171.
484. M.S.L. Lim, E.R. Johnston and C.A. Kettner, J. Med. Chem. 1993, 36, 1831.
485. T. Ishizu, A. Fujii and S. Noguchi, Chem. Pharm. Bull. 1993, 41, 235.
486. S. Koyama, H. Yu, D.C. Daigarno, T.B. Shin, L.D. Zydowsky and S.L. Schreiber, Cell 1993, 72, 945.
487. T. Maurer and H. Rueterjans, Eur. J. Biochem. 1994, 220, 111.
488. M. Cox, J. Schaller, R. Boelens, R. Kaptein, E. Rickli and M. Llinas, Chem. Phys. Lipids 1994, 67-68, 43.
489. B. Aroeti, P.A. Kosen, I.D. Kuntz, F.E. Cohen and K.E. Mostov, J. Cell Biol. 1993, 123, 1149.
490. M.D. Carr, C.J. Bauer, M.J. Gradwell and J. Feeney, Proc. Natl. Acad. Sci. USA 1994, 91, 2206.
491. K.D. Kopple, J.W. Bean, K.K. Bhandary, J. Briand, C.A. DAmbrosio and C.E. Peishoff, Biopolymers 1993, 33, 1093.
492. J.L. You, S.C.F. Milton, R.C.deL. Milton, N.S. Rangaraju and R.B. Harris, J. Protein Chem. 1993, 12, 133.
493. M.S. Prachand, S. Singh, M.M. Dhingra, U. Singh, S.K. Ghosh, V.R. Mamdapur and M.S. Chadha, Magn. Reson. Chem. 1993, 31, 944.
494. S.J. Archer, V.K. Vinson, T.D. Pollard and D.A. Torchia, Biochemistry 1993, 32, 6680.
495. V.K. Vinson, S.J. Archer, E.E. Lattmann, T.D. Pollard and D.A. Torchia, J. Cell Biol. 1993, 122, 1277.
496. W.J. Metzler, K.L Constantine, M.S. Friedrichs, A.J. Bell, E.G. Ernst, T.B. Lavoie and L. Mueller, Biochemistry 1993, 32, 13818.
497. S. Bagby, T.S. Harvey, L.E. Kay, S.G. Eagle, S. Inouye and M. Ikura, Biochemistry 1994, 33, 2409.
498. G. Mohn, P. Koehl, H. Budzikiewicz aand J.-F. Lefevre, Biochemistry 1994, 33, 2843.
499. C.H. Goerbitz and J. Krane, Acta Chem. Scand. 1993, 47, 979.
500. Y.N. Kalia, S.M. Brocklehurst, D.S. Hipps, E. Appella, K. Sakaguchi and R.N. Perham, J. Mol. Biol. 1993, 230, 323.
501. R.M.A. Knegtel, R. Boelens, M.L. Ganadu, A.V.E. George, M. Katahira, A.M.J.J. Bonvin, D. Eib, P.T. Van Der Saag and R. Kaptein, Ann. N.Y. Acad. Sci. 1993, 684, 49.
502. R.M.A. Knegtel, M. Katahira, J.G. Schilthuis, R. Boelens, D. Eib, P.T. van der Saag and R. Kaptein, NATO ASI Ser., Ser. A 1993, 246 153.
503. M.S. Lee, S.A. Kliewer, J. Provencal, P.E. Wright and R.M. Evans, Science 1993, 260, 1117.
504. M. Rico, J. Santoro, C. Gonzalez, M. Bruix, J.L. Neira and J.L. Nieto, Appl. Magn. Reson. 1993, 4, 385.
505. B.L. Golden, D.W. Hoffman, V. Ramakrishnan and S.W. White, Biochemistry 1993, 32, 12812.
506. D.S. Garrett, P.J. Lodi, Y. Shamoo, K.R. Williams, G.M. Clore and A.M. Gronenborn, Biochemistry 1994, 33, 2852.
507. H. Iwai, Y. Nakajima, S. Natari, Y. Arata and I. Shimada, Eur. J. Biochem. 1993, 217, 639.
508. S. Rebuffat, L. Conraux, M. Massias, C. Auvin-Guette and B. Bodo, Int. J. Pept. Protein Res. 1993, 41, 74.
509. B. Bersch, P. Koehl, Y. Nakatani, G. Ourisson and A. Milon, J. Biomol. NMR 1993, 3, 443.
510. N.G.J. Delaet, P. Verheyden, E.L. Hooghe-Peters, C. Bruns, D. Tourwe and G. Van Binst, Pept. Res. 1993, 6, 24.
511. H. Jaspers, A. Horvath, I. Mezo, G. Keri and G. Van Binst, Int. J. Pept. Protein Res. 1994, 43, 271.
512. Z. Huang, A. Probstl, J.R. Spencer, T. Yamazaki and M. Goodman, Int. J. Pept. Protein Res. 1993, 42, 352.

513. Y.B. He, Z. Huang, K. Raynor, T. Reisine and M. Goodman, J. Amer. Chem. Soc. 1993, 115, 8066.
514. M. Suzuki, M. Gerstein and T. Johnson, Protein Eng. 1993, 6, 565.
515. R.A. Kautz and R.O. Fox, Protein Sci. 1993, 2, 851.
516. A.T. Alexandresco, C. Abeygunawardana and D. Shortle, Biochemistry 1994, 33, 1063.
517. D. Shortle and C. Abeygunawardana, Structure 1993, 1, 121.
518. T. Szyperski, S. Scheek, J. Johansson, G. Assmann, U. Seedorf and K. Wuethrich, FEBS Lett. 1993, 335, 18.
519. P.R. Gooley, B.A. Johnson, A.I. Marcy, G.C. Cuca, S.P. Salowe, W.K. Hagmann, C.K. Esser and J.P. Springer, Biochemistry 1993, 32, 13098.
520. S.R. Van Doren, A.V. Kurochkin, Q.Z. Ye, L.L. Johnson, D.J. Hupe and E.R.P. Zuiderweg, Biochemistry 1993, 32, 13109.
521. T.C. Wong, C.M. Lee, W. Guo and D.K. Chang, Int. J. Pept. Protein Res. 1993, 41, 185.
522. M. Tallon, D. Ron, D. Halle, P. Amodeo, G. Saviano, P.A. Temussi, Z. Selinger, F. Naider and M. Chorer, Biopolymers 1993, 33, 915.
523. S. Lavielle, G. Chassaing, A. Brunissen, M. Rodriguez, J. Martinez, O. Convert, A. Carruette, C. Garret, F. Petitet *et al.* Int. J. Pept. Protein Res. 1993, 42, 270.
524. M. Zhang and T.C. Wong, Biopolymers 1993, 33, 1901.
525. J.F. OConnell, R. Bender, J.W. Engels, K.-P. Koller, M. Scarf and K. Wuethrich, Eur. J. Biochem. 1994, 220, 763.
526. X. Xu and J.W. Nelson, Proteins; Struct. Funct. Genet. 1993, 17, 124.
527. A. Wilde, C. Dempsey and G. Banting, J. Biol. Chem. 1994, 269, 7131.
528. M. Czisch, M. Schleicher, S. Hoerger, W. Voelter, T.A. Holak, Eur. J. Biochem. 1993, 218, 335.
529. C.M. Bladon, P. Bladon and J.A. Parkinson, J. Chem. Soc. Perkin Trans. 1 1993, 1687.
530. B. Gilquin, C. Roumestand, S. Zinn-Justin, A. Menez and F. Toma, Biopolymers 1993, 33, 1659.
531. J. M. Reisman, V.L. Hsu, I. Jariel-Encontre, C. Lecou, M.H. Sayre, D.R. Kearns and J. Parello, Eur. J. Biochem. 1993, 231, 865.
532. P. Viglino, F. Fogolari, S. Formisano, N. Bortolotti, G. Damante, R. Di Lauro and G. Esposito, FEBS Lett. 1993, 336, 397.
533. B. Leiting, R. De Francesco, L. Tomei, R. Cortese, G. Otting and K. Wuethrich, EMBO J. 1993, 12, 1797.
534. X. Qian, S.N. Gozani, H. Yoon, C. Jeon, K. Agarwal and M.A. Weiss, Biochemistry 1993, 32, 9944. X. Qian, C. Jeon, H. Yoon, K. Agarwal and M.A. Weiss, Nature 1993, 365, 277.
535. F.J. Moy, Y.C. Li, P. Rauenbuehler, M.E. Winkler, H.A. Scheraga and G.T. Montelione, Biochemistry 1993, 32, 7334.
536. J.A. Jarvis, S.L.A. Munro and D.J. Craik, Biochemistry 1994, 33, 33.
537. M.H. Li, P. Fan, B. Brodsky and J. Baum, Biochemistry 1993, 32, 7377.
538. J. Poznanski, A. Ejchart, K.L. Wierzchowski and M. Ciurak, Biopolymers 1993, 33, 781.
539. L. Chiche, A. Heitz, A. Padilla, D. Le-Nguyen and B. Castro, Protein Eng. 1993, 6, 675.
540. D. Le Nguyen, A. Heitz, L. Chiche, M. El Hajji and B. Castro, Protein Sci. 1993, 2, 165.
541. E.V. Curto, M.A. Jarpe, E.J. Blalock, D. Borovsky and R.N. Krishna, Biochem. Biophys. Res. Commun. 1993, 193, 688.
542. S. Ramaprasad and C. Compadre, Spectrosc. Lett 1993, 26, 639.
543. F.J. Blanco, M.A. Jimenez, J. Herranz, M. Rico, J. Santoro and J.L. Nieto, J. Amer. Chem. Soc. 1993, 115, 5887.
544. A. Martinez, C. Abeygunawardana, J. Haavik, T. Flatmark and A.S. Mildvan, Biochemistry 1993, 32, 6381.
545. V.S. Ananthanarayanan, A. Saint-Jean, B.V. Cheesman, D.W. Hamilton and A.D. Bain, J. Biomol. Struct. Dyn. 1993, 11 509.
546. B.J. Stockman, A. Euvrard and T.A. Scahill, J. Biomol. NMR 1993, 3, 285.
547. J.P.L. Cox, P.A. Evans, L.C. Packman, D.H. Williams and D.N. Woolfson, J. Mol. Biol. 1993, 234, 483.
548. M. Tessari, M.T. Foffani, S. Mammi and E. Peggion, Biopolymers 1993, 33, 1877.
549. A.P. Hansen, A.M. Petros, R.P. Meadows, D.G. Nettesheim, A.P. Mazar, E.T. Olejniczak, R.X. Xu, T.M. Pederson, J. Henkin and S.W. Fesik, Biochemistry 1994, 33, 4847.
550. R. Bhaskaran, A.I. Arunkumar and C. Yu, Biochim. Biophys. Acta 1994, 1199 115.
551. R. Jayakumar, A.B. Mandel and P.T. Manohuran, J. Chem. Soc. Chem. Commun. 1993, 853.
552. W. Lee, M.J. Jablonsky, D.D. Watt and N.R. Krishna, Biochemistry 1994, 33, 2468.
553. K.I. Nutz, W.M.F. Fabian and H. Sterk, Magn. Reson. Chem. 1993, 31, 481.
554. G. Gupta, G.M. Anantharamaiah, D.R. Scott, J.H. Eldrige and G. Myers, J. Biomol. Struct. Dyn. 1993, 11, 345.
555. R.C. Hoffman, S.J. Horvath and R.E. Klevit, Protein Sci. 1993, 2, 951.

556. M. Billeter, D. Neri, G. Otting, Y.Q. Qian and K. Wuethrich, J. Biomol. NMR 1992, 2, 257.
557. G.W. Vuister, F. Delaglio and A. Bax, J. Biomol. NMR 1993, 3, 67.
558. T.S. Harvey and W.F. van Gunsteren, Tech. Protein Chem.IV 6th 1992, 615.
559. H.M. Sulzbach, P.v.R. Schleyer and H.F. Schaefer III, J. Am. Chem. Soc. 1994, 116, 3967.
560. A. Kumar, T.L. Janes and G.C. Levy, Isr. J. Chem. 1992, 32, 257.
561. A.M.J.J. Bonvin, J. Rullmann, C. Antoon, R.M.J.N. Lamerichs, R. Boelens and R. Kaptein, Proteins: Struct. Funct. Genet. 1993, 15, 385.
562. M. Dellwo, D.M. Schneider and A.J. Wand, J. Magn. Reson. Ser. B 1994, 103, 1.
563. T. Nakai, A. Kidera and H. Nakamura, J. Biomol. NMR 1993, 3, 19.
564. K.L. Li, C.A. Tihal, M. Guo and R.E. Stark, Biochemistry 1993, 32, 9926.
565. R.E. Bruccoleri, Mol. Simul. 1993, 10, 151.
566. R.C. Hoffman, R.X. Xu, R.E. Klevit and J.R. Herriott, J. Magn. Reson., Ser. B 1993, 102, 61.
567. V.N. Maiorov and G.M. Crippen, J. Mol. Biol. 1994, 235, 625.
568. R. Abagyan and M. Totrov, J. Mol. Biol. 1994, 235, 983.
569. M.W. MacArthur and J.M. Thornton, Proteins; Struct., Funct., Genet. 1993, 17, 232.
570. D.F. Mierke, M. Kurz and H. Kessler, J. Am. Chem. Soc. 1994, 116, 1042.
571. R.B. Altman, R. Pachter and O. Jardetzky, Appl. Magn. Reson. 1993, 4, 441.
572. R. Bernstein, A. Schnuchel, C. Cieslar and T.A. Holak, J. Magn. Reson. Ser. B 1993, 102, 116.
573. P.J. Connolly, A.S. Stern and J.C. Hoch, J. Am. Chem. Soc. 1994, 116, 2675.
574. M.G. Clore, M.A. Robien and A.M. Gronenborn, J. Mol. Biol. 1993, 231, 82.
575. R. Powers, G.M. Clore, D.S. Garnett and A.M. Gronenborn, J. Magn. Reson. Ser. B 1993, 101, 325.
576. M. Nilges, Proteins: Struct., Funct., Genet. 1993, 17, 297.
577. A.P. Hinck, E.S. Eberhardt and J.L. Markley,Biochemistry 1993, 32, 11810.
578. A. Pastore, R. De Francesco, M.A.C. Morelli, D. Nalis and R. Cortese, Protein Eng. 1992, 5, 129.
579. J.G. Omichinski, G.M. Clore, O. Schaad, G. Felsenfeld, C. Trainor, E. Appella, S.J. Stahl and A.M. Gronenborn, Science 1993, 261, 438.
580. A.M. Bokman, J. Jimenez-Barbero and M. Llinas, J. Biol. Chem. 1993, 268, 13858.
581. B.O. Smith, K.A. Downing, T.J. Dudgeon, M. Cunningham, P.C. Driscoll and I.D. Campbell, Biochemistry 1994, 33, 2422.
582. A.S. Politou, M. Gautel, M. Pfuhl, S. Labait and A. Pastore, Biochemistry 1994, 33, 4730.
583. H.M. Weir, P.J. Kraulis, C.S. Hill, A.R.C. Raine, E.D. Laue and J.O. Thomas, EMBO J. 1993, 12, 1311.
584. M.J. Scanlon, M.R. Price and S.J.B. Tendler, Int. J. Oncol. 1993, 2, 271.
585. U.K. Nowak, A. Cooper, D. Saunders, R.A.G. Smith and C.M. Dobson, Biochemistry 1994, 33, 2951.
586. D.A. Horita, P.J. Hajduk, P.F. Goetinck and L.E. Lerner, J. Biol. Chem. 1994, 269, 1699.
587. J. Qin, G.N. La Mar, Y. Dou, S.J. Admirael and M. Ikeda-Saito, J. Biol. Chem. 1994, 269, 1083.
588. H. Chao, P.L. Davies, B.D. Sykes and F.D. Sonnichsen, Protein Sci. 1993, 2, 1411.
589. C. Freund, A. Ross, A. Plueckthun and T.A. Holak, Biochemistry 1994, 33, 3296.
590. N.J. Murray and M.P. Williamson, Eur. J. Biochem. 1994, 219, 915.
591. J.S. Hu and A.G. Redfield, Biochemistry 1993, 32, 6763.
592. V. Tsik Aris, E. Detsikas, M. Sakarellos-Daitsiotis, C. Sakarellos, E. Vatzaki, S.J. Tzartos, M. Marraud and M.T. Cung, Biopolymers 1993, 33, 1123.
593. B.J. Lee, H. Sakashita, T. Ohkubo, M. Ikehara, T. Doi, K. Morikawa, Y. Kyogoku, T. Osafune, S. Iwai and E. Ohtsuka, Biochemistry 1994, 33, 57.
594. R.B. Gregory, M. Gangoda, R.K. Galpin and W. Su, Biopolymers 1993, 33, 513.
595. L.K. Nicholson, T. Asakura, M. Demura and T.A. Cross, Biopolymers 1993, 33, 847.
596. M. Sone, H. Yoshimizu, H. Kurosu and I. Ando, J. Mol. Struct. 1994, 317, 111.
597. T.L. South and M.F. Summers, Protein Sci. 1993, 2, 3.
598. H. Tamamura, M. Kuroda, M. Masuda, A. Otaka, S. Funakoshi, H. Nakashima, N. Yamaoto, M. Waki, A. Matsumoto *et al.*, Biochim. Biophys. Acta 1993, 1163, 209.
599. R. de Lorimer, M.A. Moody, B.F. Haynes and L.D. Spicer, Biochemistry 1994, 33, 2055.
600. C.R. Middaugh, J.A. Thomson, C.J. Burke, H. Mach, A.M. Naylor, M.J. Bogusky, J.A. Ryan, S.M. Pitzenberger, H. Je and J.S. Cordingly, Protein Sci. 1993, 2 900.
601. A. Otter, P. Bigler, J.M. Stewart and G. Kotovych, Biopolymers 1993, 33, 769.
602. Z. Pagani, P.M. Paglialunga, I. Torrini, G. Lucente, E. Gavuzzo, F. Mazza, G. Pochetti, M. Paci, M. Sette *et al* Biopolymers 1993, 33, 437.
603. G. Mueller, M. Gurrath, M. Kurz and H. Kessler, Proteins: Struct. Funct. Genet. 1993, 15, 235.
604. K. Burgess, K.-K. Ho, B.M. Pettitt, J. Am. Chem. Soc. 1994, 116, 799.

605. A.L. Lomize, J.L. Flippen-Anderson, C. George and H.I. Mosberg, J. Am. Chem. Soc. 1994, 116, 429.
606. K.E. Kover, D. Jiao, S. Fang and V.J. Hruby, Magn. Reson. Chem. 1993, 31, 1072.
607. J.T. Davis, S. Hirani, C. Bartlett and B.R. Reid, J. Biol. Chem. 1994, 269, 3331.
608. W. Hu, K.C. Lee and T.A. Cross, Biochemistry 1993, 32, 7035.
609. H.X. Zhou, P.C. Pingching, D.E. Wemmer and N.R. Kallenbach, J. Am. Chem. Soc. 1994, 116, 1139.
610. A. Abe, S.O. Kamoto, N. Kimura, K. Tamura, H. Onigawara and J. Watanabe, Acta Polym. 1993, 44, 54.
611. H. Kessler, H. Matter, G. Gemmecker, H.J. Diehl, C. Isernia and S. Mronga, J. Pept. Protein Res. 1994, 43, 47.
612. A. Bharadwaj, M. Singh, K. Bhandary, E.L. Becker and V.S. Chauhan, Pept. Res. 1993, 6, 298.
613. E.A. Gallo and S.H. Gellman, J. Amer. Chem. Soc. 1993, 115, 9774.
614. A. Perczel, M. Hollosi, P. Sandor and G.D. Fasman, Int. J. Pept. Protein Res. 1993, 41, 223.
615. M. Pentano, F. Formaggio, M. Crisma, M.G. Bonora, S. Mammi, E. Peggion, C. Toniolo, W.H.J. Boesten, Q.B. Broxterman *et al*, Macromolecules, 1993, 26, 1980.
616. M. Crisma, G. Valle, G.M. Bonora, C. Toniolo and G. Cavicchioni, Int. J. Pept. Protein Res. 1993, 41, 553.
617. K.E. Kover, D. Jiao, S. Fang and V.J. Hruby, J. Org. Chem. 1994, 59, 991.
618. M. Marraud, V. Dupont, V. Grand, S. Zerkout, A. Lecoq, G. Boussard, J. Vidal, A. Collet and A. Aubry, Biopolymers 1993, 33, 1135.
619. P. Pristovsek, J. Kidric, J. Mavri and D. Hadzi, Biopolymers 1993, 33, 1149.
620. A. Gupta and S.V. Chauhan, Int. J. Pept. Protein Res. 1993, 41, 421.
621. S. Ma and A.F. Spatola, Int. J. Pept. Protein Res. 1993, 41, 204.
622. Y. Inai, T. Ito, T. Hirabayashi and K. Yokota, Biopolymers 1993, 33, 1173.
623. A.H. van Oijen, S. Behrens, D.F. Mierke, H. Kessler, J.H. van Boom and R.M.J. Liskamp, J. Org. Chem. 1993, 58, 3722.
624. T.P. Burkholder, E.W. Huber and G.A. Flynn, Bioorg. Med. Chem. Lett. 1993, 3, 231.
625. V. Dupont, A. Lecoq, J.P. Mangeot, A. Aubry, G. Boussard and M. Marraud, J. Am. Chem. Soc. 1993, 115, 8898.
626. J.A. Raynaud, J.P. Grivet, D. Sy and Y. Trudelle, Biochemistry 1993, 32, 4997.
627. A. Calcagni, M. Kajtar-Peredy, G. Lucente, G. Luisi, F. Pinnen, L. Radics and D. Rossi, Int. J. Pept. Protein Res. 1993, 42, 84.
628. K. Uma, R. Kishore and P. Balaram, Biopolymers 1993, 33, 865.
629. S.W. Fesik, J. Biomol. NMR 1993, 3, 261.
630. K. Wakamatsu, Asahi Garasu Zaidan Josei Kenkyu Seika Hokoku 1993, 67.
631. S.J. Landry, Methods, 1993, 5, 233.
632. G. Lippens, D. Van Belle, S.J. Wodak, K. Hallenga, N.R. Nirmala, P. Hill, K.C. Russell, D.D. Smith and V.J. Hruby, Biochemistry 1993, 32, 9423.
633. S.J. Stradley, J. Rizo and L.M. Gierasch, Biochemistry 1993, 32, 12586.
634. S. Song, D. Vander Velde, C.W. Gunn and R.H. Himes, Biochemistry 1994, 33, 693.
635. M. Zhang and H.J. Vogel, Biochemistry 1994, 33, 1163.
636. L.T. Kakalis and I.M. Armitage, Biochemistry 1994, 33, 1495.
637. E.M. Meiering, M. Bycroft, M.J. Lubienski and A.R. Ferscht, Biochemistry 1993, 32, 10975.
638. A. Wonacott, R. Cooke, F.R. Hayes, M.M. Hann, H. Jhoti, P. McMeekin, A. Mistry, P. Murray-Rust, O.N.P. Singh and M.P. Weir, J. Med. Chem. 1993, 36 3113.
639. C.A. Lepre, J.W. Cheng and J.M. Moore, J. Am. Chem. Soc. 1993, 115, 4929.
640. D. Morikis, R. Bruschweiler and P.E. Wright, J. Am. Chem. Soc. 1993, 115, 6238.
641. D.A. Slatter, B.A. Levine, V.B. Patchall, A.C.L. Wang, M. Reedy and G.M. Strasburg, Biochem. Soc. Trans. 1993, 21, 265S.
642. V.J. Basus, G. Song and E. Hawrot, Biochemistry 1993, 32, 12290.
643. P. Berthault and B. Perly, Supramol. Chem. 1993, 2, 225.
644. M. Wada, M. Sakurai, Y. Inoue, Y. Tamura and Y. Watanabe, J. Am. Chem. Soc. 1994, 116, 1537.
645. T.G. Myers, K.E. Thummel, T.F. Kalhorn and S.D. Nelson, J. Med. Chem. 1994, 37, 860.
646. K.G. Rice, P. Wu, L. Brand and Y.C. Lee, Curr. Opin. Struct. Biol. 1993, 3, 669.
647. I. Carmichael, D.M. Chipman, C.A. Podlasek and A.S. Serianni, J. Amer. Chem. Soc. 1993, 115, 10863.
648. A.J. Duben, M. Hricovini and I. Tvaroska, Carbohydr. Res. 1993, 247, 71.
649. L.M.J. Kroon-Battenburg, J. Kroon, B.R. Leeflang and J.F.G. Vliegenthart, Carbohydr. Res. 1993, 245, 21.
650. L. Poppe, J. Amer. Chem. Soc. 1993, 115, 8421.
651. J.T. Vasquez, E.Q. Morales and J.I. Padron, An. Quin. 1993, 89, 123.

652. S.C. Rao, A.J. Ratcliffe and B. Fraser-Reid, J. Chem. Soc. Perkin Trans. 1 1993, 1207.
653. A. Perczei, E. Kollat, M. Hollosi and G.D. Fasman, Biopolymers 1993, 33, 665.
654. A. Imberty, S. Perez, M. Hricovini, R.N. Shah and J.P. Carver, Int. J. Macromol. 1993, 15, 17.
655. J. Liu, F. Kong and L. Cao, Carbohydr. Res. 1993, 240, 295.
656. D.C. Lankin, S.T. Nugent and S.N. Rao, Carbohydr. Res. 1993, 244, 49.
657. S. Brunel, C. Ibarra, E. Moraga, A. Blasko and C.A. Bunton, Carbohydr. Res. 1993, 244, 1.
658. P. Fernandez and J. Jimenez-Barbero, Carbohydr. Res. 1993, 248, 15.
659. G.I. Eneva, S.L. Spassor and M.A. Haimova, J. Mol. Struct. 1993, 291, 191.
660. C. Meyer, S. Perez, C. Herve du Penhoat and V. Michon, J. Am. Chem. Soc. 1993, 115, 10300.
661. J.M. Duker and A.S. Serianni, Carbohydr. Res. 1993, 249, 281.
662. R.J. Voll, F.R. Fronczek, D. Vargas and E.S. Younathan, Carbohydr. Res. 1992, 230, 213.
663. C.L. Perrin and K.B. Armstrong, J. Amer. Chem. Soc. 1993, 115, 6825.
664. D. Keglevic, B. Kojic-Prodic, Z. Banic, S. Tomic and V. Puntarec, Carbohydr. Res. 1993, 241, 131.
665. V. Pozsgay, B. Coxon and H. Yeh, Bioorg. Med. Chem. 1993, 1, 237.
666. M. Elofsson, S. Roy, B. Walse and J. Kihlberg, Carbohydr. Res. 1993, 246, 89.
667. S. Hanessian, H. Hori, Y. Tu and Y. Boulanger, Tetrahedron 19934, 50, 77.
668. D.R. Bundle, H. Baumann, J.-R. Brisson, S.M. Gagne, A. Zdanov and C. Miroslaw, Biochemistry 1994, 33, 5183.
669. C. Mukhopadhyay, K.E. Miller and C.A. Bush, Biopolymers 1994, 34, 21.
670. H. Baumann, J.R. Brisson, F. Michon, R. Pon and H.J. Jennings, Biochemistry 1993, 32, 4007.
671. B. Molloy, M.J. Forster and C. Jones, Biochem. J. 1993, 293, 849.
672. M. Zsiska and B. Meyer, Carbohyr. Res. 1993, 243, 225.
673. J. Liu, A.L. Waterhouse and N.J. Chatterton, Carbohydr. Res. 1993, 245,
674. A. Helander, L. Kenne, S. Oscarson, T. Peters and J.R. Brisson, Carbohydr. Res. 1992, 230, 299.
675. U. Spohr, M. Bach And R.G. Spiro, Can. J. Chem. 1993, 71 1928.
676. A. Wei and Y. Kishi, J. Org. Chem. 1994, 59, 88.
677. K. Linek, J. Alfoldi and M. Durindova, Chem. Pap. 1993, 47, 247.
678. V. Durier, K. Mazeau, C. Gey, H. Driguez and F.R. Taravel, New J. Chem. 1993, 17, 843.
679. V. Durier, G. Buisson, E. Duee, H. Driguez and F.R. Taravel, J. Chem. Soc. Perkin Trans. 2 1993, 1931.
680. T. Weimar, B. Meyer and T. Peters, J. Biomol. NMR 1993, 3, 399.
681. J.H. Davis, Cholesterol Membr. Models 1993, 67-135.
682. K.S. Hamilton, H.C. Jarrell, K.M. Briere and C.W.M. Grant, Biochemistry 1993, 32, 4022.
683. J.M. Leenhouts, V. Chupin, J. de Gier and B. de Kruijff, Biochim. Biophys. Acta 1993, 1153, 257.
684. W. Scnepp and C. Schmidt, Ber. Bunsen-Ges. Phys. Chem. 1993, 97, 1399.
685. D.B. Fenske and P.R. Cullis, Biophys. J. 1993, 64, 1482.
686. R.L. Thurmond, D. Otten, M.F. Brown and K. Beyer, J. Phys. Chem. 1994, 98, 972.
687. P.L. Yeagle, A.R. Dentino, F.T. Smith, P. Spooner and A. Watts, Biochemistry 1993, 32, 12197.
688. C.R. Sanders II, J.E. Schaff and J.H. Prestegard, Biophys. J. 1993, 64, 1069.
689. S. Urano, M. Matsuo, T. Sakanaka, I. Uemura, M. Koyama, I. Kumadaki and K. Fukuzawa, Arch. Biochem. Biophys. 1993, 303, 10.
690. T.O. Matsunaga, N. Collins, V. Ramaswami, S.H. Yamamura, D.F. OBrien and V.J. Hruby, Biochemistry 1993, 32, 13180.
691. H.G. Ang, C.H. Koh, L.L. Koh and W.L. Kwik, J. Organomet. Chem. 1993, 452, 181.
692. O.J. Curnow, R.P. Hughes, E.N. Mairs and A.L. Rheingold, Organometallics 1993, 12, 3102.
693. J. Kuemmerton, I. Lange, W. Milius, A. Sebald and A. Blaschette, Organometallics 1993, 12, 3541.
694. R.P. Hughes, D.S. Tucker and A.L. Rheingold, Organometallics 1993, 12, 3069.
695. Y.T. Hsiao, P.L. Kuo, H.C. Lai, H.C. Cheng, C.Y. Cheng and S.L. Wang, Organometallics 1993, 12, 1094.
696. I. Toth and B.E. Hanson, Organometallics 1993, 12, 1506.
697. R. Ditchfield, R.P. Hughes, D.S. Tucker, E.P. Bierwagen, J. Robbins, D.J. Robinson and J.A. Zakutansky, Organometallics 1993, 12, 2258.
698. L. Girard and M.C. Baird, J. Organomet. Chem. 1993, 444, 143.
699. C.H. Lai and C.H. Cheng, Organometallics 1993, 12, 3561.
700. Y. Takeuchi, I. Zicmane, G. Manuel and R. Boukherroub, Bull. Chem. Soc. Jpn. 1993, 66, 1732.
701. P. Szczecinski and J. Zachara, J. Organomet. Chem. 1993, 447, 241.
702. A. Blasko, B. Garcia and T.C. Bruice, J. Org. Chem. 1993, 58, 5738.

703. G. Erker, M. Aulbach, M. Knickmeier, D. Wingbermuehle, C. Kreuger, M. Nolte and S. Werner, J. Amer. Chem. Soc. 1993, 115, 4590.
704. C.H. Lai, C.H. Cheng, W.C. Chou and S.L. Wang, Organometallics 1993, 12, 3418.
705. C. Mallet, R.P. Thummel and C. Hery, Inorg. Chim. Acta 1993, 210, 223.
706. J.H. Cameron and E.L. Scott, J. Chem. Soc. Dalton Trans. 1993, 397.
707. S. Jurga, K. Jurga, E.C. Reynhardt and P. Katowski, Z. Naturforsch., A: Phys. Sci. 1993, 48, 563.
708. B.V.S. Murthy, K.P. Ramesh and J. Ramakrishna, Ber. Bunsen-Ges. Phys. Chem. 1993, 97, 603.
709. L.G. Marzilli, M. Iwamoto, E. Alessio, L. Hansen and M. Calligaris, J. Am. Chem. Soc. 1994, 116, 815.
710. S. Wache and R. Taube, J. Organomet. Chem. 1993, 456, 137.
711. M. Baudler and B. Wingert, Z. Anorg. Allg. Chem. 1993, 619, 1977.
712. Y. Yoshida, T. Adachi, K. Matsumura and K. Baba, Angew. Chem. 1993, 105, 1730.
713. M. Fujiwara, Y. Ichiki and H. Wakita, Polyhedron 1993, 12, 2989.
714. P. Szczecinski, J. Organomet. Chem. 1993, 458, 119.
715. R.K. Gilpin, J. Chromatogr. 1993, 656, 217.

12
Nuclear Magnetic Resonance Spectroscopy of Living Systems

BY M. J. W. PRIOR

1 General reviews and new methodologies

1.1. General applications - A reveiw, with 20 references, on human disease diagnosis by ^{31}P-NMR spectroscopy **in vivo** has been produced.[1] A review of the application of NMR spectroscopy to clinical and metabolic studies has been produced.[2] A review of the use of NMR spectroscopy in the study of pathological states has been produced.[3]

The ability of ^{1}H NMR spectroscopy to distinguish different neural cell types has been reviewed.[4] A review with 3 references on the magnetic resonance spectroscopic examination of brain has been produced and the effects of data collection have been discussed.[5] A review with 63 references on the application of NMR spectroscopy to the study of brain biochemistry and pathology has been produced.[6]

The study of bioeneretics **in vivo**, with emphasis on the use of ^{31}P NMR to examine the bioenergetics of muscle, has been reviewed with 92 references.[7] The study of tissues by NMR spectroscopy for the examination of abnormal metabolism (in cancer) has been reviewed with 15 references.[8] The potential and limitations of ^{31}P NMR spectroscopy **in vivo** has been reviewed with 139 references.[9]

A chapter on **in vivo** spectroscopy has been published.[10] A review with 130 references of ^{19}F NMR spectroscopy **in vivo** has been poduced.[11] A review, with 101 references, of techniques, probes, drugs and drug metabolites used in ^{19}F NMR spectroscopy has been produced.[12] The most common nuclei used for biological and medical applications of NMR spectroscopy have been reviewed with 26 references.[13]

1.2 Metabolite editing, quantitation and localisation techniques - The practical applications of localisation techniques in NMR spectroscopy has been breifly reviewed and the general artifacts associated with the techniques, particularly chemical shift imaging, has been discussed.[14] A comparison of the methods of chemical shift imaging has been made using 24 references. The techniques based around spin-echo Fourier imaging have been classified and the performance of each technique has been compared as a guide to select the most appropriate technique.[15] A discussion of the technique of depth resolved surface coil spectroscopy (DRESS) has been produced which covers the history, problems and improvements of the technique.[16] The use of the proton localisation technique (SPACE) has been described.[17] The application of cyclic and slice selective J cross polarisation for heteronuclear editing and localisation has been investigated and the application of the techniques in combination for NMR spectroscopy and imaging **in vivo** has been discussed.[18]

A review of metabolite imaging in the study of tumour metabolism has been produced.[19] The use of multiple quantum techniques for spectral editing **in vivo** has been discussed in 7 chapters.[20] A method has been reported for ^{13}C-editing of localised ^{1}H spectra obtained **in vivo** without gradient coils[21] which exploits differences of coil inhomogeneity for localisation and editing. The effects of spin-spin coupling upon the flip angle in a lactate editing sequence has been described and illustrated with lactate spectra from rat brain and pig heart.[22] The potential use of ^{13}C NMR to measure the metabolic flux through the pentose phosphate pathway in human studies has been described and illustrated in a cell free system. The ratios of glutamine, alanine or lactate labelling from [^{13}C-2]-glucose is use as the indicator.[23] Discrimination of lactate and N-acetyl aspartate has been achieved in a spectroscopic imaging sequence used on the human brain.[24] Several editing techniques for the detection of g-aminobutyric acid by in vivo NMR spectroscopy have been reviewed with respect to their application in human brain studies.[25] A review, with 35 references, of the significant contributions to the development of methods for metabolite editing, efficiency of water suppression and editing of lipid resonances by double quantum coherence transfer spectroscopy has been produced.[26]

A method for solvent suppression using sequences of shaped pulses has been developed by analysis of the derivatives of $M_+(\Delta_\omega)$, where M_+ is the transverse magnetisation and Δ_ω is the resonant offset frequency. The sequence has been used for examination of deoxyhaemoglobin inside erythrocytes.[27] The use of a non-linear least-squares method for the analysis of spectral information from chemical shift imaging data has been advocated.[28] A method for the removal of residual water signals from spectra after data acquisition has been described.[29]

The measurement of metabolic fluxes using several magnetisation transfer techniques has been examined. The methods have been described with the problems of practical application and the interpretation of the experimental data has been discussed.[30] The potential of the application of two-dimensional ^{31}P-^{1}H correlation spectroscopy to the study of ^{31}P-metabolites has been discussed.[31] Inaccuracies which may arise in the mesasurement of T2 relaxation of ATP resonances by the Hahn 90°-180° spin echo experiment has been examined and shown to be the result of J-coupling.[32]

Simple methods for reducing signal attenuation due to motion during spin-echo experiments have been described which are based on the coaddition of spectral and not FID information.[33] A method for localised shimming using gradients and noise pulses has been reported.[34] The quantitation of ^{14}N- labelled metabolites in isolated amphibian muscle and other tissues has been described.[35]

1.3 Intracellular Ions and Indicators - A review with 40 references has been produced on the measurement of intracellular lithium by NMR spectroscopy.[36] The measurement of cation fluxes in cells and organs using NMR spectroscopy of ^{87}Rb has been reviewed with 99 references and the physics of ^{87}Rb-NMR spectroscopy has been examined theoretically.[37] The use of Lanthanide complexes of aminophosphonates as shift reagents for ^{7}Li and ^{23}Na in biological systems has been described.[38]

The simultaneous determination of pH and intracellular magnesium concentrations from the three 31P-NMR resonances of ATP has been described and discussed. The results show that the determination of pH in healthy tissue may be more accurate by this method compared to conventional ^{31}P NMR measurements of pH. Furthermore, the results suggest that previous ^{31}P NMR measurements of free intracellular Mg^{2+} may have been overestimated.[39] A comparison of the measurement of pH by ^{31}P NMR and a dye-based fibre-optic method has been performed. Results for the two techniques were close for extracellular measurements.[40] The use of imidazol-1-ylalkanoate esters as extrinsic indicators of intracellular pH in ^{1}H NMR spectra has been described.[41]

A new fluorinated indicator of intracellular ion concentrations (FCrown-1) has been produced which is selective for the determination of Na^{+} over K^{+}, Ca^{2+} and Mg^{2+} ions.[42] The presence of one fluorine resonance for the free indicator and another for the bound enables the determination of intracellular concentrations of Na^{+}. The modification of the existing fluorinated intracellular Ca^{2+} indicators 4FBAPTA and 5FBAPTA, to produce trifluoromethyl substituted versions which have a greater NMR sensitivity and higher Ca^{2+} affinity, has been performed.[43]

The simultaneous measurement of oxygen tension and temperature in tissues perfused with a perfluorocarbon has been described. The technique involves the measurement of the relaxation of two perfluorocarbon resonaces to unambiguously determine temperature and oxygen tension.[44]

2 Cells

2.1 Plants - The synthesis of malate and citrate has been investigated with ^{13}C NMR following the administration of $KH^{13}CO_2$ to sycamore cells (**Acer pseudoplantanus**) in a perfusion system. Discrimination of vacuolar from cytoplasmic metabolites was achieved in the presence of Mn^{2+} which resulted in the loss of signals from the vacuole. Uniform labelling of C-1 and C-4 malate implied randomizing of the label by mitochondrial fumarase. Similarly citrate was equally labelled in positions C-1, C-5 and C-6. Malate appeared in the cytoplasm first and then in the vacuole. Citrate accumulated steadily in the vacuole.[45] In a study of the effects of calcium deficiency on hypocotyl cells from **Cucurbita pepo** ^{31}P NMR spectroscopy was used to measure cytoplasmic and vacuolar pH and phosphorous energy metabolism. Calcium deficient cells had a lower vacuolar pH when exposed to an acidic environment. Nucleotide phosphates were present in normal and calcium deficient cells, though the amount of visible phosphorous metabolites was lower in the deficient cells. However, the calcium deficient cells showed a higher level of phosphorylcholine compared to control cells.[46] The effects of salinity have also been investigated in cultured carrot cells grown in suspension in the presence or absence of 150mM NaCl. Cell growth in 150mM NaCl had no effect on the cytoplasmic pH detected by ^{31}P NMR spectroscopy or a number of other techniques used to determine pH. However, the vacuolar pH of cells grown in the presence of 150mM NaCl was higher than the vacuolar pH of cells grown in the absence of 150mM NaCl. The difference was reported to be 0.38 to 1.05 pH units depending on the method used to determine pH.[47]

2.2 Mammalian - In experiments on isolated hepatocytes it has been shown that if rats are fasted prior to hepatocyte preparation increases the susceptability of the cells to anoxia. The concentration of ATP was reduced to 52% in hepatocytes from fasted rats.[48]

The redistribution of ^{15}N label from L-[2-^{15}N]glutamine, L-[5-^{15}N]glutamine and $^{15}NH_4Cl$ has been followed in HeLa and Chinese Hamster Ovary cells via the observation of ^{1}H resonances. Labelling patterns from L-[2-^{15}N]glutamine indicate that transamination occured to metabolites such as alanine, aspartate, glycine and pyrrolidone-5-carboxylic acid and no oxidative deamination occured to produce $^{15}NH_3$. The incubation of cells with L-[5-^{15}N]glutamine showed that the ammonia in the cultures was mostly derived from this amide. The production of L-[5-^{15}N]glutamine from $^{15}NH_4Cl$, was only observable in CHO cells which have a high activity of glutamine synthetase.[49]

Stimmulation of 3T3 cells by forskolin lead to an increase in intracellular cAMP and the export of cAMP into the medium surrounding the cells embedded in agar threads. The simmultaneous detection of the increase in cAMP, with a >50% decrease in phosphocreatine and ATP was carried out with ^{31}P NMR spectroscopy. Reperfusion with drug free media resulted in a restoration of ^{31}P metabolism.[50] The synthesis of lactate following allergen-induced histamine production in mast cells from rats infected with **Nippostrongylus brasiliensis** has been followed with ^{1}H NMR spectroscopy.[51]

2.3 Erythrocytes - A review has been produced which examines the measurement by NMR spectroscopy of anion transport across erythrocyte membranes by band 3. The various types of experiment have been described in which NMR parameters inside and outside the membrane are made to differ to enable transport to be measured.[52] Chlorpromazine has been shown, using ^{31}P NMR spectroscopy, to inhibit the transort of phosphate across the human erythrocyte membrane. The quenching of the fluorescence of eosinyl-5-maleimide-labelled band 3 by chlorpromazine suggest that this protein is responsible for transport .[53]

^{31}P NMR spectroscopy has been used to detect free intracellular Mg^{2+} in a study of hypomagnesaesic patients and normal volunteers. Results indicated that ^{31}P NMR spectroscopy detected Mg^{2+} may be used to reflect intracellular Mg^{2+} status.[54] The role of cAMP in the regulation of free intracellular Mg^{2+} has been investigated using ^{31}P NMR spectroscopy.[55] Exposure of erythrocytes to (-)-Isoproterenol or forskolin, for example, decreased free intracellular Mg^{2+} whereas HA1004, an inhibitor of cAMP-dependent kinases increased free intracellular Mg^{2+}. The results strongly suggested that b-adrenergic stimmulation and cAMP play an important role in regulation of free intracellular Mg^{2+}. ^{19}F and ^{31}P NMR spectroscopy have been used in an investigation of the elevated intracellular Ca^{2+} and suppression of free intracellular Mg^{2+} and pH in erythrocytes associated with hyperinsulinaemia, obesity and non-insulin dependent diabetes mellitus. Results indicated that insulin in physiological concentrations may participate in the regulation of divalent cations in the mature human erythrocyte, but that insulin per se cannot account for these ionic lesions in obesity and diabetes.[56]

The shift difference between ^{19}F-labelled anions which are intracellular (shifted downfield) and extracellular ^{19}F-labelled anions (shifted upfield) has been shown to be increased by extracellular protein (Xu, 1993). The chemical shift of dimethyl methylphosphonate in suspensions erythrocytes has been shown to be different for intracellular and extracellular compartments. The intensity of the down-field shift has been shown to be dependant on protein concenration and has been used to estimate cell volume in erythrocytes and consequently the concentration of intracellular ^{31}P-metabolites in cell suspensions.[58]

In a study of ^{23}Na double quantum filtered (DQF) spectra the shape and phase of the DQF intracellular Na^{+} signal from human erythrocytes was shown to vary with preparation time, in contrast to results from studies **in vitro**. Further investigation showed that the ratio of contributions of DQF to single quantum signal intensities did not vary over a 4-fold increase in intracellular Na^{+}, which indicates the efficiency of DQF signal generation would be unaffected by physiological conditions.[59] The two components T_{2f} and T_{2s} of the transverse magnetisation decay curve of ^{23}Na has been measured in erythrocytes in triple quantum filtered spectra.[60] Results indicated some contribution to the process by complexes of haemoglobin and 2,3-diphosphoglycerate. The measurement of the relaxation of ^{39}K in erythrocytes indicated a similar binding to that of ^{23}Na. The interaction of sodium with ordered structures in mammalian red blood cells has been investigated with ^{23}Na double quantum filtered NMR spectroscopy. Pulse sequences were implimented which enabled the exclusive detection of the contribution of the 2nd rank tensor (T_{21}) to the spectrum. The formation of T_{21} is related to the anisotropic motion of ^{23}Na and in erythrocytes was shown to be related to the integrity of the cytoskeleton network.[61] Lithium relaxation has been studied by ^{7}Li NMR in erythrocytes and solutions containing glycerol, spectrin, 2,3-bisphosphoglycerate, carbonmonoxy haemoglobin or ATP. The results indicated that the inner erythrocyte membrane was the major site of lithium binding.[62]

The rate of Na^+ transport across the erythrocyte membrane, driven by the $Na^+/K^+/Cl^-$ cotransporter, has been investigated using ^{23}Na NMR spectroscopy. Results indicated that transport was increased in elderly hypotensive patients, in contrast to results reported elsewhere.[63] The effects of hyperglycaemia and hyperinsulinaemia on intracellular Na^+ in erythrocytes from normal volunteers has been studied. The increase in intracellular Na^+ measured by ^{23}Na NMR spectroscopy during hyperglycaemia was not as high as the accumulation during hyperinsulinaemia.[64] In an investigation of erythrocytes from patients with myotonic muscular distrophy a decrease in water permeability and a decrease in Na^+ pumping was detected when compared to controls. Furthermore, an increase in glycolysis was detected, though lactate production remained the same in erythrocytes from patients with myotonic muscular distrophy.[65]

In an investigation of the affects of temperature on phosphorous metabolites, pH and free magnesium in erythrocytes, cells from the ground squirrel and human volunteers were compared. The phosphorous signal intensities from the cold-tolerant, ground squirrel were increased with lower temperature compared to the cold intolerant erythrocytes from humans. It was suggested that an increase in free intracellular Mg^{2+} (a relaxation agent) was responsible. Measurements of free intracellular Mg^{2+}, from β-ATP chemical shifts showed that free intracellular Mg^{2+} increase in the squirrel erthrocyte and decreased in the human erythrocyte with decreasing temperature. There was no change in pH detected by ^{31}P NMR spectroscopy in either species.[66] However, the temperature dependence of intracellular pH in erythrocytes has been measured with ^{31}P NMR and was consistent with that predicted by the alphastat hypothesis.[67]

The metabolism of [2-^{13}C]glucose in human erythrocytes has been followed by ^{13}C NMR spectroscopy. The flux though the pentose phosphate pathway, phosphofructokinase, the 2,3-diphosphoglycerate bipass and phosphoglycerate kinase have all been followed in a single experiment. Further evidence has been presented from experiments with [1-^{13}C], [2-^{13}C] and [3-^{13}C] glucose for a reverse flux of fructose-6-phosphate through transketolase and transaldolase.[68]

A review with 38 references on nuclear magnetic resonance studies of amine and nucleotide storage mechanism in platelet dense granules has been produced.[69] The metabolic changes which occur in human peripheral blood lymphocytes post stimmulation has been studied. ^{13}C-glucose utilisation and ^{13}C-lactate production indicated that 85% of utilised glucose was converted to lactate, but analysis of perfusing media indicated that lactate was also produced from substrates other than glucose and glycogen. Glucose accounted for 25% of lactate produced by quiesent cells and 67% of lactate produced by stimmulated cells. Further results indicated that there is a shift from mostly aerobic to mostly anaerobic latate production in the first 90 minutes after stimmulation, though glycolysis was increased 15-fold by 48h post stimmulation.[70]

2.4 Reproductive - The maturation of spermatazoa from different regions of the epididimal tract of the goat have been characterised. ^{31}P NMR spectroscopy was used to monitor intracellular pH and high energy phosphates and ^{13}C NMR spectroscopy was used to follow the production of ^{13}C-labelled metabolites.[71] The effects of ethanol on the energy metabolism of Sertoli cells isolated from the testes of 18-21 day old rats has been monitored by ^{31}P NMR. Cells were immobiliseed in agarose threads. Ethanol treatment decreased the β-ATP resonance to 50% and pre-treatment of the cells with 4-methylpyrazole (an inhibitor of ethanol metabolism) enhanced the reduction of β-ATP to 75% of control values. The results suggested that ethanol was toxic to Sertoli cells but this was not due to ethanol metabolism.[72]

2.5 Tumour - In an investigation of the effects of ascorbic acid on the growth of cancer cells ^{31}P NMR spectroscopy was used to monitor the energy status of vitamin C sensitive and resistant cell lines. In the sensitive cells the effects of ascorbic acid on energy metabolism could be seen by ^{31}P NMR spectroscopy before the cells became non-viable.[73] Apoptotic cell death, induced by dexamethazone, has been investigated in one dexamethazone sensitive and three resistant cell lines. The effects of treatment on the membrane potential was measured by the distribution of trifluoroacetate or trifluoroacetamide across the cell membrane using ^{19}F NMR spectroscopy. Extracellular signals were distinguished from intracellular signals by virtue of the difference in chemical shift. The results showed that the sensitive cells had a much reduced membrane potential whereas dexamethazone resistant cells were less affected or had an increased membrane potential.[74] The inhibitory effects of bioflavinoids on CEM-C7 cells has been investigated using ^{31}P NMR to monitor high energy phosphates and ^{13}C NMR to monitor glucose metabolism. Incubation of cells with 50mM of either 3',4',5'7-tetrahydroxy-flavone 4'4-dihydroxychalcone for 24 hours completely depleted cellular ATP and inhibited ^{13}C-glucose uptake.[75] The effects of 1,25-dihydroxyvitamin D-3 (1,25-$(OH)_2D_3$) on the intracellular calcium (Ca_i^{2+}) levels in oesteoblastic oesteosarcoma cells (ROS 17/2.8) has been measured with ^{19}F NMR spectroscopy and the Ca_i^{2+} indicator 5FBAPTA. The dose response to 1,25-$(OH)_2D_3$ was an inverted U, with a maximal response from 10nM 1,25-$(OH)_2D_3$ which elevated Ca_i^{2+} from 118nM to 237nM. Treatment of cells with Pb^{2+} increased Ca_i^{2+} but reduced the response to 1,25-$(OH)_2D_3$. Further measurements showed that Pb^{2+} pre-treatment interferes with $^{45}Ca^{2+}$ uptake.[76] The response of melanoma cells to melanocyte stimmulating hormone has been measured with ^{31}P NMR spectroscopy.[77]

^{31}P NMR spectroscopy has been used to monitor human neuroblastoma cells grown in a hollow fibre reactor. Administration of 2-deoxyglucose over a 14-16 hour period lead to the increase of the phosphomonoester peak and inorganic phosphate peak and the depletion of β-ATP.[78] The incorporation of perfluorotripropylamine (FTPA) into calcium alginate beads containing EMT6/Ro or RIF-1 murine tumour cells has been used to calculate the rate of oxygen consumption of immobilised cells. Incoporation of FTPA into gel beads had no effect on the energy metabolism of the cells monitored by ^{31}P NMR spectroscopy. The linear relationship between $1/T_1$ of the trifluoromethyl group of FTPA and oxygen tension was used to calculate the oxygen concentration within the beads. The calculation of the rate of oxygen consumption was based on the assumption that oxygen is the limiting nutrient for cell metabolism.[79]

The incubation of L929 fibroblast and RMD4 thymoma cells with antibody-targeted liposomes containing a ^{19}F-labelled compound resulted in the labelling of cells within minutes. It was suggested that this may be useful in the study the kinetics of internalisation of liposomes.[80]

Examination of several breast cancer cell types by ^{1}H NMR spectroscopy has shown differences in the lipid spectra of whole cells which have been confirmed in two dimensional correlation spectroscopy (2D COSY) spectra of membrane enriched fractions.[81] Signals from neutral lipids have been detected using ^{1}H NMR spectroscopy in melanoma cells grown in the presence of oleate. Lipid signals correlated with the presence of cytoplasmic lipid droplets observed with electron microscopy.[82]

The detection of intracellular ^{23}Na in Ehrlich asctes tumour cells has been performed using dysprosium(III) bis(tripolyphosphate) to shift extracellular ^{23}Na signals. Comparison with a related, drug-resistant strain (EHR2/DNR+) revealed a slightly higher concentration of intracellular ^{23}Na in the EHR2/DNR+ cells. The ^{23}Na NMR spectroscopy results for intracellular and extracellular Na^+ concentrations were confirmed with flame photometry .[83] The chemical shift of some boron containing compounds has been shown to be altered in the presence cancer cells compared to their chemical shift in the absence of these cells or in the presence of cells disrupted by sonication.[84]

2.6 Prokaryote cells - The rate of ATP synthesis and consumption has been measured in the green algae **Chlamydomonas reinhartii** using the ^{31}P NMR spectroscopy saturation transfer method. Evidence from the labelling of inorganic phosphate in these experiments suggests that the signals observed form this compound originate from the chloroplast.[85]

The metabolism of [2-^{13}C] glucose and [1-^{13}C] xylose has been studied in a newly isolated strain of **Klebsiella planticola.**[86] The transportation of xylose across the cell membrane of **Zymomonas mobilis** has been followed using ^{13}C NMR spectroscopy in conjunction with the extracellular shift reagent dysprosium diethylenetriaminepentaacetic acid.[87] The effects of glucose metabolism on the hydrolysis of arginine in **Micoplasma fermentans** has been studied by ^{13}C NMR spectroscopy in a continuous perfusion system. The results show the formation of lactate from glucose and the formation of citrulline from arginine. The deimination of arginine to citrulline was unaffected by the presence of glucose.[88]

The observation of **Escherichia coli** by ^{31}P NMR spectroscopy under anaerobic and aerobic conditions at cell densities above 25g dm^{-3} has been performed.[89] Nucleotide triphosphate levels and pH were lower under anaerobic conditions. Furthermore, cells were observed to maintain a transmembrane pH gradient (DpH) with the addition of extracellular acetate up to 10g dm^{-3}, but the

DpH collapsed when extracellular acetate was 20g dm^{-3}. In a study of **Zymomonas mobilis** by NMR spectroscopy **in vivo** and in perchloric acid extracts, ^{31}P NMR spectra of whole cells showed a slower rate of glucose metabolism and decreased levels nucleotide triphosphates in the presence of 10% ethanol.[90] The marine microorganisms of the genus **Alteromonas** have been investigated using ^{31}P NMR spectroscopy. Differences in the ^{31}P-containing compounds were detected in different species of the genus. The ^{31}P NMR spectra of **Alteromonas** was compared to that of other taxonomic groups.[91] Changes in the ^{31}P NMR spectra and results from the assay of extracts during glucose fermentation in **Brochothrix thermosphacta** indicated a strong correlation between the reduced pH, high nucleotide phosphate level and apparent reduction in pyruvate kinase activity. The evidence supports the hypothesis that the synergistic effects of pH and ATP on pyruvate kinase is an important regulatory factor in these cells.[92] The measurement of intracellular pH in **Propionibacterium acnes** using ^{31}P NMR spectroscopy has been compared to pH measurements with a fluorescent dye.[93] Both methods were in good agreement, and the bacteria maintained a pH gradient over a range of extracellular pH values of 5.0 - 7.4.

Measurements of the relaxation constants T_1, T_{2f} and T_{2s} of ^{23}Na in the halophilic bacterium Ba_1 isolated from the dead sea have been performed. Evidence from the T_{2f} and T_{2s} measurements suggest that the microenviroment in the cytosol remained constant when the bacteria were grown in two different media concentrations.[94]

2.7 Yeasts - A study of **Aspergillus terreus** has been reported using ^{31}P NMR spectroscopy in a 10mm diameter air lift reactor designed for mycelial/pellet forming organisms grown in suspension.[95] Signals from extracellular and intracellular inorganic phosphate, glycerol-3-phosphorylethanolamine, glycerol-3-phosphorylcholine, sugar phosphates and polyphosphates were observed. The intracellular pH was reported to be 6.2 and the polyphosphate signal disappeared when the respiratory gas was exchanged for N_2. ^{31}P NMR spectroscopy has been used to make a comparison between the metabolism of **Saccharomyces bayanus** grown in suspension culture or grown immobilsed in a gel matrix.[96] Though no differences in the concentration of phosphorous-containing compounds were detected, the pH of the immobilised cells remained constant whilst the pH of the suspended cells fell during the incubation period. The cells grown in suspension also produced a higher biomass and larger amounts of ethanol compared to those cells immobilised in agar.

The influence of glucose on the metabolism of [1-^{13}C] 2-deoxy-D-glucose has been followed in **Saccharomyces cerevisiae**. A combination of ^{1}H and ^{13}C NMR spectroscopy was used to distinguish several derivatives of deoxyglucose and trehalose. The results indicated the existance of a equilibrium which between deoxyglucose and deoxyglucose-6-phosphate which limited the production of deoxyglucose-6-phosphate. In the presence of glucose the formation of disaccharides of deoxyglucose and deoxyglucose-glucose where favoured.[97] The effects of deoxyglucose the on the metabolism of glucose was also examined in further investigations when cells were incubated with [1-^{13}C]glucose. The intracellular levels of [1-^{13}C]glucose metabolites and the influence of [1-^{13}C]glucose with or without deoxyglucose has been studied in detail.[98]

^{31}P NMR spectroscopy has been used to study the effects of hypoosmotic shock and growth stage on the cellular pH and polyphosphate metabolism in **Neurospora crassa**. The ratio of polyphosphates to orthophosphate in vacuoles increased when cells grew from early log phase to the stationary phase; cytoplasmic and vacuolar pH increased during the same period. Hypoosmotic shock caused hydrolysis of polyphosphates and an increase in orthophosphate along with an increase in vacuolar and cytoplasmic pH. The cells in early log phase showed a greatest response to hypoosmotic shock, compared to cell in mid-log phase, or cells in stationary phase which showed little response.[99] A ^{31}P NMR spectroscopy study has been performed on yeast which were expressing rat cDNA for ketohexokinase. Exposure of the cells to 5mM or 25mM fructose caused a cessation of growth, a fall in pH and a large accumulation of fructose-1-phosphate at the expense of cellular orthophosphate and nucleotide triphosphates. The cessation of growth caused by 1mM fructose lead to an accumlation of fructose-1-phosphate but did not result in the depletion of orthophosphate or nucleotide triphosphate.[100] A novel method for the estimation of free intracellular ADP has been develvoped in yeast by selectively labelling phosphoglycerate kinase with 5-fluorotryptophan. It was demonstrated, **in vitro**, that the two fluorine resonances from labelled phosphoglycerate kinase had chemical shifts which were dependent on the amount of ADP and Mg-ADP. The estimated amount of free intracellular ADP measure by this method was much less than the amount found in cell free extracts.[101]

3 Tissue studies

A review with 27 references on the use of NMR spectroscopy for the characterisation of tissue and organ oxygen metabolism has been produced.[102]

3.1 Liver - The metabolism of enflurane has been monitored **in vivo** in the liver of rats using ^{19}F NMR spectroscopy. The major metabolite (difluoromethoxy difluoroacetate) was also detected in the liver and urine.[103] ^{13}C NMR spectroscopy has been used to detect 1-(o-chlorophenyl)-1-(p-chlorophenyl)-2,2-dichloro-3-^{13}C-propane (200mg kg^{-1}) in the liver of an intact animal via a surface coil.[104] The absolute quantification of ^{31}P liver metabolites has been performed using an external standard and a electrically driven surface spoiling gradient.[105] ^{1}H NMR spectroscopy has been used to examine the effects of CCl_4 on the cellular lipid metabolism of the rat liver.[106]

The effects of the peroxisome proliferator, perfluorodecanoic acid, on the carbohydrate metabolism of the rat liver have been monitored by ^{13}C NMR spectroscopy. The intensity of the signals for [1-^{13}C]glucose and [3-^{13}C]alanine after the infusion of these substrates was greater in perfluorodecanoic acid treated rats. There was a near complete inhibition of [1-^{13}C]glucose metabolism, though some gluconeogenesis and glycogenesis occured following [3-^{13}C]alanine administration, suggesting that there was an inhibition of glucose transport or glucokinase activity. Labelling of glutamate from [3-^{13}C]alanine indicated that the ratio of pyruvate dehydrogenase activity to pyruvate carboxylase was higher in the treated rats.[107] The effects of pregnancy and feeding regime on glycogen formation in the mouse liver have been monitored using ^{13}C NMR spectroscopy.[108]

The effects of ammonium ion on the intracellular pH of the perfused rat liver have been monitored by ^{31}P NMR spectroscopy. Perfusion with 0.5mM and 10mM NH_4Cl caused a decrease in intracellular pH of 0.06 $\pm$ 0.03 and 0.11 $\pm$ 0.03 units, respectively, and a reduction in the concentration of ATP by 11% and 22, respectively. Further experiments, which examined the role of HCO_3^-, indicated that the intracellular acidification following perfusion with NH_4^- was due to the utilization of HCO_3^- for the synthesis of urea.[109] The roles of the Na^+-H^+ antiport and the Na^+-HCO_3^- symport have been investigated in perfused rat liver using ^{31}P NMR spectroscopy. The livers displayed a cytosolic pH of 7.18, even under acid load (35mM isobutyric acid). In the presence or absence of amiloride, a Na^+-H^+ antiport inhibitor, the pH remained constant even under acid load, until the the external pH was less than 7.0. Livers perfused in the absence HCO_3^- had a intracellular pH reduced by 0.2pH units.[110] Thulium 1,4,7,10-tetraazacyclododecane-1.4.7.10 tetra(methylene phosphonate) ($TmDOTP^{5-}$) has been assessed for use as a shift reagent **in vivo** for ^{23}Na NMR spectroscopy of the liver.[111] The infusion of 80mM $TmDOTP^{5-}$ allowed the resolution of intracellular ^{23}Na NMR spectroscopy signals from extracellular ^{23}Na NMR spectroscopy signals with no effect on the ^{31}P NMR spectra and no effect on the intracellular pH. $TmDOTP^{5-}$ compared very favourably to dysprosium(III) bis(tripolyphosphate) which was required to be infused at a higher concentration and gave less well resolved peaks. The use of $TmDOTP^{5-}$ gave sufficient resolution of intracellular and extracelular peaks that T_1 and T_2 for intracellular and extracelular ^{23}Na could be measured.

The administration of indocyan green to rats has been shown to cause a decrease in liver ATP levels which correlated with a decrease in bile secretion.[112] The response of normal and reperfused livers to glucagon stimmulation has been monitored with ^{31}P NMR spectroscopy **in vivo**. Ischaemia lead to the loss of ATP afer 30min, though ATP levels were recovered upon reperfusion after a 20min period of ischaemia. After 40 and 60 minutes of ischaemia, reperfusion for 60min lead to a 76% and 48% recovery of ATP, respectively. Glucagon increased blood flow in the normal liver measured using ^{19}F NMR spectroscopy to monitor the washout of trifluoromethane. The effects of glucagon on blood flow were diminished in the reperfused liver.[113]

^{31}P NMR spectroscopy has been used to monitor the recovery of ATP signals detected in the perfused liver following 18h of cold ischaemia. In livers perfused for 90min with adenosine or adenine the concentration of ATP rose to 139% and 82% of baseline values, respectively. Livers perfused without adenosine or adenine recovery 60% of ATP levels.[114] Measurements of intracellular pH and intracellular Na^+ of livers following preservation indicated affects upon the Na^+-K^+ and Na^+-H^+ exchange systems.[115] The effects of perfusion of pig liver with cooled Euro-

Collins solution compared with perfusion with oxygenated cooled Euro-collins solution were examined with ^{31}P NMR spectroscopy. The β-ATP resonance was detected longer when the oxygenated medium was used.[116] ^{31}P NMR spectroscopy has been used to monitor changes of intracellular pH and ATP concentrations in a study of the comparative efficiency of UW and HTK preservation solutions.[117]

Mouse liver has been examined by ^{31}P NMR spectroscopy following infection with the parasitic worm **Hymenolepis microstoma** for 130 or 265 days. The ratio of inorganic phosphate to β-ATP was found to be increased and the ratio of inorganic phosphate to phosphocreatine was found to be decreased after both periods. The intracellular pH was lowered in the 130 day infection group. Changes in the phosphorous metabolite concentrations were measured in liver extracts.[118] The effects on mouse liver of infection with **Hymenolepsis microstoma** and **Mesocestoides vogae** have been examined with ^{31}P NMR spectroscopy in the early and late stages of infection.[119]

3.2 Kidney - The effects of hyperglycaemia on the kidney have been monitored with ^{31}P, ^{19}F and ^{23}Na tripple quantum filtered NMR spectroscopy. No alteration in pH, Mg^{2+} and ATP were observed, though a 50% increase in Na^{+} was detected following perfusion with 25mM glucose. The level of Ca^{2+} was 390nM and 765nM when the concentration of perfused glucose was 5mM and 25mM, respectively. The increase in Na^{+} and Ca^{2+} during hyperglycaemia may indicate a role in renal hypertrophy and hypertension associated with diabetes mellitus.[120] ^{31}P NMR spectroscopy has been used to study the mechanism of the augmention of cellular ATP recovery caused by thyroxine treatment of ischaemic kidneys. Saline treatment resulted in 57.9% and 66.1% recovery of ATP following 30min and 120min reperfusion, respectively. In thyroxine treated kidneys the recovery of ATP was 70.2% and 84.0% after 30min and 120min reperfusion, respectively. Measurements of lactate dehydrogenase release, used to indicate cellular integrity, indicated that the recovery of ATP was a result of effects on synthesis as opposed to improvements in cell physiology.[121] The effects of perfusion with ATP-$MgCl_2$ on the recovery of the kidney following a 40min period of ischaemia has been observed with ^{31}P NMR spectroscopy. Although the ATP-$MgCl_2$ treated kidney did not have a higher ATP level during reperfusion, the recovery from acid pH was slower and a greater blood flow and urinary output was observed; an acidic pH has been proposed to protect cells from reperfusion injury.[122]

3.3 Heart - The effects of adenosine on regionally ischaemic myocardium have been investigated with ^{31}P NMR spectroscopy **in vivo**. During ischaemia the increase of the inorganic phosphate peak and the decrease in the phopshocreatine peak were less in those hearts pre-treated with 25mg kg^{-1} adenosine.[123] The role of adenosine in the mechanism of pre-conditoning of hearts has been studied using adenosine antagonists on perfused rat heart observed with ^{31}P NMR spectroscopy.[124] Preconditioning, a protocol whereby hearts are subjected to short, intermittent periods of ischaemia prior to a prolong (30-60min) ischaemic period, has been shown in the past to be beneficial to the recovery of cardiac function. The benefits of preconditioning on intracellular pH where not affected by treatment with the adenosine antagonist BW-A1433U, PD-115,199 or CPDPX. The changes in intracellular pH during ischaemia are largely due to anaerobic glycolysis which may be affected by the reduction in glucose uptake, implied by the reduction in the uptake of 2-deoxyglucose, observed when hearts were treated with the adenosine antagonists BW-A1433U, PD-115,199. The effects of inhibition of the ATP-sensitive potassium channels on the protective effects of preconditioning have been investigated in the isolated perfused rat heart.[125] ^{19}F and ^{31}P NMR spectroscopy was used to determine the intracellular Ca^{2+}, pH and phosphorous metabolites. The K^+-channel blockade (using glibenclamide) resulted in a more rapid depletion of high energy phosphates during preconditioning, K^+-channel blockade during preconditioning did not effect the changes observed in the intracellular ions, and did not abolish the protective effects of preconditioning observed during the 30min period of ischaemia.

The effects of trimetazidine upon the preservation of isolated arrested rat hearts has been measured with ^{31}P NMR spectroscopy to assess cellular ATP and pH.[126] The effects of cotinine (a product of nicotine oxidisation in the liver) on an ischaemis-reperfusion model in the perfused rat heart has been investigated with ^{31}P NMR spectroscopy.[127] The investigation of the benefits of a-tocopherol pre-treatment (pre-treatment for 7 days prior to the isolation of rat hearts) on the recovery of phosphorous metabolites following ischamia has been carried using ^{31}P NMR spectroscopy.[128] Tocopherol pre-treatment resulted in a more rapid recovery of high energy phosphates depleted by 30min of global ischaemia.The benefits of treatment with diltiazem (a calcium antagonist) or nicorandil (a novel coronary dilator) on high energy phosphates detected during ischaemia and reperfusion in isolated rat hearts has been followed by ^{31}P NMR spectroscopy.[129] Neither drug exhibited any beneficial effects on phosphorous metabolites during ischaemia, but both drugs improved the recovery of ATP during reperfusion.

The relationship between the pressure rate product of the heart and the flux of PCr to ATP or the flux of inorganic phosphate to ATP have been studied using the ^{31}P NMR spectroscopy saturation transfer method. The flux of PCr to ATP/pressure rate product was the same before ischaemia and during reperfusion.[130] Treatment of hearts with the drugs propranolol and diltiazem has been shown to have no effect on the value of the flux of PCr to ATP/pressure rate product. The flux of inorganic phosphate to ATP/pressure rate product was lower upon reperfusion than that observed before ischaemia, though the flux of inorganic phosphate to ATP/pressure rate product was improved by treatment with propranolol or dilitiazem.[131] Saturation transfer has been combined with localisation method to analyse the transmural profile of the creatine kinase system in the canine heart **in vivo**. Although intraleft ventricular PCr/ATP levels were constant, a 61% variation was observed in the first order rate constant and the forward creatine kinase rate.[132] The net fluxes of phosphate between intracellular and extracellular compartments of inorganic phosphate and the organic phosphate pools have been measured in the isolated rat heart perfused with or without phosphate containing buffer.[133]

The effects of lidocaine on contracture, intracellular sodium and intacellular pH in the ischaemic rat heart has been investigated using ^{23}Na NMR spectroscopy in conjunction with the shift reagent Thulium 1,4,7,10-tetraazacyclododecane-1.4.7.10 tetra(methylene phosphonate) ($TmDOTP^{5-}$), and ^{31}P NMR spectroscopy. In lidocaine treated hearts the time to contracture was reduced, intracellular Na^+ was lower, the depletion of ATP was reduced and the development of acidosis was reduced. The effects of lidocaine on the accumulation of intracellular Na^+ occur between the first 5-10min of ischaemia, whereas the effects of lidocaine on pH were observed after 9min of ischaemia.[134] ^{31}P NMR spectroscopy has been used to observe the effects of inhibition of metabolism by cyanide and iodoacetate on the isolated ferret heart. The decrease in intracellular pH following inhibition of glycolysis with iodoacetate was attributed to the release of protons from the breakdown of ATP, whilst acidosis caused by cyanide was attributed to the increase in lactate.[135] ^{31}P NMR spectroscopy has shown that the level of free intracellular Mg^{2+} is depleted by 43% in hearts perfused with 1% ethanol.[136] There was no effect on intracellular pH, though the level of phosphocreatine (PCr) and ATP were depleted and the level of inorganic phosphate was increased.

The role of glycolysis in intracellular Ca^{2+} homeostasis has been assessed using ^{31}P NMR spectroscopy to monitor intracellular phosphorous metabolism and ^{19}F NMR spectroscopy to monitor intracellular Ca^{2+}. Treatment with the glycolytic inhibitor iodoacetate, in the presence of pyruvate to maintain oxidative metabolism, showed no effect on cardiac function or ^{31}P metabolites. However, with the addition of isoproterenol, which stimmulates Ca^{2+} entry, there was an increase in mean left ventricular end-diastolic pressure, a reduction in ATP and PCr and a doubling of the diastolic Ca^{2+} concentration.[137] A new intracellular Ca^{2+} indicator, 1,2-bis-{2-[1-(hydroxycarbonyl)ethyl-(hydroxycarbonylmethyl)]amino-5-fluorophenoxy}ethane (DiMe-5FBAPTA), has been developed which has a higher affinity for Ca^{2+} than the parent compound 1,2-bis(2-amino-5-fluorophenoxy)ethane-N,N,N',N'-tetraacetic acid (5FBAPTA). The intracellular concentration of DiMe-5FBAPTA required to give adequate signal to noise caused a reduction in developed pressure which was similar to reduction observed when 5FBAPTA was used, however the results indicated that the perturbations in intracellular Ca^{2+} caused by DiMe-5FBAPTA were less than those observed when 5FBAPTA was used.[138] In a study of the roles of intracellular calcium and intracellular pH in shaping myocardial contractile response to acute respiratory alkalosis, ^{31}P NMR spectroscopy was used to measure intracellular pH; intracellular Ca^{2+} was measured by microinjected fura 2 salt.[139] The results indicated that the biphasic changes in intracellular Ca^{2+}, coupled with an out-of-phase change of intracellular pH, underlie the biphasic response of myocardial contractility to hypocapnia.

Intracellular Mg^{2+} homeostasis has been shown to be involved in the functional recovery of preconditioned hearts, using ^{31}P NMR spectroscopy to measure intracellular Mg^{2+}, pH and phosphorous metabolites.[140] ^{31}P NMR spectroscopy was used to monitor high energy phosphates and intracellular pH in a study of the effects of magnesium added to secondary cardioplegia on post ischaemic myocardial metabolism and contractile function.[141] The effects of perfusion of hearts with low (0.3mM) Mg^{2+} was shown with ^{31}P NMR spectroscopy to cause a fall in intracellular

Mg^{2+} and intracellular pH, and an increase of intracellular inorganic phosphate which reflected the fall in cellular phosphorylation potential.[142] When the extracellular Mg^{2+} concentration was returned to 1.2mM neither cardiac function nor phosphorous metabolism returned to normal values indicating irreversible myocyte damage.

Increases in the concentration of Mg^{2+} (from 1.2 to 4.8mM) in the perfusate of isolated heart[143] resulted in a 23% increase in the cardiac output, reflecting an 100% increase in stroke volume, despite a reduction in the heart rate by 35%. Perfusion of hearts in the Langendorff mode with increased extracellular Mg^{2+} resulted in increased PCr at the expense of inorganic phosphate and perfusion of hearts in the working mode with increased extracellular Mg^{2+} resulted in increased PCr from 6.3 to 9.0mM, while ATP fell by 17%. The role of intracellular Mg^{2+} in temporary post ischaemic cardiac dysfunction (stunning) has been studied by ^{31}P NMR spectroscopy, in conjunction with measurements of phosphorous metabolites and pH. The results indicated that intracellular Mg^{2+} was not involved in stunning.[144] The effects of prolonged application of isoprenaline on the concentration of intracellular free Mg^{2+} in the perfused rat heart has been studied with ^{31}P NMR spectroscopy.[145] Isoprenaline caused a rise in intracellular Mg^{2+} which could be accounted for by the observed decrease in ATP. Application of propranolol to perfused hearts did not cause a rise in intracellular Mg^{2+}.

The effects of preconditoning on perfused rat hearts has been measured using ^{31}P, ^{19}F and ^{23}Na NMR spectroscopy to detect intracellular phosphates, intracellular calcium and intracellular sodium, respectively.[146] Intacellular Na^{+} was distinguished from extracellular Na^{+} by the use of the shift reagent thulium 1,4,7,10-tetraazacyclododecane-N,N',N'',N'''-tetramethylenephosphonate. The use of preconditioning had no extra beneficial effect than hearts arrested with 16mM $MgCl_2$ (Mg^{2+} prevents Ca^{2+} entering via calcium channels), though, preconditioning attenuated the rise in intracellular calcium, intracellular sodium and intracellular pH. Using ^{31}P, ^{23}Na and ^{39}K NMR spectroscopy the changes in phosphorous metabolites and intracellular ions have been studied during ischaemia.[147] In early ischaemia there was a rise in the concentration of Na^{+} and inorganic phosphate, a decrease in the concentration of PCr and ATP and a fall in pH. After 15min ischaemia a second, larger increase in intracellular Na^{+} and a decrease of intracellular K^{+} was observed with a second increase in inorganic phosphate. After an even longer period of ischaemia a second inorganic phosphate resonance was observed which was suggested to be in a compartment where the pH was in equilibrium with the pH of the perfusing medium. The compartmental distribution of Na^{+} in hearts during hyperthermic ischaemic preservation has been investigated with ^{23}Na NMR spectroscopy.[148] The visibility of Na^{+} has been assessed as a marker of heart recovery from hypothermic ischaemic preservation.

The beneficial effects of the treatment of working perfused rat hearts with dihydrolipoic acid during and following hypoxia has been been studied with ^{31}P NMR spectroscopy. Addition of dihydrolipoic acid during hypoxia accelerated the recovery of aortic flow and attenuated the rise in inorganic phosphate. The intracellular pH of treated hearts was 7.15 during hypoxia and a higher

PCr content was also observed. Saturation transfer experiments indicated that the flux of PCr to ATP decreased in the control group during hypoxia, whereas the flux from PCr to ATP increased or remained constant in the treated hearts. The results suggested an increase in the synthesis of ATP under the influence of dihydrolipoic acid.[149] In a study of the effects of dietary creatine and cyclocreatine (a synthetic creatine analoque) on the high energy phosphate metabolism of hearts during and following ischaemia, ^{31}P NMR spectroscopy has been used to show that pretreatment with cyclocreatine offers protection against ischaemic damage.[150]

An increased tolerance to hypoxia in cardiomyopathic hamsters has been observed in ^{31}P NMR spectra and measurements of left ventricular pressure in isolated isovolumetrically beating heart preparations.[151] Cardiac function in perfused hearts isolated from rats with experimental renal failure was reduced, and the basal phosphocreatine level was reduced, when measured by ^{31}P NMR spectroscopy; the ratio of PCr/ATP was also reduced.[152] The alteration of cardiac energy state during development of hypertension in the Lyon rat has been studied with ^{31}P NMR spectroscopy in the isolated perfused heart.[153] Hearts from 12 week and 21 week old hypertensive Lyon rats were compared to hearts from age matched normotensive and hypotensive rats. All groups showed increases in inorganic phosphate and decreases in phosphocreatine in response to Ca^{2+}-induced work which were greatest in the hypertensive rats; the hypertensive rats, also, showed a work-related decrease in the concentration of ATP. The relationship between left ventricular developed pressure and the change in Pi/PCr indicated that oxidative metabolism is maximally activated in the young hypertensive rat. The mechanism of regulation of intracellular pH after an acid load has been measured in the ferret heart using ^{31}P NMR spectroscopy to detect the chemical shift of 2-deoxy-D-glucose 6-phosphate.[154] Comparison of the results obtained from experiments where hearts were perfused with HCO_3^--buffered solution or perfused with HEPES solution showed that the Na^+-H^+ antiport, and a mechanism requiring external Na^+ and HCO_3^-, contribute 50% each to H^+ efflux at pH 6.90 during recovery from an acid load.

A method has been developed for gating ^{31}P NMR spectra of isolated hearts from the pulses used for pacing of the heart beat. These have allowed spectra to be obtained during defined periods of the cardiac cycle.[155] An assessment of the use of ^{31}P NMR spectra to predict the recovery of hearts from an ischaemic episode has shown that only the absence of a peak corresponding to phosphomonoesters during ischaemia could predict the subsequent recovery of cardiac function.[156] The effect of severe pressure overload (left ventricular hypertrophy) on myocardial high energy phosphate metabolism has been studied by ^{31}P NMR spectroscopy.[157] The transmural distribution of myocardial ATP, phosphocreatine, and inorganic phosphate were quantitated by comparison against ATP levels determined in biopsies. The results showed that phosphocreatine, ATP and PCr/ATP were reduced in severe left ventricular hypertrophy but changes were not the result of persistent abnormalities of myocardial perfusion.

The metabolism of [1-^{13}C]glucose in the perfused heart from (streptozotocin-induced) diabetic rats has been studied by ^{13}C NMR spectroscopy. The effect of diabetes-induced cardiomyopathy was perceived as a lack of incorporation of ^{13}C-label into lactate or glutamate. Perfusion with insulin, however, caused the appearance of [3-^{13}C]lactate and perfusion with insulin and dichloroacetate caused the appearance of label in the 2, 3 and 4 positions of glutamate. Cardiac function was also impaired.[158] ^{13}C NMR spctroscopy has been used to examine the activity of the tricarboxylic acid cycle in post-ischaemic rat hearts by following the labelling of glutamate from ^{13}C-labelled substrate. Though the hearts showed persistent contractile dysfunction, measurements of the labelling glutamate suggested that the rate of flux through the tricarboxylic acid cycle was increased.[159] Measurements of the rate of glycogenolysis has been measured in the isolated perfused rat heart following the labelling of intracellular glycogen from a 50min infusion of D-[1-^{13}C]glucose.[160] A bolus of glucagon was used to stimmulate the metabolism of ^{13}C-labelled glycogen at a rate of 25mmol min^{-1} g wet weight^{-1}. The rate of metabolism was 0.52mmol min^{-1} g wet weight^{-1} if the 50min infusion of ^{13}C-label glucose was followed by a 50min infusion of unlabelled glucose prior to the treatment with glucagon. In the canine myocardium, the uptake and metabolism of [2-^{13}C]acetate has been followed by ^{13}C NMR spectroscopy to show the synthesis of glutamate under normal condition.[161] However, under high work loads [2-^{13}C]acetate was not incorporated into glutamate indicating a work-load dependent regulation of substrate access to the mitochondrial acetyl-CoA pool. The visibility of glycogen in ^{13}C NMR spectra of isolated perfused rat hearts was shown to be 100% from experiments of total labelling of cardiac glycogen from [1-^{13}C]glucose followed by measurements **in vivo** and **in vitro.**[162] Measurements of the chemical shift of the C-3 carbon of glycerol 3-phosphate has been used to measure intracellular pH. Measurements made in isolated rat hearts perfused under several conditions agreed well with measurements made with ^{31}P NMR spectroscopy.[163]

The distribution of trifluoroacetic acid, measured by ^{19}F NMR spectroscopy, has been used to estimate the membrane potential of perfused hearts. Measurements made using the anion exchange inhibitor, 4-acetamido-4'-isothiocyanostilbene-2,2'-disulfonic acid, suggest that trifluoroacetate uptake occurs partly via the Cl^--HCO_3^- exchanger. The results suggest that trifluoroacetate distribution in hearts reflects the chloride potential and not the membrane potential.[164] The measurement of cardiac lipolysis and oxidation of triacylglycerols by ^{1}H NMR spectroscopy has been carried out on the isolated perfused hearts of rats fed on a diet containing 20% rapeseed oil.[165] The ratio of the number of methyl, g-methylene or methylene protons next to carbon double bonds, compared to the number of methylene protons was used to assess the triglyceride content of the hearts. ^{31}P NMR spectroscopy has been used to characterise the phosphorylated compounds present in the heart of vertebrate ectotherms.[166] Resonances were detected in the region of the ^{31}P spectra between the peaks of phosphocreatine and inorganic phosphate which were characterised **in vitro**. The pattern of these compounds was different for different species.

3.4 Brain - A reveiw of the principles of NMR spectroscopy applied to the study of cerebral metabolism has been produced.[167] Current NMR spectroscopy studies of brain metabolism **in vivo** has been reviewed.[168]

The self-diffusion of brain metabolites has been measured by a technique developed from the stimmulated-echo localization method.[169] The results for water, N-acetyl aspartate, creatine and choline were 0.75 x 10^{-5}, 0.27 x 10^{-5}, 0.27 x 10^{-5} and 0.28 x 10^{-5} cm^2 s^{-1}. ^{31}P NMR spectroscopy has been used to monitor the ATP and PCr in piglet brain during ischeamia to show that interpulse repition time (1 or 25 seconds) has no effects on the changes in ^{31}P metabolites observed in the spectra of the brain.[170] The results indicate no changes in the relaxation rates of ATP and PCr during ischaemia. Measurements of T_1 for inorganic phosphate show that there was no difference in control or ischaemic brains. Furthermore, no changes in the linewidths of inorganic phosphate, ATP or PCr indcate that T_2^* is also not affected by ischaemia. ^{31}P NMR spectroscopy has been used to examine the brain following a single 6.5Gy dose of γ-irradiation.[171] The effects of x-ray on normal brain tissue has been examined with ^{31}P NMR spectroscopy and the results were compared to changes observed with histological techniques and magnetic resonance imaging.[172]

^{13}C NMR spectroscopy has been used for the study of glucose metabolism in the brain.[173] The production of γ-aminobutyric acid has been followed by ^{13}C NMR spectroscopy in the rat brain.[174] The metabolism of ^{13}C-labelled glucose has been followed with gradient-selected heteronuclear multiple quantum coherence spectroscopy and cemical shift imaging.[175] The accumulation of [1-^{13}C]glucose and the labelling of metabolites were seen within 5min. ^{15}N NMR spectroscopy[176] has been used to follow the synthesis and utilization of glutamate in the brains of hyperammonaemic rats infused with $^{15}NH_4^+$ followed by $^{14}NH_4^+$. The rate of glutamine synthesis **in vivo** was measured to be 4.8mmol h^{-1} g of brain^{-1} if ^{15}N enrichment of ammonia is equal to that of ammonia in the blood or 13.0mmol h^{-1} g of brain^{-1} if ^{15}N enrichment is equal to that measured for the whole brain. These rates represent 2-5% of optimum glutamate synthetase activity expected from measurements **in vitro**.

An age difference in the response of rat brains to hypoxia has been observed with ^{31}P NMR spectroscopy. The age group 9 to 13 days showed an severe reduction in ATP and PCr compared to younger and older rats which may be explained by known changes in the energy metabolism that occur in the 9 to 13 day period.[177] In a study of the effects of global ischaemia, ^{31}P NMR spectroscopy was used to monitor cellular energy metabolism whilst blood flow was measured by laser doppler flowmetry.[178] The loss of high energy phosphorous metabolites was observed along with a decrease in intracellular pH during ischaemia. Recovery of intracellular pH and PCr levels were observed during 60min of reperfusion, though the observed incomplete recovery of ATP was attributed to a loss of total adenylate. Differences in the recovery of brain phosphorous metabolites was related to the differences observed in the reperfusion flow. The administration of cyanide as a model of hypoxia has been used to observe the effects on phosphorous metabolites and the flux of phosphocreatine to ATP studied by ^{31}P NMR spectroscopy.[179] Low doses of cyanide (1-3mg kg^{-1}) did not affect the levels observed in the ^{31}P spectra, though the rate of flux from PCr to ATP was increased by 75%. Higher doses of cyanide (4-6mg kg^{-1}) caused a transient decrease of 30-

40% of PCr, 100% increase in inorganic phosphate and a 0.2 units decrease in intracellular pH. The flux of PCr to ATP was decreased 50-80% in brains treated with the higher dose of cyanide.

The contribution of lactic acid to the fall in brain intracellular pH normally observed during ischaemia has been studied with ^{31}P and ^{1}H NMR spectroscopy.[180] When insulin-induced hypoglycemia was used to inhibit the production of lactic acid during ischaemia a decrease of intracellular pH was still observed, showing that the production of lactic acid during ischaemia does not necessarily govern intracellular pH. The beneficial effects of taurine on brain buffering capacity have been shown with ^{1}H NMR and ^{31}P NMR spectroscopy **in vivo.**[181] The effects of treatment with the calf blood extract solcoseryl on brain subjected to ischaemia have been examined with ^{31}P NMR spectroscopy, ^{1}H NMR imaging and hystological examination.[182] The expected rise in inorganic phosphate and a fall in intracellular pH was observed during ischaemia and a return to control levels was observed in ^{31}P NMR spectra following 60min reperfusion. The solcoseryl-treated group had a faster recovery from acidified intracelllular pH following reperfusion and there was no increase in intracellular pH 48h after ischaemia in the treated group. No differences were observed between the two groups with ^{1}H NMR imaging following ischaemia, though histological examination one week later showed reduced neuronal cell death in the hippocampal CA1 region of the brains of the solcseryl-treated animals.

The effects of hypercarbia on cerebral glucose metabolism has been examine in newborn and 1 month old piglets; brain lactate production was measured by ^{1}H NMR spectroscopy.[183] The metabolism of lactate has been assessed in the rabbit brain following electric shock.[184] Levels of lactate, detected by ^{13}C-edited ^{1}H NMR spectroscopy, were elevated following electric shock and underwent complete turn over within 1h, whilst blood lactate underwent less than 50% turn over during the same 1h period. The levels of lactate in the brain of rats infused with lactate have been detected with ^{1}H NMR spectroscopy, and showed an increase during infusion.[185] Brain lactate and pH have been examined in collagenase-induced cerebral oedema using ^{1}H NMR and ^{31}P NMR spectroscopy.[186] Though lactate was elevated for 48h following injection of collagenase the pH was unaffected. The increase in lactate correlated with the increase in brain water content.

A model of cerebral hypoxia in the Gerbil has been developed, and ^{31}P and ^{1}H NMR spectroscopy have been used to examination changes in cerebral metabolism under different degrees of hypoxia. No changes in cerebral lactate or phosphorous metabolites were observed until blood p_aO_2 was below 40mm Hg and 50% of animals showed an increase in lactate prior to any changes in ^{31}P metabolites. The result suggest that the reduction of O_2 supply is the critical factor in ischaemia.[187] The energy reserves and utilization rates in developing piglet brain have been measured with ^{1}H NMR and ^{31}P NMR spectroscopy in newborn and 1 month old piglets, and following complete ischaemia as a result of cardiac arrest.[188] Hyperglycemia had no effect on the cerebral energy utilization, though hypoglycemia decreased this rate. The results also indicated that cerebral energy utilization rates decrease after the first few minutes of ischaemia. ^{31}P NMR spectroscopy has been used to monitor changes in the rat brain subjected to a short period of ischaemia to determine the effects of blood glucose concentrations on high energy phosphate metabolism.[189] The decrease in tissue pH and the increase in inorganic phosphate during ischaemia was greatest in those rats with the highest blood glucose. The rate of decline in PCr and ATP was similarly dependent. Post-ischaemic hyperglycemia delayed the recovery of pH, and in those rats which were constantly hyperglycemic, the recovery of PCr and ATP was delayed. Insulin was found to increase the rate of recovery of pH, ATP and PCr in hyperglycemic rats. In a model of acute allergic encephlomyelitis increases observed in the ^{1}H NMR spectra assigned to the peak for choline containing compounds, compared to the creatine plus phosphocreatine peak, was associated with an increase in choline, betaine and phosphorylcholine.[190] **In vitro**, a reduction in N-acetylaspartate, aspartate and N-acetylaspartyl glutamate was also detected, but there was no evidence of demyelination or neronal loss within the brain. In contrast to suggestion elsewhere the results indicate that changes in the ^{1}H NMR spectra are related to inflamation.

The effects of halothane, enfluane, sofluane or sevoflurane anaesthesia have been studied with ^{31}P NMR spectroscopy.[191] It was shown that halothane can reduce PCr levels to 75%, decrease intracellular pH and increase inorganic phosphate compared to unanaesthetised controls. The acute and chronic administration of alocohol to rats has been followed by ^{31}P NMR spectroscopy.[192] The results showed a decrease in high energy phosphates in both groups. In the chronically treated rats there were changes in the ratio of phosphomonoesters to phosphodiesters and the ratio of phosphodiesters to β-ATP. Decreases in the ratios of PCr/ATP and PCr/inorganic phosphate have been observed with ^{31}P NMR spectroscopy, along with an observed fall in intracellular Mg^{2+}, following acute alcohol administration to anaethetisd rats.[193] No changes in pH were observed over a 70min period. The effects of methyl mercury administration on the rat brain have been examined with ^{31}P NMR spectroscopy and ^{1}H NMR imaging.[194] Phosphorous spectra from rats detected a decreased in the level of PCr and an increase in the level of inorganic phosphate, whereas no differences were seen by ^{1}H NMR imaging.

The effects of acute and chronic hyponatremia on the pH buffering capacity of the brain has been monitored using ^{31}P NMR spectroscopy to measure brain pH in response to ventillation with 20% CO_2. Brain perfusion was measured using ^{19}F NMR spectroscopy to monitor the washout of trifluoromethane. The results indicated that the impairment of brain buffering capacity is unrelated to water content, perfusion, tissue catabolism or arginine vasopressin V_1 receptor activation.[195] The shift reagent dysprosium (III) triethylenetetraminehexacetate has been tested for use in the dog brain **in vivo.**[196] Two peaks were observed with cemical shifts of 9.0ppm and 0.4ppm, and the peak at 0.4ppm split into two peaks (one at 0.0ppm and one at 3.0ppm) when the blood-brain barrier was disrupted. ^{31}P NMR spectroscopy has been used to measure the relative concentrations of brain phosphorous metabolites, the pseudo first-order rate constant of the creatine kinase forward reaction and creatine kinase forward flux in relation to brain activity.[197] The brain activity was determined from measurements with electroencephalography. The administration of sodium thiopental or bicuculline was used to decrease or increase electroencephalography intensities, respectively. When electroencephalography intensities were affected no effects were observed on the intensities of the phosphorous metabolites except a decrease in the level of phosphocreatine during high brain activity. The creatine kinase forward flux increased with biculline treatment.

The fluorinated neuroleptic melperone (4'-fluoro--4-(4-methylpiperidino)butyrophenone) has been detected **in vivo** in the rat brain.[198] The half life of the drug in the brain has been reported to be 4.3h. ^{19}F NMR spectroscopy has been applied to the brains of mice to detected α-(1,1-difluoroethyl)-α-methyl-γ-butyrolactone, a fluorinated derivative of the anticonvulsent γ-butyrolactone.[199] Both the drug and some metabolites were detected, despite broad and complex spectra **in vivo**.

Measurements of changes in intracellular Ca^{2+}, measured by ^{19}F NMR spectroscopy and the intracellular Ca^{2+} indicator 5FBAPTA, during mild hypoxia, mild hypoglycemia or a combination of the two insults have been studied in superfused cortical slices.[200] Measurements, using ^{31}P NMR spectroscopy, showed a 40% decrease in PCr with hypoxia and a slight decrease in PCr and ATP with hypoglycaemia. The sequential insult of hypoxia and hypoglycaemia (in either order) caused a 75% fall in PCr and a 100% increase in the intracellular Ca^{2+}. The simmultaneous exposure of the tissues to hypoxia and hypoglycaemia had the same effect on PCr as the sequential insult but there was a 5 fold increase in intracellular Ca^{2+}. The results indicate a metabolic adaption of the tissues. The role of Ca^{2+} in cellular damage following hypoxia in the cerebral cortex has been assessed using ^{31}P NMR and ^{1}H NMR spectroscopy to monitor PCr, ATP, lactate, glutamate and N-acetyl aspartate.[201] The concentration of PCr recovered to 58% following 15min of aglycaemic hypoxia and recovered to 80% following 15min of aglycemic hypoxia in the absence of Ca^{2+}. Further experiments with several Ca^{2+}-channel antagonists revealed that neuronal damage may be mediated by Ca^{2+} influx through L-type voltage-gated Ca^{2+}-channels. Double quantum filtered ^{23}Na NMR spectroscopy has been used to examine the changes in intracellular Na^{+} during severe hypoxia.[202] Although 95% of the extracellular $^{23}Na^{+}$ signal was not observed in double quantum filtered ^{23}Na NMR spectra, 40% of the signals appeared to come from $^{23}Na^{+}$ interacting with membrane surfaces. The administration of veratidine increased the double quantum filtered ^{23}Na signal to $130 \pm 5\%$, though 30min hypoxia in the presence of 10nM glucose had no effect on the double quantum filtered ^{23}Na signal intensity. Aglycaemic hypoxia caused the loss of phosphocreatine within 7min, though no effects were seen on the level of double quantum filtered ^{23}Na signals until 15min; reaching a peak value of 125% after 25min. Neither 10mM extracellular Mg^{2+} nor 4.7μM tetrodoxin had any effect on the increase of double quantum filtered ^{23}Na signal intensity following aglycaemic hypoxia.

The detection of glutamate and N-acetylaspartate by ^{1}H NMR spectroscopy, using a semi-selective spin-echo sequence, in perfused brain cortical slices has been compared to those values obtained from total glutamate and total N-acetylaspartate found in perchlorc acid tissue extracts.[203] The ratio of glutamate to N-acetylaspartate was 20-30% less in control slices and slices treated with cyanide in the presence of glucose compared to the amounts of glutamate and N-acetylaspartate detected in extracts. Inhibiton of respiration in the absence of glucose, or treatment with nigericin or carbonyl cyanide-*m*-fluorohenylhydrazone caused an increase in the ^1H NMR observed ratio of glutamate to N-acetylaspartate in slices, but the amount of glutamate and N-acetylaspartate in extracts remained constant. The results suggest that 22% of intracellular glutamate is not detectable by spin-echo ^1H NMR spectroscopy.

Rat hippocampal slices have been examined by ^{31}P NMR spectroscopy in superfusion to determine the effect of extracellular Ca^{2+} on cellular high energy phosphates during simmulated ischaemia.[204] The depeletion of PCr and a decrease in pH were rapid when superfusion was halted. The recovery of PCr was dependent on the length of the non-perfusion period, though pH recovered in all experiments. Superfusion with low Ca^{2+} and high Mg^{2+} improved the recovery of cellular high energy phosphates. ^{31}P NMR and ^{13}C NMR spectroscopy has been used to examine the effects of depolarisation (induced by 40mM K^+) on superfused hippocampal slices.[205] Depolarization caused a 30-50% decrease in the ratio of PCr/inorganic phosphate and ATP/ inorganic phosphate and a 0.3unit decrease in the pH. A return to baseline values was observed after 30min exposure to control conditions. Depolarisation also caused a 20 fold increase in [1-^{13}C]glucose consumption and [3-^{13}C]lactate production which may account for the robust nature of hippocampal cells under long term depolarising (seizure) conditions.

3.5 Skeletal muscle - ^{31}P NMR spectroscopy has been used to follow the changes in phosphorous metabolites of control and mdx mice (a model of Duchenne muscular dystrophy) during and following exercise.[206] During 1Hz stimmulation of sciatic nerve, isometric tension did not decrease in either group. The ratio of PCr to PCr plus inorganic phosphate was lower during exercise and recovery in the mdx mice, though the ratio of ATP/PCr and the half-time of PCr recovery was the same in both groups. The recovery of intracellular pH in the mdx strain was prolonged compared to the control mice. The skeletal muscle of transgenic mice, expressing the brain and heart isoenzymes of creatine kinase, have been studied with ^{31}P NMR spectroscopy.[207] In transgenic mice, the creatine kinase catalysed reaction was 200% of the control values, though no differences in the ^{31}P NMR spectra existed between the two groups. ^{31}P NMR spectroscopy has been used in a study of peripheral arteriopathy induced by sodium lurate.[208] The beneficial effects of treatment with devocarnitine propil were determined by ^{31}P NMR spectroscopy and examination of muscle necrosis.

The resynthesis of phosphocreatine detected by ^{31}P NMR spectroscopy in rat muscle following electrical stimmulation has been shown to be a conveient measure of oxidative ATP formation.[209] The effects of creatine depletion on the contractile and metabolic profile of rat muscle has been studied with ^{31}P NMR spectroscopy.[210] The depletion of high energy phosphates, acheived by feeding rats with the creatine analogue β-guanidinopropionic acid, resulted in the shift of the speed related contractile properties of the muscle toward a slow-type associated with an improved resistance to fatigue.

In a study of the role of free-radical production in the development of halothane-induced malignant hyperthermia in the pig, ^{31}P NMR spectroscopy was used to detect changes in the muscle high energy phosphorous metabolism.[211] The development of maligant hyperthermia was associated with a fall in the level of PCr and a rise in inorganic phosphate, but C-phenyl-N-tert-butyl nitrone (a free radical spin-trap) had no effect on the halothane induced malinant hyperthermia. The effects of starvation on the high energy phosphate metabolism and muscle contraction in frogs has been studied using ^{31}P NMR spectroscopy.[212] Muscle contraction was decreased in frogs starved for 2 months and was reduced to 33% of contraction in controls after 5 months starvation. The lack of change in the concentration of PCr or ATP and a minor change in the ratio of PCr to inorganic phosphate suggested that the creatine kinase activity is decreased in starved frogs. Treatment of rats with triiodothyroxine has been shown, with ^{31}P NMR spectroscopy, to caused alkalinisation of skeletal muscle at rest and a decrease in the acidification of the exercising muscle.[213]

The use of a superfused skeletal muscle preparation of the extensor digitorum longus muscle has been assessed with ^{31}P NMR spectroscopy.[214] The levels of ATP and PCr detected in 20mg or 45mg muscles was constant over a 2h period. However, metabolic heterogeneity within the 45mg muscle preparation was indicated by a heterogenous pH measured from the shift of the inorganic phosphate peak. This variation in pH reflected a muscle core depleted of glycogen. Studies of the rabbit muscle post mortem have followed the changes in phosphorous containing compounds using ^{31}P NMR spectroscopy.[215]

3.6 Smooth muscle - The role of intracellular Ca^{2+} in organ failure during sepsis has been investigated with ^{19}F NMR spectroscopy and the intracellular Ca^{2+} indicator 1,2-bis(2-amino-5-fluorophenoxy)ethane-N,N,N',N'-tetraacetic acid (5FBAPTA). Sepsis caused a 2-fold increase in the intracellular levels of calcium and this increase was reduced by the treatment with sodium dantrolene (drug which decreases the release of Ca^{2+} from the sarcoplasmic reticulum).[216]

3.7 Tumour - The use of ^{31}P NMR spectroscopy to measure the oxygenation status of tumours has been reviewed with 19 references.[217] A study of the regional distribution of ATP, glucose, lactate and tissue pH, measured by bioluminescence and fluoroscopic imaging, has produce results which are relevant to the interpretation of volume selective NMR spectroscopy of tumour and areas of oedema.[218] A review with 95 references on the application of ^{31}P NMR spectroscopy to the study of tumour phospholipid metabolism has been produced.[219] The authors concluded that the results of many studies show there is no clear correlation between metabolites detected by ^{31}P NMR spectroscopy and cancer cell function or responsiveness to treatment.

An assessment of ^{31}P NMR spectroscopy as a method for the determination of tumour pH from the inorganic phosphate (Pi) resonance has been performed.[220] Acid extractable Pi, tumour exudate, interstitial fluid (from a micropore chamber within the tumour) and blood plasma were examined. The extracellular volumes were also determined. The results indicated that for those tumours with an extracellular volume not exceeding 55%, at least 65% of Pi detectected by ^{31}P

NMR spectroscopy is intracellular, and that pH determined from ^{31}P NMR spectroscopy is a measure of intracellular pH. The artefacts associated with ^{31}P NMR spectra of experimental tumours has been assessed by comparison of results with data from tumour extracts.[221] Despite using localised ^{31}P NMR spectroscopy it has been shown that signals from surrounding tissues may still contaminate tumour spectra. The ratio of phosphocreatine to γ-nucleotide triphosphate was measured fairly accurately when image selected **in vivo** spectroscopy was used but 1-D chemical shift imaging data from tumours was contaminated with phosphocreatine signals from the skin. In pH measurements, the affects of artifacts due to the localisation techniques were less than the biological variability in the actual tumour pH. These results indicate that phantom studies alone are not sufficient to assess the accuracy of localisation techniques. The absolute concentrations of phosphorous metabolites in RIF-1 tumours has been determined using a combination of ^{1}H, ^{2}H and ^{31}P NMR spectroscopy.[222] A decrease in the ^{31}P-metabolites, except inorganic phosphate, was observed in untreated tumours whereas treated tumours had no changes in metabolite concentrations.

^{31}P NMR spectroscopy has been used to examine the metabolism of transplanted squamous cell carcinoma VII after treatent with ^{60}Co γ-irradiation.[223] Changes were observed following radiation therapy and during tumour regrowth. The effects of the chemotherapeutic combination of N-(phosphonacetyl)-L-aspartate, 6-methylmercaptopurine and 6-aminonicotinamide with irradiation on the phosphorous metabolites of tumours has been observed by ^{31}P NMR spectroscopy.[224] The combination therapy, which caused tumour regression, was seen to deplete high energy phosphates at 10 hours after treatment. ^{31}P NMR spectroscopy has been used to examine the changes in phospholipid precursors and catabolites which occur with tumour growth following therapy.[225] The effects of growth and treatment with *cis*-dichlorodiamine platinum on the high energy phosphorus metabolism of human bladder carcinoma grown in nude mice has been folowed with ^{31}P NMR spectroscopy.[226] Changes in the phosphorous metabolites were observed during growth and detected in treated tumours as early as 6 hours after administration of *cis*-dichlorodiamine platinum. No changes in ^{1}H NMR images were observed following treatment.

Changes in tumour high energy metabolites has been observed when tumour blood flow has been affected.[227] The role of oxygen and glucose in the changes observed in ^{31}P NMR spectra has been examined in isolated perfused tumours. At lower perfusion rates (similar to those **in vivo**), the ratio of nucleotide triphosphates to inorganic phosphate was proportional to the flow rate and mostly affected through glucose deleivery. The metabolism of human adenocarcinoma cells (DU 145) perfused with different glucose concentrations and examined at different growth stages has been compared to the metabolism of the tumour line grown in nude mice.[228] Using ^{31}P NMR spectroscopy to examine cells it was shown that those cells grown in 5.5mM glucose and harvested during the log phase of growth had metabolite levels which reflected those metabolites detected by ^{31}P NMR spectroscopy in tumours grown in nude mice.

A difference in the ratio of choline to creatine has been detected in the ^{1}H NMR spectra of excised metastatic colon cancers compared with the ^{1}H NMR spectra of normal mucosa.[229] The ^{1}H NMR spectra of normal brain contralateral to implanted glioma cells has been compared to the ^{1}H NMR spectra of the tumour arising from the implanted glioma cells.[230] Localised ^{1}H NMR spectra from normal brain and brain areas contralateral to implanted tumours had resonances for N-acetylaspartate, total creatine, choline-containing compounds, myo-inositol, glutamate and glucose. However, ^{1}H NMR spectra from the tumour region had a near absence of N-acetylaspartate, a reduced level of total creatine and glucose, and an increase in choline-containing compounds. There was no consistent relationship between the changes observed by ^{1}H NMR spectroscopy and the tumour development. Primary human prostatic adenocarcinomas and their metastases have been examined by ^{1}H NMR spectroscopy **in vivo** and the results compared to ^{1}H NMR spectra of tumour extracts.[231]

A homonuclear editing sequence has been employed to detect and quantify the [3-^{13}C]lactate resonance in intercerebral C6 rat gliomas infused with [1-^{13}C]glucose.[232] The effects of hyperglycaemia on tumour lactate were also studied. The role of ^{19}F NMR spectroscopy in new anticancer drug development has been discussed.[233] The advantages and pitfalls are illustrated by results from the study of two fluorinated anti-cancer agents; a β,β-difluoro analoque of the alkylating agent chlorambucil and a fluorinated derivative of the nitroimidazole misonidozole, Ro 07-0741. The attempted detection of ^{2}H-labelled porphyrin photosensitisers by ^{2}H NMR spectroscopy has been reported.[234] The photosensitisers, which absorb red light (630nm), were injected directly into the tumour but they could not be detected when the tumour was examined in a solenoid coil.

3.8 Skin - The changes detected by ^{1}H NMR spectroscopy in the water signals from skin in diseases, aging and after ultra violet irradiation has been reviewed with 11 references.[235] Skin wounds in rats have been monitored using ^{31}P NMR spectroscopy in combination with radio frequency coil with a shallow sensitive profile.[236] Decreases in the ratio of phosphocreatine to inorganic phosphate was observed to correlate with severity of skin injury.

3.9 Lung - ^{31}P NMR spectroscopy has been used to monitor changes in the metabolism of perfused rat lung exposed to hypoxia and compliment activation.[237]

3.10 Reproductive system - Examination of the vas deferens of the blue crab[238] (**Callinectes sapidus**) revealed a chemical shift difference between the α-ATP and β-ATP resonances of 9.8ppm which implies a free intracellular magnesium concentration of less than 100μM. The crab vas deferens may, therefore, represent a new model for a low magnesium system. An investigation of the the aldose reductase activity of the rat testes has been performed using ^{19}F NMR spectroscopy to follow the uptake and metabolism of 3-fluoro-2-deoxy-D-glucose.[239]

3.11 Bladder - ^{19}F NMR spectroscopy has been employed to detect the *m*-(trifluoromethyl)-containing penicillin V analogue, LY242072 in the bladders of anaesthetised rats with occluded urethras.[240]

3.12 Eye - The ability of three preservation fluids to preserve pig corneas has been assessed by ^{31}P NMR spectroscopy.[241] ^{31}P NMR has been used to assess the damage caused by near UV irradiation to pigmented lenses from grey squirrels.[242] A fall in ATP was detected with increasing radiation dose. The presence of fructose 3-phosphate in the isolated rat, pig or rabbit lens was detected by ^{31}P NMR spectroscopy when the tissue was perfused with fructose, but not when it was perfused with glucose or sorbitol.[243] Fructose 3-phosphate may play a role in protein cross-linking in the lens during cataractogenesis. A decrease in the energy requirement of the perfused toad (**Bufo marinus**) retina during exposure to light has been indicated by results from ^{31}NMR spectroscopy.[244] In saturating levels of light (6.0 x 10^{11} or 1.5 x 10^{13} quant s^{-1} cm^{-2}) phosphocreatine and phosphomonoesters rose whilst nucleotide triphosphates and pH remained constant. The levels of pyridine nucleotides and nucleotide diphosphates fell during exposure.

3.13 Whole animal - The metabolism of [3-^{13}C]alanine has been thoroughly investigated using ^{13}C NMR spectroscopy following the administrsation of the substrate to last instar larvae of the insect **Manduca sexta**. Differences were observed for the incorporation of label into glutamate in the starved insects compared to the incorporation of label in control and parasitised larvae. The ^{13}C-label was shown to be incorporated in many sites.[245] Studies of the glucose metabolism in the beetle (**Tribolium confusum**) have utilized ^{13}C NMR spectroscopy to monitor the metabolism of D-[1-^{13}C]glucose and ^{31}P NMR spectroscopy to monitor the accumulation of phosphorylated intermediates.[246] ^{13}C NMR spectroscopy detected glycogen, trehalose, triacylglycerol, alanine and proline and ^{31}P NMR spectroscopy detected increases in arginine phosphate, sugar phosphates and uridine diphosphoglucose.

The effects of urethane anaesthesia on the phosphorous metabolism of loaches has been studied with ^{31}P NMR spectroscopy.[247] A decrease in the inorganic phosphate peak was observed at the same time as a decrease in the intracellular pH. The changes in high energy phosphates of the slug (**Incilaria bilineata**) in response to mechanical stimmulation or exposure to NaCl have been followed with ^{31}P NMR spectroscopy.[248] Mechanical stimmulation or contact with NaCl caused mucus secretion and a ^{31}P NMR-detected fall in arginine phosphate and a rise in inorganic phosphate. The β-ATP concentration depended on the amount of mucus secreted.

The kinetics of phosphoarginine-adenosine triphosphate exchange has been measured in the microscopic nematode parasite, **Stienernema carpocapsae**. Saturation transfer and inversion recovery techniques were employed to measure the kinetic constants for the forward and reverse reactions as 0.37 s^{-1} and 1.45 s^{-1}, respectively.[249] The effects of body temperature on tail muscle pH and phosphorous metabolism has been studied in newts and lungless red back salamanders using ^{31}P NMR spectroscopy.[250] In both organisms the pH of tail muscle and the ratio of phosphocreatine to inorganic phosphate decreased with increasing body temperature. Furthermore, in both species the ratio of β-ATP to inorganic phosphate rose with increasing body temperature. However, the calculated α-imidazole for skeletal muscle fell with increasing body temperature in red back salmanders, except when salamanders were supplied with extra O_2 at higher temperatures, then the level of α-imidazole was comparable to that found in newts.

4 Clinical studies

The use of localization in clinical NMR spectroscopy has been reviewed, with 80 references, comparing the merits of various techniques.[251] A review, with 82 references, has been produced on the techniques of **in vivo** magnetic resonance imaging and spectroscopy including their application to measurement of activity related changes of regional blood flow in the brain.[252] The use of NMR spectroscopy to study metabolic changes, non-invasively, in human has been reviewed without references.[253] A comparison of localisation technique for NMR spectroscopy **in vivo** has been reported[254]; the techniques of topical magnetic resonance, image selected **in vivo** spectroscopy and chemical shift imaging have been compared. ^{31}P and ^{1}H NMR spectroscopy for clinical diagnosis has been reviewed with 9 references.[255] A reveiw with 79 references has been produced on ^{13}C NMR spectroscopy in humans covering coils, methods of acquisition, spectroscopic editing and infusion studies.[256]

The use of ^{31}P NMR spectroscopy for the examination of normal and diseased heart has been described.[257] A review, with 8 references, has been produced on the use of ^{31}P NMR spectroscopy in the study of hypertrophic cardiomyopathy, cardiomegaly and cardiac insufficiency.[258] The use of ^{31}P NMR for studies of the myocardium has been reviewed with 45 references.[259] A review, with 75 references, on the role of NMR in the characterisation of secondary mechanism following traumatic brain injury has been produced[260]; the role of high energy phosphates, pH, lactic acid, excitatory amino acids and magnesium along with pharmacotherapies has been discussed. The role of nuclear magnetic resonance in the study of nephrotoxicity over the past 12 years has been reviewed with many references.[261] The potential of NMR spectroscopy techniques for the study of xenobiotic nephrotoxicity has been emphasised.

The biomedical implication of measurements of intracellular pH has been reviewed with 155 references.[262] Two non-invasive methods (NMR spectroscopy and an optical method) for the determination of mitochondrial (dis)function have been compared.[263] The prospect of analysing the ischaemic penumbra in stroke has been reviewed with 51 references.[264]

The utilization and recovery of ATP in human skeletal muscle has been measured using ^{31}P NMR spectroscopy and a combination of electrical stimmulation and muscle ischaemia techniques.[265] These techniques allowed the separate measurements of the basal metabolic rate of muscle, the rate of utilisation of phosphocreatine and the recovery of phosphocreatine following depletion. The capacity for oxidative phosphorylation was measured to be 0.4mM s^{-1}. Further measurements of skeletal muscle energetics have also been studied in relation to the utilization and recovery ATP following twitch stimulation under anaerobic conditions.[266] The results demonstrated a variation in the contractile cost and oxidative capacity between individual subjects which could be measured by ^{31}P NMR spectroscopy. In a separate study, the depletion of phosphocreatine as a result of voluntary isometric contraction and its subsequent recovery under ischaemic conditions was measured using ^{31}P NMR spectroscopy.[267] Depletion of phosphocreatine was accompanied by an increase in inorganic phosphate, though no change in ATP was observed. The rate of change in PCr and Pi during exercise was related to the severity of exercise and there was no recovery of phosphorous metabolites during a 5 minute ischaemic period immediately following exercise. It was concluded that glycolysis and glycogenolysis was halted during this ischaemic period despite the presence of potent activators and that other regulators play an important role in the control of glycolysis and glycogenolysis in ischaemic muscle.

In a study of muscle fatigue, using ^{31}P NMR spectroscopy to measure the pH and the concentration of PCr, ATP and Pi, the extent of muscle acidosis as a result of exercise was determined to be less when muscle glycogen was depleted prior to the onset of 5min bouts of exercise.[268] In glycogen depleted muscle, fatigue correlated most closely with the concentration of ADP (r=0.84), though in control subjects fatigue correlated most closely with the concentration of PCr (r=0.83). The effects of theophylline on muscle fatique has been studied using ^{31}P NMR spectroscopy.[269] Volunteers recieving 300mg of sustained release theophylline increased their maximal power output by 19% and increased the endurance of forearm musculature and delayed the onset of metabolic acidosis.

In studies of muscle metabolism using ^{31}P NMR spectroscopy the presence of a split or broadened inorganic phosphate peak, denoting a heterogeneous pH environment, has been observed many times. Using a method of localisation which allows several, sepatate regions to be observed simultanously ^{31}P NMR spectra have been obtained from single muscles; the soleus, medial and gastrocnemeus muscles.[270] The results showed that single muscles have a broadened Pi peak probably due to different pH values of different fibre types. The value of T_1 for phosphocreatine and β-ATP wre measured in human cardiac muscle and skeletal muscle.[271] The values of T_1 for PCr and ATP in cardiac muscle were reported to be 3.98s and 1.86s, respectively, and were not significantly different to the values of T_1 for PCr and ATP in skeletal muscle. The values of T_2 for α-ATP β-ATP and γ-ATP have been determined in normal muscle to be 74ms, 75ms and 93ms, respectively.[272] These values, measured using the sequence, 90°-TE/2-2662-TE/2-acq, are shorter than values generally reported elsewhere.

In a ^{13}C NMR spectroscopy study of skeletal muscle in non-insulin dependent diabetes melitus the rate of glycogenolysis and the total glcogen content of muscle were found to be 50% of normal values.[273] Studies **in vitro** indicated that the protein glycogenin could have a role in the cause of non-insulin dependent diabetes melitus. Natural abundance ^{13}C NMR spectra of gastrocnemius muscle have been used to determine the change in muscle glycogen following a mixed meal.[274] The level of muscle glycogen following overnight fasting was 83.3mM and rose to 100.2mM 4.9h after a meal. Glycogen levels then fell to 90.6mM at 7h postprandially. Muscle glycogen was calculated to have stored 26 to 35% of adsorbed carbohydrate. The use of gated proton decoupling in ^{13}C NMR spectroscopy has been compared to the use of ^{1}H-detected ^{13}C spectroscopy for the determination of muscle glycogen.[275] In both cases the increase in signal to noise was found to be 2.5-fold for **in vivo** measurements, making the techniques equivilently useful.

The lipid signals from calf muscles have been examined using a double spin-echo localisation technique.[276] Spectra showed peaks for methyl, methylene, olefinic and other fatty acids resonances, creatine plus phosphocreatine, choline plus carnitine, taurine and histidine. Variation was observed in the lipid spectra from different muscle groups. Some changes were observed in three individuals with myopathia, myositis or irradiation damage. ^{1}H NMR spectra of the soleus muscle obtained with the double spin-echo technique have revealed two lipid resonances, separated by 0.2ppm.[277] Results of lineshape analysis and relaxation measurements suggested two compartments of stored lipid, for example fat cells within muscle and lipid stored within muscle cells.

Changes in glucose concentrations during photic stimmulation have been measured in $12cm^3$ volumes centred on the calcarine fissure and encompassing the primary visual cortex.[278] The change in glucose concentration was used to determine the change in the rate of cerebral glucose metabolism. A comparison has been made of the accumulation of lactate in the visual cortex of two patients with mitochondrial encephalomyopathy and the accumulation in normal volunteers.[279] Lactate accumulation was increased in test subjects following visual stimmulation. Localised ^{1}H NMR spectroscopy with short stimmulated echo times has been used to examine a number of diseases including stroke, multiple sclerosis and AIDS-related leukoencephalopathies.[280] Examination of the data using principle component analysis distinguished between the demilination process of ischaemic lesions and the demilination process of leukoencephalopathies.

Localised ^{1}H NMR spectroscopy has been used to measure γ-aminobutyric acid in the occipital lobe of healthy volunteers and epileptic patients recieving vigabatrin, a drug known to elevate γ-aninobutyric acid.[281] The concentration of γ-aninobutyric acid was 1.1 $\mu mol\ cm^{-3}$ of brain in controls and 3.7 $\mu mol\ cm^{-3}$ of brain in patients. The absolute concentrations of N-acetyl aspartate, total creatine, choline and *myo*-inositol have been determined in 22 subjects using the compartmentation model previously established by the same authors.[282] Image-selected localised ^{1}H NMR spectroscopy has been used to determine the concentration of N-actetyl aspartate (11.5mM), choline (2.0mM) and lactate (0.6mM) in the white matter of normal volunteers.[283] Further, the ratio of choline, creatine and lactate to N-acetylaspartate was measured in patients with chronic infarction and magnetic resonance imaging-oberserve white matter hyperintensities. A change in the value of T_2 for creatine in the white matter of patients with cerebral infarcts was reported. The values of T_1 and T_2 have been determined for N-acetylaspartate, creatine plus phosphocreatine and choline containing compounds detected in the brain.[284] There were no differences detected for the values of T_1 or T_2 for these compounds in stroke patients compared to control subjects. A method for the determination of absolute quantities of metabolites in the human brain, using water as the internal standard, has been used to measure the concentrations of N-acetylaspartate (11.6mM), total creatine (7.6mM) and choline (1.7mM) in the occipital lobe of the brain in 5 healthy volunteers.[285]

A decrease in N-actetyl aspartate has been detected in the centrum semiovale and the mesial cortex of the brains of patients with HIV-associated cognitive impairment.[286] The presence of sycllo-inositol has been indentified in ^{1}H NMR spectra of mammalian brain **in vivo** and **in vitro.**[287] Its presence was noted as a singlet peak appearing at a chemical shift of 3.35ppm. Changes in the lipid peak at 1.2ppm in ^{1}H NMR spectra of patients following electroconvulsive therapy have been observed.[288] A similar increase in this peak was also observed in epileptic subjects who were temporarily off medication; the signal disappeared after remedication. The results suggest a role for free fatty acids in seizure. ^{19}F NMR spectroscopy has been applied to the determination of fluoxitine, and other fluorinated neuroleptics, in the brains of patients under steady state treatment.[289] The concentration of fluoxitine was 1.3 to 5.7 $\mu g\ cm^{-3}$ in patients recieving 40 mg day^{-1}. Signals from the other fluorinatd neurolepics were more difficult to obtain.

The effects of fish oil (ω3 fatty acids) on the ^{31}P NMR spectra of liver have been examined.[290] In normal volunteers the ratio of phosphodiesters to ATP was elevated 1 to 3 days after fish oil but this ratio returned to baseline levels after seven days. However, when spectra were recorded with a longer repitition time between scans the PME to ATP level remained high. The results were supported by evidence from studies in the rat, and suggest that there is a shift from phospholipase C to the phospholipase A1/A2 pathway of lipid breakdown after fish oil consumption. The formation of glycogen in the liver of normal volunteers has been observed using ^{13}C NMR spectroscopy following the infusion of [1-^{13}C]glucose under hyperglycaemic, hyperinsulinaemic clamp conditions and following oral intake of glucose as a bolus.[291] ^{13}C NMR signals from 99% enriched [1-^{13}C]glucose could be detected 8min after the onset of infusion but labelling of glycogen did not occur until about 60min. In addition, the formation of liver glycogen could be detected following infusion with non-labelled glucose or 6.6% labelled [1-^{13}C]glucose.

The prostate gland has been examined by localised ^{1}H NMR spectroscopy to determine the presence of citrate, a normal secretion of this gland. Results obtained with a 170mm coil were compared to results obtained by other authors using a transrectal probe.[292] ^{19}F NMR spectroscopy has been used to monitor the presence of sodium flomoxef in the normal heart, liver and kidney.[293] However, signals from the lung were only detected when a cancer associated massive atelectasis was present. The relaxation times T_1 and T_2 for water and lipids in human haemopoietic bone marrow of 14 normal volunteers and 10 patients with acute leukaemia and bone marrow carcinosis were determined using double spin-echo localised spectroscopy **in vivo.**[294] The reduced lipid signals in patients with acute lukaemia prevented measurement of these relaxation parameters, however the T_2 of water was determined to be prolonged in the lukaemia patients, though T_1 was unaltered.

5 Plants

Changes in the value of T_1 for inorganic phosphate have been observed in excised cortical tissue and N_2-fixing root nodules of **Bradyrhizobium japonicum** in the soya bean plant (**Glycine max**) when samples were perfused with N_2 or Ar in place of O_2. The change in the value of T_1 was associated with an acidic change in pH to 5.5units. There was no observed change in the value of T_2. Soya bean tissue did not exhibit the same change in the value of T_1 for inorganic phosphate. It was suggested that the changes observed in T_1 were the result of physical alterations in the environment of inorgaic phosphate in the cortical tissue.[295] In a futher study using selected combinations of perfusion with O_2, N_2 or air the changes in the observed value of T_1 were investigated in more detail.[296] The results suggested that cortical cells respond differently to the selected gas perfusions compared to the response of the inner nodule to the same perfusion regime. Furthermore, the boundary of T_1 appeared to change sharply at the interface of the inner nodule and inner cortex.

^{31}P and ^{23}N NMR spectroscopy has been used to investigate maize (**Zea mays**) and **Spartina angelica** roots during exposure to NaCl in the range of 0 - 500mM. A small cytoplasmic alkalinisation (0.1-0.2 pH units) and a larger vacuolar alkalinisation (0.6 pH units) was observed in maize root tips exposd to >200mM NaCl which suggested the role of a tonolast Na^+/H^+ antiport which exceeded the rate of the tonoplast H^+ pumps. **S. angelica** root tips accumulated sodium more slowly than maize roots with no change in cytoplasmic pH and little change in vacuolar pH. The results suggest that NaCl tolerance in **S. angelica** root tips is based partly on its ability to prevent the net influx of sodium.[297] ^{31}P NMR spectroscopy has been used to measure the phosphorous contents of the root of maize plants grown for up to 12 days in the presence of 0.1mM, 10mM or 0.5mM phosphate.[298] There was a 10-fold difference in the vacuolar inorganic phosphorous contents between these groups and the vacuolar inorganic phosphorous content of seeds grown without phosphate was too low to be measured accurately after 7 days.

$^{15}NH_4Cl$ was used to label the pathway of nitrogen assimilation in **Wolffia arrhiza** in a comparison with a previous study on **Lemma gibba.**[299] Excess $^{15}NH_4^+$ was detected in the cells of **W. arrhiza** in contrast to **L. gibba** where no $^{15}NH_4^+$ was detected; it was suggested that this is due to the lower rate of glutamate and aspartate resynthesis in **W. arrhiza**. Addition of α-ketoglutarate prevented the build-up of excess $^{15}NH_4^+$ in **W. arrhiza** and increased $^{15}NH_4^+$ assimilation in both species. Inhibition of glutamine synthetase inhibited $^{15}NH_4^+$ assimilation in both species, whereas inhibition of glutamate synthase lead to the incorporation of $^{15}NH_4^+$ solely into the glutamine-amide position suggesting that the gutamine synthetase/glutamate synthase pathway is the major route of $^{15}NH_4^+$ assimilation in these species. The metabolism of $^{15}NO_3^-$ and

$^{15}NH_4^+$ has been followed in plants labelled with either of the substrates administered in pulses.[300] Illumination of the samples allowed a high biomass density to be used for direct observation of ^{15}N assimilation. The uptake of nitrate and ammonia by maize root tips has been studied using ^{14}N and ^{15}N NMR spectroscopy **in vivo**; the results were also confirmed **in vitro.**[301] Metabolic inhibitors and N-starvation was used to reveal that raising glutamine and/or asparaginine resulted in the suppression of nitrogen whereas depletion of glutamine and/or asparaginine resulted in the increased incorporation of nitrogen in to maize metabolites. The results have been discussed in relation to the control of nitrogen absorption during growth with differing levels of nitrogen supply.

^{13}C NMR spectroscopy has been used to monitor the mobilization of lipid in the ground nut seed (**Arachis hypogaea L.**).[302] The radicle, hypocotyl and plumule exhibited time dependent ^{13}C NMR spectra. The results have been discussed in terms of the mechanisms of transformation of storage lipid. Using a pulsed field gradient method the self diffusion coefficient for water in soya beans has been measured to be greater than the literature values quoted for the measurement of the self diffusion of water made by other methods.[303] The water content was determined to be 0.1g to 1.5g H_2O/g. The use of ^1H NMR spectroscopy for the study of seasonal changes in plants, with particular reference to the development of the wine grape bud, has been described.[304]

Sugar beet root has been examined by ^{23}Na and ^{31}P NMR spectroscopy and ^1H NMR imaging.[305] The intracellular ^{23}Na was shown to be 9.2mM when root were exposed to 180-mM NaCl. A broadening of vacuolar-associated inorganic phosphate was detected by ^{31}P NMR spectroscopy when the root were exposed to NaCl. The strong water signal detected by ^1H NMR imaging in the epidermal and vicinal cells, and around the vascular bundles was reduced by NaCl treatment. The effects of treatment with 6mM Ca^{2+} on barley roots exposed to 100mM NaCl has been monitored with ^{31}P NMR spectroscopy.[306] In the absence of 6mM Ca^{2+} there was an alkalinisation of the vacuolar pH which was decreased by Ca^{2+} treatment. There was a correlation between the extent of alkalinisation and the accumulation of Na^+ within the root. which was consistent with the involvement of th Na^+/H^+ antiporter in Na+ transport .

References

1. Sanuki, Eiichi *Nichidai Igaku Zasshi* 1993, **52(10)**, 587-92.
2. Canese, R.; Podo, F. *Ann. Ist. Super. Sanita* 1991, **27(4)**, 675-708.
3. Ruiz-Cabello, Jesus; Cohen, Jack S. *Int. Rev. Cytol.* 1993, **145**, 1-63.
4. Kauppinen, Risto A. *Trends Neurosci.* 1993, **16(10)**, 384-6.
5. Kusuoka, Hideo; Fukuchi, Kazuki; Nishimura, Tsunehiko *Mol. Med. (Tokyo)* 1993, **30(11)**, 1494-7.
6. Kauppinen, Risto A.; Williams, Stephen R.; Busza, Albert L.; Van Bruggen, Nicholas *Trends Neurosci.* 1993, **16(3)**, 88-95.

7. Radda, George K.; Taylor, Doris J. *New Compr. Biochem.* 1992, **23(Molecular Mechanisms in Bioenergetics)**, 463-82.

8. Ross, Brian D. *NMR Biomed.* 1992, **5(5)**, 215-19.

9. Rudin, M.; Sauter, A. *NMR* 1992, **28(In-Vivo Magn. Reson. Spectrosc. III)**, 161- 88.

10. Berliner, Lawrence J.; Reuben, Jacques; Editors (Plenum: New York, N.Y.). 1992, 334 pp.

11. Selinsky, Barry S.; Burt, C. Tyler *Biol. Magn. Reson.* 1992, **11**, 241-76.

12. Prior, M. J. W.; Maxwell, R. J.; Griffiths, J. R. *NMR* 1992, **28(In-Vivo Magn. Reson. Spectrosc. III)**, 101-30.

13. Leibfritz, Dieter, Offermann, Werner; Stelten, Johannes; Koschorek, Frank; Jans, Arnold W. H.; Kuhn, Winifried; Soboll, Sybille *Nova Acta Leopold* 1992, **60(266, Nuklear- Magnetische Resonanzverfahren in Biologie und Medizin)**, 57-69.

14. Brown, Truman R. *NMR Biomed.* 1992, **5(5)**, 238-43.

15. Rumpel, Helmut; Pope, James M. *Concepts Magn. Reson.* 1993, **5(1)**, 43-55.

16. Bottomley, Paul A. *NMR* 1992, **27(In-Vivo Magn. Reson. Spectrosc. II)**, 67-102.

17. Doddrell, David M.; Brereton, Ian M.; Crozier, Stuart *Magn. Reson. Microsc.* 1992, 547- 62.

18. Kunze, C.; Kimmich, R.; Demco, D. E. *J. Magn. Reson. Ser. A* 1993, **101(3)**, 277-84.

19. Budinger, Thomas F. *Radiat. Res. [Proc., Int. Congr.],* 9th 1991 **(Pub. 1992), 2**, 841-8.

20. van Dijk, J. E. *Gov. Rep. Announce, Index (U.S.)* 1992, **92(11)**, Abstr. No. 228,501.

21. Koestler, Herbert; Kimmich, Rainer *J. Magn. Reson. Ser. B.* 1993, **102(2)**, 177-82.

22. Reddy, Ravinder; Subramanian, V. H.; Clark, B. J.; Leigh, John S. *J. Magn. Reson., Ser. B* 1993, **102(1)**, 20-25.

23. Bolo, N. R. *Energy Res. Abstr.* 1992, **17(7)**, Abstr. No., 19577.

24. Adalsteinsson, Elfar; Spielman, Daniel M; Wright, Graham A.; Pauly, John M.; Meyer, Craig H.; Macovski, Albert *Magn. Reson. Med.*, 1993, **30(1)**, 124-30.

25. Wilman, Alan H.; Allen, Peter S. *J. Magn. Reson., Ser. B* 1993, **101(2),** 165-71.

26. Freeman, D.; Hurd, R. *NMR* 1992, **27(In-Vivo Magn. Reson. Spectrosc. II),** 199- 222.

27. Fetler, Bayard K.; Simplaceanu, Virgil; Vander Ven, Ned S.; Lowe, Irving J.; Ho, Chien *J. Magn. Reson. Ser. B* 1993, **101(1)**, 17-27.

28. Watanabe, Hidehiro (Cl. A61B5/055), 19 Oct. 1993. Appl. 92/65,078, 23 Mar, 1992; 15 pp.

29. Deriche, Mohamed; Hu, Xiaoping *J. Magn. Reson. Ser A* 1993, **101(2)**, 229-32.

30. Rudin, Markus; Sauter, Andre *NMR* 1992, **27(In-Vivo Magn. Reson. Spectrosc. II)**, 257-93.

31. Navon, Gil; Kushnir, Tammar; Askenasy, Nadir; Kaplan, Ofer *NMR* 1992, **27(In-Vivo Magn. Reson. Spectrosc. II)**, 237-55.

32. Jung, Wulf Ingo; Straubinger, Klaus, Bunse, Michael; Widmaier, Stefan; Schick, Fritz; Kueper, Klaus; Dietze, Guenther; Lutz, Otto *Magn. Reson. Med.* 1993, **30(1)**, 138-41.

33. Ziegler, A.; Decorps, M. *J. Magn. Reson., Ser. B* 1993, **102(1)**, 26-34.

34. Mitsumori, Fumiyuki *J. Magn. Reson., Ser. A* 1993, **102(2)**, 228-30.

35. Wray, S.; Wilkie, D. R. *NMR Biomed.* 1992, **5(3)**, 137-41.

36. Hughes, Mark S. *Lithium Cell* 1991, 175-84.

37. Allis, Jonathan L. *Annu. Rep. NMR Spectrosc.*, 1993, **26**, 211-46.

38. Ramasamy, R.; Castro, M. M. C. A.; De Freitas, D. M.; Geraldes, C. F. G. C. *Biochimie* 1992, **74(9-10)**, 777-83.

39. Williams, Gerald D.; Mosher, Timothy J.; Smith, Michael B. *Anal. Biochem.* 1993, **214(2)**, 458-67.

40. Jayasundar, Rama; Hall, Laurance D.; Bleehan, Norman M. *NMR Biomed.* 1992, **5(6)**, 360-3.

41. Gil, M. S.; Cruz, F.; Cerdan, S.; Ballesteros, P. *Bioorg. Med. Chem. Lett.* 1992, **2(12)**, 1717-22.

42. Smith, Gerry A.; Kirschenlohr, Heide L.; Metcalfe, James C.; Clarke, Sonia D. *J. Chem. Soc. Perkin Trans. 2* 1993, **(6),** 1205-9.

43. Clarke, Sonia D.; Metcalfe, James C.; Smith, Gerry A. *J. Chem. Soc., Perkin Trans.* 2 1993, **(6)**, 1187-94.

44. Mason, Ralph P.; Shukla, Himu; Antich, Peter P. *Magn. Reson. Med.* 1993, **29(3)**, 296-302.

45. Gout, Elizabeth; Bligny, Richard; Pascal, Nadine; Douce, Roland *J. Biol. Chem.*, 1993, **268(6)**, 3986-92.

46. Allan, Andrew C.; Shachar-Hill, Yair; Rubery, Philip H. *Planta* 1993, **189(3)**, 306-11.

47. Reuveni, M. *Biol. Plant.* 1993, **35(1)**, 1-9.

48. Gasbarrini, Antonio; Borle, Andre B.; Farghali, Hassan; Caraceni, Paola; Thiel, David Van *Biochim. Biophys. Acta* 1993, **1178(1)**, 9-19.

49. Street, James C.; Delort, Anne Marie; Braddock, Peta S. H.; Brindle, Kevin M. *Biochem. J.* 1993, **291(2)**, 485-92.

50. Floegel, Ulrich; Leibrfritz, Dieter *Biol. Chem. Hoppe-Seyler* 1993, **374(3)**, 211-16.

51. Yoshizaki, Kazuo; Arizono, Naoki; Hayano, Takashi; Watari, Hiroshi *Magn. Reson. Med.* 1993, **29(6)**, 732-6.

52. Kuchel, Philip W.; Chapman, Bogdan E.; Xu, Arron S. L. *Prog. Cell Res.* 1992, **2 (Band 3 Proteins: Anion Transporters, Binding Proteins and Senescent Antigens)**, 105-19.

53. Yamaguchi, Takeo; Morimune, Kousuke; Kimoto, Eiji *Fukuoka Daigaku Rigaku Shuho*, 1992, **22(2)**, 153-7.

54. Rude, Robert K.; Stephen, Anthony; Nadler, Jerry *Magnesium Trace Elem.*, 1991-1992 (Pub. 1993), **10(2-4)**, 117-21.

55. Matsuura, Toshifumi; Kanayama, Yoshiharu; Inoue, Takatoshi; Takeda, Tadanao; Morishima, Isao *Biochim. Biophys. Acta* 1993, **1220(1)**, 31-6.

56. Barbagallo, M.; Gupta, R. K.; Resnick, L.M. *Diabetologia*, 1993, **36(2)**, 146-9.

57. Xu, Arron S. L.; Waldeck, A. Reginald; Kuchel, Philip W. *NMR Biomed.*, 1993, **6(2)**, 136-43.

58. Barry, Judith A.; McGovern, Kathy Ann; Lein, Yeong Hau H.; Ashmore, Brian; Gillies, Robert J. *Biochemistry* 1993, **32(17)**, 4665-70.

59. Tauskela, Joseph S.; Shoubridge, Eric A. *Biochim. Biophys. Acta*, 1993, **1158(2)**, 155-65.

60. Shinar, Hadassah; Eliav, Uzi; Navon, Gil *Isr. J. Chem.*, 1992, **32(2-3)**, 299-304.

61. Shinar, Hadassah; Knubovets, Tatyana; Eliav, Uzi; Navon, Gil *Biophys. J.*, 1993, **64(4)**, 1273-9.

62. Rong, Qinfen; Espanol, Maryceline; Mota de Freitas, Duarte; Geraldes, Carlos F. G. C. *Biochemistry*, 1993, **32(49)**, 13490-8.

63. Cacciafesta, Mauro; Ferri, Claudio; Scuteri, Angelo; Piccirillo, Gianfranco; Bellucci, Catia; Persichino, Loredana; Ariani, Alarico; Marigliano, Vincenzo; Delfini, Maurizio; Miccheli, Alfredo *Cell. Mol. Biol. (Paris)*, 1992, **38(8)**, 877-84.

64. Barbagallo, Mario; Gupta, Raj K.; Resnick, Lawrence M. *Am. J. Hypertens*, 1993, **6(4)** 264-7.

65. Kuwabara, Takeo *Niigata Igakkai Zasshi* 1992, **106(6)**, 563-72.

66. Marjanovic, M.; Gregory C.; Ghosh, P.; Willis, J.S.; Dawson, M.J. *J. Physiol. (London)* 1993, **470**, 559-74.

67. Perng, Wann Cherng; Price, William S.; Hsu, Kang; Hwang, Lian Pin *Eur. J. Clin. Chem. Clin. Biochem.*, 1993, **31(7)**, 413-18.

68. Schrader, Michael C.; Eskey, Clifford J.; Simplaceanu, Virgil; Ho, Chien *Biochim. Biophys. Acta*, 1993, **1182(2)**, 162-78.

69. Holmsen, Holm; Ugurbil, Kamil *Platelet Amine Storage Granule*, 1992, 51-71.

70. Bental, Michal; Deutsch, Carol *Magn. Reson. Med.* 1993, **29(3)**, 317-26.

71. Govil, Girjesh; Phadke, Ratna S.; Srivastava, Sudha *Spectrochim. Acta, Part A* 1992, **48A(11-12)**, 1759-66.

72. Farghali, H.; Williams, D. S.; Caraceni, P.; Borle, A. B.; Gasbarrini, A.; Gavaler, J.; Rilo, H. S.; Ho, C.; Van Thiel, D. H. *Endocrinology* 1993, **133(6)**, 2749-55.

73. Kao, Ting Lin; Meyer, Walter J. III; Post, Jan F. M. *Cancer Lett. (Shannon, Irel.)* 1993, **70(1-2)**, 101-6.

74. Adebodun, Foluso; Post, Jan F. M. *J. Cell. Physiol.* 1993, **154(1)**, 199-206.

75. Post, Jan F.M.; Varma, Rajender S. *Cancer Lett. (Shannon, Irel.)*, 1992, **67(2-3)**, 207-13.

76. Schanne, Francis, A. X.; Gupta, Raj K.; Rosen John F. *Biochem. Biophys. Acta,* 1992, **1180(2)**, 187-94.

77. Degani, Hadassa; DeJordy, John O.; Soloman, Yoram *J. Recept. Res.*, 1993, **13(1-4)**, 55-68.

78. Shikata, Takuma; Nishida, Masashi; Hosoi, Hajime; Matsumura, Takafumi; Kawakatsu, Hidekazu; Sugimoto, Tohru; Umeda, Masahiro; Nishikawa, Hiroyasu; Sawada, Tadashi *Igaku no Ayumi* 1993, **166(9)**, 623-4.

79. McGovern, K. A.; Schoeniger, J. S.; Wehrle, J. P.; Ng C. E.; Glickson, J. D. *Magn. Reson. Med.* 1993, **29(2)**, 196-204.

80. Vion-Dury, Jean; Machy, Patrick; Confort-Gouny, Sylviane; Leserman, Lee; Cozzone, Patrick *J. Magn. Reson. Med.* 1993, **29(2)**, 252-5.

81. Le Moyec, L.; Tatoud, R.; Eugene, M.; Gauville, C.; Primot, I.; Charlemagne, D.; Calvo, F. *Br. J. Cancer*, 1992, **66(4)**, 623-8.

82. Callies, R.; Sri-Pathmanathan, R. M.; Ferguson, D. Y. P.; Brindle, K. M. *Magn. Reson. Med.* 1993, **29(4)**, 546-50.

83. Hansen, Lars Lindgaard; Rasmussen, Jette; Friche, Ellen; Jaroszewski, Jerzy W. *Anal. Biochem.* 1993, **214(2)**, 506-10.

84. Yamamoto,Yoshinori; Takamatsu Satoshi; Nakamura, Hiroyuki *J. Magn. Reson., Ser. B* 1993, **101(2)**, 198-200.

85. Hentrich, S.; Hebeler, M.; Grimme, L. H.; Leibfritz, D.; Mayer, A. *Eur. Biophys. J.* 1993, **22(1)**, 31-9.

86. Rossi, C.; Lepri, A.; Picchi, M.P.; Bastianoni, S.; Medaglini, D.; Vanassina, M.; Cresta, E. *Bull. Magn. Reson.*, 1992, **14(1-4, Proc. Int. Soc. Magn. Reson. Meet., 11th 1992)**, 197-201.

87. Schoberth, S. M.; De Graaf, A. A. *Anal. Biochem.* 1993, **210(1)**, 123-8.

88. Olsen, Lyn D.; Renshaw, Carolyn A.; Rotten, Shlomo; Boal, Jila H. *FEMS Microbiol. Lett.* 1993, **108(1)**, 47-51.

89. Chen, Ruizhen; Bailey, James E. *Biotechnol. Bioeng.* 1993, **42(2)**, 215-21.

90. Strohhaecker, Joachim; de Graaf, Albert A.; Schoberth, Siegfried M.; Wittig, Rolf M.; Sahm, Hermann *Arch. Microbiol.* 1993, **159(5)**, 484-90.

91. Ivanova, E. P.; Isakov, V. V.; Mikhailov, V. V.; Sokolova, S. V.; Gorshkova, N. M.; Fedosov, Yu V.; Kiprianova, E. A. *Mikrobiologiya* 1993, **62(2)**, 260-6.

92. Singh, Satya P.; Rogers, Peter J. *J. Gen. Appl. Microbiol.* 1992, **38(6)**, 511-21.

93. Futsaether, C. M.; Kjeldstad, B.; Johnsson, A. *Can. J. Microbiol.* 1993, **39(2)**, 180-6.

94. Sakhnini, Asad; Gilboa, Haggai *Biophys. Chem.* 1993, **46(1)**, 21-9.

95. Lyngstad, Magnus; Grasdalen, Hans. *J. Biochem. Biophys. Methods* 1993, **27(2)**, 105-16.

96. Taipa, M. A.; Cabral, J. M. S.; Santos, H. *Biotechnol. Bioeng.* 1993, **41(6),** 647-53.

97. Wietzerbin, J.; Herve, M.; Un, S.; Neumann, J. M.; Namane S.; Trans-Dinh, S. *Biochimie* 1993, **75(9)**, 825-30.

98. Herve, M.; Wietzerbin, J.; Lebourguais, O.; Tran Dinh Son *Biochimie* 1992, **74(12)**, 1103-15.

99. Yang, Yun Chung; Bastos, Margarita; Chen, Kuang Yu *Biochim. Biophys. Acta* 19193, **1179(2)**, 141-7.

100. Doyle, Timothy C.; Spickett, Corinne, M.; Donaldson, Iain A.; Ratcliffe, R. George *Yeast* 1993, **9(8)**, 867-73.

101. Williams, Simon Peter; Fulton, Alexandra M.; Brindle, Kevin M. *Biochemistry* 1993, **32(18)**, 4895-902.

102. Vink, Robert *Adv. Exp. Med. Biol.*, 1992, **316 (Oxygen Transport to Tissue XIII)**, 187-93.

103. Preece, N. E.; Challands, J.; Williams, S. C. R. *NMR Biomed.*, 1992, **5(2)**, 101-6.

104. Lanens, D.; Muller, H. J.; Van de Vyver, F.; De Cock-Bunning, T.; Spanoghe, M.; Van der Linden, A.; Mulder, G. J.; Dommisse, R.; Lugtenburg, *J. Magn. Reson. Imaging*, 1992, **10(6)**, 975-81.

105. De Bisschop, Elke; Lluypaert, Robert; Annaert, Guido; Coremans, Jean; Osteaux, Michel *NMR Biomed.*, 1992, **5(6)**, 341-6.

106. Casu, M.; Gibbons, W. A.; Lai, A.; Dessi, M. A. *Recent Adv. Cell. Mol. Biol. World Congr.*, 1992, **6**, 187-91.

107. Goecke, Carol M.; Jarnot, Bruce M.; Reo, Nicholas V. *Chem. Res. Toxicol.*, 1994, **7(1)**, 15-22.

108. Changani, K. Kumar; Williams, Steve C. R.; Iles, Richard A. *Biochem. Soc. Trans.*, 1993, **21(4)**, 453S.

109. Zange, Jochen; Gronczewski, Jacek; Jans, Arnold W. H. *Biochem. J.*, 1993, **293(3)**, 667-73.

110. Durand, Thierry; Gallis, Jean Louis; Masson, Serge; Cozzone, Patrick J.; Canioni, Paul *Am. J. Physiol.*, 1993, **265(1, Pt. 1)**, G43-G50.

111. Bansal, N.; Germann, M. J.; Seshan, V.; Shires, G. T., III; Malloy, C. R.; Sherry, A. D. *Biochemistry*, 1993, **32(21)**, 5638-43.

112. Sakai, Motonari; Nakashima, Toshiaki; Nakajima, Toshikazu; Yama, Hiroyoshi; Sakamoto, Yoshikuni; Shima, Toshihide; Okanoue, Takeshi; Kashima, Kei; Yoshizaki, Kazuo *Yakuri to Chiryo*, 1992, **20(Suppl. 10)**, S2493-S2497.

113. Walsh, Thomas R.; Detre, John A.; Koretsky, Alan P.; Simplaceanu, Elena; Halow, Jessica M.; Rao, Prakash; Makowka, Leonard; Ho, Chien *Biochim. Biophys. Acta*, 1993, **1181(1)**, 7-14.

114. Palombo, John D.; Bowers, John L.; Clouse, Melvin E.; McCullough Ann; Forse, R. Armour; Bistrian, Bruce R. *Am. J. Clin. Nutr.*, 1993, **57(3)**, 420-7.

115. Orii, Takashi; Ohkohchi, N.; Satomi, S.; Taguchi, Y.; Mori, S.; Miura, I. *Transplant. Proc.*, 1993, **25(1, Book 2)**, 1657-8.

116. Negishi, Chikashi *Nippon Igaku Hoshasen Gakkai Zasshi*, 1993, **53(5)**, 551-8.

117. Carbonneau, M. A.; Delmas-Beauvieux, M. C.; Gallis, J. L.; Perromat, A.; Canioni, P.; Clerc, M. *J. Liq. Chromatogr.*, 1993, **16(6)**, 1385-417.

118. Blackburn, B. J.; Buist, R. J.; Hudspeth, C.; Novak, M. *Int. J. Parasitol.*, 1993, **23(1)**, 95-103.

119. Buist, R. J.; Novak, M.; Hudspeth, C.; Blackburn, B. J. *Bull. Magn. Reson.*, 1993, **15(1-2)**, 55-9.

120. Dowd, Terry L.; Gupta Raj K. *J. Biol. Chem.*, 1993, **268(2)**, 991-6.

121. Boydstun, Ivy I.; Thulin, Gunilla; Zhu, Xiao Hong; Avison, Malcolm J.; Gaudio, Karen M.; Siegel, Norman J. *Pediatr. Res.*, 1993, **33(6)**, 595-7.

122. Martin, Louis F.; Peter, Anastasius O.; Fehr, David M.; Landis, J. Richard; Cotter, John; Briggs, Richard W. *Am. J. Surg.*, 1992, **164(2)**, 132-9.

123. Marcovitz, Pamela M.; Aisen, Alex M.; Fencil, Laura E.; Swanson, Scott D.; Buda, Andrew J. *Am. Heart J.*, 1992, **124(5)**, 1205-12.

124. Murphy, Elizabeth; Fralix, Teresa A.; London, Robert E.; Steenbergen, Charles *Am. J. Physiol.*, 1993, **265(4, Pt. 2)**, C1146-C1155.

125. Fralix, Teresa A.; Steenbergen, Charles; London, Robert E.; Murphy, Elizabeth *Cardiovasc. Res.*, 1993, **27(4)**, 630-7.

126. Aussedat, Jean; Ray, Andre; Kay, Laurence; Verdys, Monique; Harpey, Catherine; Rossi, Andre *J. Cardiovasc. Pharmacol.*, 1993, **21(1)**, 128-35.

127. Yamada, Shigeyuki *Igaku to Seibutsugaku,* 1993, **126(6)**, 265-8.

128. Kotegawa, Masshiko; Sugiyama, Masayasu; Shoji, Takachika; Haramaki, Nobuya; Ogura, Ryohei *J. Mol. Cell. Cardiol.* 1993, **25(9)**, 1067-74.

129. Kawasaki, Shun Yoh *Shiga Ika Daigaku Zasshi,* 1992, **7**, 63-77.

130. Kobayashi, Akira; Okayama, Yoshimi; Yamazaki, Noboru *Mol. Cell. Biochem.*, 1993, **119(1-2)**.

131. Okayama, Yoshimi; Kobayashi, Akira; Fujise, Yutaka; Yamazaki, Noboru Jpn. *Circ. J.*, 1993, **57(6)**, 521-32.

132. Robitaille, P. M. L.; Abduljalil, A.; Rath, D.; Zhang, H.; Hamlin, R. L. *Magn. Reson. Med.*, 1993, **30(1)**, 4-10.

133. Polgreen, K. E.; Kemp, G. J.; Clarke, K.; Radda, G. K. *Biochem. Soc. Trans.*, 1993, **21(2)**, 180S.

134. Butwell, Nina Beth; Ramasamy, Ravichandran; Lazar, Istvan; Sherry, A. Dean; Malloy, Craig R. *Am. J. Physiol.*, 1993, **264(6, Pt. 2)**, H1884-H1889.

135. Smith, G. L.; Donoso, P.; Bauer, C. J.; Eisner, D. A. *J. Physiol. (London)*, 1993, **472**, 11-22.

136. Jelicks, Linda A.; Gupta, Raj K. *Magnesium Trace Elem.*, 1993, **10(2-4)**, 136-41.

137. Nakamura, Kazuhiro; Kusuoka, Hideo; Ambrosio, Giuseppe; Becker, Lewis C. *Am. J. Physiol.*, 1993, **264(3, Pt. 2)**, H670-H678.

138. Kirschenlohr, Heide L.; Grace, Andrew A.; Clarke, Sonia D.; Shachar-Hill, Yair; Metcalfe, James C.; Morris, Peter G.; Smith, Gerry A. *Biochem. J.*, 1993, **293(2)**, 407-11.

139. Kusuoka, Hideo; Backx, Peter H.; Camilion de Hurtado, C.; Azan-Backx, Michelle; Marban, Eduardo; Cingolani, Horacio E. *Am. J. Physiol.*, 1993, **265(5, Pt. 2)**, H1696-H1703.

140. Bradamante, Silvia; De Jong, Jan Willem; Piccinini, Francesco *Biochem. Biophys. Res. Commun.*, 1993, **196(2)**, 872-8.

141. Tian, G.; Biro, G. P.; Xiang, B.; Butler, K. W.; Deslauriers, R. *Basic Res. Cardiol.*, 1992, **87(4)**, 356-65.

142. Altura, Burton M.; Barbour, Randall L.; Dowd, Terry L.; Wu. Fann; Altura, Bella T.; Gupta, Raj K. *Biochim. Biophys. Acta*, 1993, **1182(3)**, 329-32.

143. Barbour, Randall L.; Atura, Burton M.; Reiner, Seth D.; Dowd, Terry L.; Gupta, Raj K.; Wu, Fan; Altura, Bella T. *Magnesium Trace Elem.*, 1993, **10(2-4)**, 99-116.

144. Schreur, Joop H. M.; de Beer, Ron; Van Echteld, Cees, J. A.; Ruigrok, Tom J. C. *J. Mol. Cell. Cardiol.*, 1993, **25(9)**, 1015-24.

145. Nishimura, H.; Matsubara, T.; Ikoma, Y.; Nakayama, S.; Sakamoto, N. *Br. J. Pharmacol.*, 1993, **109(2)**, 443-8.

146. Steenbergen, Charles; Perlman, Michael E.; London, Robert E.; Murphy, Elizabeth *Circ. Res.*, 1993, **72(1)**, 112-25.

147. Ishibashi, Takaharu; Nakazawa, Miklo; Imai, Shoichi *Mol. Cell. Biochem.*, 1993, **119(1-2)**, 109-20.

148. Askenasy, Nadir; Vivi, Antonio; Tassini, Maria; Navon, Gil *Magn. Reson. Med.*, 1992, **28(2)**, 249-63.

149. Assadnazari, H.; Zimmer, G.; Freisleben, H. J.; Werk, W.; Leibfritz. D. *Arzneim-Forsch*, 1993, **43(4)**, 425-32.

150. Osbakken, Mary; Ito, Kinji; Zhang, Danning; Ponomarenko, Ihor; Ivanics, Tamas; Jahngen, Edwin G. E.; Cohn, Mildred *Cardiology*, 1992, **80(3-4)**, 184-95.

151. Monomura, Shinichi; Negai, Yasushi; Ogawa, Toshiaki; Bessho, Motoaki; Yamashita, Hiroshi; Serizawa, Takashi *J. Mol. Cell. Cardiol.*, 1993, **25(5)**, 551-62.

152. Raine, A. E. G.; Seymour, A. M. L.; Roberts, A. F. C.; Radda, G. K.; Ledingham, J. G. G. *J. Clin. Invest.*, 1993, **92(6)**, 2934-40.

153. Lortet, S.; Heckmann, M.; Aussedat, J.; Ray, A.; Vincent, M.; Sassard, J.; Zimmer, H. G.; Rossi, A. *Acta Physiol. Scand.*, 1993, **149(3)**, 311-21.

154. Grace, Andrew A.; Kirschenlohr, Heide L.; Metcalfe, James C.; Smith, Gerry A.; Weissberg, Peter L.; Cragoe, Edward J., Jr.; Vandenberg, Amie I. *Am. J. Physiol.*, 1993, **265 (1. Pt. 2)**, H289-H298.

155. Tanaka, Kunio; Honda, Hajime; Akita, Nobuyuki *NMR Biomed.*, 1992, **5(6)**, 329-34.

156. Jeffrey, F. M. H.; Storey, C. J.; Malloy, C. *Res. Cardiol.*, 1992, **87(6)**, 548-58.

157. Zhang, Jianyi; Markle, Helimut; Hendrich, Kristy; Garwood, Micahel; From, Arthur H. L.; Ugurbil, Kamil; Bache, Robert J. *J. Clin. Invest.*, 1993, **92(2)**, 993-1003.

158. Chatham, John C.; Forder, John R. *J. Mol. Cell. Cardiol.*, 1993, **25(10)**, 1203-13.

159. Weiss, Robert G.; Kalil-Filho, Roberto; Herskowitz, Ahvie; Chacko, V. P.; Litt, Marc; Stern, Michael D.; Gerstenblith, Gary *Circulation*, 1993, **87(1)**, 270-82.

160. Laughlin, Maren R.; Petit, William A., Jr.; Barrett, Eugene J. *J. Mol. Cell. Cardiol.*, 1993, **25(2)**, 175-83.

161. Robitaille, Pierre Marie L.; Rath, Dipti P.; Abduljalil, Amir M.; O'Donnell, J. Michael; Jiang, Zhongchen; Zhang, Huzeng; Hamlin, Robert L. *J. Biol. Chem.*, 1993, **268(35)**, 26296-301.

162. Garlick, Pamela B.; Pritchard, Ronald D. *NMR Biomed.*, 1993, **6(1)**, 84-8.

163. Chacko, V. P.; Weiss, R. G. *Am. J. Physiol.*, 1993, **264(3, Pt. 1)**, C755-C760.

164. Ramasamy, Ravichandran; Zhao, Piyu; Gitomer, William L.; Sherry, A. Dean; Malloy, Craig R. *Am. J. Physiol.*, 1992, **263(6, Pt. 2)**, H1958-H1962.

165. Madden, Michael C.; Van Winkle, W. Barry; Kirk, Katharine; Pike, Martin M.; Pohost, Gerald M.; Wolkowicz, Paul E. *Biochim. Biophys. Acta*, 1993, **1169(2)**, 176-82.

166. Thebault, Marie T.; Raffin, Jean P.; Pennec, Jean P.; Pichon, Roger *Biochem. Biophys. Res. Commun.*, 1993, **197(1)**.

167. Bachelard, Herman; Badar-Goffer, Ronnitte *J. Neurochem.*, 1993, **61(2)**, 412-29.

168. Lee, Byung Chul; Naritomi, Hiroaki *Inkan Kwahak*, 1992, **16(3)**, 156-64.

169. Merboldt, Klaus Dietmar; Hoerstermann Dirk; Haenicke, Wolfgang; Bruhn, Harald; Frahm, Jens **Magn. Reson. Med.**, 1993, *29(1)*, 125-9.

170. Corbett, Ronald J. T.; Laptook, Abbot R. *J. Neurochem.*, 1993, **61(1)**, 144-9.

171. Semionova, N. A.; Yushmanov, V. E. *Radiobiologiya,* 1993, **33(3)**, 388-91.

172. Matoba, Munetaka *Kanazawa Ika Daigaku Zasshi*, 1993, **18(2)**, 204-16.

173. Mason, Graeme F.; Behar, Kevin L.; Martin, Margaret A.; Shulman, Robert G. *Adv. Exp. Med. Biol.*, 1993, **331(Frontiers in Cerebral Vascular Biology)**, 29-34.

174. Leo, G. C.; Shank, R.P.; Zielke, H. R. *Bull, Magn. Reson.*, 1993, **15(1-2)**, 8-13.

175. Inubushi, Toshiro; Morikawa, Sigehiro; Kito, Kouichi; Arai, Toshiyuki *Biochem. Biophys. Res. Commun.*, 1993, **191(3)**, 866-72.

176. Kanamori, Keiko; Ross, Brian D. *Biochem. J.*, 1993, **293(2)**, 461-8.

177. Jensen, Frances; Tsuji, Miles; Offutt, Martin; Firkusny, Igor; Holtzman, David *Dev. Brain Res.*, 1993, **73(1)**, 99-105.

178. Kloiber, O.; Miyazawa, T.; Hoehn-Berlage, M.; Hossmann, K.A. *NMR Biomed.*, 1993, **6(2)**, 144-52.

179. Holtzmann, David; Offutt, Martin; Tsuji, Miles; Neuringer, Leo J.; Jacobs, Danny *J. Cereb. Blood Flow Metab.*, 1993, **13(1)**, 153-61.

180. Nagai, Y.; Naruse, S.; Weiner, M. W. *NMR Biomed.*, 1993, **6(1)**, 1-6.

181. Nakada, Tsutomu; Hida Kazutoshi; Kwee, Ingrid L. *Neurosci. Res., (Shannon. Irel.)*,1992, **15(1-2)**, 155-23.

182. Nagayama, Tetsuya *Kagoshima Daigaku Igaku Zasshi*, 1992, **44(3)**, 299-316.

183. Corbett, Ronald J. T.; Laptook, Abbot R.; Sterett, Rick; Tollefsbol, Greg; Garcia, Damian *Pediatr. Res.*, 1993, **34(3)**, 370-8.

184. Petroff, Ognen A. C.; Novotny, Edward J.; Avison, Malcolm; Rothman, Douglas L.; Alger, Jeffry R.; Ogino, Takashi; Shulman, Gerald I.; Prichard, James W. *J. Cereb. Blood Flow Metab.*, 1992, **12(6)**, 1022-9.

185. Dager, S. R.; Marro, K. L.; Richards, T. L.; Metzger, G. D. *Biol. Psychiatry*, 1992, **32(10)**, 913-21.

186. Mun-Bryce, Sheila; Kroh, Franklin O.; White, Jonathan; Rosenberg, Gary A. *Am. J. Physiol.*, 1993, **265(3, Pt. 2)**, R697-R702.

187. Allen, Kathryn; Busza, Albert L.; Crockard, H. Alan; Gadian, David G. *NMR Biomed*, 1992, **5(1),** 48-52.

188. Corbett, Ronald J. T.; Laptook, Abbot R.; Garcia, Damian; Ruley, Joan I. *J. Cereb. Blood Flow Metab.*, 1993, **13(2)**, 235-46.

189. Tyson, Randy; Peeling, James; Sutherland, Garnette *Brain Res.*, 1993, **608(2)**, 288-98.

190. Brenner, R. E.; Munro, P. M. G.; Williams, S. C. R.; Bell, J. D.; Barker, G. J.; Hawkins, C. P.; *Magn. Reson. Med.*, 1993, **29(6)**, 737-45.

191. Fujibayashi, Testuo; Sugiura, Yoshihiro; Harada, Jun; Goto, *Yukio Masui to Sosei*, 1992, **28 (Suppl. 28, Masuiyaku Taisha to Zoki Shogai XIII)**, 123-7.

192. Denays, R.; Chao, S. L.; Mathur-Devre, R.; Jeghers, O.; Fruhling, J.; Noel, P.; Ham, H.R. *Magn. Reson. Med.*, 1993, **29(6)**, 719-23.

193. Altura, Burton M.; Altura Bella T.; Gupta, Raj K. *Magnesium Trace Elem.*, 1993, **10(2-4)**, 122-35.

194. Mitsumori, Fumiyuki; Nakano, Atsuhiro *Environ. Res.*, 1993, **62(1)**, 81-8.

195. Adler, Sheldon; Verbalis, Joseph G.; Williams, Donald *Am. J. Physiol.*, 1993, **264(6, Pt. 2)**, F968-F974.

196. Eleff, Scott M.; Mclennan Ian J.; Hart, Graeme K.; Maruki, Yuichi; Traystman, Richard J.; Koehler, Raymond C. *Magn. Reson. Med.*, 1993, **30(1)**, 11-17.

197. Sauter, Andre; Rudin, Markus *J. Biol. Chem.*, 1993, **268(18)**, 13166-71.

198. Guenther, Ulrich; Albert, Klaus *NMR Biomed.*, 1993, **6(1)**, 27-31.

199. Canney, Daniel J.; Covey, Douglas F.; Evers, Alex S. *Biochem. Pharmacol.*, 1993, **45(4)**, 949-59.

200. Badar-Goffer, Ronnitte S.; Thatcher, Nicola M.; Morris, Peter G.; Bachelard, Herman S. *J. Neurochem.*, 1993, **61(6)**, 2207-14.

201. Brooks, Keith J.; Kauppinen, Risto A. *Neurochem. Int.*, 1993, **23(5)**, 441-50.

202. Brooks, Keith J.; Pirttila, Tiina R. M.; Kauppinen, Risto A. *NeuroReport*, 1993, **4(2)**, 139-42.

203. Pirttila, Tiina R. M.; Hakumaki, Juhana M.; Kauppinen, Risto A. *J. Neurochem*, 1993, **60(4)**, 1274-82.

204. Schanne, Francis A. X.; Gupta, Raj K.; Stanton, Patric K. Biochim, *Biophys. Acta*, 1993, **1158(3)**, 257-63.

205. Ting, Yen-Ling Teresa; Degani, Hadassa *Brain Res.*, 1993, **610(1)**, 16-23.

206. Dunn, J. F.; Tracey, L; Radda, G. K. *Proc. R. Soc. London, Ser. B*, 1993, **251(1332)**, 201-6.

207. Brosnan, M. Julia; Raman, Sasikala P.; Chen, Lihong; Koretsky, Alan P. *Am. J. Physiol.*, 1993, **264(1, Pt. 1)**.

208. Aureli, T.; Capuani, G.; Di Cocco, M. E.; Miccheli, A.; Corsico, N.; Nardone, A.; Arrigoni Martelli, E.; Conti, F. *Bull, Magn. Reson.*, 1993, **15(3)**, 169-71.

209. Sanderson, A. L.; Thompson, C. H., Kemp, G. J.; Radda, G. K. *Biochem. Soc. Trans.*, =1993, **21(4)**, 442S.

210. Ohira, Y.; Wakatsuki, T.; Endo, N.; Nakamura, K.; Asakura, T.; Ikeda, K.; Tomiyoshi, T.; Nakojoh, M. *Guanidino Compd. Biol. Med.*, 1992, 175-9.

211. Gareau, Paula J.; Janzen, Edward G.; Towner, Rheal A.; Stewart, Wendy A. *Free Radical Res. Commun.*, 1993, **19(1)**, 43-50.

212. Chiba, Atsushi; Takashima, Masahiro; Hamaguchi, Masamitsu; Shiko, Chichibu. *Comp. Biochem, Physiol., A: Comp Physiol*, 1993, **106A(4)**, 725-9.

213. Thompson, C. H.; Anderson, Y.; Jagasia, D. H.; Radda, G. K.; Rajagopalan, B. *Clin. Sci.*, 1993, **84(6)**, 645-9.

214. Meynial-Denis, Dominique; Mignon, Michele; Foucat, Loie; Bonnet, Yves; Bielieki, Guy; Renou, Jean Pierre; Lacourt, Paulette; Lacourt, Andre; Arnal, Maurice *Biochem. J.* 1993, **293(2)**, 399-405.

215. Ito, Eriko *Yokohama Igaku*, 1993, **44(1)**, 67-74.

216. Song, Sheng Kwei; Karl, Irene E.; Ackerman, Joseph J. H.; Hotchkiss, Richard S. *Proc. Natl. Acad. Sci. U.S.A.*, 1993, **90(9)**, 3933-7.

217. Adams, G. E.; Bremner, J. C. M.; Stratford, I. J.; Wood, P. J. *BJR Suppl.*, 1992, **24(Radiation Science - of Molecules, Mice and Men)**, 137-41.

218. Hossmann, K. A.; Linn, F.; Okada, Y. *NMR Biomed.*, 1992, **5(5)**, 259-64.

219. Ruiz-Cabello, Jesus; Cohen, Jack S. *NMR Biomed.*, 1992, **5(5)**, 226-33.

220. Stubbs, M.; Bhujwalla, Z. M.; Tozer, G. M.; Rodrigues, L. M.; Maxwell, R. J.; Morgan, R.; Howe, F. A.; Griffiths, J. R. *NMR Biomed.*, 1992, **5(6)**, 351-9.

221. Howe, F. A.; Stubbs, M.; Rodrigues, L. M. Griffiths, J. R. *NMR Biomed.*, 1993, **6(1)**, 43-52.

222. Shungu, Dikoma C.; Bhujwalla, Zaver M.; Li, Shi Jiang; Rose, Lucy M.; Wehrle, Janna P.; Glickson, Jerry D. *Magn. Reson. Med.*, 1992, **28(1)**, 105-21.

223. Hori, Hiroshi; Itoh, Youko; Aihara, Mariko; Ayakawa, Yoshio; Miyata, Nobuki *Nippon Igaku Hoshasen Gakkai Zasshi*, 1993, **53(7)**, 847-55

224. Koutcher, Jason A.; Alfieri, Alan A.; Stolfi, Robert L.; Devitt, Mary L.; Colofiore, Joseph R.; Nord, L. Dee; Martin, Daniel S. *Cancer Res.*, 1993, **53(15)**, 3518-23.

225. Carpinelli, G.; Canese, R.; Giannini M.; Proietti, E.; Belardelli, F.; Podo, F., *Acta. Med. Rom.*, 1992, **30(3-4)**, 295-9.

226. De Certaines, Jacques D.; Albrectsen, Jens; Larsen, Vibeke A.; Xie, Xion; Rygaard, Joergen; Henriksen, Ole *In Vivo*, 1992, **6(6)**, 611-16.

227. Eskey, Clifford J.; Koretaky, Alan P.; Domach, Michael M.; Jain, Rakesh K. *Proc. Natl. Acad. Sci. U.S.A.*, 1993, **90(7)**, 2646-50.

228. Macdonald, Jeffrey M.; Kurbanewicz, John; Dahiya, Rajvir; Espanol, Maryceline T.; Chang, Lee Hong; Goldberg, Beth; James, Thomas L.; Narayan, Perinchery *Magn. Reson. Med.*, 1993, **29(2)**, 244-8.

229. Kyomasu, Yoshinori *Kyorin Igakkai Zasshi*, 1992, **23(4)**, 613-24.

230. Gyngell, M. L.; Hoehn-Berlage, M.; Kloiber, O.; Michaelis, T.; Ernestus, R. I.; Hoerstermann, D.; Frahm. J. *NMR Biomed.*, 1992, **5(6)**, 335-40.

231. Kurhanewicz, John; Dahiya, Rajvir; Macdonald, Jeffrey M.; Chang, Lee Hong; James, Thomas L.; Narayan, Perinchery *Magn. Reson. Med.*, 1993, **29(2)**, 149-57.

232. Schupp, Daniel G.; Merkle, Hellmut; Ellermann, Jutta M.; Ke, Yong; Garwood, Michael *Magn. Reson. Med.*, 1993, **30(1)**, 18-27.

233. Workman, Paul; Maxwell, Ross J.; Griffiths, John R. *NMR Biomed.*, 1992, **5(5)**, 270-2.

234. Ramaprasad, Subbaraya; Liu, Yohai H.; Pandey, Ravindra K.; Shiau, Fuu Y.; Smith, Kevin M. *Proc. SPIE-Int. Soc. Opt. Eng.*, 1993, **1881(Proceedings of Optical Methods for Tumor Treatment and Detection: Mechanism and Techniques in Photodynamic Therapy II, 1993)**, 325-33.

235. Kato, Kenzo *Fragrance J.*, 1993, **21(8)**, 16-24.

236. Schweizer, Martin P.; Olsen, Jay I.; Selby, Jane; Saffle, Jeffrey R.; Chick, Leland R.; Maer, Mel; Sylvester, Jessie; Jensen, Peter; Nagel, Timothy L. J. *Trauma*, 1992, **33(6)**, 828-34.

237. Glovsky, M. M.; Farrow, N.; Ross, B.; Ward, P.A.; Morganroth, M. *Int. Arch. Allergy Immunol.*, 1992, **99(2-4)**, 429-34.

238. Burt, C. Tyler; Kleps, Robert A. *Comp. Biochem. Physiol. A: Comp. Physiol.*, 1992, **103A(4)**, 715-18.

239. Nakada, Tsutomu; Kwee, Ingrid L. *Magn. Reson. Med.*, 1993, **29(4)**, 543-5.

240. Campbell, G. Douglas; Ramaprasad, Subbaraya; Olsen, Keith M.; Tryka, A. Francine; Komoroski, Richard A.; Blaszczak, Larry C.; Par, Thomas R., Jr. *J. Pharm. Sci.*, 1993, **82(1)**, 48-51.

241. Igarashi, Hiromasa; Yoshida, Akitoshi; Tanaka, Kunio *Magn. Reson. Med.*, 1993, **29(1)**, 107-9.

242. Thomas, Dwaine M.; Papadopoulou, Olga; Mahendroo, Prem P.; Zigman, Seymour *Exp. Eye Res.*, 1993, **57(1)**, 59-65.

243. Lal, Sundeep; Szwergold, Benjamin S.; Kappler, Francis; Brown, Truman *J. Biol. Chem.*, 1993, **268(11)**, 7763-7.

244. Apte, D. V.; Ebrey, T. G.; Dawson, M. Joan *J. Physiol.*, 1993, **464**, 291-306.

245. Thompson, Stewart N.; Lee, Robert W. K. *Biochim. Biophys. Acta*, 1993, **1157(3)**, 259- 69.

246. Nishina, M.; Matsushita, K.; Hori, E.; Takahashi, M.; Kato, K. *Entomol. Exp. Appl.*, 1993, **66(3)**, 269-74.

247. Chiba, Atsushi; Chichibu, Shiko Comp. Biochem. Physiol. C: *Comp. Pharmacol. Toxicol.*, 1993, **106C(1)**, 87-91.

248. Chiba, Atsushi; Chichibu, Shiko Comp. Biochem. Physiol. C: *Comp. Pharmacol. Toxicol.*, 1993, **105C(2)**, 179-83.

249. Thomson, S.N.; Platzer, E.G.; Lee, R.W.K. *Magn. Reson. Med.*, 1992, **28(2)**, 311-17.

250. Johnson, Douglas C.; Burt, C. Tyler; Perng, Wann Cherng; Hitzig, Bernard, M. *Am. J. Physiol.*, 1993, **265(5. Pt. 2)**, R1162-R1167.

251. Bolinger, Lizann; Lenkinski, Robert E. *Biol. Magn. Reson.* 1992, **11**, 1-53.

252. Shulman, R. G.; Blamire, A. M.; Rothman, D. L.; McCarthy, G. *Proc. Natl. Acad. Sci. USA* 1993, **90(8)**, 3127-33.

253. Luyten, Peter *Chem. Mag. (Rijswijk, Neth.)* 1993, **(4),** 171, 172, 175 (Neth).

254. Ordidge, R. J.; Helpern, J. A. *NMR* 1992, **27(In-Vivo Magn. Reson. Spectrosc. II)**, 103-17.

255. Mashima, Yasuoki *Igaku no Ayumi* 1993, **165(13)**, 889-92.

256. Beckmann N. *NMR*, 1992, **28(In-vivo Magn. Reson. Spectrosc. III)**, 73- 100.

257. Conway, Michael A.; Radda, George K. *Trens. Cardiovasc. Med.*, 1991, **1(7)**, 300-4.

258. Okamoto, Shinya; Konishi, Tokuji; Nakao, Takeshi *Igaku no Ayumi* 1993, **165(10)**, 731-4.

259. Ingwall, Joanne S.; Weiss, Robert G. *Trends Cardiovasc. Med.* 1993, **3(1)**, 29-37.

260. Vink, Robert *Mol. Chem. Neuropathol.* 1993, **18(3)**, 279-97.

261. Ross, Brian D.; Bellinger, Jennifer C. *Renal Dispos. Nephrotoxic. Xenobiot.* 1993, 75- 93.

262. Malhotra, Deepak; Shapiro, Joseph I. *Concepts Magn. Reson.* 1993, **5(2)**, 123-50.

263. Chance, Britton *Methods Toxicol.* 1993, **2 (Mitochondrial Dysfunction)**, 98-101.

264. Prichard, James W. *Res. Publ. - Assoc. Res. Nerv. Ment. Dis.*, 1993, **71(Mol. Cell. Approaches Treat. Neurol. Dis.)**, 153-74.

265. Blei, Michael L.; Conley, Kevin E.; Kushmerick, Martin J. *J. Physiol. (London)* 1993, **465**, 203-22.

266. Blei, Michael L.; Conley, Kevin E.; Odderson Ib R.; Esselman, Peter C.; Kushmerick, Martin J. *Proc. Natl. Acad. Sci. U.S.A.* 1993, 90(15), 7396-400.

267. Quistorff, Bjoern; Johansen, Lars; Sahlin, Kent *Biochem. J.* 1993, **291(3)**, 681-6.

268. Bertocci, Loren A.; Fleckenstein, James L.; Antonio, Jose *J. Appl. Physiol.* 1992, **73(1)**, 75-81.

269. Marsh, Gregory D.; McFadden, Robin G.; Nicholson, Richard L.; Leasa, David J.; Thompson, R. Terry *Am. Rev. Respir. Dis.* 1993, **147(4)**, 876-9.

270. Vandenborne, K.; Walter, G.; Leigh, J. S.; Goelman, G. *Am. J. Physiol.* 1993, **265(5, Pt. 1)**, C1332-C1339.

271. Sakuma, Hajime; Nelson, Sarah J.; Vigneron, Daniel B.; Hartiala, Jaakko; Higgins, Charles B. *Magn. Reson. Med.* 1993, **29(5)**, 688-91.

272. Jung, Wulf Ingo; Straubinger, Klaus; Bunse, Michael; Schnick, Fritz; Kueper, Klaus; Dietze, Gunther; Lutz, Otto *Magn. Reson. Med.*, 1992, **28(2)**, 305-10.

273. Bailey, J. Martyn; Lomako, J. P.; Lomkako, Wei; Whelan, W. J. *Biochem. Soc. Trans.* 1993, **21(2)**, 124S.

274. Taylor, Roy; Price, Thomas B.; Katz, Lee D.; Shulman, Robert R.; Shulman, Gerald I. *Am. J. Physiol.* 1993, **265(2, Pt. 1)**, E224-E229.

275. Saner, M.; McKinnon, G.; Boesiger, P. *Magn. Reson. Med.*, 1992, **28(1)**, 65-73.

276. Bongers, Hilmar; Schick, Fritz; Skalej, Martin; Jung, Wulf Ingo; Stevens, Andreas *Magn. Reson. Imaging*, 1992, **10(6)**, 957-64.

277. Schick, Fritz; Eismann, Beate; Jung, Wulf Ingo; Bongers, Hilmar; Bunse, Michael; Lutz, Otto *Magn. Reson. Med.* 1993, **29(2)**, 158-67.

278. Chen, Wei; Novotny, Edward J.; Zhu, Xiao Hong; Rothman, Douglas L.; Shulman, R. G. *Proc. Natl. Acad. Sci. USA.* 1993, **90(21)**, 9896-900.

279. Kuwabara, Takeo; Watanabe, Hiroyuki; Tsuji, Shoji; Tanaka, Keiko; Ohkubo, Masaki; Ito, Takeshi; Sakai, Kunio; Yuasa, Tatsuhiko *Igaku no Ayumi* 1993, **165(8)**, 491-2.

280. Confort-Gouny, Sylviane; Vion-Dury, Jean; Nicoli, Francois; Dano, Pierre; Gastaut, Jean Louis; Cozzone, Patrick *J. C.R. Acad. Sci., Ser III*, 1992, **315(7)**, 287-93.

281. Rothman, Douglas L.; Petroff, Ognen A. C.; Behar, Kevin L.; Mattson, Richard H. *Proc. Natl. Acad. Sci. USA.* 1993, **90(12)**, 5662-6.

282. Kreis, R.; Ernst, T.; Ross, B. D. *J. Magn. Reson., Ser. B* 1993, **102(1)**, 9-19.

283. Sappey-Marinier, Dominique; Calabrese, Giovanna; Hetherington, Hoby P.; Fisher, Sarah N. G.; Deicken, Raymond; Van Dyke, Craig; Fein, George; Weiner, Michael W. *Magn. Reson. Med.* 1992, **26(2)**, 313-27.

284. Gideon, Peter; Henriksen, Ole *Magn. Reson. Imaging*, 1992, **10(6)**, 983-6.

285. Christiansen, P.; Henriksen, O.; Stubgaard, M.; Gideon, P.; Larsson, H. B. W. *Magn. Reson. Imaging* 1993, **11(1)**, 107-18.

286. Meyerhoff, D. J.; MacKay, S.; Bachman, L.; Poole, N.; Dillon, W. P.; Weiner, M. W.; Fein, G. *Neurology* 1993, **43(3, Pt.1)**, 509-15.

287. Michaelis, Thomas; Helms, Gunther; Merboldt, Klaus Dietmar; Haenicke, Wolfgang; Bruhn, Harald; Frahm, Jens *NMR Biomed.* 1993, **6(1)**, 105-9.

288. Woods, Bryan T.; Chiu, Tak Ming *Adv. Exp. Med. Biol.*, 1992, **318(Neurobiol. Essent. Fatty Acids)**, 267-74.

289. Karson, Craig N.; Newton, Joseph E.O.; Mohanakrishnan, Parekkat; Sprigg, J.; Komoroski, Richard A. *Psychiatry Res.*, 1992, **45(2)**, 55-104.

290. Dagnelie, Pieter C.; Bell, Jimmy D.; Cox, I. Jane; Menon, David K.; Sargentoni, Janet; Coutts, Glyn A.; Williams, Steve C. R. *NMR Biomed.* 1993, **6(2)**, 157-62.

291. Beckmann, N.; Fried, R.; Turkalj, I.; Seelig, J.; Keller, U.; Stalder, G. *Magn. Reson. Med.* 1993, **29(5)**, 586-90.

292. Schick, Fritz; Bongers, Hilmar; Kurz, Siegfried; Jung, Wulf Ingo; Pfeffer, Markus; Lutz, Otto. *Magn. Reson. Med.* 1993, **29(1)**, 38-43.

293. Ito, Kengo; Narita, Hiroyuki; Kato, Takashi; Fukatzu, Hiroshi; Tadokoro, Masanori; Asai, Hideaki; Ishiguchi, Tsuneo; Ishigaki, Takeo; Sakuma, Sadayuki *Nippon Igaku Hoshasen Gakkai Zasshi* 1993, **53(1)**, 104-6.

294. Schick, F.; Bongers, H.; Jung, W. I.; Eismann, B.; Skalej, M.; Einsele, H.; Lutz, O.; Claussen, C. *Appl. Magn. Reson.* 1992, **3(6)**, 947-63.

295. Pfeffer, Phillip E.; Rolin, Dominique B.; Kumosinski, Thomas F.; MacFall, Janet S.; Schmidt, Julian H. *Plant Physiol.,* 1992, **100(4)**, 1682-90.

296. MacFall, Janet S.; Pfeffer, Phillip E.; Rolin, Dominique B.; MacFall, James R.; Johnson, G. Allan *Plant Physiol.*, 1992, **100(4)**, 1691-7.

297. Spickett, Corinne M.; Smirnoff, Nicholas; Ratcliffe, R. George *Plant Physiol.* 1993, **102(2)**, 629-38.

298. Lee, R. B.; Ratcliffe, R. G. *J. Exp. Bot.* 1993, **44(260)**, 587-98.

299. Monselise, Edna Ben Izhak; Kost, Daniel *Planta* 1993, **189(2)**, 167-73.

300. Walter, L.; Altenburger, R.; Callies, R.; Grimme, L. H.; Leibfritz, D.; Mayer, A. *Isotopenpraxis* 1992, **28(1)**, 73-80.

301. Lee, R.B.; Purves, J.V.; Ratcliffe, R.G.; Saker, L.R. *J. Exp. Bot.*, 1992, **43(256)**, 1385- 96.

302. Joshi, Syamasundar; Lakshminarayana, M.R.; Nagana, Gowda G.A.; Khetrapal, C.L. J. *Agric. Food Chem.*, 1993, **41(3)**, 481-2.

303. Watanabe, H.; Yoshida, M. *Adv. Food Eng.* 1992, 79-83.

304. Gardea, A.A.; Daley, L.S.; Kohnert, R.L.; Soeldner, A.H.; Ning, L.; Lombard, P.B.; Azarenko, A.N. *Spectroscopy (Eugene, Oreg.)*, 1993, **8(1)**, 27-35.

305. Kano, Hiromi; Ishida, Nobuaki; Takagishi, Hidejiro; Shirata, Kazuto; Koizumi, Mika *Nippon Sakumotsu Gakkai Kiji* 1993, **62(1)**, 95-104.

306. Martinez, Vicente; Lauchli, Andre *Planta* 1993, **190(4)**, 519-24.

13
Nuclear Magnetic Resonance Imaging

BY STEVE C. R. WILLIAMS

1. Introduction

As in the author's previous three annual reports[1-3] the aim of this chapter is once again to cover the whole field of nuclear magnetic resonance imaging, commonly abbreviated to MRI. The format of last year's contribution has also been adhered to where papers whose abstracts have appeared in either Chemical Abstracts NMR Selects or in a literature search of the MEDLINE database are reported, but, due to the limited length of this article, many specialised clinical case studies have been omitted. Again the emphasis of this review has remained on giving a comprehensive report on a smaller number of important articles instead of briefly cataloguing a large quantity of publications.

Journals that concentrate on the developments and applications within the field include the Journal of Magnetic Resonance, Magnetic Resonance in Medicine, the Journal of Magnetic Resonance Imaging and the similarly named Magnetic Resonance Imaging as well as more clinical publications such as Radiology and the Journal of Computer Assisted Tomography. NMR in Biomedicine and Magnetic Resonance in Chemistry have also broadened their remit to include papers relating to imaging. Theoretical aspects of the field are often documented in Medical Physics and hardware developments make a regular appearance in the Review of Scientific Instruments. Reading the abstracts from the Annual International Meeting of the recently renamed Society of Magnetic Resonance is once again recommended as a means of deriving concise, up to date information on developments within the subject.

2. Basic Principles and Education

Journals that concentrate on the educational side of magnetic resonance include Concepts in Magnetic Resonance where fundamental aspects of the technique are presented, and Magnetic Resonance Quarterly which contains articles that review various facets of the field. A monthly magazine entitled Diagnostic Imaging regularly features current trends and applications in clinical MRI. Marketing literature from the major MRI equipment manufacturers such as Siemens, General Electric, and Philips often provides both eye-catching and informative documentation on advances in their respective products. Pharmaceutical companies such as Schering and Squibb who manufacture MRI contrast agents also produce educational literature relating to their diagnostic media.

The book by Professor Peter G. Morris[4] continues to be my primary source of reference for the basic physics of MRI but for the lay reader the text of Young[5] and for the more practical aspects, Hutchinson and Foster[6] are highly recommended. The publication by Chen and Hoult[7] also presents some refreshing insights into numerous technological aspects of biomedical NMR and a more recent, excellent book by Callaghan[8] entitled "Principles and Applications of NMR Microscopy", despite the title, gives a comprehensive account of the physics involved in the whole field of MRI.

Several review articles concerning MRI have been documented this year. The bulk of these papers tend to be restricted to specialised applications and hence shall be dealt with in later sections of this chapter. Educational papers during the course of this review period have included reports on numerous interactive teaching modules linked to picture archiving and communication data bases. The primary intention of such systems is to assist in the instruction and interpretation of MR images in a more palatable form than lectures, articles, and/or textbooks alone. One group have described an interactive atlas of joint anatomy on MR images[9]. The program presented complete sets of labelled MR images of the joints by using a variety of common pulse sequences in standard imaging planes. The images could be viewed in a self-paced, structured sequence or in any arbitrary, self-directed way. Interactive features included a quiz mode that challenged the user to name the anatomic structures, one at a time. Another paper reports the use of a software package for teaching anatomy of the upper abdomen from CT and MRI data[10]. The program was created on a Macintosh computer and consisted of a series of windows, each of them containing images, texts, schematic representations and sound. Each window contained buttons which could be activated with the mouse and allowed different kinds of function to be displayed. These functions included obtaining more specific information and schematic representations. Multiple buttons were also hidden in such images which, when activated, allowed the name of the selected organ to be displayed.

3. Safety

Safety related issues continue to be an important consideration during MRI studies. The bulk of reports are concerned with the potential adverse, physiological response to both established and novel contrast agents but, due to the limited size of this chapter, such reports have been omitted. The author does however recommend an article by Carr[11] which gives a comprehensive review of the potential toxic side effects of magnetopharmaceuticals. He reports that the overall safety profile of the such compounds with current FDA approval is good with regard to acute toxicity. For gadopentetate dimeglumine, the agent with the greatest clinical experience, the overriding majority of adverse events consists of transient and relatively minor symptoms and reactions. Clinical trials of the newer agents reflect a similar experience but on a much smaller scale. Severe adverse reactions can, however, occur after the administration of magnetopharmaceuticals. These often idiosyncratic and anaphylactoid reactions may cause permanent injury or death. In the two reported cases to date, the patients had a history of reactive airway disease. The incidence of severe reactions is estimated at 1 in 350,000 to 450,000 for gadopentetate dimeglumine. The rate is considerably lower than that for both ionic and nonionic iodinated contrast agents. Magnetopharmaceuticals should therefore be regarded as safe but not innocuous. The small risk has to be weighed against the additional diagnostic information that is obtained with such agents. Radiologists should be aware of potential complications and have on site equipment and skilled personnel for management should an adverse event occur. The benefits to the patient of reduced morbidity and mortality from improved diagnosis with magnetopharmaceuticals will ensure continued use of current agents as well as development of new compounds.

Although much has been published concerning the various safety risks associated with MRI, little mention is made of the acoustic noise generated during the rapid switching of magnetic field gradients whilst scanning. One group have reported that the existing data are now out of date with regard to currently used imaging protocols and technology[12]. Out of concern for patient safety, measurements were made of the acoustic noise generated during various scan protocols on a high field (1.5T) Siemens MR imager. These data were compared with published values and it was shown that for certain imaging protocols, the exposure to acoustic noise falls outside safety guidelines unless ear protection is used. One paper has also, rather worryingly, reported that exposure to a high magnetic field strength (1.5T) can be potentially teratogenic in mammals[13]. Two teratologic end points namely, crown-rump length and craniofacial perimeter were measured in a sham control and two experimental groups of C57BL/6J mice (N = 12, 11, 12) placed in different magnet locations. A T_2 weighted spin echo technique of 36 minutes duration was used on each dam. Crown-rump length was chosen because it is a standard measure of embryotoxicity. Craniofacial perimeter was evaluated because of its relationship to the anterior neural plate, the same region in the developing embryo that gives rise to the eyes. Magnetic resonance imaging fields were found to produce crown-rump and craniofacial perimeter measures smaller than control animals ($p < 0.05$) when exposed at the isocentre ($p = 0.038$ and 0.008 respectively). The paper therefore confirmed magnetic resonance imaging related teratogenicity in the C57BL/6J mouse and demonstrated the need for further experimental investigation to ascertain the clinical safety of magnetic resonance imaging for related issues in humans.

One script has appraised whether an electric current can be induced in intracardiac catheters, thermistor wires and pacing electrodes in patients during MRI[14]. From the authors detailed analysis of such issues in a sheep heart model, they derived the conclusions that pacing catheters and thermistor wires should be safe if well insulated and disconnected from external electric connections. However, current induced in epicardial pacing wires may be a hazard, and precautions should be taken. External wire tips must be separated, insulated, and coiled to lie along the axis of the magnetic field. Electrocardiography is required, and defibrillation equipment should be readily available.

In conclusion, safety considerations for MRI will continue to require careful attention and detailed scientific investigation. As new insights are gained into the hazards of time-varying and static magnetic fields, more sophisticated assessment of the true risk will be possible. At the present time, for established clinical indications, the risk of mismanagement of disease appears to outweigh the minimal or unknown risks of an MRI scan. Other than the mechanical hazards of implanted or external metallic or electrical devices, and provided that we can come to some standardised procedure for screening our subjects[15,16], MRI seems to be an extraordinarily safe and compatible environment for patients undergoing evaluation of suspected diseases.

4. New Instrumentation

Despite major advances that regularly occur within this area, the sensitive commercial nature of such improvements tends to limit the number of conventional publications relating to advances in hardware design. One recent manuscript has reviewed methods of coil design for MRI gradient systems[17]. The conflicting requirements for satisfactory gradient coil performance were described and analyzed. Design methods were classified into either discrete wire or current density techniques. The advantages and disadvantages of each group of methods were described, and optimization protocols were summarized. These include the target field approach, and constrained inductance and power minimization. One paper has reported how a 12cm diameter local gradient coil with a solenoid radiofrequency coil (2.2cm i.d., 1.4 cm long) was interfaced to a recalibrated clinical magnetic resonance imaging system. Three-dimensional gradient-echo imaging and phase-contrast angiography were then performed on the fingers of five healthy volunteers. Data sets (256 x 256 x 28) were acquired with spatial resolution of 100 x 100 x 500 μm in less than 7 minutes. The angiograms clearly delineated blood vessels that were less than even 1mm in diameter[18]. The temporal and spatial dependence of eddy currents associated with rapid gradient switching have also been quantitatively characterized for a 40cm bore superconducting magnet by measuring the offset frequency of the proton FID obtained from a small spherical sample. The measurements were made with the pre-emphasis unit deactivated. The data obtained at each location were well fitted to a sum of four exponentially decaying components. The shortest decay was, in each case, associated with the decay of the current in the gradient coil itself, while the other decays were associated with three eddy currents with decay times ranging from 13 to 480 ms. Each of the three eddy currents generated an essentially uniform gradient for at least 4 cm on either side of the isocentre. However, when the eddy currents were generated by either the vertical transverse gradient or the longitudinal gradient, the eddy current gradients were also accompanied by a field shift[19].

Enormous effort continues to be spent on the optimisation of radiofrequency coil design and operation. The benefits from such advances include improved sensitivity and hence, higher potential spatial resolution within a fixed scan time. Phased-array coil designs have been extended from initial applications involving spinal anatomy[20] to include, for example, high resolution MRI of the temporal lobes. On the basis of measurements in a phantom, the signal-to-noise ratio of the array is at least 1.67 times better than that obtained with a standard quadrature head coil in the region of the hippocampi. The higher sensitivity of this phased array permitted use of imaging pulse sequences for which the head coil was inadequate[21]. Further application of the same design principle has also afforded high resolution images of pelvic anatomy[22]. Further improvements in image quality within the pelvic region has been obtained by the use of endorectal probes for local reception of the MR signal. Two papers have shown that such hardware shows promise for improved staging of tumours in both the prostate[23] and rectum[24]. Other less sophisticated surface coil designs have been used to improve visualization of the pancreas[25], lachrymal glands[26], wrist and skin[27] as well as the stress fracture of the great toe sesamoid in a ballerina[28].

High resolution images of atherosclerotic lesions have also been obtained with a custom-built surface coil on a 1.5T clinical system. The paper reported the characterization of the composition and morphology of an atherosclerotic plaque removed at carotid endarterectomy. Initial results have shown a good correlation with histology[29]. One recent manuscript has also documented the minor adaptation of a clinical MR system for imaging the development of structural pathologies in the rat brain[30]. Another custom-designed radiofrequency coil tuned to the magnetic field of a 1.5T system provided the technology necessary to delineate discreet regions of the rat brain with clarity. Adult, male rats were imaged, placed on a thiamine-deficient pellet diet for approximately 6 weeks, and then reimaged. Treatment associated enlargement of the lateral ventricles identified in the images was verified by subsequent histological analysis. A more sophisticated way of measuring CNS pathology in the rat has involved the use of an implanted radiofrequency coil to obtain high resolution in vivo MR images at 1.9T of rat spinal cords that have been injured using a standardized weight drop technique. The signal-to-noise ratio and motion artifact suppression of these images was superior to that achieved in earlier attempts at this field strength using an external surface coil. The high quality and spatial resolution provided by this technique afford the possibility for longitudinal studies of experimental spinal cord injury before and after treatment, as well as detailed correlation of in vivo MRI contrast, histopathological findings, and functional deficit, in a controlled setting[31]. Another related text compared the performance of an inductively coupled implanted coil with optimized surface coils for MRI of the cat spinal cord. The implanted coil was inductively coupled to an external 12 x 12cm square surface coil that was mounted on a 14cm diameter cradle . Two similar cradles were prepared with transmit-only 12 x 12cm surface coils and either a receive-only 6 x 6cm square surface coil or a receive-only quadrature coil pair (two 4 x 6 cm coils overlapped slightly to minimize their mutual inductance) with the same surface area (6 x 6cm). On the spinal cord images, the quadrature coil exhibited a factor of 1.65 increase in SNR relative to the single-turn surface coil, whereas the implanted coil achieved a factor of 2.19 increase in SNR which was reflected as a dramatic improvement in the clarity of the images[32].

5. New Techniques and Applications

During the course of this review period, very few new methods have been embraced into the routine clinical setting. Two rapid imaging methods namely *f*ast *s*pin *e*cho (FSE) and *e*cho *p*lanar *i*maging (EPI) have become more popular because of the potential for increased patient throughput and the neccessity for high temporal resolution in functional imaging studies.

Numerous papers have reported a comparison between FSE and *c*onventional *s*pin *e*cho (CSE) techniques for different anatomical locations and pathologies in man. The size, number, and location of white matter lesions were compared in proton density weighted and T_2

weighted images from HIV positive patients. Image quality, motion artifact, CSF flow artifact, and gray-white matter differentiation were also taken into consideration and it was concluded that FSE and CSE MR of the brain produced images of similar quality and showed white matter lesions equally well[33]. These results therefore support the replacement, in this case, of slower, CSE pulse sequences with more rapid FSE regimes. One group has, however shown that when we compare CSE with FSE images of suspected meniscal tears of the knee FSE images showed only 65% of the meniscal tears seen on the CSE images[34]. The results therefore suggest that the fast spin echo sequence should not be used for the diagnosis of meniscal tears. One other text has shown that a combination of *s*hort inversion *t*ime *i*nversion *r*ecovery (STIR) preparation period with FSE for the detection of musculoskeletal abnormalities gave comparable results to the much longer times required for an equivalent STIR-CSE sequence[35].

FSE methods have also been used with some success for rapid measurement of tissue temperature in thermal interventional therapy. Various fast T_1weighted imaging protocols were compared to determine the sensitivity and resolution of signal intensity for temperatures within the range of 36-66 ^{o}C in gel phantoms and in vitro porcine liver specimens. The results showed that a T_1weighted FSE sequence with a TR of 100 msec had better temperature sensitivity and resolution than other sequences with comparable temporal resolutions. The longer imaging times required for FSE sequences with a TR of 300 msec did not improve temperature sensitivity[36]. FSE collection methods have also been extended to 3D data acquisition which, when compared with more conventional gradient echo techniques showed increased signal-to-noise ratios, reduced artifacts due to both susceptibility and poor field homogeneity in times that are clinically feasible[37].

This year saw some of the first reports of *e*cho *p*lanar *i*maging (EPI) at a high magnetic field strength of 3.0T from the pioneers of the technique at Nottingham. The results presented were obtained on volunteers using an in-house designed and constructed system. They showed the feasibility of obtaining snapshot images comprising of up to 256 x 256 pixels, corresponding to a spatial resolution of 0.75 x 0.75 mm^2 with a slice thickness of 2.5 mm but some susceptibility artifacts are apparent at the bone-air interfaces as expected at 3.0 T[38]. The same group have also reported the estimation of foetal weight *in utero* by using EPI to determine foetal volume which closely correlated with actual birthweight ($R = 0.97$). The median difference (expressed as a percentage of actual birthweight) between actual and EPI-estimated birthweights was 3.0% (range 0.6-9.9); this discrepancy was significantly smaller than that found for ultrasonographic estimates (6.5% [1.7-17.8]; $p < 0.01$)[39]. The normal heart has also been evaluated with EPI techniques. When spin echo EPI and gradient echo EPI methods were compared, delineation of cardiac structures was superior with SE EPI, due to significantly higher contrast between myocardial and intraluminal signal intensity. The higher and more homogeneous intraluminal signal intensity with GRE EPI did, however allow better assessment of intracardiac flow with this method[40]. One paper has also reported the use of interleaved echo planar imaging on a standard MRI system[41]. The data acquisition was divided into two to eight repetitions rather than one to two, as implemented by dedicated EPI systems. Interleaving allowed the use of a lower sampling bandwidth with a significant increase in signal-to-noise ratio. The method also had the advantages of relative ease of implementation, no need for postprocessing to remove image distortion, and no need for shimming on a case-by-case basis. The interleaved EPI method was applied to breathhold T_2-weighted abdominal imaging and in vivo liver-lesion contrast as measured in a 35 patient study which showed increased contrast for the interleaved EPI by an average factor of 1.21 (+/- 0.34) over CSE imaging.

One other technique that has found increased application in MR research is the use of *m*agnetization *t*ransfer *c*ontrast (MTC) imaging. MTC can, for example, be generated in magnetic resonance images of water-containing macromolecular structures. Results obtained with the pulsed off-resonance saturation technique for achieving MTC have been presented, allowing comparison of this technique with others[42]. The MTC effect has also been demonstrated in newborn piglet brain and bovine muscle tissue, as well as in human subjects. MTC increases nonlinearly with the duty cycle of the off-resonance irradiation, approaching a maximum value corresponding to continuous irradiation. However, approximately half of the available contrast can be obtained at a given saturation offset frequency, with only a 16% irradiation duty cycle. Examples of MTC in various human tissues demonstrated the specificity of this technique at power levels well within U.S. Food and Drug Administration guidelines. When this technique was measured in the forearm muscles of 10 volunteers at 0.1 T in order to ascertain wheter there were exercise-induced changes in MTC it was shown that there was a 26% increase in signal intensity of active flexor muscles after exercise when imaged with ordinary gradient echo sequence. Despite this marked intensity increase, the postexercise values of MTC did not differ from the preexercise ones[43].

Applications of MTC techniques to study pathology have, however, been disappointing. MTC techniques based on a series of on-resonance binomial pulses preceding a conventional spin echo sequence have been used to investigate hepatic neoplasms in 18 patients. It was shown that the mean lesion-to-liver contrast for neoplasms worsened with T_1-MTC

(0.87 vs. 1.04), but improved slightly with proton density-MTC (1.10 vs. 1.22) when compared to the corresponding non-MTC spin echo images. However, these differences were not statistically significant ($p = 0.19$ and $p = 0.16$, respectively). Therefore, MTC imaging did not offer significant improvement in contrast between hepatic neoplasms and normal liver[44]. Likewise, another paper reported that MTC methods were inferior to spin echo techniques for distinguishing necrotic from viable tumors in rat livers[45]. Hence, evaluation of its role in increasing diagnostic specificity within the liver will require further investigation.

6. NMR Imaging of Nuclei other than Proton

During this review period, very little has been reported on the application of MR imaging to nuclei other than proton. In one text concerned with measuring water transport in concrete, background signal from enclosed water was eliminated by using deuterated water. The distribution was futher validated by the use of deuterium MRI[46].

No papers reported the direct observation of oxygen but one script discussed the non-invasive determination of tumor oxygen tension and local variation with growth by measuring ^{19}F nuclear magnetic resonance spin-lattice relaxation rate ($R_1 = 1/T_1$) of perfluorocarbon (PFC) emulsion discretely sequestered in a tumor. ^{19}F MRI was used to indicate the distribution of perfluorocarbon and subsequent ^{19}F spectroscopic relaxometry was used to measure pO_2 in the tumor. Repeated measurements were made over a period of 3 weeks which showed the variation in local pO_2 during tumor growth. Given the critical role of oxygen tension in tumor response to therapy this method has provided new insight into tumor physiology, the potential efficacy of various therapeutic approaches, and may possibly provide a clinical technique for assessing individual tumor oxygenation[47].

^{19}F MRI has also been used to determine "wash-in" and "wash-out" curves of the inert, diffusible gas CHF_3 from 0.4 cm^3 voxels in the cat brain in order to calculate cerebral blood flow in five individual animals. The average value from the "wash-in" curve for the parietal cortex was 83 ml / 100g / min with a standard deviation of +/- 10% thus showing potential for imaging CBF in humans[48]. Both fluorinated anaesthetics halothane and isoflurane have also been detected in mammalian tissue by ^{19}F MR techniques. Halothane, a hydrophobic probe showed differences in the partitioning between normal rat tissues and adenocarcinoma where, in the latter, two slow-exchanging resonances were observed at a chemical shift separation of 0.3 ppm[49]. Uptake of isofluorane in rabbit brain appeared mainly in two adjacent slices of cerebral cortex. Both slices showed an increase in anaesthetic content at similar rates. A biphasic first-order uptake pattern with an initial fast component and a slower rate for the duration of the uptake was observed by a single pulse technique. However, when using ^{19}F chemical shift imaging methods, due to constraints on temporal resolution, only the slower kinetic process was observed. They also noted that anaesthetic levels in arterial blood reached equilibrium significantly earlier than in brain, suggesting some restricted diffusion into brain tissue for isoflurane[50].

^{19}F MR methods have also been used to monitor the anti-depressant drug, fluoxetine (and its metabolite norfluoxetine) in vivo in human brain. A quadrature birdcage head coil, developed for operation at 60.1 MHz (1.5T), yielded a signal from the head 2 to 4 times stronger than that derived from surface coils. It was used to measure the in vivo ^{19}F spin-lattice relaxation time (T_1) of fluoxetine for five patients by the inversion recovery technique. The individual T_1s varied from 149 to 386 ms, which was attributed in part to interindividual differences based on the reproducibility of T_1 measurements from a phantom. Low resolution 4-dimensional spectroscopic imaging confirmed that the single in vivo ^{19}F resonance for fluoxetine arose primarily from brain. The in vivo signal arose approximately equally from fluoxetine and the active metabolite norfluoxetine, as demonstrated by the in vitro ^{19}F MR spectrum of the lipophilic extract of a small section of brain obtained at autopsy. In vitro quantitation of frozen samples from three brain regions yielded combined fluoxetine/norfluoxetine concentrations of 12.3 to 18.6 μg/ml, which is higher than typically determined in vivo, and suggested that the fluorinated drugs may not be 100% visible in vivo[51]. Selective 19F MRI was also used to shed further light on the distribution and metabolism of the antineoplastic agent 5-fluorouracil (5-FU) and its major catabolite alpha-fluoro-beta-alanine (FBAL). A chemical shift imaging sequence was used in which a chemical shift selective (CHESS) saturation pulse to suppress either the 5-FU or the FBAL resonance before the other component of the two-line 19F MR spectrum was measured. This approach yielded 5-FU and FBAL MR images free of chemical shift artifacts in both read-out and slice-selection direction. In phantom experiments, selective 5-FU and FBAL images with a spatial resolution of 15 x 15 x 20 mm^3 (4.5 ml) were obtained in 30 min from a model solution, whose drug and catabolite concentrations were similar to those estimated in the liver of tumor patients undergoing chemotherapy with 5-FU[52].

During the course of this review period, the author only found three papers published in the field of sodium MRI. One script reported the use ^{23}Na magnetic resonance imaging to study the distribution of saline in excised muscle from rabbits for 6 hr following arterial brine injection. Immediately after injection, the distribution was clearly heterogeneous. After 6 hr, a steady state was achieved but the observed brine distribution was not fully homogeneous. Ham processed in various ways (tumbling, cooking) was also investigated. Tumbling increased the homogeneity of distribution but only cooking afforded full homogenization. It was also noted that the concentrations of NMR-visible sodium in the muscles were appreciably lower than the expected values calculated from the volume of injected brine and it was presumed that some of the sodium signal was invisible with the conventional MRI methods due to its short spin-spin relaxation value[53]. The two further papers in this field reported the use of ^{23}Na MRI to investigate changes in sodium during and after cerebral ischaemia in a gerbil model of stroke. It was shown that the sodium signal in the brain decreased by approximately 20% within 4 minutes of the onset of ischaemia, and subsequently increased between 4 and 8 minutes after the onset of reperfusion. The authors inferred that these changes may be reflecting the redistribution of tissue sodium between intra- and extracellular environments resulting from energy failure and recovery[54,55].

Phosphorus magnetic resonance spectroscopic imaging has previously been used to demonstrate localized metabolic abnormalities within the epileptogenic region in patients with temporal lobe epilepsy, including alkalosis, increased inorganic phosphate level, and decreased phosphomonoester levels. A recent report on the use of ^{31}P CSI has extended this pioneering work to the analysis of a further 8 patients with frontal lobe epilepsy. The same group has shown interictal alkalosis in the epileptogenic region compared to the contralateral frontal lobe in all patients (7.10 +/- 0.05 vs 7.00 +/- 0.06, $p < 0.001$). Seven of these patients exhibited decreased phosphomonoester levels in the epileptogenic frontal lobe compared to the contralateral frontal lobe (16.0 +/- 6.0 vs 23.0 +/- 4.0, $p < 0.01$). In contrast to previous findings in temporal lobe epilepsy, inorganic phosphate level was not increased in the epileptogenic region. Based on values derived from normal control subjects, 5 patients had elevated pH in the seizure focus and 2 patients had decreased phosphomonoesters while none had abnormalities in the contralateral frontal lobe[56]. These data suggest that ^{31}P spectroscopic imagng methods may play an important role in the presurgical evaluation of patients with frontal lobe epilepsy. Another team have investigated whether pH heterogeneity in skeletal muscle during exercise, observed with ^{31}P nuclear magnetic resonance, represented muscle fibre type heterogeneity. Localized spectra were simultaneously acquired from the soleus, medial, and lateral gastrocnemius using a multivolume localization method termed the Hadamard spectroscopic imaging (HSI) technique. They showed that the contamination of nonselected regions to the localized volumes was <5%. HSI-localized spectra were obtained from the calf muscles of untrained subjects and a small group of athletes. Two plantar flexion exercise protocols were implemented: a "maximal" high frequency protocol and a "steady-state" protocol at low contraction frequency (0.25 Hz). pH heterogeneity was observed in localized spectra of single muscles during both exercise protocols, as indicated by the large Pi line width. During maximal exercise the Pi line width was up to three times wider than the phosphocreatine line width, covering an entire pH unit. During the steady-state exercise, in three subjects the Pi peak clearly resolved into two distinct peaks, one at low pH and one at high pH. As pH heterogeneity was observed in localized spectra of single muscles during both exercise protocols, it most likely reflected the metabolic heterogeneity between fibre types[57]. This may be further explained by the variation in chemical composition between different muscle groups in the human calf where there is an inverse relationship between the concentration of taurine and carnosine.

Despite the fact there were no reports on the direct MR imaging of any other nuclei during the course of this review period, indirect methods involving, in particular ^{13}C, has recently been achieved in vivo.The metabolism of [1-^{13}C]-glucose was initially studied in situ in cat brain using gradient-enhanced proton-detected heteronuclear spectroscopy. Proton detection of [1-^{13}C]-glucose, [3-^{13}C]-lactate, 4-[^{13}C]-glutamine, 4-[^{13}C]-glutamate and the combined signals 2-[^{13}C]-glutamate/glutamine and 3-[^{13}C]-glutamate/glutamine was achieved, despite the fact that some of the associated proton resonances were close to the water signal. Subsequent two-dimensional [^{1}H-1^{13}C]-spectra and spectroscopic images not only demonstrated the possibility of in situ spectral assignment with ^{1}H sensitivity and ^{13}C resolution, but also showed that such methods could be used to study the spatial dependence of glucose metabolism[58].

7. Proton Chemical Shift Imaging

In vivo proton spectroscopic imaging methods have, in the main, been concentrated on the production of metabolite maps from mammalian brain in a wide range of disease states. Two papers have reported the use of such methods in the investigation of epilepsy. Proton spectroscopic images with a $4cm^3$ effective voxel size were obtained from 8 patients suffering with unilateral temporal lobe epilepsy and it was shown that there was a siginificant reduction in the N-acetyl aspartate (NAA) signal (21%) in the hippocampal seizure focus when compared with the contralateral hippocampal formations. The focal NAA reductions were consistent with subsequent pathological findings of mesial temporal sclerosis with selective neuron loss and gliosis in the surgically resected epileptogenic foci. It was encouraging that proton MRSI correctly localized the seizure focus in all 8 cases whereas MR imaging correctly localized 7 of 8 cases and single photon emission computed tomography only correctly identified 2 of 5 cases[59]. In a related status epilepticus model of selective neuronal injury in rats, three days after injection of kainate, the same group showed a reduction of NAA as determined by 3D proton MRSI in the piriform cortex, amygdala, and hippocampus which correlated well with neuronal injury determined from histology. Changes in NAA signal intensity, without consistent MRI changes in hippocampus again indicated the greater sensitivity of MRSI for detection of neuronal injury[60]. These results from both studies are consistent with the hypothesis that a reduction in NAA measured by such methods may be a sensitive marker of neuronal injury in vivo for this disease state.

Despite the fact that NAA reduction may be a sensitive marker, it is however, non-specific. For example in 10 patients that had suffered a cerebral infarction, it has been reported that the NAA peak within the lesion was reduced below the level in contralateral normal brain in all but two. An elevation of the lactate signal was also reliably detected but clearance of lactate occurs despite the potential contribution of lactate-producing leukocytes in the subacute stage ($>$ 60 hours after the insult). Further, delayed loss of N-acetyl-aspartate signal in later examinations also suggested that late death of viable cells may continue to occur within the first 2 weeks after damage[61]. Further spectroscopic imaging studies on both normal and postmortem changes in rat brain showed choline, creatine + phosphocreatine, N-acetyl aspartate (NAA), lactate (Lac), and three small resonances in the amino acid region which included resonances from aspartate + NAA (Asp), glutamine + NAA (Gln), and glutamate + GABA (Glu). The spectroscopic images in normal brain demonstrated higher NAA and Gln and lower Glu in cortex relative to basal ganglia. The major postmortem changes were an increase of Lac, Glu and Cho and a decrease of NAA and Asp. The rise in Lac was significantly higher in cortex than in basal ganglia[62]. Elevated lactate and reduced NAA was also seen in a group of asphyxiated full-term neonates. Proton spectroscopic images were obtained from a 2cm thick slice selected through the basal ganglia using a PRESS localisation technique (TE 272, TR 2000) in conjunction with a standard 32 x 32 phase encoding regime resulting in $1cm^3$ volumes. Lactate resonances were seen only in the five patients with grade 3 postasphyxial encephalopathy. Lactate was distributed diffusely ($n = 4$), or localized in areas of infarction ($n = 1$). It was also noted that the ratio of N-acetyl-aspartate to choline was significantly lower in patients with an adverse outcome when compared to survivors without handicaps, both neonatally and at 3 months of age[63].

Two papers have reported the use of proton spectroscopic imaging methods to evaluate the spatial distribution and rate of metabolism of alcohol in the brain. Both texts reported a heterogeneous intracranial distribution of alcohol. The former quotes that, one hour after ingestion, the relative ethanol signal in cerebral spinal fluid, gray matter, and white matter were 1.00, 0.72, and 0.37, respectively[64]. The second paper noted relatively high ethanol concentrations in the cerebellum and ventricular system whereas lower levels were observed in both white and grey matter[65].

Despite the information obtainable from such techniques, the time required to collect a conventional metabolite map from the human brain renders the technique impractical in most clinical environments. Two papers published during the review period have suggested novel approaches to reduce this overhead. One method involved the collection of multiple spin echoes from the human brain within a single repetition time which resulted in a three-fold reduction in measurement time whilst preserving SNR and spatial resolution at the minor expense of reduced spectral resolution[66]. The other script reported the implementation of a one-shot imaging method in combination with time-incremented chemical shift encoding to generate the third dimension on rat brain at 4.7T which yielded spectra from volumes as small as as 3.5μl and within measuring times of as little as 3.8 min[67]. One other common problem with proton chemical shift imaging technqiues is the ringing artifact present at low spatial resolution. This can result in substantial contamination of each voxel with signal from other locations within the object thus rendering the detection of subtle changes in chemical composition impossible. One group have addressed this issue by collecting data points in high k-space with less signal averages than those at lower k-space[68]. Using an adapted reconstruction technique, they were able to circumvent noisy, high spatial frequency problems that would arise from a direct Fourier transform which augurs well for future applications in man.

8. Solid State and Non-Medical NMR Imaging

Most non-medical applications of NMR imaging tend to be focused on the analysis of rock cores for the petrochemical industry. Reservoir rock analysis by MRI requires separation of the response into brine and crude oil components. Tests on preserved core from a North Sea chalk reservoir showed that spin-lattice relaxation time distributions could be used to distinguish the two fluids. NMR estimates of oil and water saturations for 1.5" diameter core examined in a benchtop spectrometer operating at 10MHz closely match fluid contents determined by distillation[69]. Proton chemical shift imaging of a waterflood experiment on a chalk sample from a North Sea hydrocarbon field has also been performed and it was shown that the method worked well, giving separate maps of water and oil distributions within the object. From difference images derived during waterflood experiments, detailed information about fluid movements were also gained[70]. Recent developments in the industry have led to increased interest in the application of borehole NMR as a probe of petrophysical properties and it has been shown that T_1 values varied with pore surface area in sandstone cores. The spin-lattice relaxation time was also shown to be strongly influenced by the invasion of submicron-sized clay particles commonly found in drilling muds[71]. Solid state MRI has been used to visualise irreducible water in reservoir cores. One paper has described the use of a simple, robust method for producing one and two-dimensional images of brine in a sandstone and carbonate sample from which, spatially resolved pore surface relaxation estimates were derived[72]. One further non-medical MRI text has concentrated on measuring the uptake, partitioning, and release of ingredients such as water, oil, surfactant, and ions in order to improve the design and manufacture of detergents and personal products. MRI was used to study molecular mobility in three relevant cases: ion diffusion (e.g., fluoride) in concentrated dispersions, oil transport through powders, and water ingress into porous powders (zeolite). The script presented results to illustrate the potential of multiple pulse and gradient echo MRI methods for dealing with the short T_2 scenarios that are a common problem in quantitative imaging of water in many solid-containing composites[73].

9. NMR Microscopy

One obvious requirement for MR microscopy is the use of higher field strength systems which afford increased SNR and thus allow improved spatial resolution within a given unit time. However, this is not a "free lunch" since an increase in static field invariably leads to an increase in magnetic susceptibility effects which often causes dramatic image distortion in heterogeneous systems. One method of overcoming such magnetic susceptibility effects whilst maintaining high spatial resolution involves the use of *stray field imaging* (STRAFI) techniques. One group has shown that microscopic MR images of plant tissue in a soil sample can be acquired at an isotropic resolution of 150 μm with a distortion of less than 10 μm despite the fact that the NMR signal from the sample was spread over about 15 kHz in the ^1H spectrum. This was achieved by using the stray, linear 5 kG/cm magnetic flux density gradient present at the edge of the magnet[74]. STRAFI has also been used to visualise the hard tissue component of extracted teeth. Combining this method with conventional MR imaging regimes on an NMR microscope allowed the depiction of both the soft tissue and outer contour of teeth at an isotropic resolution of 100 μm[75].

NMR microscopy has also been used to monitor the formation of the gel layer in hydrating, hydrophilic polymer tablets. Such tablets are used in the controlled delivery of drugs, where it has previously been shown that the rate and extent of the swelling of the outer gel layer can critically influence the kinetics of drug release. Tablets were hydrated in distilled water at 37 ^{o}C and then imaged at discrete time intervals using a NMR microscope operating at 500MHz. The growth of the gel layer was clearly observed in time sequences of radial and axial sections. Diffusion- and T_2-weighted images provided evidence that the water in the gel layer was more strongly bound close to the dry core of the tablet than at the more fully hydrated outer surface. In images of tablets containing diclofenac, disruption of the gel layer was shown to occur primarily from the flat surfaces of the tablet, whilst the distribution of particles could be seen in tablets doped with insoluble calcium phosphate[76]. NMR microscopy has also been used to interrogate autopsy tissue at high resolution. Two papers published during the review period have concentrated on measuring the morphological features that can be observed from such a technqiue. One of these texts interrogated the three-dimensional infarct structure that can be quantified at microscopic levels with an isotropic resolution of 400μm from formalin-fixed human hearts in a 2 T system[77]. The other paper from the same group reported results from over 200 human brains including a few formalin-fixed sections of the hippocampus imaged at a high magnetic field strength of 7 T. The postmortem MR images proved to be an invaluable aid that complemented the standard pathologic examination of the brain and spinal cord. The compelling advantages of this postmortem radiographic procedure included the three-dimensional aspects of the images; the ability to detect mineral (ie, iron) deposits; small focal lesions such as hemorrhages or infarcts; and the ability to evaluate the extent of cerebral oedema. For the same reasons, as well as its archival potential for documenting the topographic distribution of pathologic processes, this technique shows great promise for forensic cases[78].

Probably the most exciting paper in this field during the course of the year was the report in Science of the use of MR microscopy to detect embryonic cell lineages and movements. It is well known that key events in vertebrate embryogenesis are difficult to observe in many species. MR microscopy was used for the first time to follow cell movements and lineages in developing frog embryos. A single cell was injected at the 16-cell stage with a contrast agent, based on the gadolinium chelate, gadolinium-diethylenetriamine pentaacetic acid-dextran. The labeled progeny cells could be followed uniquely in three-dimensional magnetic resonance images, acquired from the embryo over several days. The results show that external ectodermal and internal mesodermal tissues extend at different rates during amphibian gastrulation and neurulation. Future prospects within the field of developmental biology seem limitless[79].

10. Functional Imaging

Neurofunctional imaging by rapid MRI methods continues at apace. There are, currently, two primary approaches to the detection of a cerebral haemodynamic change in response to the presentation of a stimulus. The first involves tracking a bolus of contrast agent as it washes through the brain whereas the more common technique involves an indirect detection of the endogenous tissue contrast caused by a change in the oxygenation state of blood, i.e. we collect MR images that are sensitive to differences between diamagnetic haemoglobin and paramagnetic deoxyhaemoglobin. As the proportion of oxygenated blood arriving at the site of neuronal activity increases then so does the related proton MR signal in response to an increase in the T_2* of the spins. This latter phenomenon is often termed *bl*ood *o*xygenation *d*ependant contrast (BOLD) imaging. Results from some of the earliest studies on human brain have recently been reviewed and the specific advantages of MRI for this purpose have been discussed[80]. Another related text has compared *f*unctional *MRI* (FMRI) with *p*ositron *e*mission *t*omography (PET) and compared the advantages and disadvantages of both modalities[81].

Most of the FMRI measurements to date, have been performed with echo-planar imaging (EPI) or conventional gradient-recalled techniques (commonly termed FLASH). One paper has reported that the T_2* sensitivity of conventional FLASH methods could be improved drastically on a conventional whole body instrument by delaying the gradient echo until the subsequent TR period without increasing the total imaging time. When the method was applied to bolus tracking through the brain of healthy volunteers during visual stimulation, an average increase of regional cerebral blood volume (rCBV) in the visual cortex of 10.9% ($n = 9$, $p = .001$) was observed[82]. Further echo-shifted FLASH methods from the same group were used to demonstrate susceptibility bolus tracking in the cat brain in an imaging time of as little as 153 ms[83].

One team has made a comprehensive, empirical optimization of the parameters that can affect the quality of an FMR image derived from a gradient echo regime using the conventional hardware available on a 1.5T clinical system. Visual activation experiments were performed while imaging parameters were manipulated in a serial fashion in order to test their effect on the percentage change in signal intensity (PCSI). The parameters tested were section thickness, echo time (TE), field of view (FOV), flip angle, radio-frequency (RF) spoiling, number of readout points, and number of signals averaged. It was shown that the PCSI for visual activation experiments was typically in the 1-5% range. The best results were achieved with a small section thickness, long TE, and large FOV. Variations in other acquisition parameters had a negligible influence on the PCSI[84]. One problem that is often overlooked in functional imaging is the fact that FMRI is sensitive to signal fluctuations due to physiological motion and system instability. One paper has addressed this issue by using navigator echoes to monitor and compensate for signal fluctuations in a gradient echo sequence. The technique involved acquisition of a "navigator" signal before the application of the phase-encoding and readout gradients with subsequent correction of the phase of the acquired imaging data[85]. This method only takes into consideration in-plane motion and one would envisage the need, in the long term, to acquire three-dimensional FMR data in order to interrogate motion artifacts more fully.

FMRI has been used to interrogate the activation of the human primary visual cortex during visual recall. During recollection of a visual stimulus, focal increases in signal related to changes in blood flow were detected in V1 and V2 cortex in five of seven subjects. These experiments showed that the same areas of the early visual cortex that are excited by visual stimulation are also activated during mental representation of the same stimulus. Therefore, it can be inferred that some of the processes used in topographically mapped cortical areas during visual perception may also be utilized during visual recall[86]. A recent paper has also reported some of the first FMR images of the human auditory cortex. Five volunteers were scanned whilst listening to auditory stimuli including nonspeech noise, meaningless speech sounds, single words, and narrative text. Imaged regions included the lateral aspects of both hemispheres. Signal changes in the superior temporal gyrus and superior temporal sulcus were observed bilaterally in all subjects. Speech stimuli were associated with significantly more widespread signal changes than was the noise stimulus, while no consistent differences were observed between responses to different speech stimuli. Considerable intersubject variability in the topography of signal changes was also observed[87]. Six normal volunteers underwent FMRI using EPI at 4T while performing a covert letter fluency task. All 6 subjects showed consistent activation in the frontal operculum and premotor and primary motor cortices. Activation was also detected in the supramarginal gyrus and the posterior part of the superior temporal gyrus[88]. Hence FMRI can be used to investigate cerebral activity noninvasively during the performance of complex cognitive tasks.

FMRI has been applied to numerous motor activation paradigms including the study of finger movement-related responses which were identified in three discrete sites of mesial frontal cortex in 7 normal subjects. During imagination of the same movements there was a differential response with rostral areas more active than caudal areas. From such studies one could conclude that humans have multiple motor areas in mesial frontal cortex that subserve different functions in motor planning and execution[89]. The very impressive group from Wisconsin have also reported FMRI of complex hand movements. Six right-handed healthy subjects were scanned whilst performing self-paced simple and complex finger movements with the right and left hands. Some subjects also performed the tasks at a fixed rate (2 Hz) or imagined performing the complex task. Functional changes occurred in the contralateral primary motor cortex during simple, self-paced movements; in the contralateral (and occasionally ipsilateral) primary motor cortex, the supplementary motor area (SMA), the premotor cortex of both hemispheres, and the contralateral somatosensory cortex during complex, self-paced movements; with less intensity during paced movements, presumably due to the slower movement rates associated with the paced (relative to self-paced) condition; and in the SMA and, to a lesser degree, the premotor cortex during imagined complex movements[90]. High spatial resolution FMR images of the motor cortex have also been obtained by improving SNR using a dedicated surface coil, comprehensive shimming of the main magnetic field and optimisation of all the imaging parameters in a first order, flow rephased, gradient-echo sequence (FLASH) with fat-suppression and reduced bandwidth (16-28 Hz/pixel). Optimal imaging parameters were found to be TR = 80-120 ms, TE = 60 ms, flip angle = 40 degrees, matrix = 128 x 128, FOV = 150-250 mm, slice-thickness = 2-5 mm, NEX = 1. The total single scan time for one image was less than16 s and in a 3D version of the experiment, a slab of 32 mm thickness with 16 partitions was collected. The motor cortex stimulation was achieved by a complex finger touching exercise mentioned earlier and during stimulation and an increase in signal of the order of10-20% was detected. The authors argue that this is higher than most other related studies at a comparable field strength of 1.5 T because of the reduced partial volume effects and optimized SNR for measurements at their small voxel sizes[91].

FMRI has also become a clinically useful tool for presurgical mapping of the sensory motor cortex. Preliminary experiments have been performed in two women both of whom had intractable, simple partial motor seizures due to tumors located in or near the sensory motor cortex. They subsequently underwent invasive cortical mapping which involved direct cortical stimulation and sensory-evoked-potential recording in order to localize the affected sensory motor area prior to tumor resection. The results derived from FMRI matched well those obtained from cortical functional mapping with invasive techniques[92].

In another clinical study a 4-year-old boy suffering from frequent partial motor seizures of his right side has also been investigated by FMRI. FLASH images (TE = 60 msec) acquired every 10 seconds over intervals of 10 minutes were obtained during both a quiescent phase and during clinical seizures. Functional MRI revealed sequential activation associated with specific structurally abnormal gyri within the left hemisphere with each of five consecutive clinical seizures, and also during a period that was not associated with a detectable clinical seizure[93].

From these encouraging results we can conclude that these new procedures may offer the potential to replace some of the more established but invasive techniques currently practiced.

References

1. S.C.R. Williams, "NMR Imaging", A Specialist Periodical Report on NMR, Volume 21, The Royal Society of Chemistry, Cambridge, 1991, Chapter 13, pp 415-454
2. S.C.R. Williams, "NMR Imaging", A Specialist Periodical Report on NMR, Volume 22, The Royal Society of Chemistry, Cambridge, 1992, Chapter 13, pp 416-453
3. S.C.R. Williams, "NMR Imaging", A Specialist Periodical Report on NMR, Volume 23, The Royal Society of Chemistry, Cambridge, 1993, Chapter 13 , pp 407-438
4. P.G. Morris, "NMR Imaging in Medicine and Biology", Clarendon Press, Oxford, 1986
5. S.W. Young, "Magnetic Resonance Imaging : Basic Principles", 2nd edn. Raven Press, London, 1988
6. J.M.S. Hutchinson and M.A. Foster, "Practical NMR Imaging", IRL Press, Oxford, 1987
7. C-N. Chen and D.I. Hoult, "Biomedical Magnetic Resonance Technology", Adam Hilger, London, 1989
8. P.T. Callaghan, "Principles of NMR Microscopy", Clarendon Press, Oxford, 1991.
9. F.S. Chew and W.E. Palmer, AJR-Am-J-Roentgenol. 1994 Jul; 163(1): 211-3
10. P. Pavone, C. Catalano, A. Laghi, M. Di-Girolamo, G. Albertini-Petroni, V. Campanella and R. Passariello, Radiol-Med-Torino. 1993 Nov; 86(5): 662-7
11. J.J. Carr, Neuroimaging-Clin-N-Am. 1994 Feb; 4(1): 43-54.
12. M. McJury, A. Blug, C. Joerger, B. Condon and D. Wyper, Br-J-Radiol. 1994 Apr; 67(796): 413-5
13. D.A. Tyndall, Oral-Surg-Oral-Med-Oral-Pathol. 1993 Nov; 76(5): 655-60

14. C.J. Peden, A.G. Collins, P.C. Butson, J.G. Whitwam and I.R. Young, Crit-Care-Med. 1993 Dec; 21(12): 1923-8
15. R.D. Boutin, J.E. Briggs and M.R. Williamson, AJR-Am-J-Roentgenol. 1994 Jan; 162(1): 189-94
16. A.D. Elster, K.M. Link and J.J. Carr, AJR-Am-J-Roentgenol. 1994 Jan; 162(1): 195-9
17. R. Turner, Magn-Reson-Imaging. 1993; 11(7): 903-20
18. S.J. Blackband, I. Chakrabarti, P. Gibbs, D.L. Buckley and A. Horsman, Radiology. 1994 Mar; 190(3): 895-9
19. Q. Liu, D.G. Hughes and P.S. Allen, Magn-Reson-Med. 1994 Jan; 31(1): 73-6
20. D. Kidd, J.W. Thorpe, A.J. Thompson, B.E. Kendall, I.F. Moseley, D.G. MacManus, W.I. McDonald and D.H. Miller, Neurology. 1993 Dec; 43(12): 2632-7
21. C.E. Hayes, J.S. Tsuruda and C.M. Mathis, Radiology. 1993 Dec; 189(3): 918-20
22. R.N. Troiano and S. McCarthy, Semin-Ultrasound-CT-MR. 1994 Feb; 15(1): 38-48
23. C.M. Tempany, X. Zhou, F.A. Zerhouni, M.D. Rifkin, L.E. Quint, C.W. Piccoli, J.H. Ellis and B.J. McNeil, Radiology. 1994 Jul; 192(1): 47-54
24. M.D. Schnall, E.E. Furth, E.F. Rosato and H.V. Kressel, Radiology. 1994 Mar; 190(3): 709-14
25. Y. Takehara, K. Ichijo, N. Tooyama, N. Kodaira, H. Yamamoto, M. Tatami, M. Saito, H. Watahiki and M. Takahashi, Radiology. 1994 Jul; 192(1): 73-8
26. P.A. Rubin, J.R. Bilyk, J.W. Shore, F.C. Sutula and H.M. Cheng, Ophthalmology. 1994 Feb; 101(2): 235-43
27. J. Maurer, H. Requardt, F. Muller, H.J. Steinkamp, N. Hosten, Langer and R. Felix, Rofo-Fortschr-Geb-Rontgenstr-Neuen-Bildgeb-Verfahr. 1994 Apr; 160(4): 353-60
28. E.M. Burton and B.H. Amaker, Pediatr-Radiol. 1994; 24(1): 37-8
29. C. Yuan, J.S. Tsuruda, K.N. Beach, C.E. Hayes, M.S. Ferguson, C.E. Alpers, T.K. Foo and D.E. Strandness, J-Magn-Reson-Imaging. 1994 Jan-Feb; 4(1): 43-9
30. R.J. Pentney, J.J. Alletto, M.A. Acara, C.A. Dlugos and R.J. Fiel Alcohol-Clin-Exp-Res. 1993 Dec; 17(6): 1301-8
31. S.D. Szeglowski and J.P. Hornak, Magn-Reson-Med. 1993 Dec; 30(6): 750-3
32. E.D. Wirth 3d, T.H. Mareci, B.L. Beck, J.R. Fitzsimmons and P.J. Reier, Magn-Reson-Med. 1993 Nov; 30(5): 626-33
33. E.M. Olson, J.F. Healy, W.H. Wong, D.C. Youmans and J.R. Hesselink, AJR-Am-J-Roentgenol. 1994 May; 162(5): 1199-204
34. D.A. Rubin, J.B. Kneeland, J. Listerud, S.J. Underberg-Davis and M.K. Dalinka, AJR-Am-J-Roentgenol. 1994 May; 162(5): 1131-5

35. R.C. Smith, R.T. Constable, C. Reinhold, T. McCauley, R.C. Lange and S. McCarthy, J-Comput-Assist-Tomogr. 1994 Mar-Apr; 18(2): 209-13
36. R. Matsumoto, R.V. Mulkern, S.G. Hushek and F.A. Jolesz, J-Magn-Reson-Imaging. 1994 Jan-Feb; 4(1): 65-70
37. C. Yuan, U.P. Schmiedl, E. Weinberger, W.R. Krueck and S.D. Rand, J-Magn-Reson-Imaging. 1993 Nov-Dec; 3(6): 894-9
38. P. Mansfield, R. Coxon and P. Glover, J-Comput-Assist-Tomogr. 1994 May-Jun; 18(3): 339-43
39. P.N. Baker, I.R. Johnson, P.A. Gowland, J. Hykin, P.R. Harvey, A. Freeman, V. Adams, B.S. Worthington and P. Mansfield, Lancet. 1994 Mar 12; 343(8898): 644-5
40. C.P. Davis, G.C. McKinnon, J.F. Debatin, D. Wetter, A.C. Eichenberger, S. Duewell and G.K. von-Schulthess, Radiology. 1994 Jun; 191(3): 691-6
41. K. Butts, S.J. Riederer, R.L. Ehman, R.M. Thompson and C.R. Jack, Magn-Reson-Med. 1994 Jan; 31(1): 67-72
42. J.C. McGowan-3rd, M.D. Schnall and J.S. Leigh, J-Magn-Reson-Imaging. 1994 Jan-Feb; 4(1): 79-82
43. K.T.Mattila, M.E. Komu, S.K. Koskinen and P.T. Niemi, Acta-Radiol. 1993 Nov; 34(6): 559-62
44. M.D. Hollett, A.M. Aisen, H.N. Yeung, I.R. Francis and R.L. Bree, Magn-Reson-Imaging. 1994; 12(1): 1-8
45. K.C. Li, R.B. Jeffrey, S.C. Ning, A. Kandil, G.M. Hahn, B. Pike, G. Glover and J. Kosek, Invest-Radiol. 1993 Oct; 28(10): 896-902
46. J. Link, J. Kaufmann and K. Schenker, Magn-Reson-Imaging. 1994; 12(2): 203-5
47. R.P. Mason, P.P. Antich, E.E. Babcock, A. Constantinescu, P. Peschke and E.W. Hahn, Int-J-Radiat-Oncol-Biol-Phys. 1994 Apr 30; 29(1): 95-103
48. J. Pekar, L. Ligeti, T. Sinnwell, C.T. Moonen, J.A. Frank and A.C. McLaughlin, J-Cereb-Blood-Flow-Metab. 1994 Jul; 14(4): 656-63
49. C.T. Burt, R.R. Moore and M.F. Roberts, NMR-Biomed. 1993 Sep-Oct; 6(5): 289-96
50. P.N. Venkatasubramanian, Y.J. Shen and A.M. Wyrwicz, NMR-Biomed. 1993 Nov-Dec; 6(6): 377-82
51. R.A. Komoroski, J.E. Newton, D. Cardwell, J. Sprigg, J. Pearce and C.N. Karson, Magn-Reson-Med. 1994 Feb; 31(2): 204-11
52. G. Brix, M.E. Bellemann, H.J. Zabel, P. Bachert and W.J. Lorenz, Magn-Reson-Imaging. 1993; 11(8): 1193-201
53. J.P. Renou, S.P. Benderbous, G. Bielicki, L. Foucat and J.P. Donnat, Magn-Reson-Imaging. 1994; 12(1): 131-7
54. K.L. Allen, A.L. Busza and S.C.R. Williams, Acta-Neurochir. 1994; 60: 220-223

55. K.L. Allen, A.L. Busza, S.R. Williams and S.C.R. Williams, Magn-Reson-Imaging. 1994; 12(6): 895-901
56. P.A. Garcia, K.D. Laxer, J. van-der-Grond, J.W.Hugg, G.B. Matson and Weiner-MW, Ann-Neurol. 1994 Feb; 35(2): 217-21
57. K. Vandenborne, G. Walter, J.S. Leigh and G. Goelman, Am-J-Physiol. 1993 Nov; 265(5 Pt 1): C1332-9
58. P.C. van-Zijl, A.S. Chesnick, D. DesPres, C.T. Moonen, J. Ruiz-Cabello and P. van-Gelderen, Magn-Reson-Med. 1993 Nov; 30(5): 544-51
59. J.W. Hugg, K.D. Laxer, G.B. Matson, A.A. Maudsley and M.W. Weiner, Ann-Neurol. 1993 Dec; 34(6): 788-94
60. T. Ebisu, W.D. Rooney, S.H. Graham, M.W. Weiner and A.A.Maudsley, J-Cereb-Blood-Flow-Metab. 1994 May; 14(3): 373-82
61. G.D. Graham, A.M. Blamire, D.L. Rothman, L.M. Brass, P.B. Fayad, O.A. Petroff and J.W. Prichard, Stroke. 1993 Dec; 24(12): 1891-6
62. T. Higuchi, E.J. Fernandez, A.A. Maudsley and M.W. Weiner, NMR-Biomed. 1993 Sep-Oct; 6(5): 311-7
63. F. Groenendaal, R.H. Veenhoven, J. van-der-Grond, G.H. Jansen, T.D. Witkamp and L.S. de-Vries, Pediatr-Res. 1994 Feb; 35(2): 148-51
64. D.M. Spielman, G.H. Glover, A. Macovski and A. Pfefferbaum, Alcohol-Clin-Exp-Res. 1993 Oct; 17(5): 1072-7
65. G. Schauss, H. Schild, R. Urban, J. Bunke, T. Riepert, P. Grebe, W. Kersjes, K. Schunk and M. Thelen, Rofo-Fortschr-Geb-Rontgenstr-Neuen-Bildgeb-Verfahr. 1994 Jun; 160(6): 493-9.
66. J.H. Duyn and C.T. Moonen, Magn-Reson-Med. 1993 Oct; 30(4): 409-14
67. D.G. Norris and W. Dreher, Magn-Reson-Med. 1993 Nov; 30(5): 641-5
68. X. Hu, M. Patel and K. Ugurbil, J-Magn-Reson-B. 1994 Jan; 103(1): 30-8
69. S. Davies, A. Hardwick, D. Roberts, K. Spowage and K.J. Packer, Magn-Reson-Imaging. 1994; 12(2): 349-53
70. D. Olsen, Magn-Reson-Imaging. 1994; 12(2): 215-6
71. C. Straley, D. Rossini, L.M. Schwartz, M.E. Stromski, M. Hrovat and S. Patz, Magn-Reson-Imaging. 1994; 12(2): 313-5
72. J.J. Attard, P.J. McDonald, S.P. Roberts and T. Taylor, Magn-Reson-Imaging. 1994; 12(2): 355-9
73. E.G. Smith, J.W. Rockliffe, P.J. McDonald, A. Lonergan, M.R. Halse, B. Leone and J.H. Strange, Magn-Reson-Imaging. 1994; 12(2): 231-4
74. P. Kinchesh, E.W. Randall and K. Zick, Magn-Reson-Imaging. 1994; 12(2): 305-7

75. M.A. Baumann, D. Gross, V. Lehmann and K. Zick, Schweiz-Monatsschr-Zahnmed. 1993; 103(11): 1407-14
76. R. Bowtell, J.C. Sharp, A. Peters, P. Mansfield, A.R. Rajabi-Siahboomi, M.C. Davies and C.D. Melia, Magn-Reson-Imaging. 1994; 12(2): 361-4
77. J.C. Hsu, G.A. Johnson, W.M. Smith, K.A. Reimer and R.E. Ideker, Circulation. 1994 May; 89(5): 2133-40
78. O.B. Boyko, S.R. Alston, G.N. Fuller, C.M. Hulette, G.A. Johnson and P.C. Burger, Arch-Pathol-Lab-Med. 1994 Mar; 118(3): 219-25
79. R.E. Jacobs and S.E. Fraser, Science. 1994 Feb 4; 263(5147): 681-4
80. R. Turner, Am-J-Physiol-Imaging. 1992 Jul-Dec; 7(3-4): 136-45
81. J.J. Neil, Curr-Opin-Neurol. 1993 Dec; 6(6): 927-33
82. C.T. Moonen, F.A. Barrios, J.R. Zigun, J. Gillen, G. Liu, G. Sobering, R. Sexton, J. Woo, J. Frank and D.R. Weinberger, Magn-Reson-Imaging. 1994; 12(3): 379-85
83. G. Liu, G. Sobering, J. Duyn and C.T. Moonen, Magn-Reson-Med. 1993 Dec; 30(6): 764-8
84. R.M. Thompson, C.R. Jack, K. Butts, D.P. Hanson, S.J. Riederer, R.L. Ehman, R.W. Hynes and N.J. Hangiandreou, Radiology. 1994 Mar; 190(3): 873-7
85. X. Hu and S.G. Kim, Magn-Reson-Med. 1994 May; 31(5): 495-503
86. D. Le-Bihan, R. Turner, T.A. Zeffiro, C.A. Cuenod, P. Jezzard and V. Bonnerot, Proc-Natl-Acad-Sci-U-S-A. 1993 Dec 15; 90(24): 11802-5
87. L. Rueckert, I. Appollonio, J. Grafman, P. Jezzard, R. Johnson, D. Le-Bihan and R. Turner, J-Neuroimaging. 1994 Apr; 4(2): 67-70
88. J.R. Binder, S.M. Rao, T.A. Hammeke, F.Z. Yetkin, A. Jesmanowicz, P.A. Bandettini, E.C. Wong, L.D. Estkowski, M.D. Goldstein and V.M. Haughton; Ann-Neurol. 1994 Jun; 35(6): 662-72
89. J.M. Tyszka, S.T. Grafton, W. Chew, R.P. Woods and P.M. Colletti, Ann-Neurol. 1994 Jun; 35(6): 746-9
90. S.M. Rao, J.R. Binder, P.A. Bandettini, T.A. Hammeke, F.Z. Yetkin, A. Jesmanowicz, L.M. Lisk, G.L. Morris, W.M. Mueller and L.D. Estkowski, Neurology. 1993 Nov; 43(11): 2311-8
91. L.R. Schad, F. Wenz, M.V. Knopp, K. Baudendistel, E. Muller and W.J. Lorenz, Magn-Reson-Imaging. 1994; 12(1): 9-15
92. C.R. Jack, R.M. Thompson, R.K. Butts, F.W. Sharbrough, P.J. Kelly, D.P. Hanson, S.J. Riederer, R.L. Ehman, N.J. Hangiandreou and G.D. Cascino, Radiology. 1994 Jan; 190(1): 85-92

93. G.D. Jackson, A. Connelly, J.H. Cross, I. Gordon and D.G. Gadian, Neurology. 1994 May; 44(5): 850-6

14
NMR of Paramagnetic Species

BY R. R. SHARP

1 Introduction

Nearly half of the naturally-occurring elements possess common oxidation states which are paramagnetic, and organic and other non-metal radical species play a conspicuous role in chemical reactions. Thus paramagnetic species comprise a very broad segment of chemistry. This review will not attempt exhaustive coverage of studies in which NMR has been used for routine characterization of paramagnetic molecules. Rather attention will be directed toward novel and /or fundamental approaches to the study of paramagnetic centers in molecules that are likely to be of general interest to the broader community of chemical and biochemical NMR scientists. Two recent general reviews of paramagnetic NMR have appears[1,2]. Several reviews of more restricted coverage (e.g., paramagnetic metallo-enzymes) are noted in the relevant sections below.

As has been the custom with past reviews of this series, the focus is on areas traditionally considered to involve chemical or biological NMR, thus excluding a relatively large body of research associated with metals, alloys and superconductors. In departure from past custom however, this review will cover two areas of solid-state NMR which have recently generated a great deal of interest in Chemistry, namely, NMR studies of paramagnetic fulleride radical anions and of the superconducting salts of the alkali metal fullerides, and NMR studies of the properties of organic metals and organic ferromagnets. The interest shown by physical, organic, inorganic and materials chemists in these areas seems to warrant their inclusion in the review, accompanied by a brief background to provide the non-specialist with some perspective on these fields.

2 Nuclear Spin Relaxation in Paramagnetic Solutions

The past decade has seen considerable theoretical activity directed toward generalizing the traditional Solomon-Bloembergen-Morgan (SBM) theory of paramagnetic relaxation in solution to describe the effects of physical interactions other than the electronic Zeeman interaction on the precessional motion of the electron spin. SBM Theory, *ca* 1955-62, has provided the fundamental physical paradigm for interpreting the spin relaxation enhancements produced by paramagnetic solutes. The results of the theory are summarized in the following equations:

$$T_{1,2p}^{-1} = \left(T_{1,2p}^{-1}\right)_{dip} + \left(T_{1,2p}^{-1}\right)_{sc} \quad (1a)$$

$$\left(T_{1p}^{-1}\right)_{dip} = C_{dip}\{3\,j(\omega_I) + j(\omega_I-\omega_S) + 6\,j(\omega_I+\omega_S)\} \quad (1b)$$

$$\left(T_{2p}^{-1}\right)_{dip} = \frac{1}{2}C_{dip}\{4\,j(0) + 3\,j(\omega_I) + j(\omega_I-\omega_S) + 6\,j(\omega_I+\omega_S) + 6\,j(\omega_S)\} \quad (1c)$$

$$\left(T_{1p}^{-1}\right)_{sc} = 2C_{sc}\{j(\omega_S)\} \quad (1d)$$

$$\left(T_{2p}^{-1}\right)_{sc} = C_{sc}\{j(0) + j(\omega_S)\} \quad (1e)$$

$$C_{dip} = \frac{S(S+1)}{10}\left(\frac{\mu_o}{4\pi}\cdot\frac{\gamma_I\, g\, \beta_o}{r_{IS}^3}\right)^2; \qquad C_{sc} = (A/\hbar)^2\, S(S+1)/3 \quad (1f)$$

$$j(\omega) = \left(1+\omega^2\tau_c^{\,2}\right)^{-1} \quad (1g)$$

$$\left(\tau_c^{-1}\right)_{dip} = \tau_S^{-1} + \tau_R^{-1} + \tau_e^{-1}; \qquad \left(\tau_c^{-1}\right)_{sc} = \tau_S^{-1} + \tau_e^{-1} \quad (1h)$$

These expressions describe the dipolar and scalar parts of the intramolecular T_1 and T_2 relaxation rates (in SI units) of a nuclear spin I produced by an electron spin S on a paramagnetic solute molecule. r_{IS} is the I-S interspin distance, γ_I is the nuclear gyromagnetic ratio, $\omega_{I,S}$ are the Larmor frequencies of the I(S) spins, g is the electron g-value, β_o is the Bohr magneton, A is the scalar hyperfine coupling constant, and μ_o is the magnetic permeability of free space. $(\tau_c)_{dip,sc}$ are correlation times for the magnetic dipole and scalar interactions, expressed in terms of the electron spin relaxation time τ_S, the molecular reorientational correlation time τ_R, and the chemical exchange correlation time τ_e. A detailed discussion of SBM Theory is given by Banci et al[3].

The physical situation which has prompted the development of new theory can be summarized as followed. In the Zeeman (SBM)-limit, the precessional motion of the unpaired electron spin is assumed to be driven by its Zeeman interaction with the external magnetic field. For electron spins $S \geq 1$, however, which commonly occur for transition metal ions, the precessional motion of the spin is driven both by the Zeeman interaction and by the zero-field splitting (zfs) interaction. The latter results from the coupling, mediated by the spin -orbit coupling Hamiltonian, of the electron spin with the electrical potential of the crystal field of the transition metal ion. The zfs interaction vanishes for spins S=1/2 but is non-vanishing for $S \geq 1$ (except in special cases of cubic or higher site symmetry). When the zfs energy is larger than the Zeeman energy, the spin's precessional motion occurs about molecule-fixed axes, rather than about the external field axis. The functional dependence of the NMR relaxation times on magnetic field strength, on molecular structure, and on the physical parameters of the electron spin Hamiltonian differ markedly in the ZFS- and Zeeman-limit situations.

The profoundly different behavior encountered in the ZFS- and Zeeman-limits is illustrated in Figure 1, which is taken from Reference 100. When the precessional motion of the electron spin is quantized along molecule-fixed axes ($\omega_S/\omega_D<1$), nuclear spin relaxation rates depend on the position of the nuclear spin in the molecular coordinate frame, specifically in the coordinate frame which diagonalizes the ZFS tensor: axial positions are more strongly relaxed than equatorial positions by a factor of as much as four. Physically, this difference in behavior arises because the nuclear spins in axial and equatorial positions in the molecule experience different local dipolar fields, the dipolar field being twice as large along the z-quantization axis of the precessional motion as in the transverse (xy) plane. Since the nuclear spin relaxation rate is proportional to the mean-square dipolar coupling energy, axial positions of the nuclear spin are inherently up to four times more relaxing than equatorial positions at a given I-S interspin distance. In contrast, in the Zeeman-limit situation, which is described by SBM Theory, nuclear spin relaxation times do not depend on the angular orientation of the nuclear spin in the molecular coordinate frame. Other differences in Zeeman and zfs-limit behavior occur. One of these is that the definition of τ_R differs in the Zeeman- and zfs-limits; in the former $\tau_R=\tau_R^{(2)}$, while in the latter $\tau_R=\tau_R^{(1)}$, where $\tau_R^{(\ell)}$ is the reorientational correlation time for an ℓth rank molecule-fixed tensor. Since $\tau_R^{(1)}=3\tau_R^{(2)}$, the definition of the τ_R differs in the two limits.

Bertini et al[3], Sharp[4-8], and Fukui et al[9] have presented plots (field dispersion profiles or fdp) of the functional dependence of R_{1p} on the polarizing magnetic field strength for specified values of the polar angle θ between the I-S interspin vector and the principal axis of the ZFS tensor (as is shown for S=1 in Figure 1). The Zeeman-limit behavior, where the fdp does not depend on θ, is shown for comparison. The intermediate regime ($\omega_S/\omega_D\approx1$), where the Zeeman and ZFS interactions are of comparable magnitude and where the direction of the spin quantization axis is field-dependent, is a regime of transitional behavior in R_{1p}, where R_{1p} rises with increasing field strength (B_o) for equatorial nuclear positions and falls with increasing B_o for axial nuclear positions. This transitional behavior occurs in a range of field strengths where the Zeeman and ZFS interactions are comparable in magnitude, and thus the functional form of the profile in this regime determines the ZFS parameters D and E, as well as the g-value, of the electron spin Hamiltonian. In general, NMR relaxation data measured in the ZFS-limit depend strongly on molecular structure parameters (r_{IS},θ,ϕ of the nuclear spin), while data measured in the intermediate regime depend strongly on both the molecular structure parameters (r_{IS},θ,ϕ) as well as on the parameters of the electron spin Hamiltonian (S, ω_S, the zfs parameters D and E, and τ_S).

Sharp[6] has presented a detailed theory of these phenomena, including analytical formulae which parallel the mathematical form of the SBM equations for R_{1p} but are appropriate to the ZFS-limit. The zfs-limit expression for $\left(T_{1p}^{-1}\right)_{dip}$ corresponding to Equation 1-b, for example, is

$$\left(T_{1p}^{-1}\right)_{dip} = \frac{4}{3}\left(\frac{\mu_o}{4\pi}\frac{\gamma_I\, g\, \beta_o}{r_{IS}{}^3}\right)^2 \left\{\frac{S(S+1)}{3}\left[1+P_2(\cos\theta)\right] j(\omega_I) + (2S+1)^{-1}\left[1 - 2^{-1}P_2(\cos\theta)\right] \sum_m |c^{+}_{S,m}|^2\, j(\omega_m)\right\}. \tag{2}$$

where θ is the polar angle of the nuclear spin in the molecule-fixed coordinate system which diagonalizes the zfs tensor (Equation 2 assumes a uniaxial zfs tensor), $P_2(\cos\theta)$ is the second order Legendre polynomial. $c^{+}_{S,m}$ is the raising operator for the S spin manifold, and ω_m is the 1Q transition frequency for the m→m+1 transition. In the zfs-limit, $|\omega_m|=2\pi cD(2m+1)$, where c is the speed of light and D is the zfs D-parameter in cm^{-1}. Modified expressions for $T_{2,dip}$, $T_{1,sc}$ and $T_{2,sc}$ corresponding to the zfs-limit have also been given. A detailed analytical theory[6] which bridges the zfs-limit and intermediate regimes has also been developed and implemented in the computer program PARELAX. Bayburt and Sharp[10] have used the theory in an experimental investigation of the outer-sphere charge-transfer complexes which form between [$Mn^{III}acac_3$] and the halogenated solvents $CHCl_3$ and CH_2Cl_2. Systematic discussions of the functional forms of the fdp in the ZFS-, intermediate, and Zeeman-limit regimes, and the origin of the fdp in terms of the underlying physical phenomena, are given in References[4,5]. Fukui et al[9] and Sharp[7,8] have analyzed the effects of rhombicity (the E/D ratio) in the ZFS, which, when present, acts to depress the R_{1p}, substantially so when the rhombicity is greater than about 0.05.

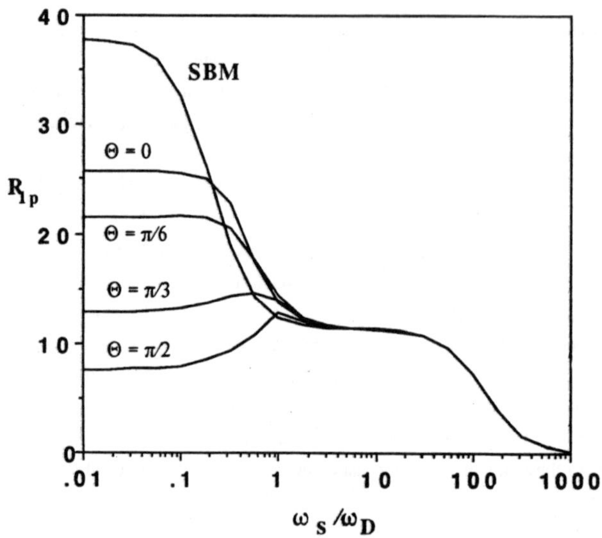

Figure 1. Field dispersion profiles of the $R_{1p}(=T_{1p}^{-1})$ paramagnetic nuclear spin relaxation rate as a function of the ratio of the Zeeman energy to the zero field splitting energy for a paramagnetic spin is S=1. Θ is the polar angle of the interspin I-S vector in the coordinate system which diagonalizes the zfs tensor. The intermediate regime of field strenths, where the 1Q Zeeman and zfs transition energies are comparable, occurs in the vicinity of $\omega_S/\omega_D=1$. For $\omega_S/\omega_D<<1$, the precessional motion of the electron spin is quantized along molecule-fixed axes. SBM is calculated from Zeeman-limit theory.

Bertini, Luchinat, Banci and their co-workers have over the past decade developed relaxation theory which includes effects of ZFS interactions as well as hyperfine interactions and Heisenberg exchange couplings; Reference 3 contains a recent review of this work. In recent model studies, Bertini et al have reported[11-14] fdp's for solvent proton R_{1p}'s induced by $Mn(H_2O)_6^{2+}$, $Fe(H_2O)_6^{3+}$, $Ti(H_2O)_6^{3+}$, and VO^{2+} in pure water and in glycerol/water mixtures. The use of glycerol/water mixtures permitted independent variation of the solution viscosity which was useful for separating the reorientational contribution to τ_c in Equation 1-h. The functional dependence of the solvent ^{1}H fdp was measured over a wide range (5 orders of magnitude) of magnetic field strength as a function of viscosity and temperature. The results were analyzed using SBM Theory and provided values for the solvent proton chemical exchange rate constant and the metal-^{1}H hyperfine coupling constant. An unusual aspect of the VO^{2+}/H_2O system is that hyperfine interactions to ^{51}V proved to be important in the analysis of the data; hyperfine interactions caused the apparent T_2 of the electron spin as measured by NMR relaxation to be much shorter than the value measured by ESR, for this reason producing deviations between experiment and the predictions of SBM Theory. A modified form of SBM theory which accounts for hyperfine coupling interactions in the electron spin Hamiltonian was used in the analysis.

Earlier theoretical work has been based on the stochastic Liouville (SL) equation, an approach developed by Westlund, Kowalewski and their co-workers which is described in previous reviews in this series. The SL approach provides a very general theoretical description, but the generality of the SL approach has been limited in its application by an important omission, namely the dependence of R_{1p} on molecular structure, i.e., on the polar angles of the IS interspin vector in the molecular coordinate frame, was not included; instead the theory assumes that the nuclear spin lies on the principal axis of the ZFS-tensor. Recent applications of the SL approach have involved proton NMR relaxation enhancements produced by Ni(II), an S=1 ion, in simple model systems. Westlund et al[15] have presented a detailed analysis of molecular motions and ^{1}H spin relaxation in solutions containing the Ni^{2+} aqua-cation. A molecular dynamics calculation was used in this study to simulate the motions of water molecules in the Ni^{2+} coordination sphere, and the results of the MD simulation were then used to calculate the time correlation functions needed in SL relaxation theory. In another model study, Ribbing et al[16] have re-examined an earlier detailed analysis of NMR relaxation data for the model S=1 complex bis(aqua)bis(malonato)Ni(II). The earlier analysis concluded, on the basis of computer fits to solvent proton relaxation data, that the zfs D-parameter for this complex is extremely large, *ca* 85 cm^{-1}, a conclustion that has now been re-examined through *ab initio* calculations of the D parameter. These calculations used a limited CI expansion of the wave function using a Slater determinant basis set constructed using open shell SCF functions. Calculated values for D were found to lie in the range 10-20 cm^{-1}, rather than the much higher value (85 cm^{-1}) that was previously inferred from ^{1}H relaxation data, and the discrepancy was attributed to an inappropriate choice of physical parameters used in the earlier simulation procedure.

In a research note, Werbelow and Thevand[18] have described interesting phenomena in paramagnetic relaxation which may result from the interference of two interactions of identical symmetries as, for example, among dipole-dipole, electric quadrupole, and chemical shift anisotropy interactions. It is pointed out that these interactions can couple even and odd spin orders.

Also, in non-English literature that is not available to the reviewer except in abstract form, Lim et al[18] have used transition metal ion-induced solvent ^1H_2O chemical shifts and relaxation enhancements as a basis for the precision measurement of the proton gyromagnetic ratio.

3 New Experimental Techniques in Studies of Paramagnetic Species

3.1. Electrochemically-generated Paramagnetic species.- Sandifer et al[19] have proposed an interesting NMR method for the measurement of the molar atomic susceptibilities of electrochemically-generated paramagnetic species in solution. For this pupose they constructed an electrochemical cell which fits into a capillary insert within a 5 mm NMR tube (Figure 2). The cell contains a working electrode (Ag) surrounded by a shrinkable Teflon shroud (TS) with a counter electrode (Au) wound around the outside of the shroud. The paramagnetic species (in their experiments, $Fe(CN)_6^{3-}$) was electrochemically generated from ferrocyanide during NMR observation (without sample spinning) of proton resonances in the solution. The bulk susceptibility difference (χ_P-χ_D) between the inner (P) and outer (D) compartments was measured using the Evans method, which is based on the fractional NMR frequency shift $\Delta f/f$ of the resonance of a diamagnetic reference compound (tert-butanol in these experiments) in the P and D compartments:

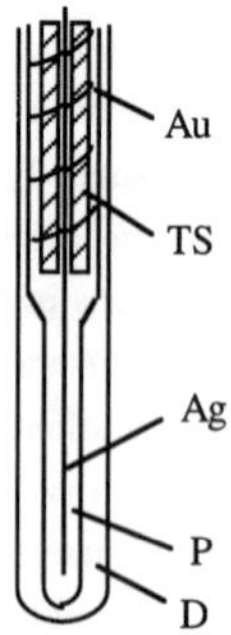

Figure 2. NMR Electrochemical cell

$$\Delta f/f = \left(\frac{4\pi}{3}\right)(\chi_P - \chi_D) \qquad (3)$$

The molar atomic susceptibility of $Fe(CN)_6^{3-}$ was then calculated from χ_P, and the result agreed well with literature values.

3.2. NMR of Paramagnetic Species in the Solid State.- Brough et al[20] have explored solid-state NMR techniques for studying paramagnetic solids by CP-MAS NMR, using the lanthanide acetates, $Ln(O_2CCH_3)_3{\cdot}4H_2O$ (La=Nd, Sm, Eu, Y) and $Pr(O_2CCH_3)_3{\cdot}H_2O$, along with their deuterated analogs. The central question addressed by these authors is the physical origin of the paramagnetic chemical shift and of the relatively large spectral linewidths encountered in lanthanide-containing solids. The authors concluded that the hyperfine shifts are largely contact (scalar) shifts, while the excess paramagnetic linewidths arise from a combination of spin relaxation and magnetic susceptibility effects along with, in protonated samples, residual linewidth due to inadequate ^1H spin decoupling. The authors also made some interspin distance estimates from measured ^{13}C linewidths. The calculated distances were, in the authors' view, in surprisingly good agreement with X-ray structure data considering the approximations employed (neglect of the diamagnetic chemical shift anisotropy and neglect of other than nearest-neighbor I-S dipolar interactions). This study lead to optimistic conclusions concerning whether useful chemical and structural information can be extracted from the CP-MAS spectra of unknown paramagnetic solids.

In connection with these CP-MAS studies, Raleigh et al[21] have proposed a relatively simple multi-frequency ^1H decoupling strategy suitable for paramagnetic solids. A major part of the excess linewidth observed in the studies described above resulted from inadequate ^1H spin decoupling due to the broad range of proton chemical shifts which occur paramagnetic solids. Even a minor deviation of the ^1H irradiation frequency from exact resonance causes a substantial decrease in the decoupling efficiency and results in substantial line broadening in CP-MAS ^{13}C spectra. Since the excess linewidth varies as ΔB_o^2, the spread of ^1H frequencies that are induced by paramagnetic metal centers can be quite large, particularly in lanthanide-containing solids, since contact and pseudo-contact chemical shifts produced by lanthanide ions are generally relatively large. The use of a simple two-frequency ^1H irradiation scheme dramatically improved the spectral quality of the CP-MAS spectra of two lanthanide salts, $Eu(CH_3CO_2)_3{\cdot}4H_2O$ and $NdNa(EDTA){\cdot}8H_2O$.

In a ^{13}C/MAS study of the weakly paramagnetic complex $Ni^{II}(Ac_2CC(CNNH)_2)$, Basato et al[22] reported narrow lines, sufficiently well resolved that specific isotopic enrichment permitted safe signal attribution. The spin moment for Ni^{II} (an S=1 ion) is much smaller than that of the majority of lanthanide ions, and excessive magnetic susceptibility broadening was not encountered.

Lin et al[23] have also studied hydrated lanthanide salts by solid-state NMR, in this case using the ^2H solid echo NMR technique to study molecular motions in $NdCl_3{\cdot}6D_2O$. The anisotropic hyperfine field of Nd^{3+} broadens the powder pattern of the ^2H resonance, and molecular motions partially narrow the hyperfine broadening of the spectrum. Thus these spectra provide information on molecular motions in the solid state. ^2H NMR spectra were collected using a shift-compensated quadrupole-echo pulse sequence, and the measured powder patterns were simulated using computer programs modified to incorporate the effects of molecular motions. This approach to the study of molecular motions in the solid state is particularly well suited to lanthanide salts and complexes, for which the electron spin highly localized on the metal ion and for which electron spin relaxation is extremely rapid (10^{-12}-10^{-13} s, except for Gd^{3+} and Eu^{2+}).

4 NMR of Paramagnetic Transition Metal Complexes

4.1. Ligand Chemical Exchange Kinetics in Transition Metal Complexes.- NMR has long been used to study fast the chemical exchange reactions which equilibrate solution species, both in paramagnetic and diamagnetic solution. The dependence of the rate constant on temperature and pressure are expressed, in classical reaction rate theory, as

$$k_{ex} = \frac{k_B T}{h} \left(\frac{\Delta S^{\ddagger}}{R} - \frac{\Delta H^{\ddagger}}{RT} \right) \qquad (4\text{-a})$$

$$\ln k_{ex} = \ln k_{ex,0} - \frac{(\Delta V^{\ddagger})\, P}{RT} + \frac{(\Delta \beta^{\ddagger})\, P^2}{2\, RT} \qquad (4\text{-b})$$

where $k_{ex,0}$ is the chemical exchange rate constant k_{ex} at zero pressure, $\Delta V^{\ddagger}$ is the activation volume, $\Delta\beta^{\ddagger}$ is the compressibility coefficient of activation, and other symbols have their common meanings. Klingler and Rathke[24] have studied solvent chemical exchange reactions of the type

$$[M(MeNH_2)_5H_2{}^{17}O]^{3+} + H_2O = [M(MeNH_2)_5H_2O]^{3+} + H_2{}^{17}O \qquad M{=}Cr, Co, Rh$$

through variable temperature and variable pressure ^{17}O NMR linewidth measurements. Analysis of their data provided values of k_{ex}, $\Delta H^\ddagger$, $\Delta S^\ddagger$, and $\Delta V^\ddagger$ for these reactions. These results were used to distinguish associative *vs* dissociative activation modes for the reaction. Dellavia and co-workers[25,26] have likewise studied the ligand exchange kinetics of dimethylformamide (DMF) solvent onto $[Ti^{III}(DMF)_6]^{3+}$ and of DMSO solvent onto $[V^{III}(DMSO)]^{3+}$ using variable temperature and pressure measurements of ^{1}H and ^{17}O NMR relaxation times. Rate constants and activation parameters were determined in these systems as part of a systematic study of the trivalent high spin first row transition metal ions under pressure.

Another model kinetic study is that of Inada et al[27], who studied the solution structure and chemical exchange reactions of Cu(II)-ethylenediamine complexes in aqueous and (neat) ethylenediamine (en) solvents. EXAFS was used to determine the solution structure of the Cu(II) coordination sphere, which contained two long axial Cu-N bonds and four short equatorial Cu-N bonds. Temperature-dependent NMR relaxation data were used to measure k_{ex} vs temperature as well as $\Delta H^\ddagger$ and $\Delta S^\ddagger$ for the reaction. Their results provided a basis for a mechanistic proposal in which the chemical exchange of ethylenediamine proceeds via five steps in a dissociative interchange mechanism: (1) first, an axial Cu-N bond breaks, followed by (2) monodentate association of a new en ligand in the open position, (3) rapid axial-equatorial interconversion, (4) dissocation of the leaving en at the new axial position, and (5) ligation at the open axial position by the second N atom of the new en ligand. Similar mechanistic conclusions have been reached by Fabian[28], who investigated the substitution reactions of NH_3 with a variety of Cu(II) complexes of mono-, bi-, tri- and tetra-dentate ligands using temperature jump and NMR line-broadening techniques. The results of this study likewise suggest the involvement of a dissociative interchange mechanism in which the rate-determining step is axial coordination of the entering ligand.

A number of other NMR studies of ligand exchange reactions of transition metal complexes and solvates have been published[29-36]. Stout et al[29] have used ^{13}C NMR to measure k_{ex}, $\Delta H^\ddagger$ and $\Delta S^\ddagger$ for the chemical exchange reactions between the CO_3^{2-} anion and the actinide cations U(VI), Np(VI), and Pu(VI) in aqueous carbonate solutions and have proposed a kinetic mechanism for these reactions. Kostromina at el[30,31] have used ^{31}P NMR to study Cu(II) complexation with phosphorous and phosphonic acids. Nazmutdonova et al[32] have use ^{19}F NMR and ESR to study the ligand exchange kinetics and to measure the activation parameters (k_{ex}, $\Delta H^\ddagger$ and $\Delta S^\ddagger$) for the chemical exchange reaction of F^- with VOF_5^{3-} and FeF_6^{3-}.

A number of solvation studies of transition metal cations in mixed solvent systems have been reported. Reynolds and Davis[37] have investigated the composition of the solvation shell of chloropentaamminecobalt(III) in water-dimethyl sulfoxide mixtures using NMR linewidth and thermodynamic measurements, studying whether a species with a unique DMSO molecule in the Co(III) coordination sphere is present, a hypothesis which their analysis supports. Devyatov et al[38] have used ^{1}H NMR, spectrophotometry and computer modeling in a study of the heterosolvates which are formed in solutions of Co(II), Ni(II), and Cu(II) in aqueous solvent mixtures.

4.2. Transition Metal Porphyrins and Tetraazamacrocyclic Complexes.- A number of experimental NMR studies of porphyrin complexes of iron[39-46] and nickel[47,48] have appeared. The chemical shifts of ^1H and ^{13}C resonances of the porphyrin are predominantly contact (scalar) shifts, which have been very useful for mapping spin density in the ligand. Such maps provide insight into the identity and structure of the HOMO, as well as providing information concerning electron withdrawing and releasing effects of substituents. Iron-porphyrins have been studied in a wide range of oxidation states, the formal oxidation states ranging from Fe^0 to Fe^V. Sinyakov and Shulga[39] have produced highly reduced Fe-porphyrin anions by interaction of the Fe^{III}-porphyrin precursor with a sodium mirror. The electronic configurations of these anions is not clear at present (e.g., $Fe^{II}P^-$ or Fe^IP?), and ^1H NMR has been used to probe the electronic structures, spin states, and paramagnetism of these species. The most highly reduced Fe^0-porphyrin appears to be diamagnetic, while the Fe^I-porphyrin and Fe^{II}-porphyrin species are paramagnetic with spins that depend on the solvent; these species exhibit substantial Fe→Porph charge transfer. Highly oxidized Fe-porphyrin complexes have also been studied. Yamaguchi et al[40] have prepared $O{=}Fe^{IV}$-porph· and $O{=}Fe^V$-porph· species, which they have characterized by ^2H NMR. Kadish et al[41] have electrochemically generated a series of cationic Fe-octaethyl-tetraphenyl-porphyrin (OETPP) species $[(OETPP)Fe^{IV}(C_6H_5)]^{n+}$ (n=1..3), among which the di-cationic species is paramagnetic. ^1H NMR was used to characterize the spin states and paramagnetism of these species.

Iron oxophlorins and verdohemes are compounds, important in the oxidative degradation of hemes, in which the porphyrin ring is oxidized at a single meso position. Proton NMR spectra have been analysed for two octaethyloxaporphyrin complexes of Fe(II) and Fe(III) (verdoheme analogs)[42] and of Fe^{III} octaethyloxophlorins[43]. The oxophlorins form interesting dimers, $[Fe^{III}(OEPO)]_2$, which are stabilized by pairwise Fe-O interactions between the iron of one heme and the ring oxygen of the other. The long-range orbital overlap in these dimers mediates weak antiferromagnetic coupling ($|J|<5$ cm^{-1}) between the two heme iron atoms. Synthetic Fe^{III} chlorins, in which the porphyrin is oxidized at one or more positions of the pyrroline rings rather than at the meso positions, have also been studied by NMR, in this case to model a peroxo-Fe^{III}-chlorin reaction intermediate in the catalytic cycle of cytochrome d[44].

Also, a number of non-porphyrin tetraaza-macrocyclic complexes of Co^{II}, Ni^{II}, and Cu^{II} have been sythesized and characterized in part by ^1H and ^{13}C NMR[49-52]. Cameron and Scott[53] have synthesized complexes of Co^{II} and Ni^{II} with a novel quinquedentate macrobicyclic ligand with all nitrogen coordination, and have used variable temperature ^1H NMR spectroscopy to study conformational equilibria in the complex.

4.3. NMR Measurement of Singlet-Triplet Gaps in Exchange-Coupled Complexes.- Transition metal centers which contain two or more transition metal ions frequently exhibit Heisenberg exchange coupling: the electron spins on individual metal ions couple to produce a total spin S_T, subject to the triangle rule of angular momentum coupling, $S_T=|S_1-S_2|, |S_1-S_2+1|,..,|S_1+S_2|$, thus forming ladders of coupled spin states. For ferromagnetic coupling, the lowest energy state of the ladder is that with highest spin, while for diamagnetic coupling (which is the more common situation for transition metal complexes) the lowest energy state is that of lowest spin. The energy spacing of the spin ladder is often of order k_BT at room temperature, and thus the thermal populations of the spin states are in general strongly temperature-dependent. Recent reports[54-56] have demonstrated that the energy level spacing of spin ladders can, in favorable situations, be measured from temperature-dependent NMR contact shift measurements of ligand resonances.

Cotton et al[54] have synthesized dinuclear chromium complexes of the formula $Cr_2(O_2CR)_4L_2$ (L=MeOH, H_2O, pyr, MeCN; R=Me, Et), in which a pair of acetate(propionate)-bridged Cr(II) ions (S=2) form a strongly antiferromagnetically coupled spin ladder with an S_T=0 ground state and an S_T=1 first excited state. The singlet-triplet spacing is >k_BT at 300 K so that only the two lowest states of the spin ladder are appreciably populated. At very low temperatures the complex is diamagnetic (S_T=0), but it becomes weakly paramagnetic as temperature rises above about 200 K. The ^1H chemical shifts of the ligand protons exhibit a strong temperature dependence as the thermal population of the S_T=1 state increases. The chemical shift $\Delta\delta$ (assumed to be a contact shift) of the paramagnetic complex relative to that of the diamagnetic complex is given by

$$\Delta\delta = \left(\frac{2\,g\,\beta_o\,A}{(\gamma_I/2\pi)\,k_BT}\right)\left(1 + 3^{-1}e^{E/kT}\right)^{-1} \tag{5}$$

where E is the singlet-triplet energy gap. The measured temperature dependence of the contact shift from 180 to 330 K followed this expression closely, yielding a value E=500 cm^{-1} for the S-T energy gap. Using this method, the S-T gap energy was measured for a series of complexes in which the Cr-Cr bond distance was varied systematically by changing the ligand L, and a close correlation was found between the gap energy and the Cr-Cr bond length. In companion studies, the S-T gap energy was measured by ^{31}P NMR in quadruply-bonded dimolybdenum and ditungsten complexes[55] and in the dinuclear molybdenum complex[56] $Mo_2Cl_4(OAc)_2(PMe_3)_2$.

4.4. Bioinorganic Model Complexes.- Ueyama et al[57] have examined peptide complexes which model the reduced form of rubredoxin, an Fe/S protein with a single iron liganded by four cysteines. These authors reported ^2H NMR evidence that the cysteine sulfurs which ligate the iron form two hydrogen bonds with amide N-H protons in these complexes. The model peptide complexes used included $(NEt_4)_2$[Fe(Z-cys-pro-val-cys-OMe)$_2$], (Z=benzyloxycarbonyl), in which an Fe^{II} ion is liganded by four cysteine sulfurs. Large isotropic shifts of four of the amide N-^{2}H peaks indicated the existence of N-H···S···N-H hydrogen bonded units in the complex. ^{19}F chemical shift and spin relaxation studies on model fluorinated pentapeptide complexes of Fe^{II} support this conclusion that amide protons can hydrogen bond with cysteinyl sulfur[58]. In another study, Ueyama et al[59] prepared peptide complexes which model the $(Fe_2S_2)^{2+}$ centers of plant ferredoxins, attempting to achieve in the models redox potentials which mirror the relatively high potential (*ca* -0.42 v) of the native protein. They used ^1H NMR to show that the isotropic shifts of the peptide model complex in DMSO are similar to that of unfolded ferredoxin (which retains its $(Fe_2S_2)^{2+}$ center) in a mixed H_2O/DMSO solvent.

Oxygen-bridged diiron centers occur in several widely studied proteins, including hemerythrin, ribonucleotide reductase, purple acid phosphatase, uteroferrin, and methane monooxygenase. Inorganic complexes which model this center and which have at least partially been characterized by NMR are reported by Mashuta et al[60] and Menage et al[61]. Mashuta et al[60] have synthesized a mixed valent Fe^{II}-Fe^{III} antiferromagnetic complex (S=1/2 ground state) with the heptadentate ligand $BIMP^-$,

BIMP⁻:

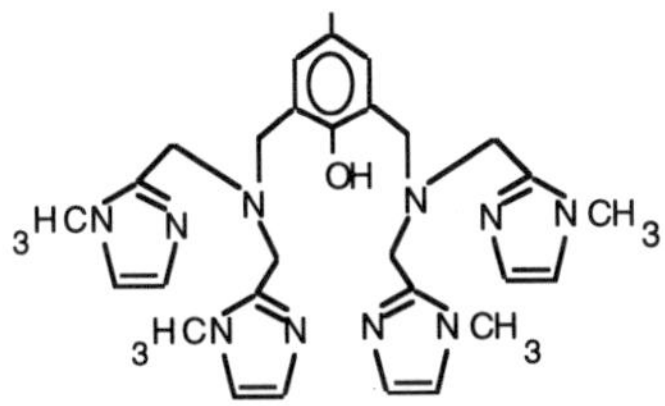

which models the diiron center of ribonucleotide reductase. Using ^1H NMR, they studied the degree of valence trapping in the complex. Well-resolved proton resonances spanning 350 ppm were observed in acetonitrile, indicating that electrons are transferred rapidly between Fe^{II} and Fe^{III}, although the data could not distinguish inter- from intra-molecular electron transfer.

Lachicotte et al[62] have used ^{13}C, ^1H, and ^{19}F NMR to characterize aqua and hydroxy bridged diiron^{II} and dicobalt^{II} model complexes of reduced non-heme diiron sites in proteins. Their NMR studies provided a direct spectroscopic distinction between bridging hydroxo- and aqua- ligands and confirmed that the solid state structures of these complexes were maintained in solution. Beer and Lippard[63] have examined the ^{13}C NMR spectra of di-μ-oxo-bridged di-Fe^{III} and tetra-Fe^{III} complexes that had bridging ^{13}C-enriched acetate ligands with the objective of finding a useful method for determining the number of inequivalent carboxylates in complexes of these types.

4.5. Other Transition Metal Complexes.- Other transition metal complexes for which NMR has been used to probe the structure, liganding, and stability with respect to ligand exchange reactions include acetylsalicylate complexes[64] of Cu^{II}, hydrazine complexes[65] of V^{II} and V^{III}, sulfosalicylate complexes[66] of Fe^{III}, Cu^{II}, Ni^{II}, and Co^{II}, heteronuclear tartrate complexes[67] of Fe^{III} and V^{IV} or V^{V}, Co^{II} β–diketonato complexes[68], acetate and perfluorocarboxylic acid complexes[69,70] of Co^{II} and Ni^{II}, the formation of Heisenberg exchange coupled dimers[71] in solutions of [N,N'-*o*-phenylenebis(salicylaldiminato)]Co^{II}, and the Cu^{II} complex[72] of the ligand N,N'-di(2-acetylcyclohexenyl)ethylenediamine. Tsuchiya et al[73] have used ^{13}C NMR relaxation to probe the site of ligation of Cu^{II} to poly(N-vinylpyrrolidone).

Zil'berman et al[74] have used NMR to study the lability and chemical exchange kinetics of Ni^{II} with amino acids in aqueous media. Zhao et al [75] used ^1H and ^{31}P NMR to study the interaction of $[Cu^{II}(dien)Cl]^+$ with 5'-nucleotides, and Steinkopf et al[76] have used ^1H and ^{15}N NMR to locate the binding sites of Cu^{II} to purines. Tonkovic[77] has used ^{13}C NMR studied the complexation of Fe^{III} with fructose. Gyurcsik et al [78] have used ^{13}C NMR relaxation to study the equilibrium structures of Cu^{II} complexes of N-D-gluconylglycine, and Araki and Tajima[79] have used ^{13}C line broadening to study site specificity in the Cu^{II} interactions with aldopentopyranoses in DMSO. Gaggelli et al[80] have used ^{1}H NMR relaxation techniques to characterize the solvent coordination and the ligand conformation of a Cu^{II} complex of a formamidine pesticide, amitraz. Gajda et al[81] have used ^{1}H NMR and ESR to study Cu^{II}, Ni^{II}, and Co^{II} complexes of carcinine, a peptide-type derivative of histamine, and Vincens et al[82] have used ^{2}H NMR to study Cr^{III} complexes of nicotinic acid.

5 Electron Transfer and Electron Self-Exchange Reactions

Electron transfer or electron self exchange between the paramagnetic and diamagnetic forms of a solute frequently produces line broadening in the NMR resonances of the diamagnetic molecule that depends on the rate of the self-exchange reaction. The classical theory of chemical exchange phenomena NMR in diamagnetic systems also describes, with minor variations, the NMR effects of electron self-exchange. The rapid transfer of the paramagnetic spin from paramagnetic to diamagnetic molecules accelerates the decay of the transverse magnetisation (i.e., shortens T_2) of the NMR lines of the diamagnetic molecule, a phenomenon which can be observed directly in a time-domain pulsed NMR experiment, or equivalently, as line-broadening in the transformed spectrum. The very large characteristic linewidths and chemical shifts of paramagnetic species are very large, and thus very small concentrations of a paramagnetic species, which itself may not be directly observable, may significantly influence the NMR spectrum of a much larger concentration of diamagnetic species if the two are in equilibrium. Alternatively, chemical exchange reactions that are too slow to study by line-width measurements may be amenable to saturation-transfer techniques, where a resonance of the oxidized (reduced) species is satuated and intensity changes in the reduced (oxidized) form are monitored as a function of time. Both line broadening and saturation transfer methods have been used to measure the rates of electron self-exchange reactions.

The elementary reaction steps of catalytic cycles frequently involve transfers of electrons, protons and hydrogen atoms. Marcus Theory provides a theoretical basis for interpreting the kinetics of e^- transfer reactions, but less is known about factors that control rates of H^+ and H atom transfers between metal complexes. Protasiewicz and Theopold[83] have used a combination of line broadening and saturation transfer experiments in ^{1}H and ^{13}C NMR to make a direct experimental comparison of the self-exchange rates of electrons, protons, and hydrogen atoms in a single redox reaction, which involved the equilibration of $Tp'Mo(CO)_3H$ and $Tp'Mo(CO)_3$ (Tp'=hydridotris(3,5-dimethylpyrazolyl)borate)). Rates of degenerate e^- transfer were found to be very fast on the NMR timescale - only a single set of broadened ^{1}H NMR peaks was observed for the oxidized and reduced species. Proton transfer rates were found to be too slow for measurement by NMR line broadening techniques, but these rates were measurable using saturation transfer experiments, for which the characteristic time-scale is seconds rather than milliseconds. Degenerate H-atom transfer was monitored by ^{13}C NMR at -34°C, but proved to be too slow for direct measurement ($k_{ex}<9.0\times10^{-3}$ $M^{-1}s^{-1}$). In summary, very large differences in the rates of e^-, H^+, and H· transfer were observed, the first being very fast, the second slow, and the third immeasurably slow in NMR experiments.

Anderson et al[84] have studied electron and atom self-exchange rates of reactions of the type

$$M(II)(cp)_2 + M(IV)(cp)_2X^+ = M(IV)(cp)_2X^+ + M(II)(cp)_2$$

where M is Ru or Os, X is a halogen, and cp is the cyclopentadienyl anion. The transition states of these reactions appear to involve a bridge halogen atom, but other mechanistic details, such as whether halogen transfer occurs as X^+, X, or X^-, remain unclear. Mechanisms other than X^+ transfer must involve a counter-transfer of electrons, which, if it occurs, may involve either an inner sphere or outer sphere mechanism, and if 2 e^- are transferred, this may occur in a concerted or sequential fashion. The authors have characterized the reaction kinetics experimentally through computer analyses based on the Bloch equations of ^1H NMR line broadening as a function of temperature, pressure, and solvent. They report rate constants, $\Delta H^‡$, $\Delta S^‡$, and $\Delta V^‡$ values for the reactions, but the mechanistic questions posed above are still largely unresolved.

Burri and Fischer[85] have proposed a set of multi-pulse FT NMR experiments to measure reversible electron self-exchange rates of radical anions and cations, or of the migration of triplet excitation energy, following pulsed optical formation of paramagnetic species in solution. Their study involved acetonitrile-d_3 solutions which contained as solutes anthracene and N,N-dimethyl-p-toluidine. Pulsed illumination of this system produced anthracene anion and toluidine cation radical pairs, which subsequently recombined to form anthracene triplet species. The authors studied the self-exchange kinetics of the triplet excitation energy between excited and ground-state anthracene molecules, as well as the decay rate of the anthracene triplet excitation.

Warren et al[86] have studied outer sphere electron transfer between $[Co^{II}(en)_3]^{2+}$ and various anionic Co^{III} oxidants, such as $[Co^{III}(edda)(oxalato)]^-$ ($edda^{2-}$=1,2-diaminoethane-N,N'-diacetate). To probe the structure of the outer-sphere complex which mediates electron transfer, they used $[Cr^{III}(en)_3]^{3+}$, a paramagnetic analog of $[Co^{II}(en)_3]^{2+}$, and monitored paramagnetic relaxation increments of protons on the diamagnetic $[Co^{III}(edda)(oxalato)]^-$ counterion as a function of the $[Cr^{III}]/[Co^{III}]$ concentration ratio. Using SBM Theory, they then deduced Cr^{III}-^1H distance information for the ion pair intermediate.

Electron transfer kinetics of Cu^I/Cu^{II} redox reactions have been studied by Rorabacher et al[87-89] using proton NMR linewidth and T_1 data. In these reactions Cu^{II} species are typically 6- or 5-coordinate and Cu(I) is typically tetrahedral, and thus electron transfer in the Cu^I/Cu^{II} couple is expected involve large reorganization of the ligands. The authors have examined the accuracy of Marcus Theory in describing these reactions. For this purpose, they measured rates of electron self-exchange in equilibrium solutions of Cu^IL and $Cu^{II}L$, where L is a macrocyclic thioether ligand (I, II, III or IV). They also studied the rates of the redox reactions which transfer electrons between the $Cu^{I/II}L$ couple and other redox couples in the solution, e.g., in the reaction $Ru^{II}(NH_3)_5py$ + $Cu^{II}L = Ru^{III}(NH_3)_5py + Cu^IL$. The rates of these reactions are well suited to the NMR time-scale, and in each case at least one resolvable proton peak of the diamagnetic Cu^IL species exhibited sufficient exchange broadening to allow kinetic studies. From studies of the cross- and self-exchange reactions, the authors measured individual rate constants for $Cu^{II}L$ reduction, for Cu^IL

oxidation, and for $Cu^{I/II}$ self-exchange. Marcus Theory predicts simple relationships among these quantities that were tested against the data and generally confirmed experimentally.

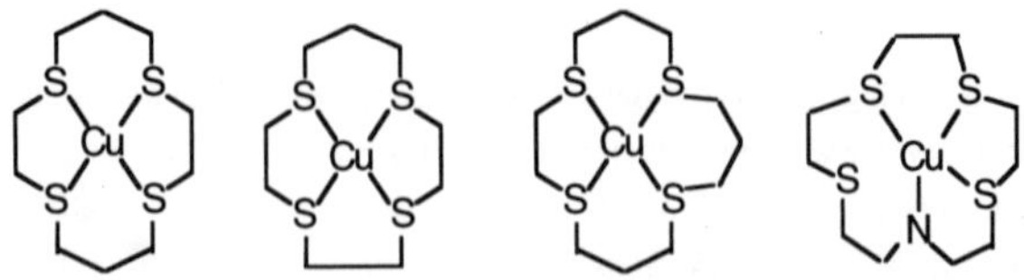

I II III IV

The $Cu^{I/II}$ redox couple is central to the biological activity of copper electron transport proteins. Among this class of proteins are plastocyanin and pseudoazurin, which are type 1 blue copper proteins. Plastocyanin is found in the chloroplasts of green plants and blue-green algae, where it acts as an aqueous electron shuttle between two membrane-bound redox-active species, cytochrome f and the P_{700}^+ chlorophyll radical cation of Photosystem I. Pseudoazurin (MW≅14000 Da), another blue copper protein, donates electrons to nitrite reductase complex of denitribying bacteria. Because of their small size and ready availability from plant material, these proteins have been studied extensively as models of biological electron transport. The rate constant for $Cu^{I/II}$ electron self-exchange between the oxidized and reduced forms of plastocyanin has been studied by proton NMR[90], using as a redox-sensitive probe four ^1H resonances on two histidine ligands (his-37 and his-87). These resonances, which were observable only in the reduced enzyme, exhibited a concentration-dependent broadening in the presence of the oxidized form of the enzyme. Analysis of the electron exchange kinetics was simplified by the demonstration that exchange was in the slow-exchange limit, where $k_{ex}[PCu]_{tot} << (1/T_{1,ox} - 1/T_{1,red})$. Electron exchange reactions of the pseudoazurin of Anabaena variabilis was studied in a similar manner[91] using as a redox-sensitive probe the ^1H NMR resonance of his-6. Kinetic analyses of the electron self- and cross-exchange (between plastocyanin and pseudoazurin) reactions showed good agreement between experiment and Marcus Theory.

Iron-sulfur electron-transport proteins are ubiquitous in living systems, where they occur in forms containing from 1-4 Fe atoms, usually with bridging S^{2-} ligands, typically in structures derived from the 4Fe/4S cubane complex. Fe/S proteins are of central importance in the biological electron transport reactions which occur in the inner membranes of chloroplast and mitochondria. Bertini et al[92] have used NMR to measure the electron self-exchange reactions of a soluble high-potential iron protein (HiPIP I) isolated from Ectothiorhodospira vacuolata. For this purpose, hyperfine-shifted ^1H resonances of cysteine ligands of the iron centers in partially-oxidized samples were used. Intermolecular e^--self-exchange between oxidized and reduced forms of the protein was slow on the NMR timescale, and separate resonances, without significant exchange broadening, were observed for the oxidized and reduced forms of the proteins. Although the kinetics were too slow to study by linewidth measurements, the kinetic time-scale was suitable for saturation transfer experiments. Saturating the signals of the oxidized species, intensity variations in the peaks of the reduced species were monitored (or vice versa). The fractional change in intensity was analysed using the equation $\eta=k'/(R_1-k')$, where $k'=k[ox]$ is a pseudo-first order rate constant for self-exchange, and R_1 is the longitudinal relaxation rate of the reduced species. Based on their data, the authors proposed a model in which e^- transfer occurs via a hydrophobic patch near the Fe/S center.

Brunel et al[93] have studied non-physiological e^--self exchange kinetics in solutions of horse-heart myoglobin. They used protein bound by $P(CH_3)_3$, which ligands both the Fe(II) and Fe(III) forms. The methyl ^1H NMR resonances of the ligand occur in an unobscured upfield region, -3.3 ppm for Fe^{II} and -1.2 ppm for Fe^{III}, where line broadening studies could conveniently be carried out. The authors measured the self-exchange rate constant as a function of ionic strength, temperature and pH, and, using Marcus Theory, calculated the ligand reorganization energy for e^- transfer between the Fe^{II} and Fe^{III} species.

6 Chemically- and Photochemically-Induced Dynamic Nuclear Polarization (CIDNP and PCIDNP)

Chemically-induced dynamic nuclear polarization is observed experimentally as a large intensity enhancement in the NMR spectra of product molecules which are formed in free radical reactions following bond cleavage, which results in the formation of a pair of free radicals (spin doublets). The spin wave functions of these doublet radicals are correlated in phase. The radical pair frequently remain in close proximity in solution after their formation due to the cage effect and ultimately recombine with each other. The product of this recombination may be an electronic singlet state if recombination is very rapid, or, alternatively, a spin triplet if recombination is slow enough to permit the differential precession of the electron spins to generate a significant phase difference. The product in this latter case is a triplet species which is relatively long-lived and which possesses a large electronic magnetic moment.

$$\text{A-B}^* \text{ (singlet,}\uparrow\downarrow) \rightarrow \text{A}\cdot(\uparrow) + \text{B}\cdot(\downarrow) \rightarrow \text{A-B (singlet }\uparrow\downarrow) + \text{A-B (triplet }\uparrow\uparrow)$$

The magnetic moment of the electron spin of the triplet product species will in general produce large polarization of the nuclear spins on the same molecule with correspondingly large NMR intensity enhancements.

A number of refinements and new theoretical approaches to the radical pair recombination theory (now more than two decades old) have been reported. Purtov and Doktorov[94] have developed a Green's function method to calculate nuclear and electron spin polarization in the product molecules, and they have used their theory to calculate the influence of CIDNP kinetics on the form of the polarized spectrum. Aizawa et al[95] and Krissinel'[96] have studied the effect of coulomb forces on the diffusional motion and recombination rate in reactions involving radical ion intermediates. Shkrob[97] has described circumstances in which the equality in CIDNP signals arising from protons which constitute a group of magnetically equivalent nuclei in a radical can be violated if there exists a correlation between the hyperfine coupling constant and the abstracted nucleus. Hwang et al.[98] have reported a nuclear isotope effect on CIDNP intensities. These authors found that ^1H enhancement factors in radical pairs generated by α-cleavage of $^{12}C{=}0$ and $^{13}C{=}0$ labeled ketones deviate from the ratios expected on the basis of the relative abundances of the respective isotopomers. This unexpected enhancement or suppression of CIDNP intensity was ascribed to the presence of a second hyperfine coupled nucleus, which affects the electron spin dephasing in the intermediate radical pair.

New experimental approaches for CIDNP studies have also been suggested. Goez[99] has described coherence transfer techniques (1D COSY and 1D relay coherence transfer (RCT) generated by selective pulses) to facilitate the identification of photo-CIDNP product species when hyperfine coupling constants in the intermediate radicals are very small. These techniques were illustrated with the photosystem 9,10-anthraquinone and triallyl amine. Another novel approach has been proposed by Mikhailov et al[100]. Their experiment involves the use of a switched high magnetic field to modulate the radical pair recombination probability. This technique, which has been analysed theoretically but not yet demonstrated experimentally, may provide information on the magnetic resonance parameters, as well as the kinetic parameters, of the intermediate radical pair.

The photocleavage of cyclic precursor molecules, such as cyclodecanone, may result in biradical species in which the two doublet intermediates are joined, in this case, by a flexible polymethylene chain. The CIDNP exit pathways of such systems are rather complex, depending on magnetic hyperfine interactions within the biradical, and these depend in turn on the chain length, as well as on the magnetic field strength and on the diffusional properties. Experimental examples and theoretical analyses of this type of system have been reported[101,102]. Closs et al[101] have reported theoretical and experimental studies of the dependence of the exchange coupling parameter $|2J|$ on the length of the intervening polymethylene chain in biradicals produced by photolysis of cyclic ketone precursor molecules. The $S \rightarrow T_-$ mixing efficiency, which is proportional to the observed CIDNP amplitude of peaks of the aldehyde photolysis product, can be influenced by varying the magnetic field strength during the photolysis act and is highest in the $S \rightarrow T_-$ crossing region. This phenomenon permits a measurement of the exchange coupling parameter $|2J|$. Experimentally, the cyclic ketone precursor was photolysed in an external magnet with a field strength in the range 0-1.8 T, after which the sample was transferred rapidly into the probe of a 60 MHz high resolution spectrometer, for measurement of the CIDNP intensity of the aldehyde peak of the photolysis product. A pronounced maximum in the CIDNP intensity was observed in the vicinity of the level crossing, the field strength of which increased in a regular manner with the length of the intervening polymethylene chain in a homologous series of cyclic ketone precursor molecules. $|2J|$ fell off rapidly with increasing length of the intervening polymethylene chain in the bi-radical as expected, although the quantitative agreement between independent ESR and CIDNP measurements was not very good.

CIDNP provides a very powerful approach to elucidation of the mechanisms of radical reactions and the identification of radical intermediates. Quantitative CIDNP studies have been used by Lee[103] to differentiate the roles of competing exit pathways in the thermal decomposition of diacyl peroxide. PhotoCIDNP probes the mechanisms of photochemically-initiated radical reactions, which may be initiated through the action of photosensitive dyes. Recently reported applications photoCIDNP are in mechanistic studies of the photoreactions of triallylamine and triethylamine with carbonyl-containing sensitizers such as anthraquinone[104], of the photolysis of benzoquinone and its methyl derivatives[105,106] and of benzaldehyde[107], of methylfurans sensitized by chloranil[108], of acetone and 3-hydroxy-3-methyl--2-butanone[109], of $Et_3GeCOPh$ sensitized by $PhCH_2Br$[110], and of $(Me_2CHO)_2PS_2\cdot$[111]. Welsh et al[112] have investigated the direct and sensitized photodecomposition of triarylsulfonium salts using a combination of optical absorption and steady-state photo-CIDNP techniques. The direct and sensitized pathways proceeded *via* different mechanisms. The direct and the anthracene-sensitized reactions proceeded *via* a singlet (diphenylsulfinyl radical cation - phenyl) radical pair,

$$[\phi_3S^+X^-]^1 \rightarrow [\phi_2S + \phi\cdot^+ + X^-]^1 \xrightarrow[RH]{} \phi\text{-}S\text{-}\phi\text{-}\phi + \phi\text{-}S\text{-}\phi + \phi\text{-}R + \phi + H\text{-}X,$$

while the acetone-sensitized pathway proceeded *via* a triplet excited state of the salt,

$$[\phi_3S^+X^-]^3 \rightarrow [\phi\text{-}\overset{\phi}{\underset{\phi}{S}}\cdot + X^-]^3 \xrightarrow[H^+]{} \phi\text{-}S\text{-}\phi + \phi + H\text{-}X.$$

The photodissociation of aromatic acetones has also been studied[113] and was found to be atypical of other carbonyl compounds.

General biological applications of photo-CIDNP have been reviewed by Hore and Broadhurst[114]. Novel applications of photo-CIDNP in these areas have been directed toward probing the structures and binding properties of biological macromolecules, rather than to detailed mechanistic studies of photochemical reactions. Serikawa et al[115] have used photoCIDNP to study the binding of the GAL4 protein to its DNA binding site. They have demonstrated the interaction of a specific tryptophan (Trp^{37}) of the protein with a specific UAS_G nucleotide sequence of DNA. Muszkat et al[116], have used photoCIDNP to elucidate the conformational basis for the immunological diversity of two synthetic multichain co-polymers, each of which contains a terminal tetrapeptide fragment, (i) TyrGluTyrGlu or (ii) TyrTyrGluGlu. The proton signals of the 3,5-aromatic protons of the tyrosines of (ii) exhibited signals indicative of free, rapidly rotating, non-interacting aromatic rings, while those of (i) gave a single, broad signal due to strong Tyr^1-Tyr^3 hydrophobic interaction. This structural difference was implicated as a basis for immunological diversity of these peptides. Alexandru et al[117] have used photoCDNIP to assist in the assignments of the proton NMR spectrum of lactalbumin. Haard et al[118] have used photoCIDNP to probe the surface exposure of the aromatic residues of a glycoprotein.

PhotoCIDNP has also been used to study the reaction mechanisms of photosensitizing drugs, e. g., sulfonamides and porphyrins (this area is reviewed by Marko and Vermeersch[119], and to characterize radical intermediates in model biological systems, specifically, in the photoreactions of pyrimidine dimers[120], in the mechanisms of oxidation of NADH analogs[121], and in the photooxidation of amino acids sensitized by vitamin B_2[122].

6.1. Dynamic Nuclear Polarization (DNP).- DNP experiments involve NMR detection of nuclear spins in free radical-containing samples during microwave saturation of the esr transitions. Enhancement of the NMR intensities is produced by a physical mechanism which is similar to that of the Nuclear Overhauser Effect (NOE). In DNP experiments, the electron spin polarization is produced, not by radical pair recombination as in CIDNP, but rather by esr saturation of the electron spin transitions during NMR detection of the nuclear resonances. At magnetic field strengths currently used for NMR spectroscopy, it is normally necessary to use a microwave horn, as opposed to a resonant cavity as in ESR spectroscopy, for the irradiation of the electron spin levels. Li et al[123] report the construction details of such a DNP spectrometer operating at 1.94 T (54 GHz for electrons). In their experiments, the electron spin polarization achieved resulted in enhancements of the ^1H and ^{13}C NMR signal intensities by 1-2 orders of magnitude.

Several groups have recently attempted DNP experiments in the solid state. Becerra et al[124] have reported DNP experiments at 5 K in polymer samples using a gyrotron 140 GHz microwave source. They studied samples of polystyrene doped with the free radical BDPA (α,γ-bisdiphenylene-β-phenyallyl) and observed ^1H enhancements of ~10 and ^{13}C enhancements of ~30. In another solid-state study, Wind et al[125] have reported substantial ^{13}C NMR enhancements due to DNP in a polymer (bisphenol A polycarbonate) sample doped with trianisylamine and trianisylaminium perchlorate. Lock et al[126] have extended this technique to the detection of nuclear resonances in surfaces. They detected the DNP-enhanced ^{13}C NMR signal, narrowed by magic-angle spinning, of a vapor deposited diamond film. Their samples contained a concentration of $2x10^{18}$ unpaired electron spins/g which, when saturated at the esr transition frequency, directly polarized the ^{13}C spin system to a level sufficient for NMR observation. The very large intensity enhancements provided by DNP experiments appear to provide exciting new approaches in future studies of surface species and possibly for paramagnetic centers in biological membranes as well.

7 Lanthanide-Containing Systems

Ions of the Lanthanide Series have partially-filled 4f subshells. For these ions, Russell-Saunders coupling is not appropriate and J, but not L or S, is a good quantum number. The angular momenta of these ions are generally considerably larger than for the d-block ions, and the electronic magnetic moments are correspondingly large. For this reason, the majority of these ions (with the exception of Sm, for which J=0 in the ground electronic state) generally induce relatively large molar chemical shifts in ligand nuclei. These ions have commonly been used as "lanthanide shift reagents" (LSR) to aid assignments of ^1H and ^{13}C NMR spectra. The chemical shifts induced by the lanthanides result predominantly from dipolar rather than scalar hyperfine couplings; i.e., they are principally pseudo-contact rather than contact chemical shifts. With chiral ligands, LSR have also been used for chiral differentiation of diastereoisomers[127-131].

While their paramagnetic moments generally induce large chemical shifts, the lanthanide ions, with the exceptions of Gd^{3+} and Eu^{2+}, have relatively little effect on NMR linewidths or nuclear spin relaxation times. Gd^{3+} and Eu^{2+} have half-filled f-shells (f^7), for which the orbital state of the free ion is an electronic singlet (^8S). These ions, atypically of the rest of the series, are very strong relaxation-enhancing agents (LRA) of nuclear spins. Thus Gd^{3+} complexes have been used widely as relaxation-enhancing contrast agents for MRI (this topic is discussed further in the following section). They are also being used, though not as widely, in chemical and industrial applications of NMR. In the latter area, Gogolashvili et al[132] have used Gd-LRA complexes in the

^1H NMR analysis of petroleum-water emulsions; addition of the Gd-LRA to the emulsion caused a selective decrease of the relaxation times of the hydrocarbon protons, thereby improving phase identification and the accuracy of the analysis. Ren et al[133] have used ^{13}C NMR for the quantitative analysis of lanthanide mixtures. Using a suitable aqueous complexing agent, these authors found that each lanthanide complex in the mixture produced at least one unique, well-resolved ^{13}C peak, suitable for quantitation purposes.

The difference in the quantization situation (L-S *vs* J) between the d-block ions on the one hand and the lanthanide ions on the other requires an appropriate modification in the traditional relaxation theory for solutions of paramagnetic ions (SBM Theory). First, S(S+1) should be replaced by J(J+1) in Equation 1. Secondly, Bertini et al[134] have proposed expressions in which the spin angular momentum is projected out from the total angular momentum. These authors have made detailed measurements of the magnetic field dependence of T_1 of the solvent protons in solutions of aqueous lanthanide salts (Pr, Sm, Dy, Ho. Er. and Yb) at proton frequencies in the range 0.01-600 MHz. The data are virtually field-independent up to field strengths of *ca* 10 T, where T_1^{-1} rises rapidly. The rise in T_1^{-1} at high field was interpreted as due to the Curie relaxation mechanism, in which nuclear spin relaxation is produced by diffusion of the solvent nuclei in the local Curie field of individual paramagnetic ions. The measured relaxation efficiencies of the lanthanide aquacations were interpreted in terms of the authors' improved relaxation theory, which yielded spin relaxation times in the range 0.1-1.0 ps.

It should be borne in mind that the above studies concerned lanthanides which are weakly relaxing with respect to solvent protons, not Gd^{3+}, which is much more highly efficient as a relaxer of proton spins and which is widely used as a relaxation reagent. Studies of the spin relaxation mechanisms of the Gd^{3+} aqua-cation, as well as of Gd(III) in two complexes, [Gd(propane-1,3-diamine-N,N,N'N'-tetraacetate)$(H_2O)_2$] (V), and [Gd(N,N'-bis[(N-methylcarbamoyl)methyl]-3-azapentane-1,5-diamine-3,N,N'triacetate)(H_2O))] (VI, R=$CH_2CONHCH_3$), have been reported by Powell et al[135]. This group measured electron spin relaxation times of Gd(III) at three field strengths from ESR (X-, S-, and Q-bands) linewidths. Their results bear directly on the use of Gd^{3+} as a LRA in macromolecular Gd(III) complexes, since, when molecular reorientation is slow, the NMR relaxation rates are inversely proportional to the electron spin relaxation rate of Gd^{3+}. They analyzed their data in terms of the relaxation matrix of the ^8S Gd ion, obtaining values for the physical parameters which are responsible for the relaxation (*viz*, the rms fluctuation of the zfs tensor, the correlation time of the fluctuation, and its activation energy). The kinetics and mechanism of the solvent chemical exchange reactions in the $Gd(H_2O)_8^{3+}$ cation and for the Gd^{III}-PDTA (PDTAH=1,3-propylenediamine tetraacetate) complex have also been characterized in detail through variable field, variable temperature, and variable pressure ^{17}O NMR studies[136]. Rate constants and activation parameters ($\Delta H^{\ddagger}$, $\Delta S^{\ddagger}$, and $\Delta V^{\ddagger}$) have been determined for these reactions.

(V) 

(VI) 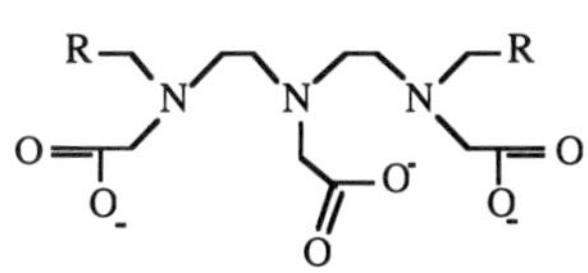

Four other studies of lanthanide spin relaxation in simple salt solutions have focused on the chemical and dynamic properties, rather than the spin physics, of these systems. Fratiello et al[137,138] have used ^{15}N NMR to study contact ion pairs of $^{15}NO_3^-$ with Nd^{3+} in aqueous solvent mixtures with acetone, Freon-12, and methanol. At low temperatures (*ca* -100°C), chemical exchange is slow on the NMR timescale in the acetone and Freon-12 solvent mixtures, and separate NMR signals were observed for uncomplexed NO_3^- in the diamagnetic solvent and in four mixed aquanitrato- species $Nd(NO_3)_n{}^{+3-n}$ and $Pr(NO_3)_n{}^{+3-n}$, where n=1 and 2 plus an undetermined combination of 4, 5, and 6, but not 3, which is a neutral species. Fleitas et al[139] have studied the concentration and temperature dependence of solvent proton and deuteron T_1's in aqueous solutions of Ce(III) and Pr(III), and Devyatov et al[140] have used ^1H NMR relaxation to study La(III) ions in mixed water/dimethylformamide solvents. Their analysis concerns the structures and dynamics of the first and second hydration spheres of the cations, and in the latter study, the formation equilibria of heterosolvate species. Arsenin et al[141] have studied proton relaxation in aqueous solutions of hafnium hydroxychloride at 0°C, interpreting their results in terms of coordination properties of the Hf^{3+} aqua-cation.

Lanthanide ions typically exhibit higher coordination numbers, often 8 or 9, than do the d-block ions. Also, the coordination properties tend to reflect simple coulombic interactions between the metal ion and the coordinating atoms of the ligands, rather than covalency involving the partially-filled f-shell, which is rather well screened from the ligand orbitals. Thus the coordination properties of the lanthanides vary relatively little across the series, in marked contrast to the behavior of the d-block ions[142]. Galera et al[143] have studied water coordination of the La-aqua-cations using Monte carlo simulations and have concluded, based on their use of two different pair potentials, that the hydration number of lanthanide ions drops from 9 to 8 with increasing atomic number across the series, in agreement with the results of variable pressure NMR experiments.

There has recently been considerable activity in the synthesis and properties of new macrocyclic ligands for the lanthanide ions. DOTA (H_4DOTA=1,4,7,10-tetraaazacyclododecane-tetraacetic acid) and DOTP (H_4DOTP=1,4,7,10-tetraaazacyclododecane-N,N',N",N"'-tetra(methylene phosphonate)) are octadentate ligands, mono- and penta-anionic respectively, with mixed O/N coordination, a 4-fold symmetry axis and a single axial water molecule (see below), which have been used fairly widely as paramagnetic relaxation agents in MRI. The solution structures of $Ln(DOTP)^{5-}$ complexes were determined from an analysis of the pseudocontact ^{13}C, ^1H, and ^{31}P chemical shifts, which are very large (up to 900 ppm for ^1H)[144,145]. The chelate structures are very rigid and virtually identical across the lanthanide series, with a single thermally-accessible conformation at room temperature. The $La(DOTA)^-$ complexes are structurally much less rigid and exhibit inter-converting Λ and Δ isomers in solution at room temperature[146,147]. The populations of these isomers are markedly temperature-dependent; the kinetics of the intramolecular isomerization reactions have been studied through variable temperature ^{13}C NMR studies in the Nd^{3+} complex and through 2D-EXSY experiments on the Yb^{3+} complex. The situation becomes even more complex for La(THP) (THP=1,4,7,10-tetrakis(2-hydroxypropyl)-1,4,7,10-tetraazacyclododecane)) complexes, where ^{13}C NMR studies of the La, Eu, and Lu chelates indicate a mixture in solution of five diastereoisomers[148]. La(THED) (THED=1,4,7,10-tetrakis(2-hydroxyethyl)-1,4,7,10-tetraazacyclododecane) represents an intermediate case, where the complex structure is rigid on the NMR timescale at -40°C but is increasingly non-rigid at higher temperatures[149].

DOTA: $R = COO^-$
DOTP: $R = CH_2PO_3^{2-}$
THP: $R = CH_2CHOHCH_3$
PDTA: $R = CH_2CH_2OH$

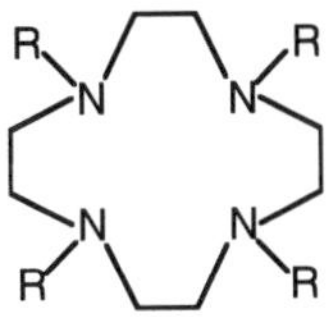

DTPA (DTPA=diethylenetriaminepentaacetic acid) (II, $R = CH_2COO^-$) and its derivatives are octadentate ligands (3N/5O coordination) which have been rather widely used as macrocyclic ligands for Gd^{3+} in MRI studies (see below). La(DTPA) complexes, like La(DOTA)$^-$ and La(DOTP)$^{5-}$, possess a single inner-sphere water molecule, but they lack the local 4-fold symmetry axis that is characteristic of the DOTA-family of ligands[144-147]. Proton NMR spectra demonstrate that La(DTPA) is structurally non-rigid in aqueous media[150], and ^{13}C NMR spectra indicate that four diastereoisomers are in rapid equilibrium at ambient temperatures[151]. The spin relaxation properties, both nuclear and electron, and the solvent chemical exchange kinetics of the inner-sphere water molecule of the Gd(DTPA) complex have been studied by means of variable temperature, pressure and magnetic field ^{17}O NMR studies[152]. TTHA (triethylenetetraminehexacetate) (I) forms trianionic nonadentate complexes with lanthanide ions and has been used as a shift reagent for Na^+ in MRI studies (see below). TTHA complexes of eleven complexes have been characterized with respect to the paramagnetic shifts that they induce in ^{23}Na, ^{35}Cl, ^{2}H, and ^{17}O resonances[153].

TTHA:

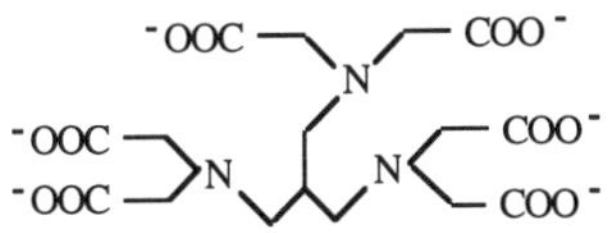

Gd^{3+} complexes of two DOTA-like ligands containing a polyhydroxy(benzyloxy)-propionamide substituent have likewise been synthesized and characterized structurally[154]. These complexes, like La(DOTA), are octadentate with a single axial inner-sphere water molecule. Their ^{1}H spin relaxation properties have been characterized through measurements of the magnetic field T_1^{-1} dispersion profile, and were found to fall off at low field, due to a shortening of the Gd^{3+} electron spin relaxation times in this complex.

Non-macrocyclic ligands of lower dentisity have also been studied. These ligands, which are generally much more labile than the macrocycles, include PDTA (1,2-propanediaminetetraacetate[155]) (penta- or hexadentate), ethylenediamine[156] (bidentate), for which the ligand exchange kinetics have been studied in detail, acetylacetonates and pyridine[157], pyridoxal (N-benzoylglycyl)hydrazone[158] (tridentate), tridentate Schiff base ligands[159], polyethlene glycol ethers[160,161], 2,6-diacetylpyridine bis(2-pyridylhydrazone)[162] (pentadentate), 4-N-(4'-antipyrylmethylidene)aminoantipyrine[163] (tridentate).

The use of lanthanide shift and relaxation reagents as aids in spectral assignment has been well established for two decades and specific applications will not be reviewed comprehensively here, but certain new approaches to the use of these agents will be described[164-169]. Lehmann and Kleinpeter[164] have systematized the analysis of LIS data of highly flexible molecules through the development of new computer algorithm, in which starting conformations for flexible molecules were generated by random variation of up to ten independent and freely-rotating dihedral angles and an R factor (which measures goodness of fit) calculated for each of the conformations generated. A steepest descent (STEED) algorithm was then used to find the minimum of the R value hypersurface, which corresponds to the predicted preferred conformation. Results from the computer-aided LIS analysis were found to be in good agreement with the known or reported preferred conformations of five highly flexible molecules, and the favored conformations were also very close to those predicted using the molecular mechanics program PIMM. The authors report that their method for analysing LIS data of flexible molecules is user-friendly, suitable for routine application, and accurate. In a related approach, Hofer and Wurz[165] have also used computer simulation of LIS data, in conjunction with force-field calculations and X-ray derived structures, to study the solution conformations of tetrahydrofuran-lignans. Less quantitative conformational studies of the amino acids[166,167] and Gly-Gly and Gly-Ala dipeptides[168,169] have also been reported. Specialized applications of LSR for studies of and spectral assignments in compounds bearing specific functional groups have also been reported. Compounds studied include primary aliphatic amines[170], conjugated unsaturated carboxylates[171], 1-butanol and cyclohexene[173], borneal-isoborneal mixtures[172], and compounds bearing a single hydroxyl functional group[174].

8 Paramagnetic Contrast Agents in Magnetic Resonance Imaging (MRI)

Probably the most widely used paramagnetic relaxation contrast agent in MRI is the lanthanide ion gadolium. The high potency of Gd^{3+} as a relaxation-enhancing agent stems from the high spin angular momentum (S=7/2) and the orbital singlet electronic ground state ($^8A_1$) of this ion. The $^8A_1$ ground state couples very weakly with the thermal lattice, so that the electron spin relaxation time of Gd^{3+} is very long relative to that of other lanthanide ions. The long electron spin relaxation time of Gd^{3+} allows the electron spin of this ion to couple very efficiently to the magnetic moments of near-by nuclear spins, thus producing a very high characteristic relaxation efficiency, particularly in macromolecular complexes where the molecular reorientation is relatively slow. Among the 3d-block transition ions, Mn^{2+} and, to a lesser degree, Fe^{3+}, have similar properties to Gd^{3+} and have also been used as relaxation-enhancing contrast agents in MRI. For these ions, SBM relaxation theory normally provides a suitable basis for calculating relaxation enhancements, and Sosnovsky[175] has proposed an equation suitable for numerical estimates of this type. Strategies for using these ions medically have focussed on macromolecular delivery systems in which the complexed metal ion is exposed to coordination by the solvent but is sequestered in a form which is both non-toxic and stable at the required dose.

The use of Gd chelate complexes in MRI have been reviewed by Tweedle[176]. Synthetic chelating ligands of the widest current interest include members of the DOTA and EDTA/DTPA families (see above), a variety of which have been synthesized and, in a number of cases, evaluated in animal studies[177-195]. Metal-porphyrins[196-198] such as Mn(III)tetraphenylporphyrin sulfonate (MnTPPS), and tetraazamacrocyclic complexes[199] also have attractive properties as contrast agents, as have Fe(III)- and Gd(III)-dextran macromolecular complexes[200-202]. An alternative, non-metal, paramagnetic contrast agent system involves ionic nitroxyl radicals[203]. These radicals have lipophilic tails and accumulate preferentially in hepatic tissue, which make them potentially of interest as organ-specific contrast agents for liver. However, experimental studies indicate that they are weakly relaxing relative to metal chelates and that they are also relatively toxic, which obviously limits their interest medically unless physiologically more benign compounds in this group can be identified.

The delivery of paramagnetic contrast agents to target organs in a safe efficient manner is a major problem in the use of these agents. One approach to this problem is compartmentalization of the contrast agent in a liposome. In this situation, the paramagnetic ion (or radical) is sequestered from the aqueous phase by a lipid boundary. The equilibration of protons in the intraliposomal and extraliposomal spaces by diffusion of the water solvent across the lipid membrane is required if the contrast agent is to affect the spin relaxation properties of the aqueous phase. Thus the problem involves the solution of the spin relaxation (Bloch) equations in a system in which diffusional equilibration between compartmentalized spaces is occurring. Barsky et al[204] have analyzed this problem and have presented a simple analytical expression for the average relaxation time in liposome-containing systems. They have also presented measurements supporting the theory.

A number of liposomal systems with potential MRI utility have been prepared[205-207]. Unger et al[205] have prepared two types of manganese-based liposome, one containing trapped $MnCl_2$, the other containing trapped alkylated manganese complexes. Experimental characterization of the *in-vitro* relaxivity, stability and toxicity of these complexes and comparison of the *in-vivo* images in rats with liver tumors indicated that the latter type of complex (alkylated manganese complexes) has superior MRI properties. A number of liposoluble compounds in which long acyl chains have been attached to the chelating agent EDTA have been synthesized as MRI contrast agents by Unger and Shen[206]. Such complexing agents can potentially be administered alone or as part of a liposomal delivery system. Kunimasa et al[207] studied mannan-coated liposomes (mannan is a polysaccharide) for the delivery of Gd-DTPA in mice. An accumulation of Gd-DTPA-containing liposomes were found in the liver and lung, suggesting that this delivery system may lower the required dose of Gd-DTPA in these target organs. Clay- and zeolite-enclosed Mn^{2+} and Gd^{3+} systems have also been proposed[208] as suitable contrast agents for the gastrointestinal tract. Aqueous suspensions of these materials are strongly relaxing, quite stable, and can readily be formulated as oral preparations.

Iron oxide (ferrite) nanoparticles (also termed superparamagnetic particles) are a new class of MRI contrast agents with very potent relaxation-enhancing properties, especially of the T_2 relaxation time. A variety of iron oxide particle-containing suspensions of potential MRI utility have been prepared by Pouliquen et al.[209], and the properties of these suspensions have been studied systematically through X-ray diffraction analysis, NMR relaxivity, static magnetic susceptibility and thermomagnetic cycling measurements. Aoke[210] has used ferrite suspensions to image liver tumors grafted into rabbits and found the contrast enhancement to be considerably better than with Gd-DTPA or X-ray CT scans. The influence of suspended magnetic particles, α-Fe_2O_3 and Fe_3O_4, on water proton T_1's and T_2's have been studied by Jang and Yu[211], who observed large frequency shifts and deviations in the relaxation behavior from that predicted by SBM Theory. By measuring the decay time of the stimulated echo envelope, it was shown that the local inhomogeneous magnetic fields of the suspended particles are responsible for the non-classical relaxation behavior.

In animal studies, iron oxide-containing particles have been used to label neural cell suspensions which were then transplanted surgically into the cerebrum[212] or striatum[213] of adult rat brains. Using rat brain fetal cells labelled with iron oxide particles covalently coupled to the lectin wheat germ agglutinin, strong MRI contrast enhancements were observed in images obtained up to three weeks after the transplant without noticeable adverse physiological effects on the labeled cells. In another study[213], neural cell suspensions were labeled with reconstituted Sendai viral envelopes containing iron oxide particles, and cell migration and cell survival were monitored *in vivo* for periods up to two months after the cell graft.

Contrast-enhancement in these applications results from the spin relaxation-enhancing properties of dissolved or suspended paramagnetic species. In addition to enhancing NMR relaxation rates, paramagnetic species also enhance the static magnetic susceptibility of solutions in which they are dissolved. Weisskoff and Kiihne[214] have exploited this phenomenon by devising an imaging experiment based on variations in the absolute magnetic susceptibility of the medium. When used with paramagnetic contrast agents, this technique responds to the degree to which the presence of the paramagnetic solute perturbs the strength of the local static magnetic field. They have demonstrated the feasibility of this approach using Gd(DTPA), Dy(DTPA), and human red blood cells loaded with superparamagnetic iron oxide particles. Dy(DTPA) is a very poor relaxation contrast agent, but it has a significant influence on the static magnetic susceptibility. In a different approach, Lurie et al[215] have patented an electron-nuclear double resonance imaging technique which has the potential of mapping the distribution of paramagnetic material in heterogeneous samples. However, this technique is technically difficult, requiring simultaneous ESR and NMR irradiation of the sample. It also requires saturation of the electron spin levels, which is normally not feasible for transition metal ion-containing systems because of their rapid electron spin relaxation times.

8.1. ^{31}P MRI.- ^{31}P NMR resonances can be used to monitor the rich chemistry of phosphorylated metabolytes. However these resonances are very sensitive to the line broadening effects of low concentrations of paramagnetic metal ions, especially Fe(III). Dixon et al[216] have studied the mechanism of ferritin-induced ^{31}P NMR line broadening of ATP, and have suggested a new 2D zero quantum (2D-ZQ) technique suitable for the observation of ATP in iron-overloaded liver. Ferritin is the principal storage site of iron in the liver. Up to 4500 Fe ions can be stored per ferritin complex, which consists of 24 polypeptide subunits (480 kDa total) which form a hollow sphere, inside which Fe is stored as a Fe(III) oxyhydroxy-phosphate complex. Ferritin-bound iron is sequestered from the aqueous phase and does not broaden ^{31}P resonances by a direct dipolar relaxation mechanism, as free Fe^{3+} does, but rather ferritin produces, by virtue of the large paramagnetic susceptibility of its core, microscopic magnetic field inhomogeneity within the suspension. The normally-observed single quantum (SQ) resonances of ^{31}P (as well as other nuclear spins) are uniformly broadened by these magnetic susceptibility effects, but ZQ transitions, the line widths of which are insensitive to magnetic field inhomogeneity, are not. Thus the authors have used a 2D-ZQ technique to minimize spectral broadening in the ^{31}P spectrum of Fe-overloaded liver tissue. They speculate that a technique of this type could be useful for iron quantitation.

8.2. ^{23}Na MRI.- ^{23}Na is a highly sensitive NMR nucleus (approx. 0.1 times the proton sensitivity at given field) which is present at relatively high concentrations (*ca* 150 mM) in the cytosol and in the extracellular space. The intracellular to extracellular sodium distribution is one of the primary determinants of action potentials necessary for the electrical function of organs such as brain, heart and skeletal muscle. ^{23}Na imaging[217-221] offers the possibility of monitoring sodium movements in these tissues in response to physiological stimuli (see Reference 217 for a review of this field). To obtain large paramagnetic shifts of monatomic cations like ^{23}Na, a shift reagent in the form of a multiply-charged, membrane-impermeant anion is needed. Eleff et al.[218] have used the lanthanide shift reagent [Dy(TTHA^{3-})] to shift the extracellular ^{23}Na resonance with respect to the intracellular resonance in dog brain. Upon infusion of the shift reagent and disruption of the blood-brain barrier, three ^{23}Na peaks were observed at 9, 3 and 0 ppm, corresponding to Na^+ in blood, in the interstitial space, and in the cytosol respectively. Subsequent cardiac arrest produced an intensity decrease in the 3 ppm signal and an increase in the 0 ppm signal, consistent with the influx of Na^+ into the cell. These experiments suggest ways in which paramagnetic shift reagents will provide new opportunities for monitoring ion movements in tissue by MRI. Another paramagnetic complex that has been used as a ^{23}Na shift reagent is the penta-anionic thulium complex, $TmDOTP^{5-}$ (see above)[219,220]. After infusion of $TmDOTP^{5-}$ in the liver, well-resolved intra- and extra-cellular ^{23}Na peaks were observed and used to measure relative Na^+ concentrations and ^{23}Na relaxation times in the two compartments[219].

The use of such highly charged anionic shift reagents presents the potential for damage to cell membranes. The damage caused by two anionic shift reagents, $[Dy(PPP)_2]^{7-}$ (PPP=tripolyphosphate) and $Dy(TTHA)^{3-}$, on the cell structures of dog and human erythrocytes, of dog kidney cortical tubules, and of rat hepatocytes has been evaluated by Boulanger et al[222]. Much more severe effects were found with the heptavalent than with the trivalent ion: exposure to concentrations of $[Dy(PPP)_2]^{7-}$ as low as 10 μM caused very serious cell damage after 60 min. This study assesses qualitatively the membrane permeation of these and other agents and illustrates the potential for cell damage which they represent.

9 Free Radicals and Spin Labels

Lipophyllic free radicals, particularly nitroxyl free radicals of the TEMPO family, have historically been used rather widely as ESR probes of the motions in, and long-range order of, bilayer membranes including native biological membranes. Recent investigations have indicated that combined NMR/ESR approaches can be more informative in spin label studies than ESR measurements alone. NMR is useful, for example, for measuring accurate hyperfine coupling constants (hfcc's) to protons or ^{13}C, particularly for coupling constants which are too small to measure directly by ESR. NMR measurements of the chemical shifts of nuclear resonances on the radical species (δ_R) relative to that on the parent diamagnetic molecule (δ_d) provides the magnitude and sign of the electron-nuclear hyperfine coupling constant A:

$$A = -\Delta\delta\,(4\,\gamma_I kT)/(\gamma_S\, g\beta B_o), \qquad \Delta\delta = \delta_R - \delta_d \qquad (6)$$

NMR assignments of hyperfine coupling constants are also highly unambiguous.

Bolton et al[223] have used the NMR approach in their investigation of the potential as spin labels of the isoindolin-2-yloxyls family of stable free radicals (VII), which exhibit much smaller ESR linewidths than do members of the TEMPO family. Bales et al[224] used ^1H NMR to measure very accurately the solvent- and temperature-dependence of the ^1H hfcc's in the nitroxide spin label CTPO (4-protio-3-carbamoyl-2,2,5,5,-tetraperdeuteromethyl-3-pyrroline-1-yloxy, (VIII). Subsequently Bales et al[225] have used ^1H NMR to measure hfcc's in doxyl-labelled n-alkyl chains, using a series of compounds where the attachment point of the doxyl radical along the alkyl chain was varied. Simulations of ESR spectra based on hfcc's measured from NMR spectroscopy were found to be in excellent agreement with experimental ESR results.

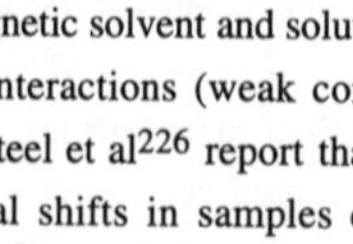

VII

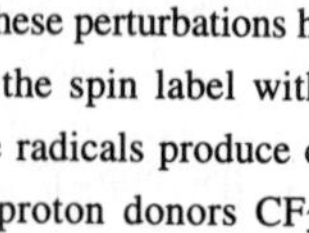

VIII

The presence of free radicals in solution in general perturbs both the chemical shifts and spin relaxation times of diamagnetic solvent and solute molecules. These perturbations have been used to probe non-convalent interactions (weak complexation) of the spin label with surrounding molecules[226-228]. Bonesteel et al[226] report that nitroxide free radicals produce concentration-dependent NMR chemical shifts in samples containing the proton donors CF_3CH_2OH and C_6F_5OH. Association constants of the H-bond-stabilized donor/radical complexes were measured, and it was found that the results of an NMR analysis differed significantly from an analysis based solely on ESR hyperfine coupling constants. Belorizky et al[227] studied solute-solute ion pairing between the $\cdot ON(SO_3)_2^{2-}$ radical anion and the Me_3P^+ cation by monitoring ^{31}P hyperfine contact shifts and the shortening of ^{31}P T_2 relaxation times.

10 Fullerenes and Fullerides

The recent interest in the chemisty and physics communities in the richness and novelty of fullerene chemistry has been peaked by the recent finding that certain alkali metal-fulleride salts are metallic conductors at ambient temperatures, and that they exhibit a transition to a superconducting state at relatively high temperatures, *ca* 20-30 K, which is much higher than for any other organic superconducting material. Doped fullerites are the first 3-dimensional organic superconductors. NMR, along with X-Ray and neutron diffraction techniques, has played a central role in elucidating the structural, electronic, and magnetic properties of fulleride materials. This section reviews briefly recent NMR work on paramagnetic fulleride anions with the objective of providing some perspective on this field for readers with backgrounds outside of solid state physics. More detailed recent reviews of the NMR properties of alkali metal fullerides have been written in English by Tycko[229] and in Russian by Loktev[230].

10.1. The C_{60}^- Radical Anion.- Buckminsterfullerene (C_{60}) is icosohedral and possesses a triply degenerate LUMO. Thus C_{60} exhibits substantial electron affinity for up to 6 electrons and is known to form alkali metal salts of the general formula A_nC_{60}, n=1-6. The compounds A_6C_{60} possess a fully occupied LUMO and are diamagnetic insulators. The compounds A_1C_{60} are ionic compounds containing A^+ cations and C_{60}^- radical anions. Above about 300 K, these compounds form high temperature NaCl lattices, which are paramagnetic insulators. The low temperature (T<300 K) phases of Rb_1C_{60} and Cs_1C_{60} are non-cubic; the electronic and magnetic properties of these phases are not well characterised. K_1C_{60} forms a high temperature NaCl-like phase, but, in contrast to Rb_1C_{60} and Cs_1C_{60}, it does not form a thermodynamically-stable low temperature phase, but rather phase-separates into K_3C_{60} and a metal-depleted phase of C_{60}. Tycko[229] and Tycko et al[231] have studied the A_1C_{60} compounds using solid-state ^{13}C, ^{133}Cs, and ^{87}Rb NMR. The measured ^{89}Rb and ^{133}Cs chemical shifts exhibited Curie Law behavior, and hyperfine coupling constants from the unpaired electron of C_{60}^- to the metal nuclei have been calculated from the Knight shifts. The calculated hfcc's were much smaller than those characteristic of bulk metals, indicating that the alkali metal cations in the fulleride salts are almost entirely ionized, and their valence s-orbitals participate only slightly in the conduction band. The ^{13}C resonances of C_{60}^- in the high temperature phases are relatively narrow (10 ppm), clearly the result of rapid rotation of C_{60}^- in the solid, and they are substantially contact shifted. The phase transition which occurs near 300 K profoundly alters the structural, electronic, and nmr properties of these materials and is accompanied by a marked decrease in the paramagnetic susceptibility of the compounds. The ^{13}C NMR spectra are much broader in the low temperature phase, and a contact shift exhibiting Curie temperature dependence was not observed. The properties of these low temperature phases are not fully understood, but it does appear that well-defined C_{60}^- radical anions are not present.

Paramagnetic C_{60}^- anion radicals can also be formed by electrochemical reduction in solution phases. Khaled et al[232] generated the -1, -2, and -3 anions of C_{60} in a 4:1 toluene-acetonitrile solvent and studied the structural properties (particularly the dynamic Jahn-Teller distortion) by ESR spectroscopy. The -2 anion was found to be diamagnetic. Zuzok et al[233] electrochemically-generated two solid compounds comprised of C_{60}^- with the organic cations, N-methyl-N-ethyl-morpholine$^+$ and acridizine$^+$. The ^{13}C NMR resonance of C_{60}^- was observed by MAS to be a single broad resonance at 142 ppm, essentially unshifted from the peak of neutral C_{60}. Relatively little rotational narrowing of the ^{13}C resonance was observed in these compounds. Dyakonov et al[234] studied the electronic and magnetic properties of C_{60}^- tetraphenylphosphonium radical anion salts using dynamic nuclear polarization (DNP) techniques. In these studies, the ESR transition of the C_{60}^- radical was saturated with microwaves while the Overhauser intensity enhancements and the spin relaxation times of the ^{13}C resonance of C_{60}^- and of the ^{13}C and ^1H resonances of the phenyl groups of the counterion were observed by NMR. The authors concluded that charge transfer from tetraphenylphosphonium to C_{60}^- are incomplete in these solids, leading to large nuclear Overhauser effects, and that the electronic magnetic moments are largely localized on specific counterions as opposed to occupying conduction bands. Schaffner and Fischer[235] studied the radical ion reactions of C_{60} with the donor N,N-dimethyl-1-naphthylamine (D),

$$C_{60} + D \;=\; C_{60}^- + D^+ \;\rightarrow\; C_{60}(T) \;+\; D.$$

in a benzonitrile/benzene solvent. Very fast back reaction between C_{60}^- and D^+ was observed and attributed to fast electron spin relaxation of C_{60}^-.

10.2. Superconducting Alkali Metal Fullerides of the Formula A_3C_{60}.- Alkali metal fullerides of the formula A_3C_{60} have a half-filled LUMO. These compounds are metallic conductors in the range 30-300 K. Below *ca* 30 K, they undergo phase transitions to a superconducting state, for which the transition temperatures are in the range $T_c \cong 18$-33 K, a range that is much higher than that of any other known organic superconductor. Not surprisingly, the A_3C_{60} compounds have generated intense interest. A great deal of electronic, magnetic, and structural information on these species has recently been derived from ^{13}C, ^{87}Rb, and ^{133}Cs NMR[229,231,236-242]. At temperatures above T_c, where the fullerides are metallic conductors, the NMR chemical shifts appear to be dominated by the Knight shift, which is caused by the hyperfine interaction of the nuclear spins and the electronic spins of conduction electrons. The NMR relaxation behavior in the metallic phases is governed by the scalar hyperfine (contact) interaction of nuclear spins with conduction electrons, for which mechanism the product $(T_1T)^{-1}$ is predicted to be temperature independent as given by

$$\frac{1}{T_1\,T} = \frac{\pi\,k_B}{\hbar}\,A_c^2\;N(E_f)^2 \qquad (7)$$

where A_c is the scalar part of the hyperfine coupling constant and $N(E_f)$ is the density of electronic states at the Fermi surface. In this situation, the Knight shifts κ of the metal (in this case ^{13}C) nuclei are related to the T_1 relaxation times by the Korringa relation,

$$\frac{\kappa^2}{T_1\,T} = \frac{\hbar}{4\pi k_B}\,(\gamma_S/\gamma_I)^2. \qquad (8)$$

Approximate temperature-independence of the $(T_1T)^{-1}$ product has been observed for the ^{13}C spins in the metallic phases as expected, and this product falls off dramatically with decreasing temperature in the superconducting phase[229]. However, the Korringa relationship was not accurately obeyed, suggesting that the electron-nuclear magnetic coupling is substantially dipolar in nature, a finding supported by *ab initio* calculations[243], which found that the states at the Fermi level have largely p^{π}-character. NMR chemical shift and spin relaxation data have been used to measure the electron-nuclear hyperfine coupling constants and the density of electronic states at the Fermi level, as well as the electron spin magnetic susceptibility and its temperature dependence, in these compounds.

11 Organic Conductors and Organic Ferromagnets

The reviews of this series have not traditionally covered NMR in metals and alloys, areas normally considered as part of solid state physics, and the present review will continue this tradition. However the field of organic electrical conductors has stimulated a great deal of interest in the synthetic organic and materials communities in recent years, and it seems appropriate here to report on the use of NMR to probe electronic structure and chemical environment in these materials.

11.1. Organic Metallic Conductors- $[DMTM(TCNQ)_2]$ (N,N-dimethylthiomorpholinium bis(tetracyanoquinodimethane)) is a quasi-one-dimensional organic conductor which becomes a metallic conductor at temperatures below its phase transition at $T_c = 272K$. Rachdi et al[244] reported the high resolution ^{13}C NMR spectrum of $[DMTM(TCNQ)_2]$, selectively ^{13}C-enriched on the cyano groups. Extremely large ^{13}C Knight shifts, -500 to -1400 ppm, probably the largest ever reported for ^{13}C, were observed in the temperature range 210-360 K, with corresponding short T_1 values (1.6 to 6.5 ms) as predicted by the Korringa relation (Equation 10.2). Kolbert et al[245] used magic angle spinning ^1H NMR to measure ^1H Knight shifts in the same material, from which they measured the hyperfine coupling constants and spin densities on the protons, measurements which they then used to characterize the degree of charge distribution associated with the transition. Relaxation and Knight shifts of both ^{13}C and ^1H in powders and single crystals have been used to study the nature of the molecular motions which accompany the transition in $[DMTM(TCNQ)_2]$[245-247].

Ishida et al[248] have used ^{63}Cu NMR to study the magnetic properties of the organic conductor $(diMe\text{-}DCNQI)_2Cu$. T_1 and Knight shift measurements were used to measure the hyperfine coupling to the ^{63}Cu nuclei to probe the degree of charge and spin distribution into the Cu 3d orbitals. ^{63}Cu T_1^{-1} data was found to obey a T_1T=constant law and the Knight shift was constant in the range 0.33-90 K, consistent with Equations 7 and 8. The authors concluded that the conduction electrons couple strongly to the copper orbitals within the 1-D conductive columns through p^{π}-d^{π} orbital overlap.

11.2. Organic Ferromagnets- Paramagnetic NMR is also playing an important role in studies of the properties of organic ferromagnets. Examples of these materials are provided by metallocene$^+$(TCNE)$^-$ salts, the solid lattices of which consist of stacks of alternating TCNE$^-$ radical anion and metallocene (cp·M·cp)$^+$ radical cations (cp=cyclopentadienyl, M=Cr, Mn, Fe, or Ni).

TCNE$^-$:
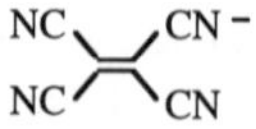

The unpaired electron spins in these compounds are known to couple ferromagnetically, but the physico-chemical mechanism of the coupling is unclear, and a better understanding of this point is needed for the rational design of organic ferromagnets. Bluemel et al[249] have conducted detailed ^1H and ^{13}C NMR studies of the coupling mechanism in metallocene-(TCNE) salts with the objective of testing competing theoretical models, specifically the McConnell-Kollmar-Kahn (MKK) model, in which the ferromagnetic coupling results from the correlation of electron spins on adjacent radicals, *vs* the McConnell-Miller-Epstein (MME) model, in which excitation occurs to a triplet state with the electron configuration $[(cp)_2Fe]^{2+}[TCNE]^{2-}$. The authors differentiated these mechanisms by measuring the sign of the unpaired electron spin density on the ligand, using for this purpose the signs of the ^{13}C and ^1H NMR contact shifts. The sign of the spin density in the atomic orbital relates directly to the sign of the contact shift, shifts to high(low) frequency corresponding to positive(negative) spin density. These studies provided compelling support for the MKK mechanism.

12 Metallo-enzymes and metal-nucleotide complexes

NMR studies of paramagnetic metallo-proteins comprise a large subfield of NMR, which has recently been reviewed by several authors[250-253]. The paramagnetic proteins studied fall into the following classes: (1) heme proteins including hemoglobin monomer[254], myoglobin[255-269], peroxidases[270-275] and cytochromes[276-292], (2) iron-sulfur proteins (ferredoxins)[293-300], (3) superoxide dismutases[301-304], (4) concanavalin A[305], (5) carboxypeptidase A[306,307], (6) carbonic anhydrase[308-310], (7) serum albumin[311], (8) ribonucleotide reductase[312], (9) sulfite reductase[313], (10) protein kinases[314,315], and (11) the manganese water-oxidizing enzyme of the photosynthetic membrane[316]. Also, investigations of the binding of transition metal ions and transition metal complexes to RNA[317] and DNA[318,319] have appeared. A detailed discussion of work on metallo-enzymes is beyond the scope of this review, and the reader is referred for this purpose to several excellent recent reviews of this area[250-253]. This section has the more limited objective of providing an overview of current NMR methodological approaches which are specifically relevant to the characterization of paramagnetic centers in metallo-enzymes.

Spectral assignments of paramagnetic metallo-proteins are complicated by the fact that resonances close to the metal center are frequently strongly shifted and broadened by hyperfine interactions with the unpaired electron spin. Cross-peaks between hyperfine-shifted resonances near the paramagnetic center and other resonances in the diamagnetic envelope of the protein can be particularly difficult to identify in phase-sensitive COSY spectra because the large linewidths of the hyperfine-shifted peaks act to suppress the cross-peaks in which they are involved. When amplitudes are very low, the cross-peaks can readily be confused with spectral artifacts. The use of magnitude COSY (MCOSY) spectra has provided a partial remedy[254,262,263,272], although Turner[283], in a detailed comparison of MCOSY and phase-sensitive COSY sensitivities, has concluded that the apparent increase in signal to noise ratio of MCOSY spectra is mis-leading and that the cross-peaks in phase sensitive COSY spectra are generally more sensitive than are MCOSY cross peaks. Sadek et al[320] have proposed the use of a modified TOCSY sequence to establish scalar connectivites. These authors used a 10 ms MLEV-17 spin-locking sequence and a data acquisition process carefully tailored to eliminate spectral artifacts. Their method was examined experimentally in studies using a ferredoxin from Clostridium pasteurianum that contains two $Fe_4S_4(S\text{-}cys)_4$ clusters (6200 Da). TOCSY spectra of the liganding cystein proton resonances (8 β–CH_2 and 4 α-CH) show very clear cross-peaks with superior sensitivity, thus providing unambiguous assignments. Bertini et al[321] have used a combination of difference NOE, NOESY, EXSY, and TOCSY ^{1}H NMR to assign completely the hyperfine-shifted protons of the four cysteine residues of selenium-substituted $[Fe_4Se_4]^{3+}$ cores of two high potential iron-sulfur proteins (HiPIP) from bacterial sources.

Bertini et al[266] have also used a pulse sequence that they term NOE-NOESY to assign hyperfine-shifted resonances in CN^--ligated myoglobin. In this method, one of the hyperfine resonances was pre-saturated by irradiation for a few to several hundred ms to allow NOE's which involve that resonance to build up prior to the start of a NOESY sequence. Difference NOESY spectra obtained with and without pre-saturation show only connectivities involving signals that experience NOE's from the pre-saturated signal. This technique facilitates spectral assignments of dipolar connectivities between the strongly hyperine-shifted resonances of the heme and its axial ligands and the diamagnetic protein matrix. This technique was demonstrated by spectral simulations and by experiments on met-myoglobin-CN^-. This study also made use of the Curie temperature dependence of the hyperfine-shifted resonances to facilitate assignments as well as to shift these peaks away from regions of severe spectral overlap.

12.1. Heme Proteins.- Heme proteins represent the most widely studied class of paramagnetic proteins. This class includes (1) the oxygen-binding proteins myoglobin and hemoglobin, (2) peroxidases, which couple the reduction of H_2O_2 to the oxidation of a variety of strongly oxidizing substrates which subsequently function as non-enzymatic oxidizing agents in oxidative metabolism, and (3) cytochromes, which function as electron transfer proteins in the respiratory and photosynthetic electron transport chains. Complete sequential NMR assignments of myoglobin and peroxidase enzymes have not yet been achieved, and the relatively high molecular weight of the peroxidases (34-46 kDa) suggests that for these enzymes complete assignments will not be attempted in the near future. However the paramagnetic heme moiety acts as an *in situ* shift reagent which separates the hyperfine-shifted resonances from the diamagnetic envelope as well as broadens them. A great deal of progress has been made in assigning the hyperfine-shifted resonances of the heme prosthetic group and of near-by amino acid resonances using 2-D COSY[259,267,254,260,262,263,272,274,277], NOESY[266,267,254,260,262,263,272-274,276,277], and TOCSY[268,274] as well as 1-D NOE-difference[254,260,262,263,270,272,274,275,276] techniques. Assignments have also been facilitated through the use of mutagenized proteins[260,261,267] as well as partially deuterated[273,268] or synthetic[259] hemes. These studies have provided a great deal of structural and magnetic information concerning the heme pocket, though complete spectral assignments have not yet been achieved.

The hyperfine shifts of nuclear spins other than those of the heme prosthetic group and its coordinating axial ligands are generally dipolar (pseudocontact) in origin and described by an expression of the form

$$\delta_{ps} = \frac{\mu_o}{8\pi N_o r_{IS}^3}\left[\frac{2\chi_z}{3} - \frac{\chi_x}{3} - \frac{\chi_y}{3}\right]\left(1\text{-}3\cos^2\theta\right) + (\chi_y\text{-}\chi_x)\sin^2\theta\cos(2\phi), \qquad (9)$$

where χ_i are the principal components of the magnetic susceptibility (MS) tensor, θ and ϕ are polar angles of the nuclear spin in the principal axis system of the MS tensor, N_o is Avogadro's number, and μ_o is the electron spin magnetic moment. The rhombic part of δ_{ps} (that due to χ_y-χ_x) is generally significant for heme proteins due to effects of axial-asymmetry in the spin density in the HOMO of the heme. Pseudocontact shift measurements have been used in conjunction with X-ray structural data to determine the magnitude and orientation of the magnetic susceptibility tensor, as well as to probe the nature of structural differences in the heme pockets of related proteins. This approach has also been used to characterize rotations of the heme moiety with respect to the surrounding protein matrix, as well as to detect tilts and rotations of the magnetic susceptibility tensor associated with bonding distortions in the axial ligation. Rajarathnam et al[262] have evaluated the accuracy of this approach in a study of the heme pocket of elephant myoglobin-CN^-. An analysis based on even a minimal data set comprised of five dipolar chemical shifts from resolved, strongly shifted and strongly relaxed resonances was found to determine the orientation of the magnetic axes with respect to the heme moiety with surprisingly high accuracy, ±2° in the tilt of the z-axis with respect to the plane of the heme, ±10° in the projection of the z-axis on the plane of the heme, and ±10° in the rhombicity. The analysis of more extensive data sets did not materially differ from the analysis based on only five resonances.

Another detailed analysis of this type was carried out by Turner et al[285,286], who analyzed ^{13}C and ^{1}H pseudocontact shifts in horse heart cytochrome c. Except for the heme and its axial histidine ligands, all paramagnetic shifts were found to be dipolar in origin and well described by an empirically-determined magnetic susceptibility tensor in conjunction with X-ray structural data. The principal components, χ_i, of the MS tensor were calculated independently from low temperature ESR data. Analysis of ^{13}C and ^{1}H NMR hyperfine shift data in two redox states of the protein indicated that a redox-dependent conformational shift of the protein with respect to the heme moiety occured. Conformational changes of this type are believed to be signficant with respect to the electron transport mechanism of the cytochrome[285].

Santos and Turner[285] have reported extensive (though not complete) assignments of the natural abundance ^{13}C spectrum (as well as the ^{1}H spectrum) of horse heart cytochrome c. These authors used the DEPT technique to distinguish methine, methylene and methyl ^{13}C peaks, and then used ^{13}C-^{1}H heteronuclear correlation spectroscopy to assign specific ^{13}C resonances to the resonances of the directely-bonded protons, the assignments of which were relatively well-known.

Of the 346 ^{13}C resonances expected for the protein, 55 methyl and 27 methine carbons were assigned in the primary structure.

Saturation transfer experiments have also provided a method for establishing connectivities between unassigned and previously assigned resonances of heme proteins, thereby facilitating the assignment of the more poorly understood data set. Yamamoto et al[264] have used this approach to assign the heme proton resonances of the low spin forms of three species of myoglobin (equine, mollusc and shark). The binding of a strong field ligand such as N_3^- or CN^- to a high spin heme iron in myoglobin converts the iron to its low spin form. In the presence of substoichiometric quantities of N_3^-, the high and low spin forms of the protein are in relatively slow chemical exchange equilibrium, and separate sets of peaks are observable for the two populations of protein. In an equilibrium mixture of the high and low spin forms, the transfer of magnetisation from selectively saturated heme resonances of the high spin form of the protein to the corresponding resonances of the low spin form was observed. In this way, saturation transfer and 2-D (EXSY) experiments were used to demonstrate connectivities between the unassigned spin system (the low spin heme) and the previously-assigned resonances of the high spin heme moiety. In a related study, Yamamoto[265] has used SCN^- as a mediator ligand in saturation transfer experiments of the acidic and alkaline forms of equine ferric myoglobin. As in Reference 264, assigments of the heme resonances were made from saturation transfer connectivities. In addition, the pseudo-first order rate constant $k_{ex}=\tau^{-1}$ for ligand exchange was determined from the measured saturation transfer ratio,

$$\frac{I_{ST}}{I_{eq}} = \frac{T_1^{-1}}{T_1^{-1}+\tau^{-1}}, \qquad (10)$$

where I_{ST} and I_{eq} are the peak intensities under saturating and unsaturated conditions, T_1 is the intrinsic spin relaxation time, and τ is the chemical exchange lifetime of the thiocyanate ligand.

A novel and potentially very powerful technique for assigning ^{15}N and ^1H resonances by establishing connectivities between resonances of the diamagnetic and paramagnetic forms of a protein has been proposed by Guiles et al[277]. Their method transfers assignments from the ^{15}N-labelled diamagnetic protein, for which assignments were known, to the isomorphous paramagnetic form of the ^{15}N-labelled protein by overlaying the ^{15}N-^1H heteronuclear correlation spectra. The ^{15}N peaks, unlike the proton peaks, are relatively insensitive to pseudocontact shifts, which provides a link between the two spectra. This approach has permitted a complete assignment of ^1H and ^{15}N resonances for rat ferricytochrome b_5 as well as an accurate determination of the orientation and magnitude of the magnetic susceptibility tensor.

Kao and Lecompte[268] have proposed a novel method for measuring the zero field splitting parameter D of the ferric heme moiety of metaquomyoglobin from measurements of the dipolar part of the paramagnetic chemical shift. Their analysis is based on the fact that the dipolar part of δ_{ps} is composed of two contributions, one (δ_{dip}') arising from magnetic susceptibility anisotropy as described in Equation 9 above and the other (δ_{dip}'') arising from effects of the zero field splitting interaction. The latter can be written

$$\delta_{dip}'' = -(28/9)\,(g\beta_o/k_BT)^2\,[3\cos^2\theta - 1]\,r_{IS}^{-3}\,D. \qquad (11)$$

δ_{dip}'' can be used to measure the zfs parameter D when the competing terms, $\delta_{con}+\delta_{dip}'$, are negligible or can be measured independently. The authors describe the application of this method to the analysis of chemical shift data of metaquomyoglobin (S=5/2), using as a diamagnetic reference carbonmonoxymyoglobin. With the aid of X-ray structure data, they have fit the chemical shifts of 37 ^{1}H resonances to the functional form of Equation 11, from which they derived the value D=9.66±0.08 cm^{-1}. (It was not clear to this reviewer however that the terms δ_{dip}' and δ_{dip}'' had been separated experimentally).

Peroxidase enzymes are heme proteins of relatively high molecular weight (34-45 kDa) which catalyze a reduction of H_2O_2 coupled to the one electron oxidation of a substrate S, which subsequently carries out non-enzymatic oxidations of terminal substrates. Different peroxidases differ in their function and in the nature of the substrate S. Peroxidases contain a 5-coordinate, high spin Fe^{III} in a protoporphyrin IX, with the fifth ligand a proximal histidine. A second, distal histidine on the opposite side of the heme has been implicated in the reaction with H_2O_2. When coordinated by CN^-, the enzyme contains a low spin, 6-coordinate Fe^{III}. Linewidths of the native protein are relatively broad but the CN^--ligated species has a very short electron spin relaxation time and relatively sharp NMR lines, and it has thus been studied extensively by NMR. Banci et al[272-274] have studied various peroxidases, including lignin- and manganese-peroxidase, for which they used ^{15}N-labeled CN^- as a heme ligand. The ^{15}N shift of the ligated CN^- consists of both contact and pseudocontact terms. The former measures the strength of the Fe^{III}-CN^- coordination, the latter measures the orientation of the coordinated cyanide ligand within the magnetic susceptibility tensor. Their analysis indicated that the z-axis of the heme magnetic susceptibility tensor is largely determined by the orientation of the cyanide ligand, which is in turn influenced by the position of the distal histidine.

Mondal et al[258], in a study of Mn(III)-reconstituted cyanometmyoglobin, have likewise used an ^{15}N-labeled CN^- ligand, in this case to measure, by means of ^{15}N T_1 data, the distance of the CN^- ligand to the metal. They concluded that bound CN^- ion does not coordinate directly to the metal but rather is 6-7 Å distant. From solvent ^{1}H T_1 measurements they likewise concluded that solvent water does not directly coordinate the metal ion. These conclusions are in apparent contradiction with the more generally-accepted picture of CN^--ligation of the iron as the agent of the high-spin to low-spin transition of the protein.

The orientation and magnitude of the MS tensor differs significantly for different heme proteins. These differences may be due both to perturbations of the axial ligation, as described above, as well as from movements of the heme prosthetic group within its binding pocket. Lee et al[278] have analyzed the pattern of hyperfine shifts in ferricytochrome b_5 from rat, beef and chicken and have concluded that the differences observed within this homologous set of proteins reflect a shift of the heme within a relatively rigid protein matrix, rather than a change in orientation of the magnetic axes of the heme due to alterations in the ligation.

Pseudocontact and contact shifts are predicted to exhibit Curie temperature dependence ($\delta \propto 1/T$), which facilitates the identification and assignment of hyperfine-shifted peaks, and which may also be useful for separating a hyperfine-shifted peak from the diamagnetic background in cases of spectral overlap. An unexpected anti-Curie temperature dependence has been reported for the hyperfine shifts of two of the four heme methyl ^{1}H resonances of horse heart cytochrome oxidase[287], which, for one pair of methyls, decreased with increasing temperature (Curie behavior), and for the other increased with increasing temperature (anti-Curie behavior). This unexpected difference in qualitative behavior was attributed to a very small energy difference (*ca* 3 kJ·mol^{-1}) between the heme MO's which are derived from the d_{xz} and d_{yz} iron orbitals. Because of the very small energy separation, even small temperature differences can cause large changes in the Boltzmann populations of the molecular orbitals of the heme and thus produce a significant redistribution of spin density. Similar behavior was observed[284] for the heme methyls of ferricytochrome c 551.

12.2 Ligand Resonances as Probes of the Paramagnetic Center.- The electron spin magnetic moment of paramagnetic proteins often produces large relaxation enhancements of nuclear spins on ligand species which can be used to investigate the nature of the metal involvement in ligand or substrate binding. When the metal is coordinated by one or more exchangeable solvent molecules, solvent proton relaxation provides a very sensitive probe of the chemical and magnetic environment of the bound water molecule, as well as of the kinetics of its chemical exchange reactions. Since, in these experiments, solvent or ligand resonances are observed rather than the resonances of the protein, the technique offers extremely high sensitivity, even for very high molecular weight systems, including including the largest enzymes (several hundred kDa) as well as redox-active paramagnetic metal centers located on biological membranes. The observed paramagnetic enhancement of the nuclear spin relaxation rate of the ligand is given by

$$T_{1p}^{-1} = \frac{[L]_M/[L]_{dia}}{T_{1m} + \tau_{ex}} \qquad (12)$$

where $[L]_M$ and $[L]_{dia}$ are molar concentrations of the ligand (which may be the solvent) bound to the paramagnetic metal ion and in the diamagnetic solvent phase, T_{1m} is the nuclear spin relaxation time in the metal coordination sphere, and τ_{ex} is the ligand lifetime in the metal coordination sphere. Thus paramagnetic relaxation enhancement experiments potentially provide information on the ligand chemical exchange kinetics and on the magnetic and chemical environment of the metal coordination sphere.

Sharp[316] has used ^{1}H relaxation enhancement techniques to study manganese redox chemistry associated with the water oxidation reaction that is catalyzed by the oxygen-evolving center of the photosynthetic (thylakoid) membrane. Photosynthetic water oxidation occurs at a center containing four manganese ions located on the inner (lumenal) face of the thylakoid membrane. During the catalytic cycle this center passes through five intermediate oxidation states (the S-states, S_0,..,S_4), in a process driven by electron transfers coupled to the primary photochemistry of Photosystem II. ^{1}H relaxation enhancements depend in a very sensitive manner on the manganese redox state, Mn(II) being a very efficient relaxer of protons, Mn(III) a very inefficient relaxer, and Mn(IV) a relaxer of intermediate efficiency. In NMR relaxation experiments, the catalytic cycle was driven by single turnover flashes applied to suspensions of oxygen-evolving thylakoid membranes, and flash-induced changes in ^{1}H relaxation times were monitored to probe the underlying manganese redox chemistry of the S-state cycle. This article discusses the observed relaxation changes and their interpretation in terms of the underlying manganese redox chemistry of the catalytic cycle of O_2 evolution.

Solvent proton T_1 relaxation enhancements have also been used to probe high affinity Ca^{2+} binding sites on protein kinase C (PKC)[314]. In these experiments, Gd^{3+} was used as a paramagnetic probe ion, and the competition of Ca^{2+} and Gd^{3+} for a very tight Ca^{2+} binding site was studied by titrating the Gd^{3+}-bound enzyme with Ca^{2+}. These experiments demonstrated the presence of a single tight Gd^{3+}/Ca^{2+} binding site on the β-isoenzyme of PKC and a single tight binding site plus a larger class of lower affinity binding sites on the α-isoenzyme. $Co(NH_3)_4ATP$, a substitution-inert analog of ATP, was also used to probe the properties of ternary (Gd^{3+})-enzyme-[Co-ATP] complexes. The authors concluded that the tight Gd^{3+}/Ca^{2+} binding site lies very close to the site of substrate binding.

A different substitution-inert ATP analog, β,γ-bidentate Cr(III)ATP, has been used to probe the conformation of the substrate at the active site of yeast phosphoglycerate kinase[315]. In this study T_1's of ^{1}H, ^{31}P, and ^{13}C nuclei of the substrate, 3-phospho-D-glycerate, were used to determine distances between the Cr(III) ion and nuclei of the substrate in the ternary metal-enzyme-substrate complex. The results were found to support a hinge-bending mechanism of the enzyme for bringing the two substrates together.

13 NMR Studies of Solvent and Ion Movements in Intact Cells and Tissues

NMR provides a highly sensitive probe of the permeability of biological membranes toward the solvent or other uncharged species. Solvent permeability through cell membranes is a rather rapid process (solvent molecules reside in the cell for times of the order of a few milliseconds), which can be studied by traditional NMR methods for rapid reactions. In this approach, the external water signal is broadened or shifted with respect to the internal water signal by adding an impermeant paramagnetic ion (e.g., Mn^{2+}) to the external medium. Solvent transport across the membrane broadens the internal peak, a phenomenon which can be analysed using the Bloch equations modified for chemical exchange. The water permeability of the erythrocyte (RBC) membrane has been studied extensively in this manner, and the RBC membrane now represents a standard system with respect to mechanistic studies of the solvent transport mechanism. The principal permeation pathway in RBC's has been linked to an integral membrane protein based on the observation that transport is inhibited by sulfhydryl binding agents such as *p*-chloromercuribenzene sulfonate (PCMBS). The characteristics of this inhibition and the possible involvement of known integral membrane proteins in the permeation mechanism has recently been studied by Benga et al[322-324], and by Lahajnar[325,326].

Manganese absorbed into plant tissue also causes pronounced broadening of the solvent proton resonance as well as of the ^{31}P resonances of ATP and other phosphorylated metabolites. Manganese is an essential element for photosynthesis and is stored in rather high concentrations in the leaves. The uptake and compartmentalization of Mn^{2+} by plant tissue (maize roots) has recently been studied by ^{31}P NMR by Quiquampoix et al[327]. Mn^{2+} localization in vacuoles, which are the primary storage site, and in the cytoplasm, as well as metabolic effects produced by anoxia and metabolic inhibitors and reflected in pH changes are characterized in this study.

References

1 L. Banci, M. Piccioli, A. Scozzafava, *Coord. Chem. Rev.,* 1992, **120**, 1.
2 C. Luchinat, Z. Xia, *Coord. Chem. Rev.*, 1992, **120**, 281.
3 L. Banci, I. Bertini, and C. Luchinat, 'Nuclear and Electron Relaxation', VCH Publ. New York, 1991.
4 R. R. Sharp, *J. Chem. Phys.*, 1993, **98**, 2507.
5 R. R. Sharp, *J. Chem. Chem.*, 1993, **98**, 912.
6 R. R. Sharp, *J. Magn. Reson.*, 1992, **100**, 491.
7 R. R. Sharp, *J. Chem. Phys.*, 1993, **98**, 6092.
8 J.-M. Bovet, R. R. Sharp, *J. Chem. Phys.*, 1993, **99**, 18.
9 H. Fukui, K. Miura, H. Matsuda, *J. Magn. Reson.,* 1990, **88**, 311.
10 T. M. Bayburt, R. R. Sharp, *J. Phys. Chem.,* 1993, **97**, 4558.
11 I. Bertini, Z. Xia, C. Luchinat, *J. Magn. Reson.*, 1993, **101**(2), 198.
12 I. Bertini, F. Capozzi, C. Luchinat, Z. Xia, *J. Phys. Chem.* 1993, **97**(6), 1134.
13 I. Bertini, C. Luchinat, Z. Xia, *Inorg. Chem.*, 1992, **31**(14), 3152.
14 I. Bertini, Z. Xia, C. Luchinat, *J. Magn. Reson.*, 1992, **99**(2), 235.
15 P. O. Westlund, T. P. Larsson, O. Teleman, *Molec. Phys.* 1993, **78**(6), 1365.
16 C. Ribbing, M. Odelius, J. Kowalewski, L. Pettersson, *Theor. Chem. Acta* 1994, **87**(4-5), 307.
17 L. Werbelow, A. Thevand, *J. Magn. Reson.*, Ser. A, 1993, **101**(3), 317.
18 A. R. Lim, K. S. Ryu, S. H. Choh, *Sae Mulli* (Korean), 1992, **32**(2), 239.
19 M. E. Sandifer, M. Zhao, S. Kim, D. A. Scherson, *Anal. Chem.* 1993, **65**(15), 2093.
20 A. R. Brough, C. P. Grey, C. M. Dobson, *J. Am. Chem. Soc.,* 1993, **115**(16), 7318.
21 D. P. Raleigh, C. P. Grey, N. Soffe, C. M. Dobson, *J. Magn. Reson.,* 1992, **97**(1), 162.
22 M. Basato, G. Favero, A. C. Veronese, A. Grassi, *Inorg. Chem.*, 1993, **32**(5), 763.
23 T. H. Lin, J. A. DiNatale, R. R. Vold, *J. Am. Chem. Soc.,* 1994, **116**(5), 2133.
24 Klingler and Rathke, *Inorg. Chem.*, **31**, 804, (1992).
25 I. Dellavia, L. Helm, A. E. Merbach, *Inorg. Chem.*, 1992, **31**(11), 2230.
26 I. Dellavia, P. Y. Sauvageat, L. Helm, Y. Ducommun, A. E. Merbach, *Inorg. Chem.*, 1992, **31**(5), 792.
27 Y. Inada, K. Ozutsumi, S. Funahashi, S. Soyama, T. Kawashima, M. Tanaka, *Inorg. Chem.* 1993, **342**(14), 3010.
28 I. Fabian, *Conf. Coord. Chem.* 1993, 14th, 349.
29 B. E. Stout, G. R. Choopin, J. C. Sullivan, *Transuranium Elem. Symp.* 1992, 225.
30 N. A. Kostromina, V. B. Chernogorenko, Sh. B. Tasybaeva, *Ukr. Khim. Zh.* 1993, **59**(2), 115.
31 N. A. Kostromina, V. B. Chernogorenko, Sh. B. Tasybaeva, P. B. Tikhonova, *Ukr. Khim. Zh.* 1991, **57**(6), 563.
32 G. A. Nazmutdinova, V. G. Shtyrlin, A. V. Zakharov, Yu. I. Sal'nikov, *Koord. Khim.* 1993, **19**(9), 701.
33 Shan, Q. Liu, Z. Zu, Y. He, *Bopuxue Zazhi* 1993, **10**(3), 303.
34 E. Gaggelli, A. Maccotta, G. Valensin, *J. Inorg. Biochem.* 1992, **48**(3), 173.
35 V. G. Shtyrlin, Ya. E. Zil'berman, O. V. Kukushkina, A. V. Zakharov, *Zh. Neorg. Khim.* 1992, **37**(3), 642.
36 M. K. Lee, C. J. Yoon, Y. S. Choi, *J. Korean Chem. Soc.*, 1992, **36**(3), 345.
37 W. L. Reynolds, S. Davis, *Croat. Chem. Acta* 1992 **64**(4), 663.
38 F. V. Devyatov, V. F. Safina, L. G. Lazareva, Yu. I. Sal'nikov, *Zh. Neorg. Khim.* 1993, **38**(6), 1085.
39 G. N. Sinyakov, A. M. Shulga, *J. Mol. Struct.* 1993, **295**(1-3), 1.
40 K. Yamaguchi, Y. Watanabe, I. Morishima, *J. Chem. Soc., Chem. Commun.* 1992, (23), 1721.
41 K. M. Kadish, E. Van Caemelbecke, F. D'Souza, C. J. Medforth, K. M. Smith, A. Tabard, R. Guilard, *Organometallics* 1993, **12**(7), 2411.
42 A. L. Balch, L. Latos-Grazynski, B. C. Noll, M. M. Olmstead, L. Szterenberg, N. Safari, *J. Am. Chem. Soc.* 1993, **115**(4), 1422.
43 A. L. Balch, L. Latos-Grazynski, B. C. Noll, M. M. Olmstead, E. P. Zovinka, *Inorg. Chem.* 1992, **31**(11), 2248.
44 S. Ozawa, Y. Watanabe, I. Morishima, *Inorg. Chem.,* 1994, **33**(2), 306.
45 S. P. Gavrish, Y. D. Lampieda, *Neorg. Khim.* 1993, **38**(8), 1304

46 K. Sakata, M. Hashimoto, H. Nakamura, *Synth. React. Inorg. Met.-Org. Chem.* 1994, **24**(1), 1.
47 D. Coucouvanis, S. G. Jonasdottir, D. Christodoulou, C. G. Kim, J. W. Kampf, *Inorg. Chem.*, 1993, **32**(14), 2987.
48 M. Shakir, S. P. Varkey, *Polyhedron,* 1994, **13**(5), 791.
49 J. H. Cameron, E. L. Scott, *J. Chem. Soc., Dalton Trans.*, 1993, (3), 397.
50 K. R. Rodgers, R. A. Reed, Y. O. Su, T. G. Spiro, *Inorg. Chem.* 1992, **31**(13), 2688.
51 M. F. Isaac, Q. Lin, U. Simonis, D. J. Suffian, D. L. Silson, F. A. Walker, *Inorg. Chem.* 1993, **32**(19), 4030.
52 J. Lisowski, L. Latos-Grazynski, L. Szterenberg, *Inorg. Chem.* 1992, **31**(10), 1933.
53 P. J. Chmielewski, L. Latos-Grazynski, *Inorg. Chem.* 1992, **31**(25), 5231.
54 F. A. Cotton, H. Chen, L. M. Daniels, X. Feng, *J. Am. Chem. Soc.* 1992, **114**(23) 8980.
55 F. A. Cotton, J. Su, Z. S. Sun, Ho. Chen, *Inorg. Chem.* 1993, **32**(22) 4871.
56 F. A. Cotton, J. L. Eglin, H. Bo, C. A. James, *Inorg. Chem.* 1993, **32**(10) 2104.
57 N. Ueyama, W. Y. Sun, A. Nakamura, *Inorg. Chem.*, 1992, **31**(20), 4053.
58 W. Y. Sun, N. Ueyama, A. Nakamura, *J. Chem. Soc., Dalton Trans.*, 1993, (12), 1871.
59 N. Ueyama, S. Ueno, A. Nakamura, K. Wada, H. Matsubara, S. Kumagai, S. Sakakibara, T. Tsukihara, *Biopolymers,* 1992 **32**, 1535.
60 M. S. Mashuta, R. J. Wevv, J. K. McCusker, E. A. Schmitt, K. J. Oberhausen, J. F. Richardson, R. M. Buchanan, D. N. Hendrickson, *J. Am. Chem. Soc.* 1992, **114**(10), 3815.
61 S. Menage, Y. Zang, M. P. Hendrich, L. Que, Jr., *J. Am. Chem. Soc.*, 1992, **114**(20), 7786.
62 R. Lachicotte, A. Kitaygorodskiy, K. S. Hagen, *J. Am. Chem. Soc.*, 1993, **115**(19), 8883.
63 R. H. Beer, S. J. Lippard, *Inorg. Chem.*, 1993, **32**(6), 1030.
64 D. A. Davlatshoeva, A. M. Glebov, V. V. Ginzburg, Z. N. Yusupov, Sh. T. Yusupov, *Zh. Neorg. Khim.*, 1993, **38**(5), 856.
65 W. Tsagkalidis, C. Woitha, D. Rehder, *Inorg. Chem. Acta*, 1993, **205**(2), 239.
66 N. I. Porphyr'eva, *Koord. Khim.,* 1993, **19**(9), 711.
67 A. N. Glebov, R. S. Mavlyautdinov, O. Yu. Tarasov, *Zh. Neorg. Khim.* 1993, **38**(8), 1320.
68 A. P. Gulya, G. V. Novitskii, S. G. Shova, M. D. Mazus, I. Sandu, *Koord. Khim.*, 1993, **19**(3), 227.
69 V. V. Matveev, B. N. Chernyshov, *Koord. Khim.*, 1993, **19**(4), 316.
70 V. V. Matveev, B. N. Chernyshov, V. A. Shcherbakov, *Koord. Khim.*, 1993, **19**(3), 223.
71 Y. Ohashi, M. Nakamura, *Chem. Lett.*, 1993, (8), 1389.
72 J. N. Fernandez-G., R. G. Enriques, A. Tobon-Cervantes, M. I. Bernal-Uruchurtu, R. Villena-I., W. F. Reynolds, J. P. Yang, *Can. J. Chem.*, 1993, **71**(3), 358.
73 K. Tsuchiya, K. Takegoshi, K. Hikichi, *Polym. J. (Tokyo)*, 1994, **26**(4), 479.
74 Ya. E. Zil'berman, V. G. Shtyrlin, A. V. Zakharov, Z. A. Saprykova, *Zh. Neorg. Khim.*, 1992, **37**(2), 388.
75 G. Zhao, Z. Wu, Y. Guan, W. Tang, *Wuji Huaxue Xuebao*, 1993, **9**(1), 7.
76 S. Steinkopf, Q. Liu, N. A. Froeystein, E. Sletten, *Acta Chem. Scand.*, 1192, **46**(5), 446.
77 M. Tonkovic, *Carbohydr. Res.,* 1994, **254**, 277.
78 B. Gyurcsik, T. Gajda, L. Nagy, K. Burger, *J. Chem. Soc., Dalton Trans.*, 1992, (19), 2787.
79 K. Araki, H. Rajima, *J. Inorg. Biochem.,* 1993, **52**(2), 89.
80 E. Gaggelli, N. Gaggelli, G. Valensin, A. Franchi, *Polyhedron*, 1993, **12**(19), 2355.
81 T. Gajda, B. Henry, J. J. Delpuech, *J. Chem. Soc., Dalton Trans.*, 1992, (15), 2313.
82 M. Vicens, J. J. Fiol, A. Terron, V. Moreno, D. M. L. Goodgame, R. N. Sheppard, *Inorg. Chim. Acta*, 1992, **192**(1), 139.
83 J. D. Protasiewicz, K. H. Theopold, *J. Am. Chem. Soc.* 1993, **115**(13), 5559.
84 K. A. Anderson, K. Kirchner, H. W. Dodgen, J. P. Hunt, S. Wherland, *Inorg. Chem.* 1992, **31**(12), 2605.
85 J. Burri, H. Fischer, *Chem. Phys.*, 1992, **161**(3) 429.
86 R. M. L. Warren, A. Tatehata, A. G. Lappin, *Inorg. Chem.*, 1993, **32**(7), 1191.
87 N. E. Meagher, K. L. Juntunen, C. A. Salhi, L. A. Ochrymowyca, D. B.Rorabacher, *J. Am. Chem. Soc.* 1992, **114**(26), 10411.
88 A. M. Q. Vande Linde, B. C.Westerby, L. A. Ochrymowycz, D. B. Rorabacher, *Inorg. Chem.* 1993, **32**(3), 251.

89 G. H. Leggett, B. C. Dunn, A. M. Q. Vande Linde, L. A. Ochrymowyca, D. B. Rorabacher, *Inorg. Chem.*, 1993, **32**(26), 5911.
90 C. Dennison, P. Kyritsis, W. McFarlane, G. A. Sykes, *J. Chem. Soc., Dalton Trans.*, 1993, (13), 1959.
91 C. Dennison, T. Kohzuma, W. McFarlane, S. Suzuki, A. G. Sykes, *J. Chem. Soc., Dalton Trans.*, 1994, (4), 437.
92 I. Bertini, A. Gaudemer, C. Luchinat, M. Piccioli, *Biochemistry* 1993, **32**(47), 12887.
93 C. Brunel, A. Bondon, G. Simonneaux, *Biochim. Biophys. Acta* 1992, **1101**(1), 73.
94 P. A. Purtov, A. B. Doktorov, *Chem. Phys.* 1993, **178**(1-3), 47.
95 T. Aizawa, T. Sakata, S. Itoh, K. Maeda, T. Axumi, *Chem. Phys. Lett.* 1992, **195**(1) 16.
96 E. B. Krissinel, Chem. Phys. 1993, **169**(2), 207.
97 I. A. Shkrob, *Chem. Phys. Lett.* 1993, **210**(4-6), 432.
98 K. C. Hwang, H. D. Roth, N. J. Turro, K. M. Welsh, *J. Phys. Org. Chem.* 1992, **5**(4), 209.
99 M. Goez, *J. Magn. Reson., Ser.* A 1993, **102**(2), 144.
100 S. A. Mikhailov, P. A. Purtov, A. B. Doktorov, *Chem. Phys.* 1992, **166**(1-2), 35.
101 G. L. Closs, M. D. E. Forbes, P. Piotrowiak, *J. Am. Chem. Soc.* 1992, **114**(9), 3285.
102 K. Maeda, N. Suzuki, T. Azumi, *J. Phys. Chem.* 1993, **97**(38), 9562.
103 S. G. Lee, J. Chem. Soc., *Perkin Trans.* 2, 1993, (7), 1361.
104 M. Goez, I. Sartorius, *J. Am. Chem. Soc.* 1993, **115**(24), 11123.
105 Q. X. Meng, K. Suzuki, K. Maeda, M. Terazima, T. Azumi, *J. Phys. Chem.* 1993, **97**(7), 1265.
106 Q. Meng, Y. Yamakage, K. Maeda, T. Azumi, *Z. Phys. Chem. (Munich)* 1993, **180**(1-2), 95.
107 Y. Yamakage, Q.-X. Meng, K. Maeda, T. Azumi, *Chem. Phys. Lett.* 1993, **204**(5-6), 411.
108 L. Cheng, G. Xu, B. Yan, H. Yuan, *Ganguang Kexue Yu Kuang Huaxue* 1992, **10**(3), 279.
109 Yu. P. Tsentalovich, A. A. Frantsev, A. B. Doktorov, A. V. Yurkovskaya, R. Z. Sagdeev, *J. Phys. Chem.* 1993, **97**(35), 8900.
110 M. B. Taraban, V. I. Maryasova, T. V. Leshina, D. Pfeifer, *Main Group Met. Chem.* 1991, **14**(1) 33.
111 R. M. Gainullin, A. V. Il'yasov, B. M. Odintsov, *Khim. Fiz.* 1992, **11**(4) 486.
112 K. M. Welsh, J. L. Dektar, M. A. Garcia-Garibaya, N. P. Hacker, N. J. Turro, *J. Org. Chem.* 1992, **57**(15), 4179.
113 S. Grimme, H. Dreeskamp, *J. Photochem. Photobiol. A.* 1992, **65**(3), 371.
114 P. J. Hore, R. W. Broadhurst, *Prog. Nucl. Magn. Reson. Spectrosc.* 1993, 25(4), 345.
115 Y. Serikawa, M. Shirakawa, Y. Kyogoku, *FEBS Lett.* 1992, **299**(3), 205.
116 K. A. Muszkat, B. Schechter, M. Sela, *Mol. Immunol.* 1992, **29**(9) 1049.
117 A. T. Alexandrescu, R.W.Broadhurst, C.Wormald,C.L.Chyan,J. Baum, C.M.Dobson, *Eur. J. Biochem.* 1992, **210**, 699.
118 K. Haard, J. P. Kamerling, J. F. G. Vliegenthart, *Carbohydr. Res.* 1992, **236,** 315.
119 J. Marko, G. Vermeersch, *Trends Photochem. Photobiol.* 1991, **2**, 387.
120 R. R. Rustandi, H. Fischer, *J. Am. Chem. Soc.* 1993, **115**(6), 2537.
121 N. E. Polyakov, M. B. Taraban, A. I. Kruppa, N. I. Avdievich, V. V. Mokrushin, P. V. Schastnev, T. V. Leshina, V. Lusis, D. Muceniece, et al. *J. Photochem. Photobiol.*, A 1993, **74**(1) 75.
122 L. Cheng, G. Xu, B. Yan, H. Yuan, *Ganguang Kexue Yu Kuang Huaxue* 1992, **10**(3), 279.
123 L. Li, J. Qiu, S. Hu, W. Liu, C. Ye, *Sci. China, Ser. A*, 1992, **35**(10), 1214.
124 L. R. Becerra, G. J. Gerfen, R. J. Temkin, K. J. Singel, R. G. Griffen, *Phys. Rev. Lett.* 1993, **71**(21), 3561.
125 R. A. Wind, N. Zumbulyadis, R. H. Young, Y. Hung, L. Li, R. H. D. Nuttall, G. Maciel, *Solid State Nucl. Magn. Reson.* 1992, **1**(2), 55.
126 H. Lock. G. E. Maciel, C. E. Johnson, *J. Mater. Res.* 1992, **7**(10), 2791.
127 K. Roberts, R. Rothchild, H. Wyss, *Spectrosc. Lett.* 1993, **26**(10), 1901.
128 R. Rothchild, K. Sanders, K. S. Venkatasubban, *Spectrosc. Lett.* 1993, **26**(4), 597; *ibid.*, **26**(10), 1997.
129 T. J. Wentzel, C. A. Morin, A. A. Brechting, *J. Org. Chem.*, 1992, **57**(13) 3594.

130 R. Benshafrut, R. Rothchild, M. E. Sullivan, K. S. Venkatasubban, S. Kunisi, *Spectrosc. Lett.*, 1992, **25**(6), 841.
131 T. A. Powers, S. A. Evans. *Heteroat. Chem.*, 1992, **3**(1) 41.
132 E. L. Gogolashvili, N. G. Dzyubenko, N. P. Huz'mina, B. Ya. Margulis, L. I. Martynenko, *Zh. Prikl. Khim. (S.-Peterburg)*, 1993, **66**(1), 138.
133 J. Ren, F. Jing, S. Zhang, F. Pei, W. Wang, *Yingyong Huaxue*, 1993, **10**(3), 72.
134 I. Bertini, F. Capozzi, C. Luchinat, G. Nicastro, Z. Xia, *J. Phys. Chem.*, 1993, **97**(24), 6351.
135 D. H. Powell, A. E. Merbach, G. Gonzalez, E. Bruecher, K. Micskei, M. F. Ottaviani, K. Koehler, A. von Zelewsky, O. Y. Grinberg, Y. S. Levedev, *Helv. Chim. Acta*, 1993, **76**(5), 2129.
136 K. Micskei, H. D. Powell, L. Helm, E. Brucher, A. E. Merbach, *Magn. Reson. Chem.*, 1993, **31**(11), 1011.
137 A. Fratiello, V. Kubo-Anderson, S. Azimi, F. Laghaei, R. D. Perrigan, F. Reyes, *J. Solution Chem.*, 1992, **21**(10), 1015.
138 A. Fratiello, V. Kubo-Anderson, S. Azimi, F. Laghaei, R. D. Perrigan, *J. Solution Chem.*, 1992, **22**(6), 519.
139 N. Fleitas, C. Cabal, V. I. Chizhik, *Rev. Cubana Quim.* 1992, **6**(2), 24; *ibid.* 38.
140 F. V. Devyatov, V. F. Safina, S. G. Vul'fson, Yu. I. Sal'nikov, *Koord. Khim.*, 1993, **19**(2), 163.
141 K. I. Arsenin, L. A. Malinko, I. A. Sheka, R. V. Tikhonova, I. Ya. Pishchai, A. N. Antishdo, *Zh. Neorg. Khim.*, 1992 **37**(1) 196.
142 E. N. Rizkalla, *Radiochim. Acta*, 1993, **61**(3-4), 181.
143 S. Galera, J. M. Lluch, A. Olliva, J. Bertran, F. Foglia, L. Helm, A. E. Merbach, *New J. Chem.*, 1993, **17**(12), 773.
144 S. Aime, M. Botta, E. Terreno, P. L. Anelli, F. Uggeri, *Magn. Reson. Med.*, 1993, **30**(5), 583.
145 C. F. G. C. Geraldes, A. D. Sherry, G. E. Kiefer, *J. Magn. Reson.*, 1992, **97**(2), 290.
146 S. Hoeft, K. Roth, *Chem. Ber.*, 1993, **126**(4), 869.
147 S. Aime, M. Botta, G. Ermondi, *Inorg. Chem.*, 1992, **31**(21), 4291.
148 K. O. A. Chin, J. R. Morrow, C. H. Lake, M. R. Churchill, *Inorg. Chem.*, 1994, **33**(4), 656.
149 J. R. Morrow, K. O. A. Chin, *Inorg. Chem.*, 1994, **32**(15), 3357.
150 C. F. G. C. Geraldes, A. M. Urbano, M. A. Hofnagel, J. A. Peters, *Inorg. Chem.* 1993, **32**(11), 2426.
151 E. N. Rizkalla, G. R. Choppin, W. Cacheris, *Inorg. Chem.*, 1993, **32**(5), 582.
152 G. Gonzalez, D. H. Powell, V. Tissieres, A. E. Merbach, *J. Phys. Chem.*, 1994, **98**(1), 53.
153 S. Hua, R. Song, G. Huang, *Gaodeng Xuexiao Huaxue Xuebao*, 1992, **13**(1), 81.
154 S. Aime, P. L. Anelli, M. Botta, F. Fedeli, M. Grandi, P. Paoli, F. Uggeri, *Inorg. Chem.* 1992, **31**(12), 2422.
155 R. Song, C. Zhang, F. Li, S. Hua, *Chin. J. Chem.*, 1993, **11**(1), 59.
156 J. H. Forsberg, T. J. Dolter, A. M. Carter, D. Singh, S. A. Aubuchon. A. T. Timperman, A. Ziaee, *Inorg. Chem.*, 1992, **31**(26), 5555.
157 A. K. Trikha, K. Dilbagi, *Lanthanide Actinide Res.*, 1991, **3**(4-6) 367.
158 G. Singh, P. A. Kumar, T. R. Rao, *J. Chem. Res.*, Synop., 1994, (4), 123.
159 R. Goyal, V. Gupta, A. Sharma, B. R. Patel, R. K. Mehta, *Proc. Natl. Acad. Sci., India, Sect. A*, 1992, **62**(3), 319.
160 K. Yao, Leshen, L. Cai, L. Shen, Q. Tian, *Polyhedron*, 1992, **11**(17), 2245.
161 D. Feng, B. Wang, *Polyhedron*, 1992, **11**(8), 937.
162 M. Sakamoto, Y. Sadaoka, A. Matsumoto, *Nippon Kagaku Kaishi*, 1993, (5), 465.
163 M. K. M. Nair, P. K. Radhakrishnan, *Polyhedron*, 1993, **12**(10), 1227.
164 J. Lehmann, E. Kleinpeter, *Magn. Reson. Chem.*, 1993, **31**(1) 68; *ibid.*, **31**(4) 367.
165 O. Hofer, G. Wurz, *Monatsh. Chem.*, 1992, **123**(1-2), 105.
166 R. Song, G. Huang, S. Hua, *Jilin Daxue Ziran Kexue Xuebao*, 1991, (3), 111.
167 J. Wang, X. Wang, Y. Ouyang, Y. You, A. Dai, *Wuli Huaxue Xuebao*, 1992, **8**(5), 647.
168 J. Ren, F. Pei, W. Wang, J. Ni, *Yingyong Huaxue*, 1992, **9**(4), 41.
169 J. Ren, C. Hiu, F. Pei, W. Wang, J. Ni, *Wuli Huaxue Xuebao*, 1992, **8**(2) 270.
170 I. V. Komarov, V. N. Nemykin, N. B. Subbotin, *Appl. Magn. Reson.*, 1993, **4**(3), 377.
171 J. Huskens, H. Van Bekkum, J. A. Peters, *J. Chem. Soc., Dalton Trans.*, 1992, (7), 1139.
172 H. S. Amoli, B. Madda, *Indian J. Chem., Sect. A*, 1993, **32A**(11), 996.

173 R. Benshafrut, R. Rothchild, *Spectrosc. Lett.*, 1992, **25**(3), 433.
173 A. D. Wright, G. M. Koenig, O. Sticher, *Phytochem. Anal.*, 1992, **3**(2), 73.
175 G. Sosnovsky, *Appl. Magn. Reson.*, 1992, **3**(1), 131.
176 M. F. Tweedle, *Invest. Radiol.*, 1992 **27**(suppl. 1), 2; *J. Alloys Compd.*, 1992, **180**, 317.
177 O. Clement, A. Muhler, V. Vexler, Y. Berthezene, R. C. Brasch, *Invest. Radiol.*, 1992, **27**(8), 612.
178 K. Micskei, L. Helm, E. Brucher, A. E. Merbach, *Inorg. Chem.*, 1993, **32**(18), 3844.
179 P. Gowland, P. Mansfield, P. bullock, M. Stehling, B. Worthington, J. Firth, *Magn. Reson. Med.*, 1992, **26**(2), 241.
180 G. Schuhmann-Giampieri, H. Schmitt-Willich, W.-R. Press, C. Hegishi, H. J. Weinmann, U. Speck, *Radiology,* 1992 **183**(1) 59.
181 M. Sun, M. Juo, C. Tang, L. Li, H. Mei, T. Shen, X. Chen, Shanghai, *Yike Daxue Xuebao,* 1992, **19**(1), 67.
182 R. Kravtzoff, E. Urvoase, C. Chambon, C. Ropars, *Adv. Exp. Med. Biol.*, 1992, **326**, 347.
183 H. Mollier, A. Naini, C. Dumone, D. Audo, G. Serratrice, M. Vincens, M. Vidal, S. Grand, J. F. Lebas, *Bull. Soc. Chim., Fr.* 1992, **129**(4), 376.
184 J. Kenney, U. Schmiedl, K. Maravilla, F. Starr, M. Graham, A. Spence, J. Nelson, *Magn. Reson. Med.,* 1992, **27**(1), 68.
185 A. Muhler, M. Saeed, R. C. Brasch, C. B. Higgins, *Radiology*, 1992, **183**(2), 523.
186 M. Saeed, M. F. Wendland, K. K. Yu, C. B. Higgins, *J. Am. Coll. Cardiol.,* 1992, **20**(7), 1634.
187 R. A. Wallace, **U.S. US 5,138,040**, 2 Jul 1991; **U.S. US 5,077,037**, 31 Dec 1991; (patents).
188 T. Kasokat, K. Urich, *Arzneim.-Forsch.*, 1992, **42**(6), 869.
189 X. Wan, T. C. Fu, R. E. London, *Magn. Reson. Med.,* 1992, **27**(1), 135.
190 P. Wedeking, K. Kumar, M. F. Tweedle, *Magn. Reson. Imaging,* 1992, **10**(4), 641.
191 G. E. Jackson, L. L. Du Toit, *Polyhedron*, 1992, **11**(11), 1315.
192 M. Periasamy, D. White, L. DeLearie, D. Moore, R. Wallace, W. Lin, J. Dunn, W. Hirth, W. Cacheris, et al., *Invest. Radiol.,* 1991, **26**(suppl. 1), S217.
193 G. Schuhmann-Giampieri, *Arzneim.-Forsch.,* 1993, **43**(9), 1020.
194 D. L. White, R. G. Eason, **PCT Int. Appl. WO 93, 03,351 09** Aug 1991.
195 I. Heuze, B. Robert, J. Y. Petit, L. Welin, A. De Kersaint-Gilly, F. Resche, J. Mantoura, Y. Lajat, *Pharmacol. Res.*, 1993, **27**(1), 89.
196 M. Hoehn-Berlage, D. Norris, K. Bockhorst, R. I. Ernestus, O. Kloiber, P. Bonnekoh, D. Leibfritz, K. A. Hossman, *Magn. Reson. Med.,* 1992, **27**(2), 201.
197 I. Sakata, et al., **Jpn. Kokai Tokkyo Koho JP 04 59,779**, 26 Feb 1992 (patent).
198 H. Mollier, A. Naini, C. Dumone, D. Audo, G. Serratrice, M. Vincens, M. Vidal, S. Grand, J. F. Lebas, *Bull. Soc. Chim. Fr.,* 1992, **129**(4), 376.
199 C. F. G. C. Geraldes, E. Brucher, S. Cortes, S. H., Koenig, A. D. Sherry, *J. Chem. Soc., Dalton Trans.,* 1992, (16), 2517.
200 D. L. White, K. P. Aicher, A. A. Tzike, J. Kucharczyk, B. L. Engelstad, M. E. Moseley, *Magn. Reson. Med.* 1992 **24**(1), 14.
201 P. Niemi, T. Reisto, M. Kormano, *Invest. Radiol.,* 1991, **26**(suppl. 1), S48.
202 Y. Hashiguchi, S. Seri, K. Kubomura, Y. Abe, K. Kamimura, C. Fujimoto, T. Iguchi, K. Iwai, *Nippon Kagaku Kaishi*, 1993, (5), 521; *ibid.*, 528.
203 B. Gallez, R. Demeure, R. Debuyst, D. Leonard, F. Dejehet, P. Dumont, *Magn. Reson. Imaging,* 1992, **10**(3), 445.
204 D. Barsky, B. Putz, K. Schulten, R. L. Magin, *Magn. Reson. Med.,* 1992, **24**(1), 1.
205 E. C. Unger, D. Shen, G. Wu, *Invest. Radiol.,* 1993, **28**(10), 933.
206 E. C. Unger, D. Shen, **PCT Int. Appl. WO 92 21,017**, 26 Nov 1992.
207 J. Kunimasa, K. Inui, R. Hori, Y. Kawamura, K. Endo, *Chem. Pharm. Bull.*, 1992, **40**(9), 2565.
208 K. J. Balkus, Jr. **PCT Int. Appl. WO 93 25,895**, 15 Jun 1992.
209 D. Pouliquen, H. Perroud, F. Calza, P. Jallet, J. J. Le, Jeune, *Magn. Reson. Med.,* 1992, **24**(1) 75.
210 F. Aoki, *Nippon Ika Daigaku Zasshi*, 1992, **59**(1), 29.
211 E. K. Jang, I. Yu, *Colloids Surf., A*, 1993, **72**, 229.
212 N. Hawrylak, P. Ghosh, J. Broadus, C. Schlueter, W. T. Greenough, P. C. Lauterbur, *Exp. Neurol.*, 1993, **121**(2), 181.
213 A. B. Norman, S. R. Thomas, R. G. Pratt, S. Y. Lu, R. B. Norgren, *Brain Res.*, 1992, **594**(2), 279.

214 R. M. Weisskoff, S. Kiihne, *Magn. Reson. Med.*, 1992, **24**(2), 375.

215 D. J. Lurie, J. R. Mallard, I. Nicholson, **PCT Int. Appl. WO 94 03,824**, 05 Aug 1992.

216 R. M. Dixon, J. L. Allis, B. Rajagopalan, *J. Magn. Reson., Ser. B*, 1994, **103**(1), 68.

217 S. K. Miller, G. A. Elgavish, *Biol. Magn. Reson.*, 1992, **11**, 159.

218 S. M. Eleff, I. J. Mclennan, G. K. Hart, Y. Maruki, R. J. Traystman, R. C. Koehler, *Magn. Reson. Med.*, 1993, **30**(1), 11.

219 N. Bansal, M. J. Germann, V. Seshan, G. T. Shires II, C. R. Malloy, A. D. Sherry, *Biochemistry*, 1993, **32**(21), 5638.

220 M. S. Albert, W. Huang, J.-H. Lee, J. A. Balschi, C. S. Springer, Jr., *NMR Biomed.*, 1993, **6**(1), 7.

221 P. Tu, R. Zhang, N. Zhao, *Shengwu Wuli Xuebao*, 1993, **9**(1), 51.

222 Y. Boulanger, A. Fleser, R. Amarouche, H. Ammann, M. Bergeron, P. Vinay, *NMR Biomed.*, 1992, **5**(1) 1.

223 R. Bolton, D. G. Gillies, L. H. Sutcliffe, X. Wu, *J. Chem. Soc., Perkin Trans. 2*, 1993, (11), 2049.

224 B. L. Bales, D. Mareno, F. L. Harris, *J. Magn. Reson., Ser. A*, 1993, **104**(1), 37.

225 B. L. Bales, R. A. Blum, D. Mareno, M. Peric, H. J. Halpern, *J. Magn. Reson.* 1992, **98**(2), 299.

226 J. A. K. Bonesteel, B. Borah, R. D. Bates, Jr., *J. Magn. Reson.* 1992, **98**(3), 475.

227 E. Belorizky, P. H. Fries, W. Gorecki, M. Jeannin, C. Roby, *J. Phys. Chem.* 1994, **98**(17), 4517.

228 J. K. Bonesteel, I. Al-Bala'a, R. D. Bates, Jr., *Bull. Magn. Reson.*, 1993, **15**(1-2), 18.

229 R. Tycko, *J. Phys. Chem. Solids,* 1993, **54**(12), 1712.

230 V. M. Loktev, *Fiz. Nizk, Temp. (Kiev)*, 1992, **18**(3), 217.

231 R. Tycko, G. Dabbagh, D. W. Murphy, Q. Zhu, J. E. Fischer, *Phys. Rev. G: Condens. matter,* 1993, **48**(12), 9097.

232 M. M. Khaled, R. T. Carlin, P. C. Trulove, G. R. Eaton, S. S. Eaton, *J. Am. Chem. Soc.*, 1994, **116**(8), 3465.

233 R. Zuzok, P. Wzietek, M. Fourmigue, P. Batail, P. Auban-Senzier, S. Roth, D. Jerome, *Synth. Met.*, 1993, **56**(2-3), 3235.

234 V. Dyakonov, G. Roesler, H. Klos, B. Gotschy, G. Denninger, A. Hirsch, *Synth. Met.* 1993, **56**(2-3), 3214.

235 E. Schaffner, H. Fischer, *J. Phys. Chem.*, 1993, **97**(50), 13149.

236 V. A. Stenger, C. Recchia, J. Vance, C. H. Pennington, D. R. Buffinger, R. P. Ziebarth, *Phys. Rev. B: Condens. Matter,* 1993, **48**(13), 9942.

237 G. Zimmer, M. Helmle, M. Mehring, F. Rachdi, J. Reichenbach, L. Firley, P. Bernier, *Europhys. Lett.*, 1993, **24**(1), 59.

238 Y. Maniwa, K. Mizoguchi, K. Kume, K. Tanigaki, T. W. Ebbesen, S. Saito, J. Mizuki, J. S. Tsai, Y. Kubo, *Solid State Commun.*, 1992, **82**(10), 783.

239 Y. Maniwa, T. Saito, A. Ohi, K. Mizoguchi, K. Kume, K. Kikuchi, I. Ikemoto, S. Suzuki, Y. Achiba, *et al*, *J. Phys. Soc. Jpn.*, 1994, **63**(3), 1139.

240 G. Quirion, C. Bourbonnais, E. Barthel, P. Auban, D. Jerome, J. M. Lambert, A. Zahab, P. Bernier, C. Fabre, A. Rassat, *Europhys. Lett.*, 1993, **21**(2), 233.

241 M. Kosaka, K. Tanigaki, T. W. Ebbesen, *Appl. Phys. Lett.*, 1993, **63**(18), 2561.

242 M. Baenitz, M. Kraus, S. Gaertner, H. M. vieth, H. Werner, R. Schloegl, W. Kraetschmer, M. Kanowski, K. Lueders, *Springer Ser. Solid-State Sci.*, 1993, **114**, 475.

243 V. P. Antropov, I. I. Maxzin, O. K. Andersen, A. I. Liechtenstein, O. Jepsen, *Phys. Rev. B: Condens. Matter,* 1993, **47**(18), 12373.

244 F. Rachdi, T. Nunes, M. Ribet, P. Bernier, M. Helmele, M. Mehring, M. Almeida, *Phys. Rev. B: Condense. Matter* 1992, **45**(14) 8134.

245 A. C. Kolbert, G. Zimmer, F. Rachdi, P. Bernier, M. Almeida, M. Mehring, *Phys. Rev. B: Condens. Matter* 1992, **46**(2), 674.

246 G. Zimmer, A. C. Kolbert, M. Mehring, F. Rachdi, P. Bernier, M. Almeida, *Phys. Rev. B: Condens. Matter* 1993, **47**(2) 763.

247 A. C. Kolbert, G. Zimmer, F. Rachdi, P. Bernier, M. Almeida, M. Mehring, *Phys. Rev. B: Condens. Matter* 1993, **46**(2) 674.

248 K. Ishida, Y. Kitaoka, H. Masuda, K. Asayama, T. Takahashi, K. Kanoda, A. Kobayashi, R. Kato, H. Kobayashi, Physica B (Amsterdam) 1993, 186-8, 1059.

249 J. Bluemel, N. Hebendanz, P. Huydeczek, F. H. Koehler, W. Strauss, *J. Am. Chem. Soc.*, 1992, **114**(11) 4223.

250 A. V. Xavier, D. L. Turner, H. Santos, *Meth. Enzymol.* 1993, **227**(Metallobiochemistry, Pt. D), 1.
251 H. Thomann, M. Bernardo, Methods Enzymol. 1993, 227(Metallobiochemistry, Pt. D), 118.
252 G. W. Canters, C. W. Hilbers, M. Van De Kamp, S. S. Wijmenga, *Meth. Enzymol.* 1993, **227**(Metallobiochemistry, Pt. D), 244.
253 L. Banci, I. Bertini, C. Luchinat, L. Messori, P. Turano, *Appl. Magn. Reson.* 1993, **4**(4), 461.
254 I. Bertini, S. Ciurli, A. Dikiy, C. Luchinat, *J. Am. Chem. Soc.*, 1993, **115**(25), 12020.
254 S. L. Alam, J. D. Satterlee, *Biochemistry* 1994, **33**(13), 4008.
255 L. Banci, I. Bertini, S. Marconi, R. Pierattelli, *Eur. J. Biochem.* 1993, **215**(2), 431.
256 B. R. Van Dyke, D. A. Bakan, K. A. M. Glover, J. C. Hegenauer, P. Saltman, B. A. Springer, S. G. Sligar, *Proc. Natl. Acad. Sci, U.S.A.* 1992, **89**(17) 8016.
257 J. Qin, G. N. La Mar, F. Ascoli, M. Brunori, *J. Mol. Biol.* 1993, **231**(4), 1009.
258 M. S. Mondal, S. Mazumdar, S. Mitra, *Inorg. Chem.* 1993, **32**(23), 5362.
259 G. N. La Mar, N. L. davis, R. d. Johnson, W. S. Smith, J. B. Hauksson, D. L. Budd, F. Dalichow, K. C. Langry, I. K. Morris, K. M. Smith, *J. Am. Chem. Soc.* 1993, **115**(10), 3869.
260 J. Qin, G. N. La Mar, Y. Dou, S. J. Admiraal, M. Ikeda-Saito, *J. Biol. Chem.* 1994, **269**(2), 1083.
261 M. Ikeda-Saito, H. Hori, L. A. Andersson, R. C. Prince, I. J. Pickering, G. N. George, C. R. Sanders, II, R. S. Lutz, E. J. McKelvey, R. Mattera, *J. Biol. Chem.* 1992, **267**(32), 22843.
262 K. Vyas, K. Rajarathnam, L. P. Yu, H. Mizukami, *J. Biol. Chem.* 1993, **268**(20), 14826.
263 J. Qin, G. N. La Mar, *J. Biomol. NMR* 1992, **2**(6), 597.
264 Y. Yamamoto, Y. Inoue, T. Suzuki, *Magn. Reson. Chem.* 1993, **31**, S8-S16.
265 Y. Yamamoto, *Biochem. Biophys. Res. Commun.* 1993, **196**(1), 348.
266 I. Bertini, A. Dikiy, C. Luchinat, M. Piccioll, D. Tarchi, *J. Magn. Reson., Ser. B* 1994, **103**(3), 278.
267 K. Rajarathnam, G. N. La Mar, M. L. Chiu, S. G. Sligar, *J. Am. Chem. Soc.* 1992, **114**(23), 9048.
268 Y. H. Kao, J. T. J. Lecomte, *J. Am. Chem. Soc.* 1993, **115**(21), 9754.
269 K. A. Keating, G. N. La Mar, F. Y. Shiau, K. M. Smith, *J. Am. Chem. Soc.*, 1992, **114**(16), 6513.
270 L. B. Dugad, H. M. Goff, *Biochim. Biophys. Acta* 1992, **1122**(1), 63.
271 R. Gilmour, C. F. Goodhew, G. W. Pettigrew, S. Prazeres, I. Moura, J. J. G. Moura, *Biochem. J.* 1993, **294**(3), 745.
272 L. Banci, I. Bertini, E. A. Pease, M. Tien, P. Turano, *Biochemistry* 1992, **31**(41), 10009.
273 L. Banci, I. Bertini, I. C. Kuan, M. Tien, P. Turano, A. J. Vila, *Biochemistry* 1993, 3**2**(49), 13483.
274 L. Banci, I. Bertini, T. Bini, M. Tien, P. Turano, *Biochemistry* 1993, **32**(22), 5825.
275 L. B. Dugad, X. Wang, C. C. Wang, G. S. Lukat, H. M. Goff, *Biochemistry*, 1992, **31**(6), 1651.
276 M. Rivera, C. Barillas-Mury, K. A. Christensen, J. W. Little, M. A. Wells, F. A. Walker, *Biochemistry* 1992, **31**(48), 12233.
277 R. D. Guiles, V. J. Basus, S. Sarma, S. Malpure, K. M. Fox, I. D. Kuntz, L. Waskell, *Biochemistry* 1993, **32**(32), 8329.
278 K.-B. Lee, G. N. La Mar, K. E. Mansfield, K. M. Smith, T. C. Pochapsky, S. G. Sligar, Biochim. Biophys. Acta 1993, 1202(2), 189.
279 R. M. Greene, S. F. Betz, S. Hilgen-Willis, D. S. Auld, J. B. Fencl, G. J. Pielak, J. Inorg. Biochem. 1993, 51(3), 663.
280 J. C. Ferrer, J. G. Guillemette, R. Bogumil, S. C. Inglis, M. Smith, A. G. Mauk, *J. Am. Chem. Soc.* 1993, **115**(16), 7505.
281 H. Akutsu, M. Hirasawa, *FEBS Lett.* 1992, **308**(3) 264.
282 R. Timkovich, M. Cai, Biochemistry, 1993, 32(43), 11516.
283 D. L. Turner, *J. Magn. Reson., Ser. A.* 1993, **104**(2), 197.
284 L. Banci, I. Bertini, M. T. Cambria, F. Capozzi, A. Dikiy, *Eur. J. Biochem.* 1994, **219**, 663.
285 H. Santos, D. L. Turner, Eur. J. Biochem. 1992, **206**, 721.
286 D. L. Turner, R. J. P. Williams, *Eur. J. Biochem.*, 1993, **211**(3), 555.
287 D. L. Turner, Eur. J. Biochem. 1993, **211**, 563.

288 T. G. Myers, K. E. Thummel, T. F. Kalhorn, S. D. Nelson, *J. Med. Chem.* 1994, **37**(6), 860.
289 Q. Yi, J. E. Erman, J. D. Satterlee, *Biochemistry* 1993, **32**(41), 10988.
290 Q. Yi, J. E. Erman, J. D. Satterlee, *J. Am. Chem. Soc.* 1994, **116**(5), 1981.
291 S. Tuzi, K. Shinzawa, T. Erata, A. Naito, S. Yoshikawa, H. Saito, *Eur. J. Biochem.* 1992, **208**(3), 713.
292 M. Hayer-Hartl, H. Schaegger, G. Von Jagow, K. Beyer, *Eur. J. Biochem.* 1992, **209**(1), 423.
293 L. Banci, I. Bertini, Capossi, F., P. Carloni, S. Ciurli, C. Luchinat, M. Piccioli, *J. Am. Chem. Soc.*, 1993, **115**(9), 3431.
294 H. Cheng, K. Grohmann, W. Sweeney, *J. Biol. Chem.* 1992, **267**(12), 8073.
295 L. Banci, I. Bertini, P. Carloni, C. Luchinat, P. L. Orioli, J. Am. Chem. Soc. 1992, 114(27), 10684.
296 L. Banci, I. Bertini, S. Ciurli, S. Ferretti, C. Luchinat, M. Piccioli, *Biochemistry*, 1993, **32**(36), 9387.
297 L. Blanchard, F. Payan, M. Qian, R. Haser, M. Noailly, M. Bruschi, F. Guerlesquin, *Biochim. Biophys. Acta* 1993, **1144**(2), 125.
298 H. Cheng, B. Xia, G. H. Reed, J. L. Markley, *Biochemistry* 1994, **33**(11), 3155.
299 S. C. Busse, G. N. La Mar, L. P. Yu, J. B. Howard, E. T. Smith, Z. H. Zhou, M. W. W. Adams, *Biochemistry* 1992, **31**(47), 11952.
300 I. Bertini, F. Capozzi, C. Luchinat, M. Piccioli, M. Vincens Oliver, *Inorg. Chem. Acta* 1992, **198-200**, 483.
301 L. Banci, I. Bertini, C. Luchinat, M. Piccioli, A. Scozzafava, *Gazz. Chim. Ital.* 1993, **123**(2), 95.
302 L. J. Ming, J. B. Lynch, R. C. Holz, L. Que, Jr. Inorg. Chem. 1994, 33(1), 83.
303 M. Sette, M. Paci, A. Desideri, F. Rotilio, *Biochemistry*, 1992, **31**(49) 12410.
304 M. Sette, M. Paci, A. Desideri, G. Rotilio, *Eur. J. Biochem.* 1993, **213**(1), 391.
305 A. L. Petrescu, R. Georgescu, M. Rotaru, F. Lehaci, S. M. Petrescu, Stud. Cercet. Biochim. 1992, 35(2), 101.
306 J. M. Moratal, A. Donaire, J. Salfado, J. Castells, R. Domingo, M. J. Martinez Ferrer, V. Soler, *An. Quim.* 1991, **87**(6) 718.
307 J. M. Moratal, J. Castells, A. Donaire, J. Salgado, H. R. Jimenez, *J. Chem. Soc., Dalton Trans.* 1992, (23), 3317.
308 A. Bencini, I. Bertini, T. Bini, *Inorg. Chem.* 1993, **32**(15), 3312.
309 L. Banci, L. B. Dugad, G. N. La Mar, K. A. Keating, C. Luchinat, R. Pierattelli, *Biophys. J.* 1992, **63**(2), 530.
310 I. Bertini, C. Luchinat, R. Pierattelli, A. J. Vila, *Eur. J. Biochem.* 1992, **208**(3), 607.
311 P. J. Sadler, A. Tucker, J. H. Viles, Eur. J. Biochem. 1994, 220(1), 193.
312 T. E. Elgren, L. J. Ming, L. Q. Jr., *Inorg. Chem.* 1994, **33**(5), 891.
313 J. Kaufman, L. M. Siegel, L. D. Spicer, *Tech. Protein Chem. IV, [Pap. Protein Soc. Symp.]* 1993, 577.
314 M. C. Maurer, J. J. Sando, C. M. Grisham, *Biochemistry*, 1992, **31**(33), 7714.
315 J. D. Gregory, E. H. Serpersu, *J. Biol. Chem.* 1993, **268**(6), 3880.
316 R. R. Sharp, Manganese Redox Enzymes 1992, 177 (Ed. V. L. Pecoraro) VCH Publ.
317 G. Ott, L. Arnold, S. Limmer, *Nucleic Acids Res.* 1993, **21**(25), 5859.
318 E. A. Kesicki, M. A. DeRosch, L. H. Freeman, C. L. Walton, D. F. Harvey, W. C. Trogler, *Inorg. Chem.* 1993, **32**(25), 5851.
319 N. A. Froeystein, J. T. Davis, B. R. Reid, E. Sletten, *Acta Chem. Scand.* 1993, **47**(7) 649.
320 M. Sadek, R. T. C. Brownlee, S. D. B. Scrofani, A. G. Wedd, *J. Magn. Reson., Ser. B.* 1993, **101**(3), 309.
321 I. Bertini, S. Ciurli, A. Dikiy, C. Luchinat, *J. Am. Chem. Soc.*, 1993, **115**(25), 12020.
322 G. Benga, T. Borza, O. Popescu, D. Porutiu, H. Matei, *Comp. Biochem. Physiol., G: Comp. Biochem.* 1993, **104B**(3), 589.
323 G. Benga, H. Matei, T. Borza, D. Porutiu, C. Lupse, *Comp. Biochem. Physiol., B: Comp. Biochem.* 1993, **106B**(2), 281.
324 G. Benga, G. B. Ralston, T. Borza, B. E. Chapman, C. H. Gallagher, P. W. Kuchel, *Comp. Biochem. Physiol., B: Comp. Biochem.* 1994, **107B**(1), 45.
325 G. Lahajnar, *Croat. Chem. Acta* 1992, **65**(1), 191.
326 G. Lahajnar, *J. Mol. Struct.* 1993, **294**, 139.
327 H. Quiquampoix, B. C. Lughman, R. G. Ratcliffe, *J. Exp. Bot.* 1993, **44**(269), 1819.

15
NMR of Liquid Crystals and Micellar Solutions

BY ALI KHAN

1 Introduction

This chapter covers the contribution of n.m.r. to the study of the structures, conformations and dynamics of liquid crystals and micellar solutions formed by amphiphilic systems during the last two years (June, 1992 - May, 1994). The liquid crystalline phases are characterized by fast molecular motions and long-range order. Thus the intermolecular interactions are averaged, while dipolar and quadrupolar n.m.r. splittings can be observed from intramolecular couplings. The partial averaging of static effects gives rise to complex n.m.r. relaxation behaviour, while the static effects are sensitive to the type of molecular alignment. Here, "the liquid crystals" include both the lyotropic and the thermotropic liquid crystalline mesophases. The two classes of liquid crystals, namely, the lyotropic and thermotropic mesophases, are similar in that macroscopic structure corresponds to some space group, describing the symmetry properties of the crystalline phase. Microscopically, the two classes are also equal, since they are disordered giving rise to a liquid character of the phases. However, the thermotropics are usually considered as close packed assemblies of more or less rigid molecules while the lyotropics are regarded as built up by molecular aggregates being arranged in space. The "micellar solutions" include normal micelles, reversed micelles and microemulsions. In isotropic solution phases, the absence of long-range order and fast molecular motions average to zero static nuclear effects, such as direct dipole-dipole couplings or nuclear quadrupolar couplings; thus a n.m.r. spectrum usually consists of narrow lines. Nuclei such as ^{1}H, ^{2}H, ^{13}C, ^{31}P, ^{19}F etc. are employed extensibly in this study. Inorganic counterions, notably $^{23}Na^{+}$, $^{133}Cs^{+}$, $^{7}Li^{+}$, $^{19}F^{-}$ and a limited number of organic counterions as well as some paramagnetic ions are also used.

The format of this chapter is similar to that used in previous volumes[1]. The review articles published are included in a section followed by theoretical models and new n.m.r. experimental techniques. The thermotropic mesomorphism and the lyotropic polymorphism constitute the largest sections of the chapter. Each of these topics is divided systematically and is described under the appropriate section. Polymer/protein-surfactant interactions in solution which is, currently, a popular research area is also included. Finally, the water (solvent) and ions that bring about important changes in the phase behaviour and aggregate microstructure by their association at the interface, is reviewed.

2 Reviews

Many review papers dealing with natural and synthetic amphiphilic systemshave been published. A review article[2] has been published describing the principles and applications of nuclear spin relaxation to the study of the microstructure and dynamics of lyotropic liquid crystals. Specific results for lamellar, hexagonal, cubic and nematic liquid crystalline phases are presented. The self-assembly of mixed surfactant systems is reviewed[3], with special reference to the thermodynamic and related properties of the micellar aggregates. The application of n.m.r. to the study of molecular dynamics of the nematic phase of EBBA is also reviewed[4]. Depending on the fatty acid acyl chain lengths, phospholipids can form monomers, micelles and bilayers with water and the systems produce ideal interfaces to examine the kinetics of surface-active enzymes, e.g., water soluble phospholipases[5]. N.m.r. methods applied to investigate the phase equilibria of the phosphatidylcholine-cholesterol-water system[6], membrane structure with respect to phospholipid orientational order and dynamics[6-8], membrane-protein[9], and membrane-peptide[10-12] interactions are reviewed. Properties such as phase behaviour, bilayer stability, the rate of membrane fusion in the presence of protein/peptide as well as the pressure induced unfolding of protein and high pressure membrane-n.m.r. are discussed. A database, LIPIDAT, is also available[13] containing data, such as, lipid phase transition temperature, enthalpy change, etc. Finally. the structural and motional properties of liposomes are presented in another review paper[14].

3 Theory and Methods

A Monte Carlo dynamics method[15] has been developed to study the equilibrium and dynamic properties of lipid-water systems in the context of a diamond lattice realisation. The model is able to predict the transition of phases and various bilayer properties. The conformations of hydrocarbon chains and surfactants in various media are modelled by computer simulation[16,17] and the results are in agreement with observations made by n.m.r. The spin-echo attenuation corresponding to pulsed-gradient n.m.r. self-diffusion experiments was stimulated by the Brownian dynamics for spins confined to spheres and between planes[18] and the results are applicable to experimentally determined n.m.r. restricted self-diffusion. The PGSE method is also applied in a highly concentrated emulsion system in which the emulsion is considered as belonging to the general class of porous systems. The water diffusion behaviour is found to exhibit diffraction-like effects in the emulsion system. The method can be used to obtain the mean droplet size of the emulsion[19]. Several approximate models are developed to monitor the properties of phospholipid membranes in water with or without added substrate(s). Computer simulations based on Langevin dynamics and a mean-field based on the Marcelja model are applied to unsaturated phospholipid membranes[20]. Molecular dynamics simulations are reported of a fluid phase DPPC lipid bilayer in water and of neat hexadecane and compared with n.m.r. spin-lattice relaxation data. The calculated fast reorientation correlation times and the microscopic lateral diffusion of the lipids show excellent agreement with the experimental results.[21] A thermodynamic model based on the Meyer-Overton hypothesis is also presented[22] for anesthetic actions on membrane. A model that can be used to study the solute interactions with a water-phosphatidylcholine interface is available[23]. Non-bilayer structures are important components in membrane fusion events. A ^{31}P n.m.r. method suitable to detect the non-bilayer structure is described[24]. Another method[25] has been developed to measure the exact fractional component of binary mixtures of phospholipids at the air-water interface. Orientational constraints as three-dimensional structural constraints from chemical shift anisotropy have been used to evaluate models for the channel conformation of gramicidine A. The results are tested for gramicidine in sodium dodecyl sulphate micelles[26]. A theoretical description[27] has been given of nuclear spin relaxation mediated by magnetic resonance imaging contrast agents and transport processes in liposome systems. A simple formula for average relaxation times in liposome systems is presented and the resulting relaxation times agree well with observations.

4 Thermotropic Mesomorphism

Many crystalline organic compounds that form liquid crystals upon heating have elongated molecules and this class of crystalline solids are very rich in mesomorphism. The vast majority of liquid crystalline substances that yield smectic and nematic liquid crystals are based on the structure

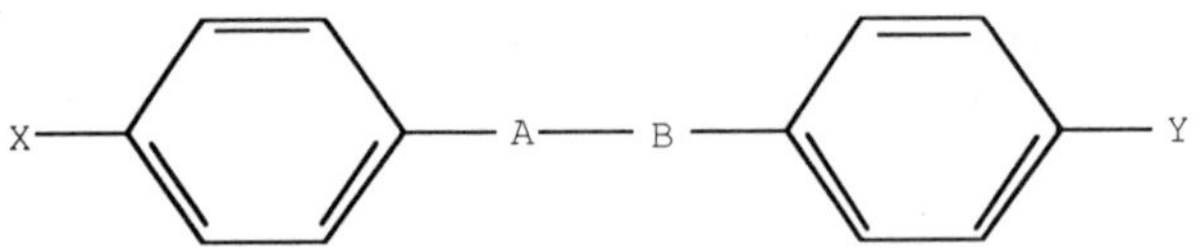

where they possess: (a) two or more aromatic rings, usually benzene rings; (b) one or more bridging groups, A-B, that bind the rings together, e.g. -CH=N-; and (c) two terminal groups, X and Y, usually on the long axis of the molecule, e.g. X=CH_3O-, (OR) and Y=C_4H_9, (R). In addition to smectic and nematic liquid crystals, one may obtain cholesteric mesophases by adding chiral solutes to the nematic phases. Recently discotic mesophases consisting of disc-like rather than rod-like molecules have been discovered. This article reports the papers published on theses liquid crystalline phases, but does not include (i) plastic phases with cubic symmetry and liquid crystals from the synthetic polymeric compounds, and (ii) solute molecules dissolved and oriented in the thermotropic liquid crystal solvents for structure determinations. The sub-topics include: (4.1) relaxation studies, probing mainly re-orientational dynamics and (4.2) bandshape analysis, giving information on molecular ordering.

4.1 Relaxation Studies. - Relaxation studies are performed in various thermotropic liquid crystals by ^{1}H, ^{2}H and ^{13}C n.m.r. methods. Rotational reorientations and self-diffusion are shown to be the essential relaxation mechanisms in the smectic A phase of 4-pentylphenyl 4'-cyanobenzoyloxy-benzoate[28] and of a mixture containing CBOOA(4-cyanobenzylidene-4'-octyloxyaniline) and DOBCA (4-dodecyloxybenzylidene-4'-cyanoaniline)[29], but for the nematic phase, the order director fluctuation mechanism dominates the relaxation process[28]. Using an improved field-cycling n.m.r. spectrometer, frequency dependent ^2H T_1 measurements in the range between 500 Hz and 7 MHz for the various nematogens (PAA-d_6, PAA-d_9, etc.) show that the relaxation results do not support the Pincus-Blinc model, i.e., $T_1 \approx \upsilon^{1/2}$, where T_1 is the relaxation times of the atom considered and υ is the Larmer frequency[30]. The deuteron relaxation times, T_{1Z} and T_{1Q} are measured in the nematic phase of 5OB-d_{15} as a function of temperature at two magnetic fields and the data are discussed in terms of conformation motions of the alkyl chain[31]. Broad band Jeener-Broekaert sequence in combination with the inversion recovery technique are shown to yield spectral densities with high precision for 1 CB Me and aromatic deuterons and the relaxation of the alkyl chain C_1 deuterons in the nematic liquid crystals for the binary mixtures of 1 CB and 5 CB. The data are interpreted in terms of three correlation times[32]. The Zeeman (T_{1Z}) and the quadrupolar (T_{1Q}) deuteron relaxation times determined by the Jeener-Broekaert experiment for the system 8 OBCAB-d_2 and its mixture with compounds of the n TPCHB series show a connection between the presence of the nematic reentrant behaviour and the slowing down of both D(parallel) and D(perpendicular) elements of the rotational diffusion tensor[33]. Similar results are also obtained in the nematogen 2-fluorenyl-d9 4-tetradecyloxybenzoate[34], but the relaxation times, T_{1Z} in HB(4.4'-diheptylazobenzene), HAB(4.4'-diheptylazoxybenzene), OAB(4.4'-dioctylazoxybenzene) show a discontinuity at the isotropic-nematic transition for all deuterons[35]. ^2H relaxation times measured for the system 5 CB confined in cylindrical microcavities reveal that in the isotropic phase, the relaxation is not effected by the confinement, but in the nematic phase, the relaxation rate is effected by order director fluctuation and orientational rotation[36]. The results of the high field ^{13}C T_1 times of the ring carbons of 5 CB indicate that order director fluctuations account for ≈ 50% of the relaxation rate in the nematic phase. The authors suggest that the high field rotating frame ^{13}C relaxation technique may be a rich source of information on critical fluctuations[37].

4.2 Bandshapes-Order Parameters. - The orientational order of a compound exhibiting nematic, smectic A_d and reentrant nematic mesophase is studied through the analysis of ^1H spectra of the core molecule and ^2H spectra of the chain and the results are compared with different theoretical models[38]. The orientational order determined by the n.m.r. method in the nematic phase of 1-OCB(4-methoxy-4'-cyanobiphenyl and PAA(4.4'-dimethoxyazoxybenzene) is compared with that obtained by the neutron diffraction isotope labelling technique[39]. The conformation of ethoxy tail flanking in the cyanobiphenyl core is also reported in the liquid crystalline phase as well as in isotropic solution[40]. The conformation of the tail is elucidated in terms of RIS(rotational isomeric state) simulation scheme. A technique is described to align the director of the nematic phase of 5 CB and 8 CB along the magnetic field.

The ^2H quadrupolar coupling constant and ^1H-^2H dipolar coupling constant are measured by following the ^2H n.m.r. lineshape as a function of temperature and the applications of the results are illustrated by testing the Emsley-Luckhurst-Stockley theory[41]. The odd-even behaviour observed for the liquid crystal dimers is claimed to be a sensitive function of the geometry of the link joining the mesogenic groups to the flexible spacer and the prediction is found to be in good accord with methylene linked cyanobiphenyl dimers[42]. In a lecture published[43], the author has compared experimentally determined orientational order parameters in liquid crystalline phases with different theoretical models. The nematic phase of 2,3,4-tri-n-hexyloxycinnamic acid is found to consist of an uniaxial symmetry as revealed by the powder-pattern ^{2}H spectra obtained by selectively deuterated ethylenic bond[44]. The mesomorphic properties of homologs of octaalkyloxy series are reported. The mesophase consists of either an uniaxial- (discotic A) or a biaxial (discotic B) symmetry depending on the length of side chains[45]. The ^{2}H quadrupolar splitting and ^{13}C chemical shifts measured in the nematic and isotropic solution of cyanobiphenylpiperidines are used to obtain the orientational order parameter of the liquid crystals[46].

The effect of molecular structure on molecular dynamics and ordering is studied by the solid-state ^{13}C n.m.r. Introduction of a branched alkyl chain into a S_C molecule increases the molecular tilt angle and the S_A - S_C transition enthalpy as well as broadens the ^{13}C spectra of the core carbons. The ordering of the molecular core is probably affected through the interlayer interaction[47].

The ^{2}H n.m.r. lineshapes recorded in the nematic liquid crystals confined to cylindrical cavities reveal a structural transition from an escaped-twisted to a planar bipolar configuration as the curvature is varied. The method allows the determination of polar and azimuthal energies[48]. Reports on confinement and orientational studies of liquid crystals also reveal a homeotropic-to-planar anchoring transition in the nematic phase and an orientational wetting transition in the isotropic phase[49], a pronounced 1st order nematic-isotropic transition compared to a weak discontinuity of the non-confined system of 5 CB[50], a perpendicular alignment of the nematic director with respect to the pore axis when the pore surface is treated with lecithin against the parallel alignment of the director for the untreated membrane[51]. Based on the ^{2}H n.m.r. experiments the measurements are reported of the saddle-splay surface elastic constant in a nematic liquid crystal. These measurements are made from observations of nematic director-field configurations and a configuration transition in submicrometer-sized cylindrical cavities of membrane. Detailed experimental technique is given and the results are discussed with theories[52].

There are several reports that deal with the studies of mesomorphic behaviour of mesogenic compounds,[53-56] many of which are newly synthesized[57-62]. Characterization of phases, phase transition temperature as well as the orientational order of liquid crystalline phases of these compounds are studied by n.m.r. methods.

5 Lyotropic Polymorphism

Surfactants (both synthetic and biological) are often rich in polymorphism. The water-soluble surfactants form normal micellar aggregates with water followed by different liquid crystalline phases with increased surfactant concentrations. Insoluble amphiphiles can not form micellar solution phase but they often form mesophases. Addition of other components, such as, another amphiphile, a polymer, a protein, a polar or a non-polar substrate, etc. to a binary aqueous system may result to a very complex phase behaviour. The identification of phases and the determination of phase diagrams constitute a very important task for the surfactant systems. The following section covers the reported phase equilibria for the surfactant systems during the last two years.

5.1 Phase Diagrams and Phase Structures. - The binary (composition vs. temperature) and isothermal multi-component phase diagrams for a variety of surfactant and lipid sysyems are described in this section. Water deuteron, ^{31}P, etc., n.m.r. in combination with other techniques such as polarizing optical microscopy, calorimetry and SAXS are used to determine the equilibrium phase diagrams. Multinuclear ^{1}H, ^{13}C, ^{31}P, ^{2}H , and counterions such as ^{23}Na, ^{133}Cs, etc.,(for ionic surfactants) are also used to obtain phase structures. Theoretical models are used to explain experimentally determined phase diagrams for many systems.

The phase behaviour for the following surfactant systems has been published: Didodcyldimethylammonium chloride (DDAC), sulphate (DDAS), hydroxide (DDAOH), acetate (DDAAc) with water and DDAS-water-dodecane[63,64], octadecylpyridinium bromide (ODPB)-water[65], cetyltrimethylammonium bromide (CTAB)/cetylpyridinium bromide (CPB)-water/non-aqueous solvents[66], and octylamine-water[67] The important findings of phase equilibria studies are that both DDAC and DDAS are practically insoluble with water while DDAOH and DDAAc dissolve easily forming micellar solution phase, $L_{1.}$ These surfactants also form lamellar liquid crystalline phase, D, but the stability of the D phase at maximum water is much less stable with SO_4^{-2} counterion compared with other monovalent ions. Moreover, DDAS system also forms a L_1, a normal hexagonal, E and a bicontinuous cubic, I_b, phase with an oil. Both the CPB and ODPB systems display a succession of conventional phase sequence with water - L_1-to-E-to-I_b-to-D. But organic solvents with low cohesion energy form only D phases with both CTAB and CPB.

Na 2-ethylhexyloxypropionate-water system[68] is shown to form three liquid crystalline phases - a E phase at the highest water content, an extensive D phase at the highest surfactant content and an intermediate phase. The E phase is separated from the D phase by a reentrant micellar solution. The coexistence of two lamellar phases is identified for the system Na dodecyl-5-p-benzenesulphonate-water[69]. The authors have attributed the coexistence of phases to the presence of an interbilayer attractive component caused by a specific Na^+ - headrgoup interaction. The coexistence of two lamellar phases is also indicated in the aqueous system of a surfactant containing thermotropic and lyotropic mesogenic groups[70]. The phase diagram of the mixed SDS (sodium dodecyl sulphate)-DDAB-water system[71] is also published. This system is rich with polymorphism but is very complex. The existence of spontaneously formed stable vesicles at high dilution, concentration-dependent phases - E, D, I_b, and the coexistence of 2 and 3 liquid crystalline phases are identified. The mesomorphism and the phase transition temperatures are also determined for the system K laurate-decylammoniumhydrochloride-water[72].

The nonionic surfactant system $C_{22}EO_6$-H_2O forms two high-temperature bicontinuos cubic phases and a low-temperature fluid intermediate phase in addition to the usual E and D phases. The intermediate phase appears to consist of a lamellar-like structure[73]. The partial phase diagram for the system $C_{16}EO_6$-water has also been determined by microscopy and ^{2}H n.m.r. methods. Besides the conventional D and E phases, the system exhibits a wide range of intermediate behaviour including a defected lamellar phase, a nematic phase of rod micelles and a cubic or an intemediate phase[74]. The solubilization of a thermotropic mesogen, 5 CB, to the lyotropic mesophases formed for the system $C_{16}EO_8$-water as well as the solubilization of the nonionic surfactant into the nematic of 5 CB are reported[75]. A double tail nonionic surfactant with a double branched lipophilic moiety is found to form two lamellar phases with water[76]. Two surfactants of the type 4-n-alkylpyridine-N-oxide show a limited swelling capability with water and they form only lamellar phases[77]. The phase diagram has been determined for the system C_{16}-C_{18} saturated monoglyceride-water[78]. The transition enthapies between the lamellar and the gel phase, the surfactant hydration and the ordering of water are also studied. Lamellar, hexagonal and isotropic phases are identified for the system n-alkyl-β-D-glucopyranosides with water and benzene. The details of the phase diagrams depend on the sugar homolog and the solvent[73].

^{19}F n.m.r. study reveals the formation of an isotropic phase followed by an anisotropic phase with nematic nature in the ammonium perfluorooctanoate-water system[79]. Using ^{133}Cs n.m.r. the authors reported a significant isotope effect on the phase behaviour and the aggregate structure when H_2O is replaced by ^2H_2O for the system Cs perfluorooctanoate[80].

The thermal behaviour of di-, tri-, and tetravalent metal soaps (carboxylates) is discussed with regard to the phase behaviour, transport and other related physical properties, and thermal decomposition[81]. In many cases, liquid crystalline phases are observed between the solid and the liquid indicating the presence of aggregate structures in the melt[82]. Multinuclear, ^1H, ^{13}C and ^{207}Pb n.m.r. lineshapes and the relaxation studies of lead dioctanoate and didecanoate soaps also indicate the existence of aggregates in the liquid phase. Multifield ^{13}C T_1 times data further support the view that the liquid phase of the molten carboxylates consists of small aggregates[83]. An incorporation of phospholipid to the metal soaps does not lead to the destabilization of the original aggregate structure of the soap[84].

The phase behaviour of mixed bilayers consisting of PO (palmitoyloleoyl) PC-POPE[85], POPC-DMPA (dimyristoylglycerophosphate)[86], DMPC-DMG (glycerol)[87], and DMPC-DS (distearoyl) PC[88], glycosphingolipid-SOPC[89,90,77] with water are reported. For these systems, the existence of phases such as gel, lamellar and hexagonal liquid crystals, isotropic solution phases as well as the transition temperatures between the phases at different mixing ratios are determined. Depending on temperature and water contents, DOPE forms either a D or a F phase[91]. It is observed that the ^2H n.m.r. of headgroup deuterated phospholipid can be employed to sense and to quantify inhomogeneities in the lateral distribution of charged membrane components in mixtures of charged and uncharged phospholipds[86]. The effects of cholesterol and androstenol on the phase behaviour of aqueous dispersions of SEPC-d_{35}-lecithin[92], and DPPC-melittin complex[93] are also available. Cholesterol and Cholesterol 3-sulphate, a component of intercellular lipid are found to exert a stabilizing effect on membranes composed of different lipids[94]. The phase behaviour of an equimolar dispersions of bovine brain sphingomyelin, cholesterol and palmitic acid with an equivalent dispersion in which bovine brain ceramide is substituted for sphingomyelin is reported. While the sphingomyelin dispersions remain in a homogeneous fluid lamellar phase, those containing ceramide display complex polymorphism[95]. The first moment calculated from the ^2H n.m.r. spectra over the temperature range for the aqueous DMPC-d_{27} bilayers with added tricyclohexaprenol and isoarborinol indicate that the former substrate almost abolishes the phase transition of DMPC, but the latter has no effect on the lipid chain order in the liquid crystalline phase of DMPC[96].

Upon heating, aqueous dispersions of diacyl PE with chain length, $n \leq 17$, exhibit a single transition from the crystalline form to a lamellar phase in contrast to PE homologs with $n \geq 18$ which exhibit crystals-to-lamellar-to-hexagonal transitions with increased temperatures[97]. The similarities and differences of phase behaviour between PE and PC lipid systems are also explained. On addition of retinoic acid and retinol to DEPE membranes, a reduction of transition temperature is recorded for both the lamellar and the hexagonal phases[98]. The transition temperature between the lamellar-hexagonal phases of PE membranes is also found to be dependent on both the heating rate and the lipid composition. It is further suggested that the conversion to non-bilayer phase is not caused by the formation of a cubic phase[99]. The thermotropic phase behaviour is presented for the system POPG in the presence of the surfactant dimethyldihexylammonium bromide. The surfactant increases the transition temperature from the gel state to the liquid crystalline state and the conformation of the PG headgroup is also characterized[100]. The formation of a lamellar liquid crystalline phase and a gel phase is identified for the system DOPC-glycerol-water. On sonication, the lamellar dispersions are found to transform to unilamellar vesicles of variable diameters depending on sample compositions[101]. The chain length dependence of lipid bilayer properties is also examined near the liquid crystals to gel phase transition[102].

6 Isotropic Micellar Solution Phases

Isotropic micellar solutions are formed in two- or multi-component surfactant systems. In normal micellar solutions, the aggregates adopt a positive curvature (the aggregate interface is curved towards oil) and in most cases the solution phase contains an excess aqueous part whereas the reversed micelles which contain an excess non-aqueous component(s) have a negative curvature. A microemulsion phase which contains oil as one of the components, can have either a positive or a negative curvature. But for a balanced surfactant, the microemulsions obtained with a few percent of surfactant, may have solution structures varying from the normal-to-bicontinuous-to-reversed type microstructures. Due to rapid and isotropic motions, narrow lines in the n.m.r. spectra are normally obtained in these systems. Where broad lines do appear their origin may be due to the presence of rather large non-spherical micellar aggregates. This section is presented as follows: normal micelles, solubilization and mixed micelles, and reversed micelles and microemulsions.

6.1 Normal Micelles. - The ^2H n.m.r. relaxation studies of selectively deuterated Na dodecylsulphate (SDS) in formamide indicate a moderate growth of micellar size with increased surfactant concentrations. Fraction of surfactant monomers vs. composition derived from the relaxation data are compared with the prediction of the Poisson-Boltzmann cell model[103]. Three types of micellar aggregates are detected in aqueous solutions of SDS - spherical micelles at high dilution (above CMC), ellipsoidal and rodlike micelles at high surfactant concentrations. Mechanisms leading to the transformation of aggregate-shape are also discussed[104]. The NOESY experiment performed in aqueous solutions of Na 7-dodecen-1-yl sulphate demonstrates that the alkyl chain undergoes a conformational change upon the formation of micelles[104]. In another study[105], evidence is provided for two binding sites on micellization of 1-methyl-4,4' bipyridinium reporter ion in SDS. Like the SDS system, micellization of dodecyltrimethylammonium bromide (DoTAB)[106], Triton X-114[107], and linear alkyl Na benzenesulphonate and Na oleate[108] with water is studied by the ^1H n.m.r. chemical shift and relaxation techniques. The micellar size, shape and conformation of surfactant molecules are obtained and the results are discussed. The ^{14}N relaxation and the self-diffusion coefficients are measured in the micellar solution and the cubic phase for the system cetylpyridinium chloride-Na salicylate-water. Both the relaxation rates and the self-diffusion are found to vary smoothly without having any discontinuity at the phase boundary between the solution and the liquid crystalline phases. From the results it is concluded that the micellar solution consists of a bicontinuous-type microstructure[109].

The micellar properties of two new types of surfactants are reported. The gemini surfactants form small aggregation with high CMC as well as submicellar complexes with water[110]. The contrafacial surfactants appear to undergo a stepwise aggregation rather than a cooperative micellization[111]. Like monopolar surfactants, the micellization of bipolar surfactants and their mixtures with monopolar ones can be described in terms of Israelachvili's model[112].

Multi-frequency relaxation studies[113,114] are carried out on a range of aqueous solutions of anionic surfactants such as Na octanoate, Na decanoate, etc. Data are analysed within the framework of the two-step model. The correletion times associated with slow and fast motions and the order parameters are discussed with regard to the surfactant chain length. Some studies are also subjected to a three-step model and the data are presented showing how the dimensions of the non-spherical micelles vary with temperature[113]. The structural changes of ((dodecyloxy)-18-crown-6) micelles as a function of the degree of K^+ chelation and temperature[115], the surfactant properties of β-diphosphoramides[116], the micellization of double tail surfactants containing one hydro- and one fluoro carbon tails[117], and the diffusion in a micellar solution with respect to the structural changes are also reported[118,119]. In another study[120], mixed surfactant systems containing Na perfluorooctanoate and various types of hydrocarbon surfactants are studied by the ^1H and ^{19}F n.m.r. If both surfactants in the mixture are anionic, the system shows positive deviation from ideal behaviour and for long alkyl chains, two types of micelles coexist at certain concentrations. For mixtures containing nonionic or cationic surfactants, negative deviation from ideal behaviour is observed.

N.m.r. methods are employed to study the surfactant aggregate properties at solid surface-water interfaces by adsorbing surfactants on solid surfaces. ^{13}C n.m.r. chemical shifts measured for four cationic and two anionic surfactants adsorbed on appropriately charged solid surfaces are compared with the shifts of the corresponding micelles in solution and concluded that for most surfactants, the surface aggregate does resemble to the micellar state[121]. DDAC is adsorbed as a complete bilayer on solid silica surface. ^2H n.m.r. T_1, T_2 and bandshape results determined from adsorbed bilayer are analysed to obtain the molecular mobility and ordering of the surfactant molecules[122]. Similar studies are also performed on adsorbing SDS on Al_2O_3 surface[123]. SDS is found to have a bilayer conformation on the solid surface. The spin relaxation data are interpreted by a motional model and the order parameter is calculated from the ^2H bandshapes. The adsorption of nonionic surfactants $C_{12}EO_6$ and $C_{12}EO_{25}$ from aqueous solutions on hydrophilic surfaces is also reported. The

adsorbed amounts and the layer thickness are obtained. N.m.r. results indicate that the EO segments in the adsorbed layers experience conformational restrictions[124].
^1H and ^{17}O n.m.r. relaxation measurements are carried out to obtain the state of water in skim milk and caseinate solutions at different pH[125]. The results suggest that the aggregation of caseins by Ca^{2+} and colloidal Ca phosphate is mainly responsible for the excess hydration in skim milk micelles. A biosurfactant is synthesized from Bacillus licheniformis JF-2. The surfactant exhibits a cmc and measures a very low interfacial tension with water[126].

6.2 Solubilization and Mixed Micelles. - Analysis of self-diffusion coefficients of surfactant molucules measured by the PGSE technique suggests the growth of micelles at different mixing ratios of surfactant at total surfactant concentrations studied for the two mixed surfactant systems Na decylsulphonate-octyltrimethylammonium bromide (OTAB), Na octylsulphate-OTAB with water[127]. Mixed micelles are also reported to form when anionic dyes Remozols are added to the aqueous micellar solutions of ionic and zwitterionic surfactants[128], Na oleate and Na laurate to the Na octylbenzenesulphonate (SOBS)[129]. Solubilization and aggregation numbers are obtained for the micellar mixtures of SDS and DoTAB with tetraethylene glycol (TEG) and tetraethylene glycol dimethyl ether (TEGDM). TEG is hardly incorporated in either SDS or DoTAB micelles while TEGDM is solubilized easily by SDS micelles, but not by DoTAB micelles[130]. The effect of Na toluene sulphonate on the structure and microviscosity of micelles of SDS, SOBS are reported[131]. 2D-COSY experiments are performed to obtain the binding sites in the SDS micelles with added phenyl alkanols[132]. Anabaseine, a pyridine compound, binds to SDS micelles in solution and brings about a conformational change of surfactant molecules[133].
Solubilization of peptides[134-138] and proteins[26] in SDS micellar phase is determined. N.m.r. methods are used to obtain the binding sites and molecular conformations. The cyclopeptides become partially immobilized by interaction with micelles and the conformation of peptide molecules are derived[134]. Bacterioopsin polypeptide is shown to adopt a right-handed α-helical structure[135,136]. The peptides from uteroglobin interact with SDS micelles and change their conformation to an α-helix as well as the cmc of SDS is lowered[137]. The ^1H n.m.r. spectral analysis and chemical shifts are employed to obtain information about the interaction of the drug PAV (papavarine) with ionic and nonionic surfactants at different pH. The interaction is described between the heterocycle of PAV and the polar headgroups of surfactants[138].
Mixed surfactant systems composed of Na perfluorooctanoate and various hydrocarbon surfactants are studied by ^1H and ^{19}F n.m.r.[120,139]. Depending on the charge of the surfactant, length of the alkyl chain, 1 or 2 types of mixed micelles or large aggregates are detected. Both the positive or negative deviations from the ideal behaviour are also observed. Nonideal, unfavoured mixing of the hydro- and fluorocarbon surfactants are also observed for the system NH_4 perfluorooctanoate-NH_4 decanoate[140]. Micellar properties of a hybrid hydro- and fluoro carbon double tail surfactant are monitored by ^{19}F n.m.r. It is found that the rotation of the fluorocarbon chain along the C-C bonds is restricted and the exchange rate between the monomers and the micellar states is lower compared to that of the common ionic surfactants[141]. In another study[142], a triple chain surfactant consisting of 2 hydro- and 1 fluoro carbon chains forms bilayer assembly in fluorocarbon media.

Field-dependent ^{13}C and NOE are measured in methanolic and aqueous solutions of laurocholate, caprylic acid and capryloyl glycerol. The alkyl chain dynamics in mixed micelles are analysed in both aqueous and non-aqueous solvents with a customary two-site model[143]. The concentration-dependent ^{1}H n.m.r. spectra are recorded for the aqueous solutions of taurine-conjugated bile salts and the results are used to show the role of hydrophobicity in the association process[144]. The solubilization of bilrubin in submicellar and micellar aqueous solutions of Na deoxycholate is investigated and observed that the cholate interacts preferentially with the bilrubin molecules having the left-handed chirality[145].

Solubilization and aggregation behaviour of mixed micelles of nonionic surfactants-additives are studied by the PGSE self-diffusion and ^{1}H n.m.r. methods[146-148]. A micellar growth is indicated for the system dimethyldodecylamine oxide-H_2O as a function of surfactant concentration. Upon addition of hydrophobic molecules, TMS or dodecane, the spherical micelles are induced having a narrow size distribution[149]. It is inferred that the Ph group of the surfactant Triton X-102 and Igepal CA 720 is incorporated inside the β-cyclodextrin when the dextrin is added to the aqueous solutions of the surfactants[147]. An addition of benzene to the micellar solutions of DoTAB leads to the growth of micelles and the C_6H_6 molecules are shown to solubilize near the micelle-water interface[146]. The theoretical model that can predict the active agent distribution between the hydrophobic core and the hydrophilic interface of nonionic surfactants is applied to several active agents and the results are compared with those obtained by the n.m.r. methods[150].

6.3 Reversed Micelles and Microemulsions.- Multi-component self-diffusion coefficients measured by the PGSE n.m.r. method show that the lecithin[151] and DPPC[152,153] form reversed micelles in organic solvents such as chloroform, benzene, toluene, etc., and on adding water, the growth of micelles is also observed. The amino group of the drug, diamino diphenyl sulphone is reported to interact with the headgroup of the DPPC in the reversed micellar system[152]. The structure of gels prepared by mixing the appropriate amounts of lecithin, cyclohexane and water appears not to build up by the chain entanglement. Instead, a marked stiffening of the triglyceride moiety and phosphate group is observed without having any loss of chain mobility[154]. Organogels are also prepared through the H bondings between a variety of phenols and the headgroup of the surfactant Aerosol OT (AOT) in a non-polar solvent[155]. The solubilization of water and formation of reversed micelles by $C_{12}EO_6$ in cyclohexane and decane are also reported[156].

Multi-component n.m.r. self-diffusion coefficients (in combination with other n.m.r. methods for some systems) are also used to study the microstructure of microemulsions prepared in 3- and 4 component systems with DDAB[157], SDS[158], Na tetradecylbenzene sulphonate[159], $C_{12}E_5$[160], dodecyl polyglucosides[161], octyl diglucoside[162], mixed surfactants - Na oleate-monoalkyl poly(oxyethylene)ethers[163] and Na oleate-pluronic copolymer[164]. A bicontinuous type of microemulsion is identified in the DDAB system with high volume fraction of both surfactant and water, and upon dilution with oil, there is a gradual transition to discrete particle structure. Based on the water content, transitions between O/W-to-bicontinuous-to-W/O type structures are also reported in 4 component microemulsions containing Na oleate- copolymer[164], dodecyl polyglucosides[161], octyl diglucoside[162] and Na oleate[163]. Ionic microemulsions stabilized by SDS and alcohol are characterized in terms of Winsor type I, II, and III systems. The composition of phases are discussed at equilibrium with relation to structural aspects[158]. At a constant surfactant to oil wt ratio, the microemulsion near the phase boundary of the $C_{12}EO_5$ system consists of discrete oil-swollen micelles over the entire dilution range[160]. The ternary and quaternary microemulsions based on perfluoropolyether oil and an anionic surfactant with a perfluoropolyether hydrophobic chain consists of W/O type droplets over the entire microemulsion region. The effects of oil chain length and alcohols are also reported[165,166].

^{23}Na spin relaxation data obtained in a microemulsion of the AOT system are interpreted in terms of closed droplet structure as a function of the H_2O/AOT ratio. However the existence of small reversed micelles coexisting with the classical microemulsion droplets is invoked for the quantitative analysis of the relaxation data[167]. The n.m.r. self-diffusion in the L_3 phase of the AOT-NaCl-H_2O system is examined. The results indicate that the L_3 phase is simultaneously brine- and surfactant continuous and the solution microstructure is topologically related to that of bilayer continuous cubic phases[168]. ^{17}O relaxation results obtained in the AOT reversed micelles hosting hydophilic enzymes indicate that the enzyme shows maximum activity when all the water molecules present inside the core are strongly bound to the micellar interface[169]. The morphology and structure of reversed micelles composed of AOT and Brij-96 are examined by adding albumin[170], and acetone in the presence of chlorophyll[171]. The ternary microemulsion of the $C_{14}EO_9$ system is claimed to have a lamellar-like structure at a specific water to oil ratio[172].

7 Lyotropic Mesophases

As described earlier, surfactants are often rich in mesomorphism. They form liquid crystals with water as well as in multicomponent systems. The usual mesophases - lamellar, hexagonal (normal and reversed) and cubic liquid crystals have been studied extensively. The so-called "intermediate" liquid crystals are attracting the interest of increasing number of research scientists. Multi-nuclear and multi-field n.m.r. methods are often employed to obtain the structures, conformations and dynamics of alkyl chain and headgroup of amphiphile molecules. We begin the section with mesophase structures followed by vesicles and dispersions/liposomes. Vesicles are used extensively as models for biological membranes. The vesicles prepared mechanically are thermodynamically unstable. Spontaneous formation of vesicles which are probably thermodynamically stable are reported for some surfactant systems. Natural systems, mainly biological membranes, are also included. Polymer-surfactant interactions which are also included here, are becoming important topics for n.m.r. studies as shown by many publications in this review. Water and ions, important constituents in amphiphilic systems, are the concluding part of this review article.

7.1 Alkyl chains and Headgroups in Liquid Crystalline Phases.- The ^{2}H n.m.r. splittings of α-deuterated surfactants in combination with SAXS data are used to obtain the intrinsic structural defects in the lamellar phase for the system SDS-decanol-water as functions of sample composition and temperature. It is found that the bilayer defects are promoted by a small [alcohol]/[surf.] ratio[173]. The curvature defects in the oriented lamellar phase (D)[174] and the static ^{2}H-^{2}H dipole interaction in the oriented hexagonal phase (E)[175] for the same SDS system are, respectively, studied by the ^{23}Na spin relaxation and ^{2}H n.m.r. methods. The results are discussed by theoretical models. Two intermediate phases are identified for the system dodecyltrimethylammonium chloride (DoTAC)-water. One intermediate phase is composed of noncylindrical aggregates and the other, an uniaxial phase[176]. The ribbon phase characterized in the liquid crystals of the dodecyl-1, 3-propylenebisamine-HCl-water system[177] is shown to consist of a centered rectangular structure from the n.m.r. and SAXS results in combination with predictions from a cell model. It is proposed that the undulations stabilize the dilute D phase of the $C_{12}EO_5$-water system. An addition of alcohols affects the undulating membrane drastically. The results obtained under the different experimental conditions are discussed with reference to the molecular geometry and the bending properties of the membrane[178]. A continuous decrease of the orientational order parameter is measured in the liquid crystalline phase from the ^{2}H splitting-values with increased temperatures for the system $C_{12}EO_6$ methyl ether[146,179,180]

The ^{2}H splittings are also used to obtain the conformational and orientational characteristics of the surfactant molecules by incorporating additive molecules to the liquid crystalline phases for the systems Na octanoate-decanol-water[181], K laurate-decanol-water-perylene[182], SDS-decanol-formamide[183], C_{10} ammonium chloride-phosphoric acid monodecylester-water[184], and perfluorinated ionomer membranes[185]. In the E phase, the 2nd C-C bond of C_8COO^- is flexible compared with the D and F (reversed hexagonal) phases. The alcohol chain is, however, rigid in the E phase. In the F phase, both C_8COO^- and $C_{10}OH$ chains are much extended and wobbled, thus showing small order parameters. The lamellar surface is found to be flat. The low order parameter of surfactant and alcohol molecules in combination with decreased compatibility between these two components are considered to be the main factors for the less stability of the D phase in formamide compared to water as solvent[183]. The ^{2}H n.m.r. lineshapes are simulated to match the experimental spectra recorded in the liquid crystalline materials for the systems Na and Cs tetradecanoate-water. The method is also applicable to determine tieline and tri-triangle end points in multi-phase liquid crystalline samples[186].

The diffusion of water as a solvent calculated with computer simulation in liquid crystals is found to be in agreement with the values obtained by the PGSE method of the AOT-water system[187,188]. The obstruction factors for water diffusion parallel and perpendicular to the lamella normal are calculated from the water self-diffusion coefficient measured by the PGSE technique in the oriented D phase of the Cs perfluorooctanoate (CsPFO)-water system. The results support the defected nature of the D phase. The structural models for the defects are presented[189]. The diffusion behaviour measured in the liquid crystals for the systems CsPFO-water, DMPC-CHAPSO-water are compared with simulation results. It is concluded that for the fluorinated surfactant, disk normals parallel to the field conforms with published results but disk normals perpendicular to the field results a new liquid crystalline phase for the DMPC system[190]. An alternative method is described to measure the lateral diffusion of phospholipids along the plane of a spherical bilayer on a solid support [191].

The n.m.r. diffusion results show that the bicontinuous structure remains unperturbed by the radical polymerization of acrylamide in the bicontinuous cubic phase of the dodecyldimethylammino oxide-water system[192]. The lipid diffusion coefficient in the D phase of the monoolein-glycerol system is found to decrease continuously with increased glycerol content and the decrease is explained in terms of changes of the viscosity in the interbilayer regions[193]. The diffusion coefficient of lipid molecules along with the ^{31}P n.m.r. lineshapes are used to show that the cubic phase of the lauroyl lyso PC-water system is composed of short non-spherical micelles[194]. The PGSE method is also used to obtain the microstructure of cubic phases reported previously for many lipid-water systems[195].

^{13}C MAS n.m.r. experiments are carried out to probe the headgroup, glycerol backbone and acyl chain orientational order and dynamics of membranes composed of different biological lipids[196-199]. The results reveal that the dipalmitoyl glycerol has a distinct motionally-averaged structure in egg PC bilayers that are similar to those of the DPPC[197]. The backbone and the headgroup carbons of many membranes show similar high frequency motion-independent of headgroup size and charge[196]. The specific changes in the ^2H quadrupolar splittings of the choline headgroup deuterated PC bilayers are observed on insertion of the negatively charged cardiolipin. The observations are explained in terms of the electrostatic effects[198]. The results obtained in an oriented bilayer containing CHAPSO show an average structure possessing a high degree of order[200].

The temperature dependence of the mean orientational order parameter in the vicinity of the liquid crystalline phase to the gel phase transition[102], the acyl chain conformational ordering[201,202], bilayer thickness[203], headgroup area[204], and chain incompatibilty in mixed bilayers[205] are studied by recording the ^2H n.m.r. spectra of appropriately deuterated membranes in water. The measured order parameters in the liquid crystal to gel transition are explained by the Landau expansion[102]. The ^2H splitting-values in combination with the FTIR data lead to a better understanding of a chain order parameter in PC and PE membranes[201]. When saturated and unsaturated chains in a bilayer are to pack side by side, the system shows a complex chain conformation disorder, instability and marked polymorphism. The hydratin can partly stabilize the interactions[205].

A new ^2H technique is used to measure the hydration forces of frozen PC membranes[206,207]. The data are used to estimate the magnitude of freezing-induced membrane stresses and discuss their effects on the response of cells and organelles to freezing and thawing. The influence of fatty acid carbonyl groups on phospholipid bilayers of DPPC and DHPC (dihexadecyl PC) are studied over a wide range of water concentrations. The two lipids have identical hydration forces between their bilayers in the lamellar liquid crystalline phase. It is concluded that the carbonyl groups are not essential for lipid structure and hydration[208].

The ^2H n.m.r. technique is also used to gain information on the orientational order of acyl chain perdeuterated lipids in lamellar phases for a number of lipid-water systems[209]. Similar studies are also reported in the non-lamellar liquid crystal aggregates[210]. The results in combination with predictions from the theoretical models describe intermolecular interactions in the lamellar phase and curvature effects, order and dynamics of lipid molecules in the hexagonal phase.

Several phospholipids are used to obtain the structures, dynamics and thermodynamic behaviour of PC bilayer containing cholesterol[211-214]. At high cholesterol content, the ^2H lineshape analysis indicates the presence of a new phase. Moreover, the presence of cholesterol leads to a more ordered bilayer structure[212]. The ^2H relaxation data obtained in the oriented bilayer with cholesterol can be explained in terms of a simple rotational diffusion model for the dynamics of membrane lipid molecules[214]. An addition of cholesterol also increases the bilayer thickness of the PC liquid crystals. However, when peptide is incorporated into the liquid crystals, the bilayer is perturbed. The increase or decrease of the bilayer thickness depends on the degree of mismatch between the hydrophobic length of the peptide and the acyl chain of the lipid[213]. The relaxation study shows that the molecular order parameter is decreased while the conformational order parameter remains almost unaltered by incorporating peptides in the PC bilayers[215,216]. The ^2H n.m.r. results obtained with mixtures of glycosylceramide and DMPC containing cholesterol suggest that the average acyl chain backbone conformation of glycosphingolipid in phospholipid membrane is closely related to that predicted from single x-ray studies[217]. The solid-state multinuclear n.m.r. lineshape analysis performed in the aligned liquid crystalline materials containing the peptide, melittin does not indicate the formation of any non-bilayer liquid crystalline structure, but an increase of ^1H T_1-values of the acyl chain methylene protons is measured[218]. The binding of cytochrome c to cardiolipin bilayers brings about a change in the molecular dynamics around the double bonds in the acyl chains of bilayers[219]. The ^2H (of appropriately deuterated lipid molecules) and ^{31}P n.m.r. are employed to provide information about the morphological structure adopted by the reconstituted phospholipid-coat protein of bacteriophage M13 system. It is found that the protein in the predominant β-sheet conformation causes a fraction of lipids to be trapped and the trapped lipids are arranged in a nonbilayer structure[220]. The properties of the liquid crystalline bilayers are also investigated by incorporating doxorubicin[221], amlodipine and nimodipine[222], and aromatic amphiphiles[223] into the liquid crystals.

7.2 Alkyl Chains and Headgroups in Vesicles. - The ^2H n.m.r. relaxation rates are measured for lipid vesicles containing paramagnetic ions. The results are discussed by comparing the correlation time of fluctuations with the correlation time for rotation of the vesicle. The diameter of the vesicles can also be determined by the method[224]. The temperature dependence of vesicle-size prepared with the hydrogenated castor oil is also determined[225]. ^{19}F n.m.r. in combination with photon correlation spectroscopy are used to study some critical features of unilamellar vesicles containing fluorinated steroids. It is observed that on increasing temperature, there is a decrease of hydrophilicity resulting the coalescence of vesicles[226]. Large unilamellar vesicles (LUV) are prepared from several phospholipids containing glycerol. The ^2H n.m.r. spectra recorded in vesicle phase display the powder pattern similar to those in the liquid crystalline phase. The powder pattern spectra arise due to reducing the rate of phospholipid reorientations. The order parameter profile is presented by integration of the dePaked spectra[227]. Lyso-phospholipids are shown to inhibit the fusion of LUVs between LUVs of N-methyl-DOPE and the fusion between LUVs and Sendai virus. The ^{31}P n.m.r. data show that lyso-phospholipid stabilizes the lamellar phase and destabilizes the nonbilayer structure. Curvature effects are used to rationalize the observed results[228]. Conditions are obtained to prepare vesicles with bipolar lipids extracted from Sulfolobus solfataricus. Like unipolar lipids, the self-assembly of a bipolar lipid results from the balance of interaction free energy, entropy and molecular geometry. The calculated results agree closely to those obtained from proton n.m.r. measurements[229]. AC2 is a synthetic chemical analog of the DMPC. The property of AC2 is studied by the n.m.r. methods in the sonicated form. On sonication, the lipid is found to form vesicles which, however, are not thermodynamically stable; because small sonicated vesicles fuse forming LUVs. The dynamics of acyl chain and headgroup of AC2 are also studied[230].

The relaxation behaviour of phospholipid vesicles in supported and unsupported membranes is studied by the ^{31}P n.m.r. as well as by the computer simulation technique. The results obtained by the two methods are compared and discussed[231,232]. It is demonstrated that the addition of physiological amounts of aracidonic acid into either phosphatidylcholine or plasmenylcholine membrane bilayers modifies the motional dynamics of the membrane and the aracidonic acid. However, the effect on the aracidonic acid is more pronounced in phosphatidylcholine vesicles[233]. Upon prolonged incubation, the vesicles composed of DMPC undergo spontaneous fragmentation forming discoidal-type particles at low p^H. A model for the formation of discoidal complex is presented. An addition of high content of cholesterol into the vesicles can prevent the fragmentation process[234].

Diffusion ordered 2D n.m.r. (DOSY) is used to characterize large unilamellar phospholipd vesicles. Entrapped sucrose is used as an n.m.r. active label for the vesicles. It is shown that the fraction of sucrose entrapped in the aqueous core of a vesicle and the tracer diffusion coefficient of vesicles can be determined simultaneously. The method is also sufficiently sensitive to obtain the size-distribution of vesicles[235,236]. ^{13}C T_2^{-1} rates are measured in lipid vesicles. It is shown that the lipid lateral diffusion coefficient in vesicles can be estimated from the analysis of the spectral density functions[237]. γ-irradiation of DPPC vesicles produces new molecular species which are identified by the ^{31}P and ^{1}H n.m.r. analysis. The fraction of formation of new species is also measured vs the radiation dose. The molecular and structural changes resulting from irradiation are also discussed[238].

Spontaneous vesiculation is observed in model membrane composed of POPC-POPS with or without cholesterol. The properties of vesicles are investigated as a function of temperature. The coexistence of small daughter vesicles with large unilamellar vesicles are identified and the mechanism of their coexistence is explained by the theories of shape fluctuations[239]. There is a partial orientation of multilamellar vesicles (MLV) at high magnetic field. The orientation effect is measured with ^{2}H and ^{31}P n.m.r. on MLVs of DPPC with cholesterol. MLVs are found to adopt an ellipsoidal shape. A method is also proposed to prevent the orientation of vesicles at high field[240]. The ^{2}H quadrupole splittings and ^{31}P CSAs (chemical shift anisotropies) are measured in giant unilamellar vesicles composed of mixed phospholipds - POPC + DMPG and cholesterol. By varying experimental conditions with added Ca^{2+} or perchlorate ions, it is possible to resolve the differences in lipid conformations and the reorientational order of the two monolayers of a lipid bilayer in giant vesicles[241]. Multi-field ^{13}C relaxation studies are made to measure the dynamics of cholesterol incorporated into DMPC vesicles. The data are discussed by the model-free approach of Lipari and Szabo[242].

The interactions of DMPC small vesicles with amphiphilic peptides[243], membrane-bound conformations of a signal peptide[244], melittin-induced transition of vesicles[245], perturbation of vesicles by insertion of α-tocopherol and ubiquinone-10[246,247], and the effect of dantrolene on the phospholipid membranes with or without halothane[248] are investigated by multi-nuclear n.m.r. methods. Well-defined complexes are formed by the association of one amino acid residue with two lipid molecules. In the presence of peptide, fusion of small unilamellar vesicles are also observed at certain concentration[243]. Part of the peptide primarily interacts with the membrane surface[244]. Below the transition temperature, the membrane is induced by melittin and the system forms discoidal objects[245]. It is demonstrated that α-tocopherol is preferentially partitioned in the most fluid domains of the membrane[247]. It is found that the relaxing agent, dantrolene cures the membrane that is destructured by halothane[248]. Amphotericin B is found to aggregate in vesicles and the extent of aggregation depends on the lipid/drug concentration ratio[249]. Doxorubicin is also found to induce a disordering of acyl chains in bacterial plasma membranes and their total lipid extracts[250].

The computer modelling studies indicate that the aromatic portion of the alkylbenzene amine lies in close proximity to the phospholipid ester alkyl cains while the N-alkyl groups may either project into the region occupied by the phospholipid headgroups or fold back into the more hydrophobic environment in lipid vesicles. The hypothesis is validated by the 2D NOE experiments[251]. It is shown from the results of n.m.r. paramagnetic relaxation experiments that both polyethylene oxide and polypropylene oxide do not interact with small unilamellar vesicles made with phospholipids; they are located exclusively in the bulk water phase. However, the low molecular weight polypropylene oxide distributes itself between the aqueous and the lipid phase[252]. High concentrations of polyethylene oxide induce large, unilamellar, DPPC vesicles to leak and fuse when the bilayers contain amphipathic perturbent molecules. In this paper[253], the authors attempt to resolve if fusion occurs independently of leakage or if fusion can be described as local bilayer rupture followed by resealing.

Mixtures of anionic and cationic surfactants with long alkyl chains produce stable vesicles spontaneously with water and the size of the vesicles can be controlled by varying the surface charge of the aggregates. These systems also provide a new way to study the changes in surfactant organization[71,254,255].

7.3 Alkyl Chains and Headgroups in Liposomes/Dispersions. - In an attempt to isolate the effects of oleoylgalactosylceramide fatty acid unsaturation on glycolipid behaviour in membranes, ^{2}H n.m.r. spectral analysis of appropriately deuterated acyl chain is performed by incorporating the lipid in several phospholipd liposomes (DMPC, DSPC, POPC). The results indicate that the effects of acyl chain unsaturation as well as the overall dynamics of oleoylgalactosylceramide in fluid phospholipid membranes are very similar to that of glycerolipids with camparable acyl chains[256]. Hydrophobic interactions between the Me_{10} group of methacryloyloxydecyl dihydrogen phosphate (MDP) and the acyl chains of phospholipids are indicated from the ^{1}H n.m.r. chemical shifts measured in liposomes composed of DDPC + MDP, DLEA + MDP + cholesterol[257].

The influence of various parameters, such as, Ca^{2+} ions, and lipid concentrations are examined by ^{31}P n.m.r.on the formation of nonbilayer phases in cardiolipin-phospholipid liposomes. In particular, an increase of the cardiolipin concentration favours the induction of isotropic and hexagonal phases[258]. Human myelin lipoprotein disrupts the multilamellar phospholipid bilayers into small lipoprotein particles[259]. Stratum corneum lipid mixed with phospholipid liposomes has different phase characteristics, being in a liquid crystalline state at low temperature and transforming apparantly to a reversed hexagonal phase at high temperature. The consequences of the penetration of liposomes through the stratum corneum of the skin are discussed[260]. The membrane surface charge is manipolated by the simultaneous addition of the cationic DDAB and the anionic DMPG to the neutral DMPC in aqueous mixtures. The ^{2}H quadrupole splittings are found not to respond directly to the membrane surface charge. Instead, the splitting values measured for a given ternary system can be reproduced by summing the individual effects of the charged lipids in binary mixtures, weighted according to their appropriate mole fractions[261]. The autoperoxidation of liposomes is also studied by n.m.r. The lamellar structure is converted to an isotropic solution upon autooxidation[262]. But the addition of PE tends to inhibit the formation of hydroperoxide. The reversed hexagonal phase is reported to be more sensitive to hydroperoxidation than the lamellar phase in phospholipid aqueous dispersions[263,264].

Solid-state ^{13}C n.m.r. experiments are performed on oriented phospholipid lamellar dispersions containing surfactants, Triton X-100, CHAPSO, etc., to obtain chemical shift anisotropies and ^{13}C-^{31}P dipolar couplings. Taking spectral parameter with pure DMPC and DMPC with detergents, and spectral parameter vs. bilayer order together, it is suggested that the average conformation and dynamic of DMPC in the presence of detergents are similar to those of pure DMPC lamellar phase[265]. Similar experiments are also carried out in aqueous dispersions of mixed bovin brain sulfatide and phospholipid as functions of temperature and composition. In the liquid crystalline state, the hydrocarbon chains of sulfatide are packed less orderly than those of phospholipid and there are two conformers for the ceramide residues of natural lipids[266]. In another study[267], the characteristic effects of sulfatide and its derivatives are also reported on the stability of small liposomes. Results obtained by the multinuclear n.m.r. in an aqueous dispersion of egg PC and triolein show that the spectra obtained with MAS exhibit resolution comparable to that of sonicated vesicles. Moreover, an incorporation of triolein produces a bilayer which remains in the liquid crystalline state[268].

The hypothesis states that there are several conformations present in liquid crystalline phospholipid dispersions. The interconversion between these conformers takes place at rates that are fast on the n.m.r. time scale but slow on the IR time scale and the component IR spectra are recorded at order-disorder phase transition[269]. The influence of cytochrome c binding to cardiolipin bilayers on the motional characteristics of each component has been analyzed by the multinuclear MAS n.m.r[270]. Synthetic pulmonary surfactant-associated protein SP-B interacts with phospholipid bilayers in aqueous dispersions. The ^{2}H n.m.r. data obtained in dispersions show that the protein causes only small perturbations of the spectra in the gel and the liquid crystalline states. The conformations of protein in liposomes are also considered[271].

Physical and chemical properties of vitamin E are studied in aqueous phospholipid dispersions by multinuclear n.m.r. methods[272-274]. No specific interaction is identified between the OH group of α-tocopherol and the polar interfacial region of the bilayer and at the same time the OH group is not located deep in the hydrophobic region[273]. α-trienol is found to be more potent in protecting against free-radical-induced oxidation stress than α-tocopherol. The antioxidant potency of α-trienol appears to be related to its effects on the molecular properties of the membranes[274]. α-tocopherol is found to broaden gel-to-liquid crystalline phase transition, but the membrane retains the normal lamellar bilayer structure. A model for α-tocopherol-lipid interaction is also constructed using computer-aided graphics[272].

The drug amantadine upon adding to DOPC membrane adopts two populations within the bilayer - one site is close to the bilayer surface and the other is much deeper. However, no evidence of bilayer perturbation was observed with either the protonated or the deprotonated forms of amantadine[275]. The NOE difference experiments suggest that part of the nicotine acetylcholine receptor antagonist molecules partition themselves into the hydrophobic interior of phospholipid liposomes[276]. Aqueous mixtures of phospholipid and peptide are used to obtain the components' structural and orientational informations from n.m.r. experiments[277-279] Melittin is found to adopt a helical conformation and the helix is oriented with the hydrophobic face directed towards the membrane interior[277]. In phosphatidylcholine bilayers containing cardiolipin or phosphatidylinositol, P thionin causes an increase in membrane permeability and induces a fusion of sonicated vesicles. At high thionin concentration, nonbilayer structures are also detected[278]. The conformation of gramicidin A in phospholipid dispersions is found to be dependent on the acyl chain length of lipid molecules and at certain chain length, gramicidin A may adopt a channel conformation[279]. The interaction of myelin protein with single bilayers on a solid support is studied by ^{2}H n.m.r. and observed that the protein strongly interacts with the binary lipid systems (DMPC + DMPG), but not with the pure DMPC bilayers. The interaction is described in terms of the electrostatic effect[280]. Moreover, phospholipase A_2 is found to bring about a destruction of egg PC liposomes. An addition of lipids that can inhibit the destruction of liposomes is also reported[281]. Various plant sterols that have the ability to order the soybean lecithin bilayers are reported[282].

7.4 Alkyl Chains and Headgroups in Natural Systems. - Using ^{2}H and ^{31}P n.m.r. techniques, the effects of temperature variation and the addition of phenethyl alcohol are investigated on lipid acyl chain order and on the macroscopic lipid organization of membrane system derived from cells of the E. coli. Increased temperature affects the membrane integrity by forming the non-bilayer structures and the membrane fluidity is also increased by the addition of phenethyl alcohol[283]. The bilayer structure of phospholipid membrane is found to be little changed by the incorporation of hydroxyphenyl derivatives of fatty acids[284]. Benzylpyrimidines are also found to cause a partial inactivation of the E. coli membrane due to a binding as well as a change in membrane mechanism of membrane action[285]. ^{31}P n.m.r. measurements as a function of temperature indicate that the pore-forming domain of Colcin A stabilizes the lipid bilayer structure of the E. coli membrane at low p^H and also it has a large effect on acyl chain order. The results are indicative of a surface localization of the protein[286]. Model lipid systems are established with compositions similar to those of normal erythrocytes, malaria-infected erythrocytes or malaria parasite membranes. The order parameter determined by the deuteron n.m.r. of the infected and parasite membranes are reduced markedly relative to that of the normal erythrocytes. An addition of stearoylarachidonoylglycerol (SAG) induces the formation of nonbilayer phases in all lipid systems. But the addition of chloroquine to the SAG-embedded membrane prevents the formation of nonbilayer structures. The membrane-stabilizing action of chloroquine is discussed[287]. It is observed from the ^{1}H and ^{31}P n.m.r. results that dibucaine bound to the erythrocyte membrane increases the mobility of the acyl chains of phospholipids , but it decreases the mobility and/or changes the structure of the polar headgroup. The results further suggest that dibucaine locates across the polar and the hydrophobic areas of the lipid phase of the membrane[288]. It is demonstrated that the coat protein in its membrane-embedded state has a very high α-helical content with virtually no β-sheet structures present. This finding suggests that the structures of coat protein are similar whether the protein is intact in phage or embedded in the membrane [289]. The extracted pulmonary surfactant C (SP-C) is combined with DPPC-d_4 membrane. The ^{2}H splitting of α–^{2}H spectrum in liquid crystalline state is insensitive to temperature, but is decreased with increasing protein concentration due to the fact that the charges on SP-C are located near the bilayer surface. But the β–^{2}H splitting is found to show reversed trends with respect to temperature and protein concentration suggesting that the torsion angle about the C_{α}-C_{β} bond may be sensitive to steric interactions between the lipid headgroup and the protein[290]. Bacteriorhodopsin (BR) is complexed with DMPC vesicles. The ^{2}H n.m.r. quadrupole splitting of appropriately deuterated lipid molecules and the ^{31}P n.m.r. chemical shift anisotropy (CSA) of the phosphate group are measured with increased BR contents. The variation of splitting and CSA vs. protein concentration indicate that BR induces a net positive charge to the lipid headgroups at the protein-lipid interface[291]. A model system of the inner mitochondrial membrane is constracted. The ^{31}P T_1 times for each lipid phosphate in the membrane and the chemical shift anisotropy of the ^{31}P n.m.r. are measured as a function of temperature in the absence and presence of bound cytochrome c. It is found that the strongest interaction of protein in the mitochondrial membrane occurs with the cardiolipin component[292]. In order to establish the phase preference of the E. coli membrane, a detailed experimental procedure is described to adjust the ratio of bilayer to non-bilayer lipids. The ^{31}P n.m.r. results obtained under different experimantal conditions are discussed in terms of the preferred membrane structure[293]. Solid-state ^{31}P n.m.r. experiments are performed on model phospholipid membrane to measure spectral lineshapes, T_{1Z} and T_{1E} relaxation times. The results are interpreted with respect to the membrane dynamics[294].

The ^{2}H n.m.r. techniques are used to obtain the complete orientational profile of intact membranes of A. laidlawii strain B grown on mixtures of perdeuterated palmitic acid and nondeuterated fatty acids of varying acyl chains length and unsaturation. It is concluded that the modulation of the membrane lipid headgroup composition has the potential to play an important role in maintaining the membrane order[295]. In another study, it is observed that the membrane lipid composition in several strains of A. laidlawii is regulated upon changing in the growth conditions. The authors have studied the phase behaviour with two lipids isolated from A. laidlawii strain A-EF-2 and the results are discussed in terms of a theoretical model[296].

The extraordinarily effective barrier property of Stratum corneum has also been attributed to the highly ordered arrangement of the constituent lipids in the bilayer configuration[297]. It is shown from the ^{1}H n.m.r. measurements that the addition of very small amounts of Ca^{2+} ions and spermine causes a decrease in bilayer fluidity of the microsome membranes[298].

8 Polymer/Protein-Surfactant Systems

The interactions between polymers and surfactants are studied in aqueous solutions by the n.m.r. methods. Few references, 26, 134-138, have already been quoted under section 6.2. The ^{13}C n.m.r. chemical shifts are measured for the complex formed by polyvinylpyrrolidone (PVP) and SDS with water. The data are used to show that the structure of the complex involves the docking of PVP with a micellelike aggregate of the surfactant[299]. The interactions between hydroxypropyl Me cellulose and SDS are also detected at high dilution from the ^1H chemical shift measurements. The self-diffusion results suggest that neither the size nor the shape of the SDS micelles change significantly by the addition of polysaccharide[300]. An associative phase separation is reported for the system - hydrophobically modified polymer-SDS-water. The phase separation is dependent on the added Na_2SO_4 and at certain concentration of the electrolyte, the two-phase region disappears giving rise to one isotropic solution phase. The interaction between the polymer and surfactant occurs in the single phase and at a certain polymer concentration, the interaction leads to the formation of a gel phase[301]. The binding of PSS (poly(Na 4-styrenesulphonate)) to mixed micelles composed of hexadecylphosphocholine (with deuterated headgroup) and CTAB is followed by the ^2H n.m.r. lineshape analysis as a function of added PSS. The observed increase and then the decrease of linewidths are correlated with the precipitation and resolubilization of the colloidal dispersion by PSS[302]. A viscoelastic micellar solution is formed for the system CTAB-Na salicylate-water. Diffusion-ordered 2D n.m.r. experiments are performed in the viscoelastic system with added polymer polyvinyl Me ether. The addition of polymer reduces the aggregate size - the long rod in viscoelastic solution is broken up into relatively small and probably spherical micelles[303].

Structures of solid dispersions in the system polyethylene oxide-griseofulvin with added surfactants are investigated by the ^{13}C CP/MAS solid-state n.m.r. The structure of the dispersions is strongly dependent on the chemical nature of surfactants used (SDS, DTAB, Brij). The charged surfactants interact with the polymer forming aggregates while no interaction is observed with the nonionic surfactant[304,305]. The self-diffusion behaviour of surfactant ($C_{12}EO_6$ or $C_{12}EO_8$) is studied in solutions and in agarose gells containing NaSCN. In the salt-free solutions of $C_{12}EO_6$ containing agarose, the micelles display an increase in size with increasing temperature, but on addition of salt to the system, no significant growth of micelles is observed[306]. The self-diffision coefficient of surfactant measured for the system $C_{12}EO_8$-ethylhydroxyethyl cellulose-water indicates that the micelle diffusion is influenced due to obstruction by the polymer network as well as by the association of micelles to the polymer chains. The separation between the two effects can be achieved by adjusting the experimental parameters. The results are discussed in terms of a theoretical model[307].

There are several reports that deal with interactions between surfactants and proteins in aqueous systems. The surfactants used are of biological lipids and the proteins are: myelin basic protein (MBP)[308], obalbumin (OA)[309], serum albumin[310], mitochondrial cytochrome c reductase[311,312], and coat protein[313]. The conformational changes induced in the protein molecules by the association with lipid micelles depend on acyl chain length of the surfactant as well as on the size and fluidity of the micelles[308]. Egg phospholipids do not affect the protein structure of OA, but lyso-phospholipids can bring about a conformational change of the protein. The ^{31}P and ^{13}C n.m.r. spectra and T_2 relaxation times of complexes indicate that the molecular motion of acyl chains and glycerol carbons of lyso-lipids are restricted by the interaction with the protein[309]. Lipid-protein interactions in complexes of fatty acids and serum albumin are studied by the ^{13}C n.m.r. The native fatty acids (as non-perturbing probes) determine microscopic properties of fatty acids in individual binding sites of albumin, rather than the average properties reflecting multiple binding sites[310]. Neutral phospholipids are shown to have lower affinity for protein than cardiolipin[312]. The ^{31}P n.m.r. lineshape measurements suggest that the protein binding induces less restricted isotropic-like motions in the lipid phosphate. The motional characteristics of the headgroup deuterons are also accessible to a dynamic n.m.r. approach using the Carr-Purcell-Meiboom-Gill multiple pulse experiment and the dynamics of the cardiolipin headgroup in the protein-lipid complex are reported[311]. The complex formed by SDS and gelatin with water is also studied by the ^{13}C and PGSE n.m.r. diffusion methods. Gelatin interacts with micellelike aggregates of surfactant through electrostatic and hydrophobic attractions leading to a considerable reduction of surfactant diffusion coefficient compared to the diffusion of gelatin-free solution phase[314].

9 Water and Ions

The relative degree of counterion binding in aqueous micellar solutions of Na decyl phosphate and SDS has been determined by measuring the ^{23}Na n.m.r. relaxation rates on addition of Cl^- salts of Na^+, K^+, Rb^+, Cs^+, Me_4N^+ and Et_4N+ ions. The wide variations of counterion binding observed for different counterions are explained in terms of the headgroup charge and the hydration effect at the interface[315]. The competition of ion binding is also studied by following the linewidths and/or chemical shifts of $^{79}Br^-$ or $^{35}Cl^-$ ion for the micelles of cetyltrimethylammonium surfactant containing several inorganic anions. The results are discussed by the pseudo-phase ion-exchange model and/or by the Poisson-Boltzmann equation[316]. The ^{31}P n.m.r. linewidth measurements are used to monitor the counter ion binding to the bilayer vesicles composed of a series of identical and mixed chain dialkyl phosphate in the presence of Na^+, K^+, Rb^+, Ca^{2+}, and Me_4N^+ ions. The degree of counterion binding is reported to decrease in the order $Na^+ > K^+ > Me_4N^+$. The ordering of alkyl chain in the core of the bilayer is used to explain the observed counterion effects[317].

There are two reports that describe the effects of organic counterions on the properties of surfactant aggregates with water. The ^{13}C chemical shifts of the counterions butyl- and pentylammonium for the micellar solutions of alkylammonium decylsulphonate show a downfield shift in comparison with those of the corresponding chlorides. Moreover, T_1 relaxation times of the $C_{10}SO_3^-$ ions have irregularities caused by collision between the decyl chain and the terminal Me group of the counterions. Multi-field ^{13}C relaxation studies are reported to obtain the dynamical behaviour and conformations of octylphosphate anion in micellar solution, and hexagonal and lamellar liquid crystals formed from pyridinium octylphosphate with water. The conformer populations, the rate of conformational changes and the reorientation correlation times are derived. The data are discussed by a rotation jump model[318].

The effects of counterions on the property of water in reversed micellar solutions are examined by following the H_2O ^1H n.m.r. chemical shifts. The results obtained in the reversed micelles of the metal bis(2-ethylhexyl)sulphosuccinate in the presence of several inorganic counterions suggest that the effect of the cation on the encased water depends on the charge density of the respective cation[319]. The solubilized states of water are also studied in the reversed micellar solution for the system butyldodecyldimethylammonium bromide-chlorobenzene-water. The state of water solubilized in the reversed micelles of cetyltrimethylammonium bromide in organic solvents are examined by water ^1H and surfactant ^{13}C n.m.r. chemical shift measurements. It is found that the structure of the solubilized water is the same over the whole volume of the aqueous droplet[320]. The water proton spectral features indicate that there are three types of water - the water bound directly to the ionic headgroups of the surfactant followed, in succession, by the water held together with the hydrated ionic groups by H-bonds and the bulk-like water[321]. The water proton relaxation rates which are proportional to the correlation time of molecular motions is found to increase sharply at the critical gel concentration for the sol-gel transition in aqueous system of egg lecithins. The results are interpreted on the ground that the bound water is strongly related to the sol-gel transition[322]. The conformational and dynamic behaviour of the headgroup to hydration are examined by measuring the ^2H (appropriately deuterated headgroup) quadrupole splittings and T_1^{-1} relaxation rates of DOPC. The hydration dependence reveals a concerted change in headgroup conformation together with an increase in the mobility. The spectroscopy parameters are linearly correlated with the activity of the interbilayer water. By linking spectroscopy results with thermodynamics, the repulsive hydration force between apposing bilayers is calculated[323].

The effect of inorganic cations, such as Ca^{2+}, Mg^{2+}, Ba^{2+}, Sr^{2+} on lipid organization of biological membranes is also reported[324,325]. The cations are found to decrease the molecular surface area of cardiolipin in the order $Ba^{2+}>Sr^{2+}>Mg^{2+}\approx Ca^{2+}$. The ^{31}P n.m.r. demonstrates a comparable sequence for the ability of cations to promote a reversed hexagonal phase formation in dispersions of the E. coli cardiolipin[325]. Dispersions are prepared by mixing several phospholipids with water. The association of cations with individual phospholipids are examined by different ^{31}P n.m.r. experiments. Considering only interactions with Ca^{2+} ions, the phospholipids are divided into three groups - high interaction potential, intermediate interaction potential and essentially no interaction potential[324]. More references on "Water and Ions" are given under sections 5.1, 6.1, 6.2, 6.3 and 7.

10 References

1. A. Khan, (G.A. Webb, ed.), *Nucl. Magn. Reson.*, (Specialist Periodic Reports), Royal Society of Chemistry, London, 1993; Vol. 22, pp 497.
2. B. Halle; P. O. Quist; I. Furo, *Liq. Cryst.*, 1993, **14**, 227.
3. R. Zana, *Surfact. Sci. Ser.*, 1992, **46**, 337.
4. M. Ricco; M. P. Fontana, *NATO ASI Ser., Ser. B*, 1992, **290**, 273.

5. M. F. Robert, *Biomembr. Struct. Funct. State of the Art, Proc. Indo-U.S. Workshop,* 1992, 273.
6. J. H. Davis, *Cholesterol. Membr. Models,* 1993, 67.
7. E. Oldfield, *Portland Press Res. Monogr.,* 1992, **3**, 119.
8. M. Bloom; C. Morroson; E. Sternin; J. L. Thewalt, *Pulsed Magn. Reson: NMR, ESR, Opt.,* 1992, 274.
9. A. Watts, *Phospholipids Handb.,* 1993, 687.
10. R. M. Epand; R. F. Epand, *Struct. Biol. Membr.,* 1992, 573.
11. P. V. Bubovski; L. A. Vasilenko, *Biol. Memb.,* 1992, **9**, 1189.
12. J. Jonas, *Colloq. INSERM,* 1992, **224**, 123.
13. T. D. Caffery; J. Hogan, *Chem. Phys. Lipids,* 1992, **61**, 1-109.
14. D. B. Fenske, *Chem. Phys. Lipids,* 1993, **64**, 143.
15. M. Milik; J. Skolnick; A. Kolinski, *J. Phys. Chem.,* 1992, **96**, 4015.
16. S. Karaborni; J. P. O´Connel, *Surfactants Solution,* 1991, **11**, 83.
17. S. Karaborni; J. P. O'Connel, 1993, **30**, 235.
18. B. Balinov; B. Jönsson; P. Linse; O. Söderman, *J. Magn. Reson., Ser A,* 1993, **104**, 17.
19. B. Balinov; O. Söderman; J. C. Revey, *J. Phys. Chem.,* 1994, **98**, 393.
20. L. L. Pearce; S. C. Harvey, *Biophys. J.,* 1993, **65**, 1084.
21. R. M. Vanable; Y. Zhang; B. J. Hardy; R. W. Pastor, *Science,* 1993, **262**, 223.
22. L. Ma; T. F. Taraschi; N. Janes, *Bull. Magn. Reson.,* 1992, **14**, 293.
23. C. R. Sanders; J. P. Schwonek, *Biophys. J.,* 1993, **65**, 1207.
24. P. L. Yeagle, *Methods Enzymol.,* 1993, **220**, 68.
25. F. R. Rana; A. J. Mautone; R. A. Dluhy, *Appl. Spectrosc.,* 1993, **47**, 1015.
26. W. H. W. Mai; C. Wang; T. A. Cross, *Prioein Sci.,* 1993, **2**, 532.
27. D. Barsky; B. Putz; K. Schulten; R. L. Magin, *Magn.Reson.Med.,* 1992, **24**, 1-13.
28. P. J. Sebastiao; A. C. Ribeiro; H. T. Nguyen; F. Noack, *Z. Naturforsch., A: Phys. Sci.,* 1993, **48**, 851.
29. P. J. Sebastiao; G. M.H.; A. C. Ribeiro; D. Guillon; M. Vilfan, *Liq. Cryst,.* 1992, **11**, 621.
30. K. H. Schweikert; F. Noack, *Mol. Cryst. Liq. Cryst. Sci TecnolL. Sect. A,* 1992, **212**, 33.
31. R. Y. Dong; G. M. Richards, *J. Chem. Soc., Faraday Trans.,* 1992, **88**, 1885.
32. G. L. Hoatson; T. Y. Tse; R. L. Vold, *J. Magn. Reson.,* 1992, **98**, 342.
33. D. Imberdelli; B. Wazynska; A. Golemma; G. Chidichimo; R. Dabrowski, *Mol. Phys.,* 1993, **79**, 1275.
34. J. M. Goetz; G. L. Hoatson; R. L. Vold, *J. Chem. Phys.,* 1992, **97**, 1306.
35. D. Catalano; M. Geppi; C. A. Veracini; C. Forte, *Mol. Cryst. Liq. Cryst. Sci. Technol., Sect. A,* 1992, **221**, 7.
36. N. Vrbancic; M. Vilfan; R. Blinc; J. Dolinsek; G. P. Crawford; J. W. Doane, *J. Chem. Phys.,* 1993, **98**, 3540.
37. J. S. Lewis; E. Tomechuk; E. Bock, *Liq. Cryst.,* 1993, **14**, 1507.
38. J. L. Figueirinhas; C. Cruz; A. C. Ribeiro; H. T. Nguyen, *Mol. Cryst. Liq. Cryst. Technol. Sect. A,* 1992, **212**, 263.
39. R. W. Date; I. W. Hamley; G. R. Luckhurst; J. M. Seddon; R. M. Richardson, *Mol. Phys.,* 1992, **76**, 951.
40. A. Abe; N. Kimura; M. Nakamura, *Makromol. Chem. Theory Simul.,* 1992, **1**, 401.
41. S. K. Heeks; G. R. Luckhurst, *J. Chem. Soc., Faraday Trans.,* 1993, **89**, 3783.
42. P. J. Barnes; A. G. Douglass; S. K. Heeks; G. R. Luckhurst, *Liq.. Cryst.,* 1993, **13**, 603.
43. G. R. Luckhurst, *Ber. Bunsen-Ges. Phys. Chem.,* 1993, **97**, 1169.
44. S. M. Fan; I. D. Fletcher; B. Guendogan; N. J. Heaton; G. Kothe; G. R. Luckhurst; K. Praefcke, *Chem. Phys. Lett.,* 1993, **204**, 517.

45. N. Spielberg; M. Sarkar; Z. Luz; R. Poupko; J. Billards; H. Zimmermann, *Liq. Cryst.*, 1993, **15**, 311.
46. B. M. Sheik-Ali; C. L. Khetrapal; R. G. Weiss, *J. Phys. Chem.*, 1994, **98**, 1213.
47. A. Yoshizawa; A. Yokoyama; H. Kikuzaki; T. Hirai, *Liq. Cryst,*, 1993, **14**, 513.
48. R. J. Ondris-Crawford; G. P. Crawford; S. Zumer; J. W. Doane, *Phys. Rev. Lett.*, 1993, **70**, 194.
49. G. P. Crawford; R. Ondris-Crawford; S. Zumer; J. W. Doane, *Phys. Rev. Lett.*, 1993, **70**, 1838.
50. S. Kralj; G. Lahajnar; A. Zidansek; N. Vrbancic-Kopac; M. Vilfan; R. Blinc; M. Kosec, *Phys. Rev. E: Stat. Phys., Plasmas, Fluids, Relat. Interdiscip. Top.*, 1993, **48**, 340.
51. G. Iannacchione; D. Finotello, *Liq. Cryst.*, 1993, **14**, 1135.
52. G. P. Crawford; D. W. Allender; J. W. Doane, *Phys. Rev. A*, 1992, **45**, 8693.
53. M. Cavazza; C. Forte; G. Galli; M. Geppi; F. Pietra; C. A. Veracini, *Liq. Cryst.*, 1993, **15**, 275.
54. G. Rizi; S. K. Chosh, *Nuovo Cimento Soc. Ital. Fis. D*, 1993, **15D**, 669.
55. B. M. Fung; M. L. Magnuson; T. H. Tong; M. S. Ho, *Liq. Cryst.*, 1993, **14**, 1495.
56. H. Yoshida, *Phys. Lett. A*, 1993, **172**, 267.
57. X. Yang; X. Hua; G. Li, *Huadong Huagong Xueyuan Xuebao*, 1991, **17**, 722.
58. C. J. Booth; J. W. Goodby; J. P. Hardy; O. C. Lettington; K. J. Toyne, *J. Mater. Chem.*, 1993, **3**, 821.
59. J. P. Bayle; M. S. Ho; B. M. Fung, *New J. Chem.*, 1993, **17**, 345.
60. A. Mori; K. Katahira; K. Kida; H. Takeshita, *Chem. Lett.*, 1992, **9**, 1767.
61. Q. Wang; S. Y. Fan; H. N. C. Wong; Z. Li; B. M. Fung; R. J. Twieg; H. T. Nguyen, *Tetrahedron*, 1993, **49**, 619.
62. J. P. Bayle; B. M. Fung, *Liq. Cryst.*, 1993, **15**, 87.
63. A. Khan; C. Kang, *Progr. Colloid Polym. Sci.*, 1993, **93**, 146.
64. C. Kang; A. Khan, *J. Colloid Interface Sci.*, 1993, **156**, 218.
65. T. A. Bleasdale; G. J. T. Tiddy, *Surfactant Sci. Ser.*, 1992, **44**, 125.
66. X. Auvray; T. Perche; C. Petipas; R. Anthore; M. J. Marti; I. L. A. Rico, *Langmuir*, 1992, **8**, 2671.
67. T. Wärnheim; U. Henriksson, *Surfact. Sci. Ser.*, 1992, **44**, 221.
68. K. C. Gounden; P. Ganguli; G. J. T. Tiddy, *J.Phys.Chem.*, 1992, **96**, 4627
.69. B. A. Timimi; K. S. Narayan; G. J. T. Tiddy, *J. Phys. Chem.*, 1993, **97**, 6767.
70. S. Fuller; J. Hopwood; A. Rahman; N. Shinde; G. J. Tiddy; G. S. Attard; O. Howell; S. Sproston, *Liq. Cryst.*, 1992, **12**, 521.
71. E. Marques; A. Khan; M. d. G. Miguel; B. Lindman, *J. Phys. Chem.*, 1993, **97**, 4729.
72. R. Bartolino; G. Chidichimo; A. Golemme; F. P. Nicoletta, *NATO ASI Ser., Ser. B*, 1992, **290**, 427.
73. S. S. Funari; M. C. Holmes; G. J. T. Tiddy, *J. Phys. Chem.*, 1992, **96**, 11029.
74. S. S. Funari; M. C. Holmes; G. J. T. Tiddy, *J. Phys. Chem.*, 1994, **98**, 3015.
75. J. Corcoran; S. Fuller; A. Rahman; N. Shinde; G. T. Tiddy; G. S. Attard, *J. Mater. Chem.*, 1992, **2**, 695.
76. K. Kratzat; C. Schmidt; H. Finkelmann, *J. Colloid Interf. Sci.*, 1994, **163**, 190.
77. H. Morgans; G. Williams; G. J. T. Tiddy; A. R. Katritzky; G. P. Savage, *Liq. Cryst.*, 1993, **15**, 899.
78. W. G. Morley; G. J. T. Tiddy, *J. Chem. Soc. Faraday Trans.*, 1993, **89**, 2823.
79. S. Ristori; G. Gebel; G. Martini, *Colloids Surf., A*, 1993, **80**, 113.
80. N. Boden; K. W. Jolley; M. H. Smith, *J. Phys. Chem.*, 1993, **97**, 7678.
81. M. S. Akanni; E. K. Okoh; H. D. Burrows; H. A. Ellis, *Themochim. Acta*, 1992, **208**, 1.

82. H. D. Burrows, *NATO ASI Ser., Ser. C,* 1992, **379**, 333.
83. G. Feio; H. D. Burrows; C. F. G. S. Geraldes; T. J. T. Pinheiro, *J. Chem. Soc., Faraday Trans.,* 1993, **89**, 3117.
84. H. D. Burrows; C. F. G. C. Geraldes; M. d. G. M. Miguel; T. J. T. Pinheiro; J. J. C. C. Pinto, *Thermochim. Acta,* 1992, **206**, 203.
85. G. Bryant; J. M. Pope; J. Wolfe, *Eur. Biophys. J.,* 1992, **21**, 223.
86. F. M. Marassi; S. Djukic; P. M. Macdonald, *Biochim. Biophys. Acta,* 1993, **1146**, 219.
87. T. Heimburg; U. Wuerz; D. Marsh, *Biophys. J.,* 1992, **63**, 1369.
88. M. B. Sankaram; T. E. Thompson, *Biochemistry,* 1992, **31**, 8258.
89. M. R. Morrow; D. Singh; D. Lu; C. W. M. Grant, *Biochim. biophys. Acta,* 1992, **1106**, 85.
90. D. Lu; D. Singh; M. R. Morrow; C. W. M. Grant, *Biochemistry,* 1993, **32**, 290.
91. K. Gawrisch; V. A. Parsegian; D. A. Hajduk; M. W. Tate; S. M. Gruner; N. L. Fuller; R. P. Rand, *Biochemistry,* 1992, **31**, 2856.
92. T. P. McMullen; R. N. A. H. Lewis; R. N. McElhaney, *Biophys. J.,* 1994, **66**, 741.
93. M. Monette; M. R. Van calsteren; M. Lafleur, *Biochim. Biophys. Acta,* 1993, **1149**, 319.
94. N. Kitson; M. Monck; K. Wong; J. Thewalt; P. Cullis, *Biochim. Biophys. Acta,* 1992, **1111**, 127.
95. J. Thewalt; N. Kitson; C. Araujo; A. MacKay; M. Bloom, *Biochem. Biophys. Res. Commun.,* 1992, **188**, 1247.
96. M. A. Krajewski-Bertrand; A. Milon; Y. Nakatani; G. Ourisson, *Biochim. Biophys. Acta.,* 1992, **1105**, 213.
97. R. N. A. S. Lewis; R. N. McElhaney, *Biophys. J.,* 1993, **64**, 1081.
98. A. Ortiz; F. J. Aranda; J. Villalain; J. C. Gomez-Fernandez, *Biochim. Biophys. Acta,* 1992, **1112**, 226.
99. R. M. Epand; C. T. Lemay, *Chem. Phys. Lipids,* 1993, **66**, 181.
100. R. K. G. Habiger; J. Seelig, *Springer Proc. Phys.,* 1992, **66**, 49.
101. L. B. A. Johansson; B. Kalman; G. Wikander; Å. Fransson; K. Fontell; B. Bergenstål; G. Lindblom, *Biochim. Biophys. Acta,* 1993, **1149**, 285.
102. M. R. Morrow; J. P. Whitehead; D. Lu, *Biophys. J.,* 1992, **63**, 18.
103. A. Ceglie; G. Colafemmina; M. D. Monica; U. Olsson; B. Jönsson, *Langmuir,* 1993, **9**, 1449.
104. J. Zhao; B. M. Fung, *Langmuir,* 1993, **9**, 1228.
105. J. A. Zoltewicz; L. B. Bloom, *J. Phys. Chem.,* 1992, **96**, 5156.
106. Y. S. Lee; K. W. Woo, *Bull. Korean Chem. Soc..,* 1993, **14**, 392.
107. A. Zhelyaskova; A. Derzhanski, *J. Mol. Liq.,* 1992, **53**, 155.
108. S. Das; R. G. Bhirud; N. Nayyar; K. S. Narayan; V. V. Kumar, *J. Phys. Chem.,* 1992, **96**, 7454.
109. M. Monduzzi; U. Olsson; O. Söderman, *Langmuir,* 1993, **9**, 2914.
110. F. M. Menger; C. A. Littau, *J. Am. Chem. Soc.,* 1993, **115**, 10083.
111. D. G. Barrett; S. H. Gellman, *J. Am. Chem. Soc.,* 1993, **115**, 9343.
112. A. Relini; F. Cavagnetto; A. Gliozzi, *Springer Proc. Phys.,* 1992, 189.
113. P. J. Bratt; D. G. Gillies; A. M. L. Krebber; L. H. Sutcliffe, *Magn. Reson. Chem.,* 1992, **30**, 1000.
114. N. Mahieu; P. Tekely; J. C. Boubel; J. M. Cases; D. Canet, *J. Phys. Chem.,* 1993, **97**, 9513.
115. S. Ozeki; H. Seki, *J. Phys. Chem.,* 1992, **96**, 10074.
116. N. Laakel; P. Rubini; L. Rodehuser; J. J. Delpuech, *New J. Chem.,* 1992, **16**, 809.
117. W. Guo; Z. Li; B. M. Fung; E. A. O'Rear; J. H. Harwell, *J. Phys. Chem.,* 1992, **96**, 6738.
118. T. Kato; T. Terao; M. Tsukada; T. Seimiya, *J. Phys. Chem.,* 1993, **97**, 3910.
119. P. Neogi, *Langmuir,* 1994, **10**, 1410.

120. W. Guo; B. M. Fung; S. D. Christian; E. K. Guzman, *ACS Symp. Ser.*, 1992, **50**, 244.
121. E. Söderlind; P. Stilbs, *Langmuir,* 1993, **9**, 1678.
122. E. Söderlind; M. Björling; P. Stilbs, *Langmuir,* 1994, **10**, 890.
123. E. Söderlind, *Langmuir,* 1994, **10**, 1122.
124. M. R. Bohmer; L. K. Koopal; R. Janssen; E. M. Lee; R. K. Thomas; A. R. Rennie, *Langmuir,* 1992, **8**, 2228.
125. F. Mariette; C. Tellier; G. Brule; P. Marchal, *J. Diary Res.*, 1993, **60**, 175.
126. S. C. Lin; M. A. Minton; M. M. Sharma; G. Georgiou, *Appl. Environ. Microbiol.*, 1994, **60**, 31.
127. T. Kato; H. Takeuchi; T. Seimiya, *J. Phys.Chem.*, 1992, **96**, 6839.
128. L. N. Guo; I. Arnaud; M. Petit-Ramel; R. Gauthier; C. Monnet; P. Le Perchec; Y. Chevalier, *J. Colloid Interf. Sci.*, 1994, **163**, 334.
129. N. Navyer; R. G. Bhirud; K. S. Narayan; A. P. Singh; V. V. Kumar, *J. Colloid Interf. Sci.*, 1993, **160**, 496.
130. D. G. Marangoni; A. P. Rodenhiser; J. M. Thomas; J. C. T. Kwak, *Langmuir,* 1993, **9**, 438.
131. V. A. Gaevoi; V. S. Kuts; V. I. Kovalev, *Teor. Eksp. Khim.*, 1991, **27**, 606.
132. J. J. Chung; J. B. Kang; K. H. Lee, *J Korean Chem. Soc.*, 1994, **38**, 194.
133. J. A. Zoltewicz; L. B. Blum, *J. Phys. Chem.*, 1992, **96**, 2007.
134. M. D. Bruch; J. Rizo; L. M. Gierasch, *Biopolymers,* 1992, **32**, 1741.
135. A. L. Lomize; K. V. Pervushin; S. Arsenev, *J. Biomol. NMR,* 1992, **2**, 361.
136. K. V. Pervushin; A. S. Arsenev, *FEBS Lett.*, 1992, **308**, 190.
137. M. Tessari; M. T. Foffani; S. Mammi; E. Peggion, *Biopolymers,* 1993, **33**, 1877.
138. V. E. Yushmanov; J. R. Perussi; H. Imasato; M. Tabak, *Biochim. Biophys. Acta,* 1994, **1189**, 74.
139. W. Guo; E. K. Guzman; S. D. Heavin; Z. Li; B. M. Fung; S. D. Christian, *Langmuir,* 1992, **8**, 2368.
140. R. M. Clapperton; R. H. Ottewill; B. T. Ingram, *Langmuir,* 1994, **10**, 51.
141. W. Guo; B. M. Fung; E. A. O´Rear, *J. Phys. Chem.*, 1992, **96**, 10068.
142. H. Kuwahara; M. Hamada; Y. Ishikawa; T. Kunitake, *J. Am. Chem. Soc.*, 1993, **115**, 3002.
143. D. Wang; X. Xu; N. Deng; G. Wang; S. Mao; R. E. Stark, *J. Phys. Chem.*, 1992, **96**, 8187.
144. R. D. Stevens; A. A. Ribeoro; L. Lack; P. G. Killenberg, *J. Lipid Res.*, 1992, **33**, 21-29.
145. M. D´Alagni; M. Delfini; L. Galantini; E. Giglio, *J. Phys. Chem.*, 1992, **96**, 10520.
146. Y. S. Lee; K. W. Woo, *Bull. Korean Chem. Soc.*, 1993, **14**, 599.
147. V. K. Smith; T. T. Ndou; I. M. Warner, *Appl. Spectrosc.*, 1992, **46**, 659.
148. J. K. McCulloch; D. Fornasiero; J. M. Perera; B. S. Murray; G. W. Stevens; F. Grieser, *J. Colloid Interf. Sci.*, 1993, **157**, 180.
149. G. Orädd; G. Lindblom; L. B. A. JohanssonN; G. Wikander, *J. Phys. Chem,* 1992, **96**, 5170.
150. M. S. Goldenberg; L. A. Bruno; E. L. Rennwantz, *J. Colloid Interf. Sci.*, 1993, **158**, 351.
151. A. S. Sadaghiani; A. Noori; A. Khan, *J. Surf. Sci. Technol.*, 1991, **7**, 163.
152. G. Datta; P. S. Parvathanathan; U. R. K. Rao; K. U. Deniz, *Physiol. Chem. Phys. Med. NMR,* 1992, **24**, 51.
153. G. Datta; P. S. Parvathanathan; K. U. Deniz, *J. Surf. Sci. Technol.*, 1991, **7**, 283.
154. D. Capitani; E. Rossi; A. L. Segre; M. Giustini; P. L. Luisi, *Langmuir,* 1993, **9**, 685.
155. M. Tata; V. T. John; Y. Y. Waguespack; G. L. Mcpherson, *J. Phys. Chem.*, 1994, **98**, 3809.
156. T. Kawai; N. Shindo; K. Konno, *J. Surf. Sci. Technol.*, 1992, **8**, 145.
157. R. Skurtveit; U. Olsson, *J. Phys. Chem.*, 1992, **96**, 8640.
158. S. Backlund; G. Lundsten; M. Ruths; M. Sjöberg; T. Wänheim, *Prog. Colloid Polym. Sci.*, 1992, **88**, 36.

159. K. P. Datema; J. A. Bolt-Westerhoff; A. Jaspers; J. G. R. Daane; L. A. M. Rupert, *Magn. Reson. Chem.*, 1992, **30**, 760.
160. U. Olsson; P. Schurtenberger, *Langmuir,* 1993, **9**, 3389.
161. K. Fukuda; O. Söderman; K. Shinoda; B. Lindman, *Langmuir,* 1993, **9**, 2921.
162. W. O. Parker; C. Genova; G. Carignano, *Colloids Surf., A,* 1993, **72**, 275.
163. T. Tamura; M. Fujitso; K. Itoh, *Yukagaku,* 1993, **42**, 661.
164. M. Fujitsu; T. Tamura, *Yukagaku,* 1994, **43**, 131.
165. M. Monduzzi; A. Chittofrati; M. Visca, *Langmuir,,* 1992, **8**,, 1278-84.
166. A. Sanguineti; A. Chittofrati; D. Lenti; M. Visca, *J. Colloid Interface Sci.,* 1993, **155**, 402.
167. P. H. Kenez; G. Carlström; I. Furo; B. Halle, *J. Phys. Chem.,* 1992, **96**, 9524.
168. R. Strey; W. Jahn; M. Skouri; G. Porte; J. Marignan; U. Olsson, *NATO ASI Ser., Ser. C,* 1992, **369**, 351.
169. M. Varshney; T. K. Jain; A. Maitra, *J. Surf. Sci. Technol.,* 1992, **8**, 243.
170. Y. E. Shapiro; E. G. Pykhteeva; A. V. Levashov; N. L. Klyachko, *Biol. Membr.,* 1993, **10**, 397.
171. A. Ceglie; G. Colafemmina; M. Della Monica; G. Palazzo, *Colloids Surf., A,* 1993, **72**, 285.
172. S. Wang; Z. Chen; C. Hao; S. Wu; F. Jia; L. Gao; D. Chen, *Huaxue Xuebao,* 1993, **51**, 736.
173. P. O. Quist; K. Fontell; B. Halle, *Liq. Cryst.,* 1993, **16**, 236.
174. P. O. Quist; B. Halle, *Phys. Rev. E: Stat. Phys., Plasma, Fluids, Relat. Interdisp. Top.,* 1993, **47**, 3374.
175. P. O. Quist, *J. Magn. Reson., Ser. A,* 1993, **101**, 131.
176. C. Kang; O. Söderman; P. O. Eriksson; J. S. V. Holstein, *Liq. Cryst.,* 1992, **12**, 71.
177. H. Hagslätt; O. Söderman; B. Jönsson, *Liq. Cryst.,* 1992, **12**, 667.
178. M. Jonströmer; R. Strey, *J. Phys. Chem.,* 1992, **96**, 5993.
179. W. Schnepp; S. Disch; C. Schmidt, *Liq. Cryst.,* 1993, **14**, 843.
180. W. Schnepp; C. Schmidt, *Ber. Bunsen-Ges. Phys. Chem.,* 1993, **97**, 1399.
181. Y. Sasanuma; M. Nakamura; A. Abe, *J. Phys. Chem.,* 1993, **97**, 5155.
182. D. Catalano; A. Lenzi; C. A. Veracini, *Liq. Cryst.,* 1993, **14**, 1457.
183. R. Guo, *Gaodeng Xuexiao Huaxue Xuebao,* 1991, **12**, 1526.
184. M. Acimis; A. Alaman, *Spectrosc. Lett.,* 1992, **25**, 959.
185. M. Rankothge; M. G. Haryadi; J. Hook; L. Van Gorkom, *Solid States Ionics,* 1994, **67**, 241.
186. J. C. Blackburn; P. K. Kalpatrick, *Langmuir,* 1992, **8**, 1679.
187. G. Celebre; L. Coppola; G. A. Ranieri, *J. Chem. Phys.,* 1992, **97**, 7781.
188. L. Coppola; C. L. Mesa; G. A. Ranieri; M. Terenzi, *J. Chem. Phys.,* 1993, **98**, 5087.
189. M. C. Holmes; P. Sotta; Y. Hendrikx; B. Deloche, *J. Phys. II,* 1993, **3**, 1735.
190. J. Chung; J. H. Prestegard, *J. Phys. Chem.,* 1993, **97**, 9837.
191. T. Koechy; T. M. Bayerl, *Phys. Rev. E: Stat. Phys. Plasmas, Fluids, Relat. Interdisp. Top.,* 1993, **47**, 2109.
192. P. Ström, *J. Colloid Interface Sci.,* 1992, **154**, 184.
193. G. Orädd; G. Wikander; G. Lindblom; L. B. A. Johansson, *J. Chem. Soc. Faraday Trans.,* 1994, **90**, 305.
194. G. Lindblom; L. B. A. Johansson; G. Wikander; P. O. Eriksson; G. Arvidson, *Biophys. J.,* 1992, **63**, 723.
195. P. O. Eriksson; G. Lindblom, *Biophys. J.,* 1993, **64**, 129.
196. F. Adebodun; J. Chung; B. Montez; E. Oldfield; X. Shan, *Biochemistry,* 1992, **31**, 4502.
197. S. O. Smith; I. Kustanovich; S. Bhamidipati; A. Salmon; J. A. Hamilton, *Biochemistry,* 1992, **31**, 11660.
198. T. J. T. Pinheiro; A. A. Duralski; A. Watts, *Biochemistry,* 1994, **33**, 4896.
199. W. Guo; J. A. Hamilton, *Biochemistry,* 1993, **32**, 9038.

200. Y. Aubin; Y. Ito; J. C. Paulson; J. H. Prestegard, *Biochemistry,* 1993, **32**, 13405.
201. M. A. Davies; W. Hubner; A. Blume; R. Mendelsohn, *Biophys. J.,* 1992, **63**, 1059.
202. T. M. Alam, *J. Bliophys,* 1993, **64**, 1681.
203. A. Leonard; J. C. Maillet; J. Dufourcq; E. J. Dufourc, *Prog. Colloid Polym. Sci.,* 1992, **89**, 315.
204. J. F. Nagle, *J. Biophys.,* 1993, **64**, 1476.
205. L. Di; D. M. Small, *J. Lipid Res.,* 1993, **34**, 1611.
206. J. Wolfe; Z. Yan; J. M. Pope, *Biophys. Chem.,* 1994, **49**, 51.
207. Z. Yan; J. Pope; J. Wolfe, *J.Chem. Soc., Faraday Trans.,* 1993, **89**, 2583.
208. K. Gawrisch; D. Ruston; J. Zimmerberg; V. A. Parsegian; R. P. Rand; N. Fuller, *Biophys. J.,* 1992, **61**, 1213.
209. M. Jansson; R. L. Thurmond; J. A. Barry; M. F. Brown, *J. Phys. Chem.,* 1992, **96**, 9532.
210. R. L. Thurmond; G. Lindblom; M. F. Brown, *Biochemistry,* 1993, **32**, 5394.
211. Krajewski-Bertrand.M.A.; Y. Nakatani; G. Ourisson; E. j. Dufourc; A. Milon, *J. Chim. Phys.-Chim. Biol.,* 1992, **89**, 237.
212. T. H. Huang; C. W. B. Lee; S. K. Das Gupta; A. Blume; R. G. Griffin, *Biochemistry,* 1993, **32**, 13277.
213. F. A. Nezil; M. Bloom, *Biophys. J.,* 1992, **61**, 1176.
214. T. P. Trouard; T. M. Alam; J. Zajicek; M. K. Brown, *Chem.Phys.Lett.,* 1992, **189**, 67-75.
215. R. C. Prosser; J. H. Davis; C. Mayer; K. Weisz; G. Kothe, *Biochemistry,* 1992, **31**, 9355.
216. P. V. Dubovskii; O. V. Esipova; I. A. Vasilenko; E. N. Zvonkova, *Biol. Membr.,* 1992, **9**, 184.
217. K. S. Hamilton; H. C. Jarrell; K. M. Briere; C. W. M. Grant, *Biochemistry,* 1993, **32**, 4022.
218. R. Smith; F. Separovic; F. C. Bennett; B. A. Cornell, *Biophys. J.,* 1992, **63**, 469.
219. P. J. Bratt; I. H. Davis; D. C. Doetschman; A. Watts, *J. Chem. Soc., Perkin Trans. 2,* 1993, **11**, 2113.
220. J. C. Sanders; T. W. Poile; C. J. A. M. Wolfs; M. A. Hemminga, *Biochim. Biophys. Acta,* 1992, **1110**, 218.
221. F. A. D. Wolf; K. Nicolay; B. D. Kruijff, *Biochemistry,* 1992, **31**, 9252.
222. H. D. Baeuerle; J. Seelig, *Springer Proc. Phys.,* 1992, **66**, 76.
223. C. R. Sanders; J. E. Schaff; J. H. Prestegard, *Biophys. J.,* 1993, **64**, 1069.
224. C. Tilcock; Q. F. Ahkong; S. H. Koenig; R. D. B. III; M. Davis; G. Kabalka, *Magn. Reson. Med.,* 1992, **27**, 44.
225. T. Horiuchu; K. Tajima, *Yukagaku,* 1992, **41**, 1197.
226. J. M. Devoiselle; J. Vion-Dury; S. Confort-Gouny; D. Coustaut; P. J. Cozzone, *J. Pharm. Sci.,* 1992, **81**, 249.
227. D. B. Fenske; P. R. Cullis, *J. Biophys.,* 1993, **64**, 1482.
228. P. L. Yeacle; F. T. Smith; J. E. F. Young T.D., *Biochemistry,* 1994, **33**, 1820.
229. F. Cavagnetto; A. Relini; Z. Mirghani; A. Gliozzi; D. Bertoia; A. Gampacorta, *Biochim. Biophys. Acta,* 1992, **1106**, 273.
230. X. Qiu; C. Pidgeon, *Biochemistry,* 1994, **33**, 960.
231. C. Dolainsky; M. A; T. M. Bayerl, *J. Chem. Phys.,* 1993, **98**, 1712.
232. C. Dolainsky; T. Koechy; C. Naumann; T. Brumm; S. J. Johnson; T. M. Bayerl, *Springer Proc. Phys.,* 1992, **66**, 34.
233. X. Chen; X. Han; R. W. Gross, *Biochim. Biophys. Acta,* 1993, **1149**, 241.
234. H. S. Han; H. Kim, *J. Biochem.,* 1994, **115**, 26.
235. D. P. Hinton; C. S. Johnsopn, *Chem. Phys. Lipids,* 1994, **69**, 175.
236. D. P. Hinton; C. S. Johnson, *J. Phys. Chem..,* 1993, **97**, 9064.
237. J. F. Ellena; L. S. Lepore; D. S. Cafiso, *J. Phys. Chem.,* 1993, **97**, 2952.

238. G. Erriu; S. Onnis; N. Zucca; M. Casu; A. Lai; C. Giori; G. Schianchi, *Nuovo Cimento Soc. Ital. Fis., D,* 1993, **15D**, 1451.
239. F. A. Nezil; S. Bayerl; M. Bloom, *Biophys. J.*, 1992, **61**, 1413.
240. T. Brumm; A. Moeps; C. Dolainsky; S. Brueckner; T. M. Bayerl, *Biophys. J.*, 1992, **61**, 1018.
241. F. M. Marassi; R. R. Shivers; P. M. Macdonald, *Bichemistry,* 1993, **32**, 9936.
242. J. Higinbotham; P. H. Beswick; R. J. Malcolmson; D. Reed; J. A. Parkinson; I. H. Sadler, *Chem. Phys. Lipids,* 1993, **66**, 1.
243. J. A. Raynaud; J. P. Grivet; D. Sy; Y. Trudelle, *Biochemistry,* 1993, **32**, 4997.
244. Z. Wang; J. D. Jones; J. Rizo; L. M. Gierasch, *Biochemistry,* 1993, **32**, 13991.
245. D. Pott; E. J. Dufourc, *Bull. Magn. Reson.*, 1993, **15**, 121.
246. J. Salgado; J. Villalain; J. C. Gomez-Fernandez, *Eur. Biophys. J.*, 1993, **22**, 151.
247. J. C. Gomez-Fernandez; J. Villalain; F. J. Aranda, *Vitm. E Health Dis.*, 1993, 223.
248. S. Gaillard; E. J. Dufourc; M. Bonnet; J. P. Renou, *Eur.Biophys.J.*, 1992, **21**, 13.
249. A. R. Balakrishnan; K. R. K. Easwaran, *Biochemistry,* 1993, **32**, 4139.
250. F. A. d. Wolf; R. W. H. M. Staffhorst; W. H. M. Rutger; H. P. Smits; M. F. Onwezen; B. d. Kruijff, *Biochemistry,* 1993, **32**, 6688.
251. E. A. Coats; M. Wiese; H. L. Chi; H. P. Cordes; J. K. Seydel, *Quant. Struct.-Act. Relat.*, 1992, **11**, 364.
252. R. Labonte; Z. Gao; J. C. T. Kwak, *Colloids Surf., A,* 1993, **78**, 271.
253. D. Massenburg; B. R. Lentz, *Biochemistry,* 1993, **32**, 9172.
254. M. Ambuehl; F. Bangerter; P. L. Luisi; P. Skrabal; H. J. Watzke, *Langmuir,* 1993, **9**, 36.
255. O. Regev; C. Kang; A. Khan, *J. Phys. Chem.*, 1994, **98**, 6619.
256. D. Singh; J. H. Davis, *Biochim. Biophys. Acta.*, 1992, **1107**, 23.
257. S. Fujisawa; Y. Komoda, *Dent. Mater. J.*, 1993, **12**, 69.
258. A. Schanck; M. Deleers, *Biochem. Biophys. Res. Commun.*, 1993, **195**, 654.
259. M. Roux; F. A. Nezil; M. Monck; M. Bloom, *Biochemistry,* 1994, **33**, 307.
260. A. Blume; M. Jansen; M. Ghyczy; J. Gareiss, *Int. J. Pharm.*, 1993, **99**, 291.
261. F. M. Marassi; P. M. Macdonald, *Biochemistry,* 1992, **31**, 10031.
262. V. Misik; K. Ondrias; P. Balgavy, *Gen. Physiol. Biophys.*, 1992, **11**, 317.
263. J. W. Wang; T. Miyazawa; K. Fujimoto; Z. W. Wang; T. Nozawa, *Int. Congr. Ser. - Excerpta Med.*, 1992, **998**, 257.
264. J. Y. Wang; T. Miyazawa; K. Fujimoto; Z. Y. Wang; T. Nozawa, *FEBS Lett.*, 1992, **310**, 106.
265. C. R. Sanders.II, *Biophys. J.*, 1993, **64**, 171.
266. L. M. Chi; C. H. Hsieh; W. G. Wu, *J. Chin. Chem. Soc(Taipei),* 1992, **39.**, 35.
267. H. Noda; M. Kurono; N. Ohishi; K. Yagi, *Biochim. Biophys. Acta,* 1993, **1153**, 127.
268. K. L. Li; C. A. Tihal; M. Guo; R. E. Stark, *Biochemistry,* 1993, **32**, 9926.
269. W. Huebner; H. H. Mantsch; F. Paltauf; H. Hauser, *Biochemistry,* 1994, **33**, 320.
270. P. J. R. Spooner; A. Watts, *Biochemistry,* 1992, **31**, 10129.
271. M. R. Morrow; J. Perez-Gil; G. Simatos; C. Boland; J. Stewart; D. Absolom; V. Sarin; K. M. W. Keough, *Biochemistry,* 1993, **32**, 4397.
272. G. Govil; R. S. Phadke; S. Srivastava, *Lipid-Soluble Antioxid.*, 1992, 27.
273. K. Fukuzawa; W. Ikebata; A. Shibata; I. Kumadaki; T. Sakanaka; S. Urano, *Chem. Phys. Lipids,* 1992, **63**, 69.
274. Y. J. Suzuki; M. Tsuchiya; S. R. Wassall; Y. M. Choo; G. Govil; V. E. Kagan; L. Packer, *Biochemistry,* 1993, **32**, 10692.
275. K. C. Duff; A. J. Cudmore; J. P. Bradshaw, *Biochim. Biophys. Acta,* 1993, **1145**, 149.
276. R. A. Goodnow; C. J. Turner, *J. Liposome Res.*, 1993, **3**, 123.

277. C. E. Dempsey; G. S. Butler, *Biochemistry,* 1992, **31**, 11973.
278. S. E. Gasanov; L. P. Vernon; T. F. Aripov, *Arch. Biochem. Biophys.*, 1993, **301**, 367.
279. D. V. Greathouse; J. F. Hinton; K. S. Kim; R. E. Koeppe, *Biochemistry,* 1994, **33**, 4291.
280. H. M. Reini; T. M. Bayerl, *Biochim. Biophys. Acta,* 1993, **115**, 127.
281. V. V. Chupin; M. V. Anikin; G. A. Serebrennikova; Y. M. Margolin; V. M. Kreines; I. M. Ust'yantseva; V. G. Bulgakov; A. N. Shalnev, *Biol. Membr.*, 1992, **9**, 349.
282. M. A. Krajewski-Bertrand; A. Milon; M. A. Hartmann, *Chem. Phys. Lipids,* 1992, **63**, 235.
283. J. A. Killian; C. H. Fabrie; W. Baart; S. Morein; B. D. Kruijff, *Biochim. Biophys. Acta,* 1992, **1105**, 253.
284. E. Y. Lurie; P. V. Dubovski; A. P. Kaplun; O. E. Oksynoid; V. I. Shvets, *Biol. Membr.*, 1992, **9**, 1204.
285. J. K. Seydel; V. M. Albores; E. A. Coats; H. P. Cordes; B. Kunz; M. Wiese, *Quant. Struct.-Act. Relat.*, 1992, **11**, 205.
286. V. Geli; M. C. Koorengevel; R. A. Demel; C. Lazdunski; J. A. Killian, *Biochemistry,* 1992, **31**, 11089.
287. R. Zidovetzki; I. W. Sherman; M. Cardenas; D. B. Borchardt, *Biochem. Pharmacol.*, 1993, **45**, 183.
288. M. Wakita; Y. Kuroda; T. Nakagawa, *Chem. Pharm. Bull.*, 1992, **40**, 1361.
289. J. C. Sanders; P. I. Haris; D. Chapman; C. Otto; M. A. Hemminga, *Biochemistry,* 1993, **32**, 12446.
290. M. R. Morrow; S. Taneva; G. A. Simatos; L. A. Allwood; K. M. W. Keough, *Biochemistry,* 1993, **32**, 11338.
291. P. Gale; A. Watts, *Biochim. Biophys. Acta,* 1992, **1106**, 317.
292. T. J. T. Pinheiro; A. Watts, *Biochemistry,* 1994, **33**, 2459.
293. A. G. Rietveld; J. A. Killian; W. Dowhan; B. d. Kruijff, *J. Biol.Chem.*, 1993, **268**, 12427.
294. E. J. Dufourc; C. Mayer; J. Stohrer; G. Kothe, *J.Chim.Phys.Phys.-Chim.Biol.*, 1992, **89**, 243-52.
295. M. A. Monck; M. Bloom; M. Lafleur; R. N. A. H. Lewis; R. N. McElhaney; P. R. Cullis, *Biochemistry,* 1992, **31**, 10037.
296. G. Lindblom; J. B. Hauksson; L. Rilfors; B. Bergenstål; Å. Wieslander; P. O. Eriksson, *J. Biol. Chem.*, 1993, **268**, 16198.
297. W. Abraham, *Proc. Int. Symp. Controlled Release Bioact. Mater., 19th,* 1992, 135.
298. D. J. McCormac; J. F. Todd; G. Paliyath; J. E. Thompson, *Plant Physiol.Biochem,* 1993, **31**, 1.
299. K. Chari, *J.Colloid Interf.Sci.*, 1992, **151**, 294.
300. A. Hammarström; L. O. Sundelöf, *Colloid polym. Sci.*, 1993, **271**, 1129.
301. J. J. Effing; L. J. McLennan; J. C. T. Kwak, *J. Phys. Chem.*, 1994, **98**, 2499.
302. P. M. Macdonald; D. Staring; Y. Yue, *Langmuir,* 1993, **9**, 381.
303. K. F. Morris; C. S. Johnson; T. C. Wong, *J. Phys. Chem.*, 1994, **98**, 603.
304. M. Alden; J. Tegenfeldt; E. Sjökvist, *Int. J. Pharm.*, 1992, **83**, 47.
305. M. Alden; J. Tegenfeldt; E. S. Särs, *Int. J. Pharm.*, 1993, **94**, 31.
306. M. H. G. M. Penders; S. Nilsson; L. Piculell; B. Lindman, *J. Phys. Chem.*, 1993, **97**, 11332.
307. K. Zhang; M. Jonströmer; B. Lindman, *J. Phys. Chem.*, 1994, **98**, 2459.
308. G. L. Mendz; I. M. Jamie; J. W. White, *Biophys. Chem.*, 1992, **45**, 61.
309. Y. Mine; K. Chiba; M. Tada, *J. Agric. Food Chem.*, 1993, **41**, 157.
310. J. A. Hamilton, *News Physiol. Sci.*, 1992, **7**, 264.
311. P. J. R. Spooner; A. A. Duralski; S. E. Rankin; T. J. T. Pinheiro; A. Watts, *Biophys. J.*, 1993, **65**, 106.
312. M. Hayer-Hartl; H. Schägger; G. V. Jagow; K. Beyer, *Eur. J. Biochem.*, 1992, **209**, 423.

313. P. A. McDonnel; S. J. Opella, *J. Magn. Reson., Ser B,* 1993, **102**, 120.
314. D. D. Miller; W. Lenhart; B. J. Antalek; A. J. Williams; J. M. Hewitt, *Langmuir,* 1994, **10**, 68.
315. L. S. Romsted; C. O. Yoon, *J. Am. Chem. Soc.,* 1993, **115**, 989.
316. A. Elasko; C. A. Bunton; C. G.; D. C. McKenzie, *J. Phys. Chem.,* 1993, **97**, 11324.
317. A. Wagenaar; L. Streefland; D. Hoekstra; J. B. F. N. Engberts, *J. Phys. Org. Chem.,* 1992, **5**, 451.
318. C. Chachaty; T. Bredel, *J. Chem. Soc., Faraday Trans.,* 1992, **88**, 1893.
319. X. X. Zhu; E. Bardez; L. Dallery; B. Larrey; B. Valeur, *New J. Chem.,* 1992, **16**, 973.
320. O. A. El seoud; M. I. El seoud; J. A. Mickiewicz, *J. Colloid Interf. Sci.,* 1994, **163**, 87.
321. T. Kawai; K. Hamada; K. Konno, *Bull. Chem. Soc. Japan,* 1993, **66**, 2804.
322. T. Matsumoto, *Nihon Reoroji Gakkaishi,* 1993, **21**, 18.
323. A. S. Ulrich; A. Watts, *Biophys. Chem.,* 1994, **49**, 39.
324. E. Merchant; T. Glonek, *Lipids,* 1992, **27**, 551.
325. J. A. Killian; M. C. Koorengevel; J. A. Boustra; G. Gooris; W. Dowhan; B. de Kruijff, *Biochim. Biophys. Acta,* 1994, **1189**, 225.

Author Index

In this index the number in parenthesis is the Chapter number of the citation and this is followed by the reference number or numbers of the relevant citations within that Chapter.